# Environmental Remediation

## Assemblies Cost Book

## 1998

**Senior Editor**
**Delta Technologies Group, Inc.**

Richard R. Rast, President

**Technical Editors**
**Delta Technologies Group, Inc.**

Tim Burks
Kevin L. Klink, PE
Jacqueline Crenca Rast, PE
R. Douglas Taylor, PE
C. Tim Wallace
Laura Williams

**Contributing Editors**
**R.S. Means Company**

Jeannene D. Murphy
Jesse R. Page
Phillip R. Waier, PE

**R.S. Means Company**
**President**

Durwood S. Snead

**Vice President and**
**General Manager**

Roger J. Grant

**Vice President,**
**Sales and Marketing**

John M. Shea

**Vice President,**
**Operations**

Andrew J. Centauro

**Manager, Engineering**
**Operations**

John H. Ferguson, PE

**Production Manager**

Michael Kokernak

**Production Coordinator**

Marion E. Schofield

Underlying portions of this work are individually owned and copyrighted by Delta Technologies Group, Inc. and ECHOS, L.L.C. and may not be reproduced, translated or appropriated without their prior written authorization.

All rights reserved. No part of this work may be reproduced, translated or appropriated in any form or by any means (including electronic, mechanical or other such as photocopying, recording or any information storage and retrieval system) without prior written authorization.

Copyright 1998 Delta Technologies Group, Inc. and ECHOS, L.L.C.
All rights reserved.

# ECHOS

**ENVIRONMENTAL**

**COST HANDLING**

**OPTIONS AND**

**SOLUTIONS**

Delta Technologies Group, Inc.
7340 E. Caley Ave., Suite 120
Englewood, CO 80111
(303) 771-3103

**Published by**
**R.S. Means Company, Inc.**
**and Delta Technologies**
**Group, Inc.**

# Preface

## THE BOOK

The ECHOS *Environmental Remediation Assemblies Cost Data* is "the cost source book" for environmental restoration activities beginning with initial site investigation and continuing through studies, design, remediation, and long-term monitoring and operation. Containing over 4,000 assembly cost items, this publication is updated and expanded biannually to reflect the latest cost and technology information in the rapidly changing environmental field.

The ECHOS *Environmental Remediation Assemblies Cost Data Book* provides you with the procedures, forms, data, and descriptive information needed to prepare complete cost estimates for almost any type of environmental restoration project, ranging from simple underground storage tank removals to complex multimedia/multicontaminant hazardous waste sites listed on the US EPA's National Priority List.

Cost information is provided for performing work at various OSHA-dictated safety levels depending on the level of contamination and potential for exposure for on-site work crews. Location factors are supplied by zip code, allowing you to adjust your estimate to local conditions. For ease of use, the cost data is organized around different phases of work and remediation technologies that may be encountered at the particular site. Assembly numbering follows the recently introduced US Government Interagency Code of Accounts, an evolving standard for organizing environmental restoration costs.

## THE DATA

The ECHOS research staff is constantly gathering, monitoring, and developing construction and environmental restoration cost information throughout the US. In doing so, the ECHOS database reflects the most current trends in both procedures and unit costs for environmental restoration activities. This book is the result of over seven years of research in environmental restoration cost and the cost information used in this book has been successfully used on over 1,500 environmental restoration projects in every state in the US.

This data is received by us from sources we believe to be reliable, but no warranty, guaranty or representation is made by ECHOS as to the correctness or sufficiency of any information, prices, or representation contained in the ECHOS *Environmental Remediation Assemblies Cost Data Book* and ECHOS assumes no responsibility or liability in connection therewith.

Material costs are determined through contact with product manufacturers, dealers, supply houses, distributors, and contractors. Labor costs are based on crews and productivity factors determined by ECHOS environmental engineering and construction experts. Productivity rates for different safety levels are based on research conducted by the ECHOS research staff combined with research and published reports provided by the US EPA. Equipment costs are based on rental rates for different equipment items in some cases and purchase and annual cost of ownership in other cases. These differences are noted in the equipment item descriptions throughout the book.

## REGULATORY ENVIRONMENT

There are numerous state and federal laws and regulations that govern the practice of environmental restoration activities. The two primary laws that set the standards for this book are the Comprehensive Environmental Response Compensation and Liability Act (CERCLA) and its amendments (commonly referred to as the SUPERFUND Law), and the Resource Conservation and Recovery Act (RCRA). The environmental restoration technologies and processes used in this book are primarily designed to be used on projects that are regulated by these laws but the data can be used for other unregulated projects.

## ABOUT ECHOS

ECHOS is a joint venture between Delta Technologies Group, Inc., experts in environmental restoration cost estimating and technology application, and R. S. Means Company, Inc. leading publisher of construction cost information in North America. Through this collaboration, exhaustive cost research, and seasoned technical experts, all of the practical tools necessary to assemble or analyze restoration costs have been established.

# Table of Contents

## How to Use This Book

| | |
|---|---|
| Estimating Forms and Instructions | 1-1 |
| How to Interpret the Cost Pages | 1-14 |
| List of Abbreviations | 1-16 |
| Localization Information | 1-18 |

## Cost Data for Site Studies -- RI/FS and RFI/CMS

| | |
|---|---|
| Tasks | 2-1 |
| Professional Labor | 2-4 |
| Groundwater Monitoring Wells | 2-6 |
| Sampling | 2-9 |
| Analysis | 2-21 |
| Miscellaneous | 2-38 |

## Cost Data for Remediation

| | |
|---|---|
| Air Sparged Hydrocyclone | 3-1 |
| Air Sparging | 3-5 |
| Air Stripping | 3-8 |
| Bioremediation, Water (Ex Situ) | 3-15 |
| Bioslurping | 3-23 |
| Bulk Material Storage | 3-28 |
| Capping | 3-33 |
| Carbon Adsorption (Gas) | 3-44 |
| Carbon Adsorption (Liquid) | 3-49 |
| Chemical Precipitation | 3-54 |
| Coagulation / Flocculation | 3-61 |
| Commercial Disposal (Incineration) | 3-65 |
| Composting | 3-69 |
| Decontamination Facilities | 3-75 |
| Dewatering (Sludge) | 3-82 |
| Discharge to POTW | 3-85 |
| Drum Removal | 3-92 |
| Ex Situ Vapor Extraction | 3-95 |
| Excavation, Buried Waste | 3-99 |
| Extraction Wells | 3-104 |
| Field Sampling | 3-117 |
| Free Product Removal (French Drain) | 3-127 |
| Groundwater Monitoring Wells | 3-136 |
| Heat-Enhanced Vapor Extraction | 3-141 |
| In Situ Biodegradation (Bioventing) | 3-146 |

## Cost Data for Remediation (continued)

| | |
|---|---|
| In Situ Biodegradation (Land Treatment) | 3-152 |
| In Situ Biodegradation (Saturated Zone) | 3-155 |
| In Situ Solidification | 3-162 |
| In Situ Vitrification | 3-164 |
| Incineration (On Site) | 3-167 |
| Infiltration Gallery | 3-170 |
| Injection Wells | 3-179 |
| Land Farming (Ex Situ) | 3-185 |
| Landfill Disposal | 3-191 |
| Low Level Radioactive Soil Treatment | 3-197 |
| Low Temperature Thermal Desorption | 3-203 |
| Media Filtration | 3-206 |
| Monitoring | 3-209 |
| Neutralization | 3-232 |
| Oil / Water Separation | 3-235 |
| Ordnance & Explosive Waste Remediation | 3-240 |
| Passive Water Treatment | 3-244 |
| Permeable Barriers | 3-249 |
| Piping | 3-253 |
| Sampling and Analysis | 3-262 |
| Slurry Walls | 3-285 |
| Soil Flushing | 3-288 |
| Soil Vapor Extraction | 3-293 |
| Soil Washing | 3-298 |
| Solidification / Stabilization | 3-304 |
| Solvent Extraction | 3-309 |
| Storage Tank Installation | 3-311 |
| Thermal & Catalytic Oxidation | 3-317 |
| Transportation | 3-320 |
| Ultraviolet Oxidation | 3-330 |
| UST Closure | 3-338 |
| Well Drilling & Installation | 3-341 |

**Environmental Remediation: Assemblies Cost Book**

# Table of Contents

## Cost Data for Remediation -- Ex Situ Treatments

| | |
|---|---|
| Air Stripping | 3-8 |
| Bioremediation, Water (Ex Situ) | 3-15 |
| Bioslurping | 3-23 |
| Carbon Adsorption (Gas) | 3-44 |
| Carbon Adsorption (Liquid) | 3-49 |
| Chemical Precipitation | 3-54 |
| Coagulation / Flocculation | 3-61 |
| Composting | 3-69 |
| Dewatering (Sludge) | 3-82 |
| Ex Situ Vapor Extraction | 3-95 |
| Incineration (On Site) | 3-167 |
| Land Farming (Ex Situ) | 3-185 |
| Low Level Radioactive Soil Treatment | 3-197 |
| Low Temperature Thermal Desorption | 3-203 |
| Media Filtration | 3-206 |
| Neutralization | 3-232 |
| Oil / Water Separation | 3-235 |
| Passive Water Treatment | 3-244 |
| Soil Washing | 3-298 |
| Solidification / Stabilization | 3-304 |
| Solvent Extraction | 3-309 |
| Thermal & Catalytic Oxidation | 3-317 |
| Ultraviolet Oxidation | 3-330 |

## Cost Data for Remediation -- In Situ Treatments

| | |
|---|---|
| Air Sparging | 3-5 |
| Heat-Enhanced Vapor Extraction | 3-141 |
| In Situ Biodegradation (Bioventing) | 3-146 |
| In Situ Biodegradation (Land Treatment) | 3-152 |
| In Situ Biodegradation (Saturated Zone) | 3-155 |
| In Situ Solidification | 3-162 |
| In Situ Vitrification | 3-164 |
| Permeable Barriers | 3-249 |
| Soil Flushing | 3-288 |
| Soil Vapor Extraction | 3-293 |

## Cost Data for Remediation -- Soil Treatments

| | |
|---|---|
| Commercial Disposal (Incineration) | 3-65 |
| Composting | 3-69 |
| Dewatering (Sludge) | 3-82 |
| Ex Situ Vapor Extraction | 3-95 |
| Heat-Enhanced Vapor Extraction | 3-141 |
| In Situ Biodegradation (Bioventing) | 3-146 |
| In Situ Biodegradation (Land Treatment) | 3-152 |
| In Situ Solidification | 3-162 |
| In Situ Vitrification | 3-164 |
| Incineration (On Site) | 3-167 |
| Land Farming (Ex Situ) | 3-185 |
| Low Level Radioactive Soil Treatment | 3-197 |
| Low Temperature Thermal Desorption | 3-203 |
| Soil Flushing | 3-288 |
| Soil Vapor Extraction | 3-293 |
| Soil Washing | 3-298 |
| Solidification / Stabilization | 3-304 |
| Solvent Extraction | 3-309 |

## Cost Data for Remediation -- Air Treatments

| | |
|---|---|
| Carbon Adsorption (Gas) | 3-44 |
| Thermal & Catalytic Oxidation | 3-317 |

## Cost Data for Remediation -- Groundwater Treatments

| | |
|---|---|
| Air Sparged Hydrocyclone | 3-1 |
| Air Sparging | 3-5 |
| Air Stripping | 3-8 |
| Bioremediation, Water (Ex Situ) | 3-15 |
| Bioslurping | 3-23 |
| Carbon Adsorption (Liquid) | 3-49 |
| Chemical Precipitation | 3-54 |
| Coagulation / Flocculation | 3-61 |
| Commercial Disposal (Incineration) | 3-65 |
| Free Product Removal (French Drain) | 3-127 |
| In Situ Biodegradation (Saturated Zone) | 3-155 |
| Incineration (On Site) | 3-167 |
| Media Filtration | 3-206 |
| Oil / Water Separation | 3-235 |

**Environmental Remediation: Assemblies Cost Book**

©1998 by ECHOS. All rights reserved.

# Table of Contents

## Cost Data for Remediation -- Groundwater Treatment (Continued)

| | |
|---|---|
| Permeable Barriers | 3-249 |
| Slurry Walls | 3-285 |
| Ultraviolet Oxidation | 3-330 |

## Cost Data for Remediation -- Disposal Methods

| | |
|---|---|
| Commercial Disposal (Incineration) | 3-65 |
| Discharge to POTW | 3-85 |
| Infiltration Gallery | 3-170 |
| Injection Wells | 3-179 |
| Landfill Disposal | 3-191 |
| Low Level Radioactive Soil Treatment | 3-197 |

## Cost Data for Remediation -- Transfer Technologies

| | |
|---|---|
| Air Sparging | 3-5 |
| Air Stripping | 3-8 |
| Carbon Adsorption (Gas) | 3-44 |
| Carbon Adsorption (Liquid) | 3-49 |
| Ex Situ Vapor Extraction | 3-95 |
| Heat-Enhanced Vapor Extraction | 3-141 |
| Low Temperature Thermal Desorption | 3-203 |
| Soil Flushing | 3-288 |
| Soil Vapor Extraction | 3-293 |
| Soil Washing | 3-298 |

## Cost Data for Remediation -- Destruction Technologies

| | |
|---|---|
| Bioremediation, Water (Ex Situ) | 3-15 |
| Commercial Disposal (Incineration) | 3-65 |
| Composting | 3-69 |
| In Situ Biodegradation (Bioventing) | 3-146 |
| In Situ Biodegradation (Land Treatment) | 3-152 |
| In Situ Biodegradation (Saturated Zone) | 3-155 |
| In Situ Vitrification | 3-164 |
| Incineration (On Site) | 3-167 |
| Land Farming (Ex Situ) | 3-185 |
| Passive Water Treatment | 3-244 |
| Thermal & Catalytic Oxidation | 3-317 |

## Cost Data for Remediation -- Destruction Technologies (Continued)

| | |
|---|---|
| Bioslurping | 3-23 |
| Chemical Precipitation | 3-54 |
| Coagulation / Flocculation | 3-61 |
| Free Product Removal (French Drain) | 3-127 |
| Low Level Radioactive Soil Treatment | 3-197 |
| Media Filtration | 3-206 |
| Oil / Water Separation | 3-235 |
| Passive Water Treatment | 3-244 |
| Solvent Extraction | 3-309 |

## Cost Data for Remediation -- Isolation / Containment Technologies

| | |
|---|---|
| Capping | 3-33 |
| Extraction Wells | 3-104 |
| In Situ Solidification | 3-162 |
| In Situ Vitrification | 3-164 |
| Slurry Walls | 3-285 |
| Solidification / Stabilization | 3-304 |

## Cost Data for Remediation - Remediation Evaluation / Support

| | |
|---|---|
| Analysis | 3-21 |
| Bulk Material Storage | 3-28 |
| Decontamination Facilities | 3-75 |
| Dewatering (Sludge) | 3-82 |
| Drum Removal | 3-92 |
| Excavation, Buried Waste | 3-99 |
| Extraction Wells | 3-104 |
| Field Sampling / Mobile Laboratory | 3-117 |
| Groundwater Monitoring Wells | 3-136 |
| Monitoring | 3-209 |
| Neutralization | 3-232 |
| Ordnance & Explosive Waste Remediation | 3-240 |

**Environmental Remediation: Assemblies Cost Book**

# Table of Contents

## Cost Data for Remediation - Remediation Evaluation / Support
### (Continued)

| | |
|---|---|
| Piping | 3-253 |
| Professional Labor | 2-4 |
| Sampling | 2-9 |
| Storage Tank Installation | 3-311 |
| Transportation | 3-320 |
| UST Closure | 3-338 |
| Well Drilling & Installation | 3-341 |

## Cost Data for Site Work

| | |
|---|---|
| Access Roads | 4-1 |
| Arterial Roads / Divided Highways | 4-3 |
| Bridges | 4-5 |
| Cleanup and Landscaping | 4-8 |
| Clear and Grub | 4-10 |
| Communications | 4-12 |
| Demolition - Bridges | 4-14 |
| Demolition - Catch Basins / Manholes | 4-15 |
| Demolition - Curbs | 4-16 |
| Demolition - Fencing | 4-17 |
| Demolition - Pavements | 4-18 |
| Demolition - Pipes | 4-19 |
| Demolition - Sidewalks | 4-20 |
| Excavation, Cut and Fill | 4-21 |
| Excavation, Trench / Channel | 4-23 |
| Fencing | 4-24 |
| Gas Distribution | 4-25 |
| Heating / Cooling Distribution System | 4-27 |
| Lighting - Interstate, Roadway, Parking | 4-31 |
| Load and Haul | 4-32 |
| Materials Plant | 4-34 |
| Overhead Electrical Distribution | 4-35 |
| Parking Lots | 4-37 |
| Railroad Tracks and Crossings | 4-39 |
| Restriping Roadways / Parking Lots | 4-41 |
| Resurfacing Roadways / Parking Lots | 4-42 |
| Retaining Wall, CIP Concrete | 4-44 |
| Sanitary Sewer | 4-45 |
| Sidewalks | 4-48 |
| Sprinkler System | 4-49 |
| Storm Sewer | 4-50 |

## Cost Data for Site Work
### (Continued)

| | |
|---|---|
| Structures - Culverts | 4-57 |
| Treatment Plants / Lift Stations | 4-59 |
| Underground Electrical Distribution | 4-60 |
| Water Distribution | 4-62 |
| Water Storage Tanks | 4-65 |

## Contractor Costs

| | |
|---|---|
| Contractor Costs/General Conditions | 5-1 |
| Other General Conditions | 5-4 |
| State Sales Tax | 5-5 |

**Environmental Remediation: Assemblies Cost Book**

©1998 by ECHOS. All rights reserved

# How to Use This Book
## Estimating Forms and Instructions

Creating a cost estimate for environmental restoration, whether it be an RI/FS, RFI/CMS, or a Remedial Action, is a multi-step process. This section explains how to create a cost estimate using the ***Environmental Remediation: Assemblies Cost Book.*** Blank forms are included at the end of the section.

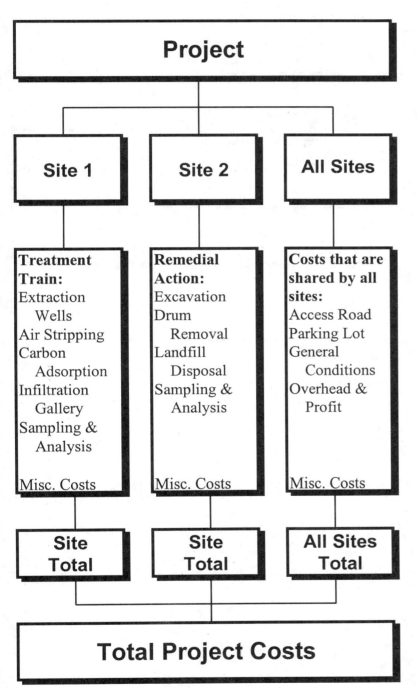

### Step #1
*Describe your project and estimate -- Project Name or ID, Location, Description, Estimator's Name, Preparation Date.*

### Step #2
*Identify your site(s) -- Projects may involve only one site or there may be many sites. For multiple-site projects, assign a Site ID to each site and then establish a "site" (shown here as "All Sites") for costs that pertain to all sites.*

### Step #3
*Identify the technologies, treatments, processes, etc. that are to be included in your estimate for each site.*

### Step #4
*Estimate the cost of each technology, treatment, process, etc. for each site. The estimates are prepared by selecting and quantifying assemblies from this book.*

### Step #5
*Identify any miscellaneous costs that were not included in the estimates.*

### Step #6
*Compute the total costs for each site (technologies, treatments, processes, and miscellaneous costs).*

### Step #7
*Sum the costs for all sites and adjust for location and escalation. Add design costs (if applicable), overhead and profit, and contingencies.*

---

**Environmental Remediation: Assemblies Cost Book**

1998 by ECHOS. All rights reserved.

# How to Use This Book
## Estimating Forms and Instructions

## *Step #1 -- Project Description*

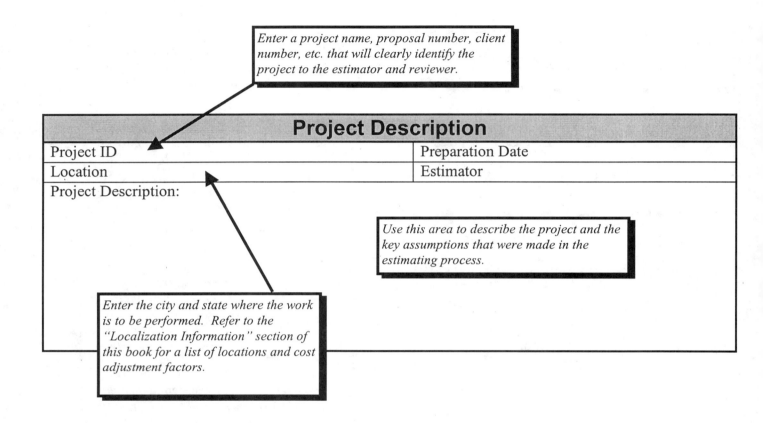

**Environmental Remediation: Assemblies Cost Book**

# How to Use This Book
## Estimating Forms and Instructions

## Step #2 -- Description of Site(s)

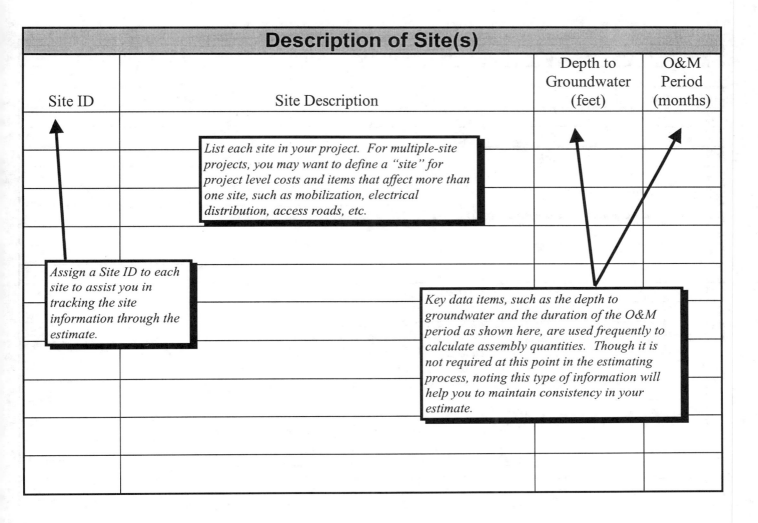

**Environmental Remediation: Assemblies Cost Book**

**How to Use This Book**
**Estimating Forms and Instructions**

# *Step #3 -- Treatment Trains*

| Treatment Trains | | | | | |
|---|---|---|---|---|---|
| Site ID | | Site ID | | Site ID | |
| Treatment Train: | | Treatment Train: | | Treatment Train: | |
| 1. | | 1. | | 1. | |
| 2. | | 2. | | 2. | |
| 3. | | 3. | | 3. | |
| 4. | | 4. | | 4. | |
| 5. | | 5. | | 5. | |
| 6. | | 6. | | 6. | |
| 7. | | 7. | | 7. | |
| 8. | | 8. | | 8. | |
| 9. | | 9. | | 9. | |
| 10. | | 10. | | 10. | |
| 11. | | 11. | | 11. | |
| 12. | | | | | |
| 13. | | | | | |
| 14. | | | | | |
| 15. | | | | | |
| 16. | | | | | |
| 17. | | | | | |
| 18. | | | | | |
| 19. | | | | | |
| 20. | | | | | |

*Enter the Site IDs that you established in Step #2. Use additional forms if needed.*

*List the technologies, treatments, processes, etc. that are to be estimated for each site. Be sure that you define a* complete treatment train *(including sampling and analysis, disposal, and treatment of any residuals). Also list any required* site work and utilities.

*For assistance in determining a treatment train, refer to the cost sections of this book. For help in getting started, refer to the Table of Contents which contains a categorized list of technologies, processes, treatments, site work, and utilities.*

---

**Environmental Remediation: Assemblies Cost Book**

Page 1-4

©1998 by ECHOS. All rights reserved.

# How to Use This Book
## Estimating Forms and Instructions

# *Step #4 -- Assembly Cost Estimates*

**IMPORTANT**
1. *Complete this step for each technology, process, treatment, etc., that you identified for the site.*
2. *Repeat the process for each site.*

3. *Enter the Site ID that you established in Step #2 and the technology that you identified in Step #3.*
4. *Locate the technology/process in this book.*

| Cost Estimate | Site ID _____<br>Technology/Process _____ | | | | |
|---|---|---|---|---|---|
| Assembly | Safety Level | Unit Cost | Quantity | Unit of Measure | Total Cost |
| | | | | | |
| | | | | | |
| | | | | | |
| | | | | | |
| | | | | | |
| | | | | | |
| | | | | | |
| | | | | | |
| | | | | | |

5. *Identify the items of work, equipment, and materials that are required for the technology or process. Note that many technologies have both capital costs (construction and start up) and O&M (operations and maintenance) costs.*
6. *Enter the assembly numbers or descriptions in this space.*

7. *Enter the safety level for which the item of work is to be performed, then enter the unit cost that corresponds to that safety level.*
8. *Enter the quantity and unit of measure for each assembly, making sure that the quantity you enter is consistent with the unit of measure for the assembly.*
9. *Multiply the quantity by the unit cost and enter the result.*

**Notes:**

*This space is provided for important comments and assumptions that affect your cost estimate.*

**RI/FS and RFI/CMS:** *For professional labor, use one form for each task. For sampling and analysis, use one form per media, per round (i.e., Groundwater, Round 1). Use separate forms for the installation of monitoring wells and other costs (geophysical investigations, pilot tests, etc.).*

## Environmental Remediation: Assemblies Cost Book

Page 1-5

©1998 by ECHOS. All rights reserved.

# How to Use This Book
## Estimating Forms and Instructions

## Step #5 -- Misc. Costs for the Site(s) (Optional)

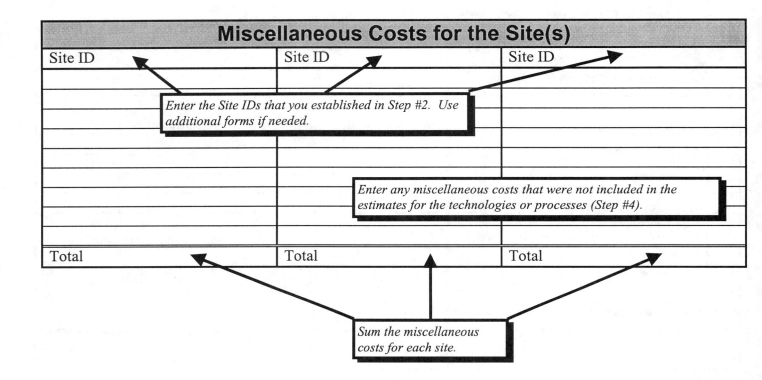

**Environmental Remediation: Assemblies Cost Book**

## How to Use This Book
### Estimating Forms and Instructions

# *Step #6 -- Site Totals*

| Site Totals | | | | | |
|---|---|---|---|---|---|
| **Site ID** | | **Site ID** | | **Site ID** | |
| Process | Cost | Process | Cost | Process | Cost |
| | | | | | |
| | | | | | |
| | | | | | |
| | | | | | |
| | | | | | |
| | | | | | |
| | | | | | |
| | | | | | |
| | | | | | |
| | | | | | |
| | | | | | |
| | | | | | |
| | | | | | |
| | | | | | |
| Misc. Costs | | Misc. Costs | | Misc. Costs | |
| Total | | Total | | Total | |

*Using as many forms as necessary, enter the following for each site:*
- *The Site ID that you established in Step #2*
- *The technologies and processes that you identified in Step #3*
- *The cost for each technology that you estimated in Step #4*
- *The miscellaneous costs that you estimated in Step #5*

*Calculate the total costs for each site.*

**IMPORTANT:** *If you estimated Overhead and Profit using assemblies, do not enter that amount on this form. Overhead and Profit, which can be estimated either by the assembly method or by a simple percentage, should be entered on the form for Step #7.*

---

## Environmental Remediation:  Assemblies Cost Book

Page 1-7

©1998 by ECHOS.  All rights reserved.

# How to Use This Book
## Estimating Forms and Instructions

# *Step #7 -- Project Summary*

| Project Summary | | |
|---|---|---|
| | + | |
| *1. Using one line per site, enter the Site ID (Step #2) and the total cost (Step #6).* | | |
| | + | |
| | + | |
| | + | |
| | + | |
| | + | |
| | + | |
| | + | |
| Total for All Sites | $ | *2. Compute the sum for all sites.* |
| Location Multiplier | X | 3 |
| Site Total, Adjusted for Location | $ | 4 |
| Escalation to ___/___/___ | + | 5 |
| Escalated Costs | $ | 6 |
| Design (___%) | + | 7 |
| Overhead and Profit (___%) | + | 8 |
| Contingencies (___%) | + | 9 |
| Project Total | $ | 10 |

**Calculate Total Project Costs:**

3. *Find your location (Step #1) in the "Location Information" section of this book. If your location is not listed, consider nearby locations, the state average, or a combination of locations. Enter the location multiplier on this form.*
4. *Multiply the Site Total by the location multiplier and enter the result.*
5. *Enter the escalation date and the dollar amount of the escalation.*
6. *Subtotal (Escalated Costs).*
7. *If your project does not include design, enter zeros here. Otherwise, enter a percentage and calculate the dollar value using the Escalated Costs.*
8. *If you estimated overhead and profit using the assembly method, enter the dollar amount here and calculate the percentage. Otherwise, enter a percentage here and calculate the dollar amount.*
9. *Using a percentage, estimate contingencies for your project. Enter the percentage and the dollar amount on the form.*
10. *Compute the total project cost by summing the Escalated Costs, Design Costs (if any), Overhead & Profit, and Contingencies.*

---

## Environmental Remediation: Assemblies Cost Book

# How to Use This Book
## Estimating Forms and Instructions

### Project Description

| Project ID | | Preparation Date | |
|---|---|---|---|
| Location | | Estimator | |

Project Description:

### Description of Site(s)

| Site ID | Site Description | Depth to Groundwater (feet) | O&M Period (Months) |
|---|---|---|---|
| | | | |
| | | | |
| | | | |
| | | | |
| | | | |
| | | | |
| | | | |
| | | | |
| | | | |
| | | | |
| | | | |

**Environmental Remediation: Assemblies Cost Book**

Page 1-9

©1998 by ECHOS. All rights reserved.

## How to Use This Book
## Estimating Forms and Instructions

| Treatment Trains | | |
|---|---|---|
| Site ID | Site ID | Site ID |
| Treatment Train: | Treatment Train: | Treatment Train: |
| 1. | 1. | 1. |
| 2. | 2. | 2. |
| 3. | 3. | 3. |
| 4. | 4. | 4. |
| 5. | 5. | 5. |
| 6. | 6. | 6. |
| 7. | 7. | 7. |
| 8. | 8. | 8. |
| 9. | 9. | 9. |
| 10. | 10. | 10. |
| 11. | 11. | 11. |
| 12. | 12. | 12. |
| 13. | 13. | 13. |
| 14. | 14. | 14. |
| 15. | 15. | 15. |
| 16. | 16. | 16. |
| 17. | 17. | 17. |
| 18. | 18. | 18. |
| 19. | 19. | 19. |
| 20. | 20. | 20. |

**Environmental Remediation: Assemblies Cost Book**

# How to Use This Book
## Estimating Forms and Instructions

| Cost Estimate | Site ID _____ Technology/Process _____ | | | | |
|---|---|---|---|---|---|
| Assembly | Safety Level | Unit Cost | Quantity | Unit of Measure | Total Cost |
|  |  |  |  |  |  |
|  |  |  |  |  |  |
|  |  |  |  |  |  |
|  |  |  |  |  |  |
|  |  |  |  |  |  |
|  |  |  |  |  |  |
|  |  |  |  |  |  |
|  |  |  |  |  |  |
|  |  |  |  |  |  |
|  |  |  |  |  |  |
|  |  |  |  |  |  |
|  |  |  |  |  |  |
|  |  |  |  |  |  |
|  |  |  |  |  |  |
|  |  |  |  |  |  |
|  |  |  |  |  |  |
|  |  |  |  |  |  |
|  |  |  |  |  |  |
|  |  |  |  |  |  |
|  |  |  |  |  |  |
|  |  |  |  |  |  |
|  |  |  |  |  |  |
|  |  |  |  |  |  |
|  |  |  |  |  |  |
|  |  |  |  |  |  |
|  |  |  |  |  |  |
|  |  |  |  |  |  |
|  |  |  |  |  |  |
|  |  |  |  |  |  |
|  |  |  |  |  |  |
|  |  |  |  |  |  |

**Notes:**

## Environmental Remediation: Assemblies Cost Book

Page 1-11

©1998 by ECHOS. All rights reserved.

# How to Use This Book
## Estimating Forms and Instructions

### Miscellaneous Costs for the Site(s)

| Site ID | Site ID | Site ID |
|---|---|---|
|  |  |  |
|  |  |  |
|  |  |  |
|  |  |  |
|  |  |  |
|  |  |  |
|  |  |  |
|  |  |  |
| Total | Total | Total |

### Site Totals

| Site ID | | Site ID | | Site ID | |
|---|---|---|---|---|---|
| Process | Cost | Process | Cost | Process | Cost |
|  |  |  |  |  |  |
|  |  |  |  |  |  |
|  |  |  |  |  |  |
|  |  |  |  |  |  |
|  |  |  |  |  |  |
|  |  |  |  |  |  |
|  |  |  |  |  |  |
|  |  |  |  |  |  |
|  |  |  |  |  |  |
|  |  |  |  |  |  |
|  |  |  |  |  |  |
|  |  |  |  |  |  |
|  |  |  |  |  |  |
|  |  |  |  |  |  |
| Misc. Costs |  | Misc. Costs |  | Misc. Costs |  |
| Total |  | Total |  | Total |  |

### Environmental Remediation: Assemblies Cost Book

# How to Use This Book
## Estimating Forms and Instructions

| Project Summary | |
|---|---|
| | + |
| | + |
| | + |
| | + |
| | + |
| | + |
| | + |
| | + |
| | + |
| | + |
| | + |
| | + |
| | + |
| | + |
| | + |
| | + |
| | + |
| | + |
| | + |
| Total for All Sites | $ |
| Location Multiplier | X |
| Site Total, Adjusted for Location | $ |
| Escalation to ___/___/___ | + |
| Escalated Costs | $ |
| Design (___%) | + |
| Overhead and Profit (___%) | + |
| Contingencies (___%) | + |
| Project Total | $ |

**Environmental Remediation: Assemblies Cost Book**

## How to Use This Book
## How to Interpret the Cost Pages

| Assembly | Description | Unit | Unit Cost by Safety Level | | | | |
|---|---|---|---|---|---|---|---|
| | | | A | B | C | D | E |
| | **Capital Costs** | | | | | | |
| 99 99 9993 | This could be one or more lines of description | EA | 52,888.24 | 50,000.79 | 47,222.22 | 33,333.33 | 22,222.22 |
| 99 99 9994 | Assembly title such as 2" diameter stainless steel piping | LF | 2.00 | 1.65 | 1.04 | 0.86 | 0.52 |
| 99 99 9995 etc. | Another title etc. | SF etc. | 4.00 | | | | |
| | **Operations & Maintenance** | | | | | | |
| 99 99 9999 | assembly description | CY | 9.99 | 9.99 | 9.99 | 9.99 | 9.99 |
| ①1 | ②2 | ③3 | ④4 | ⑤5 | ⑥6 | ⑦7 | ⑧8 |

**Column 1**  **Assembly Number**

This eight-digit number identifies the "assembly" for which the costs are shown. The assembly number is the link between this book and the ***Environmental Remediation: Cost Data – Unit Price Book***, which provides the line item(s) that are included in each assembly as well as the labor, material, and equipment costs for each line item.

**Column 2**  **Assembly Description**

This text describes the assembly. Again, for more detailed information about an assembly, refer to the ***Environmental Remediation: Cost Data – Unit Price Book***.

**Column 3**  **Unit of Measure**

A unit of measure is shown for each assembly. If you are not familiar with the abbreviations used, refer to the List of Abbreviations included in this book.

**Columns 4 - 8**  **Unit Costs at Safety Levels A, B, C, D, and E**

These unit costs represent the sum of the labor, equipment, and materials costs for the assembly. The costs have been adjusted to reflect the reduced productivity of labor and equipment that result from safety level requirements, as follows:

| Safety Level | Labor Productivity | Equipment Productivity |
|---|---|---|
| A | 37% (typically 10-40%) | 50% |
| B | 48% (typically 25-60%) | 60% |
| C | 55% (typically 25-70%) | 75% |
| D | 82% (typically 50-90%) | 100% |
| E (Non-Hazardous) | 100% (basis) | 100% (basis) |

## Environmental Remediation:  Assemblies Cost Book

## How to Use This Book
## How to Interpret the Cost Pages

### Analysis Assemblies -- Water/Liquid

| Assembly | Description | Unit Cost, by Turnaround Time/Quality Control Level | | | |
|---|---|---|---|---|---|
| | | 24-72 hr CLP | 24-72 hr Std QC | Std TAT CLP | Std TAT Std QC |
| 33021601 | Specific Conductance (EPA 120.1) | 24.00 | 20.00 | 14.00 | 10.00 |
| 33021602 | pH (EPA 150.1) | 24.00 | 20.00 | 14.00 | 10.00 |
| 33021603 | Total Dissolved Solids (EPA 160.1) | 24.00 | 20.00 | 14.00 | 10.00 |
| 33021604 | Total Suspended Solids (EPA 160.2) | 24.00 | 20.00 | 14.00 | 10.00 |
| 33021605 | Total Solids (EPA 160.3) | 24.00 | 20.00 | 14.00 | 10.00 |
| 1 | 2 | 3 | 4 | 5 | 6 |

**Column 1**

### Assembly Number

This eight-digit number identifies the analysis for which the costs are shown. The assembly number is the link between this book and the *Environmental Remediation: Cost Data – Unit Price Book*, which provides the line item(s) that are included in each assembly. Unlike the assemblies shown on the previous page, analysis assemblies are material costs only.

**Column 2**

### Assembly Description

This text describes the analysis. Again, for more detailed information about an assembly, refer to the *Environmental Remediation: Cost Data – Unit Price Book*.

**Columns 3 - 6**

### Unit Costs

Four unit costs are provided for each analysis, each reflecting the premium charged for nonstandard Turnaround Time (TAT) and Quality Control (QC) Levels, as follows:

| Turnaround Time | TAT Markups |
|---|---|
| Standard (2-3 weeks) | 0% (cost basis) |
| 14 Days | 20% |
| 4-7 Days | 50% |
| 24-72 Hours | 100% |

| Quality Control Level | QC Markups |
|---|---|
| Level 1 (Standard) | 0% (cost basis) |
| Level 2 | 10% |
| Level 3 | 25% |
| Level 4 (CLP) | 40% |

### Environmental Remediation: Assemblies Cost Book

# How to Use This Book

## List of Abbreviations

| | |
|---|---|
| ASME | American Society of Mechanical Engineers |
| BCY | Bank Cubic Yards |
| BOD | Biochemical Oxygen Demand |
| BTEX | Benzene, Toluene, Ethylbenzene, Xylene |
| CaCl2 | Calcium Chloride |
| CEC | Cation Exchange Capacity |
| CERCLA | Comprehensive Environmental Response Compensation and Liability Act |
| CF | Cubic Feet |
| CFM | Cubic Feet per Minute |
| CI | Cast Iron |
| CIP | Cast-In-Place |
| CLP | Certified Laboratory Program |
| CMP | Corrugated Metal Pipe |
| COD | Chemical Oxygen Demand |
| CPT | Cone Penetrometer Test |
| CY | Cubic Yards |
| Decon | Decontamination |
| Demob | Demobilization |
| Dia | Diameter |
| DIP | Ductile-Iron Pipe |
| DOT | Department of Transportation |
| DRM | Drums |
| EA | Each |
| EPA | Environmental Protection Agency |
| ESVCP | Extra Strength Vitrified Clay Pipe |
| FML | Fabric Membrane Liner |
| FRP | Fiberglass Reinforced Plastic |
| FT | Feet |
| GAL | Gallons |
| GPD | Gallons per Day |
| GPH | Gallons per Hour |
| GPM | Gallons per Minute |
| HDPE | High-Density Polyethylene |
| HP | Horse Power |
| ISV | In Situ Vitrification |
| KGAL | Thousand Gallons |
| KWH | Kilowatt Hours |
| LB | Pounds |
| LF | Linear Feet |
| MBH | Thousand BTUs per Hour |
| MCF | Thousand Cubic Feet |
| MI | Miles |
| MO | Months |
| Mob | Mobilization |
| MWK | Man-weeks |

**Environmental Remediation: Assemblies Cost Book**

# How to Use This Book

## List of Abbreviations

| | |
|---|---|
| **NAPL** | Nonaqueous Phase Liquid |
| **NPL** | National Priority List |
| **NPT** | National Pipe Thread |
| **O & M** | Operations and Maintenance |
| **OD** | Outside Diameter |
| **ODCs** | Other Direct Costs |
| **OH** | Overhead |
| **OSHA** | Occupational Safety and Health Administration |
| **OVA** | Organic Vapor Analyzer |
| **PE** | Polyethylene |
| **POTW** | Publicly-Owned Treatment Works |
| **PPE** | Personal Protective Equipment |
| **PPHPR** | Parts per HundredPair |
| **PSF** | Pounds per Square Foot |
| **PSI** | Pounds per Square Inch |
| **QA** | Quality Assurance |
| **QC** | Quality Control |
| **RA** | Remedial Action |
| **RCP** | Reinforced Concrete Pipe |
| **RCRA** | Resource Conservation and Recovery Act |
| **RFI/CMS** | RCRA Facility Investigation / Corrective Measures Study |
| **RI/FS** | Remedial Investigation / Feasibility Study |
| **SCBASCF** | Self-contained Breathing ApparatusStandard Cubic Feet |
| **SCFM** | Standard Cubic Feet per Minute |
| **SF** | Square Feet |
| **SVE** | Soil Vapor Extraction |
| **SY** | Square Yards |
| **TAL** | Target Analyte List |
| **TAT** | Turn Around Time |
| **TCLP** | Toxicity Characteristic Leach Procedure |
| **TN** | Tons |
| **TOC** | Total Organic Carbon |
| **UST** | Underground Storage Tank |
| **VEP** | Vapor Extraction Point |
| **VLDPE** | Very Low Density Polyethylene |
| **VOC** | Volatile Organic Compound |
| **w/** | With |
| **w/o** | Without |
| **WK** | Weeks |
| **YR** | Years |

**Environmental Remediation: Assemblies Cost Book**

# How to Use This Book

## Localization Information - By Zip Code

| Alabama | | Arkansas | | California | |
|---|---|---|---|---|---|
| 350 | 0.85 | 716 | 0.80 | 922 | 1.08 |
| 351 | 0.85 | 717 | 0.72 | 923 | 1.07 |
| 352 | 0.85 | 718 | 0.76 | 924 | 1.07 |
| 354 | 0.79 | 719 | 0.71 | 925 | 1.09 |
| 355 | 0.77 | 720 | 0.81 | 926 | 1.08 |
| 356 | 0.83 | 721 | 0.81 | 927 | 1.08 |
| 357 | 0.83 | 722 | 0.81 | 928 | 1.11 |
| 358 | 0.83 | 723 | 0.81 | 930 | 1.10 |
| 359 | 0.82 | 724 | 0.81 | 931 | 1.09 |
| 360 | 0.80 | 725 | 0.78 | 932 | 1.06 |
| 361 | 0.80 | 726 | 0.79 | 933 | 1.06 |
| 362 | 0.75 | 727 | 0.68 | 934 | 1.10 |
| 363 | 0.79 | 728 | 0.77 | 935 | 1.07 |
| 364 | 0.80 | 729 | 0.80 | 936 | 1.09 |
| 365 | 0.84 | 749 | 0.82 | 937 | 1.09 |
| 366 | 0.84 | **California** | | 938 | 1.09 |
| 367 | 0.79 | 900 | 1.11 | 939 | 1.12 |
| 368 | 0.79 | 901 | 1.11 | 940 | 1.24 |
| 369 | 0.79 | 902 | 1.11 | 941 | 1.24 |
| **Alaska** | | 903 | 1.09 | 942 | 1.11 |
| 995 | 1.26 | 904 | 1.09 | 943 | 1.17 |
| 996 | 1.26 | 905 | 1.09 | 944 | 1.18 |
| 997 | 1.26 | 906 | 1.10 | 945 | 1.17 |
| 998 | 1.25 | 907 | 1.10 | 946 | 1.19 |
| 999 | 1.30 | 908 | 1.10 | 947 | 1.33 |
| **Arizona** | | 910 | 1.08 | 948 | 1.17 |
| 850 | 0.90 | 911 | 1.08 | 949 | 1.19 |
| 852 | 0.86 | 912 | 1.08 | 950 | 1.15 |
| 853 | 0.90 | 913 | 1.10 | 951 | 1.20 |
| 855 | 0.88 | 914 | 1.10 | 952 | 1.10 |
| 856 | 0.88 | 915 | 1.10 | 953 | 1.10 |
| 857 | 0.88 | 916 | 1.10 | 954 | 1.19 |
| 859 | 0.88 | 917 | 1.10 | 955 | 1.11 |
| 860 | 0.91 | 918 | 1.10 | 956 | 1.11 |
| 863 | 0.88 | 919 | 1.07 | 957 | 1.11 |
| 864 | 0.88 | 920 | 1.07 | 958 | 1.11 |
| 865 | 0.88 | 921 | 1.07 | 959 | 1.10 |

**Environmental Remediation: Assemblies Cost Book**

# How to Use This Book

## Localization Information - By Zip Code (Cont.)

| California | |
|---|---|
| 960 | 1.09 |
| 961 | 1.09 |

| Colorado | |
|---|---|
| 800 | 0.94 |
| 801 | 0.94 |
| 802 | 0.94 |
| 803 | 0.86 |
| 804 | 0.91 |
| 805 | 0.92 |
| 806 | 0.86 |
| 807 | 0.90 |
| 808 | 0.91 |
| 809 | 0.91 |
| 810 | 0.91 |
| 811 | 0.89 |
| 812 | 0.89 |
| 813 | 0.87 |
| 814 | 0.85 |
| 815 | 0.87 |
| 816 | 0.91 |

| Connecticut | |
|---|---|
| 60 | 1.06 |
| 61 | 1.06 |
| 62 | 1.06 |
| 63 | 1.05 |
| 64 | 1.05 |
| 65 | 1.06 |
| 66 | 1.05 |
| 67 | 1.06 |
| 68 | 1.05 |
| 69 | 1.07 |

| Delaware | |
|---|---|
| 197 | 1.00 |
| 198 | 0.99 |
| 199 | 1.00 |

| District of Columbia | |
|---|---|
| 200 | 0.95 |

| District of Columbia | |
|---|---|
| 201 | 0.95 |
| 202 | 0.95 |
| 203 | 0.95 |
| 204 | 0.95 |
| 205 | 0.95 |

| Florida | |
|---|---|
| 320 | 0.85 |
| 321 | 0.89 |
| 322 | 0.85 |
| 323 | 0.80 |
| 324 | 0.74 |
| 325 | 0.85 |
| 326 | 0.84 |
| 327 | 0.87 |
| 328 | 0.87 |
| 329 | 0.89 |
| 330 | 0.88 |
| 331 | 0.88 |
| 332 | 0.88 |
| 333 | 0.88 |
| 334 | 0.84 |
| 335 | 0.85 |
| 336 | 0.85 |
| 337 | 0.86 |
| 338 | 0.84 |
| 339 | 0.83 |
| 340 | 0.88 |
| 342 | 0.84 |
| 346 | 0.85 |
| 347 | 0.87 |
| 349 | 0.84 |

| Georgia | |
|---|---|
| 300 | 0.89 |
| 301 | 0.89 |
| 302 | 0.89 |
| 303 | 0.89 |
| 304 | 0.69 |

| Georgia | |
|---|---|
| 305 | 0.76 |
| 306 | 0.81 |
| 307 | 0.68 |
| 308 | 0.81 |
| 309 | 0.81 |
| 310 | 0.82 |
| 311 | 0.82 |
| 312 | 0.82 |
| 313 | 0.83 |
| 314 | 0.83 |
| 315 | 0.76 |
| 316 | 0.78 |
| 317 | 0.81 |
| 318 | 0.80 |
| 319 | 0.80 |
| 399 | 0.89 |

| Hawaii | |
|---|---|
| 967 | 1.23 |
| 968 | 1.23 |
| 969 | 0.82 |

| Idaho | |
|---|---|
| 832 | 0.94 |
| 833 | 0.81 |
| 834 | 0.85 |
| 835 | 1.02 |
| 836 | 0.94 |
| 837 | 0.94 |
| 838 | 0.92 |

| Illinois | |
|---|---|
| 600 | 1.07 |
| 601 | 1.07 |
| 602 | 1.07 |
| 603 | 1.07 |
| 604 | 1.06 |
| 605 | 1.07 |
| 606 | 1.11 |
| 609 | 1.00 |

**Environmental Remediation: Assemblies Cost Book**

# How to Use This Book

## Localization Information - By Zip Code  (Cont.)

| Illinois | |
|---|---|
| 610 | 1.02 |
| 611 | 1.02 |
| 612 | 0.96 |
| 613 | 0.99 |
| 614 | 0.97 |
| 615 | 1.00 |
| 616 | 1.00 |
| 617 | 0.99 |
| 618 | 1.00 |
| 619 | 1.00 |
| 620 | 0.99 |
| 621 | 0.99 |
| 622 | 0.99 |
| 623 | 0.94 |
| 624 | 0.96 |
| 625 | 0.97 |
| 626 | 0.98 |
| 627 | 0.98 |
| 628 | 0.98 |
| 629 | 0.96 |

| Indiana | |
|---|---|
| 424 | 0.91 |
| 460 | 0.92 |
| 461 | 0.95 |
| 462 | 0.95 |
| 463 | 0.98 |
| 464 | 0.98 |
| 465 | 0.90 |
| 466 | 0.90 |
| 467 | 0.91 |
| 468 | 0.91 |
| 469 | 0.90 |
| 470 | 0.88 |
| 471 | 0.88 |
| 472 | 0.89 |
| 473 | 0.90 |
| 474 | 0.90 |

| Indiana | |
|---|---|
| 475 | 0.93 |
| 476 | 0.93 |
| 477 | 0.93 |
| 478 | 0.94 |
| 479 | 0.91 |

| Iowa | |
|---|---|
| 500 | 0.92 |
| 501 | 0.92 |
| 502 | 0.92 |
| 503 | 0.92 |
| 504 | 0.82 |
| 505 | 0.80 |
| 506 | 0.84 |
| 507 | 0.84 |
| 508 | 0.86 |
| 509 | 0.92 |
| 510 | 0.84 |
| 511 | 0.84 |
| 512 | 0.79 |
| 513 | 0.81 |
| 514 | 0.83 |
| 515 | 0.88 |
| 516 | 0.77 |
| 520 | 0.87 |
| 521 | 0.82 |
| 522 | 0.91 |
| 523 | 0.91 |
| 524 | 0.91 |
| 525 | 0.86 |
| 526 | 0.80 |
| 527 | 0.94 |
| 528 | 0.94 |

| Kansas | |
|---|---|
| 660 | 0.93 |
| 661 | 0.93 |
| 662 | 0.93 |
| 664 | 0.86 |

| Kansas | |
|---|---|
| 665 | 0.86 |
| 666 | 0.86 |
| 667 | 0.86 |
| 668 | 0.83 |
| 669 | 0.84 |
| 670 | 0.86 |
| 671 | 0.86 |
| 672 | 0.86 |
| 673 | 0.81 |
| 674 | 0.83 |
| 675 | 0.78 |
| 676 | 0.84 |
| 677 | 0.84 |
| 678 | 0.85 |
| 679 | 0.78 |

| Kentucky | |
|---|---|
| 400 | 0.90 |
| 401 | 0.90 |
| 402 | 0.90 |
| 403 | 0.86 |
| 404 | 0.86 |
| 405 | 0.86 |
| 406 | 0.89 |
| 407 | 0.75 |
| 408 | 0.75 |
| 409 | 0.75 |
| 410 | 0.93 |
| 411 | 0.96 |
| 412 | 0.96 |
| 413 | 0.75 |
| 414 | 0.75 |
| 415 | 0.84 |
| 416 | 0.84 |
| 417 | 0.75 |
| 418 | 0.75 |
| 420 | 0.90 |
| 421 | 0.89 |

**Environmental Remediation:  Assemblies Cost Book**

# How to Use This Book
## Localization Information - By Zip Code  (Cont.)

| Kentucky | |
|---|---|
| 422 | 0.89 |
| 423 | 0.90 |
| 424 | 0.91 |
| 425 | 0.74 |
| 426 | 0.74 |
| 427 | 0.88 |

| Louisiana | |
|---|---|
| 700 | 0.86 |
| 701 | 0.86 |
| 703 | 0.86 |
| 704 | 0.84 |
| 705 | 0.83 |
| 706 | 0.85 |
| 707 | 0.84 |
| 708 | 0.84 |
| 710 | 0.81 |
| 711 | 0.81 |
| 712 | 0.80 |
| 713 | 0.79 |
| 714 | 0.79 |

| Maine | |
|---|---|
| 39 | 0.82 |
| 40 | 0.91 |
| 41 | 0.91 |
| 42 | 0.91 |
| 43 | 0.81 |
| 44 | 0.93 |
| 45 | 0.80 |
| 46 | 0.85 |
| 47 | 0.82 |
| 48 | 0.85 |
| 49 | 0.80 |

| Maryland | |
|---|---|
| 206 | 0.88 |
| 207 | 0.90 |
| 208 | 0.90 |
| 209 | 0.89 |

| Maryland | |
|---|---|
| 210 | 0.92 |
| 211 | 0.92 |
| 212 | 0.92 |
| 214 | 0.91 |
| 215 | 0.88 |
| 216 | 0.70 |
| 217 | 0.89 |
| 218 | 0.79 |
| 219 | 0.86 |

| Massachusetts | |
|---|---|
| 10 | 1.04 |
| 11 | 1.04 |
| 12 | 1.01 |
| 13 | 1.01 |
| 14 | 1.06 |
| 15 | 1.08 |
| 16 | 1.08 |
| 17 | 1.10 |
| 18 | 1.11 |
| 19 | 1.10 |
| 20 | 1.17 |
| 21 | 1.17 |
| 22 | 1.17 |
| 23 | 1.10 |
| 24 | 1.10 |
| 25 | 1.07 |
| 26 | 1.08 |
| 27 | 1.10 |

| Michigan | |
|---|---|
| 480 | 0.99 |
| 481 | 1.01 |
| 482 | 1.05 |
| 483 | 0.99 |
| 484 | 0.99 |
| 485 | 0.99 |
| 486 | 0.96 |
| 487 | 0.96 |

| Michigan | |
|---|---|
| 488 | 0.97 |
| 489 | 0.97 |
| 490 | 0.93 |
| 491 | 0.93 |
| 492 | 0.94 |
| 493 | 0.88 |
| 494 | 0.92 |
| 495 | 0.88 |
| 496 | 0.88 |
| 497 | 0.92 |
| 498 | 0.94 |
| 499 | 0.94 |

| Minnesota | |
|---|---|
| 540 | 0.93 |
| 550 | 1.08 |
| 551 | 1.08 |
| 553 | 1.10 |
| 554 | 1.10 |
| 556 | 1.04 |
| 557 | 1.04 |
| 558 | 1.04 |
| 559 | 1.02 |
| 560 | 0.98 |
| 561 | 0.88 |
| 562 | 0.88 |
| 563 | 1.01 |
| 564 | 0.99 |
| 565 | 0.96 |
| 566 | 0.98 |
| 567 | 0.94 |

| Mississippi | |
|---|---|
| 386 | 0.68 |
| 387 | 0.79 |
| 388 | 0.73 |
| 389 | 0.70 |
| 390 | 0.79 |
| 391 | 0.79 |

**Environmental Remediation:  Assemblies Cost Book**

# How to Use This Book
## Localization Information - By Zip Code  (Cont.)

| Mississippi | |
|---|---|
| 392 | 0.79 |
| 393 | 0.77 |
| 394 | 0.70 |
| 395 | 0.81 |
| 396 | 0.68 |
| 397 | 0.73 |

| Missouri | |
|---|---|
| 630 | 1.02 |
| 631 | 1.02 |
| 633 | 0.96 |
| 634 | 0.95 |
| 635 | 0.90 |
| 636 | 0.99 |
| 637 | 0.98 |
| 638 | 0.93 |
| 639 | 0.92 |
| 640 | 0.95 |
| 641 | 0.95 |
| 644 | 0.92 |
| 645 | 0.92 |
| 646 | 0.84 |
| 647 | 0.93 |
| 648 | 0.87 |
| 650 | 0.92 |
| 651 | 0.92 |
| 652 | 0.89 |
| 653 | 0.90 |
| 654 | 0.83 |
| 655 | 0.83 |
| 656 | 0.87 |
| 657 | 0.87 |
| 658 | 0.87 |

| Montana | |
|---|---|
| 590 | 0.96 |
| 591 | 0.96 |
| 592 | 0.95 |
| 593 | 0.96 |

| Montana | |
|---|---|
| 594 | 0.96 |
| 595 | 0.94 |
| 596 | 0.96 |
| 597 | 0.94 |
| 598 | 0.94 |
| 599 | 0.93 |

| Nebraska | |
|---|---|
| 680 | 0.89 |
| 681 | 0.89 |
| 683 | 0.84 |
| 684 | 0.84 |
| 685 | 0.84 |
| 686 | 0.75 |
| 687 | 0.85 |
| 688 | 0.84 |
| 689 | 0.81 |
| 690 | 0.72 |
| 691 | 0.81 |
| 692 | 0.76 |
| 693 | 0.73 |

| Nevada | |
|---|---|
| 889 | 1.03 |
| 890 | 1.03 |
| 891 | 1.03 |
| 893 | 0.97 |
| 894 | 0.98 |
| 895 | 0.98 |
| 897 | 0.98 |
| 898 | 0.96 |

| New Hampshire | |
|---|---|
| 30 | 0.97 |
| 31 | 0.97 |
| 32 | 0.96 |
| 33 | 0.96 |
| 34 | 0.83 |
| 35 | 0.85 |
| 36 | 0.81 |

| New Hampshire | |
|---|---|
| 37 | 0.81 |
| 38 | 0.94 |

| New Jersey | |
|---|---|
| 70 | 1.13 |
| 71 | 1.13 |
| 72 | 1.09 |
| 73 | 1.12 |
| 74 | 1.13 |
| 75 | 1.13 |
| 76 | 1.10 |
| 77 | 1.09 |
| 78 | 1.11 |
| 79 | 1.08 |
| 80 | 1.07 |
| 81 | 1.09 |
| 82 | 1.08 |
| 83 | 1.07 |
| 84 | 1.08 |
| 85 | 1.12 |
| 86 | 1.12 |
| 87 | 1.10 |
| 88 | 1.10 |
| 89 | 1.10 |

| New Mexico | |
|---|---|
| 870 | 0.91 |
| 871 | 0.91 |
| 872 | 0.91 |
| 873 | 0.92 |
| 874 | 0.92 |
| 875 | 0.91 |
| 877 | 0.91 |
| 878 | 0.91 |
| 879 | 0.88 |
| 880 | 0.85 |
| 881 | 0.91 |
| 882 | 0.91 |
| 883 | 0.92 |

**Environmental Remediation:  Assemblies Cost Book**

# How to Use This Book

## Localization Information - By Zip Code  (Cont.)

| New Mexico | | | New York | | | North Dakota | |
|---|---|---|---|---|---|---|---|
| 884 | 0.91 | | 135 | 0.96 | | 580 | 0.84 |
| New York | | | 136 | 0.98 | | 581 | 0.84 |
| 100 | 1.34 | | 137 | 0.96 | | 582 | 0.84 |
| 101 | 1.34 | | 138 | 0.96 | | 583 | 0.84 |
| 102 | 1.34 | | 139 | 0.96 | | 584 | 0.84 |
| 103 | 1.29 | | 140 | 1.04 | | 585 | 0.85 |
| 104 | 1.28 | | 141 | 1.04 | | 586 | 0.84 |
| 105 | 1.21 | | 142 | 1.04 | | 587 | 0.85 |
| 106 | 1.20 | | 143 | 1.04 | | 588 | 0.83 |
| 107 | 1.24 | | 144 | 1.03 | | Ohio | |
| 108 | 1.22 | | 145 | 1.03 | | 430 | 0.94 |
| 109 | 1.15 | | 146 | 1.03 | | 431 | 0.94 |
| 110 | 1.28 | | 147 | 0.93 | | 432 | 0.94 |
| 111 | 1.29 | | 148 | 0.95 | | 433 | 0.92 |
| 112 | 1.28 | | 149 | 0.95 | | 434 | 0.97 |
| 113 | 1.30 | | North Carolina | | | 435 | 0.97 |
| 114 | 1.28 | | 270 | 0.79 | | 436 | 0.97 |
| 115 | 1.26 | | 271 | 0.78 | | 437 | 0.90 |
| 116 | 1.29 | | 272 | 0.79 | | 438 | 0.90 |
| 117 | 1.26 | | 273 | 0.79 | | 439 | 0.96 |
| 118 | 1.28 | | 274 | 0.79 | | 440 | 0.96 |
| 119 | 1.27 | | 275 | 0.79 | | 441 | 1.01 |
| 120 | 0.98 | | 276 | 0.79 | | 442 | 0.99 |
| 121 | 0.98 | | 277 | 0.79 | | 443 | 0.99 |
| 122 | 0.98 | | 278 | 0.67 | | 444 | 0.96 |
| 123 | 0.99 | | 279 | 0.67 | | 445 | 0.96 |
| 124 | 1.11 | | 280 | 0.77 | | 446 | 0.95 |
| 125 | 1.13 | | 281 | 0.78 | | 447 | 0.95 |
| 126 | 1.13 | | 282 | 0.78 | | 448 | 0.93 |
| 127 | 1.10 | | 283 | 0.78 | | 449 | 0.93 |
| 128 | 0.94 | | 284 | 0.77 | | 450 | 0.93 |
| 129 | 0.95 | | 285 | 0.66 | | 451 | 0.92 |
| 130 | 0.99 | | 286 | 0.65 | | 452 | 0.92 |
| 131 | 0.99 | | 287 | 0.78 | | 453 | 0.92 |
| 132 | 0.99 | | 288 | 0.78 | | 454 | 0.92 |
| 133 | 0.96 | | 289 | 0.67 | | 455 | 0.92 |
| 134 | 0.96 | | North Dakota | | | 456 | 0.94 |

**Environmental Remediation:  Assemblies Cost Book**

# How to Use This Book

## Localization Information - By Zip Code (Cont.)

| Ohio | |
|---|---|
| 457 | 0.88 |
| 458 | 0.92 |

| Oklahoma | |
|---|---|
| 730 | 0.83 |
| 731 | 0.83 |
| 734 | 0.82 |
| 735 | 0.83 |
| 736 | 0.82 |
| 737 | 0.82 |
| 738 | 0.81 |
| 739 | 0.70 |
| 740 | 0.83 |
| 741 | 0.83 |
| 743 | 0.82 |
| 744 | 0.75 |
| 745 | 0.78 |
| 746 | 0.82 |
| 747 | 0.81 |
| 748 | 0.80 |

| Oregon | |
|---|---|
| 970 | 1.07 |
| 971 | 1.07 |
| 972 | 1.07 |
| 973 | 1.06 |
| 974 | 1.05 |
| 975 | 1.05 |
| 976 | 1.05 |
| 977 | 1.05 |
| 978 | 1.02 |
| 979 | 0.97 |

| Pennsylvania | |
|---|---|
| 150 | 1.03 |
| 151 | 1.03 |
| 152 | 1.03 |
| 153 | 1.01 |
| 154 | 1.00 |
| 155 | 0.97 |

| Pennsylvania | |
|---|---|
| 156 | 1.01 |
| 157 | 0.99 |
| 158 | 0.97 |
| 159 | 0.98 |
| 160 | 0.98 |
| 161 | 0.98 |
| 162 | 1.00 |
| 163 | 0.95 |
| 164 | 0.97 |
| 165 | 0.97 |
| 166 | 0.97 |
| 167 | 0.97 |
| 168 | 0.98 |
| 169 | 0.96 |
| 170 | 0.98 |
| 171 | 0.98 |
| 172 | 0.96 |
| 173 | 0.97 |
| 174 | 0.97 |
| 175 | 0.96 |
| 176 | 0.96 |
| 177 | 0.93 |
| 178 | 0.96 |
| 179 | 0.96 |
| 180 | 1.03 |
| 181 | 1.02 |
| 182 | 0.97 |
| 183 | 0.97 |
| 184 | 1.01 |
| 185 | 1.01 |
| 186 | 0.97 |
| 187 | 0.97 |
| 188 | 0.98 |
| 189 | 1.06 |
| 190 | 1.10 |
| 191 | 1.10 |
| 193 | 1.05 |

| Pennsylvania | |
|---|---|
| 194 | 1.07 |
| 195 | 0.99 |
| 196 | 0.99 |

| Rhode Island | |
|---|---|
| 28 | 1.06 |
| 29 | 1.06 |

| South Carolina | |
|---|---|
| 290 | 0.77 |
| 291 | 0.77 |
| 292 | 0.77 |
| 293 | 0.76 |
| 294 | 0.78 |
| 295 | 0.75 |
| 296 | 0.76 |
| 297 | 0.69 |
| 298 | 0.70 |
| 299 | 0.72 |

| South Dakota | |
|---|---|
| 570 | 0.83 |
| 571 | 0.83 |
| 572 | 0.81 |
| 573 | 0.80 |
| 574 | 0.82 |
| 575 | 0.81 |
| 576 | 0.81 |
| 577 | 0.80 |

| Tennessee | |
|---|---|
| 370 | 0.85 |
| 371 | 0.85 |
| 372 | 0.85 |
| 373 | 0.84 |
| 374 | 0.84 |
| 375 | 0.86 |
| 376 | 0.81 |
| 377 | 0.81 |
| 378 | 0.81 |
| 379 | 0.81 |

**Environmental Remediation: Assemblies Cost Book**

# How to Use This Book

## Localization Information - By Zip Code  (Cont.)

| Tennessee | |
|---|---|
| 380 | 0.86 |
| 381 | 0.86 |
| 382 | 0.71 |
| 383 | 0.77 |
| 384 | 0.78 |
| 385 | 0.70 |

| Texas | |
|---|---|
| 750 | 0.84 |
| 751 | 0.84 |
| 752 | 0.86 |
| 753 | 0.86 |
| 754 | 0.75 |
| 755 | 0.80 |
| 756 | 0.76 |
| 757 | 0.81 |
| 758 | 0.76 |
| 759 | 0.79 |
| 760 | 0.83 |
| 761 | 0.83 |
| 762 | 0.81 |
| 763 | 0.82 |
| 764 | 0.76 |
| 765 | 0.79 |
| 766 | 0.82 |
| 767 | 0.82 |
| 768 | 0.74 |
| 769 | 0.77 |
| 770 | 0.90 |
| 771 | 0.90 |
| 772 | 0.90 |
| 773 | 0.76 |
| 774 | 0.79 |
| 775 | 0.88 |
| 776 | 0.87 |
| 777 | 0.87 |
| 778 | 0.83 |
| 779 | 0.82 |

| Texas | |
|---|---|
| 780 | 0.80 |
| 781 | 0.84 |
| 782 | 0.84 |
| 783 | 0.81 |
| 784 | 0.81 |
| 785 | 0.79 |
| 786 | 0.83 |
| 787 | 0.83 |
| 788 | 0.70 |
| 789 | 0.75 |
| 790 | 0.82 |
| 791 | 0.82 |
| 792 | 0.80 |
| 793 | 0.82 |
| 794 | 0.82 |
| 795 | 0.79 |
| 796 | 0.79 |
| 797 | 0.82 |
| 798 | 0.78 |
| 799 | 0.78 |
| 885 | 0.78 |

| Utah | |
|---|---|
| 840 | 0.87 |
| 841 | 0.87 |
| 842 | 0.86 |
| 843 | 0.87 |
| 844 | 0.86 |
| 845 | 0.82 |
| 846 | 0.88 |
| 847 | 0.88 |

| Vermont | |
|---|---|
| 50 | 0.75 |
| 51 | 0.75 |
| 52 | 0.71 |
| 53 | 0.76 |
| 54 | 0.86 |
| 56 | 0.85 |

| Vermont | |
|---|---|
| 57 | 0.86 |
| 58 | 0.78 |
| 59 | 0.77 |

| Virginia | |
|---|---|
| 220 | 0.90 |
| 221 | 0.90 |
| 222 | 0.90 |
| 223 | 0.91 |
| 224 | 0.86 |
| 225 | 0.86 |
| 226 | 0.81 |
| 227 | 0.80 |
| 228 | 0.76 |
| 229 | 0.82 |
| 230 | 0.83 |
| 231 | 0.83 |
| 232 | 0.83 |
| 233 | 0.82 |
| 234 | 0.82 |
| 235 | 0.82 |
| 236 | 0.82 |
| 237 | 0.81 |
| 238 | 0.83 |
| 239 | 0.75 |
| 240 | 0.78 |
| 241 | 0.78 |
| 242 | 0.76 |
| 243 | 0.71 |
| 244 | 0.72 |
| 245 | 0.78 |
| 246 | 0.71 |

| Washington | |
|---|---|
| 980 | 1.05 |
| 981 | 1.05 |
| 982 | 1.04 |
| 983 | 1.05 |
| 984 | 1.05 |

**Environmental Remediation:  Assemblies Cost Book**

# How to Use This Book
## Localization Information - By Zip Code  (Cont.)

| Washington | |
|---|---|
| 985 | 1.04 |
| 986 | 1.04 |
| 987 | 1.05 |
| 988 | 1.01 |
| 989 | 1.02 |
| 990 | 1.00 |
| 991 | 1.00 |
| 992 | 1.00 |
| 993 | 1.01 |
| 994 | 0.99 |

| West Virginia | |
|---|---|
| 247 | 0.86 |
| 248 | 0.86 |
| 249 | 0.90 |
| 250 | 0.94 |
| 251 | 0.94 |
| 252 | 0.94 |
| 253 | 0.94 |
| 254 | 0.79 |
| 255 | 0.96 |
| 256 | 0.96 |
| 257 | 0.96 |
| 258 | 0.92 |
| 259 | 0.92 |
| 260 | 0.95 |
| 261 | 0.94 |
| 262 | 0.94 |
| 263 | 0.94 |
| 264 | 0.94 |
| 265 | 0.94 |
| 266 | 0.94 |
| 267 | 0.92 |
| 268 | 0.92 |

| Wisconsin | |
|---|---|
| 530 | 1.00 |
| 531 | 0.98 |
| 532 | 1.00 |

| Wisconsin | |
|---|---|
| 534 | 0.99 |
| 535 | 0.97 |
| 537 | 0.95 |
| 538 | 0.90 |
| 539 | 0.94 |
| 541 | 0.96 |
| 542 | 0.96 |
| 543 | 0.96 |
| 544 | 0.91 |
| 545 | 0.92 |
| 546 | 0.93 |
| 547 | 0.94 |
| 548 | 0.94 |
| 549 | 0.92 |

| Wyoming | |
|---|---|
| 820 | 0.84 |
| 821 | 0.81 |
| 822 | 0.80 |
| 823 | 0.80 |
| 824 | 0.79 |
| 825 | 0.81 |
| 826 | 0.84 |
| 827 | 0.78 |
| 828 | 0.84 |
| 829 | 0.81 |
| 830 | 0.81 |
| 831 | 0.81 |

### Canadian Factors
*(reflect Canadian currency)*

**Alberta**

| Calgary | 0.99 |
|---|---|
| Edmonton | 0.99 |

**British Columbia**

| Vancouver | 1.09 |
|---|---|
| Victoria | 1.09 |

**Manitoba**

| Winnipeg | 1.00 |
|---|---|

**New Brunswick**

| Moncton | 0.94 |
|---|---|
| Saint John | 0.97 |

**Newfoundland**

| St. John's | 0.96 |
|---|---|

**Nova Scotia**

| Halifax | 0.98 |
|---|---|

**Ontario**

| Barrie | 1.10 |
|---|---|
| Brantford | 1.12 |
| Cornwall | 1.10 |
| Hamilton | 1.11 |
| Kingston | 1.10 |
| Kitchener | 1.05 |
| London | 1.09 |
| North Bay | 1.09 |
| Oshawa | 1.10 |
| Ottawa | 1.10 |
| Owen Sound | 1.09 |
| Peterborough | 1.10 |
| Sarnia | 1.10 |
| St. Catherines | 1.05 |
| Sudbury | 1.05 |
| Thunder Bay | 1.06 |
| Toronto | 1.13 |
| Windsor | 1.07 |

**Prince Edward Island**

| Charlottetown | 0.93 |
|---|---|

**Quebec**

| Chicoutimi | 1.02 |
|---|---|
| Montreal | 1.03 |
| Quebec | 1.03 |

**Saskatchewan**

| Regina | 0.92 |
|---|---|
| Saskatoon | 0.92 |

**Environmental Remediation:  Assemblies Cost Book**

# Cost Data for RI/FS and RFI/CMS
## RI/FS and RFI/CMS

**General:** This book addresses two types of preremediation investigations and accompanying studies:
1. Remedial Investigation/Feasibility Study (RI/FS) which is regulated by the Comprehensive Environmental Response Compensation and Liability Act (CERCLA)
2. RCRA Facility Investigation/Corrective Measures Study (RFI/CMS) which is regulated by the Resource Conservation and Recovery Act (RCRA)

**RI/FS:** The RI/FS project phase provides the detailed investigation component of the remedial action process. The RI/FS represents the methodology for characterizing the nature and extent of releases of hazardous constituents to the environment and for evaluating potential remedial action alternatives. The RI/FS stage consists of the following seven interactive/iterative steps:
· RI/FS Scoping
· Development of Alternatives
· Site Characterization
· Screening of Alternatives
· Treatability Investigation
· Analysis of Remedial Alternatives
· Remedy Selection

**RFI/CMS:** The RFI/CMS project phase provides the detailed investigation component of the RCRA Corrective Action process. The RFI/CMS represents the methodology for characterizing the nature and extent of releases of hazardous constituents to the environment and for evaluating potential Corrective Action alternatives. The RFI/CMS stage consists of the following interactive/iterative steps:
· RFI/CMS Scoping
· Identify and Develop Corrective Measure Alternatives
· Facility Investigation (or Site Characterization)
· Evaluation of Corrective Measure Alternatives
· Treatability Investigation
· Analysis of Corrective Measure Alternatives
· Corrective Measure Selection

**Common Cost Components:**
1. Professional Labor Tasks (Other than Fieldwork)
2. Groundwater Monitoring Wells
3. Sampling and Analysis
4. Geophysical Investigation and Soil Gas Survey
5. Bench and Pilot Treatability Tests
6. Other Direct Costs (Travel, Photocopying, Postage, etc.)

**Other Cost Considerations:** Site work and utilities.

---

**Environmental Remediation: Assemblies Cost Book**

*Page 2-1*

1998 by ECHOS. All rights reserved

# Cost Data for RI/FS and RFI/CMS

## RI/FS and RFI/CMS

### RI/FS

#### RI/FS Scoping

Community Relations Activities
Prepare SOW for RI/FS Work Plan
Identify Preliminary set of ARARs
Identify Preliminary Alternatives
Compose RI/FS Work Plan
Compose RI/FS Supplemental Plan
    Field Sampling Plan
    Quality Assurance Project Plan
    Health and Safety Plan
Interagency Agreement (NPL)
Data Quality Objectives (NPL)

#### Development of Alternatives

Identify Treatment Technologies
Evaluate/Screen Treatment Technologies
Assemble Technologies into Actions
Identify Action-Specific ARARs

#### Site Characterization

Characterize Site Meteorology
Conduct Fieldwork Support Activities
Characterize Site Surface Features
Characterize Site Geology
Characterize Site Hydrogeology
Characterize Site Surface Hydrogeology
Characterize Populations and Land Usage
Characterize Site Ecology
Characterize Nature and Extent of Contamination
Model Contaminant Fate and Transport
Characterize Site Soils and Vadose Zone
Conduct Baseline Risk Assessment
Compose Draft RI Report
Additional Field Data, if Required
Respond to Comments/Finalize RI Report
Define Source(s) of Contamination

#### Screening of Alternatives

Screen Alternatives for Implementability
Screen Alternatives for Cost
Screen Alternatives for Effectiveness
Evaluate Action-Specific ARARs

#### Treatability Investigation

Additional Field Data, if Required
Literature Survey on Technologies
Bench-Scale Treatability Studies

#### Analysis of Remedial Alternatives

Evaluate Alternatives by Nine Criteria
Compose Draft FS Report
Implement Community Relations
Prepare Preferred Alternatives
Public Meetings and Prepare Transcript
Respond to Comments on FS Report

#### Remedy Selection

Record of Decision (RoD)/Decision Document
Compose Final FS Report
Finalize RoD or Decision Document
Update Administrative Record

---

**Environmental Remediation: Assemblies Cost Book**

# Cost Data for RI/FS and RFI/CMS

## RI/FS and RFI/CMS

### RFI/CMS

#### RFI/CMS Scoping

Data Management Plan

Data Collection Quality Assurance Plan

Pre-Investigation Evaluation of Corrective Measure Technologies

Compose RFI/CMS Report of Current Conditions

Compose RFI/CMS Supplemental Plans:

Health and Safety Plan

Community Relations Plan

Prepare Project Management Plan for RFI/CMS

#### Identify and Develop Corrective Measure Alternatives

Establish Corrective Measure Objectives

Screening of Corrective Measure Technologies

Statement of Purpose/Description of Current Situation

Identification of the Corrective Measure Alternative(s)

#### Facility Investigation (or Site Characterization)

Characterize Populations and Land Usage

Conduct Fieldwork Support Activities

Characterize Site Surface Features

Characterize Site Geology

Characterize Site Hydrogeology

Characterize Site Soils and Vadose Zone

Potential Receptors (Baseline Risk Assessment)

Characterize Site Surface Water and Sediment

Characterize Site Meteorology

Define Source(s) of Contamination

Contaminant Fate and Transport

Compose Draft RFI Report

Investigation Analysis

Finalize RFI Report

Characterize Site Ecology

Nature and Extent of Contamination

#### Evaluation of Corrective Measure Alternatives

Screen for Costs

Evaluate Corrective Measure Alternatives

#### Treatability Investigation

Additional Field Data if Required

Literature Survey on Technologies

Bench/Pilot Scale Treatabililty Studies

#### Analysis of Corrective Measure Alternatives

Evaluate Corrective Measure Alternatives by Nine Criteria

Compose Draft CMS Report

Implement Community Relations

#### Corrective Measure Selection

Response to Comments (RTC)

Compose Final CMS Report

Statement of Basis (SB)

Public Comment Period

Update Administrative Record

---

**Environmental Remediation: Assemblies Cost Book**

# Cost Data for RI/FS and RFI/CMS
## Professional Labor

| Assembly | | | Description | Unit Cost (Hourly Rate) |
|---|---|---|---|---|
| 33 | 22 | 0101 | Project Manager | 36.06 |
| 33 | 22 | 0102 | Contract Administrator | 24.04 |
| 33 | 22 | 0103 | Office Manager | 23.08 |
| 33 | 22 | 0104 | QA/QC Officer | 23.08 |
| 33 | 22 | 0105 | Project Engineer | 31.25 |
| 33 | 22 | 0106 | Project Hydrogeologist | 29.81 |
| 33 | 22 | 0107 | Staff Engineer | 25.48 |
| 33 | 22 | 0108 | Staff Hydrogeologist | 22.36 |
| 33 | 22 | 0109 | Community Relations Specialist | 17.55 |
| 33 | 22 | 0110 | Engineer | 22.91 |
| 33 | 22 | 0111 | Geologist | 18.51 |
| 33 | 22 | 0112 | Surveyor | 18.03 |
| 33 | 22 | 0113 | Health & Safety Officer | 18.03 |
| 33 | 22 | 0114 | Chemist | 22.36 |
| 33 | 22 | 0115 | Biologist | 18.51 |
| 33 | 22 | 0116 | Toxicologist | 29.81 |
| 33 | 22 | 0117 | Field Technician | 13.94 |
| 33 | 22 | 0118 | Draftsman/CADD Operator | 13.94 |
| 33 | 22 | 0119 | Word Processing/Clerical | 12.02 |
| 33 | 22 | 0123 | Senior Project Manager | 45.67 |
| 33 | 22 | 0124 | Junior Project Manager | 28.85 |
| 33 | 22 | 0125 | Senior Contract Administrator | 31.25 |
| 33 | 22 | 0126 | Junior Contract Administrator | 17.31 |
| 33 | 22 | 0127 | Senior Office Manager | 31.25 |
| 33 | 22 | 0128 | Junior Office Manager | 16.83 |
| 33 | 22 | 0129 | Senior QA/QC Officer | 31.25 |
| 33 | 22 | 0130 | Junior QA/QC Officer | 17.31 |
| 33 | 22 | 0131 | Senior Project Engineer | 43.27 |
| 33 | 22 | 0132 | Junior Project Engineer | 25.48 |
| 33 | 22 | 0133 | Senior Project Hydrogeologist | 40.87 |
| 33 | 22 | 0134 | Junior Project Hydrogeologist | 23.08 |
| 33 | 22 | 0135 | Senior Certified Industrial Hygienist | 40.87 |
| 33 | 22 | 0136 | Certified Industrial Hygienist | 29.81 |
| 33 | 22 | 0137 | Junior Certified Industrial Hygienist | 23.08 |
| 33 | 22 | 0138 | Senior Staff Engineer | 31.25 |

**Environmental Remediation: Assemblies Cost Book**

# Cost Data for RI/FS and RFI/CMS
## Professional Labor

| Assembly | Description | Unit Cost (Hourly Rate) |
|---|---|---|
| 33 22 0139 | Junior Staff Engineer | 17.31 |
| 33 22 0140 | Senior Staff Hydrogeologist | 27.88 |
| 33 22 0141 | Junior Staff Hydrogeologist | 17.31 |
| 33 22 0142 | Senior Community Relations Specialist | 23.08 |
| 33 22 0143 | Junior Community Relations Specialist | 14.90 |
| 33 22 0144 | Senior Engineer | 28.04 |
| 33 22 0145 | Junior Engineer | 12.98 |
| 33 22 0146 | Senior Geologist | 24.04 |
| 33 22 0147 | Junior Geologist | 12.98 |
| 33 22 0148 | Senior Surveyor | 23.32 |
| 33 22 0149 | Junior Surveyor | 13.70 |
| 33 22 0150 | Senior Health & Safety Officer | 23.32 |
| 33 22 0151 | Junior Health & Safety Officer | 13.70 |
| 33 22 0152 | Senior Staff Attorney | 72.12 |
| 33 22 0153 | Senior Chemist | 27.88 |
| 33 22 0153 | Staff Attorney | 43.27 |
| 33 22 0154 | Junior Staff Attorney | 24.04 |
| 33 22 0154 | Junior Chemist | 17.31 |
| 33 22 0155 | Senior Biologist | 24.04 |
| 33 22 0156 | Junior Biologist | 12.98 |
| 33 22 0157 | Senior Toxicologist | 40.87 |
| 33 22 0158 | Junior Toxicologist | 23.08 |
| 33 22 0159 | Senior Field Technician | 17.31 |
| 33 22 0160 | Junior Field Technician | 9.13 |
| 33 22 0161 | Senior Draftsman | 17.31 |
| 33 22 0162 | Junior Draftsman | 9.13 |
| 33 22 0163 | Senior Word Processing/Clerical | 17.31 |
| 33 22 0164 | Junior Word Processing/Clerical | 8.17 |

**Environmental Remediation: Assemblies Cost Book**

# Cost Data for RI/FS and RFI/CMS

## Groundwater Monitoring Wells

**Common Cost Components:**

1. Mobilization of Drill Rig and Crew
2. Drilling
3. Well Casing, Screen, Plug
4. Well Cover
5. Slug Testing
6. Well Development
7. Sample Collection During Borehole Advancement
8. Drumming of Drill Cuttings and Development Water
9. Professional Labor Hours (Drilling Supervision, Well Development, Slug Tests)
10. Decontamination

| Assembly | Description | Unit | Unit Cost by Safety Level | | | | |
|---|---|---|---|---|---|---|---|
| | | | A | B | C | D | E |
| 33 01 0101 | Mobilize/DeMobilize Drilling Rig & Crew | LS | 4,949 | 4,031 | 3,308 | 2,401 | 2,280 |
| 33 02 0303 | Organic Vapor Analyzer Rental, per Day | DAY | 184.30 | 184.30 | 184.30 | 184.30 | 184.30 |
| 33 02 0405 | Monitoring Well Slug Testing Equipment Rental | WEEK | 552.08 | 552.08 | 552.08 | 552.08 | 552.08 |
| 33 17 0808 | Decontaminate Rig, Augers, Screen (Rental Equipment) | DAY | 205.34 | 205.34 | 205.34 | 205.34 | 205.34 |
| 33 23 0101 | 2" PVC, Schedule 40, Well Casing | LF | 17.45 | 14.39 | 11.98 | 8.95 | 8.55 |
| 33 23 0102 | 4" PVC, Schedule 40, Well Casing | LF | 26.98 | 22.38 | 18.77 | 14.24 | 13.63 |
| 33 23 0103 | 6" PVC, Schedule 40, Well Casing | LF | 27.07 | 22.66 | 19.19 | 14.84 | 14.25 |
| 33 23 0104 | 8" PVC, Schedule 40, Well Casing | LF | 34.50 | 29.27 | 25.15 | 19.98 | 19.29 |
| 33 23 0105 | 10" PVC, Schedule 40, Well Casing | LF | 38.28 | 33.04 | 28.92 | 23.75 | 23.06 |
| 33 23 0106 | 12" PVC, Schedule 40, Well Casing | LF | 44.32 | 39.08 | 34.97 | 29.80 | 29.10 |
| 33 23 0121 | 2" Stainless Steel, Well Casing | LF | 31.75 | 28.14 | 25.30 | 21.73 | 21.25 |
| 33 23 0122 | 4" Stainless Steel, Well Casing | LF | 47.66 | 43.06 | 39.45 | 34.92 | 34.31 |
| 33 23 0123 | 6" Stainless Steel Well Casing, 5' Sections, Flush Threaded | LF | 370.73 | 311.96 | 265.77 | 207.74 | 199.94 |
| 33 23 0125 | 8" Stainless Steel Well Casing, 5' Sections, Flush Threaded | LF | 526.42 | 452.94 | 395.18 | 322.61 | 312.87 |
| 33 23 0127 | 10" Stainless Steel Well Casing, 5' Sections, Flush Threaded | LF | 533.65 | 460.17 | 402.41 | 329.84 | 320.10 |
| 33 23 0129 | 12" Stainless Steel Well Casing, 5' Sections, Flush Threaded | LF | 561.57 | 488.09 | 430.33 | 357.76 | 348.02 |
| 33 23 0201 | 2" PVC, Schedule 40, Well Screen | LF | 23.47 | 19.52 | 16.42 | 12.52 | 12.00 |
| 33 23 0202 | 4" PVC, Schedule 40, Well Screen | LF | 38.16 | 32.00 | 27.16 | 21.09 | 20.27 |
| 33 23 0203 | 6" PVC, Schedule 40, Well Screen | LF | 47.13 | 39.78 | 34.01 | 26.75 | 25.78 |
| 33 23 0204 | 8" PVC, Schedule 40, Well Screen | LF | 62.02 | 52.84 | 45.61 | 36.54 | 35.33 |
| 33 23 0221 | 2" Stainless Steel, Well Screen | LF | 26.91 | 23.84 | 21.44 | 18.41 | 18.01 |
| 33 23 0222 | 4" Stainless Steel, Well Screen | LF | 47.66 | 43.06 | 39.45 | 34.92 | 34.31 |
| 33 23 0223 | 6" Stainless Steel Well Screen, 5' Sec, Flush Threaded | LF | 366.05 | 307.28 | 261.09 | 203.06 | 195.26 |
| 33 23 0225 | 8" Stainless Steel Well Screen, 5' Sec, Flush Threaded | LF | 462.22 | 388.74 | 330.98 | 258.41 | 248.67 |
| 33 23 0301 | 2" PVC, Well Plug | EACH | 29.37 | 24.77 | 21.16 | 16.63 | 16.02 |
| 33 23 0302 | 4" PVC, Well Plug | EACH | 55.65 | 48.91 | 43.62 | 36.97 | 36.07 |
| 33 23 0303 | 6" PVC, Well Plug | EACH | 112.63 | 101.15 | 92.13 | 80.79 | 79.27 |
| 33 23 0304 | 8" PVC, Well Plug | EACH | 132.70 | 119.58 | 109.26 | 96.31 | 94.57 |

**Environmental Remediation: Assemblies Cost Book**

*Page 2-6*

1998 by ECHOS. All rights reserved

# Cost Data for RI/FS and RFI/CMS
## Groundwater Monitoring Wells

| Assembly | Description | Unit | Unit Cost by Safety Level | | | | |
|---|---|---|---|---|---|---|---|
| | | | A | B | C | D | E |
| 33 23 0311 | 2" Stainless Steel, Well Plug | EACH | 83.24 | 74.06 | 66.83 | 57.76 | 56.55 |
| 33 23 0312 | 4" Stainless Steel, Well Plug | EACH | 112.24 | 102.03 | 94.01 | 83.93 | 82.58 |
| 33 23 0313 | 6" Stainless Steel, Well Plug | EACH | 573.04 | 481.18 | 408.98 | 318.27 | 306.09 |
| 33 23 0314 | 8" Stainless Steel, Well Plug | EACH | 702.53 | 597.56 | 515.04 | 411.38 | 397.46 |
| 33 23 0503 | 4" Submersible Pump, 13 - 37 GPM, < 180' Head | EACH | 3,084 | 2,790 | 2,637 | 2,352 | 2,264 |
| 33 23 0506 | 2" Submersible Pump Rental, Day | DAY | 63.86 | 63.86 | 63.86 | 63.86 | 63.86 |
| 33 23 0507 | 2" Submersible Pump Rental, Week | WEEK | 191.57 | 191.57 | 191.57 | 191.57 | 191.57 |
| 33 23 0508 | 2" Submersible Pump Rental, Month | MONTH | 478.92 | 478.92 | 478.92 | 478.92 | 478.92 |
| 33 23 0509 | 4" Submersible Pump Rental, Day | DAY | 63.86 | 63.86 | 63.86 | 63.86 | 63.86 |
| 33 23 0510 | 4" Submersible Pump Rental, Week | WEEK | 191.57 | 191.57 | 191.57 | 191.57 | 191.57 |
| 33 23 0511 | 4" Submersible Pump Rental, Month | MONTH | 478.92 | 478.92 | 478.92 | 478.92 | 478.92 |
| 33 23 1101 | Hollow-stem Auger, 8" Outside Diameter Borehole for 2" Well | LF | 89.98 | 73.28 | 60.15 | 43.66 | 41.45 |
| 33 23 1102 | Hollow-stem Auger, 11" Outside Diameter Borehole for 4" Well | LF | 109.98 | 89.57 | 73.52 | 53.36 | 50.66 |
| 33 23 1103 | Hollow-stem Auger, 13 3/4" Outside Diameter Borehole for 6" Well | LF | 141.40 | 115.16 | 94.53 | 68.61 | 65.13 |
| 33 23 1105 | Hollow-stem Auger, 16" Outside Diameter Borehole for 8" Well | LF | 164.97 | 134.35 | 110.28 | 80.05 | 75.99 |
| 33 23 1106 | Split Spoon Sample, 2" x 24", During Drilling | EACH | 46.67 | 46.67 | 46.67 | 46.67 | 46.67 |
| 33 23 1107 | Coring (Fluid or Air) | LF | 102.14 | 88.16 | 78.67 | 64.96 | 62.17 |
| 33 23 1110 | Continuous Split Spoon Sample | LF | 484.76 | 392.72 | 324.27 | 233.60 | 218.99 |
| 33 23 1111 | Well Development Equipment Rental | WEEK | 478.91 | 462.93 | 456.09 | 440.65 | 434.98 |
| 33 23 1121 | Standby for Drilling | EACH | 618.63 | 503.81 | 413.56 | 300.17 | 284.95 |
| 33 23 1122 | Move Rig/Equipment Around Site | EACH | 154.66 | 125.95 | 103.39 | 75.04 | 71.24 |
| 33 23 1125 | Concrete Coring (Minimum Charge) | DAY | 2,475 | 2,015 | 1,654 | 1,201 | 1,140 |
| 33 23 1126 | Furnish 55 Gallon Drum for Drill Cuttings & Development Water | EACH | 65.19 | 65.19 | 65.19 | 65.19 | 65.19 |
| 33 23 1128 | Jackhammer Rental, per Day | DAY | 166.03 | 166.03 | 166.03 | 166.03 | 166.03 |
| 33 23 1129 | Concrete Saw Rental, per Day | DAY | 102.00 | 102.00 | 102.00 | 102.00 | 102.00 |
| 33 23 1151 | Mud Drilling, 4" Diameter Borehole | LF | 73.65 | 59.98 | 49.23 | 35.73 | 33.92 |
| 33 23 1152 | Mud Drilling, 6" Diameter Borehole | LF | 81.01 | 65.98 | 54.16 | 39.31 | 37.31 |
| 33 23 1153 | Mud Drilling, 8" Diameter Borehole | LF | 88.38 | 71.97 | 59.08 | 42.88 | 40.71 |
| 33 23 1154 | Mud Drilling, 10" Diameter Borehole | LF | 100.65 | 81.97 | 67.29 | 48.84 | 46.36 |
| 33 23 1155 | Mud Drilling, 12" Diameter Borehole | LF | 112.92 | 91.97 | 75.49 | 54.79 | 52.01 |
| 33 23 1156 | Mud Drilling, 15" Diameter Borehole | LF | 137.47 | 111.96 | 91.90 | 66.71 | 63.32 |
| 33 23 1161 | Air Rotary 6" Borehole in Unconsolidated | LF | 73.65 | 59.98 | 49.23 | 35.73 | 33.92 |
| 33 23 1162 | Air Rotary 8" Borehole in Unconsolidated | LF | 122.74 | 99.96 | 82.06 | 59.56 | 56.54 |
| 33 23 1163 | Air Rotary 10" Borehole in Unconsolidated | LF | 220.94 | 179.93 | 147.70 | 107.20 | 101.77 |
| 33 23 1164 | Air Rotary 12" Borehole in Unconsolidated | LF | 294.59 | 239.91 | 196.93 | 142.94 | 135.69 |
| 33 23 1168 | Air Rotary 6" Borehole in Consolidated | LF | 58.92 | 47.98 | 39.39 | 28.59 | 27.14 |

**Environmental Remediation: Assemblies Cost Book**

*Page 2-7*

1998 by ECHOS. All rights reserved

# Cost Data for RI/FS and RFI/CMS
## Groundwater Monitoring Wells

| Assembly | Description | Unit | A | B | C | D | E |
|---|---|---|---|---|---|---|---|
| | | | Unit Cost by Safety Level | | | | |
| 33 23 1170 | Air Rotary 10" Borehole in Consolidated | LF | 294.59 | 239.91 | 196.93 | 142.94 | 135.69 |
| 33 23 1171 | Air Rotary 12" Borehole in Consolidated | LF | 613.72 | 499.81 | 410.28 | 297.79 | 282.69 |
| 33 23 1401 | 2" Screen, Filter Pack | LF | 16.49 | 13.88 | 11.84 | 9.27 | 8.92 |
| 33 23 1402 | 4" Screen, Filter Pack | LF | 29.10 | 24.50 | 20.89 | 16.36 | 15.75 |
| 33 23 1403 | 6" Screen, Filter Pack | LF | 42.19 | 35.53 | 30.29 | 23.72 | 22.83 |
| 33 23 1404 | Gravel Pack, 8" Well | LF | 29.47 | 23.93 | 19.82 | 14.36 | 13.48 |
| 33 23 1502 | Surface Pad, Concrete, 4' x 4' x 4" | EACH | 24.21 | 21.82 | 20.74 | 18.43 | 17.61 |
| 33 23 1504 | Surface Pad, Concrete, 2' x 2' x 4" | EACH | 6.05 | 5.45 | 5.18 | 4.61 | 4.40 |
| 33 23 1811 | 2" Well, Portland Cement Grout | LF | 0.92 | 0.92 | 0.92 | 0.92 | 0.92 |
| 33 23 1812 | 4" Well, Portland Cement Grout | LF | 1.38 | 1.38 | 1.38 | 1.38 | 1.38 |
| 33 23 1813 | 6" Well, Portland Cement Grout | LF | 7.80 | 7.80 | 7.80 | 7.80 | 7.80 |
| 33 23 1814 | 8" Well, Portland Cement Grout | LF | 11.02 | 11.02 | 11.02 | 11.02 | 11.02 |
| 33 23 1815 | 10" Well, Portland Cement Grout | LF | 12.85 | 12.85 | 12.85 | 12.85 | 12.85 |
| 33 23 1816 | 12" Well, Portland Cement Grout | LF | 14.69 | 14.69 | 14.69 | 14.69 | 14.69 |
| 33 23 1822 | Well Abandonment, 2" Well | LF | 32.46 | 26.55 | 21.90 | 16.06 | 15.27 |
| 33 23 1823 | Well Abandonment, 4" Well | LF | 46.38 | 37.93 | 31.28 | 22.94 | 21.82 |
| 33 23 1824 | Well Abandonment, 6" Well | LF | 0.00 | 0.00 | 0.00 | 0.00 | 0.00 |
| 33 23 1825 | Well Abandonment, 8" Well | LF | 2.12 | 2.12 | 2.12 | 2.12 | 2.12 |
| 33 23 2101 | 2" Well, Bentonite Seal | EACH | 63.00 | 52.67 | 44.55 | 34.34 | 32.97 |
| 33 23 2102 | 4" Well, Bentonite Seal | EACH | 157.54 | 131.70 | 111.39 | 85.87 | 82.44 |
| 33 23 2103 | 6" Well, Bentonite Seal | EACH | 252.01 | 210.68 | 178.18 | 137.37 | 131.89 |
| 33 23 2105 | 8" Well, Bentonite Seal | EACH | 346.60 | 289.75 | 245.06 | 188.92 | 181.39 |
| 33 23 2213 | 8" x 7.5" Locking Manhole Cover, Watertight | EACH | 345.53 | 288.12 | 242.99 | 186.30 | 178.69 |
| 33 23 2214 | 12" x 7.5" Locking Manhole Cover, Watertight | EACH | 459.57 | 383.02 | 322.86 | 247.27 | 237.12 |
| 33 23 2215 | 4" x 4" x 5' Steel Protective Cover, Lockable | EACH | 539.95 | 448.09 | 375.89 | 285.18 | 273.00 |
| 33 23 2216 | 6" x 6" x 5' Steel Protective Cover, Lockable | EACH | 560.94 | 469.08 | 396.88 | 306.17 | 293.99 |
| 33 23 2217 | 8" x 8" x 5' Steel Protective Cover, Lockable | EACH | 620.41 | 528.55 | 456.35 | 365.64 | 353.46 |
| 33 23 2301 | 5' Guard Posts, Cast Iron, Concrete Fill | EACH | 98.83 | 83.39 | 76.77 | 61.84 | 56.37 |
| 33 23 2401 | Teflon Bailer, 3/4" Outside Diameter x 1', 60 cc | EACH | 179.77 | 179.77 | 179.77 | 179.77 | 179.77 |
| 33 23 2402 | Teflon Bailer, 3/4" Outside Diameter x 3', 180 cc | EACH | 227.74 | 227.74 | 227.74 | 227.74 | 227.74 |
| 33 23 2403 | Teflon Bailer, 1" Outside Diameter x 1', 80 cc | EACH | 191.28 | 191.28 | 191.28 | 191.28 | 191.28 |
| 33 23 2404 | Teflon Bailer, 1 7/8" Outside Diameter x 1', 350 cc | EACH | 181.13 | 181.13 | 181.13 | 181.13 | 181.13 |
| 33 23 2405 | Teflon Bailer, 1 7/8" Outside Diameter x 2', 700 cc | EACH | 198.20 | 198.20 | 198.20 | 198.20 | 198.20 |
| 33 23 2406 | Teflon Bailer, 1 7/8" Outside Diameter x 3', 1,050 cc | EACH | 261.53 | 261.53 | 261.53 | 261.53 | 261.53 |
| 33 23 2407 | Disposable Bailer, Polyethylene, 1.5" Outside Diameter x 36" | EACH | 27.60 | 27.60 | 27.60 | 27.60 | 27.60 |
| 33 23 2411 | Stainless Steel Bailer, 1 3/4" Outside Diameter x 2' | EACH | 124.23 | 124.23 | 124.23 | 124.23 | 124.23 |
| 33 23 2421 | Emptying Stand | EACH | 302.32 | 302.32 | 302.32 | 302.32 | 302.32 |
| 33 23 2422 | Suspension Cable, Teflon Coated | FT | 1.00 | 1.00 | 1.00 | 1.00 | 1.00 |
| 33 23 2423 | Hand Reel | EACH | 9.89 | 9.89 | 9.89 | 9.89 | 9.89 |

**Environmental Remediation: Assemblies Cost Book**

# Cost Data for RI/FS and RFI/CMS
## Sampling & Analysis Assemblies

**Common Cost Components:**

1. *Groundwater Samples*
2. *Surface Water Samples*
3. *Sediment Samples*
4. *Bioassays*
5. *Surface Soil Samples*
6. *Soil Borings*
7. *Subsurface Soil Samples*
8. *Air Samples*
9. *QA/QC Samples (replicates/duplicates, trip blanks, equipment blanks, field blanks, spikes and blanks)*
10. *Mobilization of Sampling Crew*
11. *Analses (consider turnaround time and quality control level)*

The Sampling and Analysis assemblies in this section are organized as follows:

| Sampling | Analysis |
|---|---|
| Field Personnel | Water/Liquid |
| Mobilization/Per Diem | Soil/Sludge/Sediment |
| Soil Sampling/Drilling | Immunoassay |
| Drummed Waste | Air/Gas |
| Surface Water/Liquid | Biological |
| Groundwater Sampling | Explosive |
| Air Sampling | Other Analyses |
| Laboratory | Radiological Water/Liquid |
| Transportation | Radiological Vegetation/Soil/Sediment |
| Sample Containers/Disposables | Radiological Air |
| Overnight Delivery | Radiological Animal Tissue/Bone |
| Packaging | Radiological Urine & Feces |
| Sample Preparation & Cleanup | Radiological Miscellaneous |
| Miscellaneous Radioactive | |

**Environmental Remediation: Assemblies Cost Book**

# Cost Data for RI/FS and RFI/CMS
## Sampling Assemblies

| Sampling Assemblies - Air Sampling | | | Unit Cost by Safety Level | | | | |
|---|---|---|---|---|---|---|---|
| Assembly | Description | Unit | A | B | C | D | E |
| 33 02 0301 | Air Monitoring Station | EACH | 718.68 | 718.68 | 718.68 | 718.68 | 718.68 |
| 33 02 0302 | Organic Vapor Analyzer Rental, per Month | MONTH | 1,920 | 1,920 | 1,920 | 1,920 | 1,920 |
| 33 02 0303 | Organic Vapor Analyzer Rental, per Day | DAY | 184.30 | 184.30 | 184.30 | 184.30 | 184.30 |
| 33 02 0309 | Bacterial Air Sampler, Monthly Rental | MONTH | 436.37 | 436.37 | 436.37 | 436.37 | 436.37 |
| 33 02 0310 | Microbial Air Sampler, Monthly Rental | MONTH | 436.37 | 436.37 | 436.37 | 436.37 | 436.37 |
| 33 02 0311 | Cotton Dust Sampler, Monthly Rental | MONTH | 480.53 | 480.53 | 480.53 | 480.53 | 480.53 |
| 33 02 0312 | Digital Dust Sampler, Monthly Rental | MONTH | 552.97 | 552.97 | 552.97 | 552.97 | 552.97 |
| 33 02 0313 | Flow Calibrator, Monthly Rental | MONTH | 212.00 | 212.00 | 212.00 | 212.00 | 212.00 |
| 33 02 0314 | Personal Low Flow Sampling Pump, Monthly Rental | MONTH | 178.43 | 178.43 | 178.43 | 178.43 | 178.43 |
| 33 02 0315 | Ambient Air Monitor, Monthly Rental | MONTH | 826.80 | 826.80 | 826.80 | 826.80 | 826.80 |
| 33 02 0316 | Ambient CO Analyzer, Monthly Rental | MONTH | 378.95 | 378.95 | 378.95 | 378.95 | 378.95 |
| 33 02 0317 | Ambient Ozone Monitor, Monthly Rental | MONTH | 1,758 | 1,758 | 1,758 | 1,758 | 1,758 |
| 33 02 0318 | Ambient Sulfur Dioxide Analyzer, Monthly Rental | MONTH | 2,690 | 2,690 | 2,690 | 2,690 | 2,690 |
| 33 02 0319 | Calibration Gas Monitor | EACH | 487.60 | 487.60 | 487.60 | 487.60 | 487.60 |
| 33 02 0320 | Chloride Analyzer, Monthly Rental | MONTH | 749.07 | 749.07 | 749.07 | 749.07 | 749.07 |
| 33 02 0321 | Gas Monitors, Monthly Rental | MONTH | 657.20 | 657.20 | 657.20 | 657.20 | 657.20 |
| 33 02 0322 | Hydrogen Sulfide Analyzer, Monthly Rental | MONTH | 249.10 | 249.10 | 249.10 | 249.10 | 249.10 |
| 33 02 0323 | Manual Air Sampling Kit, Monthly Rental | MONTH | 85.68 | 85.68 | 85.68 | 85.68 | 85.68 |
| 33 02 0324 | Manual Air Sampling Kit, Detection Tubes | EACH | 88.69 | 88.69 | 88.69 | 88.69 | 88.69 |
| 33 02 0325 | Mercury Vapor Monitor, Monthly Rental | MONTH | 1,277 | 1,277 | 1,277 | 1,277 | 1,277 |
| 33 02 0326 | Nitrogen Dioxide Analyzer, Monthly Rental | MONTH | 2,385 | 2,385 | 2,385 | 2,385 | 2,385 |
| 33 02 0327 | Personal Alarms, Monthly Rental | MONTH | 318.00 | 318.00 | 318.00 | 318.00 | 318.00 |
| 33 02 0328 | Portable Ambient Air Analyzer, Monthly Rental | MONTH | 1,745 | 1,745 | 1,745 | 1,745 | 1,745 |
| 33 02 0329 | Portable CO2 Monitor, Monthly Rental | MONTH | 485.83 | 485.83 | 485.83 | 485.83 | 485.83 |
| 33 02 0330 | Portable Combustible Gas/Oxygen Indicator, Monthly Rental | MONTH | 355.10 | 355.10 | 355.10 | 355.10 | 355.10 |
| 33 02 0331 | Portable Organic Vapor Analyzer, Monthly Rental | MONTH | 1,920 | 1,920 | 1,920 | 1,920 | 1,920 |
| 33 02 0332 | Portable Oxygen Deficiency Monitor, Monthly Rental | MONTH | 409.87 | 409.87 | 409.87 | 409.87 | 409.87 |
| 33 02 0333 | Portable Toxic Gas Analyzer, Monthly Rental | MONTH | 265.00 | 265.00 | 265.00 | 265.00 | 265.00 |
| 33 02 0334 | Solvent Vapor Indicator, Monthly Rental | MONTH | 1,920 | 1,920 | 1,920 | 1,920 | 1,920 |
| 33 02 0335 | Specific Vapor Analyzer, Monthly Rental | MONTH | 1,290 | 1,290 | 1,290 | 1,290 | 1,290 |
| 33 02 0336 | Stack Sampling Probe, Monthly Rental | MONTH | 648.37 | 648.37 | 648.37 | 648.37 | 648.37 |

## Environmental Remediation: Assemblies Cost Book

# Cost Data for RI/FS and RFI/CMS
## Sampling Assemblies

### Sampling Assemblies - Air Sampling

| Assembly | Description | Unit | Unit Cost by Safety Level | | | | |
|---|---|---|---|---|---|---|---|
| | | | A | B | C | D | E |
| 33 02 0337 | Trace-Gas Analyzer, Monthly Rental | MONTH | 1,852 | 1,852 | 1,852 | 1,852 | 1,852 |
| 33 02 0338 | Instrument Shelter, 30.0" x 20.0" x 32.0", Pine with Latex Paint | EACH | 738.96 | 722.90 | 716.02 | 700.49 | 694.80 |
| 33 02 0339 | Instrument Shelter, 18.0" x 13.0" x 20.0", Wood Painted White | EACH | 289.52 | 273.46 | 266.58 | 251.05 | 245.36 |
| 33 02 0340 | Instrument Shelter, 81.0" x 5.5.0" x 16.5", Wood Painted White | EACH | 266.47 | 250.40 | 243.52 | 228.00 | 222.30 |
| 33 02 1501 | Saturation Indicator | EACH | 106.53 | 106.53 | 106.53 | 106.53 | 106.53 |
| 33 02 1502 | Thermostat & Humidity Control Devices | EACH | 329.49 | 279.97 | 258.78 | 210.91 | 193.37 |
| 33 02 1506 | Monitoring Port with Gas Monitor | EACH | 28.81 | 22.42 | 19.69 | 13.52 | 11.26 |

### Sampling Assemblies - Sample Containers and Disposables

| Assembly | Description | Unit | Unit Cost by Safety Level | | | | |
|---|---|---|---|---|---|---|---|
| | | | A | B | C | D | E |
| 33 02 0401 | Disposable Materials per Sample | EACH | 6.87 | 6.87 | 6.87 | 6.87 | 6.87 |
| 33 02 0402 | Decontamination Materials per Sample | EACH | 6.26 | 6.26 | 6.26 | 6.26 | 6.26 |
| 33 02 2019 | 1 Gallon, 128 Oz, Glass Sample Container, Case of 4 | EACH | 12.80 | 12.80 | 12.80 | 12.80 | 12.80 |
| 33 02 2020 | 1 Liter, 32 Oz, Clear Wide Mouth Jar, Case of 12 | EACH | 44.37 | 44.37 | 44.37 | 44.37 | 44.37 |
| 33 02 2021 | 500 ml, 16 Oz, Clear Wide Mouth Jar, Case of 12 | EACH | 61.67 | 61.67 | 61.67 | 61.67 | 61.67 |
| 33 02 2022 | 250 ml, 8 Oz, Clear Wide Mouth Jar, Case of 24 | EACH | 47.09 | 47.09 | 47.09 | 47.09 | 47.09 |
| 33 02 2023 | 120 ml, 4 Oz, Clear Wide Mouth Jar, Case of 24 | EACH | 34.47 | 34.47 | 34.47 | 34.47 | 34.47 |
| 33 02 2024 | 1 Liter, 32 Oz, Boston Round Bottle, Case of 12 | EACH | 28.78 | 28.78 | 28.78 | 28.78 | 28.78 |
| 33 02 2025 | 500 ml, 16 Oz, Boston Round Bottle, Case of 12 | EACH | 19.91 | 19.91 | 19.91 | 19.91 | 19.91 |
| 33 02 2026 | 40 ml, Clear Vial, Case of 72 | EACH | 89.91 | 89.91 | 89.91 | 89.91 | 89.91 |
| 33 02 2027 | 20 ml, Clear Vial, Case of 72 | EACH | 74.92 | 74.92 | 74.92 | 74.92 | 74.92 |
| 33 02 2028 | 250 ml, 8 oz, Clear Wide Mouth Vial with Septa, Case of 24 | EACH | 58.01 | 58.01 | 58.01 | 58.01 | 58.01 |
| 33 02 2029 | 125 ml, 4 oz, Clear Wide Mouth Vial with Septa, Case of 24 | EACH | 53.53 | 53.53 | 53.53 | 53.53 | 53.53 |
| 33 02 2030 | 1 Liter, 32 oz, High-density Polyethylene Bottle, Case of 12 | EACH | 32.64 | 32.64 | 32.64 | 32.64 | 32.64 |
| 33 02 2031 | 500 ml, 16 oz, High-density Polyethylene Bottle, Case of 24 | EACH | 28.06 | 28.06 | 28.06 | 28.06 | 28.06 |
| 33 02 2032 | 250 ml, 8 oz, High-density Polyethylene Bottle, Case of 24 | EACH | 15.58 | 15.58 | 15.58 | 15.58 | 15.58 |
| 33 02 2033 | 125 ml, 4 oz, High-density Polyethylene Bottle, Case of 24 | EACH | 13.71 | 13.71 | 13.71 | 13.71 | 13.71 |
| 33 02 2034 | Custody Seals, Package of 10 | EACH | 15.60 | 15.60 | 15.60 | 15.60 | 15.60 |
| 33 02 2035 | Safe Transport Can Filled with Vermiculite, 1 Gallon, Case of 4 | EACH | 28.94 | 28.94 | 28.94 | 28.94 | 28.94 |

## Environmental Remediation: Assemblies Cost Book

# Cost Data for RI/FS and RFI/CMS
## Sampling Assemblies

### Sampling Assemblies - Sample Containers and Disposables

| Assembly | Description | Unit | A | B | C | D | E |
|---|---|---|---|---|---|---|---|
| 33 02 2036 | Documentation Package for QA Verification, Data & Benchwork | EACH | 100.00 | 100.00 | 100.00 | 100.00 | 100.00 |
| 33 02 2050 | Blue Ice Soft Packs (Equivalent to 7 Lbs Ice) | EACH | 1.73 | 1.73 | 1.73 | 1.73 | 1.73 |

### Sampling Assemblies - Overnight Delivery

| Assembly | Description | Unit | A | B | C | D | E |
|---|---|---|---|---|---|---|---|
| 33 02 2037 | Overnight Delivery, 8 oz Letter | EACH | 10.17 | 10.17 | 10.17 | 10.17 | 10.17 |
| 33 02 2038 | Overnight Delivery, 1 Lb Package | LB | 11.08 | 11.08 | 11.08 | 11.08 | 11.08 |
| 33 02 2039 | Overnight Delivery, 2 - 5 Lb Package | LB | 16.08 | 16.08 | 16.08 | 16.08 | 16.08 |
| 33 02 2040 | Overnight Delivery, 6 - 10 Lb Package | LB | 21.58 | 21.58 | 21.58 | 21.58 | 21.58 |
| 33 02 2041 | Overnight Delivery, 11 - 20 Lb Package | LB | 34.25 | 34.25 | 34.25 | 34.25 | 34.25 |
| 33 02 2042 | Overnight Delivery, 21 - 50 Lb Package | LB | 57.27 | 57.27 | 57.27 | 57.27 | 57.27 |
| 33 02 2043 | Overnight Delivery, 51 - 70 Lb Package | LB | 78.27 | 78.27 | 78.27 | 78.27 | 78.27 |
| 33 02 2044 | Overnight Delivery, 71 Lbs or Greater | LB | 171.25 | 171.25 | 171.25 | 171.25 | 171.25 |

### Sampling Assemblies - Drummed Waste

| Assembly | Description | Unit | A | B | C | D | E |
|---|---|---|---|---|---|---|---|
| 33 19 0201 | Drummed Waste, Minimum Charge for Shipment | EACH | 2,350 | 2,350 | 2,350 | 2,350 | 2,350 |
| 33 19 9921 | DOT Steel Drum, 55 Gallon | EACH | 65.19 | 65.19 | 65.19 | 65.19 | 65.19 |
| 33 19 9922 | Polyethylene Closed Head Drum, 55 Gallon | EACH | 43.63 | 43.63 | 43.63 | 43.63 | 43.63 |
| 33 19 9924 | DOT Steel Salvage Drum, 85 Gallon, 16 Gauge | EACH | 57.06 | 57.06 | 57.06 | 57.06 | 57.06 |
| 33 19 9925 | DOT Steel Salvage Drum Composite Overpack, 85 Gallon, 16 Gauge | EACH | 70.49 | 70.49 | 70.49 | 70.49 | 70.49 |
| 33 23 1126 | Furnish 55 Gallon Drum for Drill Cuttings & Development Water | EACH | 65.19 | 65.19 | 65.19 | 65.19 | 65.19 |
| 33 23 1127 | Furnish 55 Gallon Drum for Development/Purge Water | EACH | 65.19 | 65.19 | 65.19 | 65.19 | 65.19 |

### Sampling Assemblies - Groundwater Sampling

| Assembly | Description | Unit | A | B | C | D | E |
|---|---|---|---|---|---|---|---|
| 33 02 0405 | Monitoring Well Slug Testing Equipment Rental | WEEK | 552.08 | 552.08 | 552.08 | 552.08 | 552.08 |
| 33 02 0549 | Lysimeter, 1.90" Outside Diameter, PVC Tube Type - PVC Body/Teflon Filter | EACH | 565.44 | 488.90 | 428.73 | 353.14 | 342.99 |

**Environmental Remediation: Assemblies Cost Book**

# Cost Data for RI/FS and RFI/CMS
## Sampling Assemblies

### Sampling Assemblies - Groundwater Sampling

| Assembly | Description | Unit | \multicolumn{5}{c}{Unit Cost by Safety Level} |
|---|---|---|---|---|---|---|---|
| | | | A | B | C | D | E |
| 33 02 0550 | Lysimeter, 1.90" Outside Diameter, Teflon Tube Type - Teflon Body/Filter | EACH | 636.22 | 559.67 | 499.50 | 423.91 | 413.76 |
| 33 02 0551 | Lysimeter, 2.38" Outside Diameter, PVC Tube Type - PVC Body/Teflon Filter | EACH | 565.44 | 488.90 | 428.73 | 353.14 | 342.99 |
| 33 02 0552 | Lysimeter, 2.38" Outside Diameter, Teflon Tube Type - Teflon Body/Filter | EACH | 544.92 | 468.38 | 408.21 | 332.62 | 322.47 |
| 33 02 0553 | Lysimeter, 1.90" Outside Diameter, PVC Cup Type - PVC Body/Teflon Filter | EACH | 565.44 | 488.90 | 428.73 | 353.14 | 342.99 |
| 33 02 0554 | Lysimeter, 1.90" Outside Diameter, Teflon Cup Type - Teflon Body/Filter | EACH | 630.42 | 553.87 | 493.70 | 418.12 | 407.97 |
| 33 02 0555 | Lysimeter, 1.90" Outside Diameter, PVC Deep Sample Type - Teflon Body/Filter | EACH | 644.64 | 568.09 | 507.92 | 432.33 | 422.18 |
| 33 02 0556 | Lysimeter, 1.90" Outside Diameter, Teflon Deep Sample Type - Teflon Body/Filter | EACH | 741.01 | 664.46 | 604.29 | 528.70 | 518.56 |
| 33 02 0557 | Lysimeter Head Assembly, All PVC Fittings | EACH | 577.72 | 501.17 | 441.00 | 365.41 | 355.26 |
| 33 02 0558 | Lysimeter Head Assembly, Teflon & PVC Fittings | EACH | 684.13 | 607.58 | 547.41 | 471.83 | 461.68 |
| 33 02 0559 | Silica Flour, 50 Lb Pail for Media Continuum | EACH | 174.04 | 151.08 | 133.03 | 110.35 | 107.31 |
| 33 02 0560 | Vacuum Pressure Hand Pump | EACH | 46.40 | 46.40 | 46.40 | 46.40 | 46.40 |
| 33 02 0561 | Nylon Tubing, 1/4" Outside Diameter | LF | 0.47 | 0.47 | 0.47 | 0.47 | 0.47 |
| 33 02 0562 | Teflon Tubing, 1/4" Outside Diameter | LF | 1.21 | 1.21 | 1.21 | 1.21 | 1.21 |
| 33 02 1509 | Water Quality Parameter Testing Device | WEEK | 227.90 | 227.90 | 227.90 | 227.90 | 227.90 |
| 33 23 0503 | 4" Submersible Pump, 13 - 37 GPM, < 180' Head | EACH | 3,084 | 2,790 | 2,637 | 2,352 | 2,264 |
| 33 23 0505 | 6" Submersible Pump, 15 - 135 GPM, 200' < Head < 500' | EACH | 4,472 | 3,885 | 3,579 | 3,008 | 2,834 |
| 33 23 1111 | Well Development Equipment Rental | WEEK | 478.91 | 462.93 | 456.09 | 440.65 | 434.98 |
| 33 23 2401 | Teflon Bailer, 3/4" Outside Diameter x 1', 60 cc | EACH | 179.77 | 179.77 | 179.77 | 179.77 | 179.77 |
| 33 23 2402 | Teflon Bailer, 3/4" Outside Diameter x 3', 180 cc | EACH | 227.74 | 227.74 | 227.74 | 227.74 | 227.74 |
| 33 23 2403 | Teflon Bailer, 1" Outside Diameter x 1', 80 cc | EACH | 191.28 | 191.28 | 191.28 | 191.28 | 191.28 |
| 33 23 2404 | Teflon Bailer, 1 7/8" Outside Diameter x 1', 350 cc | EACH | 181.13 | 181.13 | 181.13 | 181.13 | 181.13 |
| 33 23 2405 | Teflon Bailer, 1 7/8" Outside Diameter x 2', 700 cc | EACH | 198.20 | 198.20 | 198.20 | 198.20 | 198.20 |
| 33 23 2406 | Teflon Bailer, 1 7/8" Outside Diameter x 3', 1,050 cc | EACH | 261.53 | 261.53 | 261.53 | 261.53 | 261.53 |
| 33 23 2407 | Disposable Bailer, Polyethylene, 1.5" Outside Diameter x 36" | EACH | 27.60 | 27.60 | 27.60 | 27.60 | 27.60 |
| 33 23 2421 | Emptying Stand | EACH | 302.32 | 302.32 | 302.32 | 302.32 | 302.32 |
| 33 23 2422 | Suspension Cable, Teflon Coated | FT | 1.00 | 1.00 | 1.00 | 1.00 | 1.00 |
| 33 23 2423 | Hand Reel | EACH | 9.89 | 9.89 | 9.89 | 9.89 | 9.89 |

## Environmental Remediation: Assemblies Cost Book

*Page 2-13*

1998 by ECHOS. All rights reserved

# Cost Data for RI/FS and RFI/CMS
## Sampling Assemblies

### Sampling Assemblies - Laboratory

| Assembly | Description | Unit | Unit Cost by Safety Level A | B | C | D | E |
|---|---|---|---|---|---|---|---|
| 33 02 9913 | Mobile Laboratory Trailer, 8' W x 30' L, Rental | MONTH | 7,950 | 7,950 | 7,950 | 7,950 | 7,950 |
| 33 02 9914 | Gas Chromatograph, HP5890A, Rental | MONTH | 1,855 | 1,855 | 1,855 | 1,855 | 1,855 |
| 33 02 9915 | Mass Spectrometer, HP5970B MD, Rental | MONTH | 1,617 | 1,617 | 1,617 | 1,617 | 1,617 |
| 33 02 9918 | Spectrophotometer, Monthly Rental | MONTH | 275.00 | 275.00 | 275.00 | 275.00 | 275.00 |
| 33 02 9919 | Spectrophotometer, Purchase | EACH | 1,900 | 1,900 | 1,900 | 1,900 | 1,900 |
| 33 02 9921 | Gas Chromatograph Weekly Rental | WK | 910.00 | 910.00 | 910.00 | 910.00 | 910.00 |
| 33 02 9923 | Radiological Scintillation Counter | MONTH | 2,850 | 2,850 | 2,850 | 2,850 | 2,850 |
| 33 02 9924 | pH/Temperature/Conductivity Tester Rental | MONTH | 71.00 | 71.00 | 71.00 | 71.00 | 71.00 |
| 33 02 9925 | Infrared Spectrometer Rental | MONTH | 2,650 | 2,650 | 2,650 | 2,650 | 2,650 |
| 33 02 9926 | Computer Rental | MONTH | 200.00 | 200.00 | 200.00 | 200.00 | 200.00 |
| 33 02 9927 | Recalibration/Maintenance per Instrument per Month | MONTH | 235.00 | 235.00 | 235.00 | 235.00 | 235.00 |
| 33 02 9928 | Utilities Hook-up Fee | EACH | 2,000 | 2,000 | 2,000 | 2,000 | 2,000 |
| 33 02 9929 | Laboratory Utilities Allowance | MONTH | 200.00 | 200.00 | 200.00 | 200.00 | 200.00 |
| 33 02 9930 | Laboratory Expendables Allowance | EACH | 10.00 | 10.00 | 10.00 | 10.00 | 10.00 |
| 33 02 9931 | Electric Laboratory Scales, Purchase | EACH | 195.00 | 195.00 | 195.00 | 195.00 | 195.00 |
| 33 02 9932 | Fire Extinguisher | EACH | 99.45 | 99.45 | 99.45 | 99.45 | 99.45 |

### Sampling Assemblies - Mobilization and Per Diem

| Assembly | Description | Unit | Unit Cost by Safety Level A | B | C | D | E |
|---|---|---|---|---|---|---|---|
| 33 01 0101 | Mobilize/DeMobilize Drilling Rig & Crew | LS | 4,949 | 4,031 | 3,308 | 2,401 | 2,280 |
| 33 01 0201 | Mobilize Crew, >= 500 Miles, per Person | EACH | 357.75 | 357.75 | 357.75 | 357.75 | 357.75 |
| 33 01 0202 | Per Diem | DAY | 104.94 | 104.94 | 104.94 | 104.94 | 104.94 |
| 33 01 0203 | Mobilize Crew, 250 Miles, per Person | EACH | 178.88 | 178.88 | 178.88 | 178.88 | 178.88 |
| 33 01 0204 | Mobilize Crew, 100 Miles, per Person | EACH | 71.55 | 71.55 | 71.55 | 71.55 | 71.55 |
| 33 01 0205 | Mobilize Crew, 50 Miles, per Person | EACH | 53.66 | 53.66 | 53.66 | 53.66 | 53.66 |
| 33 01 0206 | Mobilize Crew, Local, per Person | EACH | 17.89 | 17.89 | 17.89 | 17.89 | 17.89 |
| 33 02 0640 | Mobilize/Demobilize CPT Rig and Crew | EACH | 2,986 | 2,986 | 2,986 | 2,986 | 2,986 |
| 33 02 9922 | Mobilize/Demobilize Laboratory Trailer | EACH | 3,500 | 3,500 | 3,500 | 3,500 | 3,500 |

## Environmental Remediation: Assemblies Cost Book

# Cost Data for RI/FS and RFI/CMS
## Sampling Assemblies

### Sampling Assemblies - Field Personnel

| Assembly | Description | Unit | Unit Cost by Safety Level | | | | |
|---|---|---|---|---|---|---|---|
| | | | A | B | C | D | E |
| 33 02 9906 | Subcontracted Sampling per Hour, One-man Crew | HOUR | 85.25 | 85.25 | 85.25 | 85.25 | 85.25 |
| 33 02 9907 | Subcontracted Sampling per Day, One-man Crew | DAY | 680.00 | 680.00 | 680.00 | 680.00 | 680.00 |

### Sampling Assemblies - Packaging

| Assembly | Description | Unit | Unit Cost by Safety Level | | | | |
|---|---|---|---|---|---|---|---|
| | | | A | B | C | D | E |
| 33 02 2045 | 48 Quart Ice Chest | EACH | 27.62 | 27.62 | 27.62 | 27.62 | 27.62 |
| 33 02 2046 | 60 Quart Ice Chest | EACH | 46.27 | 46.27 | 46.27 | 46.27 | 46.27 |
| 33 02 2047 | 80 Quart Ice Chest | EACH | 79.85 | 79.85 | 79.85 | 79.85 | 79.85 |
| 33 02 2048 | 102 Quart Ice Chest | EACH | 135.64 | 135.64 | 135.64 | 135.64 | 135.64 |
| 33 02 2049 | 120 Quart Ice Chest | EACH | 216.05 | 216.05 | 216.05 | 216.05 | 216.05 |

### Sampling Assemblies - Sample Preparation and Cleanup

| Assembly | Description | Unit | Unit Cost by Safety Level | | | | |
|---|---|---|---|---|---|---|---|
| | | | A | B | C | D | E |
| 33 02 2001 | Sample Preparation (3005/3010/3020/3040/3050) | EACH | 10.00 | 10.00 | 10.00 | 10.00 | 10.00 |
| 33 02 2002 | Sample Preparation (3510/3520/3540/3550/5030) | EACH | 25.00 | 25.00 | 25.00 | 25.00 | 25.00 |
| 33 02 2003 | Sample Preparation, Waste Dilution (3580) | EACH | 20.00 | 20.00 | 20.00 | 20.00 | 20.00 |
| 33 02 2004 | Cleanup (3610/3611/3620/3630/3650/3660) | EACH | 25.00 | 25.00 | 25.00 | 25.00 | 25.00 |
| 33 02 2005 | Cleanup, Gel-Permeation (3640) | EACH | 25.00 | 25.00 | 25.00 | 25.00 | 25.00 |

### Sampling Assemblies - Miscellaneous Radioactive

| Assembly | Description | Unit | Unit Cost by Safety Level | | | | |
|---|---|---|---|---|---|---|---|
| | | | A | B | C | D | E |
| 33 02 0206 | Prefiltering Liquids | EACH | 20.00 | 20.00 | 20.00 | 20.00 | 20.00 |
| 33 02 0207 | Acid Digestion | EACH | 27.50 | 27.50 | 27.50 | 27.50 | 27.50 |
| 33 02 0208 | Drying and Grinding | EACH | 22.50 | 22.50 | 22.50 | 22.50 | 22.50 |
| 33 02 0209 | Dissolution by Microwave | EACH | 40.00 | 40.00 | 40.00 | 40.00 | 40.00 |
| 33 02 0210 | Urine Preparation | EACH | 25.00 | 25.00 | 25.00 | 25.00 | 25.00 |
| 33 02 0211 | Feces Preparation | EACH | 33.33 | 33.33 | 33.33 | 33.33 | 33.33 |
| 33 02 0212 | Blood Preparation | EACH | 33.33 | 33.33 | 33.33 | 33.33 | 33.33 |
| 33 02 0213 | Animal Bone/Tissue Preparation | EACH | 66.67 | 66.67 | 66.67 | 66.67 | 66.67 |
| 33 02 0214 | Gas Flow Proportional Counting | EACH | 27.50 | 27.50 | 27.50 | 27.50 | 27.50 |

**Environmental Remediation: Assemblies Cost Book**

# Cost Data for RI/FS and RFI/CMS
## Sampling Assemblies

### Sampling Assemblies - Miscellaneous Radioactive

| Assembly | Description | Unit | A | B | C | D | E |
|---|---|---|---|---|---|---|---|
| 33 02 0215 | Liquid Scintillation | EACH | 27.50 | 27.50 | 27.50 | 27.50 | 27.50 |
| 33 02 0216 | Alpha Spectroscopy | EACH | 27.50 | 27.50 | 27.50 | 27.50 | 27.50 |
| 33 02 0217 | Gamma Spectroscopy | EACH | 20.00 | 20.00 | 20.00 | 20.00 | 20.00 |
| 33 02 0218 | Fluorometric or Laser | EACH | 27.50 | 27.50 | 27.50 | 27.50 | 27.50 |
| 33 02 0219 | Alpha Scintillation Counting | EACH | 27.50 | 27.50 | 27.50 | 27.50 | 27.50 |
| 33 02 0220 | Radon De-emanation | EACH | 27.50 | 27.50 | 27.50 | 27.50 | 27.50 |
| 33 02 0221 | Beta/Gamma Coincidence | EACH | 27.50 | 27.50 | 27.50 | 27.50 | 27.50 |
| 33 02 0222 | Radioactivity Screening - Tritium Not Present | EACH | 32.50 | 32.50 | 32.50 | 32.50 | 32.50 |
| 33 02 0223 | Radioactivity Screening - Tritium Present | EACH | 10.00 | 10.00 | 10.00 | 10.00 | 10.00 |
| 33 02 0224 | Repackage Sample | EACH | 10.00 | 10.00 | 10.00 | 10.00 | 10.00 |
| 33 02 0225 | Repackage and Ship Sample | EACH | 23.33 | 23.33 | 23.33 | 23.33 | 23.33 |
| 33 02 0341 | Alpha, Beta, Gamma Detector Equipment | EACH | 194.33 | 194.33 | 194.33 | 194.33 | 194.33 |
| 33 02 2357 | Characterization of Any Unknown (Low Level) Radioactive Sample | EACH | 750.00 | 750.00 | 750.00 | 750.00 | 750.00 |

### Sampling Assemblies - Soil Sampling and Drilling

| Assembly | Description | Unit | A | B | C | D | E |
|---|---|---|---|---|---|---|---|
| 17 03 0210 | D7 with U-blade Bulldozer | HOUR | 296.12 | 242.11 | 197.86 | 144.43 | 138.38 |
| 17 03 0230 | Crawler-mounted, 1.0 CY, 215 Hydraulic Excavator | HOUR | 210.40 | 170.50 | 140.73 | 101.43 | 95.15 |
| 33 02 0601 | Drilling 2.5" Diameter Soil Borings, No Sampling | LF | 21.74 | 18.75 | 16.71 | 13.78 | 13.18 |
| 33 02 0603 | Surface Soil Sampling Equipment | EACH | 280.80 | 280.80 | 280.80 | 280.80 | 280.80 |
| 33 02 0605 | Hand Auger Rental | DAY | 13.25 | 13.25 | 13.25 | 13.25 | 13.25 |
| 33 02 0606 | Power Auger Rental | DAY | 15.96 | 15.96 | 15.96 | 15.96 | 15.96 |
| 33 02 0613 | Soil Recovery Probe, Unslotted, 7/8" x 1' | EACH | 121.64 | 121.64 | 121.64 | 121.64 | 121.64 |
| 33 02 0614 | Soil Recovery Probe, Unslotted, 7/8" x 2' or 9/8" x 1' | EACH | 140.10 | 140.10 | 140.10 | 140.10 | 140.10 |
| 33 02 0615 | Soil Recovery Probe, Unslotted, 9/8" x 2' | EACH | 151.23 | 151.23 | 151.23 | 151.23 | 151.23 |
| 33 02 0616 | Soil Recovery Probe, Liners, 7/8" per LF, Butyrate Plast | LF | 0.86 | 0.86 | 0.86 | 0.86 | 0.86 |
| 33 02 0617 | Soil Recovery Probe, Liners, 9/8" per LF, Butyrate Plast | LF | 1.41 | 1.41 | 1.41 | 1.41 | 1.41 |
| 33 02 0618 | Soil Recovery Probe, Liners, Stainless Steel, 7/8" | EACH | 6.51 | 6.51 | 6.51 | 6.51 | 6.51 |
| 33 02 0619 | Soil Recovery Probe, Liners, Stainless Steel, 9/8" | EACH | 9.54 | 9.54 | 9.54 | 9.54 | 9.54 |
| 33 02 0620 | Soil Gas Vapor Probe, Stainless Steel, Manual, Nonremovable Tip | EACH | 45.93 | 45.93 | 45.93 | 45.93 | 45.93 |

## Environmental Remediation: Assemblies Cost Book

# Cost Data for RI/FS and RFI/CMS
## Sampling Assemblies

| Sampling Assemblies - Soil Sampling and Drilling | | | Unit Cost by Safety Level | | | | |
|---|---|---|---|---|---|---|---|
| Assembly | Description | Unit | A | B | C | D | E |
| 33 02 0621 | Soil Gas Vapor Probe, Stainless Steel, Electric, Rotary, Up/Down Hammer | EACH | 226.73 | 226.73 | 226.73 | 226.73 | 226.73 |
| 33 02 0622 | Screw Auger, High-carbon Steel, 1.5" x 6" | EACH | 79.89 | 79.89 | 79.89 | 79.89 | 79.89 |
| 33 02 0623 | Screw Auger, High-carbon Steel, 1.75" x 6" | EACH | 97.33 | 97.33 | 97.33 | 97.33 | 97.33 |
| 33 02 0624 | Soil Auger Buckets, Dutch, 3" Diameter, Carbon/Stainless Steel | EACH | 129.80 | 129.80 | 129.80 | 129.80 | 129.80 |
| 33 02 0625 | Soil Auger Buckets, Mud, 2" Diameter, Carbon/Stainless Steel | EACH | 120.12 | 120.12 | 120.12 | 120.12 | 120.12 |
| 33 02 0626 | Soil Auger Buckets, Reg, 2" Diameter, Carbon/Stainless Steel | EACH | 100.46 | 100.46 | 100.46 | 100.46 | 100.46 |
| 33 02 0627 | Soil Auger Buckets, Sand, 2" Diameter, Carbon/Stainless Steel | EACH | 114.29 | 114.29 | 114.29 | 114.29 | 114.29 |
| 33 02 0628 | Soil Recovery Auger Buckets, Stainless Steel, 2.25" x 8" | EACH | 342.05 | 342.05 | 342.05 | 342.05 | 342.05 |
| 33 02 0629 | Soil Recovery Auger Buckets, Stainless Steel, 3.25" x 10" | EACH | 352.12 | 352.12 | 352.12 | 352.12 | 352.12 |
| 33 02 0630 | Soil Recovery Auger Buckets, Stainless Steel with Carbon Bits, 2.25" x 8" | EACH | 315.75 | 315.75 | 315.75 | 315.75 | 315.75 |
| 33 02 0631 | Soil Recovery Auger Buckets, Stainless Steel with Carbon Bits, 3.25" x 10" | EACH | 325.82 | 325.82 | 325.82 | 325.82 | 325.82 |
| 33 02 0632 | Soil Recovery Auger Buckets, Liners, Butyrate Plast, 3" x 10" or 2" x 8" | EACH | 5.13 | 5.13 | 5.13 | 5.13 | 5.13 |
| 33 02 0633 | Soil Recovery Auger Buckets, Liners, Stainless Steel, 3" x 10" or 2" x 8" | EACH | 17.14 | 17.14 | 17.14 | 17.14 | 17.14 |
| 33 02 0634 | Sludge Sampler, Stainless Steel, 3.25" x 12", Thread On | EACH | 378.97 | 378.97 | 378.97 | 378.97 | 378.97 |
| 33 02 0635 | Tensiometer Soil Moist Probe, Standard, 7/8" Outside Diameter x 18", Fixed | EACH | 83.69 | 83.69 | 83.69 | 83.69 | 83.69 |
| 33 02 0636 | Tensiometer Soil Moist Probe, Jet Fill, 7/8" Outside Diameter x 18", Fixed | EACH | 71.81 | 71.81 | 71.81 | 71.81 | 71.81 |
| 33 02 0637 | Tensiometer Soil Moist Probe, Quick Draw, 18", Quick Read | EACH | 412.34 | 412.34 | 412.34 | 412.34 | 412.34 |
| 33 02 0638 | Soil Test Kit, 50 Tests | EACH | 278.25 | 278.25 | 278.25 | 278.25 | 278.25 |
| 33 02 0639 | CPT Rig, Includes Labor, Sampling, Punching, Decontamination | DAY | 2,844 | 2,844 | 2,844 | 2,844 | 2,844 |
| 33 02 0641 | Standby Time for CPT Rig and Crew | HOUR | 3,015 | 3,015 | 3,015 | 3,015 | 3,015 |
| 33 02 0642 | Setup Cost per Each Hole, CPT Rig | EACH | 273.83 | 273.83 | 273.83 | 273.83 | 273.83 |
| 33 02 0643 | Decontamination Trailer Rental for CPT Rig | DAY | 92.75 | 92.75 | 92.75 | 92.75 | 92.75 |
| 33 02 0644 | Decontaminate Cone on CPT Rig | DAY | 273.83 | 273.83 | 273.83 | 273.83 | 273.83 |
| 33 02 0645 | Level "D" PPE Rental per 2-Man CPT Crew | DAY | 2,874 | 2,874 | 2,874 | 2,874 | 2,874 |
| 33 02 0646 | Level "C" PPE Rental per 2-Man CPT Crew | DAY | 3,122 | 3,122 | 3,122 | 3,122 | 3,122 |

## Environmental Remediation: Assemblies Cost Book

# Cost Data for RI/FS and RFI/CMS
## Sampling Assemblies

### Sampling Assemblies - Soil Sampling and Drilling

| Assembly | Description | Unit | Unit Cost by Safety Level | | | | |
|---|---|---|---|---|---|---|---|
| | | | A | B | C | D | E |
| 33 02 0647 | In Situ Batch Sampling for Groundwater, CPT Rig (per Sample) | EACH | 92.75 | 92.75 | 92.75 | 92.75 | 92.75 |
| 33 02 0648 | In Situ Soil Sampling, CPT Rig (per Sample) | EACH | 60.95 | 60.95 | 60.95 | 60.95 | 60.95 |
| 33 02 0649 | Hydropunch with CPT Rig, Hard Soil | FT | 10.60 | 10.60 | 10.60 | 10.60 | 10.60 |
| 33 02 0650 | Hydropunch with CPT Rig, Medium Soil | FT | 10.60 | 10.60 | 10.60 | 10.60 | 10.60 |
| 33 02 0651 | Hydropunch with CPT Rig, Soft Soil | FT | 10.60 | 10.60 | 10.60 | 10.60 | 10.60 |
| 33 02 0652 | Grout Hole After Hydropunching | FT | 4.38 | 4.38 | 4.38 | 4.38 | 4.38 |
| 33 02 0653 | CPT Rig with ROST Analytical Equipment for Obtaining VOC in Soil | DAY | 7,100 | 7,100 | 7,100 | 7,100 | 7,100 |
| 33 17 0808 | Decontaminate Rig, Augers, Screen (Rental Equipment) | DAY | 205.34 | 205.34 | 205.34 | 205.34 | 205.34 |
| 33 23 1106 | Split Spoon Sample, 2" x 24", During Drilling | EACH | 46.67 | 46.67 | 46.67 | 46.67 | 46.67 |
| 33 23 1107 | Coring (Fluid or Air) | LF | 102.14 | 88.16 | 78.67 | 64.96 | 62.17 |
| 33 23 1121 | Standby for Drilling | EACH | 618.63 | 503.81 | 413.56. | 300.17 | 284.95 |
| 33 23 1122 | Move Rig/Equipment Around Site | EACH | 154.66 | 125.95 | 103.39 | 75.04 | 71.24 |
| 33 23 1123 | Load Supplies/Equipment | LS | 1,856 | 1,511 | 1,241 | 900.52 | 854.85 |
| 33 23 1124 | Security Pass/Protocol | LS | 309.32 | 251.91 | 206.78 | 150.09 | 142.47 |
| 33 23 1125 | Concrete Coring (Minimum Charge) | DAY | 2,475 | 2,015 | 1,654 | 1,201 | 1,140 |
| 33 23 1128 | Jackhammer Rental, per Day | DAY | 166.03 | 166.03 | 166.03 | 166.03 | 166.03 |
| 33 23 1129 | Concrete Saw Rental, per Day | DAY | 102.00 | 102.00 | 102.00 | 102.00 | 102.00 |
| 33 23 1801 | 2" Well, Grout (Annular Seal) | LF | 83.99 | 68.83 | 56.92 | 41.95 | 39.94 |
| 33 23 1802 | 4" Well, Grout (Annular Seal) | LF | 145.07 | 118.89 | 98.31 | 72.46 | 68.99 |
| 33 23 1803 | 6" Well, Grout (Annular Seal) | LF | 213.78 | 175.20 | 144.88 | 106.78 | 101.67 |
| 33 23 1811 | 2" Well, Portland Cement Grout | LF | 0.92 | 0.92 | 0.92 | 0.92 | 0.92 |
| 33 23 1812 | 4" Well, Portland Cement Grout | LF | 1.38 | 1.38 | 1.38 | 1.38 | 1.38 |
| 33 23 1813 | 6" Well, Portland Cement Grout | LF | 7.80 | 7.80 | 7.80 | 7.80 | 7.80 |
| 33 23 2104 | 3/8", 1' Thick, Sodium Bentonite Pellets | CF | 275.35 | 229.42 | 193.32 | 147.97 | 141.88 |

### Sampling Assemblies - Surface Water and Liquids

| Assembly | Description | Unit | Unit Cost by Safety Level | | | | |
|---|---|---|---|---|---|---|---|
| | | | A | B | C | D | E |
| 33 02 0404 | 12' Aluminum Pole with Polyhead Sampler | EACH | 396.66 | 396.66 | 396.66 | 396.66 | 396.66 |
| 33 02 0521 | Hip Waders Rental | WEEK | 51.44 | 51.44 | 51.44 | 51.44 | 51.44 |
| 33 02 0522 | Boat with Motor, Rental | DAY | 58.83 | 58.83 | 58.83 | 58.83 | 58.83 |

## Environmental Remediation: Assemblies Cost Book

# Cost Data for RI/FS and RFI/CMS
## Sampling Assemblies

### Sampling Assemblies - Surface Water and Liquids

| Assembly | Description | Unit | Unit Cost by Safety Level | | | | |
|---|---|---|---|---|---|---|---|
| | | | A | B | C | D | E |
| 33 02 0523 | Boat without Motor, Rental | DAY | 21.73 | 21.73 | 21.73 | 21.73 | 21.73 |
| 33 02 0524 | Glass Coliwasas, Disposable, 7/8" x 42", 200 ml, Case of 12 | EACH | 187.22 | 187.22 | 187.22 | 187.22 | 187.22 |
| 33 02 0525 | Glass Coliwasas, Reusable, 7/8" x 42", 200 ml, Case of 4 | EACH | 67.42 | 67.42 | 67.42 | 67.42 | 67.42 |
| 33 02 0526 | PVC Coliwasas, 1.9" Outside Diameter x 49", 760 ml | EACH | 155.01 | 155.01 | 155.01 | 155.01 | 155.01 |
| 33 02 0527 | Bomb Sampler, 2.5" Diameter x 10", 500 ml | EACH | 524.35 | 524.35 | 524.35 | 524.35 | 524.35 |
| 33 02 0528 | Drum Thief Tubes, 3/4" x 42", 150 ml, Disposable, Case of 25 | EACH | 63.94 | 63.94 | 63.94 | 63.94 | 63.94 |
| 33 02 0529 | Drum Thief Tubes, 1/2" x 42", 75 ml, Disposable, Case of 25 | EACH | 44.75 | 44.75 | 44.75 | 44.75 | 44.75 |
| 33 02 0530 | Bottom Sampler, 3 Lb Brass, 7.5" x 4.5", No Cable Required | EACH | 179.67 | 179.67 | 179.67 | 179.67 | 179.67 |
| 33 02 0531 | Bottom Sampler, 17 Lb Stainless Steel, 6" x 6" x 6", with 100' Cable | EACH | 466.91 | 466.91 | 466.91 | 466.91 | 466.91 |
| 33 02 0532 | Bottom Sampler, 60 Lb Stainless Steel, 9" x 9" | EACH | 915.24 | 915.24 | 915.24 | 915.24 | 915.24 |
| 33 02 0533 | Water Level Indicator, 100' Tape, Electric, Light & Horn | EACH | 461.65 | 461.65 | 461.65 | 461.65 | 461.65 |
| 33 02 0534 | Water Level Indicator, 300' Tape, Electric, Light & Horn | EACH | 688.99 | 688.99 | 688.99 | 688.99 | 688.99 |
| 33 02 0535 | Water Level Indicator, 500' Tape, Electric, Light & Horn | EACH | 996.15 | 996.15 | 996.15 | 996.15 | 996.15 |
| 33 02 0536 | Water Level Indicator, Manual Steel Chalk Tape, with Weight, 100' Cable | EACH | 138.33 | 138.33 | 138.33 | 138.33 | 138.33 |
| 33 02 0538 | DO Meter, Analog, Portable, 6' Cable, Polarographic Probe | EACH | 577.70 | 577.70 | 577.70 | 577.70 | 577.70 |
| 33 02 0539 | DO Meter, Digital, Portable, 10' Cable, Small Diameter Probe | EACH | 1,114 | 1,114 | 1,114 | 1,114 | 1,114 |
| 33 02 0540 | DO Meter, Portable, Probe, 10' Cable, Quick Readings | EACH | 517.60 | 517.60 | 517.60 | 517.60 | 517.60 |
| 33 02 0541 | Conductivity Meter with Dip-type Sensor, Digital, 9V Battery, 7 x 3.2 x 2" | EACH | 977.27 | 977.27 | 977.27 | 977.27 | 977.27 |
| 33 02 0542 | Conductivity Meter, 1" Diameter Probe, 150' Cable, Also Temperature/Level | EACH | 914.25 | 914.25 | 914.25 | 914.25 | 914.25 |
| 33 02 0543 | Conductivity Meter, 10' Cable, Probe, Also Temperature/ Salinity | EACH | 870.79 | 870.79 | 870.79 | 870.79 | 870.79 |
| 33 02 0544 | Thermometer, Stainless Steel, Bi-Metal, 40 - 160 F, 12" Stem | EACH | 62.47 | 62.47 | 62.47 | 62.47 | 62.47 |
| 33 02 0545 | Thermometer, Thermocouple | EACH | 144.48 | 144.48 | 144.48 | 144.48 | 144.48 |
| 33 02 0546 | Thermometer, Digital | EACH | 38.50 | 38.50 | 38.50 | 38.50 | 38.50 |
| 33 02 0547 | pH Meter, Test Paper Kits, 200 Strips, Charts | EACH | 9.89 | 9.89 | 9.89 | 9.89 | 9.89 |
| 33 02 0548 | pH Meter, Digital, LCD, 0 - 14 pH Range | EACH | 361.81 | 361.81 | 361.81 | 361.81 | 361.81 |
| 33 02 0602 | Stainless Steel Bottom Sampler | EACH | 378.97 | 378.97 | 378.97 | 378.97 | 378.97 |

## Environmental Remediation: Assemblies Cost Book

# Cost Data for RI/FS and RFI/CMS
## Sampling Assemblies

### Sampling Assemblies - Surface Water and Liquids

| Assembly | Description | Unit | A | B | C | D | E |
|----------|-------------|------|---|---|---|---|---|
| | | | | | | | |

### Sampling Assemblies - Transportation

| Assembly | Description | Unit | A | B | C | D | E |
|----------|-------------|------|---|---|---|---|---|
| 33 01 0102 | Van or Pickup Rental | DAY | 32.69 | 32.69 | 32.69 | 32.69 | 32.69 |
| 33 01 0103 | Flatbed Truck Rental | DAY | 45.94 | 45.94 | 45.94 | 45.94 | 45.94 |
| 33 01 0104 | Car or Van Mileage Charge | MILE | 0.35 | 0.35 | 0.35 | 0.35 | 0.35 |
| 33 01 0105 | Pickup Truck Mileage Charge | MILE | 0.56 | 0.56 | 0.56 | 0.56 | 0.56 |
| 33 01 0106 | Flatbed Truck Mileage Charge | MILE | 1.01 | 1.01 | 1.01 | 1.01 | 1.01 |
| 33 07 9901 | Lease, Portable 40' W x 150' L x 18' High Emissions Containment Building | YEAR | 7,500 | 7,500 | 7,500 | 7,500 | 7,500 |
| 33 07 9902 | Buy, Portable 40' W x 150' L x 18' High Emissions Containment Building | EACH | 23,503 | 23,503 | 23,503 | 23,503 | 23,503 |
| 33 07 9903 | Assemble Emissions Containment Building, 5 Days | EACH | 12,346 | 9,591 | 8,298 | 5,629 | 4,722 |
| 33 07 9904 | Disassemble Emissions Containment Building, 3 Days | EACH | 7,402 | 5,750 | 4,975 | 3,374 | 2,831 |

**Environmental Remediation: Assemblies Cost Book**

# Cost Data for RI/FS and RFI/CMS
## Analysis Assemblies

| Sampling Assemblies - Biological | | Cost by Turnaround Time/Quality Control Level | | | |
|---|---|---|---|---|---|
| **Assembly** | **Description** | **24-72 hr CLP** | **24-72 hr Std QC** | **Std TAT CLP** | **Std TAT Std QC** |
| 33 02 1901 | Freshwater Discharge, LC50, Bioassay Analysis | 1,360 | 1,133 | 793.34 | 566.67 |
| 33 02 1902 | Saltwater Discharge, Lc50, Bioassay Analysis | 1,320 | 1,100 | 770.00 | 550.00 |
| 33 02 1903 | Drilling Mud, Lc50, Bioassay Analysis | 1,600 | 1,333 | 933.34 | 666.67 |
| 33 02 1904 | Freshwater Chronic Toxicity Bioassay Analysis | 2,760 | 2,300 | 1,610 | 1,150 |
| 33 02 1905 | Saltwater Chronic Toxicity Bioassay Analysis | 3,560 | 2,967 | 2,077 | 1,483 |
| 33 02 1906 | Ames (Full) Microbiological Analysis | 4,280 | 3,567 | 2,497 | 1,783 |
| 33 02 1907 | Ames (Screen) Microbiological Analysis | 420.00 | 350.00 | 245.00 | 175.00 |
| 33 02 1908 | Daphnia IQ Toxicity Analysis | 760.01 | 633.34 | 443.34 | 316.67 |
| 33 02 1909 | Microtox Microbiological Analysis | 780.00 | 650.00 | 455.00 | 325.00 |
| 33 02 1910 | Total Heterotrophic Plate Count | 102.00 | 85.00 | 59.50 | 42.50 |
| 33 02 1911 | Hydrocarbon Degraders | 186.00 | 155.00 | 108.50 | 77.50 |
| 33 02 1912 | Fluorescent Pseudomonas | 132.00 | 110.00 | 77.00 | 55.00 |

| Sampling Assemblies - Explosive | | Cost by Turnaround Time/Quality Control Level | | | |
|---|---|---|---|---|---|
| **Assembly** | **Description** | **24-72 hr CLP** | **24-72 hr Std QC** | **Std TAT CLP** | **Std TAT Std QC** |
| 33 02 2401 | EPA Method 8330 (11 Compounds) Nitroaromatics/Nitramines | 689.64 | 574.70 | 402.29 | 287.35 |
| 33 02 2402 | Nitroglycerine | 567.04 | 472.53 | 330.77 | 236.27 |
| 33 02 2403 | Nitrocellulose | 459.76 | 383.14 | 268.20 | 191.57 |
| 33 02 2404 | Nitroquanadine | 459.76 | 383.14 | 268.20 | 191.57 |
| 33 02 2405 | Ammonium Perchlorate | 260.53 | 217.11 | 151.98 | 108.56 |

| Sampling Assemblies - Air and Gas | | Cost by Turnaround Time/Quality Control Level | | | |
|---|---|---|---|---|---|
| **Assembly** | **Description** | **24-72 hr CLP** | **24-72 hr Std QC** | **Std TAT CLP** | **Std TAT Std QC** |
| 33 02 1801 | VOST Analysis, Air (30/5040/8240) | 540.00 | 450.00 | 315.00 | 225.00 |
| 33 02 1802 | Principal Organic Hazardous Constituents, Air (30/5040/8240) | 462.00 | 385.00 | 269.50 | 192.50 |
| 33 02 1803 | Tentative ID Compounds, GC/MS, Air (30/5040/8240) | 616.80 | 514.00 | 359.80 | 257.00 |
| 33 02 1804 | Tedlar Bag (Single Component), Air (30/5040/8240) | 240.00 | 200.00 | 140.00 | 100.00 |
| 33 02 1805 | Sorbent Tube (Single Component), Air (5040/8240) | 240.00 | 200.00 | 140.00 | 100.00 |
| 33 02 1806 | BTEX, Air (5040/8020) | 228.00 | 190.00 | 133.00 | 95.00 |

**Environmental Remediation: Assemblies Cost Book**

# Cost Data for RI/FS and RFI/CMS
## Analysis Assemblies

| Sampling Assemblies - Air and Gas | | Cost by Turnaround Time/Quality Control Level | | | |
|---|---|---|---|---|---|
| **Assembly** | **Description** | **24-72 hr CLP** | **24-72 hr Std QC** | **Std TAT CLP** | **Std TAT Std QC** |
| 33 02 1807 | BTEX/Total Volatile Petroleum Hydrocarbons, Air (5040/8020/8015) | 268.01 | 223.34 | 156.34 | 111.67 |
| 33 02 1808 | Semivolatiles, Air (10/3550/8270) | 1,276 | 1,063 | 744.34 | 531.67 |
| 33 02 1809 | Multimetal Train, Air, Various Methods | 553.20 | 461.00 | 322.70 | 230.50 |
| 33 02 1810 | Pesticides/PCBs, GC, Air, Various Methods | 360.00 | 300.00 | 210.00 | 150.00 |
| 33 02 1811 | Hydrocarbons, BTEX, Air (NIOSH 1500) | 245.59 | 204.66 | 143.26 | 102.33 |
| 33 02 1812 | Polyaromatic Hydrocarbons, Air (NIOSH 5515) | 1,110 | 925.00 | 647.50 | 462.50 |
| 33 02 1813 | Metals by ICP, Air (NIOSH 7300) | 598.80 | 499.00 | 349.30 | 249.50 |
| 33 02 1814 | Metals by Graphite Furnace, Air (NIOSH 7300) | 675.19 | 562.66 | 393.86 | 281.33 |
| 33 02 1815 | Mercury, Air (245.2) | 84.00 | 70.00 | 49.00 | 35.00 |
| 33 02 1816 | Cyanide, Air (335.2) | 103.99 | 86.66 | 60.66 | 43.33 |
| 33 02 1817 | Ion Chromatography, Air | 43.20 | 36.00 | 25.20 | 18.00 |
| 33 02 1818 | Volatile Aromatic Halocarbons, Air, TO-1 | 228.00 | 190.00 | 133.00 | 95.00 |
| 33 02 1819 | Highly Volatile Nonpolar Organics, TO-2 | 732.00 | 610.00 | 427.00 | 305.00 |
| 33 02 1820 | Volatile Nonpolar Organics, TO-3 | 222.00 | 185.00 | 129.50 | 92.50 |
| 33 02 1821 | Organochlorine Pesticides/PCBs, Air, TO-4 | 582.00 | 485.00 | 339.50 | 242.50 |
| 33 02 1822 | Aldehydes & Ketones, Air, TO-5 | 270.00 | 225.00 | 157.50 | 112.50 |
| 33 02 1823 | Phosgene in Ambient Air, TO-6 | 504.00 | 420.00 | 294.00 | 210.00 |
| 33 02 1824 | N-Nitrosodimethylamine, Air, TO-7 | 438.00 | 365.00 | 255.50 | 182.50 |
| 33 02 1825 | Phenol & Methyl Phenols, Air, TO-8 | 174.00 | 145.00 | 101.50 | 72.50 |
| 33 02 1826 | Dioxins, Air, TO-9 | 774.00 | 645.00 | 451.50 | 322.50 |
| 33 02 1827 | Total Nonmethane Organics, Air, TO-12 | 120.00 | 100.00 | 70.00 | 50.00 |
| 33 02 1828 | Polyaromatic Hydrocarbons, Air, TO-13 | 240.00 | 200.00 | 140.00 | 100.00 |
| 33 02 1829 | Canister Samples by GC or GC/MS, TO-14 | 960.00 | 800.00 | 560.00 | 400.00 |
| 33 02 1830 | Hydrocarbon Speciation, C1-C8, GC/FID, Air (TO-12/14) | 180.00 | 150.00 | 105.00 | 75.00 |
| 33 02 1831 | Hydrocarbon Speciation, C8-C22, GC/FID, Air (TO-12/14) | 180.00 | 150.00 | 105.00 | 75.00 |
| 33 02 1832 | Hydrocarbon Speciation, C1-C22, GC/FID, Air (TO-12/14) | 180.00 | 150.00 | 105.00 | 75.00 |
| 33 02 1833 | Hydrocarbon Speciation, C4-C22, GC/FID, Air (TO-12/14) | 180.00 | 150.00 | 105.00 | 75.00 |

**Environmental Remediation: Assemblies Cost Book**

# Cost Data for RI/FS and RFI/CMS
## Analysis Assemblies

| Sampling Assemblies - Immunoassay | | Cost by Turnaround Time/Quality Control Level | | | |
|---|---|---|---|---|---|
| Assembly | Description | 24-72 hr CLP | 24-72 hr Std QC | Std TAT CLP | Std TAT Std QC |
| 33 02 2501 | EPA 4010 PCP Immunoassay Field Test | 67.44 | 56.20 | 39.34 | 28.10 |
| 33 02 2502 | EPA 4015 2,4-D Immunoassay Field Test | 107.28 | 89.40 | 62.58 | 44.70 |
| 33 02 2503 | EPA 4020 PCBs Immunoassay Field Test | 85.81 | 71.51 | 50.06 | 35.76 |
| 33 02 2504 | EPA 4030 Petroleum Hydrocarbons Immunoassay Field Test | 73.56 | 61.30 | 42.91 | 30.65 |
| 33 02 2505 | EPA 4031 BTEX Immunoassay Field Test | 107.28 | 89.40 | 62.58 | 44.70 |
| 33 02 2506 | EPA 4035 PAH Immunoassay Field Test | 107.28 | 89.40 | 62.58 | 44.70 |
| 33 02 2507 | EPA 4040 Toxaphene Immunoassay Field Test | 107.28 | 89.40 | 62.58 | 44.70 |
| 33 02 2508 | EPA 4041 Chlordane Immunoassay Field Test | 107.28 | 89.40 | 62.58 | 44.70 |
| 33 02 2509 | EPA 4042 DDT Immunoassay Field Test | 107.28 | 89.40 | 62.58 | 44.70 |
| 33 02 2551 | EPA Draft 8515 TNT Field Test | 444.44 | 370.36 | 259.26 | 185.18 |
| 33 02 2552 | Dioxin Field Test | 73.56 | 61.30 | 42.91 | 30.65 |

| Sampling Assemblies - Other Analyses | | Cost by Turnaround Time/Quality Control Level | | | |
|---|---|---|---|---|---|
| Assembly | Description | 24-72 hr CLP | 24-72 hr Std QC | Std TAT CLP | Std TAT Std QC |
| 33 02 0307 | Soil Gas Investigation & Analysis | 254.40 | 212.00 | 148.40 | 106.00 |
| 33 02 0504 | Clarifier Indicator Analyses | 468.02 | 390.02 | 273.01 | 195.01 |
| 33 02 0505 | Groundwater Contamination Analysis, per Sample | 302.40 | 252.00 | 176.40 | 126.00 |
| 33 02 0506 | Tank Decontamination Analysis, per Sample | 278.33 | 231.94 | 162.36 | 115.97 |
| 33 02 0507 | Primary Drinking Water Parameters Analysis, per Sample | 819.94 | 683.28 | 478.30 | 341.64 |
| 33 02 0508 | Detection/Compliance Monitoring Analysis, per Well | 2,376 | 1,980 | 1,386 | 990.17 |
| 33 02 0509 | Compliance Monitoring 40 CFR 261, Appendix IX Constituents | 4,116 | 3,430 | 2,401 | 1,715 |
| 33 02 0510 | Analysis of Wastewater for POTW Permit | 708.70 | 590.58 | 413.41 | 295.29 |
| 33 02 0511 | Analysis of Lysimeters for Soil-Pore Liquid Monitoring | 857.59 | 714.66 | 500.26 | 357.33 |
| 33 02 0512 | Rinsate Analysis | 278.33 | 231.94 | 162.36 | 115.97 |
| 33 02 0612 | Soil Analysis | 1,215 | 1,013 | 708.82 | 506.30 |
| 33 02 1101 | Geotechnical Characteristics Analysis | 54.00 | 45.00 | 31.50 | 22.50 |
| 33 02 1102 | Soil Moisture Content ASTM D2216 | 54.00 | 45.00 | 31.50 | 22.50 |
| 33 02 1507 | Continuous Monitoring and Recording of Air Flow | 23,688 | 19,740 | 13,818 | 9,870 |
| 33 02 9901 | Magnetometer | 127.20 | 106.00 | 74.20 | 53.00 |
| 33 02 9902 | Electromagnetics | 127.20 | 106.00 | 74.20 | 53.00 |

**Environmental Remediation: Assemblies Cost Book**

# Cost Data for RI/FS and RFI/CMS
## Analysis Assemblies

| Sampling Assemblies - Other Analyses | | Cost by Turnaround Time/Quality Control Level | | | |
|---|---|---|---|---|---|
| Assembly | Description | 24-72 hr CLP | 24-72 hr Std QC | Std TAT CLP | Std TAT Std QC |
| 33 02 9903 | Ground Penetrating Radar | 120.84 | 100.70 | 70.49 | 50.35 |
| 33 02 9904 | Seismic Refraction | 6,678 | 5,565 | 3,896 | 2,783 |
| 33 02 9905 | Resistivity | 6,678 | 5,565 | 3,896 | 2,783 |
| 33 02 9908 | Double-ring Infiltrometer Test (DRIT Test) | 1,320 | 1,100 | 770.00 | 550.00 |
| 33 02 9909 | Falling Head Rigid Wall Consolidometer Tech CAE 1110-2-1906 | 540.00 | 450.00 | 315.00 | 225.00 |
| 33 02 9911 | Falling Head Permeability Test EPA-9100 | 516.00 | 430.00 | 301.00 | 215.00 |
| 33 02 9912 | Constant Head Permeability Test ASTM D2432 | 463.99 | 386.66 | 270.66 | 193.33 |
| 33 02 9916 | Unconfined Compressive Strength (Pocket Penetrometer) | 9.60 | 8.00 | 5.60 | 4.00 |
| 33 19 9543 | Initial Waste Stream Evaluation, Non-PCB | 3,300 | 2,750 | 1,925 | 1,375 |
| 33 19 9544 | Initial PCB Waste Stream Evaluation | 1,515 | 1,263 | 883.75 | 631.25 |

| Sampling Assemblies - Radiological Animal Tissue and Bone | | Cost by Turnaround Time/Quality Control Level | | | |
|---|---|---|---|---|---|
| Assembly | Description | 24-72 hr CLP | 24-72 hr Std QC | Std TAT CLP | Std TAT Std QC |
| 33 02 2201 | Tissue/Bone, Americium Isotopic, Alpha Spectroscopy | 352.01 | 293.34 | 205.34 | 146.67 |
| 33 02 2202 | Tissue/Bone, Curium Isotopic, Alpha Spectroscopy | 348.00 | 290.00 | 203.00 | 145.00 |
| 33 02 2203 | Tissue/Bone, Neptunium Isotopic, Alpha Spectroscopy | 348.00 | 290.00 | 203.00 | 145.00 |
| 33 02 2204 | Tissue/Bone, Plutonium Isotopic, Alpha Spectroscopy | 360.00 | 300.00 | 210.00 | 150.00 |
| 33 02 2205 | Tissue/Bone, Thorium Isotopic, Alpha Spectroscopy | 352.01 | 293.34 | 205.34 | 146.67 |
| 33 02 2206 | Tissue/Bone, Uranium Isotopic, Alpha Spectroscopy | 324.00 | 270.00 | 189.00 | 135.00 |
| 33 02 2207 | Tissue/Bone, Strontium - 89, 90, Gas Flow Proportional Counting | 400.01 | 333.34 | 233.34 | 166.67 |
| 33 02 2208 | Tissue/Bone, Strontium - 90, Gas Flow Proportional Counting | 373.61 | 311.34 | 217.94 | 155.67 |
| 33 02 2209 | Tissue/Bone, Technetium - 99, Gas Flow Proportional Counting | 307.99 | 256.66 | 179.66 | 128.33 |
| 33 02 2210 | Tissue/Bone, Americium Isotopic, Gamma Spectroscopy | 316.01 | 263.34 | 184.34 | 131.67 |
| 33 02 2211 | Tissue/Bone, Gamma Isotopic, Gamma Spectroscopy | 336.00 | 280.00 | 196.00 | 140.00 |
| 33 02 2212 | Tissue/Bone, Iodine - 25 (Filter or Charcoal), Gamma Spectroscopy | 360.00 | 300.00 | 210.00 | 150.00 |
| 33 02 2213 | Tissue/Bone, Iodine - 129 (Filter or Charcoal), Gamma Spectroscopy | 360.00 | 300.00 | 210.00 | 150.00 |
| 33 02 2214 | Tissue/Bone, Iodine - 131 (Filter or Charcoal), Gamma Spectroscopy | 336.00 | 280.00 | 196.00 | 140.00 |

**Environmental Remediation: Assemblies Cost Book**

# Cost Data for RI/FS and RFI/CMS
## Analysis Assemblies

| Sampling Assemblies - Radiological Animal Tissue and Bone | | Cost by Turnaround Time/Quality Control Level | | | |
|---|---|---|---|---|---|
| Assembly | Description | 24-72 hr CLP | 24-72 hr Std QC | Std TAT CLP | Std TAT Std QC |
| 33 02 2215 | Tissue/Bone, Krypton - 85, Gamma Spectroscopy | 246.00 | 205.00 | 143.50 | 102.50 |
| 33 02 2216 | Tissue/Bone, Lead - 210, Gamma Spectroscopy | 316.01 | 263.34 | 184.34 | 131.67 |
| 33 02 2217 | Tissue/Bone, Radium - 226, 228, Gamma Spectroscopy | 348.00 | 290.00 | 203.00 | 145.00 |
| 33 02 2218 | Tissue/Bone, Uranium - Total, Gamma Spectroscopy | 354.00 | 295.00 | 206.50 | 147.50 |
| 33 02 2219 | Tissue/Bone, Carbon - 14, Liquid Scintillation | 318.41 | 265.34 | 185.74 | 132.67 |
| 33 02 2220 | Tissue/Bone, Uranium Isotopic, Fluorometric | 420.00 | 350.00 | 245.00 | 175.00 |
| 33 02 2221 | Tissue/Bone, Uranium Isotopic, Laser Phosphorometer | 180.00 | 150.00 | 105.00 | 75.00 |

| Sampling Assemblies - Radiological Air | | Cost by Turnaround Time/Quality Control Level | | | |
|---|---|---|---|---|---|
| Assembly | Description | 24-72 hr CLP | 24-72 hr Std QC | Std TAT CLP | Std TAT Std QC |
| 33 02 2222 | Air, Americium Isotopic, Alpha Spectroscopy | 352.01 | 293.34 | 205.34 | 146.67 |
| 33 02 2223 | Air, Curium Isotopic, Alpha Spectroscopy | 348.00 | 290.00 | 203.00 | 145.00 |
| 33 02 2224 | Air, Neptunium - 237, Alpha Spectroscopy | 348.00 | 290.00 | 203.00 | 145.00 |
| 33 02 2225 | Air, Polonium - 210, Alpha Spectroscopy | 352.01 | 293.34 | 205.34 | 146.67 |
| 33 02 2226 | Air, Plutonium Isotopic, Alpha Spectroscopy | 360.00 | 300.00 | 210.00 | 150.00 |
| 33 02 2227 | Air, Radium - 226, Alpha Spectroscopy | 343.99 | 286.66 | 200.66 | 143.33 |
| 33 02 2228 | Air, Thorium Isotopic, Alpha Spectroscopy | 352.01 | 293.34 | 205.34 | 146.67 |
| 33 02 2229 | Air, Uranium Isotopic, Alpha Spectroscopy | 352.01 | 293.34 | 205.34 | 146.67 |
| 33 02 2230 | Air, Gross Beta - Total, Gas Flow Proportional Counting | 108.00 | 90.00 | 63.00 | 45.00 |
| 33 02 2231 | Air, Lead - 210, Gas Flow Proportional Counting | 352.01 | 293.34 | 205.34 | 146.67 |
| 33 02 2232 | Air, Strontium - 89, 90, Gas Flow Proportional Counting | 414.00 | 345.00 | 241.50 | 172.50 |
| 33 02 2233 | Air, Strontium - 90, Gas Flow Proportional Counting | 324.00 | 270.00 | 189.00 | 135.00 |
| 33 02 2234 | Air, Technetium - 99, Gas Flow Proportional Counting | 340.01 | 283.34 | 198.34 | 141.67 |
| 33 02 2235 | Air, Gamma Isotopic, Gamma Spectroscopy | 264.00 | 220.00 | 154.00 | 110.00 |
| 33 02 2236 | Air, Radium - 226, 228, Gamma Spectroscopy | 424.01 | 353.34 | 247.34 | 176.67 |
| 33 02 2237 | Air, Carbon - 14, Liquid Scintillation | 378.00 | 315.00 | 220.50 | 157.50 |
| 33 02 2238 | Air, Iron - 55, Liquid Scintillation | 348.00 | 290.00 | 203.00 | 145.00 |
| 33 02 2239 | Air, Gross Alpha - Total, Gas Flow Proportional Counting or Alpha Scintillation | 103.99 | 86.66 | 60.66 | 43.33 |
| 33 02 2240 | Air, Gross Alpha & Gross Beta - Total, Gas Flow | 103.99 | 86.66 | 60.66 | 43.33 |

**Environmental Remediation: Assemblies Cost Book**

# Cost Data for RI/FS and RFI/CMS
## Analysis Assemblies

| Sampling Assemblies - Radiological Air | | Cost by Turnaround Time/Quality Control Level | | | |
|---|---|---|---|---|---|
| Assembly | Description | 24-72 hr CLP | 24-72 hr Std QC | Std TAT CLP | Std TAT Std QC |
| 33 02 2241 | Air, Plutonium Isotopic, Plutonium - 241, Alpha Spectroscopy | 340.01 | 283.34 | 198.34 | 141.67 |
| 33 02 2242 | Air, Radium - 226, Radon De-emanation | 187.99 | 156.66 | 109.66 | 78.33 |
| 33 02 2243 | Air, Uranium - Total, Laser Phosphorometer | 144.00 | 120.00 | 84.00 | 60.00 |
| 33 02 2244 | Air, Uranium - Total, Fluorometric | 144.00 | 120.00 | 84.00 | 60.00 |

| Sampling Assemblies - Radiological Miscellaneous | | Cost by Turnaround Time/Quality Control Level | | | |
|---|---|---|---|---|---|
| Assembly | Description | 24-72 hr CLP | 24-72 hr Std QC | Std TAT CLP | Std TAT Std QC |
| 33 02 2356 | Vegetation/Soil/Sediment, Uranium - Total, Fluorometric | 120.00 | 100.00 | 70.00 | 50.00 |

| Sampling Assemblies - Radiological Urine and Feces | | Cost by Turnaround Time/Quality Control Level | | | |
|---|---|---|---|---|---|
| Assembly | Description | 24-72 hr CLP | 24-72 hr Std QC | Std TAT CLP | Std TAT Std QC |
| 33 02 2301 | Urine & Feces, Americium Isotopic, Alpha Spectroscopy | 342.00 | 285.00 | 199.50 | 142.50 |
| 33 02 2302 | Urine & Feces, Curium Isotopic, Alpha Spectroscopy | 366.00 | 305.00 | 213.50 | 152.50 |
| 33 02 2303 | Urine & Feces, Neptunium - 237, Alpha Spectroscopy | 366.00 | 305.00 | 213.50 | 152.50 |
| 33 02 2304 | Urine & Feces, Polonium - 210, Alpha Spectroscopy | 306.00 | 255.00 | 178.50 | 127.50 |
| 33 02 2305 | Urine & Feces, Plutonium Isotopic, Alpha Spectroscopy | 360.00 | 300.00 | 210.00 | 150.00 |
| 33 02 2306 | Urine & Feces, Thorium Isotopic, Alpha Spectroscopy | 330.00 | 275.00 | 192.50 | 137.50 |
| 33 02 2307 | Urine & Feces, Uranium Isotopic, Alpha Spectroscopy | 330.00 | 275.00 | 192.50 | 137.50 |
| 33 02 2308 | Urine & Feces, Gross Beta - Total, Gas Flow Proportional Counting | 192.00 | 160.00 | 112.00 | 80.00 |
| 33 02 2309 | Urine & Feces, Ruthenium - 106, Gas Flow Proportional Counting | 294.00 | 245.00 | 171.50 | 122.50 |
| 33 02 2310 | Urine & Feces, Strontium - 89, 90, Gas Flow Proportional Counting | 432.00 | 360.00 | 252.00 | 180.00 |
| 33 02 2311 | Urine & Feces, Strontium - 90, Gas Flow Proportional Counting | 432.00 | 360.00 | 252.00 | 180.00 |
| 33 02 2312 | Urine & Feces, Cerium - 144, Gamma Spectroscopy | 294.00 | 245.00 | 171.50 | 122.50 |
| 33 02 2313 | Urine & Feces, Gamma Isotopic, Gamma Spectroscopy | 324.00 | 270.00 | 189.00 | 135.00 |
| 33 02 2314 | Urine & Feces, Iodine - 131, Gamma Spectroscopy | 324.00 | 270.00 | 189.00 | 135.00 |

**Environmental Remediation: Assemblies Cost Book**

# Cost Data for RI/FS and RFI/CMS
## Analysis Assemblies

| Sampling Assemblies - Radiological Urine and Feces | | Cost by Turnaround Time/Quality Control Level | | | |
|---|---|---|---|---|---|
| **Assembly** | **Description** | **24-72 hr CLP** | **24-72 hr Std QC** | **Std TAT CLP** | **Std TAT Std QC** |
| 33 02 2315 | Urine & Feces, Radium - 228, Gamma Spectroscopy | 324.00 | 270.00 | 189.00 | 135.00 |
| 33 02 2316 | Urine & Feces, Ruthenium - 106, Gamma Spectroscopy | 294.00 | 245.00 | 171.50 | 122.50 |
| 33 02 2317 | Urine & Feces, Carbon - 14, Liquid Scintillation | 396.00 | 330.00 | 231.00 | 165.00 |
| 33 02 2318 | Urine & Feces, Promethium - 147, Liquid Scintillation | 246.00 | 205.00 | 143.50 | 102.50 |
| 33 02 2319 | Urine & Feces, Phosphorus - 32, Liquid Scintillation | 948.00 | 790.00 | 553.00 | 395.00 |
| 33 02 2320 | Urine & Feces, Tritium (Direct Counting), Liquid Scintillation | 240.00 | 200.00 | 140.00 | 100.00 |
| 33 02 2321 | Urine & Feces, Tritium (Distillation), Liquid Scintillation | 300.00 | 250.00 | 175.00 | 125.00 |
| 33 02 2322 | Urine & Feces, Mixed Fission Products, Gas Flow | 228.00 | 190.00 | 133.00 | 95.00 |
| 33 02 2323 | Urine & Feces, Plutonium Isotopic, Plutonium - 241, Alpha Spectroscopy | 366.00 | 305.00 | 213.50 | 152.50 |
| 33 02 2324 | Urine & Feces, Radium - 226, Radon De-emanation | 210.00 | 175.00 | 122.50 | 87.50 |
| 33 02 2325 | Urine & Feces, Radium - 226, 228, Radon De-emanation | 348.00 | 290.00 | 203.00 | 145.00 |
| 33 02 2326 | Urine & Feces, Uranium - Total, Laser Phosphorometer | 180.00 | 150.00 | 105.00 | 75.00 |
| 33 02 2327 | Urine & Feces, Uranium - Total, Fluorometric | 168.00 | 140.00 | 98.00 | 70.00 |
| 33 02 2328 | Urine & Feces, Uranium - Total, Radiometric, Alpha Scintillation | 198.00 | 165.00 | 115.50 | 82.50 |
| 33 02 2329 | Urine & Feces, Uranium - Total, Laser Phosphorometer | 198.00 | 165.00 | 115.50 | 82.50 |

| Sampling Assemblies - Radiological Vegetation, Sediment, and Soil | | Cost by Turnaround Time/Quality Control Level | | | |
|---|---|---|---|---|---|
| **Assembly** | **Description** | **24-72 hr CLP** | **24-72 hr Std QC** | **Std TAT CLP** | **Std TAT Std QC** |
| 33 02 2330 | Vegetation/Soil/Sediment, Americium Isotopic, Alpha Spectroscopy | 336.00 | 280.00 | 196.00 | 140.00 |
| 33 02 2331 | Vegetation/Soil/Sediment, Curium Isotopic, Alpha Spectroscopy | 331.99 | 276.66 | 193.66 | 138.33 |
| 33 02 2332 | Vegetation/Soil/Sediment, Neptunium - 237, Alpha Spectroscopy | 331.99 | 276.66 | 193.66 | 138.33 |
| 33 02 2333 | Vegetation/Soil/Sediment, Plutonium Isotopic, Alpha Spectroscopy | 343.99 | 286.66 | 200.66 | 143.33 |
| 33 02 2334 | Vegetation/Soil/Sediment, Thorium Isotopic, Alpha Spectroscopy | 336.00 | 280.00 | 196.00 | 140.00 |
| 33 02 2335 | Vegetation/Soil/Sediment, Uranium Isotopic, Alpha Spectroscopy | 336.00 | 280.00 | 196.00 | 140.00 |

**Environmental Remediation: Assemblies Cost Book**

# Cost Data for RI/FS and RFI/CMS
## Analysis Assemblies

| Sampling Assemblies - Radiological Vegetation, Sediment, and Soil | | Cost by Turnaround Time/Quality Control Level | | | |
|---|---|---|---|---|---|
| Assembly | Description | 24-72 hr CLP | 24-72 hr Std QC | Std TAT CLP | Std TAT Std QC |
| 33 02 2336 | Vegetation/Soil/Sediment, Gross Beta - Total, Gas Flow Proportional Counting | 120.00 | 100.00 | 70.00 | 50.00 |
| 33 02 2337 | Vegetation/Soil/Sediment, Ruthenium - 106, Gas Flow Proportional Counting | 348.00 | 290.00 | 203.00 | 145.00 |
| 33 02 2338 | Vegetation/Soil/Sediment, Strontium - 89, 90, Gas Flow Proportional Counting | 408.00 | 340.00 | 238.00 | 170.00 |
| 33 02 2339 | Vegetation/Soil/Sediment, Strontium - 90, Gas Flow Proportional Counting | 324.00 | 270.00 | 189.00 | 135.00 |
| 33 02 2340 | Vegetation/Soil/Sediment, Technetium - 99, Gas Flow Proportional Counting | 300.00 | 250.00 | 175.00 | 125.00 |
| 33 02 2341 | Vegetation/Soil/Sediment, Americium Isotopic, Gamma Spectroscopy | 295.99 | 246.66 | 172.66 | 123.33 |
| 33 02 2342 | Vegetation/Soil/Sediment, Gamma Isotopic, Gamma Spectroscopy | 295.99 | 246.66 | 172.66 | 123.33 |
| 33 02 2343 | Vegetation/Soil/Sediment, Iodine - 129, Gamma Spectroscopy | 295.99 | 246.66 | 172.66 | 123.33 |
| 33 02 2344 | Vegetation/Soil/Sediment, Lead - 210, Gamma Spectroscopy | 316.01 | 263.34 | 184.34 | 131.67 |
| 33 02 2345 | Vegetation/Soil/Sediment, Polonium - 210, Gamma Spectroscopy | 316.01 | 263.34 | 184.34 | 131.67 |
| 33 02 2346 | Vegetation/Soil/Sediment, Radium - 226, 228, Gamma Spectroscopy | 295.99 | 246.66 | 172.66 | 123.33 |
| 33 02 2347 | Vegetation/Soil/Sediment, Ruthenium - 106, Gamma Spectroscopy | 216.00 | 180.00 | 126.00 | 90.00 |
| 33 02 2348 | Vegetation/Soil/Sediment, Uranium - Total, Gamma Spectroscopy | 162.00 | 135.00 | 94.50 | 67.50 |
| 33 02 2349 | Vegetation/Soil/Sediment, Carbon - 14, Liquid Scintillation | 360.00 | 300.00 | 210.00 | 150.00 |
| 33 02 2350 | Vegetation/Soil/Sediment, Nickel - 63, Liquid Scintillation | 348.00 | 290.00 | 203.00 | 145.00 |
| 33 02 2351 | Vegetation/Soil/Sediment, Tritium, Liquid Scintillation | 144.00 | 120.00 | 84.00 | 60.00 |
| 33 02 2352 | Vegetation/Soil/Sediment, Gross Alpha - Total, Gas Flow Proportional Counting | 103.99 | 86.66 | 60.66 | 43.33 |
| 33 02 2353 | Vegetation/Soil/Sediment, Gross Alpha/Gross Beta - Total, Gas Flow | 120.00 | 100.00 | 70.00 | 50.00 |
| 33 02 2354 | Vegetation/Soil/Sediment, Plutonium Isotopic/Plutonium 241, Alpha Spectroscopy | 355.99 | 296.66 | 207.66 | 148.33 |
| 33 02 2355 | Vegetation/Soil/Sediment, Uranium - Total, Laser Phosphorometer | 144.00 | 120.00 | 84.00 | 60.00 |

**Environmental Remediation: Assemblies Cost Book**

# Cost Data for RI/FS and RFI/CMS
## Analysis Assemblies

### Sampling Assemblies - Radiological Water and Liquid

| Assembly | Description | 24-72 hr CLP | 24-72 hr Std QC | Std TAT CLP | Std TAT Std QC |
|---|---|---|---|---|---|
| 33 02 2245 | Liquid, Americium Isotopic, Alpha Spectroscopy | 316.01 | 263.34 | 184.34 | 131.67 |
| 33 02 2246 | Liquid, Curium Isotopic, Alpha Spectroscopy | 331.99 | 276.66 | 193.66 | 138.33 |
| 33 02 2247 | Liquid, Neptunium - 237, Alpha Spectroscopy | 331.99 | 276.66 | 193.66 | 138.33 |
| 33 02 2248 | Liquid, Polonium - 210, Alpha Spectroscopy | 316.01 | 263.34 | 184.34 | 131.67 |
| 33 02 2249 | Liquid, Plutonium Isotopic, Alpha Spectroscopy | 360.00 | 300.00 | 210.00 | 150.00 |
| 33 02 2250 | Liquid, Radium - 226, Alpha Spectroscopy | 304.01 | 253.34 | 177.34 | 126.67 |
| 33 02 2251 | Liquid, Radium - 226, 228, Alpha Spectroscopy | 400.01 | 333.34 | 233.34 | 166.67 |
| 33 02 2252 | Liquid, Thorium Isotopic, Alpha Spectroscopy | 316.01 | 263.34 | 184.34 | 131.67 |
| 33 02 2253 | Liquid, Uranium Isotopic, Alpha Spectroscopy | 316.01 | 263.34 | 184.34 | 131.67 |
| 33 02 2254 | Liquid, Iodine - 131, Beta/Gamma Coincidence | 300.00 | 250.00 | 175.00 | 125.00 |
| 33 02 2255 | Liquid, Radium - 226, 228, Beta/Gamma Coincidence | 379.99 | 316.66 | 221.66 | 158.33 |
| 33 02 2256 | Liquid, Radium - 228, Beta/Gamma Coincidence | 235.99 | 196.66 | 137.66 | 98.33 |
| 33 02 2257 | Liquid, Gross Beta - Total, Gas Flow Proportional Count | 103.99 | 86.66 | 60.66 | 43.33 |
| 33 02 2258 | Liquid, Calcium - 45, Gas Flow Proportional Counting | 348.00 | 290.00 | 203.00 | 145.00 |
| 33 02 2259 | Liquid, Cerium - 144, Gas Flow Proportional Counting | 324.00 | 270.00 | 189.00 | 135.00 |
| 33 02 2260 | Liquid, Chlorine - 36, Gas Flow Proportional Counting | 324.00 | 270.00 | 189.00 | 135.00 |
| 33 02 2261 | Liquid, Iodine - 131, Gas Flow Proportional Counting | 324.00 | 270.00 | 189.00 | 135.00 |
| 33 02 2262 | Liquid, Lead - 210, Gas Flow Proportional Counting | 324.00 | 270.00 | 189.00 | 135.00 |
| 33 02 2263 | Liquid, Radium - 226, 228, Gas Flow Proportional Count | 400.01 | 333.34 | 233.34 | 166.67 |
| 33 02 2264 | Liquid, Radium - 228, Gas Flow Proportional Counting | 304.01 | 253.34 | 177.34 | 126.67 |
| 33 02 2265 | Liquid, Ruthenium - 106, Gas Flow Proportional Counting | 348.00 | 290.00 | 203.00 | 145.00 |
| 33 02 2266 | Liquid, Strontium - 89, 90, Gas Flow Proportional Count | 355.99 | 296.66 | 207.66 | 148.33 |
| 33 02 2267 | Liquid, Strontium - 90, Gas Flow Proportional Counting | 331.99 | 276.66 | 193.66 | 138.33 |
| 33 02 2268 | Liquid, Sulfur - 35, Gas Flow Proportional Counting | 426.00 | 355.00 | 248.50 | 177.50 |
| 33 02 2269 | Liquid, Technetium - 99, Gas Flow Proportional Counting | 328.01 | 273.34 | 191.34 | 136.67 |
| 33 02 2270 | Liquid, Cerium - 144, Gamma Spectroscopy | 214.80 | 179.00 | 125.30 | 89.50 |
| 33 02 2271 | Liquid, Gamma Isotopic, Gamma Spectroscopy | 214.80 | 179.00 | 125.30 | 89.50 |
| 33 02 2272 | Liquid, Iodine - 125, Gamma Spectroscopy | 214.80 | 179.00 | 125.30 | 89.50 |
| 33 02 2273 | Liquid Iodine - 129, Gamma Spectroscopy | 214.80 | 179.00 | 125.30 | 89.50 |
| 33 02 2274 | Liquid, Ruthenium - 106, Gamma Spectroscopy | 214.80 | 179.00 | 125.30 | 89.50 |

**Environmental Remediation: Assemblies Cost Book**

# Cost Data for RI/FS and RFI/CMS
## Analysis Assemblies

| Sampling Assemblies - Radiological Water and Liquid | | Cost by Turnaround Time/Quality Control Level | | | |
|---|---|---|---|---|---|
| Assembly | Description | 24-72 hr CLP | 24-72 hr Std QC | Std TAT CLP | Std TAT Std QC |
| 33 02 2275 | Liquid, Sulfur - 35, Gamma Spectroscopy | 214.80 | 179.00 | 125.30 | 89.50 |
| 33 02 2276 | Liquid, Chlorine - 36, Liquid Scintillation | 324.00 | 270.00 | 189.00 | 135.00 |
| 33 02 2277 | Liquid, Carbon - 14, Liquid Scintillation | 342.00 | 285.00 | 199.50 | 142.50 |
| 33 02 2278 | Liquid, Iron - 55, Liquid Scintillation | 348.00 | 290.00 | 203.00 | 145.00 |
| 33 02 2279 | Liquid, Iron - 125, Liquid Scintillation | 240.00 | 200.00 | 140.00 | 100.00 |
| 33 02 2280 | Liquid, Iron - 129, Liquid Scintillation | 240.00 | 200.00 | 140.00 | 100.00 |
| 33 02 2281 | Liquid, Nickel - 63, Liquid Scintillation | 348.00 | 290.00 | 203.00 | 145.00 |
| 33 02 2282 | Liquid, Promethium - 147, Liquid Scintillation | 324.00 | 270.00 | 189.00 | 135.00 |
| 33 02 2283 | Liquid, Sulfur - 35, Liquid Scintillation | 324.00 | 270.00 | 189.00 | 135.00 |
| 33 02 2284 | Liquid, Tritium, (Direct Counting), Liquid Scintillation | 162.00 | 135.00 | 94.50 | 67.50 |
| 33 02 2285 | Liquid, Tritium, (Distillation), Liquid Scintillation | 172.01 | 143.34 | 100.34 | 71.67 |
| 33 02 2286 | Liquid, Tritium, (Electrolytic Enrichment), Liquid Scintillation | 174.00 | 145.00 | 101.50 | 72.50 |
| 33 02 2287 | Liquid, Gross Alpha - Total, Gas Flow &/or Alpha Spectroscopy | 132.00 | 110.00 | 77.00 | 55.00 |
| 33 02 2288 | Liquid, Suspended or Dissolved, Gross Alpha/Gross Beta | 192.00 | 160.00 | 112.00 | 80.00 |
| 33 02 2289 | Liquid, Suspended & Dissolved, Gross Alpha/Gross Beta | 192.00 | 160.00 | 112.00 | 80.00 |
| 33 02 2290 | Liquid, Gross Alpha & Gross Beta - Total | 132.00 | 110.00 | 77.00 | 55.00 |
| 33 02 2291 | Liquid, Plutonium Isotopic/Plutonium - 241, Alpha Spectroscopy | 427.99 | 356.66 | 249.66 | 178.33 |
| 33 02 2292 | Liquid, Total Radium, Alpha Scintillation | 168.00 | 140.00 | 98.00 | 70.00 |
| 33 02 2293 | Liquid, Radium - 226, Radon De-emanation | 187.99 | 156.66 | 109.66 | 78.33 |
| 33 02 2294 | Liquid, Radium - 226, 228, Radon De-emanation | 414.00 | 345.00 | 241.50 | 172.50 |
| 33 02 2295 | Liquid, Radon 222 | 168.00 | 140.00 | 98.00 | 70.00 |
| 33 02 2296 | Liquid, Uranium - Total, Laser Phosphorometer | 168.00 | 140.00 | 98.00 | 70.00 |
| 33 02 2297 | Liquid, Uranium - Total, Fluorometric | 120.00 | 100.00 | 70.00 | 50.00 |

| Sampling Assemblies - Soil, Sludge, and Sediment | | Cost by Turnaround Time/Quality Control Level | | | |
|---|---|---|---|---|---|
| Assembly | Description | 24-72 hr CLP | 24-72 hr Std QC | Std TAT CLP | Std TAT Std QC |
| 33 02 1701 | EP Toxicity, Metals (EPA 1310) | 350.40 | 292.00 | 204.40 | 146.00 |
| 33 02 1702 | TCLP (RCRA) (EPA 1311) | 3,274 | 2,729 | 1,910 | 1,364 |
| 33 02 1703 | Chromium (SW 7191), with Prep | 67.99 | 56.66 | 39.66 | 28.33 |

**Environmental Remediation: Assemblies Cost Book**

# Cost Data for RI/FS and RFI/CMS
## Analysis Assemblies

### Sampling Assemblies - Soil, Sludge, and Sediment

| Assembly | Description | 24-72 hr CLP | 24-72 hr Std QC | Std TAT CLP | Std TAT Std QC |
|---|---|---|---|---|---|
| 33 02 1705 | Targeted TCLP (Metals, Volatiles, Semivolatiles only) | 1,606 | 1,339 | 937.06 | 669.33 |
| 33 02 1706 | Chromium (EPA 6010) | 28.80 | 24.00 | 16.80 | 12.00 |
| 33 02 1708 | Purgeable Aromatics (SW 5030/SW 8020) | 232.01 | 193.34 | 135.34 | 96.67 |
| 33 02 1709 | Target Analyte List Metals (EPA 6010/7000S), Soil | 695.21 | 579.34 | 405.54 | 289.67 |
| 33 02 1710 | Metals (EPA 6010), per Each Metal | 27.19 | 22.66 | 15.86 | 11.33 |
| 33 02 1711 | Arsenic (SW 7060), with Prep | 49.61 | 41.34 | 28.94 | 20.67 |
| 33 02 1712 | Mercury (SW 7470), with Prep | 71.21 | 59.34 | 41.54 | 29.67 |
| 33 02 1713 | Selenium (SW 7740), with Prep | 55.99 | 46.66 | 32.66 | 23.33 |
| 33 02 1714 | BTEX Purgeable Aromatics (SW 5030/SW 8020) | 196.01 | 163.34 | 114.34 | 81.67 |
| 33 02 1715 | BTEX/Gasoline Hydrocarbons (Mod 8020)(PID/FID), with Prep | 199.99 | 166.66 | 116.66 | 83.33 |
| 33 02 1716 | Phenols (SW 3550/SW 8040) | 180.79 | 150.66 | 105.46 | 75.33 |
| 33 02 1717 | Pesticides/PCBs (SW 3550/SW 8080) | 496.01 | 413.34 | 289.34 | 206.67 |
| 33 02 1718 | Chlordane (SW 3550/SW 8080) | 511.99 | 426.66 | 298.66 | 213.33 |
| 33 02 1719 | Chlorinated Phenoxy Acid Herbicides (SW 3550/SW 8150) | 559.99 | 466.66 | 326.66 | 233.33 |
| 33 02 1720 | Volatile Organic Analysis (SW 5030/SW 8240) | 420.00 | 350.00 | 245.00 | 175.00 |
| 33 02 1721 | Base/Neutral & Acid Extractable Organics (SW 3550/SW 8270) | 1,096 | 913.34 | 639.34 | 456.67 |
| 33 02 1722 | Polynuclear Aromatic Hydrocarbons (PAH) (SW 8310), with Prep | 451.99 | 376.66 | 263.66 | 188.33 |
| 33 02 1723 | Cyanide (EPA 9010) | 115.99 | 96.66 | 67.66 | 48.33 |
| 33 02 1724 | Phenols (EPA 9065) | 104.81 | 87.34 | 61.14 | 43.67 |
| 33 02 1725 | CEC (EPA 9081) | 342.00 | 285.00 | 199.50 | 142.50 |
| 33 02 1726 | Diesel Hydrocarbon, Mod 8015, GC/FID, with Prep | 258.00 | 215.00 | 150.50 | 107.50 |
| 33 02 1728 | Gasoline Hydrocarbon, Mod 8015, GC/FID, with Prep | 208.01 | 173.34 | 121.34 | 86.67 |
| 33 02 1729 | Hydrocarbon Screen GC/FID, Mod 8015, with Prep | 316.01 | 263.34 | 184.34 | 131.67 |
| 33 02 1730 | Sludge Constituents and Characteristics Analysis Tests | 343.20 | 286.00 | 200.20 | 143.00 |
| 33 02 1731 | Purgeable Halocarbons (SW 5030/SW 8010) | 271.99 | 226.66 | 158.66 | 113.33 |
| 33 02 1732 | Total Petroleum Hydrocarbons (SW 5030/SW 8015) | 160.01 | 133.34 | 93.34 | 66.67 |
| 33 02 1733 | Acrolein, Acrylonitrile, Acetonitrile (SW 5030/SW 8030) | 271.99 | 226.66 | 158.66 | 113.33 |
| 33 02 1734 | Phthalate Esters (SW 3550/SW 8060/SW 8061) | 432.00 | 360.00 | 252.00 | 180.00 |
| 33 02 1735 | Nitroaromatics & Cyclic Ketones (SW 3550/SW 8090/SW8091) | 432.00 | 360.00 | 252.00 | 180.00 |

**Environmental Remediation: Assemblies Cost Book**

# Cost Data for RI/FS and RFI/CMS
## Analysis Assemblies

| Sampling Assemblies - Soil, Sludge, and Sediment | | Cost by Turnaround Time/Quality Control Level | | | |
|---|---|---|---|---|---|
| | | 24-72 hr CLP | 24-72 hr Std QC | Std TAT CLP | Std TAT Std QC |
| Assembly | Description | | | | |
| 33 02 1736 | Polynuclear Aromatic Hydrocarbons (SW 3550/SW 8100) | 451.99 | 376.66 | 263.66 | 188.33 |
| 33 02 1737 | Chlorinated Hydrocarbons (SW 3550/SW 8120/SW 8121) | 360.00 | 300.00 | 210.00 | 150.00 |
| 33 02 1738 | Organophosphorus Pesticides (SW 3550/SW 8140/SW 8141) | 516.00 | 430.00 | 301.00 | 215.00 |
| 33 02 1739 | Semivolatile Organics, Packed Column (SW 8250), with Prep | 648.00 | 540.00 | 378.00 | 270.00 |
| 33 02 1740 | Dioxins & Dibenzofurans (SW 3550/SW 8280) | 456.00 | 380.00 | 266.00 | 190.00 |
| 33 02 1741 | Total Organic Halides, TOX (EPA 9020/9021) | 180.00 | 150.00 | 105.00 | 75.00 |
| 33 02 1742 | Total Dissolved Sulfide (EPA 9030) | 40.01 | 33.34 | 23.34 | 16.67 |
| 33 02 1743 | Sulfate (EPA 9035/9036/9038) | 39.17 | 32.64 | 22.85 | 16.32 |
| 33 02 1744 | pH (EPA 9040/9041/9045) | 24.00 | 20.00 | 14.00 | 10.00 |
| 33 02 1745 | Specific Conductance, Conductivity (EPA 9050) | 24.00 | 20.00 | 14.00 | 10.00 |
| 33 02 1746 | Total Organic Carbon, TOC (EPA 9060) | 85.61 | 71.34 | 49.94 | 35.67 |
| 33 02 1747 | Total Recoverable Oil & Grease (EPA 9070) | 156.00 | 130.00 | 91.00 | 65.00 |
| 33 02 1748 | Oil & Grease Extraction Method (EPA 9071) | 115.99 | 96.66 | 67.66 | 48.33 |
| 33 02 1749 | Waste/Membrane Liner Compatibility Test (EPA 9090) | 258.00 | 215.00 | 150.50 | 107.50 |
| 33 02 1750 | Paint Filter Test (EPA 9095) | 54.00 | 45.00 | 31.50 | 22.50 |
| 33 02 1751 | Saturated Hydraulic Conductivity/Leachate/Intrinsic Permeability (EPA 9100) | 627.60 | 523.00 | 366.10 | 261.50 |
| 33 02 1752 | Total Coliform (EPA 9131/9132) | 62.40 | 52.00 | 36.40 | 26.00 |
| 33 02 1753 | Nitrate (EPA 9200) | 41.59 | 34.66 | 24.26 | 17.33 |
| 33 02 1754 | Chloride (EPA 9250/9251/9252) | 73.20 | 61.00 | 42.70 | 30.50 |
| 33 02 1756 | Ignitability (EPA 1010/1020) | 76.01 | 63.34 | 44.34 | 31.67 |
| 33 02 1757 | Corrosivity (EPA 1110) | 220.01 | 183.34 | 128.34 | 91.67 |
| 33 02 1758 | Reactivity, SW 846 Ch7 P7.4 | 187.99 | 156.66 | 109.66 | 78.33 |
| 33 02 1759 | Sorbent Cartridges from VOC Sampling (EPA 5040) | 337.99 | 281.66 | 197.16 | 140.83 |
| 33 02 1760 | Metals, Flame, 7000s, per Each | 27.19 | 22.66 | 15.86 | 11.33 |
| 33 02 1761 | Metals, Furnace, 7000s, per Each | 78.00 | 65.00 | 45.50 | 32.50 |
| 33 02 1762 | Chromium, Hexavalent (SW 7195), with Prep | 42.00 | 35.00 | 24.50 | 17.50 |
| 33 02 1764 | Selenium, Hydride (SW 7741), with Prep | 91.99 | 76.66 | 53.66 | 38.33 |
| 33 02 1765 | Arsenic, Hydride (SW 7061), with Prep | 91.99 | 76.66 | 53.66 | 38.33 |
| 33 02 1766 | Total Petroleum Hydrocarbons (EPA 9073) | 160.01 | 133.34 | 93.34 | 66.67 |
| 33 02 1767 | Chromium, Total (SW 7190) with Prep | 38.40 | 32.00 | 22.40 | 16.00 |

**Environmental Remediation: Assemblies Cost Book**

*Page 2-32*

1998 by ECHOS. All rights reserved

# Cost Data for RI/FS and RFI/CMS
## Analysis Assemblies

### Sampling Assemblies - Soil, Sludge, and Sediment

| | | Cost by Turnaround Time/Quality Control Level | | | |
|---|---|---|---|---|---|
| Assembly | Description | 24-72 hr CLP | 24-72 hr Std QC | Std TAT CLP | Std TAT Std QC |
| 33 02 1768 | Antimony (SW 7040) with Prep | 82.80 | 69.00 | 48.30 | 34.50 |
| 33 02 1769 | Antimony (SW 7041) with Prep | 60.00 | 50.00 | 35.00 | 25.00 |
| 33 02 1770 | Thallium (SW 7840) with Prep | 82.80 | 69.00 | 48.30 | 34.50 |
| 33 02 1771 | Thallium (SW 7841) with Prep | 60.00 | 50.00 | 35.00 | 25.00 |
| 33 02 1772 | Gasoline Group (Mod 8010/8020, Lead, EDB) | 216.00 | 180.00 | 126.00 | 90.00 |
| 33 02 1773 | Kerosene Group (Mod 8010/8020, 8100, TPH, Lead, EDB) | 216.00 | 180.00 | 126.00 | 90.00 |
| 33 02 1774 | BTEX/MTBE (Mod 8020) | 199.99 | 166.66 | 116.66 | 83.33 |
| 33 02 1775 | Purgeable Organics (SW 3550/SW 8260) | 420.00 | 350.00 | 245.00 | 175.00 |
| 33 02 2124 | Metal Analysis, Priority 17 Metals | 314.40 | 262.00 | 183.40 | 131.00 |

### Sampling Assemblies - Water and Liquids

| | | Cost by Turnaround Time/Quality Control Level | | | |
|---|---|---|---|---|---|
| Assembly | Description | 24-72 hr CLP | 24-72 hr Std QC | Std TAT CLP | Std TAT Std QC |
| 33 02 1601 | Specific Conductance (EPA 120.1) | 24.79 | 20.66 | 14.46 | 10.33 |
| 33 02 1602 | pH (EPA 150.1) | 24.00 | 20.00 | 14.00 | 10.00 |
| 33 02 1603 | Total Dissolved Solids (EPA 160.1) | 36.00 | 30.00 | 21.00 | 15.00 |
| 33 02 1604 | Total Suspended Solids (EPA 160.2) | 36.00 | 30.00 | 21.00 | 15.00 |
| 33 02 1605 | Total Solids (EPA 160.3) | 34.39 | 28.66 | 20.06 | 14.33 |
| 33 02 1606 | Total Volatile Solids (EPA 160.4) | 44.81 | 37.34 | 26.14 | 18.67 |
| 33 02 1607 | Temperature (EPA 170.1) | 24.00 | 20.00 | 14.00 | 10.00 |
| 33 02 1608 | Nitrogen/Nitrite/Nitrate (EPA 300.1/354.1) | 47.21 | 39.34 | 27.54 | 19.67 |
| 33 02 1609 | Acidity/Alkalinity (EPA 305.1) | 36.00 | 30.00 | 21.00 | 15.00 |
| 33 02 1610 | Cyanide (EPA 335.2) | 108.00 | 90.00 | 63.00 | 45.00 |
| 33 02 1611 | Ammonia Nitrogen (EPA 350.2) | 72.00 | 60.00 | 42.00 | 30.00 |
| 33 02 1612 | Phosphorus (EPA 365.3) | 67.20 | 56.00 | 39.20 | 28.00 |
| 33 02 1613 | Oil and Grease (EPA 413.2) | 132.00 | 110.00 | 77.00 | 55.00 |
| 33 02 1614 | Total Petroleum Hydrocarbons (EPA 418.1) | 160.01 | 133.34 | 93.34 | 66.67 |
| 33 02 1615 | Phenols (Method 420.1) | 104.81 | 87.34 | 61.14 | 43.67 |
| 33 02 1617 | Pesticides/PCBs (EPA 608) | 496.01 | 413.34 | 289.34 | 206.67 |
| 33 02 1618 | Volatile Organic Analysis (EPA 624) | 420.00 | 350.00 | 245.00 | 175.00 |
| 33 02 1619 | Base Neutral & Acid Extractable Organics (EPA 625) | 1,096 | 913.34 | 639.34 | 456.67 |
| 33 02 1620 | Target Analyte List Metals (EPA 6010/7000s), Water | 695.21 | 579.34 | 405.54 | 289.67 |

**Environmental Remediation: Assemblies Cost Book**

# Cost Data for RI/FS and RFI/CMS
## Analysis Assemblies

| Sampling Assemblies - Water and Liquids | | Cost by Turnaround Time/Quality Control Level | | | |
|---|---|---|---|---|---|
| | | 24-72 hr CLP | 24-72 hr Std QC | Std TAT CLP | Std TAT Std QC |
| **Assembly** | **Description** | | | | |
| 33 02 1621 | Purgeable Halocarbons (EPA 601) | 271.99 | 226.66 | 158.66 | 113.33 |
| 33 02 1622 | Purgeable Aromatics (EPA 602) | 232.01 | 193.34 | 135.34 | 96.67 |
| 33 02 1623 | Acrolein & Acrylonitrile (EPA 603) | 271.99 | 226.66 | 158.66 | 113.33 |
| 33 02 1624 | Phenols (EPA 604) | 400.01 | 333.34 | 233.34 | 166.67 |
| 33 02 1625 | Benzidines (EPA 605) | 480.00 | 400.00 | 280.00 | 200.00 |
| 33 02 1626 | Phthalate Esters (EPA 606) | 432.00 | 360.00 | 252.00 | 180.00 |
| 33 02 1627 | Nitrosamines (EPA 607) | 400.01 | 333.34 | 233.34 | 166.67 |
| 33 02 1628 | Nitroaromatics & Isophorone (EPA 609) | 432.00 | 360.00 | 252.00 | 180.00 |
| 33 02 1629 | Polynuclear Aromatic Hydrocarbons, PAH (EPA 610) | 451.99 | 376.66 | 263.66 | 188.33 |
| 33 02 1630 | Haloethers (EPA 611) | 360.00 | 300.00 | 210.00 | 150.00 |
| 33 02 1631 | Chlorinated Hydrocarbons (EPA 612) | 360.00 | 300.00 | 210.00 | 150.00 |
| 33 02 1632 | 2,3,7,8-Tetrachloro-Dibzo-P-Dioxin (EPA 613) | 438.00 | 365.00 | 255.50 | 182.50 |
| 33 02 1633 | Organophosphorus Pesticides (EPA 614) | 516.00 | 430.00 | 301.00 | 215.00 |
| 33 02 1634 | Chlorinated Phenoxy Acid Herbicide (EPA 615) | 540.00 | 450.00 | 315.00 | 225.00 |
| 33 02 1635 | Organochlorine Pesticides & PCBs (EPA 617) | 374.40 | 312.00 | 218.40 | 156.00 |
| 33 02 1636 | Triazine Pesticides (EPA 619) | 439.99 | 366.66 | 256.66 | 183.33 |
| 33 02 1637 | Carbamate & Urea Pesticides (EPA 632) | 420.00 | 350.00 | 245.00 | 175.00 |
| 33 02 1638 | Color (EPA 110.2/110.3) | 40.80 | 34.00 | 23.80 | 17.00 |
| 33 02 1639 | Hardness, Total (EPA 130.2) | 37.20 | 31.00 | 21.70 | 15.50 |
| 33 02 1640 | Odor (EPA 140.1) | 27.19 | 22.66 | 15.86 | 11.33 |
| 33 02 1641 | Settleable Solids (EPA 160.5) | 32.81 | 27.34 | 19.14 | 13.67 |
| 33 02 1642 | Total Solids (EPA 166.3) | 34.39 | 28.66 | 20.06 | 14.33 |
| 33 02 1643 | Turbidity (EPA 180.1) | 30.41 | 25.34 | 17.74 | 12.67 |
| 33 02 1644 | Metals Screen, Flame, per Each (EPA 200S) | 27.19 | 22.66 | 15.86 | 11.33 |
| 33 02 1645 | Metals Screen, Furnace, per Each (EPA 200S) | 78.00 | 65.00 | 45.50 | 32.50 |
| 33 02 1646 | Arsenic, Hydride (EPA 206.3) | 72.00 | 60.00 | 42.00 | 30.00 |
| 33 02 1647 | Boron, Colorimetric (EPA 212.3) | 28.80 | 24.00 | 16.80 | 12.00 |
| 33 02 1648 | Chromium, Hexavalent (EPA 218.5) | 52.01 | 43.34 | 30.34 | 21.67 |
| 33 02 1649 | Mercury, Cold Vapor (EPA 245.1) | 78.00 | 65.00 | 45.50 | 32.50 |
| 33 02 1650 | Ion Chromatography (EPA 300) | 43.99 | 36.66 | 25.66 | 18.33 |
| 33 02 1651 | Alkalinity (EPA 310.1/310.2) | 36.00 | 30.00 | 21.00 | 15.00 |
| 33 02 1652 | Bromide, Titrimetric (EPA 320.1) | 78.00 | 65.00 | 45.50 | 32.50 |
| 33 02 1653 | Chloride (EPA 325.2/325.3) | 40.80 | 34.00 | 23.80 | 17.00 |

**Environmental Remediation: Assemblies Cost Book**

*Page 2-34*

1998 by ECHOS. All rights reserved

# Cost Data for RI/FS and RFI/CMS
## Analysis Assemblies

| Sampling Assemblies - Water and Liquids | | Cost by Turnaround Time/Quality Control Level | | | |
|---|---|---|---|---|---|
| | | 24-72 hr CLP | 24-72 hr Std QC | Std TAT CLP | Std TAT Std QC |
| **Assembly** | **Description** | | | | |
| 33 02 1654 | Chloride, Total Residual (EPA 330.2/330.5) | 40.01 | 33.34 | 23.34 | 16.67 |
| 33 02 1655 | Fluoride SPADNS (EPA 340.1) | 300.00 | 250.00 | 175.00 | 125.00 |
| 33 02 1656 | Fluoride Electrode (EPA 340.2) | 44.81 | 37.34 | 26.14 | 18.67 |
| 33 02 1657 | Iodide (EPA 345.1) | 67.99 | 56.66 | 39.66 | 28.33 |
| 33 02 1658 | Nitrogen, Nitrate-Nitrite (EPA 350.3) | 47.21 | 39.34 | 27.54 | 19.67 |
| 33 02 1659 | Nitrogen, Kjeldahl, Total (EPA 351.2/351.3/351.4) | 91.99 | 76.66 | 53.66 | 38.33 |
| 33 02 1660 | Nitrogen, Organic (EPA 351.3) | 139.20 | 116.00 | 81.20 | 58.00 |
| 33 02 1661 | Nitrogen, Nitrate (EPA 352.1) | 47.21 | 39.34 | 27.54 | 19.67 |
| 33 02 1662 | Nitrogen, Nitrate & Nitrite (EPA 353.2) | 139.20 | 116.00 | 81.20 | 58.00 |
| 33 02 1663 | Dissolved Oxygen (EPA 360.1) | 27.60 | 23.00 | 16.10 | 11.50 |
| 33 02 1664 | Phosphorus, Ortho (EPA 365.1) | 51.19 | 42.66 | 29.86 | 21.33 |
| 33 02 1665 | Phosphorus, Total (EPA 365.2/365.4) | 64.80 | 54.00 | 37.80 | 27.00 |
| 33 02 1666 | Silica, Dissolved (EPA 370.1) | 72.00 | 60.00 | 42.00 | 30.00 |
| 33 02 1667 | Sulfate (EPA 375.2/375.3/375.4) | 40.01 | 33.34 | 23.34 | 16.67 |
| 33 02 1668 | Sulfide (EPA 376.1) | 39.17 | 32.64 | 22.85 | 16.32 |
| 33 02 1669 | Sulfite (EPA 377.1) | 39.17 | 32.64 | 22.85 | 16.32 |
| 33 02 1670 | Metals Screen, 25 Metals Listed in Method EPA 200.7 | 679.80 | 566.50 | 396.55 | 283.25 |
| 33 02 1671 | Biochemical Oxygen Demand, BOD (EPA 405.1) | 83.21 | 69.34 | 48.54 | 34.67 |
| 33 02 1672 | Chemical Oxygen Demand, COD (EPA 410.1/410.2/410.3/410.4) | 67.99 | 56.66 | 39.66 | 28.33 |
| 33 02 1673 | Total Organic Carbon, TOC (EPA 415.1/415.2) | 73.61 | 61.34 | 42.94 | 30.67 |
| 33 02 1674 | MBAS, Surfactants (EPA 425.1) | 91.20 | 76.00 | 53.20 | 38.00 |
| 33 02 1675 | NTA (EPA 430.2) | 60.00 | 50.00 | 35.00 | 25.00 |
| 33 02 1676 | Total Organic Halogens, TOX (EPA 450.1) | 180.00 | 150.00 | 105.00 | 75.00 |
| 33 02 1677 | Trihalomethanes (EPA 501.1) | 252.00 | 210.00 | 147.00 | 105.00 |
| 33 02 1688 | 1,2-Dibromo-3-Chloropropane (EPA 504) | 252.00 | 210.00 | 147.00 | 105.00 |
| 33 02 1689 | Organohalide Pesticides & Aroclors (EPA 505) | 374.40 | 312.00 | 218.40 | 156.00 |
| 33 02 1693 | Chlorinated Herbicides (EPA 515) | 540.00 | 450.00 | 315.00 | 225.00 |
| 33 02 2101 | Color (SM 204B) | 40.80 | 34.00 | 23.80 | 17.00 |
| 33 02 2102 | Total Solids (SM 209A) | 34.39 | 28.66 | 20.06 | 14.33 |
| 33 02 2103 | Total Dissolved Solids (SM 209B) | 36.00 | 30.00 | 21.00 | 15.00 |
| 33 02 2104 | Total Suspended Solids (SM 209C) | 36.00 | 30.00 | 21.00 | 15.00 |
| 33 02 2105 | Total Volatile Solids (SM 209D) | 44.81 | 37.34 | 26.14 | 18.67 |

**Environmental Remediation: Assemblies Cost Book**

# Cost Data for RI/FS and RFI/CMS
## Analysis Assemblies

| Sampling Assemblies - Water and Liquids | | Cost by Turnaround Time/Quality Control Level | | | |
|---|---|---|---|---|---|
| Assembly | Description | 24-72 hr CLP | 24-72 hr Std QC | Std TAT CLP | Std TAT Std QC |
| 33 02 2106 | Turbidity (SM 214A) | 30.41 | 25.34 | 17.74 | 12.67 |
| 33 02 2107 | Hardness, Total (SM 314B) | 37.20 | 31.00 | 21.70 | 15.50 |
| 33 02 2108 | Acidity (SM 402) | 36.00 | 30.00 | 21.00 | 15.00 |
| 33 02 2109 | Alkalinity (SM 403) | 36.00 | 30.00 | 21.00 | 15.00 |
| 33 02 2110 | Bromide (SM 405) | 78.00 | 65.00 | 45.50 | 32.50 |
| 33 02 2111 | Chlorine, Residual (SM 408D) | 40.01 | 33.34 | 23.34 | 16.67 |
| 33 02 2112 | Cyanide, Total (SM 412B/412D) | 115.99 | 96.66 | 67.66 | 48.33 |
| 33 02 2113 | Fluoride, Electrode (SM 413B) | 44.81 | 37.34 | 26.14 | 18.67 |
| 33 02 2114 | Fluoride, SPADNS (SM 413C) | 300.00 | 250.00 | 175.00 | 125.00 |
| 33 02 2115 | Nitrogen, Total (SM 420) | 91.99 | 76.66 | 53.66 | 38.33 |
| 33 02 2116 | Sulfite (SM 428A) | 39.17 | 32.64 | 22.85 | 16.32 |
| 33 02 2117 | Oil & Grease (SM 503A) | 132.00 | 110.00 | 77.00 | 55.00 |
| 33 02 2118 | Total Organic Carbon, TOC (SM 505B) | 73.61 | 61.34 | 42.94 | 30.67 |
| 33 02 2119 | Biochemical Oxygen Demand BOD (SM 507) | 83.21 | 69.34 | 48.54 | 34.67 |
| 33 02 2120 | Chemical Oxygen Demand COD (SM 508C) | 67.99 | 56.66 | 39.66 | 28.33 |
| 33 02 2121 | Surfactants (SM 512B) | 91.20 | 76.00 | 53.20 | 38.00 |
| 33 02 2122 | Fecal Coliform (SM 909C) | 62.40 | 52.00 | 36.40 | 26.00 |
| 33 02 2125 | Arsenic, Hydride (SW 7061) with Prep | 435.19 | 362.66 | 253.86 | 181.33 |
| 33 02 2126 | Arsenic, (SW 7060) with Prep | 69.60 | 58.00 | 40.60 | 29.00 |
| 33 02 2127 | Metals Screen, 25 Listed in SW 3005/SW 6010 | 679.99 | 566.66 | 396.66 | 283.33 |
| 33 02 2128 | Chromium, Hexavalent (SW 7195) with Prep | 52.01 | 43.34 | 30.34 | 21.67 |
| 33 02 2129 | Mercury, Cold Vapor (SW 7470) with Prep | 100.01 | 83.34 | 58.34 | 41.67 |
| 33 02 2130 | Cyanide (SW 9010) with Prep | 103.99 | 86.66 | 60.66 | 43.33 |
| 33 02 2131 | Purgeable Halocarbons (SW 5030/SW 8010) | 292.01 | 243.34 | 170.34 | 121.67 |
| 33 02 2132 | Purgeable Aromatics (SW 5030/SW 8020) | 232.01 | 193.34 | 135.34 | 96.67 |
| 33 02 2133 | Pesticides/PCBs (SW 3510/SW 8080) | 496.01 | 413.34 | 289.34 | 206.67 |
| 33 02 2134 | Polynuclear Aromatic Hydrocarbons, PAH (SW 3510/SW 8310) | 451.99 | 376.66 | 263.66 | 188.33 |
| 33 02 2135 | Base Neutral & Acid Extractable Organics (SW 3510/SW 8270) | 1,043 | 869.34 | 608.54 | 434.67 |
| 33 02 2136 | Carbamate & Urea Pesticides (SW 3510/SW 8318) | 492.00 | 410.00 | 287.00 | 205.00 |
| 33 02 2137 | Gasoline Group (Mod EPA 601/602, Lead, EDB) | 448.01 | 373.34 | 261.34 | 186.67 |
| 33 02 2138 | Kerosene Group (Mod EPA 601/602, 610, TPH, Lead, EDB) | 448.01 | 373.34 | 261.34 | 186.67 |

## Environmental Remediation: Assemblies Cost Book

# Cost Data for RI/FS and RFI/CMS
## Analysis Assemblies

| Sampling Assemblies - Water and Liquids | | Cost by Turnaround Time/Quality Control Level | | | |
|---|---|---|---|---|---|
| Assembly | Description | 24-72 hr CLP | 24-72 hr Std QC | Std TAT CLP | Std TAT Std QC |
| 33 02 2139 | BTEX/MTBE (Mod EPA 602) | 192.00 | 160.00 | 112.00 | 80.00 |
| 33 02 2140 | Ethylene Dibromide (EDB) (EPA 501.4) | 187.99 | 156.66 | 109.66 | 78.33 |
| 33 02 2141 | Chromium, Total (EPA 218.1) | 45.60 | 38.00 | 26.60 | 19.00 |
| 33 02 2142 | Titanium (EPA 283.2) | 43.20 | 36.00 | 25.20 | 18.00 |
| 33 02 2143 | Lead (SW 3005/SW 7421) | 100.01 | 83.34 | 58.34 | 41.67 |
| 33 02 2144 | Ethylene Dibromide (SW 8011) with Prep | 312.00 | 260.00 | 182.00 | 130.00 |
| 33 02 2145 | Dioxins (SW 8280) with Prep | 1,769 | 1,474 | 1,032 | 737.00 |
| 33 02 2146 | Selenium (SW 7740) with Prep | 88.01 | 73.34 | 51.34 | 36.67 |
| 33 02 2147 | Nonhalogenated Volatile Organics (SW 5030/SW 8015) | 264.00 | 220.00 | 154.00 | 110.00 |
| 33 02 2148 | Organophosphorous Pesticides (SW 3510/SW 8140) | 516.00 | 430.00 | 301.00 | 215.00 |
| 33 02 2149 | Chlorinated Phenoxy Herbicides (SW 3510/SW 8150) | 559.99 | 466.66 | 326.66 | 233.33 |

**Environmental Remediation: Assemblies Cost Book**

# Cost Data for RI/FS and RFI/CMS
## Miscellaneous

**Common Cost Components:**
1. Other Direct Costs (ODCs) Related to Professional Labor
2. Geophysical Investigations
3. Soil Gas Surveys
4. Bench and Pilot Treatability Tests

**Other Direct Costs:** ODCs generally include miscellaneous costs associated with the RI/FS or RFI/CMS such as travel, telephone, regular and overnight mail, reproductio costs, and research materials. ODCs may be estimated as a percentage of professiona labor costs or by preparing a detailed accounting of cost items.

**Geophysical Investigations:** Geophysical investigations are typically used for reconnaissance subsurface mapping during the initial phases of site investigation of land where buried wastes are suspected or known to be present.

**Soil Gas Surveys:** Soil gas surveys are an effective reconnaissance tool because numerous data points can be sampled over a short period of time for a relatively low cost. Soil gas surveying is applicable to most volatile organic compounds. Soil gas surveying has an advantage over geophysical techniques for detecting shallow contaminatio because identifiable compounds rather than indirect physical changes in the subsurface are measured. Moreover, soil gas techniques are sensitive to both dissolved VOCs and to pure product layers.

**Bench and Pilot Treatability Tests:** Bench scale tests are tabletop laboratory test conducted at an off-site laboratory on a sample (container) of soil, water, or air to determine the applicability (treatability) of several alternative remedial action techniques. Alternatives that pass the bench scale tests can then be pilot tested at the site. Pilot scale testing can range from small scale to full scale. Bench and pilot tests can be estimated as a percentage of estimated remedial action capital costs or by preparing a detailed accounting of cost items.

### Geophysical Investigations and Soil Gas Surveys

| Assembly | Description | Unit | Unit Cost |
|---|---|---|---|
| 33 02 0307 | Soil Gas Investigation & Analysis | DAY | 106.00 |
| 33 02 9901 | Magnetometer | DAY | 53.00 |
| 33 02 9902 | Electromagnetics | DAY | 53.00 |
| 33 02 9903 | Ground Penetrating Radar | DAY | 50.35 |
| 33 02 9904 | Seismic Refraction | DAY | 782.50 |
| 33 02 9905 | Resistivity | DAY | 782.50 |

## Environmental Remediation: Assemblies Cost Book

# Cost Data for Remediation
## Air Sparged Hydrocyclone

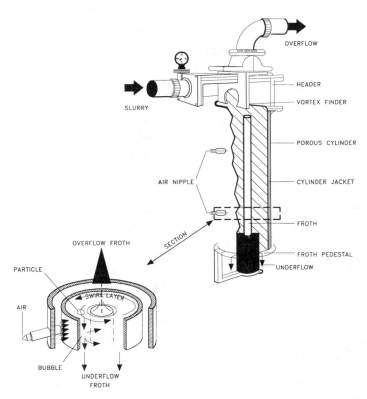

### General:
*An air sparged hydrocyclone (ASH) is a physical separation device which has applications in the recovery of metals/minerals from mine tailings, the removal of fine contaminant particles from soils, and the removal of oil and volatile organic carbon (VOC) compounds from water.*

### Common Cost Components:
1. Groundwater Extraction Wells
2. Excavation of Contaminated Soils
3. Material Screening (Materials Plant)
4. Bulk Material Storage
5. Injection wells
6. Discharge to POTW
7. Coagulation/Flocculation
8. Media Filtration
9. Chemical Precipitation
10. Dewatering (Sludge)
11. Oil/Water Separation
12. Ultraviolet Oxidation
13. Sampling and Analysis

### Typical Treatment Train:
*ASH systems require a feed/conditioning tank, reagent feed equipment, compressed air supply, pumps, piping, and instrumentation. Trailer-mounted ASH systems have 2-4 hydrocyclones consisting of rougher (upstream) and scavenger (downstream) units. Trailer-mounted ASH systems are pre-wired, pre-plumbed, and contain all operating equipment except for holding tanks (optional).*

### Other Cost Considerations:
*Electrical distribution, fencing, clearing and grubbing*

**Environmental Remediation: Assemblies Cost Book**

# Cost Data for Remediation

## Air Sparged Hydrocyclone

| Assembly | Description | Unit | Unit Cost by Safety Level A | B | C | D | E |
|---|---|---|---|---|---|---|---|
| | **Capital Costs** | | | | | | |
| 17 03 0220 | 910, 1.25 CY, Wheel Loader | HOUR | 130.20 | 104.06 | 87.23 | 61.63 | 55.86 |
| 17 03 0221 | 916, 1.5 CY, Wheel Loader | HOUR | 114.32 | 90.83 | 76.64 | 53.69 | 47.92 |
| 17 03 0222 | 926, 2.0 CY, Wheel Loader | HOUR | 154.30 | 124.14 | 103.30 | 73.68 | 67.91 |
| 17 03 0223 | 950, 3.0 CY, Wheel Loader | HOUR | 180.10 | 145.64 | 120.50 | 86.58 | 80.81 |
| 17 03 0224 | 966, 4.0 CY, Wheel Loader | HOUR | 211.58 | 171.88 | 141.48 | 102.32 | 96.55 |
| 17 03 0225 | 980, 5.25 CY, Wheel Loader | HOUR | 298.84 | 244.38 | 199.68 | 145.79 | 139.74 |
| 17 03 0345 | Standby, 910, 1.25 CY Wheel Loader | HOUR | 15.13 | 12.61 | 10.09 | 7.57 | 7.57 |
| 17 03 0346 | Standby, 916, 1.5 CY Wheel Loader | HOUR | 11.32 | 9.44 | 7.55 | 5.66 | 5.66 |
| 17 03 0347 | Standby, 926, 2.0 CY Wheel Loader | HOUR | 21.40 | 17.83 | 14.27 | 10.70 | 10.70 |
| 17 03 0348 | Standby, 950, 3.0 CY Wheel Loader | HOUR | 27.24 | 22.70 | 18.16 | 13.62 | 13.62 |
| 17 03 0349 | Standby, 966, 4.0 CY Wheel Loader | HOUR | 34.27 | 28.56 | 22.85 | 17.13 | 17.13 |
| 17 03 0350 | Standby, 980, 5.25 CY Wheel Loader | HOUR | 57.52 | 47.94 | 38.35 | 28.76 | 28.76 |
| 19 01 0207 | 4", Class 200, PVC Piping | LF | 12.33 | 9.86 | 8.72 | 6.32 | 5.50 |
| 19 01 0208 | 6", Class 200, PVC Piping | LF | 14.95 | 12.20 | 10.93 | 8.27 | 7.36 |
| 19 04 0403 | 4,000 Gallon Polyethylene Wastewater Tank, Rental | MONTH | 545.90 | 545.90 | 545.90 | 545.90 | 545.90 |
| 19 04 0405 | 6,000 Gallon, Polyethylene Aboveground Wastewater Holding Tank, Rental | MONTH | 641.30 | 641.30 | 641.30 | 641.30 | 641.30 |
| 19 04 0406 | 21,000 Gallon Steel Wastewater Holding Tank, Rental | MONTH | 1,219 | 1,219 | 1,219 | 1,219 | 1,219 |
| 33 10 9657 | 1,000 Gallon Single-wall Steel Aboveground Tank | EACH | 2,118 | 1,844 | 1,727 | 1,463 | 1,366 |
| 33 10 9658 | 2,000 Gallon Single-wall Steel Aboveground Tank | EACH | 3,457 | 2,997 | 2,777 | 2,332 | 2,183 |
| 33 10 9659 | 3,000 Gallon Single-wall Steel Aboveground Tank | EACH | 4,433 | 3,959 | 3,733 | 3,274 | 3,121 |
| 33 10 9660 | 5,000 Gallon Single-wall Steel Aboveground Tank | EACH | 6,079 | 5,494 | 5,214 | 4,646 | 4,457 |
| 33 10 9661 | 8,000 Gallon Single-wall Steel Aboveground Tank | EACH | 7,876 | 7,205 | 6,884 | 6,234 | 6,017 |
| 33 10 9662 | 10,000 Gallon Single-wall Steel Aboveground Tank | EACH | 9,147 | 8,415 | 8,065 | 7,356 | 7,119 |
| 33 10 9663 | 12,000 Gallon Single-wall Steel Aboveground Tank | EACH | 10,460 | 9,655 | 9,270 | 8,490 | 8,230 |
| 33 10 9664 | 15,000 Gallon Single-wall Steel Aboveground Tank | EACH | 11,591 | 10,697 | 10,269 | 9,402 | 9,113 |
| 33 10 9665 | 20,000 Gallon Single-wall Steel Aboveground Tank | EACH | 14,594 | 13,445 | 12,895 | 11,780 | 11,408 |
| 33 10 9666 | 30,000 Gallon Single-wall Steel Aboveground Tank | EACH | 6,427 | 5,085 | 4,444 | 3,143 | 2,709 |
| 33 13 1301 | Rental of 10-20 GPM Turnkey Mobile ASH Units | MO | 1,650 | 1,650 | 1,650 | 1,650 | 1,650 |
| 33 13 1302 | Rental of 100-400 GPM Turnkey Mobile ASH Units | MO | 8,501 | 8,501 | 8,501 | 8,501 | 8,501 |
| 33 13 1305 | Purchase of 10-20 GPM Turnkey Mobile ASH Units | EA | 20,064 | 20,050 | 20,044 | 20,030 | 20,026 |
| 33 13 1306 | Purchase of 100-400 GPM Turnkey Mobile ASH Units | EA | 160,141 | 160,113 | 160,100 | 160,073 | 160,063 |
| 33 13 1307 | Install a 10-20 GPM Unit (equip. not included) | EA | 62.31 | 48.33 | 41.89 | 28.35 | 23.68 |
| 33 13 1308 | Install a 100-400 GPM Unit (equip. not included) | EA | 124.61 | 96.66 | 83.77 | 56.70 | 47.36 |

## Environmental Remediation: Assemblies Cost Book

*Page 3-2*

1998 by ECHOS. All rights reserved

# Cost Data for Remediation

## Air Sparged Hydrocyclone

| Assembly | Description | Unit | Unit Cost by Safety Level | | | | |
|---|---|---|---|---|---|---|---|
| | | | A | B | C | D | E |
| | **Capital Costs** | | | | | | |
| 33 13 1310 | Purchase of a 2 in., 10-20 GPM ASH device | EA | 5,051 | 5,039 | 5,034 | 5,023 | 5,019 |
| 33 13 1311 | Purchase of a 6 in., 100-400 GPM ASH device | EA | 15,068 | 15,053 | 15,046 | 15,032 | 15,027 |
| 33 13 1312 | Purchase of a 15 in., 1,200-1,600 GPM ASH device | EA | 30,087 | 30,068 | 30,059 | 30,041 | 30,035 |
| 33 13 1320 | Transport of 10-20 GPM Turnkey Mobile ASH Units | MI | 1.65 | 1.65 | 1.65 | 1.65 | 1.65 |
| 33 13 1321 | Transport of 100-400 GPM Turnkey Mobile ASH Units | MI | 2.05 | 2.05 | 2.05 | 2.05 | 2.05 |
| 33 13 1322 | Startup of a 10-20 GPM Unit | EA | 37.77 | 29.12 | 25.41 | 17.04 | 13.98 |
| 33 13 1323 | Startup of a 100-400 GPM Unit Treating Solids | EA | 151.10 | 116.47 | 101.65 | 68.18 | 55.91 |
| 33 13 1324 | Startup of a 100-400 GPM Unit Treating Liquids | EA | 75.55 | 58.23 | 50.82 | 34.09 | 27.95 |
| 33 13 1330 | Decon/Breakdown of 10-20 GPM Turnkey Mobile ASH Units | EA | 62.31 | 48.33 | 41.89 | 28.35 | 23.68 |
| 33 13 1331 | Decon/Breakdown of 100-400 GPM Turnkey Mobile ASH Units | EA | 124.61 | 96.66 | 83.77 | 56.70 | 47.36 |
| 33 13 1340 | Operational Labor for a 10-20 GPM ASH Unit | HR | 0.94 | 0.73 | 0.64 | 0.43 | 0.35 |
| 33 13 1341 | Oper. Labor for a 100-400 GPM ASH Unit Treating Solids | HR | 1.89 | 1.46 | 1.27 | 0.85 | 0.70 |
| 33 13 1342 | Oper. Labor for a 100 - 400 GPM Unit Treating Liquids | HR | 0.94 | 0.73 | 0.64 | 0.43 | 0.35 |
| 33 13 1345 | Maintenance for 10-20 GPM Turnkey Mobile ASH Units | HR | 0.94 | 0.73 | 0.64 | 0.43 | 0.35 |
| 33 13 1346 | Maintenance for 100-400 GPM Turnkey Mobile ASH Units | HR | 1.89 | 1.46 | 1.27 | 0.85 | 0.70 |
| 33 13 1347 | Monthly Spare Parts Charge for 10-20 GPM ASH Units | MO | 85.01 | 85.01 | 85.01 | 85.01 | 85.01 |
| 33 13 1348 | Monthly Spare Parts Charge for 100-400 GPM ASH Units | MO | 650.06 | 650.06 | 650.06 | 650.06 | 650.06 |
| 33 13 1360 | Reagent Frother Usage | LB | 0.72 | 0.72 | 0.72 | 0.72 | 0.72 |
| 33 23 1306 | High Sump Level Switch for Avoiding Overflow | EACH | 214.65 | 214.65 | 214.65 | 214.65 | 214.65 |
| 33 26 0104 | 4" Carbon Steel Piping | LF | 23.20 | 19.86 | 18.43 | 15.21 | 14.02 |
| 33 26 0105 | 6" Carbon Steel Piping | LF | 36.37 | 30.24 | 27.46 | 21.52 | 19.44 |
| 33 27 0104 | 4" PVC, Schedule 40, Tee | EACH | 137.04 | 108.84 | 96.77 | 69.51 | 59.51 |
| 33 27 0105 | 6" PVC, Schedule 40, Tee | EACH | 205.86 | 169.47 | 153.89 | 118.72 | 105.82 |
| 33 27 0114 | 4" PVC, 90 Degree, Elbow | EACH | 101.72 | 80.57 | 71.51 | 51.07 | 43.57 |
| 33 27 0115 | 6" PVC, 90 Degree, Elbow | EACH | 135.44 | 111.27 | 100.92 | 77.56 | 68.99 |
| 33 27 0552 | 4" Carbon Steel 90-degree Elbow, Schedule 80 | EACH | 190.36 | 152.53 | 135.42 | 98.81 | 85.97 |
| 33 27 0554 | 6" Carbon Steel 90-degree Elbow, Schedule 80 | EACH | 297.54 | 245.36 | 221.75 | 171.24 | 153.54 |
| 33 27 0562 | 4" Carbon Steel Tee, Schedule 80 | EACH | 307.17 | 250.43 | 224.76 | 169.84 | 150.59 |
| 33 27 0564 | 6" Carbon Steel Tee, Schedule 80 | EACH | 452.39 | 374.65 | 339.48 | 264.23 | 237.85 |
| 33 42 0101 | Electrical Charge | KWH | 0.06 | 0.06 | 0.06 | 0.06 | 0.06 |
| 33 42 0120 | Generator Supplied Electricity | kWh | 0.08 | 0.08 | 0.08 | 0.08 | 0.08 |
| 33 42 0301 | Water | KGAL | 8.39 | 8.39 | 8.39 | 8.39 | 8.39 |

## Environmental Remediation: Assemblies Cost Book

# Cost Data for Remediation

## Air Sparged Hydrocyclone

| Assembly | Description | Unit | Unit Cost by Safety Level | | | | |
|---|---|---|---|---|---|---|---|
| | | | A | B | C | D | E |
| | **Capital Costs** | | | | | | |
| 33 42 0302 | Process Water, Supplied by Water Line | KGAL | 2.56 | 2.56 | 2.56 | 2.56 | 2.56 |

**Environmental Remediation: Assemblies Cost Book**

# Cost Data for Remediation
## Air Sparging

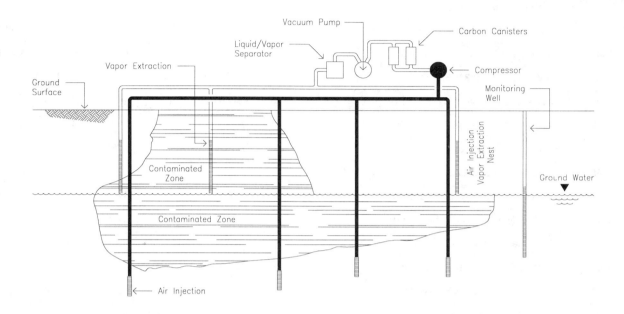

### General:

*Air Sparging is the injection of pressurized air into contaminated media which results in the volatilization of volatile organic compounds and enhanced biodegradation of contaminants susceptible to aerobic microbial degradation. Though there are other applications of air sparging, the information presented here assumes air sparging in situ groundwater via injection wells.*

### Typical Treatment Train:

*Groundwater recovery, vapor extraction, groundwater monitoring wells, disposal of drill cuttings, underground connection piping, sampling and analysis.*

### Common Cost Components:

1. Air Injection Wells (drill rig mobilization, drilling, screen, casing, decontamination)
2. Connection Piping
3. Air Compressor
4. Blowers
5. Air Flow Monitoring and Control
6. Monitoring System
7. Soil Sample Collection during Borehole Advancement
8. Containment of Drill Cuttings
9. Operations and Maintenance

### Other Cost Considerations:

*Electrical distribution, fencing, clearing and grubbing.*

**Environmental Remediation: Assemblies Cost Book**

# Cost Data for Remediation

## Air Sparging

| Assembly | Description | Unit | A | B | C | D | E |
|---|---|---|---|---|---|---|---|
| | **Capital Costs** | | | | | | |
| 33 01 0101 | Mobilize/DeMobilize Drilling Rig & Crew | LS | 4,949 | 4,031 | 3,308 | 2,401 | 2,280 |
| 33 02 0303 | Organic Vapor Analyzer Rental, per Day | DAY | 184.30 | 184.30 | 184.30 | 184.30 | 184.30 |
| 33 13 9001 | Blower 98 SCFM, 3.2 HP, 5 PSI | EACH | 5,163 | 4,617 | 4,383 | 3,854 | 3,660 |
| 33 13 9002 | Blower 170 SCFM, 10.3 HP, 10 PSI | EACH | 5,184 | 4,637 | 4,403 | 3,874 | 3,681 |
| 33 13 9003 | Blower 163 SCFM, 15.0 HP, 15 PSI | EACH | 4,927 | 4,381 | 4,146 | 3,618 | 3,424 |
| 33 13 9004 | Blower 426 SCFM, 84.0 HP, 30 PSI | EACH | 9,508 | 8,961 | 8,727 | 8,199 | 8,005 |
| 33 17 0808 | Decontaminate Rig, Augers, Screen (Rental Equipment) | DAY | 205.34 | 205.34 | 205.34 | 205.34 | 205.34 |
| 33 23 0101 | 2" PVC, Schedule 40, Well Casing | LF | 17.45 | 14.39 | 11.98 | 8.95 | 8.55 |
| 33 23 0111 | 2" PVC, Schedule 80, Well Casing | LF | 17.78 | 14.72 | 12.31 | 9.28 | 8.88 |
| 33 23 0121 | 2" Stainless Steel, Well Casing | LF | 31.75 | 28.14 | 25.30 | 21.73 | 21.25 |
| 33 23 0201 | 2" PVC, Schedule 40, Well Screen | LF | 23.47 | 19.52 | 16.42 | 12.52 | 12.00 |
| 33 23 0211 | 2" PVC, Schedule 80, Well Screen | LF | 18.70 | 15.64 | 13.23 | 10.20 | 9.80 |
| 33 23 0221 | 2" Stainless Steel, Well Screen | LF | 26.91 | 23.84 | 21.44 | 18.41 | 18.01 |
| 33 23 0301 | 2" PVC, Well Plug | EACH | 29.37 | 24.77 | 21.16 | 16.63 | 16.02 |
| 33 23 0311 | 2" Stainless Steel, Well Plug | EACH | 83.24 | 74.06 | 66.83 | 57.76 | 56.55 |
| 33 23 1101 | Hollow-stem Auger, 8" Outside Diameter Borehole for 2" Well | LF | 89.98 | 73.28 | 60.15 | 43.66 | 41.45 |
| 33 23 1106 | Split Spoon Sample, 2" x 24", During Drilling | EACH | 46.67 | 46.67 | 46.67 | 46.67 | 46.67 |
| 33 23 1121 | Standby for Drilling | EACH | 618.63 | 503.81 | 413.56 | 300.17 | 284.95 |
| 33 23 1122 | Move Rig/Equipment Around Site | EACH | 154.66 | 125.95 | 103.39 | 75.04 | 71.24 |
| 33 23 1125 | Concrete Coring (Minimum Charge) | DAY | 2,475 | 2,015 | 1,654 | 1,201 | 1,140 |
| 33 23 1126 | Furnish 55 Gallon Drum for Drill Cuttings & Development Water | EACH | 65.19 | 65.19 | 65.19 | 65.19 | 65.19 |
| 33 23 1152 | Mud Drilling, 6" Diameter Borehole | LF | 81.01 | 65.98 | 54.16 | 39.31 | 37.31 |
| 33 23 1161 | Air Rotary 6" Borehole in Unconsolidated | LF | 73.65 | 59.98 | 49.23 | 35.73 | 33.92 |
| 33 23 1401 | 2" Screen, Filter Pack | LF | 16.49 | 13.88 | 11.84 | 9.27 | 8.92 |
| 33 23 1504 | Surface Pad, Concrete, 2' x 2' x 4" | EACH | 6.05 | 5.45 | 5.18 | 4.61 | 4.40 |
| 33 23 1811 | 2" Well, Portland Cement Grout | LF | 0.92 | 0.92 | 0.92 | 0.92 | 0.92 |
| 33 23 2101 | 2" Well, Bentonite Seal | EACH | 63.00 | 52.67 | 44.55 | 34.34 | 32.97 |
| 33 23 2301 | 5' Guard Posts, Cast Iron, Concrete Fill | EACH | 98.83 | 83.39 | 76.77 | 61.84 | 56.37 |
| 33 26 0210 | 2" Stainless Steel, Schedule 40, Type 304, Connection Piping | LF | 33.47 | 29.21 | 27.39 | 23.28 | 21.77 |
| 33 26 0213 | 2" Stainless Steel, Schedule 40, Type 304, Manifold Piping | LF | 33.47 | 29.21 | 27.39 | 23.28 | 21.77 |
| 33 26 0406 | 2" PVC, Schedule 40, Connection Piping | LF | 6.14 | 4.87 | 4.33 | 3.11 | 2.66 |

**Unit Cost by Safety Level**

---

**Environmental Remediation: Assemblies Cost Book**

# Cost Data for Remediation

## Air Sparging

| Assembly | Description | Unit | Unit Cost by Safety Level A | B | C | D | E |
|---|---|---|---|---|---|---|---|
| | **Capital Costs** | | | | | | |
| 33 26 0417 | 2" PVC, Schedule 40, Manifold Piping | LF | 6.14 | 4.87 | 4.33 | 3.11 | 2.66 |
| 33 27 0102 | 2" PVC, Schedule 40, Tee | EACH | 68.73 | 53.63 | 46.80 | 32.18 | 27.06 |
| 33 27 0112 | 2" PVC, 90 Degree, Elbow | EACH | 41.36 | 32.22 | 28.30 | 19.46 | 16.22 |
| 33 27 0201 | 2" Stainless Steel, Tee | EACH | 173.32 | 140.34 | 126.22 | 94.34 | 82.65 |
| 33 27 0211 | 2" Stainless Steel, 90 Degree, Elbow | EACH | 114.28 | 93.45 | 84.53 | 64.39 | 57.01 |
| 33 27 0402 | 2" Iron Body Check Valve | EACH | 188.96 | 174.31 | 168.03 | 153.86 | 148.67 |
| 33 31 0209 | Pressure Gauge | EACH | 140.16 | 121.61 | 113.67 | 95.74 | 89.16 |
| 33 42 0101 | Electrical Charge | KWH | 0.06 | 0.06 | 0.06 | 0.06 | 0.06 |
| | **Operations & Maintenance** | | | | | | |
| 33 13 2311 | Operational Labor Cost | DAY | 1,686 | 1,306 | 1,134 | 765.60 | 636.52 |
| 33 42 0101 | Electrical Charge | KWH | 0.06 | 0.06 | 0.06 | 0.06 | 0.06 |
| 33 42 0106 | Miscellaneous Electrical Site Usage | MONTH | 249.60 | 249.60 | 249.60 | 249.60 | 249.60 |

**Environmental Remediation: Assemblies Cost Book**

# Cost Data for Remediation
## Air Stripping

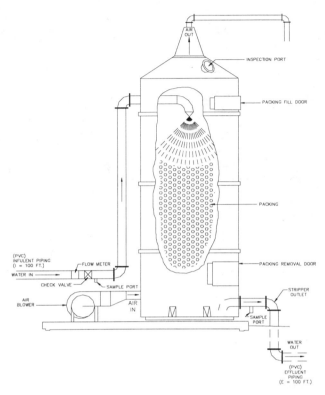

### General:
Air Stripping is a process for removing VOCs from water. The process has been widely and successfully applied to groundwater remediation involving a number of contaminants including TCE, benzene, toluene, xylene, and methylene chloride. The packed tower air stripper is the most widely used treatment system for extracted groundwater impacted by VOCs.

### Typical Treatment Train:
Groundwater Extraction, Influent Water Pretreatment, Off-Gas Treatment, Posttreatment for Effluent Water, Sampling and Analysis, Flow Equalization, Disposal.

### Common Cost Components:
1. Air Stripping Towers (single ower, parallel towers, towers in series)
2. Packing
3. Accessories (sump tanks, transfer pumps, biofouling prevention chemical feeder)
4. Influent Piping
5. Effluent Piping
6. Operations and Maintenance (packing reconditioning, electricity, blower motor maintenance)

### Other Cost Considerations:
Electrical Distribution, Fencing, Signage.

# Cost Data for Remediation

## Air Stripping

| Assembly | Description | Unit | A | B | C | D | E |
|---|---|---|---|---|---|---|---|
| | **Capital Costs** | | | | | | |
| 18 02 0320 | 4" Structural Slab on Grade | SF | 6.45 | 5.45 | 4.98 | 4.00 | 3.67 |
| 18 02 0321 | 6" Structural Slab on Grade | SF | 6.98 | 5.95 | 5.47 | 4.48 | 4.14 |
| 18 02 0322 | 8" Structural Slab on Grade | SF | 9.75 | 8.38 | 7.71 | 6.38 | 5.95 |
| 18 02 0323 | 10" Structural Slab on Grade | SF | 10.56 | 9.13 | 8.43 | 7.04 | 6.58 |
| 18 02 0324 | 12" Structural Slab on Grade | SF | 11.57 | 10.04 | 9.30 | 7.82 | 7.34 |
| 19 01 0202 | 1", Class 200, PVC Piping | LF | 7.48 | 5.84 | 5.07 | 3.48 | 2.93 |
| 19 01 0204 | 2", Class 200, PVC Piping | LF | 8.86 | 6.96 | 6.07 | 4.23 | 3.60 |
| 19 01 0206 | 3", Class 200, PVC Piping | LF | 11.81 | 9.34 | 8.20 | 5.80 | 4.98 |
| 19 01 0207 | 4", Class 200, PVC Piping | LF | 12.33 | 9.86 | 8.72 | 6.32 | 5.50 |
| 19 01 0208 | 6", Class 200, PVC Piping | LF | 14.95 | 12.20 | 10.93 | 8.27 | 7.36 |
| 19 01 0212 | 8", Class 150, PVC Piping | LF | 18.36 | 15.27 | 13.84 | 10.85 | 9.82 |
| 19 02 0101 | 4", Class 50, Bell & Spigot Sanitary Sewer, Cast-iron Pipe | LF | 8.49 | 7.49 | 7.05 | 6.08 | 5.72 |
| 19 02 0102 | 5", Class 50, Bell & Spigot Sanitary Sewer, Cast-iron Pipe | LF | 11.80 | 10.41 | 9.81 | 8.47 | 7.98 |
| 19 02 0103 | 6", Class 50, Bell & Spigot Sanitary Sewer, Cast-iron Pipe | LF | 13.25 | 11.79 | 11.16 | 9.75 | 9.23 |
| 19 02 0104 | 8", Class 50, Bell & Spigot Sanitary Sewer, Cast-iron Pipe | LF | 18.98 | 17.26 | 16.52 | 14.86 | 14.25 |
| 19 02 0105 | 10", Class 50, Bell & Spigot Sanitary Sewer, Cast-iron Pipe | LF | 27.66 | 25.51 | 24.59 | 22.51 | 21.75 |
| 19 02 0106 | 12", Class 50, Bell & Spigot Sanitary Sewer, Cast-iron Pipe | LF | 36.36 | 34.21 | 33.29 | 31.21 | 30.45 |
| 19 02 0107 | 15", Class 50, Bell & Spigot Sanitary Sewer, Cast-iron Pipe | LF | 51.76 | 49.01 | 47.83 | 45.17 | 44.19 |
| 19 04 0601 | 275 Gallon Steel Sump, Aboveground with Supports & Fittings | EACH | 1,325 | 1,153 | 1,080 | 913.28 | 852.29 |
| 19 04 0602 | 550 Gallon Steel Sump, Aboveground with Supports & Fittings | EACH | 1,598 | 1,390 | 1,301 | 1,099 | 1,026 |
| 19 04 0603 | 1,000 Gallon Steel Sump, Aboveground with Supports & Fittings | EACH | 2,118 | 1,844 | 1,727 | 1,463 | 1,366 |
| 19 04 0604 | 1,500 Gallon Steel Sump, Aboveground with Supports & Fittings | EACH | 2,697 | 2,356 | 2,209 | 1,879 | 1,758 |
| 19 04 0605 | 2,000 Gallon Steel Sump, Aboveground with Supports & Fittings | EACH | 3,457 | 2,997 | 2,777 | 2,332 | 2,183 |
| 19 04 0606 | 5,000 Gallon Steel Sump, Aboveground with Supports & Fittings | EACH | 6,079 | 5,494 | 5,214 | 4,646 | 4,457 |
| 19 04 0621 | 550 Gallon Horizontal Plastic Sump with 4" NPT Connection | EACH | 1,699 | 1,596 | 1,551 | 1,452 | 1,415 |
| 19 04 0622 | 1,000 Gallon Horizontal Plastic Sump with 4" NPT Connection | EACH | 2,179 | 2,042 | 1,984 | 1,852 | 1,803 |
| 19 04 0623 | 2,000 Gallon Horizontal Plastic Sump with 6" NPT Connection | EACH | 3,089 | 2,918 | 2,845 | 2,680 | 2,619 |
| 19 04 0624 | 4,000 Gallon Horizontal Plastic Sump with 6" NPT Connection | EACH | 4,130 | 3,927 | 3,841 | 3,645 | 3,573 |

**Environmental Remediation: Assemblies Cost Book**

# Cost Data for Remediation

## Air Stripping

| Assembly | Description | Unit | A | B | C | D | E |
|---|---|---|---|---|---|---|---|
| | **Capital Costs** | | | | | | |
| 19 04 0625 | 6,000 Gallon Horizontal Plastic Sump with 6" NPT Connection | EACH | 5,248 | 4,975 | 4,858 | 4,594 | 4,497 |
| 19 04 0626 | 8,000 Gallon Horizontal Plastic Sump with 6" NPT Connection | EACH | 5,857 | 5,516 | 5,369 | 5,039 | 4,918 |
| 19 04 0627 | 10,000 Gallon Horizontal Plastic Sump with 6" NPT Connection | EACH | 7,082 | 6,622 | 6,402 | 5,956 | 5,807 |
| 19 04 0628 | 12,000 Gallon Horizontal Plastic Sump with 6" NPT Connection | EACH | 8,308 | 7,722 | 7,442 | 6,875 | 6,686 |
| 19 04 0629 | 15,000 Gallon Horizontal Plastic Sump with 6" NPT Connection | EACH | 14,184 | 13,540 | 13,232 | 12,608 | 12,400 |
| 19 04 0631 | 20,000 Gallon Horizontal Plastic Sump with 6" NPT Connection | EACH | 15,799 | 15,084 | 14,742 | 14,048 | 13,817 |
| 19 04 0632 | 25,000 Gallon Horizontal Plastic Sump with 6" NPT Connection | EACH | 24,407 | 23,487 | 23,047 | 22,155 | 21,858 |
| 19 04 0633 | 30,000 Gallon Horizontal Plastic Sump with 6" NPT Connection | EACH | 35,112 | 34,039 | 33,525 | 32,485 | 32,138 |
| 19 04 0634 | 40,000 Gallon Horizontal Plastic Sump with 6" NPT Connection | EACH | 40,844 | 39,234 | 38,465 | 36,904 | 36,383 |
| 19 04 0635 | 48,000 Gallon Horizontal Plastic Sump with 6" NPT Connection | EACH | 59,758 | 57,612 | 56,586 | 54,505 | 53,811 |
| 19 07 0120 | 1 1/4", 60 PSI, Polyethylene Pipe | LF | 5.29 | 4.21 | 3.72 | 2.68 | 2.32 |
| 19 07 0121 | 1 1/2", 60 PSI, Polyethylene Pipe | LF | 5.38 | 4.30 | 3.81 | 2.77 | 2.41 |
| 19 07 0122 | 2", 60 PSI, Polyethylene Pipe | LF | 6.05 | 4.85 | 4.31 | 3.15 | 2.75 |
| 19 07 0123 | 3" Polyethylene 40' Joints, 60 PSI | LF | 7.90 | 6.47 | 5.82 | 4.42 | 3.94 |
| 19 07 0124 | 4" Polyethylene 40' Joints, 60 PSI | LF | 19.10 | 15.75 | 13.90 | 10.64 | 9.71 |
| 19 07 0125 | 6" Polyethylene 40' Joints, 60 PSI | LF | 25.85 | 22.22 | 20.22 | 16.68 | 15.67 |
| 19 07 0126 | 8" Polyethylene 40' Joints, 60 PSI | LF | 35.37 | 31.01 | 28.61 | 24.37 | 23.16 |
| 33 12 9901 | 1.7 Gallon Bypass Chemical Shot Feeder, In-line Mount, 125 PSIG | EACH | 847.93 | 691.64 | 624.73 | 473.67 | 418.28 |
| 33 12 9902 | 5 Gallon Bypass Chemical Shot Feeder, Floor Mount, 175 PSIG | EACH | 1,236 | 1,001 | 900.54 | 673.47 | 590.22 |
| 33 12 9903 | 12 Gallon Bypass Chemical Shot Feeder, Floor Mount, 175 PSIG | EACH | 1,626 | 1,315 | 1,182 | 881.31 | 771.02 |
| 33 12 9904 | 5 Gallon Bypass Chemical Shot Feeder, Floor Mount, 150 Lb ASME | EACH | 1,236 | 1,001 | 900.54 | 673.47 | 590.22 |
| 33 12 9905 | 10 Gallon Bypass Chemical Shot Feeder, Floor Mount, 150 Lb ASME | EACH | 1,626 | 1,315 | 1,182 | 881.31 | 771.02 |
| 33 12 9906 | 5 Gallon Bypass Chemical Shot Feeder, Floor Mount, 300 Lb ASME | EACH | 1,676 | 1,441 | 1,341 | 1,114 | 1,030 |
| 33 12 9907 | 10 Gallon Bypass Chemical Shot Feeder, Floor Mount, 300 Lb ASME | EACH | 2,084 | 1,772 | 1,639 | 1,338 | 1,228 |

**Environmental Remediation: Assemblies Cost Book**

# Cost Data for Remediation

## Air Stripping

| Assembly | Description | Unit | \multicolumn{5}{c}{Unit Cost by Safety Level} | | | | |
|---|---|---|---|---|---|---|---|
| | | | A | B | C | D | E |
| | **Capital Costs** | | | | | | |
| 33 13 0701 | Packing Reconditioning | EACH | 3,996 | 3,189 | 2,678 | 1,888 | 1,705 |
| 33 13 0704 | Install Air Stripper Tower, 1' - 3' Diameter, <= 12' High | EACH | 7,637 | 5,942 | 5,132 | 3,489 | 2,941 |
| 33 13 0705 | Install Air Stripper Tower, 1' - 3' Diameter, 13' - 20' High | EACH | 8,207 | 6,386 | 5,515 | 3,750 | 3,161 |
| 33 13 0706 | Install Air Stripper Tower, 1' - 3' Diameter, 21' - 30' High | EACH | 8,783 | 6,834 | 5,903 | 4,013 | 3,383 |
| 33 13 0707 | Install Air Stripper Tower, 1' - 3' Diameter, > 30' High | EACH | 9,447 | 7,351 | 6,348 | 4,316 | 3,638 |
| 33 13 0736 | Internal Parts for Air Stripper, < 20' High | SF | 3,222 | 3,222 | 3,222 | 3,222 | 3,222 |
| 33 13 0737 | Internal Parts for Air Stripper, >= 20' High | SF | 4,832 | 4,832 | 4,832 | 4,832 | 4,832 |
| 33 13 0738 | 1" - 3.5" Packing for Air Stripper Tower | CF | 16.13 | 16.13 | 16.13 | 16.13 | 16.13 |
| 33 13 0741 | Electrical Controls for Air Stripper | EACH | 7,171 | 6,485 | 6,175 | 5,510 | 5,277 |
| 33 13 0742 | Install Air Stripper Tower, 4' - 8' Diameter, <= 12' High | EACH | 8,358 | 6,504 | 5,617 | 3,819 | 3,219 |
| 33 13 0743 | Install Air Stripper Tower, 4' - 8' Diameter, 13' - 20' High | EACH | 8,783 | 6,834 | 5,903 | 4,013 | 3,383 |
| 33 13 0744 | Install Air Stripper Tower, 4' - 8' Diameter, 21' - 30' High | EACH | 11,553 | 8,989 | 7,764 | 5,278 | 4,449 |
| 33 13 0745 | Install Air Stripper Tower, 4' - 8' Diameter, > 30' High | EACH | 15,115 | 11,761 | 10,157 | 6,906 | 5,821 |
| 33 13 0746 | Install Air Stripper Tower, 9' - 12' Diameter, <= 12' High | EACH | 8,783 | 6,834 | 5,903 | 4,013 | 3,383 |
| 33 13 0747 | Install Air Stripper Tower, 9' - 12' Diameter, 13' - 20' High | EACH | 14,170 | 11,026 | 9,523 | 6,474 | 5,457 |
| 33 13 0748 | Install Air Stripper Tower, 9' - 12' Diameter, 21' - 30' High | EACH | 25,018 | 19,466 | 16,812 | 11,430 | 9,635 |
| 33 13 0749 | Install Air Stripper Tower, 9' - 12' Diameter, > 30' High | EACH | 35,219 | 27,404 | 23,668 | 16,091 | 13,564 |
| 33 13 0751 | 1.0' Diameter Tower, Skid Mount | EACH | 644.83 | 644.83 | 644.83 | 644.83 | 644.83 |
| 33 13 0752 | 1.5' Diameter Tower, Skid Mount | EACH | 779.10 | 779.10 | 779.10 | 779.10 | 779.10 |
| 33 13 0753 | 2.0' Diameter Tower, Skid Mount | EACH | 886.87 | 886.87 | 886.87 | 886.87 | 886.87 |
| 33 13 0754 | 3.0' Diameter Tower, Skid Mount | EACH | 1,207 | 1,207 | 1,207 | 1,207 | 1,207 |
| 33 13 0755 | 4.0' Diameter Tower, Skid Mount | EACH | 1,882 | 1,882 | 1,882 | 1,882 | 1,882 |
| 33 13 0771 | 12' Truss-mounted Towers, Aluminum Safety Ladder & Cage | EACH | 4,305 | 3,983 | 3,829 | 3,517 | 3,413 |
| 33 13 0772 | 15' - 19' Truss-mounted Towers, Aluminum Safety Ladder & Cage | EACH | 5,316 | 4,887 | 4,682 | 4,265 | 4,127 |
| 33 13 0781 | 19' - 25' Truss-mounted Towers, Aluminum Safety Ladder & Cage | EACH | 6,949 | 6,305 | 5,997 | 5,372 | 5,164 |
| 33 13 0782 | 3' - 4' Diameter Tower Blowers, Blower Weather Enclosure | EACH | 707.60 | 652.93 | 629.53 | 576.69 | 557.31 |
| 33 13 0783 | 1' - 3' Diameter Tower Blowers, Mastic Coating of Blower | EACH | 630.00 | 630.00 | 630.00 | 630.00 | 630.00 |
| 33 13 0784 | 4' Diameter Tower Blowers, Mastic Coating of Blower | EACH | 862.50 | 862.50 | 862.50 | 862.50 | 862.50 |
| 33 13 0785 | 1' - 2' Diameter Tower Blowers, PVC Coating of Blower | EACH | 6.08 | 6.08 | 6.08 | 6.08 | 6.08 |
| 33 13 0786 | 3' Diameter Tower Blowers, PVC Coating of Blower | EACH | 8.12 | 8.12 | 8.12 | 8.12 | 8.12 |
| 33 13 0787 | 4' Diameter Tower Blowers, PVC Coating of Blower | EACH | 8.13 | 8.13 | 8.13 | 8.13 | 8.13 |

**Environmental Remediation: Assemblies Cost Book**

# Cost Data for Remediation

## Air Stripping

| Assembly | Description | Unit | Unit Cost by Safety Level A | B | C | D | E |
|---|---|---|---|---|---|---|---|
| | **Capital Costs** | | | | | | |
| 33 13 0788 | 1.0' Diameter x Height, Prefabricated, Fiberglass Reinforced Plastic, Air Stripper Column/Shell Only | FT | 277.99 | 277.99 | 277.99 | 277.99 | 277.99 |
| 33 13 0789 | 1.5' Diameter x Height, Prefabricated, Fiberglass Reinforced Plastic Air Stripper Column/Shell Only | FT | 328.60 | 328.60 | 328.60 | 328.60 | 328.60 |
| 33 13 0790 | 2.0' Diameter x Height, Prefabricated, Fiberglass Reinforced Plastic, Air Stripper Column/Shell Only | FT | 348.21 | 348.21 | 348.21 | 348.21 | 348.21 |
| 33 13 0791 | 3.0' Diameter x Height, Prefabricated, Fiberglass Reinforced Plastic, Air Stripper Column/Shell Only | FT | 471.70 | 471.70 | 471.70 | 471.70 | 471.70 |
| 33 13 0792 | 4.0' Diameter x Height, Prefabricated, Fiberglass Reinforced Plastic, Air Stripper Column/Shell Only | FT | 575.05 | 575.05 | 575.05 | 575.05 | 575.05 |
| 33 13 0793 | 5.0' Diameter x Height, Prefabricated, Fiberglass Reinforced Plastic, Air Stripper Column/Shell Only | FT | 675.75 | 675.75 | 675.75 | 675.75 | 675.75 |
| 33 13 0794 | 6.0' Diameter x Height, Prefabricated, Fiberglass Reinforced Plastic, Air Stripper Column/Shell Only | FT | 805.60 | 805.60 | 805.60 | 805.60 | 805.60 |
| 33 13 0795 | 8.0' Diameter x Height, Prefabricated, Fiberglass Reinforced Plastic, Air Stripper Column/Shell Only | FT | 940.75 | 940.75 | 940.75 | 940.75 | 940.75 |
| 33 13 0796 | 10' Diameter x Height, Prefabricated, Fiberglass Reinforced Plastic, Air Stripper Column/Shell Only | FT | 1,222 | 1,222 | 1,222 | 1,222 | 1,222 |
| 33 13 0797 | 12' Diameter x Height, Prefabricated, Fiberglass Reinforced Plastic, Air Stripper Column/Shell Only | FT | 1,465 | 1,465 | 1,465 | 1,465 | 1,465 |
| 33 13 0801 | 1 Tray Aerator Unit, 135 CFM Air Flow | EA | 8,557 | 8,010 | 7,776 | 7,248 | 7,054 |
| 33 13 0802 | 2 Tray Aerator Unit, 140 CFM Air Flow | EA | 8,734 | 8,187 | 7,953 | 7,424 | 7,231 |
| 33 13 0803 | 3 Tray Aerator Unit, 210 CFM Air Flow | EA | 9,876 | 9,233 | 8,958 | 8,336 | 8,109 |
| 33 13 0804 | 4 Tray Aerator Unit, 280 CFM Air Flow | EA | 10,162 | 9,519 | 9,244 | 8,623 | 8,395 |
| 33 13 0805 | 5 Tray Aerator Unit, 350 CFM Air Flow | EA | 13,107 | 12,378 | 12,066 | 11,362 | 11,104 |
| 33 13 0806 | 6 Tray Aerator Unit, 420 CFM Air Flow | EA | 13,598 | 12,869 | 12,558 | 11,853 | 11,595 |
| 33 13 0807 | 7 Tray Aerator Unit, 490 CFM Air Flow | EA | 15,315 | 14,587 | 14,275 | 13,571 | 13,312 |
| 33 13 0808 | 8 Tray Aerator Unit, 560 CFM Air Flow | EA | 16,664 | 15,935 | 15,623 | 14,919 | 14,661 |
| 33 23 1306 | High Sump Level Switch for Avoiding Overflow | EACH | 214.65 | 214.65 | 214.65 | 214.65 | 214.65 |
| 33 26 0101 | 1" Carbon Steel Piping | LF | 6.46 | 5.28 | 4.77 | 3.63 | 3.21 |
| 33 26 0102 | 2" Carbon Steel Piping | LF | 10.50 | 8.72 | 7.95 | 6.23 | 5.60 |
| 33 26 0103 | 3" Carbon Steel Piping | LF | 18.02 | 15.20 | 13.99 | 11.26 | 10.26 |
| 33 26 0104 | 4" Carbon Steel Piping | LF | 23.20 | 19.86 | 18.43 | 15.21 | 14.02 |
| 33 26 0105 | 6" Carbon Steel Piping | LF | 36.37 | 30.24 | 27.46 | 21.52 | 19.44 |
| 33 26 0106 | 8" Carbon Steel Piping | LF | 40.77 | 34.08 | 31.15 | 24.68 | 22.36 |
| 33 26 0107 | 10" Carbon Steel Piping | LF | 54.40 | 46.05 | 42.38 | 34.30 | 31.39 |
| 33 26 0201 | 1" Stainless Steel Piping, Schedule 40, Threaded | LF | 18.67 | 16.12 | 15.03 | 12.56 | 11.65 |
| 33 26 0202 | 2" Stainless Steel Piping, Schedule 40, Threaded | LF | 33.47 | 29.21 | 27.39 | 23.28 | 21.77 |

**Environmental Remediation: Assemblies Cost Book**

# Cost Data for Remediation

## Air Stripping

| Assembly | Description | Unit | A | B | C | D | E |
|---|---|---|---|---|---|---|---|
| | **Capital Costs** | | | | | | |
| 33 26 0203 | 3" Stainless Steel Piping, Schedule 40, Threaded | LF | 57.00 | 50.61 | 47.88 | 41.71 | 39.45 |
| 33 26 0204 | 4" Stainless Steel Piping, Schedule 40, Threaded | LF | 80.15 | 71.72 | 68.12 | 59.98 | 57.00 |
| 33 26 0205 | 6" Stainless Steel Piping, Schedule 10, Type 316 | LF | 78.58 | 70.01 | 66.14 | 57.84 | 54.94 |
| 33 26 0206 | 8" Stainless Steel Piping, Schedule 40, Welded | LF | 197.22 | 183.32 | 176.89 | 163.42 | 158.80 |
| 33 26 0601 | (1 1/2", 3") Stainless Steel Double-wall Piping, with Fittings | LF | 134.61 | 117.05 | 109.53 | 92.56 | 86.33 |
| 33 26 0602 | (2", 4") Stainless Steel Double-wall Piping, with Fittings | LF | 185.98 | 163.46 | 153.81 | 132.04 | 124.05 |
| 33 26 0603 | (2 1/2", 4") Stainless Steel Double-wall Piping, with Fittings | LF | 206.86 | 182.17 | 171.60 | 147.74 | 138.99 |
| 33 26 0604 | (4", 6") Stainless Steel Double-wall Piping, with Fittings | LF | 339.64 | 280.79 | 254.18 | 197.21 | 177.23 |
| 33 26 0621 | (1 1/2", 3") PVC Double-wall Piping, with Fittings | LF | 33.15 | 27.05 | 24.43 | 18.53 | 16.36 |
| 33 26 0622 | (2", 4") PVC Double-wall Piping, with Fittings | LF | 46.22 | 37.98 | 34.43 | 26.46 | 23.55 |
| 33 26 0623 | (2 1/2", 4") PVC Double-wall Piping, with Fittings | LF | 49.22 | 40.39 | 36.61 | 28.07 | 24.94 |
| 33 26 0624 | (4", 6") PVC Double-wall Piping, with Fittings | LF | 70.61 | 58.00 | 52.59 | 40.40 | 35.93 |
| 33 26 0641 | (1 1/2", 3") Carbon Steel Double-wall Piping, with Fittings | LF | 48.16 | 40.64 | 37.42 | 30.15 | 27.48 |
| 33 26 0642 | (2", 4") Carbon Steel Double-wall Piping, with Fittings | LF | 64.71 | 55.69 | 51.83 | 43.11 | 39.91 |
| 33 26 0643 | (2 1/2", 4") Carbon Steel Double-wall Piping, with Fittings | LF | 70.81 | 61.02 | 56.83 | 47.37 | 43.90 |
| 33 26 0644 | (4", 6") Carbon Steel Double-wall Piping, with Fittings | LF | 114.39 | 100.33 | 94.32 | 80.73 | 75.75 |
| 33 29 0102 | 10 GPM, 1/2 HP, Centrifugal Pump | EACH | 1,063 | 952.32 | 905.07 | 798.39 | 759.27 |
| 33 29 0103 | 50 GPM, 100' Head, 3 HP, Centrifugal Pump | EACH | 2,658 | 2,500 | 2,432 | 2,279 | 2,223 |
| 33 29 0104 | 75 GPM, 100' Head, 5 HP, Centrifugal Pump | EACH | 2,946 | 2,754 | 2,672 | 2,487 | 2,419 |
| 33 29 0106 | 100 GPM, 150' Head, 7.5 HP, Centrifugal Pump | EACH | 3,419 | 3,191 | 3,093 | 2,873 | 2,793 |
| 33 29 0107 | 125 GPM, 150' Head, 10 HP, Centrifugal Pump | EACH | 3,815 | 3,512 | 3,382 | 3,088 | 2,981 |
| 33 29 0108 | 10 HP, 200 GPM, Centrifugal Pump | EACH | 2,941 | 2,714 | 2,616 | 2,396 | 2,316 |
| 33 29 0109 | 10 HP, 250 GPM, Centrifugal Pump | EACH | 3,032 | 2,805 | 2,707 | 2,487 | 2,407 |
| 33 29 0110 | 15 HP, 300 GPM, Centrifugal Pump | EACH | 3,441 | 3,160 | 3,040 | 2,770 | 2,670 |
| 33 29 0111 | 20 HP, 500 GPM, Centrifugal Pump | EACH | 4,544 | 4,241 | 4,111 | 3,818 | 3,710 |
| 33 29 0112 | 30 HP, 750 GPM, Centrifugal Pump | EACH | 5,213 | 4,808 | 4,635 | 4,244 | 4,100 |
| 33 29 0113 | 40 HP, 1050 GPM, Centrifugal Pump | EACH | 5,655 | 5,199 | 5,004 | 4,564 | 4,403 |
| 33 29 0114 | 60 HP, 1500 GPM, Centrifugal Pump | EACH | 9,842 | 8,823 | 8,350 | 7,363 | 7,024 |
| 33 29 0115 | 75 HP, 2000 GPM, Centrifugal Pump | EACH | 12,456 | 11,233 | 10,666 | 9,482 | 9,075 |
| 33 29 0116 | 100 HP, 3000 GPM, Centrifugal Pump | EACH | 15,575 | 14,046 | 13,338 | 11,857 | 11,348 |
| 33 29 0401 | 25 GPM, 1 1/2" Discharge, Cast-iron Sump Pump | EACH | 2,432 | 2,274 | 2,206 | 2,053 | 1,997 |
| 33 29 0402 | 75 GPM, 2" Discharge, Cast-iron Sump Pump | EACH | 3,011 | 2,820 | 2,738 | 2,552 | 2,484 |
| 33 29 0403 | 100 GPM, 2 1/2" Discharge, Cast-iron Sump Pump | EACH | 3,243 | 3,016 | 2,918 | 2,698 | 2,618 |

**Environmental Remediation: Assemblies Cost Book**

# Cost Data for Remediation

## Air Stripping

| Assembly | Description | Unit | A | B | C | D | E |
|---|---|---|---|---|---|---|---|
| | **Capital Costs** | | | | | | |
| 33 29 0404 | 150 GPM, 3" Discharge, Cast-iron Sump Pump | EACH | 3,703 | 3,422 | 3,302 | 3,032 | 2,932 |
| 33 29 0405 | 200 GPM, 3" Discharge, Cast-iron Sump Pump | EACH | 3,904 | 3,601 | 3,471 | 3,178 | 3,070 |
| 33 29 0406 | 300 GPM, 4" Discharge, Cast-iron Sump Pump | EACH | 4,817 | 4,453 | 4,297 | 3,945 | 3,816 |
| 33 29 0407 | 500 GPM, 5" Discharge, Cast-iron Sump Pump | EACH | 4,969 | 4,610 | 4,456 | 4,110 | 3,982 |
| 33 29 0408 | 800 GPM, 6" Discharge, Cast-iron Sump Pump | EACH | 12,550 | 12,113 | 11,911 | 11,488 | 11,342 |
| 33 29 0409 | 1,000 GPM, 6" Discharge, Cast-iron Sump Pump | EACH | 15,213 | 14,703 | 14,467 | 13,974 | 13,804 |
| 33 29 0410 | 1,600 GPM, 8" Discharge, Cast-iron Sump Pump | EACH | 17,561 | 16,949 | 16,666 | 16,074 | 15,870 |
| 33 29 0411 | 2,000 GPM, 8" Discharge, Cast-iron Sump Pump | EACH | 19,430 | 18,751 | 18,436 | 17,778 | 17,552 |
| 33 31 0147 | 100 CFM, 6" Pressure, 1/3 HP, Blower | EACH | 1,069 | 1,006 | 978.53 | 917.23 | 894.75 |
| 33 31 0148 | 150 CFM, 6" Pressure, 1/2 HP, Blower | EACH | 1,126 | 1,056 | 1,026 | 958.87 | 934.14 |
| 33 31 0149 | 250 CFM, 6" Pressure, 3/4 HP, Blower | EACH | 1,305 | 1,224 | 1,189 | 1,111 | 1,082 |
| 33 31 0150 | 500 CFM, 6" Pressure, 1 HP, Blower | EACH | 974.03 | 882.23 | 842.93 | 754.20 | 721.67 |
| 33 31 0151 | 750 CFM, 8" Pressure, 1.5 HP, Blower | EACH | 1,207 | 1,100 | 1,055 | 952.48 | 914.88 |
| 33 31 0152 | 1,000 CFM, 8" Pressure, 2.5 HP, Blower | EACH | 1,487 | 1,373 | 1,324 | 1,213 | 1,172 |
| 33 31 0153 | 1,500 CFM, 8" Pressure, 4 HP, Blower | EACH | 1,783 | 1,642 | 1,582 | 1,447 | 1,397 |
| 33 31 0154 | 2,000 CFM, 7" Pressure, 6 HP, Blower | EACH | 2,658 | 2,389 | 2,274 | 2,015 | 1,920 |
| 33 31 0155 | 3,000 CFM, 6" Pressure, 7.5 HP, Blower | EACH | 2,500 | 2,232 | 2,117 | 1,858 | 1,762 |
| 33 31 0156 | 4,000 CFM, 6" Pressure, 10 HP, Blower | EACH | 2,747 | 2,456 | 2,332 | 2,051 | 1,948 |
| 33 31 0157 | 5,000 CFM, 6" Pressure, 10 HP, Blower | EACH | 3,037 | 2,746 | 2,622 | 2,341 | 2,238 |
| 33 31 0158 | 8,000 CFM, 6" Pressure, 15 HP, Blower | EACH | 3,532 | 3,221 | 3,087 | 2,786 | 2,676 |
| 33 31 0159 | 11,000 CFM, 6" Pressure, 20 HP, Blower | EACH | 3,969 | 3,658 | 3,525 | 3,223 | 3,113 |
| 33 31 0160 | 15,000 CFM, 6" Pressure, 25 HP, Blower | EACH | 4,220 | 3,884 | 3,741 | 3,416 | 3,298 |
| 33 31 0161 | 18,000 CFM, 6" Pressure, 25 HP, Blower | EACH | 4,909 | 4,573 | 4,430 | 4,105 | 3,987 |
| 33 31 0162 | 20,000 CFM, 6" Pressure, 30 HP, Blower | EACH | 5,461 | 5,098 | 4,942 | 4,591 | 4,462 |
| 33 41 0201 | Blower and Motor Maintenance and Repair | EACH | 865.17 | 671.48 | 581.58 | 393.96 | 329.64 |
| 33 42 0101 | Electrical Charge | KWH | 0.06 | 0.06 | 0.06 | 0.06 | 0.06 |
| | **Operations & Maintenance** | | | | | | |
| 33 13 0701 | Packing Reconditioning | EACH | 3,996 | 3,189 | 2,678 | 1,888 | 1,705 |
| 33 41 0201 | Blower and Motor Maintenance and Repair | EACH | 865.17 | 671.48 | 581.58 | 393.96 | 329.64 |
| 33 42 0101 | Electrical Charge | KWH | 0.06 | 0.06 | 0.06 | 0.06 | 0.06 |

**Environmental Remediation: Assemblies Cost Book**

# Cost Data for Remediation
## Bioremediation, Water (Ex Situ)

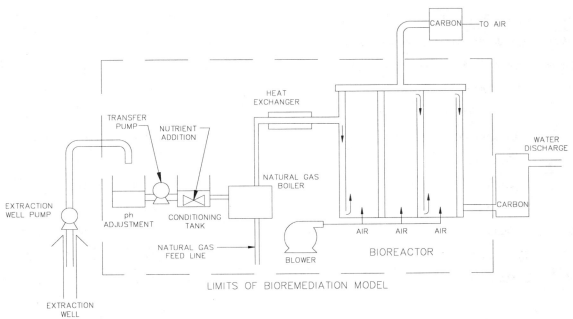

TYPICAL FIXED FILM BIOREACTOR SYSTEM

### General:

Biological water treatment processes employ microorganisms to degrade contaminants. The chief agents of biological degradation are heterotrophic bacteria and fungi. These organisms are capable of utilizing organic substrates as sources of carbon for energy and for the biosynthesis of new cellular material. Biological treatment systems are applicable to a wide range of organic contaminants, including polycyclic aromatic hydrocarbons (PAHs), phenols, gasoline, chlorinated solvents, diesel fuel, and chlorobenzene.

### Typical Treatment Train:

Extraction wells, oil/water separator, carbon adsorption (liquid), disposal (infiltration gallery, injection wells, discharge to POTW), sampling and analysis.

### Common Cost Components:

1. Mixing Tank, Structural Slab
2. Chemical Shot Feeder
3. Pumps
4. Boiler
5. Heat Exchanger
6. Bioreactor
7. Piping
8. Natural Gas, Electric
9. Nutrients
10. pH Adjustment
11. Operations and Maintenance

### Other Cost Considerations:

Electrical distribution, gas distribution, fencing.

# Cost Data for Remediation

## Bioremediation, Water (Ex Situ)

Unit Cost by Safety Level

| Assembly | Description | Unit | A | B | C | D | E |
|---|---|---|---|---|---|---|---|
| | **Capital Costs** | | | | | | |
| 18 02 0301 | Asphalt Pavement - 10" Subgrade, 9" Base, 1 1/2" Topping | SY | 61.11 | 50.94 | 43.58 | 33.57 | 31.83 |
| 18 02 0321 | 6" Structural Slab on Grade | SF | 6.98 | 5.95 | 5.47 | 4.48 | 4.14 |
| 18 02 0322 | 8" Structural Slab on Grade | SF | 9.75 | 8.38 | 7.71 | 6.38 | 5.95 |
| 18 02 0324 | 12" Structural Slab on Grade | SF | 11.57 | 10.04 | 9.30 | 7.82 | 7.34 |
| 18 02 0331 | 6" Mesh Reinforced Slab on Grade | SF | 6.01 | 5.10 | 4.67 | 3.79 | 3.49 |
| 18 02 0341 | 6" Unreinforced Slab on Grade | SF | 4.87 | 4.14 | 3.79 | 3.08 | 2.85 |
| 19 01 0202 | 1", Class 200, PVC Piping | LF | 7.48 | 5.84 | 5.07 | 3.48 | 2.93 |
| 19 01 0204 | 2", Class 200, PVC Piping | LF | 8.86 | 6.96 | 6.07 | 4.23 | 3.60 |
| 19 01 0206 | 3", Class 200, PVC Piping | LF | 11.81 | 9.34 | 8.20 | 5.80 | 4.98 |
| 19 01 0207 | 4", Class 200, PVC Piping | LF | 12.33 | 9.86 | 8.72 | 6.32 | 5.50 |
| 19 01 0208 | 6", Class 200, PVC Piping | LF | 14.95 | 12.20 | 10.93 | 8.27 | 7.36 |
| 19 01 0212 | 8", Class 150, PVC Piping | LF | 18.36 | 15.27 | 13.84 | 10.85 | 9.82 |
| 19 01 0213 | 10", Class 150, PVC Piping | LF | 22.90 | 19.37 | 17.73 | 14.32 | 13.14 |
| 19 01 0214 | 12", Class 150, PVC Piping | LF | 28.75 | 24.63 | 22.72 | 18.73 | 17.36 |
| 19 01 0310 | 5,000 Gallon Water Tank, Steel, Horizontal @ Grade | EACH | 15,945 | 13,359 | 12,252 | 9,752 | 8,835 |
| 19 01 0311 | 10,000 Gallon Water Tank, Steel, Horizontal @ Grade | EACH | 18,756 | 16,170 | 15,063 | 12,563 | 11,646 |
| 19 01 0312 | 20,000 Gallon Water Tank, Steel, Horizontal @ Grade | EACH | 28,844 | 24,853 | 23,144 | 19,286 | 17,871 |
| 19 01 0313 | 35,000 Gallon Water Tank, Steel, Horizontal @ Grade | EACH | 45,913 | 40,740 | 38,525 | 33,525 | 31,692 |
| 19 04 0401 | 550 Gallon, Stainless Steel Aboveground Wastewater Holding Tank, Rental | MONTH | 320.65 | 320.65 | 320.65 | 320.65 | 320.65 |
| 19 04 0402 | 630 Gallon, Polyethylene Aboveground Wastewater Holding Tank, Rental | MONTH | 320.65 | 320.65 | 320.65 | 320.65 | 320.65 |
| 19 04 0403 | 4,000 Gallon Polyethylene Wastewater Tank, Rental | MONTH | 545.90 | 545.90 | 545.90 | 545.90 | 545.90 |
| 19 04 0405 | 6,000 Gallon, Polyethylene Aboveground Wastewater Holding Tank, Rental | MONTH | 641.30 | 641.30 | 641.30 | 641.30 | 641.30 |
| 19 07 0120 | 1 1/4", 60 PSI, Polyethylene Pipe | LF | 5.29 | 4.21 | 3.72 | 2.68 | 2.32 |
| 19 07 0121 | 1 1/2", 60 PSI, Polyethylene Pipe | LF | 5.38 | 4.30 | 3.81 | 2.77 | 2.41 |
| 19 07 0122 | 2", 60 PSI, Polyethylene Pipe | LF | 6.05 | 4.85 | 4.31 | 3.15 | 2.75 |
| 33 11 9301 | 200 MBH Gas-fired Water Boiler | EACH | 6,218 | 5,682 | 5,425 | 4,905 | 4,731 |
| 33 11 9302 | 275 MBH Gas-fired Water Boiler | EACH | 8,169 | 7,249 | 6,809 | 5,917 | 5,620 |
| 33 11 9303 | 360 MBH Gas-fired Water Boiler | EACH | 8,247 | 7,327 | 6,887 | 5,995 | 5,698 |
| 33 11 9304 | 520 MBH Gas-fired Water Boiler | EACH | 9,546 | 8,473 | 7,959 | 6,919 | 6,572 |
| 33 11 9305 | 600 MBH Gas-fired Water Boiler | EACH | 10,154 | 9,081 | 8,567 | 7,527 | 7,180 |
| 33 11 9306 | 720 MBH Gas-fired Water Boiler | EACH | 11,730 | 10,442 | 9,826 | 8,578 | 8,161 |

**Environmental Remediation: Assemblies Cost Book**

# Cost Data for Remediation

## Bioremediation, Water (Ex Situ)

| Assembly | Description | Unit | A | B | C | D | E |
|---|---|---|---|---|---|---|---|
| | **Capital Costs** | | | | | | |
| 33 11 9307 | 960 MBH Gas-fired Water Boiler | EACH | 13,789 | 12,179 | 11,410 | 9,849 | 9,328 |
| 33 11 9308 | 1,220 MBH Gas-fired Water Boiler | EACH | 14,398 | 12,788 | 12,019 | 10,458 | 9,937 |
| 33 11 9309 | 1,440 MBH Gas-fired Water Boiler | EACH | 23,432 | 21,594 | 20,716 | 18,934 | 18,340 |
| 33 11 9310 | 1,680 MBH Gas-fired Water Boiler | EACH | 25,423 | 23,277 | 22,251 | 20,170 | 19,476 |
| 33 11 9311 | 1,920 MBH Gas-fired Water Boiler | EACH | 27,362 | 24,885 | 23,701 | 21,300 | 20,499 |
| 33 11 9312 | 2,160 MBH Gas-fired Water Boiler | EACH | 31,110 | 27,891 | 26,351 | 23,230 | 22,188 |
| 33 11 9313 | 2,400 MBH Gas-fired Water Boiler | EACH | 31,410 | 28,191 | 26,651 | 23,530 | 22,488 |
| 33 11 9314 | 7 GPM Water Heat Exchanger | EACH | 1,265 | 1,199 | 1,171 | 1,107 | 1,084 |
| 33 11 9315 | 16 GPM Water Heat Exchanger | EACH | 1,728 | 1,649 | 1,615 | 1,538 | 1,510 |
| 33 11 9316 | 34 GPM Water Heat Exchanger | EACH | 2,532 | 2,434 | 2,391 | 2,296 | 2,260 |
| 33 11 9317 | 55 GPM Water Heat Exchanger | EACH | 3,609 | 3,477 | 3,420 | 3,293 | 3,246 |
| 33 11 9318 | 74 GPM Water Heat Exchanger | EACH | 4,503 | 4,331 | 4,257 | 4,091 | 4,030 |
| 33 11 9319 | 86 GPM Water Heat Exchanger | EACH | 5,935 | 5,727 | 5,637 | 5,436 | 5,362 |
| 33 11 9320 | 112 GPM Water Heat Exchanger | EACH | 7,395 | 7,142 | 7,022 | 6,777 | 6,695 |
| 33 11 9321 | 3,000 Gallon Fixed Film Bioreactor | EACH | 62,328 | 62,328 | 62,328 | 62,328 | 62,328 |
| 33 11 9322 | 5,000 Gallon Fixed Film Bioreactor | EACH | 87,927 | 87,927 | 87,927 | 87,927 | 87,927 |
| 33 11 9323 | 9,000 Gallon Fixed Film Bioreactor | EACH | 139,125 | 139,125 | 139,125 | 139,125 | 139,125 |
| 33 11 9324 | 11,500 Gallon Fixed Film Bioreactor | EACH | 150,255 | 150,255 | 150,255 | 150,255 | 150,255 |
| 33 11 9325 | 17,500 Gallon Fixed Film Bioreactor, Short Retention | EACH | 194,775 | 194,775 | 194,775 | 194,775 | 194,775 |
| 33 11 9326 | 17,500 Gallon Fixed Film Bioreactor, Long Retention | EACH | 194,775 | 194,775 | 194,775 | 194,775 | 194,775 |
| 33 11 9327 | 18,500 Gallon Fixed Film Bioreactor | EACH | 194,775 | 194,775 | 194,775 | 194,775 | 194,775 |
| 33 11 9328 | 24,000 Gallon Fixed Film Bioreactor | EACH | 233,730 | 233,730 | 233,730 | 233,730 | 233,730 |
| 33 11 9329 | 27,000 Gallon Fixed Film Bioreactor | EACH | 267,120 | 267,120 | 267,120 | 267,120 | 267,120 |
| 33 11 9330 | 29,500 Gallon Fixed Film Bioreactor | EACH | 289,380 | 289,380 | 289,380 | 289,380 | 289,380 |
| 33 11 9331 | Expansion Tank, Pipe, & Fittings, 1 - 1,220 MBH Boiler | EACH | 4,657 | 3,871 | 3,535 | 2,776 | 2,497 |
| 33 11 9332 | Expansion Tank, Pipe, & Fittings, 1,440 - 2,400 MBH Boiler | EACH | 20,553 | 17,748 | 16,543 | 13,832 | 12,841 |
| 33 11 9902 | Light Petroleum Hydrocarbon Degraders, Microorganisms | LB | 19.88 | 19.88 | 19.88 | 19.88 | 19.88 |
| 33 11 9903 | Light Petroleum Hydrocarbon Degraders, 100 Lb Bag, Microorganisms | EACH | 1,948 | 1,948 | 1,948 | 1,948 | 1,948 |
| 33 11 9904 | Light Petroleum Hydrocarbon Degraders, 1,000 Lb Bag, Microorganisms | EACH | 17,490 | 17,490 | 17,490 | 17,490 | 17,490 |
| 33 11 9905 | Heavy Petroleum Hydrocarbon/Creosote Degraders, Microorganisms | LB | 13.25 | 13.25 | 13.25 | 13.25 | 13.25 |
| 33 11 9906 | Heavy Petroleum Hydrocarbon/Creosote Degraders, 100 Lb Bag, Microorganisms | EACH | 1,007 | 1,007 | 1,007 | 1,007 | 1,007 |

**Environmental Remediation: Assemblies Cost Book**

# Cost Data for Remediation

## Bioremediation, Water (Ex Situ)

| Assembly | Description | Unit | A | B | C | D | E |
|---|---|---|---|---|---|---|---|
| | **Capital Costs** | | | | | | |
| 33 11 9907 | Heavy Petroleum Hydrocarbon/Creosote Degraders, 1,000 Lb Bag, Microorganisms | EACH | 8,745 | 8,745 | 8,745 | 8,745 | 8,745 |
| 33 11 9951 | Bionutrients, 50 Lb Bag | EACH | 58.30 | 58.30 | 58.30 | 58.30 | 58.30 |
| 33 12 9901 | 1.7 Gallon Bypass Chemical Shot Feeder, In-line Mount, 125 PSIG | EACH | 847.93 | 691.64 | 624.73 | 473.67 | 418.28 |
| 33 12 9902 | 5 Gallon Bypass Chemical Shot Feeder, Floor Mount, 175 PSIG | EACH | 1,236 | 1,001 | 900.54 | 673.47 | 590.22 |
| 33 12 9903 | 12 Gallon Bypass Chemical Shot Feeder, Floor Mount, 175 PSIG | EACH | 1,626 | 1,315 | 1,182 | 881.31 | 771.02 |
| 33 12 9904 | 5 Gallon Bypass Chemical Shot Feeder, Floor Mount, 150 Lb ASME | EACH | 1,236 | 1,001 | 900.54 | 673.47 | 590.22 |
| 33 12 9905 | 10 Gallon Bypass Chemical Shot Feeder, Floor Mount, 150 Lb ASME | EACH | 1,626 | 1,315 | 1,182 | 881.31 | 771.02 |
| 33 12 9906 | 5 Gallon Bypass Chemical Shot Feeder, Floor Mount, 300 Lb ASME | EACH | 1,676 | 1,441 | 1,341 | 1,114 | 1,030 |
| 33 12 9907 | 10 Gallon Bypass Chemical Shot Feeder, Floor Mount, 300 Lb ASME | EACH | 2,084 | 1,772 | 1,639 | 1,338 | 1,228 |
| 33 13 0101 | 3' Diameter Electric Automatic Pressure Filter Unit | EACH | 9,960 | 8,539 | 7,759 | 6,375 | 5,978 |
| 33 13 0102 | 4' Diameter Electric Automatic Pressure Filter Unit | EACH | 14,937 | 13,345 | 12,472 | 10,923 | 10,477 |
| 33 13 0103 | 5' Diameter Electric Automatic Pressure Filter Unit | EACH | 15,593 | 14,001 | 13,128 | 11,579 | 11,133 |
| 33 13 0104 | 6' Diameter Electric Automatic Pressure Filter Unit | EACH | 20,547 | 18,788 | 17,823 | 16,110 | 15,618 |
| 33 13 0105 | 7' Diameter Electric Automatic Pressure Filter Unit | EACH | 23,444 | 21,685 | 20,720 | 19,007 | 18,515 |
| 33 13 0106 | 8' Diameter Electric Automatic Pressure Filter Unit | EACH | 29,455 | 27,301 | 26,120 | 24,023 | 23,421 |
| 33 13 0401 | 9' Diameter, 25 GPM, Waste Flow, Contact Clarifier | EACH | 125,549 | 109,509 | 101,224 | 85,638 | 80,833 |
| 33 13 0402 | 12' Diameter, 45 GPM, Waste Flow, Contact Clarifier | EACH | 143,480 | 127,441 | 119,155 | 103,569 | 98,764 |
| 33 13 0403 | 15' Diameter, 70 GPM, Waste Flow, Contact Clarifier | EACH | 148,807 | 132,768 | 124,482 | 108,896 | 104,091 |
| 33 13 0404 | 20' Diameter, 130 GPM, Waste Flow, Contact Clarifier | EACH | 158,323 | 137,909 | 127,364 | 107,527 | 101,412 |
| 33 13 0405 | 25' Diameter, 208 GPM, Waste Flow, Contact Clarifier | EACH | 162,595 | 142,181 | 131,636 | 111,799 | 105,683 |
| 33 13 0406 | 30' Diameter, 305 GPM, Waste Flow, Contact Clarifier | EACH | 172,420 | 152,006 | 141,461 | 121,624 | 115,508 |
| 33 13 0407 | 40' Diameter, 546 GPM, Waste Flow, Contact Clarifier | EACH | 300,630 | 272,561 | 258,061 | 230,785 | 222,377 |
| 33 13 0408 | 50' Diameter, 850 GPM, Waste Flow, Contact Clarifier | EACH | 386,586 | 349,161 | 329,828 | 293,460 | 282,249 |
| 33 13 0409 | 60' Diameter, 1,194 GPM, Waste Flow, Contact Clarifier | EACH | 490,608 | 445,698 | 422,498 | 378,857 | 365,403 |
| 33 13 0410 | 70' Diameter, 1,645 GPM, Waste Flow, Contact Clarifier | EACH | 596,008 | 539,871 | 510,871 | 456,320 | 439,503 |
| 33 13 2916 | Natural Gas Usage, per 1,000 CF | MCF | 5.50 | 5.50 | 5.50 | 5.50 | 5.50 |
| 33 26 0101 | 1" Carbon Steel Piping | LF | 6.46 | 5.28 | 4.77 | 3.63 | 3.21 |
| 33 26 0102 | 2" Carbon Steel Piping | LF | 10.50 | 8.72 | 7.95 | 6.23 | 5.60 |
| 33 26 0103 | 3" Carbon Steel Piping | LF | 18.02 | 15.20 | 13.99 | 11.26 | 10.26 |

**Environmental Remediation: Assemblies Cost Book**

# Cost Data for Remediation

## Bioremediation, Water (Ex Situ)

| Assembly | Description | Unit | Unit Cost by Safety Level A | B | C | D | E |
|---|---|---|---|---|---|---|---|
| | **Capital Costs** | | | | | | |
| 33 26 0104 | 4" Carbon Steel Piping | LF | 23.20 | 19.86 | 18.43 | 15.21 | 14.02 |
| 33 26 0105 | 6" Carbon Steel Piping | LF | 36.37 | 30.24 | 27.46 | 21.52 | 19.44 |
| 33 26 0106 | 8" Carbon Steel Piping | LF | 40.77 | 34.08 | 31.15 | 24.68 | 22.36 |
| 33 26 0201 | 1" Stainless Steel Piping, Schedule 40, Threaded | LF | 18.67 | 16.12 | 15.03 | 12.56 | 11.65 |
| 33 26 0202 | 2" Stainless Steel Piping, Schedule 40, Threaded | LF | 33.47 | 29.21 | 27.39 | 23.28 | 21.77 |
| 33 26 0203 | 3" Stainless Steel Piping, Schedule 40, Threaded | LF | 57.00 | 50.61 | 47.88 | 41.71 | 39.45 |
| 33 26 0204 | 4" Stainless Steel Piping, Schedule 40, Threaded | LF | 80.15 | 71.72 | 68.12 | 59.98 | 57.00 |
| 33 26 0205 | 6" Stainless Steel Piping, Schedule 10, Type 316 | LF | 78.58 | 70.01 | 66.14 | 57.84 | 54.94 |
| 33 26 0206 | 8" Stainless Steel Piping, Schedule 40, Welded | LF | 197.22 | 183.32 | 176.89 | 163.42 | 158.80 |
| 33 26 0207 | 8" Stainless Steel, Schedule 10, Type 316, Manifold Piping | LF | 126.78 | 112.88 | 106.45 | 92.98 | 88.36 |
| 33 26 0210 | 2" Stainless Steel, Schedule 40, Type 304, Connection Piping | LF | 33.47 | 29.21 | 27.39 | 23.28 | 21.77 |
| 33 26 0211 | 4" Stainless Steel, Schedule 40, Type 304, Connection Piping | LF | 80.15 | 71.72 | 68.12 | 59.98 | 57.00 |
| 33 26 0212 | 4" Stainless Steel, Schedule 40, Type 304, Manifold Piping | LF | 80.15 | 71.72 | 68.12 | 59.98 | 57.00 |
| 33 26 0301 | 1 1/2" Polypropylene Pipe Including Fittings | LF | 7.56 | 6.11 | 5.48 | 4.08 | 3.57 |
| 33 26 0302 | 2" Polypropylene Pipe Including Fittings | LF | 8.79 | 7.16 | 6.46 | 4.88 | 4.30 |
| 33 26 0303 | 3" Polypropylene Pipe Including Fittings | LF | 13.80 | 11.43 | 10.42 | 8.13 | 7.29 |
| 33 26 0304 | 4" Polypropylene Pipe Including Fittings | LF | 18.36 | 15.28 | 13.96 | 10.99 | 9.90 |
| 33 26 0601 | (1 1/2", 3") Stainless Steel Double-wall Piping, with Fittings | LF | 134.61 | 117.05 | 109.53 | 92.56 | 86.33 |
| 33 26 0602 | (2", 4") Stainless Steel Double-wall Piping, with Fittings | LF | 185.98 | 163.46 | 153.81 | 132.04 | 124.05 |
| 33 26 0603 | (2 1/2", 4") Stainless Steel Double-wall Piping, with Fittings | LF | 206.86 | 182.17 | 171.60 | 147.74 | 138.99 |
| 33 26 0604 | (4", 6") Stainless Steel Double-wall Piping, with Fittings | LF | 339.64 | 280.79 | 254.18 | 197.21 | 177.23 |
| 33 26 0621 | (1 1/2", 3") PVC Double-wall Piping, with Fittings | LF | 33.15 | 27.05 | 24.43 | 18.53 | 16.36 |
| 33 26 0622 | (2", 4") PVC Double-wall Piping, with Fittings | LF | 46.22 | 37.98 | 34.43 | 26.46 | 23.55 |
| 33 26 0623 | (2 1/2", 4") PVC Double-wall Piping, with Fittings | LF | 49.22 | 40.39 | 36.61 | 28.07 | 24.94 |
| 33 26 0624 | (4", 6") PVC Double-wall Piping, with Fittings | LF | 70.61 | 58.00 | 52.59 | 40.40 | 35.93 |
| 33 26 0641 | (1 1/2", 3") Carbon Steel Double-wall Piping, with Fittings | LF | 48.16 | 40.64 | 37.42 | 30.15 | 27.48 |
| 33 26 0642 | (2", 4") Carbon Steel Double-wall Piping, with Fittings | LF | 64.71 | 55.69 | 51.83 | 43.11 | 39.91 |
| 33 26 0643 | (2 1/2", 4") Carbon Steel Double-wall Piping, with Fittings | LF | 70.81 | 61.02 | 56.83 | 47.37 | 43.90 |
| 33 26 0644 | (4", 6") Carbon Steel Double-wall Piping, with Fittings | LF | 114.39 | 100.33 | 94.32 | 80.73 | 75.75 |
| 33 27 0102 | 2" PVC, Schedule 40, Tee | EACH | 68.73 | 53.63 | 46.80 | 32.18 | 27.06 |
| 33 27 0104 | 4" PVC, Schedule 40, Tee | EACH | 137.04 | 108.84 | 96.77 | 69.51 | 59.51 |
| 33 27 0106 | 8" PVC, Schedule 40, Tee | EACH | 336.90 | 291.78 | 272.46 | 228.85 | 212.85 |

**Environmental Remediation: Assemblies Cost Book**

*Page 3-19*

# Cost Data for Remediation

## Bioremediation, Water (Ex Situ)

| Assembly | Description | Unit | Unit Cost by Safety Level | | | | |
|---|---|---|---|---|---|---|---|
| | | | A | B | C | D | E |
| | Capital Costs | | | | | | |
| 33 27 0112 | 2" PVC, 90 Degree, Elbow | EACH | 41.36 | 32.22 | 28.30 | 19.46 | 16.22 |
| 33 27 0114 | 4" PVC, 90 Degree, Elbow | EACH | 101.72 | 80.57 | 71.51 | 51.07 | 43.57 |
| 33 27 0116 | 8" PVC, 90 Degree, Elbow | EACH | 200.21 | 172.01 | 159.94 | 132.68 | 122.68 |
| 33 27 0201 | 2" Stainless Steel, Tee | EACH | 173.32 | 140.34 | 126.22 | 94.34 | 82.65 |
| 33 27 0202 | 4" Stainless Steel, Tee | EACH | 583.41 | 504.25 | 470.36 | 393.85 | 365.80 |
| 33 27 0203 | 8" Stainless Steel, Schedule 10, Type 316, Tee | EACH | 2,283 | 1,934 | 1,772 | 1,434 | 1,317 |
| 33 27 0211 | 2" Stainless Steel, 90 Degree, Elbow | EACH | 114.28 | 93.45 | 84.53 | 64.39 | 57.01 |
| 33 27 0212 | 4" Stainless Steel, 90 Degree, Elbow | EACH | 391.27 | 338.50 | 315.91 | 264.90 | 246.20 |
| 33 27 0402 | 2" Iron Body Check Valve | EACH | 188.96 | 174.31 | 168.03 | 153.86 | 148.67 |
| 33 27 0404 | 4" Iron Body Check Valve | EACH | 354.29 | 315.96 | 299.54 | 262.49 | 248.90 |
| 33 27 0405 | 6" Iron Body Check Valve | EACH | 618.60 | 554.57 | 523.95 | 461.86 | 441.15 |
| 33 27 0406 | 8" Iron Body Check Valve | EACH | 1,094 | 980.95 | 926.96 | 817.50 | 780.98 |
| 33 27 0602 | 2" Tee, Fiberglass Reinforced | EACH | 131.59 | 109.76 | 100.41 | 79.31 | 71.57 |
| 33 27 0604 | (2", 4") Tee, Fiberglass Reinforced | EACH | 279.06 | 235.39 | 216.70 | 174.49 | 159.01 |
| 33 27 0612 | 2" 90-degree Elbow, Fiberglass Reinforced | EACH | 126.48 | 104.65 | 95.30 | 74.20 | 66.46 |
| 33 27 0614 | (2", 4") 90-degree Elbow, Fiberglass Reinforced | EACH | 258.24 | 214.57 | 195.88 | 153.67 | 138.19 |
| 33 27 0622 | 2" 45-degree Elbow, Fiberglass Reinforced | EACH | 126.48 | 104.65 | 95.30 | 74.20 | 66.46 |
| 33 27 0624 | (2", 4") 45-degree Elbow, Fiberglass Reinforced | EACH | 258.24 | 214.57 | 195.88 | 153.67 | 138.19 |
| 33 27 0632 | 2" Coupling, Fiberglass Reinforced | EACH | 102.87 | 81.04 | 71.69 | 50.59 | 42.85 |
| 33 27 0634 | (2", 4") Coupling, Fiberglass Reinforced | EACH | 221.99 | 178.32 | 159.63 | 117.42 | 101.94 |
| 33 27 0642 | 2" Termination Reducer, Fiberglass Reinforced | EACH | 109.33 | 87.50 | 78.15 | 57.05 | 49.31 |
| 33 27 0644 | (2", 4") Termination Reducer, Fiberglass Reinforced | EACH | 223.39 | 179.72 | 161.03 | 118.82 | 103.34 |
| 33 27 0902 | 2" Fiberglass Reinforced Piping | EACH | 12.46 | 10.27 | 9.34 | 7.23 | 6.46 |
| 33 27 0904 | (2", 4") Fiberglass Reinforced Double-wall Piping | LF | 25.86 | 21.50 | 19.63 | 15.41 | 13.86 |
| 33 29 0102 | 10 GPM, 1/2 HP, Centrifugal Pump | EACH | 1,063 | 952.32 | 905.07 | 798.39 | 759.27 |
| 33 29 0103 | 50 GPM, 100' Head, 3 HP, Centrifugal Pump | EACH | 2,658 | 2,500 | 2,432 | 2,279 | 2,223 |
| 33 29 0104 | 75 GPM, 100' Head, 5 HP, Centrifugal Pump | EACH | 2,946 | 2,754 | 2,672 | 2,487 | 2,419 |
| 33 29 0106 | 100 GPM, 150' Head, 7.5 HP, Centrifugal Pump | EACH | 3,419 | 3,191 | 3,093 | 2,873 | 2,793 |
| 33 29 0107 | 125 GPM, 150' Head, 10 HP, Centrifugal Pump | EACH | 3,815 | 3,512 | 3,382 | 3,088 | 2,981 |
| 33 29 0108 | 10 HP, 200 GPM, Centrifugal Pump | EACH | 2,941 | 2,714 | 2,616 | 2,396 | 2,316 |
| 33 29 0109 | 10 HP, 250 GPM, Centrifugal Pump | EACH | 3,032 | 2,805 | 2,707 | 2,487 | 2,407 |
| 33 29 0110 | 15 HP, 300 GPM, Centrifugal Pump | EACH | 3,441 | 3,160 | 3,040 | 2,770 | 2,670 |
| 33 29 0111 | 20 HP, 500 GPM, Centrifugal Pump | EACH | 4,544 | 4,241 | 4,111 | 3,818 | 3,710 |

## Environmental Remediation: Assemblies Cost Book

# Cost Data for Remediation

## Bioremediation, Water (Ex Situ)

| Assembly | Description | Unit | Unit Cost by Safety Level | | | | |
|---|---|---|---|---|---|---|---|
| | | | A | B | C | D | E |
| | **Capital Costs** | | | | | | |
| 33 33 0101 | Sodium Hydroxide Solution, 190 Lb Drummed Liquid | EACH | 62.34 | 57.81 | 55.07 | 50.64 | 49.53 |
| 33 33 0102 | Sodium Hydroxide Solution, 700 Lb Drummed Liquid | EACH | 163.04 | 158.51 | 155.77 | 151.34 | 150.23 |
| 33 33 0103 | Sodium Hydroxide, 500 Lb Container, Beads | EACH | 234.06 | 229.53 | 226.79 | 222.36 | 221.25 |
| 33 33 0104 | Sodium Hydroxide, 100 Lb Container, Flakes | EACH | 89.90 | 85.37 | 82.63 | 78.20 | 77.09 |
| 33 33 0105 | Sodium Hydroxide, 400 Lb Container, Flakes | EACH | 169.60 | 169.60 | 169.60 | 169.60 | 169.60 |
| 33 33 0106 | Magnesium Hydroxide Slurry, 55 Gallon Drums, < 40 Drums | EACH | 249.77 | 249.77 | 249.77 | 249.77 | 249.77 |
| 33 33 0107 | Magnesium Hydroxide Slurry, 55 Gallon Drums, 40 - 71 Drums | EACH | 234.26 | 234.26 | 234.26 | 234.26 | 234.26 |
| 33 33 0108 | Magnesium Hydroxide Slurry, 55 Gallon Drums, 72 Drums | EACH | 107.33 | 107.33 | 107.33 | 107.33 | 107.33 |
| 33 33 0109 | Magnesium Hydroxide, Bulk Quantity, 20 Tons | EACH | 4,452 | 4,452 | 4,452 | 4,452 | 4,452 |
| 33 33 0110 | Sulfuric Acid Solution, 220 Lb Drummed Liquid | EACH | 45.38 | 40.85 | 38.11 | 33.68 | 32.57 |
| 33 33 0111 | Sulfuric Acid Solution, 750 Lb Drummed Liquid | EACH | 93.61 | 89.08 | 86.34 | 81.91 | 80.80 |
| 33 33 0112 | Soda Ash, 50 Lb Bag, Powdered | EACH | 28.42 | 23.89 | 21.15 | 16.72 | 15.61 |
| 33 33 0113 | Soda Ash, 100 Lb Bag, Powdered | EACH | 41.14 | 36.61 | 33.87 | 29.44 | 28.33 |
| 33 33 0114 | Quicklime, 1/4" Nominal Granules, Bulk Quantity | TON | 96.49 | 96.49 | 96.48 | 96.48 | 96.47 |
| 33 33 0115 | Quicklime, 3/4" Nominal Granules, Bulk Quantity | TON | 96.49 | 96.49 | 96.48 | 96.48 | 96.47 |
| 33 33 0116 | Quicklime, Combination 1/4" & 3/4" Granules, Bulk Quantity | TON | 96.49 | 96.49 | 96.48 | 96.48 | 96.47 |
| 33 33 0117 | Hydrated Lime, Powdered, Bulk | TON | 97.55 | 97.55 | 97.54 | 97.54 | 97.53 |
| 33 33 0118 | Hydrated Lime, Powdered, 50 Lb Bag | EACH | 25.76 | 21.23 | 18.49 | 14.06 | 12.94 |
| 33 42 0101 | Electrical Charge | KWH | 0.06 | 0.06 | 0.06 | 0.06 | 0.06 |
| | **Operations & Maintenance** | | | | | | |
| 19 04 0401 | 550 Gallon, Stainless Steel Aboveground Wastewater Holding Tank, Rental | MONTH | 320.65 | 320.65 | 320.65 | 320.65 | 320.65 |
| 19 04 0402 | 630 Gallon, Polyethylene Aboveground Wastewater Holding Tank, Rental | MONTH | 320.65 | 320.65 | 320.65 | 320.65 | 320.65 |
| 19 04 0403 | 4,000 Gallon Polyethylene Wastewater Tank, Rental | MONTH | 545.90 | 545.90 | 545.90 | 545.90 | 545.90 |
| 19 04 0405 | 6,000 Gallon, Polyethylene Aboveground Wastewater Holding Tank, Rental | MONTH | 641.30 | 641.30 | 641.30 | 641.30 | 641.30 |
| 33 11 9902 | Light Petroleum Hydrocarbon Degraders, Microorganisms | LB | 19.88 | 19.88 | 19.88 | 19.88 | 19.88 |
| 33 11 9903 | Light Petroleum Hydrocarbon Degraders, 100 Lb Bag, Microorganisms | EACH | 1,948 | 1,948 | 1,948 | 1,948 | 1,948 |
| 33 11 9904 | Light Petroleum Hydrocarbon Degraders, 1,000 Lb Bag, Microorganisms | EACH | 17,490 | 17,490 | 17,490 | 17,490 | 17,490 |
| 33 11 9905 | Heavy Petroleum Hydrocarbon/Creosote Degraders, Microorganisms | LB | 13.25 | 13.25 | 13.25 | 13.25 | 13.25 |

**Environmental Remediation: Assemblies Cost Book**

# Cost Data for Remediation

## Bioremediation, Water (Ex Situ)

| Assembly | Description | Unit | \*Unit Cost by Safety Level A | B | C | D | E |
|---|---|---|---|---|---|---|---|
| | **Operations & Maintenance** | | | | | | |
| 33 11 9906 | Heavy Petroleum Hydrocarbon/Creosote Degraders, 100 Lb Bag, Microorganisms | EACH | 1,007 | 1,007 | 1,007 | 1,007 | 1,007 |
| 33 11 9907 | Heavy Petroleum Hydrocarbon/Creosote Degraders, 1,000 Lb Bag, Microorganisms | EACH | 8,745 | 8,745 | 8,745 | 8,745 | 8,745 |
| 33 11 9951 | Bionutrients, 50 Lb Bag | EACH | 58.30 | 58.30 | 58.30 | 58.30 | 58.30 |
| 33 13 2916 | Natural Gas Usage, per 1,000 CF | MCF | 5.50 | 5.50 | 5.50 | 5.50 | 5.50 |
| 33 33 0101 | Sodium Hydroxide Solution, 190 Lb Drummed Liquid | EACH | 62.34 | 57.81 | 55.07 | 50.64 | 49.53 |
| 33 33 0102 | Sodium Hydroxide Solution, 700 Lb Drummed Liquid | EACH | 163.04 | 158.51 | 155.77 | 151.34 | 150.23 |
| 33 33 0103 | Sodium Hydroxide, 500 Lb Container, Beads | EACH | 234.06 | 229.53 | 226.79 | 222.36 | 221.25 |
| 33 33 0104 | Sodium Hydroxide, 100 Lb Container, Flakes | EACH | 89.90 | 85.37 | 82.63 | 78.20 | 77.09 |
| 33 33 0105 | Sodium Hydroxide, 400 Lb Container, Flakes | EACH | 169.60 | 169.60 | 169.60 | 169.60 | 169.60 |
| 33 33 0106 | Magnesium Hydroxide Slurry, 55 Gallon Drums, < 40 Drums | EACH | 249.77 | 249.77 | 249.77 | 249.77 | 249.77 |
| 33 33 0107 | Magnesium Hydroxide Slurry, 55 Gallon Drums, 40 - 71 Drums | EACH | 234.26 | 234.26 | 234.26 | 234.26 | 234.26 |
| 33 33 0108 | Magnesium Hydroxide Slurry, 55 Gallon Drums, 72 Drums | EACH | 107.33 | 107.33 | 107.33 | 107.33 | 107.33 |
| 33 33 0109 | Magnesium Hydroxide, Bulk Quantity, 20 Tons | EACH | 4,452 | 4,452 | 4,452 | 4,452 | 4,452 |
| 33 33 0110 | Sulfuric Acid Solution, 220 Lb Drummed Liquid | EACH | 45.38 | 40.85 | 38.11 | 33.68 | 32.57 |
| 33 33 0111 | Sulfuric Acid Solution, 750 Lb Drummed Liquid | EACH | 93.61 | 89.08 | 86.34 | 81.91 | 80.80 |
| 33 33 0112 | Soda Ash, 50 Lb Bag, Powdered | EACH | 28.42 | 23.89 | 21.15 | 16.72 | 15.61 |
| 33 33 0113 | Soda Ash, 100 Lb Bag, Powdered | EACH | 41.14 | 36.61 | 33.87 | 29.44 | 28.33 |
| 33 33 0114 | Quicklime, 1/4" Nominal Granules, Bulk Quantity | TON | 96.49 | 96.49 | 96.48 | 96.48 | 96.47 |
| 33 33 0115 | Quicklime, 3/4" Nominal Granules, Bulk Quantity | TON | 96.49 | 96.49 | 96.48 | 96.48 | 96.47 |
| 33 33 0116 | Quicklime, Combination 1/4" & 3/4" Granules, Bulk Quantity | TON | 96.49 | 96.49 | 96.48 | 96.48 | 96.47 |
| 33 33 0117 | Hydrated Lime, Powdered, Bulk | TON | 97.55 | 97.55 | 97.54 | 97.54 | 97.53 |
| 33 33 0118 | Hydrated Lime, Powdered, 50 Lb Bag | EACH | 25.76 | 21.23 | 18.49 | 14.06 | 12.94 |
| 33 42 0101 | Electrical Charge | KWH | 0.06 | 0.06 | 0.06 | 0.06 | 0.06 |

**Environmental Remediation: Assemblies Cost Book**

# Cost Data for Remediation
## Bioslurping

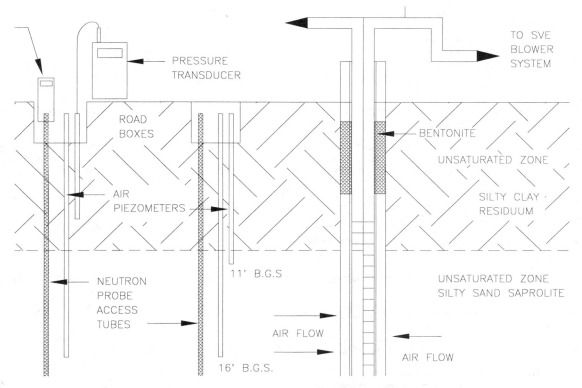

### General:

*Bioslurping is a process which combines vacuum enhanced free product recovery and soil aeration. By integrating these two technologies, bioslurping recovers free phase light non aqueous phase liquid (LNAPL) while simultaneously stimulating bioremediation of vadose zone soil by drawing air through the soil matrix. Bioslurping systems extract soil vapor, product, and groundwater in the same process stream, and a single vacuum pump can be connected to multiple extraction wells.*

### Common Cost Components:

1. Well Installation
2. Piping
3. Clearing and Grubbing

### Typical Treatment Train:

*Bioslurping is typically part of a treatment train which includes off-gas and water treatment components. Product, groundwater, and soil gas, which are extracted in the same process stream, are separated, treated as necessary, and discharged. Typically, an air/water separator is used to separate the gas and liquid streams. Soil gas vapor can be treated using gas phase carbon adsorption, catalytic oxidation, or vapor reinjection into the vadose zone (if allowed). The liquid stream is normally passed to an oil/water separator.*

### Other Cost Considerations:

*System Maintenance/Utilities, Sampling and Analysis, Treatment/Disposal of Process Residuals.*

# Cost Data for Remediation

## Bioslurping

| Assembly | Description | Unit | Unit Cost by Safety Level A | B | C | D | E |
|---|---|---|---|---|---|---|---|
| | **Capital Costs** | | | | | | |
| 17 03 1002 | 2" Diameter Contractor's Trash Pump, 75 GPM | DAY | 50.20 | 47.01 | 45.64 | 42.55 | 41.42 |
| 17 03 1003 | 3" Diameter Contractor's Trash Pump, 150 GPM | DAY | 61.72 | 57.05 | 55.05 | 50.54 | 48.88 |
| 17 03 1004 | 4" Diameter Contractor's Trash Pump, 300 GPM | DAY | 80.30 | 74.23 | 71.63 | 65.77 | 63.61 |
| 19 01 0203 | 1 1/2", Class 200, PVC Piping | LF | 7.59 | 5.95 | 5.18 | 3.59 | 3.04 |
| 19 04 0401 | 550 Gallon, Stainless Steel Aboveground Wastewater Holding Tank, Rental | MONTH | 320.65 | 320.65 | 320.65 | 320.65 | 320.65 |
| 19 04 0403 | 4,000 Gallon Polyethylene Wastewater Tank, Rental | MONTH | 545.90 | 545.90 | 545.90 | 545.90 | 545.90 |
| 19 04 0405 | 6,000 Gallon, Polyethylene Aboveground Wastewater Holding Tank, Rental | MONTH | 641.30 | 641.30 | 641.30 | 641.30 | 641.30 |
| 19 04 0407 | 21,000 Gallon, Steel Closed Stationary Aboveground Wastewater Holding Tank, Rental | MONTH | 1,219 | 1,219 | 1,219 | 1,219 | 1,219 |
| 19 04 0602 | 550 Gallon Steel Sump, Aboveground with Supports & Fittings | EACH | 1,598 | 1,390 | 1,301 | 1,099 | 1,026 |
| 19 04 0603 | 1,000 Gallon Steel Sump, Aboveground with Supports & Fittings | EACH | 2,118 | 1,844 | 1,727 | 1,463 | 1,366 |
| 19 04 0604 | 1,500 Gallon Steel Sump, Aboveground with Supports & Fittings | EACH | 2,697 | 2,356 | 2,209 | 1,879 | 1,758 |
| 19 04 0605 | 2,000 Gallon Steel Sump, Aboveground with Supports & Fittings | EACH | 3,457 | 2,997 | 2,777 | 2,332 | 2,183 |
| 19 04 0606 | 5,000 Gallon Steel Sump, Aboveground with Supports & Fittings | EACH | 6,079 | 5,494 | 5,214 | 4,646 | 4,457 |
| 19 04 0625 | 6,000 Gallon Horizontal Plastic Sump with 6" NPT Connection | EACH | 5,248 | 4,975 | 4,858 | 4,594 | 4,497 |
| 19 04 0626 | 8,000 Gallon Horizontal Plastic Sump with 6" NPT Connection | EACH | 5,857 | 5,516 | 5,369 | 5,039 | 4,918 |
| 19 04 0627 | 10,000 Gallon Horizontal Plastic Sump with 6" NPT Connection | EACH | 7,082 | 6,622 | 6,402 | 5,956 | 5,807 |
| 19 04 0628 | 12,000 Gallon Horizontal Plastic Sump with 6" NPT Connection | EACH | 8,308 | 7,722 | 7,442 | 6,875 | 6,686 |
| 19 04 0629 | 15,000 Gallon Horizontal Plastic Sump with 6" NPT Connection | EACH | 14,184 | 13,540 | 13,232 | 12,608 | 12,400 |
| 19 04 0631 | 20,000 Gallon Horizontal Plastic Sump with 6" NPT Connection | EACH | 15,799 | 15,084 | 14,742 | 14,048 | 13,817 |
| 19 04 0632 | 25,000 Gallon Horizontal Plastic Sump with 6" NPT Connection | EACH | 24,407 | 23,487 | 23,047 | 22,155 | 21,858 |
| 33 01 0101 | Mobilize/DeMobilize Drilling Rig & Crew | LS | 4,949 | 4,031 | 3,308 | 2,401 | 2,280 |
| 33 11 1301 | 3 hp Liquid Ring Vacuum Pump | EA | 6,249 | 6,247 | 6,246 | 6,244 | 6,244 |
| 33 11 1302 | 5 hp Liquid Ring Vacuum Pump | EA | 7,289 | 7,287 | 7,286 | 7,284 | 7,284 |
| 33 11 1303 | 7.5 hp Liquid Ring Vacuum Pump | EA | 8,537 | 8,535 | 8,534 | 8,532 | 8,532 |
| 33 11 1304 | 10 hp Liquid Ring Vacuum Pump | EA | 9,891 | 9,889 | 9,888 | 9,885 | 9,885 |

**Environmental Remediation: Assemblies Cost Book**

# Cost Data for Remediation

## Bioslurping

| Assembly | Description | Unit | Unit Cost by Safety Level A | B | C | D | E |
|---|---|---|---|---|---|---|---|
| | **Capital Costs** | | | | | | |
| 33 11 1305 | 15 hp Liquid Ring Vacuum Pump | EA | 15,611 | 15,609 | 15,608 | 15,606 | 15,605 |
| 33 11 1306 | Seal Water Tank for Liquid Ring Pump | EA | 531.94 | 487.69 | 467.40 | 424.55 | 409.70 |
| 33 13 1218 | 10 GPM Oil Water Separator | EA | 8,388 | 6,465 | 5,643 | 3,785 | 3,103 |
| 33 13 1219 | Emulsion Tank | EA | 0.00 | 0.00 | 0.00 | 0.00 | 0.00 |
| 33 13 2343 | Knockout Drum | EACH | 65.19 | 65.19 | 65.19 | 65.19 | 65.19 |
| 33 17 0808 | Decontaminate Rig, Augers, Screen (Rental Equipment) | DAY | 205.34 | 205.34 | 205.34 | 205.34 | 205.34 |
| 33 23 0101 | 2" PVC, Schedule 40, Well Casing | LF | 17.45 | 14.39 | 11.98 | 8.95 | 8.55 |
| 33 23 0102 | 4" PVC, Schedule 40, Well Casing | LF | 26.98 | 22.38 | 18.77 | 14.24 | 13.63 |
| 33 23 0103 | 6" PVC, Schedule 40, Well Casing | LF | 27.07 | 22.66 | 19.19 | 14.84 | 14.25 |
| 33 23 0111 | 2" PVC, Schedule 80, Well Casing | LF | 17.78 | 14.72 | 12.31 | 9.28 | 8.88 |
| 33 23 0112 | 4" PVC, Schedule 80, Well Casing | LF | 28.55 | 23.95 | 20.34 | 15.81 | 15.20 |
| 33 23 0113 | 6" PVC, Schedule 80, Well Casing | LF | 47.52 | 40.17 | 34.40 | 27.14 | 26.17 |
| 33 23 0121 | 2" Stainless Steel, Well Casing | LF | 31.75 | 28.14 | 25.30 | 21.73 | 21.25 |
| 33 23 0122 | 4" Stainless Steel, Well Casing | LF | 47.66 | 43.06 | 39.45 | 34.92 | 34.31 |
| 33 23 0123 | 6" Stainless Steel Well Casing, 5' Sections, Flush Threaded | LF | 370.73 | 311.96 | 265.77 | 207.74 | 199.94 |
| 33 23 0201 | 2" PVC, Schedule 40, Well Screen | LF | 23.47 | 19.52 | 16.42 | 12.52 | 12.00 |
| 33 23 0202 | 4" PVC, Schedule 40, Well Screen | LF | 38.16 | 32.00 | 27.16 | 21.09 | 20.27 |
| 33 23 0203 | 6" PVC, Schedule 40, Well Screen | LF | 47.13 | 39.78 | 34.01 | 26.75 | 25.78 |
| 33 23 0211 | 2" PVC, Schedule 80, Well Screen | LF | 18.70 | 15.64 | 13.23 | 10.20 | 9.80 |
| 33 23 0212 | 4" PVC, Schedule 80, Well Screen | LF | 30.01 | 25.41 | 21.80 | 17.27 | 16.66 |
| 33 23 0213 | 6" PVC, Schedule 80, Well Screen | LF | 49.72 | 42.37 | 36.60 | 29.34 | 28.37 |
| 33 23 0221 | 2" Stainless Steel, Well Screen | LF | 26.91 | 23.84 | 21.44 | 18.41 | 18.01 |
| 33 23 0222 | 4" Stainless Steel, Well Screen | LF | 47.66 | 43.06 | 39.45 | 34.92 | 34.31 |
| 33 23 0223 | 6" Stainless Steel Well Screen, 5' Sec, Flush Threaded | LF | 366.05 | 307.28 | 261.09 | 203.06 | 195.26 |
| 33 23 0301 | 2" PVC, Well Plug | EACH | 29.37 | 24.77 | 21.16 | 16.63 | 16.02 |
| 33 23 0302 | 4" PVC, Well Plug | EACH | 55.65 | 48.91 | 43.62 | 36.97 | 36.07 |
| 33 23 0303 | 6" PVC, Well Plug | EACH | 112.63 | 101.15 | 92.13 | 80.79 | 79.27 |
| 33 23 0311 | 2" Stainless Steel, Well Plug | EACH | 83.24 | 74.06 | 66.83 | 57.76 | 56.55 |
| 33 23 0312 | 4" Stainless Steel, Well Plug | EACH | 112.24 | 102.03 | 94.01 | 83.93 | 82.58 |
| 33 23 0313 | 6" Stainless Steel, Well Plug | EACH | 573.04 | 481.18 | 408.98 | 318.27 | 306.09 |
| 33 23 1101 | Hollow-stem Auger, 8" Outside Diameter Borehole for 2" Well | LF | 89.98 | 73.28 | 60.15 | 43.66 | 41.45 |
| 33 23 1102 | Hollow-stem Auger, 11" Outside Diameter Borehole for 4" Well | LF | 109.98 | 89.57 | 73.52 | 53.36 | 50.66 |

## Environmental Remediation: Assemblies Cost Book

*Page 3-25*

1998 by ECHOS. All rights reserved

# Cost Data for Remediation

## Bioslurping

| Assembly | Description | Unit | A | B | C | D | E |
|---|---|---|---|---|---|---|---|
| | **Capital Costs** | | | | | | |
| 33 23 1103 | Hollow-stem Auger, 13 3/4" Outside Diameter Borehole for 6" Well | LF | 141.40 | 115.16 | 94.53 | 68.61 | 65.13 |
| 33 23 1106 | Split Spoon Sample, 2" x 24", During Drilling | EACH | 46.67 | 46.67 | 46.67 | 46.67 | 46.67 |
| 33 23 1122 | Move Rig/Equipment Around Site | EACH | 154.66 | 125.95 | 103.39 | 75.04 | 71.24 |
| 33 23 1126 | Furnish 55 Gallon Drum for Drill Cuttings & Development Water | EACH | 65.19 | 65.19 | 65.19 | 65.19 | 65.19 |
| 33 23 1152 | Mud Drilling, 6" Diameter Borehole | LF | 81.01 | 65.98 | 54.16 | 39.31 | 37.31 |
| 33 23 1153 | Mud Drilling, 8" Diameter Borehole | LF | 88.38 | 71.97 | 59.08 | 42.88 | 40.71 |
| 33 23 1154 | Mud Drilling, 10" Diameter Borehole | LF | 100.65 | 81.97 | 67.29 | 48.84 | 46.36 |
| 33 23 1161 | Air Rotary 6" Borehole in Unconsolidated | LF | 73.65 | 59.98 | 49.23 | 35.73 | 33.92 |
| 33 23 1162 | Air Rotary 8" Borehole in Unconsolidated | LF | 122.74 | 99.96 | 82.06 | 59.56 | 56.54 |
| 33 23 1163 | Air Rotary 10" Borehole in Unconsolidated | LF | 220.94 | 179.93 | 147.70 | 107.20 | 101.77 |
| 33 23 1306 | High Sump Level Switch for Avoiding Overflow | EACH | 214.65 | 214.65 | 214.65 | 214.65 | 214.65 |
| 33 23 1401 | 2" Screen, Filter Pack | LF | 16.49 | 13.88 | 11.84 | 9.27 | 8.92 |
| 33 23 1402 | 4" Screen, Filter Pack | LF | 29.10 | 24.50 | 20.89 | 16.36 | 15.75 |
| 33 23 1403 | 6" Screen, Filter Pack | LF | 42.19 | 35.53 | 30.29 | 23.72 | 22.83 |
| 33 23 1502 | Surface Pad, Concrete, 4' x 4' x 4" | EACH | 24.21 | 21.82 | 20.74 | 18.43 | 17.61 |
| 33 23 1811 | 2" Well, Portland Cement Grout | LF | 0.92 | 0.92 | 0.92 | 0.92 | 0.92 |
| 33 23 1812 | 4" Well, Portland Cement Grout | LF | 1.38 | 1.38 | 1.38 | 1.38 | 1.38 |
| 33 23 1813 | 6" Well, Portland Cement Grout | LF | 7.80 | 7.80 | 7.80 | 7.80 | 7.80 |
| 33 23 2101 | 2" Well, Bentonite Seal | EACH | 63.00 | 52.67 | 44.55 | 34.34 | 32.97 |
| 33 23 2102 | 4" Well, Bentonite Seal | EACH | 157.54 | 131.70 | 111.39 | 85.87 | 82.44 |
| 33 23 2103 | 6" Well, Bentonite Seal | EACH | 252.01 | 210.68 | 178.18 | 137.37 | 131.89 |
| 33 26 0204 | 4" Stainless Steel Piping, Schedule 40, Threaded | LF | 80.15 | 71.72 | 68.12 | 59.98 | 57.00 |
| 33 26 0205 | 6" Stainless Steel Piping, Schedule 10, Type 316 | LF | 78.58 | 70.01 | 66.14 | 57.84 | 54.94 |
| 33 26 0407 | 4" PVC, Schedule 40, Connection Piping | LF | 13.38 | 10.65 | 9.48 | 6.84 | 5.88 |
| 33 26 0409 | 6" PVC, Schedule 40, Connection Piping | LF | 18.73 | 15.09 | 13.53 | 10.01 | 8.73 |
| 33 26 0704 | 50' Flexible, Product Discharge Hose | EACH | 159.00 | 159.00 | 159.00 | 159.00 | 159.00 |
| 33 27 0102 | 2" PVC, Schedule 40, Tee | EACH | 68.73 | 53.63 | 46.80 | 32.18 | 27.06 |
| 33 27 0104 | 4" PVC, Schedule 40, Tee | EACH | 137.04 | 108.84 | 96.77 | 69.51 | 59.51 |
| 33 27 0105 | 6" PVC, Schedule 40, Tee | EACH | 205.86 | 169.47 | 153.89 | 118.72 | 105.82 |
| 33 27 0114 | 4" PVC, 90 Degree, Elbow | EACH | 101.72 | 80.57 | 71.51 | 51.07 | 43.57 |
| 33 27 0115 | 6" PVC, 90 Degree, Elbow | EACH | 135.44 | 111.27 | 100.92 | 77.56 | 68.99 |
| 33 27 0151 | 4" x 2" Reducer, PVC Schedule 40 | EACH | 9.09 | 9.09 | 9.09 | 9.09 | 9.09 |

**Environmental Remediation: Assemblies Cost Book**

# Cost Data for Remediation

## Bioslurping

| Assembly | Description | Unit | A | B | C | D | E |
|---|---|---|---|---|---|---|---|
| | **Capital Costs** | | | | | | |
| 33 27 0201 | 2" Stainless Steel, Tee | EACH | 173.32 | 140.34 | 126.22 | 94.34 | 82.65 |
| 33 27 0202 | 4" Stainless Steel, Tee | EACH | 583.41 | 504.25 | 470.36 | 393.85 | 365.80 |
| 33 27 0204 | 6" Stainless Steel, Tee | EACH | 1,508 | 1,220 | 1,090 | 811.00 | 713.21 |
| 33 27 0212 | 4" Stainless Steel, 90 Degree, Elbow | EACH | 391.27 | 338.50 | 315.91 | 264.90 | 246.20 |
| 33 27 0213 | 6" Stainless Steel, 90 Degree, Elbow | EACH | 425.65 | 376.17 | 355.00 | 307.18 | 289.64 |
| 33 27 0231 | 4" x 2" Stainless Steel Reducer, Schedule 40, Type 304 | EACH | 27.01 | 27.01 | 27.01 | 27.01 | 27.01 |
| 33 27 0404 | 4" Iron Body Check Valve | EACH | 354.29 | 315.96 | 299.54 | 262.49 | 248.90 |
| 33 27 0405 | 6" Iron Body Check Valve | EACH | 618.60 | 554.57 | 523.95 | 461.86 | 441.15 |
| 33 27 0425 | 2" Bronze Gate Valve | EACH | 125.01 | 107.16 | 99.52 | 82.26 | 75.94 |
| 33 29 0412 | 15 GPM Submersible Sump Pump | EACH | 492.79 | 460.07 | 446.06 | 414.43 | 402.83 |
| 33 29 0413 | 42 GPM Submersible Sump Pump | EACH | 792.79 | 760.07 | 746.06 | 714.43 | 702.83 |
| 33 29 0414 | 52 GPM Submersible Sump Pump | EACH | 871.35 | 832.08 | 815.27 | 777.32 | 763.40 |
| 33 31 0209 | Pressure Gauge | EACH | 140.16 | 121.61 | 113.67 | 95.74 | 89.16 |
| 33 42 0101 | Electrical Charge | KWH | 0.06 | 0.06 | 0.06 | 0.06 | 0.06 |
| | **Operations & Maintenance** | | | | | | |
| 19 04 0401 | 550 Gallon, Stainless Steel Aboveground Wastewater Holding Tank, Rental | MONTH | 320.65 | 320.65 | 320.65 | 320.65 | 320.65 |
| 19 04 0403 | 4,000 Gallon Polyethylene Wastewater Tank, Rental | MONTH | 545.90 | 545.90 | 545.90 | 545.90 | 545.90 |
| 19 04 0405 | 6,000 Gallon, Polyethylene Aboveground Wastewater Holding Tank, Rental | MONTH | 641.30 | 641.30 | 641.30 | 641.30 | 641.30 |
| 19 04 0407 | 21,000 Gallon, Steel Closed Stationary Aboveground Wastewater Holding Tank, Rental | MONTH | 1,219 | 1,219 | 1,219 | 1,219 | 1,219 |
| 33 42 0101 | Electrical Charge | KWH | 0.06 | 0.06 | 0.06 | 0.06 | 0.06 |

**Unit Cost by Safety Level**

**Environmental Remediation: Assemblies Cost Book**

# Cost Data for Remediation
## Bulk Material Storage

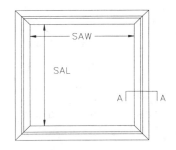

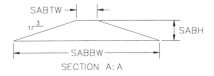
SECTION A: A

SAW = Storage Area Width
SAL = Storage Area Length
SABH = Storage Area Berm Height
SABTW = Storage Area Berm Top Width
SABBW = Storage Area Berm Bottom Width

## SOIL BERMED STORAGE AREA

### General:
Bulk material storage facilities are frequently required at remediation sites. Bulk materials stored on-site may be either waste materials or materials used for treating wastes, such as stabilization agents or dewatering additives. Common storage methods include surface impoundments, waste piles, tanks, and portable storage buildings. Temporary storage facilities must be in compliance with Resource Conservation and Recovery Act (RCRA) rules (40 CFR 264) to store the material that will be contained in them, and they must be managed in accordance.

### Typical Treatment Train:
Transportation of bulk material (hazardous or non-hazardous) to and from storage facility, sampling and analysis.

### Common Cost Components:
1. Storage Facility (soil bermed storage area, inflatable containment berms, roll-off/bulk containers/tanks, portable storage building)
2. Liners
3. Storm Drainage (inlets, piping, sumps, holding tanks)

### Other Cost Considerations:
Loading and hauling.

**Environmental Remediation: Assemblies Cost Book**

# Cost Data for Remediation

## Bulk Material Storage

| Assembly | Description | Unit | Unit Cost by Safety Level | | | | |
|---|---|---|---|---|---|---|---|
| | | | A | B | C | D | E |
| | **Capital Costs** | | | | | | |
| 17 03 0102 | Rough Grading, 12G, 1 Pass | SY | 1.28 | 1.04 | 0.85 | 0.62 | 0.59 |
| 17 03 0105 | Fine Grading, Hand | SY | 4.21 | 3.24 | 2.83 | 1.90 | 1.56 |
| 17 03 0106 | Fine Grading, 12G, 2 Passes | SY | 0.62 | 0.50 | 0.42 | 0.30 | 0.28 |
| 17 03 0423 | Unclassified Fill, 6" Lifts, Off-Site | CY | 11.08 | 9.81 | 8.83 | 7.58 | 7.40 |
| 17 03 0426 | Sand, 6" Lifts, Off-Site | CY | 15.55 | 14.34 | 13.42 | 12.22 | 12.04 |
| 17 03 0430 | Gravel, 6" Lifts | CY | 15.46 | 13.36 | 12.05 | 9.99 | 9.49 |
| 18 02 0203 | 26" x 26", 5' Deep Area Drain with Grate | EACH | 3,425 | 2,959 | 2,752 | 2,302 | 2,141 |
| 18 02 0204 | 27" x 20", 5' Deep Area Drain with Grate | EACH | 3,346 | 2,880 | 2,673 | 2,223 | 2,062 |
| 18 02 0205 | 24" Diameter, 5' Deep Area Drain with Grate | EACH | 3,131 | 2,673 | 2,470 | 2,027 | 1,869 |
| 19 02 0311 | 12' x 36" Diameter Reinforced Concrete Pipe Wet Well for Lift Station | EACH | 9,840 | 8,577 | 7,816 | 6,581 | 6,270 |
| 19 02 0312 | 24' x 60" Diameter Reinforced Concrete Pipe Wet Well for Lift Station | EACH | 39,142 | 33,897 | 30,681 | 25,554 | 24,296 |
| 19 02 0313 | 5' x 5' x 5' Reinforced Concrete Sump | EACH | 3,948 | 3,282 | 2,979 | 2,335 | 2,110 |
| 19 03 0112 | 8" Corrugated Metal Pipe, Plain | LF | 8.69 | 7.91 | 7.51 | 6.75 | 6.51 |
| 19 03 0113 | 10" Corrugated Metal Pipe, Plain | LF | 9.99 | 9.21 | 8.81 | 8.05 | 7.81 |
| 19 03 0114 | 12" Corrugated Metal Pipe, Plain | LF | 11.23 | 10.38 | 9.96 | 9.13 | 8.87 |
| 19 03 0115 | 15" Corrugated Metal Pipe, Plain | LF | 13.37 | 12.43 | 11.95 | 11.04 | 10.74 |
| 19 03 0116 | 18" Corrugated Metal Pipe, Plain | LF | 16.76 | 15.33 | 14.61 | 13.22 | 12.77 |
| 19 03 0117 | 24" Corrugated Metal Pipe, Plain | LF | 24.55 | 22.64 | 21.68 | 19.82 | 19.23 |
| 19 03 0174 | 12" Reinforced Concrete Pipe, Class 4, with Gaskets | LF | 14.33 | 13.02 | 12.29 | 11.01 | 10.65 |
| 19 03 0175 | 15" Reinforced Concrete Pipe, Class 4, with Gaskets | LF | 16.47 | 14.84 | 13.94 | 12.36 | 11.91 |
| 19 03 0176 | 18" Reinforced Concrete Pipe, Class 4, with Gaskets | LF | 23.67 | 21.05 | 19.60 | 17.05 | 16.33 |
| 19 03 0178 | 24" Reinforced Concrete Pipe, Class 4, with Gaskets | LF | 34.58 | 31.22 | 29.36 | 26.09 | 25.17 |
| 19 04 0602 | 550 Gallon Steel Sump, Aboveground with Supports & Fittings | EACH | 1,598 | 1,390 | 1,301 | 1,099 | 1,026 |
| 19 04 0603 | 1,000 Gallon Steel Sump, Aboveground with Supports & Fittings | EACH | 2,118 | 1,844 | 1,727 | 1,463 | 1,366 |
| 19 04 0604 | 1,500 Gallon Steel Sump, Aboveground with Supports & Fittings | EACH | 2,697 | 2,356 | 2,209 | 1,879 | 1,758 |
| 19 04 0605 | 2,000 Gallon Steel Sump, Aboveground with Supports & Fittings | EACH | 3,457 | 2,997 | 2,777 | 2,332 | 2,183 |
| 19 04 0606 | 5,000 Gallon Steel Sump, Aboveground with Supports & Fittings | EACH | 6,079 | 5,494 | 5,214 | 4,646 | 4,457 |
| 19 04 0625 | 6,000 Gallon Horizontal Plastic Sump with 6" NPT Connection | EACH | 5,248 | 4,975 | 4,858 | 4,594 | 4,497 |

**Environmental Remediation: Assemblies Cost Book**

# Cost Data for Remediation

## Bulk Material Storage

| Assembly | Description | Unit | Unit Cost by Safety Level | | | | |
|---|---|---|---|---|---|---|---|
| | | | A | B | C | D | E |
| | **Capital Costs** | | | | | | |
| 19 04 0626 | 8,000 Gallon Horizontal Plastic Sump with 6" NPT Connection | EACH | 5,857 | 5,516 | 5,369 | 5,039 | 4,918 |
| 19 04 0627 | 10,000 Gallon Horizontal Plastic Sump with 6" NPT Connection | EACH | 7,082 | 6,622 | 6,402 | 5,956 | 5,807 |
| 33 08 0504 | Herbicide Application | ACRE | 487.34 | 400.41 | 343.04 | 257.83 | 239.51 |
| 33 08 0505 | Soil/Bentonite Liner | SF | 0.78 | 0.77 | 0.76 | 0.75 | 0.74 |
| 33 08 0507 | Clay 10E-7, 6" Lifts, Off-Site | CY | 31.11 | 26.22 | 22.84 | 18.04 | 17.11 |
| 33 08 0541 | 20 Mil Polymeric Liner, Very Low Density Polyethylene | SF | 1.25 | 1.02 | 0.91 | 0.69 | 0.61 |
| 33 08 0542 | 30 Mil Polymeric Liner, Very Low Density Polyethylene | SF | 1.98 | 1.59 | 1.41 | 1.03 | 0.90 |
| 33 08 0543 | 40 Mil Polymeric Liner, Very Low Density Polyethylene | SF | 2.38 | 1.91 | 1.70 | 1.24 | 1.09 |
| 33 08 0544 | 60 Mil Polymeric Liner, Very Low Density Polyethylene | SF | 3.54 | 2.83 | 2.51 | 1.83 | 1.60 |
| 33 08 0545 | 80 Mil Polymeric Liner, Very Low Density Polyethylene | SF | 4.35 | 3.49 | 3.10 | 2.27 | 1.98 |
| 33 08 0546 | 100 Mil Polymeric Liner, Very Low Density Polyethylene | SF | 5.50 | 4.41 | 3.91 | 2.85 | 2.49 |
| 33 08 0551 | 36 Mil Polymeric Liner, Chlorosulfanated Polyethylene | SF | 0.75 | 0.73 | 0.72 | 0.70 | 0.70 |
| 33 08 0552 | 45 Mil Polymeric Liner, Chlorosulfanated Polyethylene | SF | 0.83 | 0.81 | 0.80 | 0.79 | 0.78 |
| 33 08 0561 | 20 Mil Polymeric Liner, PVC | SF | 0.24 | 0.23 | 0.22 | 0.21 | 0.20 |
| 33 08 0562 | 30 Mil Polymeric Liner, PVC | SF | 0.32 | 0.31 | 0.30 | 0.29 | 0.28 |
| 33 08 0563 | 40 Mil Polymeric Liner, PVC | SF | 0.38 | 0.36 | 0.36 | 0.34 | 0.34 |
| 33 08 0564 | 60 Mil Polymeric Liner, PVC | SF | 0.72 | 0.70 | 0.69 | 0.67 | 0.67 |
| 33 08 0565 | 80 Mil Polymeric Liner, PVC | SF | 0.98 | 0.96 | 0.95 | 0.93 | 0.93 |
| 33 08 0566 | 100 Mil Polymeric Liner, PVC | SF | 1.28 | 1.26 | 1.25 | 1.23 | 1.22 |
| 33 08 0571 | 40 Mil Polymeric Liner, High-density Polyethylene | SF | 2.35 | 1.88 | 1.67 | 1.21 | 1.06 |
| 33 08 0572 | 60 Mil Polymeric Liner, High-density Polyethylene | SF | 3.49 | 2.80 | 2.48 | 1.81 | 1.58 |
| 33 08 0573 | 80 Mil Polymeric Liner, High-density Polyethylene | SF | 4.32 | 3.47 | 3.08 | 2.25 | 1.97 |
| 33 10 0109 | 250 Gallon Bulk Container, Polyethylene with Metal Case | EACH | 749.95 | 749.95 | 749.95 | 749.95 | 749.95 |
| 33 10 0110 | 350 Gallon Bulk Container, Polyethylene | EACH | 294.33 | 294.33 | 294.33 | 294.33 | 294.33 |
| 33 10 0111 | 450 Gallon Bulk Container, Polyethylene | EACH | 266.77 | 266.77 | 266.77 | 266.77 | 266.77 |
| 33 10 0112 | 550 Gallon Bulk Container, Polyethylene | EACH | 324.36 | 324.36 | 324.36 | 324.36 | 324.36 |
| 33 10 0120 | 20 CY Open Top Roll-Off Container | EACH | 4,817 | 4,817 | 4,817 | 4,817 | 4,817 |
| 33 10 0121 | 25 CY Open Top Roll-Off Container | EACH | 5,000 | 5,000 | 5,000 | 5,000 | 5,000 |
| 33 10 0122 | 30 CY Open Top Roll-Off Container | EACH | 5,233 | 5,233 | 5,233 | 5,233 | 5,233 |
| 33 10 0123 | 20 CY Closed Top Roll-Off Container | EACH | 4,817 | 4,817 | 4,817 | 4,817 | 4,817 |
| 33 10 0124 | 19 CY Chassis-mounted Ejector Body | EACH | 13,959 | 13,959 | 13,959 | 13,959 | 13,959 |
| 33 10 0125 | 24 CY Chassis-mounted Ejector Body | EACH | 17,369 | 17,369 | 17,369 | 17,369 | 17,369 |

**Environmental Remediation: Assemblies Cost Book**

# Cost Data for Remediation

## Bulk Material Storage

| Assembly | Description | Unit | Unit Cost by Safety Level | | | | |
|---|---|---|---|---|---|---|---|
| | | | A | B | C | D | E |
| | **Capital Costs** | | | | | | |
| 33 10 0126 | 25 CY Horizontal Discharge Trailer | EACH | 33,596 | 33,596 | 33,596 | 33,596 | 33,596 |
| 33 10 0127 | 34 CY Horizontal Discharge Trailer | EACH | 34,349 | 34,349 | 34,349 | 34,349 | 34,349 |
| 33 10 0128 | 42 CY Horizontal Discharge Trailer | EACH | 38,653 | 38,653 | 38,653 | 38,653 | 38,653 |
| 33 10 0130 | 4,000 Gallon Tanker Trailer | EACH | 41,000 | 41,000 | 41,000 | 41,000 | 41,000 |
| 33 10 0131 | 5,000 Gallon Tanker Trailer | EACH | 46,000 | 46,000 | 46,000 | 46,000 | 46,000 |
| 33 10 0132 | 6,000 Gallon Tanker Trailer | EACH | 49,000 | 49,000 | 49,000 | 49,000 | 49,000 |
| 33 19 0333 | Tanker Trailer Drop (Daily) | DAY | 117.50 | 117.50 | 117.50 | 117.50 | 117.50 |
| 33 19 0334 | Tanker Trailer Drop (Monthly) | MONTH | 107.50 | 107.50 | 107.50 | 107.50 | 107.50 |
| 33 19 9901 | 750 Gallon, 10' x 10' x 17" Inflatable Containment Berm, Single Lined | EACH | 2,289 | 2,272 | 2,265 | 2,249 | 2,243 |
| 33 19 9902 | 1,800 Gallon, 13' x 13' x 17" Inflatable Containment Berm, Single Lined | EACH | 3,461 | 3,444 | 3,437 | 3,420 | 3,414 |
| 33 19 9903 | 3,400 Gallon, 18' x 18' x 17" Inflatable Containment Berm, Single Lined | EACH | 4,881 | 4,847 | 4,833 | 4,800 | 4,788 |
| 33 19 9904 | 6,100 Gallon, 24' x 24' x 17" Inflatable Containment Berm, Single Lined | EACH | 6,295 | 6,254 | 6,237 | 6,198 | 6,183 |
| 33 19 9905 | 10,800 Gallon, 32' x 32' x 17" Inflatable Containment Berm, Single Lined | EACH | 8,332 | 8,281 | 8,259 | 8,210 | 8,192 |
| 33 19 9906 | 21,600 Gallon, 32' x 32' x 34" Inflatable Containment Berm, Single Lined | EACH | 12,226 | 12,145 | 12,110 | 12,031 | 12,002 |
| 33 19 9907 | 53,400 Gallon, 74' x 34' x 34" Inflatable Containment Berm, Single Lined | EACH | 19,561 | 19,425 | 19,367 | 19,236 | 19,188 |
| 33 19 9908 | 7,600 Gallon, 45' x 15' x 17" Inflatable Containment Berm, Double Lined | EACH | 7,568 | 7,517 | 7,495 | 7,446 | 7,428 |
| 33 19 9909 | 11,000 Gallon, 65' x 16' x 17" Inflatable Containment Berm, Double Lined | EACH | 9,725 | 9,674 | 9,652 | 9,603 | 9,585 |
| 33 19 9910 | 11,600 Gallon, 50' x 22' x 17" Inflatable Containment Berm, Double Lined | EACH | 9,564 | 9,513 | 9,492 | 9,442 | 9,424 |
| 33 19 9911 | Air Blower to Inflate Portable Containment Berm | EACH | 160.59 | 160.59 | 160.59 | 160.59 | 160.59 |
| 33 23 1306 | High Sump Level Switch for Avoiding Overflow | EACH | 214.65 | 214.65 | 214.65 | 214.65 | 214.65 |
| 33 26 0402 | 2" PVC Piping Including Fittings & Hangers | LF | 17.85 | 14.06 | 12.38 | 8.72 | 7.42 |
| 33 29 0401 | 25 GPM, 1 1/2" Discharge, Cast-iron Sump Pump | EACH | 2,432 | 2,274 | 2,206 | 2,053 | 1,997 |
| 33 29 0402 | 75 GPM, 2" Discharge, Cast-iron Sump Pump | EACH | 3,011 | 2,820 | 2,738 | 2,552 | 2,484 |
| 33 29 0403 | 100 GPM, 2 1/2" Discharge, Cast-iron Sump Pump | EACH | 3,243 | 3,016 | 2,918 | 2,698 | 2,618 |
| 33 43 0102 | 5.58' x 5.5' x 8.33' Hazardous Material Storage Building, 4 Drum Capacity | EACH | 3,202 | 3,202 | 3,202 | 3,202 | 3,202 |
| 33 43 0104 | 10.25' x 5.5' x 8.33' Hazardous Material Storage Building, 8 Drum Capacity | EACH | 11,448 | 11,448 | 11,448 | 11,448 | 11,448 |
| 33 43 0106 | 10.0' x 12.0' x 8.67' Hazardous Material Storage Building | EACH | 9,612 | 9,467 | 9,377 | 9,235 | 9,201 |

**Environmental Remediation: Assemblies Cost Book**

# Cost Data for Remediation

## Bulk Material Storage

| Assembly | Description | Unit | Unit Cost by Safety Level | | | | |
|---|---|---|---|---|---|---|---|
| | | | A | B | C | D | E |
| | **Capital Costs** | | | | | | |
| 33 43 0107 | 12.0' x 20.0' x 8.67' Hazardous Material Storage Building | EACH | 12,949 | 12,755 | 12,636 | 12,447 | 12,401 |
| 33 43 0108 | 20.0' x 20.0' x 10.0' Hazardous Material Storage Building | EACH | 23,423 | 23,133 | 22,954 | 22,670 | 22,602 |
| 33 43 0109 | 24.0' x 20.0' x 10.0' Hazardous Material Storage Building | EACH | 29,779 | 29,417 | 29,192 | 28,838 | 28,753 |
| 33 43 0110 | 20.0' x 10.0' x 10.0' Hazardous Material Storage Building Expansion | EACH | 9,849 | 9,655 | 9,536 | 9,347 | 9,301 |
| 33 43 0111 | 24.0' x 10.0' x 10.0' Hazardous Material Storage Building Expansion | EACH | 11,949 | 11,755 | 11,636 | 11,447 | 11,401 |
| 33 43 0115 | Non-Skid Concrete Ramp | EACH | 812.55 | 695.09 | 644.81 | 531.27 | 489.64 |

**Environmental Remediation: Assemblies Cost Book**

# Cost Data for Remediation
## Capping

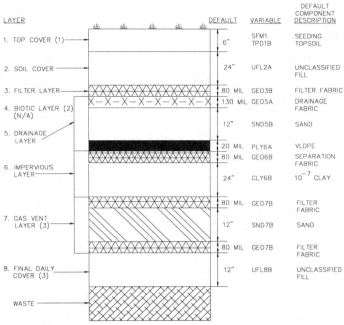

1. CERTAIN LOCATIONS DEPENDING UPON CLIMATE WILL DEFAULT TO ROCK FOR TOP COVER RATHER THAN VEGETATIVE. THE DEFAULT ROCK TOP COVER INCLUDES ONE LAYER OF SEPARATION FABRIC AND A 12" LAYER OF 10-100 LB RIP-RAP.
2. BIOTIC LAYER DEFAULTS TO NONE IN ALL CASES.
3. APPLICABLE TO LANDFILL "SITE TYPE" ONLY.

### General:

Capping is an overlying layered system of vegetative cover, natural soils, rock, synthetics, pavement, and/or polymeric liners as a means to control hydrogeologic processes. Capping a hazardous waste site is one of the most commonly used passive remediation technologies. It is a proven rather than an innovative technology. The objectives of capping include storm water runoff control, infiltration reduction, erosion control, containment of landfill gases, support for vegetation, consistency with end use, low maintenance remediation, and low life-cycle costs.

### Common Cost Components:

1. Top Cover
2. Soil Cover
3. Filter Layer
4. Biotic Layer
5. Drainage Layer
6. Impervious Layer
7. Gas Vent Layer
8. Final Daily Cover
9. Operations and Maintenance

### Typical Treatment Train:
Groundwater monitoring.

### Other Cost Considerations:
Cleanup and landscaping, fencing and signage, clearing and grubbing, monitoring wells, storm sewer, decontamination facilities, retaining wall.

**Environmental Remediation: Assemblies Cost Book**

# Cost Data for Remediation

## Capping

| Assembly | Description | Unit | Unit Cost by Safety Level | | | | |
|---|---|---|---|---|---|---|---|
| | | | A | B | C | D | E |
| | **Capping - Top Cover** | | | | | | |
| 18 01 0313 | Jointed Concrete, Place & Finish, 4" - 12" | SY | 10.99 | 9.18 | 8.38 | 6.63 | 6.02 |
| 18 01 0314 | Jointed Mesh Reinforced Concrete, Place & Finish, 4" - 12" | SY | 17.21 | 14.26 | 12.96 | 10.11 | 9.09 |
| 18 01 0315 | Jointed Reinforced Concrete, Place & Finish, 4" - 12" | SY | 28.21 | 24.00 | 22.15 | 18.08 | 16.60 |
| 18 01 0316 | Field Mix Concrete | CY | 90.38 | 85.86 | 83.82 | 79.44 | 77.89 |
| 18 02 0301 | Asphalt Pavement - 10" Subgrade, 9" Base, 1 1/2" Topping | SY | 61.11 | 50.94 | 43.58 | 33.57 | 31.83 |
| 18 05 0202 | Rock Cover, Riprap, Light (10 to 100 Lb Pieces) | CY | 26.72 | 24.15 | 22.59 | 20.09 | 19.46 |
| 18 05 0203 | Rock Cover, Riprap, Medium (10 to 200 Lb Pieces) | CY | 27.05 | 24.48 | 22.92 | 20.42 | 19.79 |
| 18 05 0204 | Rock Cover, Riprap, Heavy (25 to 500 Lb Pieces) | CY | 27.39 | 24.82 | 23.26 | 20.76 | 20.13 |
| 18 05 0205 | Rock Cover, Bank-run Gravel | CY | 32.96 | 27.35 | 23.49 | 17.98 | 16.89 |
| 18 05 0301 | Topsoil, 6" Lifts, Off-Site | CY | 32.76 | 29.51 | 27.43 | 24.25 | 23.53 |
| 18 05 0302 | Topsoil, 6" Lifts, On-Site | CY | 14.14 | 11.54 | 9.45 | 6.88 | 6.57 |
| 18 05 0402 | Seeding, Vegetative Cover | ACRE | 1,874 | 1,823 | 1,791 | 1,741 | 1,729 |
| 18 05 0405 | Sodding, Vegetative Cover | ACRE | 25,697 | 22,064 | 20,238 | 16,710 | 15,589 |
| 18 05 0413 | Watering with 3,000-Gallon Tank Truck, per Pass | ACRE | 111.81 | 90.08 | 75.73 | 54.43 | 49.85 |
| 18 05 0711 | Automated Pop-Up Sprinkler System | ACRE | 8,779 | 6,984 | 6,188 | 4,451 | 3,832 |
| 33 05 0801 | 10' Wide Grass Drainage Swale | LF | 6.61 | 5.80 | 5.24 | 4.43 | 4.27 |
| 33 05 0802 | Grass Ditching, 3' Bottom, 3' Deep, 2:1 Side Slopes | LF | 18.76 | 15.91 | 13.91 | 11.11 | 10.58 |
| 33 05 0803 | Grass Ditching, 5' Bottom, 5' Deep, 4:1 Side Slopes | LF | 57.12 | 48.23 | 42.00 | 33.26 | 31.62 |
| 33 05 0804 | Riprap Ditching, 3' Bottom, 3' Deep, 2:1 Side Slopes | LF | 23.84 | 20.88 | 18.81 | 15.90 | 15.34 |
| 33 05 0805 | Concrete Ditching, 3' Bottom, 3' Deep, 2:1 Side Slopes | LF | 68.58 | 58.45 | 53.68 | 43.86 | 40.53 |
| 33 08 0511 | Drainage Netting, 1/4" Thick High-density Polyethylene | SF | 0.23 | 0.21 | 0.20 | 0.19 | 0.18 |
| 33 08 0512 | Drainage Netting, Geotextile Fabric Heat-bonded 1 Side | SF | 0.41 | 0.39 | 0.38 | 0.36 | 0.35 |
| 33 08 0513 | Drainage Netting, Geotextile Fabric Heat-bonded 2 Sides | SF | 0.48 | 0.45 | 0.44 | 0.42 | 0.41 |
| 33 08 0521 | Geogrid, Biaxial 9.8' x 164' Roll, TM = 14,000 PSF | SF | 0.23 | 0.21 | 0.21 | 0.19 | 0.19 |
| 33 08 0522 | Geogrid, Biaxial 9.8' x 164' Roll, TM = 18,500 PSF | SF | 0.37 | 0.36 | 0.35 | 0.33 | 0.33 |
| 33 08 0523 | Geogrid, Uniaxial 4.3' x 98' Roll, TM = 50,000 PSF | SF | 0.61 | 0.59 | 0.59 | 0.57 | 0.56 |
| 33 08 0524 | Geogrid, Uniaxial 4.3' x 98' Roll, TM = 130,000 PSF | SF | 1.21 | 1.18 | 1.17 | 1.15 | 1.14 |
| 33 08 0525 | Geogrid, Nylon Geomatrix/PVC-coated Polyester | SF | 1.45 | 1.43 | 1.42 | 1.40 | 1.39 |
| 33 08 0531 | 60 Mil Geotextile, Nonwoven | SY | 1.53 | 1.31 | 1.21 | 1.00 | 0.93 |
| 33 08 0532 | 80 Mil Geotextile, Nonwoven | SY | 1.57 | 1.35 | 1.25 | 1.04 | 0.97 |
| 33 08 0533 | 105 Mil Geotextile, Nonwoven | SY | 1.92 | 1.65 | 1.52 | 1.26 | 1.17 |
| 33 08 0534 | 130 Mil Geotextile, Nonwoven | SY | 2.56 | 2.20 | 2.03 | 1.68 | 1.56 |
| 33 08 0535 | 170 Mil Geotextile, Nonwoven | SY | 2.96 | 2.60 | 2.43 | 2.08 | 1.96 |

**Environmental Remediation: Assemblies Cost Book**

*Page 3-34*

1998 by ECHOS. All rights reserved

# Cost Data for Remediation

## Capping

| Assembly | Description | Unit | Unit Cost by Safety Level A | B | C | D | E |
|---|---|---|---|---|---|---|---|
| | **Capping - Top Cover** | | | | | | |
| 33 08 0541 | 20 Mil Polymeric Liner, Very Low Density Polyethylene | SF | 1.25 | 1.02 | 0.91 | 0.69 | 0.61 |
| 33 08 0542 | 30 Mil Polymeric Liner, Very Low Density Polyethylene | SF | 1.98 | 1.59 | 1.41 | 1.03 | 0.90 |
| 33 08 0543 | 40 Mil Polymeric Liner, Very Low Density Polyethylene | SF | 2.38 | 1.91 | 1.70 | 1.24 | 1.09 |
| 33 08 0544 | 60 Mil Polymeric Liner, Very Low Density Polyethylene | SF | 3.54 | 2.83 | 2.51 | 1.83 | 1.60 |
| 33 08 0545 | 80 Mil Polymeric Liner, Very Low Density Polyethylene | SF | 4.35 | 3.49 | 3.10 | 2.27 | 1.98 |
| 33 08 0546 | 100 Mil Polymeric Liner, Very Low Density Polyethylene | SF | 5.50 | 4.41 | 3.91 | 2.85 | 2.49 |
| 33 08 0551 | 36 Mil Polymeric Liner, Chlorosulfanated Polyethylene | SF | 0.75 | 0.73 | 0.72 | 0.70 | 0.70 |
| 33 08 0552 | 45 Mil Polymeric Liner, Chlorosulfanated Polyethylene | SF | 0.83 | 0.81 | 0.80 | 0.79 | 0.78 |
| 33 08 0561 | 20 Mil Polymeric Liner, PVC | SF | 0.24 | 0.23 | 0.22 | 0.21 | 0.20 |
| 33 08 0562 | 30 Mil Polymeric Liner, PVC | SF | 0.32 | 0.31 | 0.30 | 0.29 | 0.28 |
| 33 08 0563 | 40 Mil Polymeric Liner, PVC | SF | 0.38 | 0.36 | 0.36 | 0.34 | 0.34 |
| 33 08 0564 | 60 Mil Polymeric Liner, PVC | SF | 0.72 | 0.70 | 0.69 | 0.67 | 0.67 |
| 33 08 0565 | 80 Mil Polymeric Liner, PVC | SF | 0.98 | 0.96 | 0.95 | 0.93 | 0.93 |
| 33 08 0566 | 100 Mil Polymeric Liner, PVC | SF | 1.28 | 1.26 | 1.25 | 1.23 | 1.22 |
| 33 08 0571 | 40 Mil Polymeric Liner, High-density Polyethylene | SF | 2.35 | 1.88 | 1.67 | 1.21 | 1.06 |
| 33 08 0572 | 60 Mil Polymeric Liner, High-density Polyethylene | SF | 3.49 | 2.80 | 2.48 | 1.81 | 1.58 |
| 33 08 0573 | 80 Mil Polymeric Liner, High-density Polyethylene | SF | 4.32 | 3.47 | 3.08 | 2.25 | 1.97 |
| | **Capping - Soil Cover** | | | | | | |
| 17 03 0422 | Unclassified Fill, 6" Lifts, On-Site | CY | 14.23 | 11.65 | 9.60 | 7.05 | 6.72 |
| 17 03 0423 | Unclassified Fill, 6" Lifts, Off-Site | CY | 11.08 | 9.81 | 8.83 | 7.58 | 7.40 |
| | **Capping - Filter Layer** | | | | | | |
| 17 03 0425 | Sand, 6" Lifts, On-Site | CY | 13.88 | 11.35 | 9.36 | 6.87 | 6.54 |
| 17 03 0426 | Sand, 6" Lifts, Off-Site | CY | 15.55 | 14.34 | 13.42 | 12.22 | 12.04 |
| 17 03 0430 | Gravel, 6" Lifts | CY | 15.46 | 13.36 | 12.05 | 9.99 | 9.49 |
| 33 08 0511 | Drainage Netting, 1/4" Thick High-density Polyethylene | SF | 0.23 | 0.21 | 0.20 | 0.19 | 0.18 |
| 33 08 0512 | Drainage Netting, Geotextile Fabric Heat-bonded 1 Side | SF | 0.41 | 0.39 | 0.38 | 0.36 | 0.35 |
| 33 08 0513 | Drainage Netting, Geotextile Fabric Heat-bonded 2 Sides | SF | 0.48 | 0.45 | 0.44 | 0.42 | 0.41 |
| 33 08 0521 | Geogrid, Biaxial 9.8' x 164' Roll, TM = 14,000 PSF | SF | 0.23 | 0.21 | 0.21 | 0.19 | 0.19 |
| 33 08 0522 | Geogrid, Biaxial 9.8' x 164' Roll, TM = 18,500 PSF | SF | 0.37 | 0.36 | 0.35 | 0.33 | 0.33 |
| 33 08 0523 | Geogrid, Uniaxial 4.3' x 98' Roll, TM = 50,000 PSF | SF | 0.61 | 0.59 | 0.59 | 0.57 | 0.56 |
| 33 08 0524 | Geogrid, Uniaxial 4.3' x 98' Roll, TM = 130,000 PSF | SF | 1.21 | 1.18 | 1.17 | 1.15 | 1.14 |
| 33 08 0525 | Geogrid, Nylon Geomatrix/PVC-coated Polyester | SF | 1.45 | 1.43 | 1.42 | 1.40 | 1.39 |
| 33 08 0531 | 60 Mil Geotextile, Nonwoven | SY | 1.53 | 1.31 | 1.21 | 1.00 | 0.93 |

**Environmental Remediation: Assemblies Cost Book**

*Page 3-35*

1998 by ECHOS. All rights reserved

# Cost Data for Remediation

## Capping

| Assembly | Description | Unit | Unit Cost by Safety Level | | | | |
|---|---|---|---|---|---|---|---|
| | | | A | B | C | D | E |
| | **Capping - Filter Layer** | | | | | | |
| 33 08 0532 | 80 Mil Geotextile, Nonwoven | SY | 1.57 | 1.35 | 1.25 | 1.04 | 0.97 |
| 33 08 0533 | 105 Mil Geotextile, Nonwoven | SY | 1.92 | 1.65 | 1.52 | 1.26 | 1.17 |
| 33 08 0534 | 130 Mil Geotextile, Nonwoven | SY | 2.56 | 2.20 | 2.03 | 1.68 | 1.56 |
| 33 08 0535 | 170 Mil Geotextile, Nonwoven | SY | 2.96 | 2.60 | 2.43 | 2.08 | 1.96 |
| | **Capping - Biotic Layer** | | | | | | |
| 17 03 0430 | Gravel, 6" Lifts | CY | 15.46 | 13.36 | 12.05 | 9.99 | 9.49 |
| 18 05 0202 | Rock Cover, Riprap, Light (10 to 100 Lb Pieces) | CY | 26.72 | 24.15 | 22.59 | 20.09 | 19.46 |
| 18 05 0203 | Rock Cover, Riprap, Medium (10 to 200 Lb Pieces) | CY | 27.05 | 24.48 | 22.92 | 20.42 | 19.79 |
| 18 05 0204 | Rock Cover, Riprap, Heavy (25 to 500 Lb Pieces) | CY | 27.39 | 24.82 | 23.26 | 20.76 | 20.13 |
| 33 08 0511 | Drainage Netting, 1/4" Thick High-density Polyethylene | SF | 0.23 | 0.21 | 0.20 | 0.19 | 0.18 |
| 33 08 0512 | Drainage Netting, Geotextile Fabric Heat-bonded 1 Side | SF | 0.41 | 0.39 | 0.38 | 0.36 | 0.35 |
| 33 08 0513 | Drainage Netting, Geotextile Fabric Heat-bonded 2 Sides | SF | 0.48 | 0.45 | 0.44 | 0.42 | 0.41 |
| 33 08 0521 | Geogrid, Biaxial 9.8' x 164' Roll, TM = 14,000 PSF | SF | 0.23 | 0.21 | 0.21 | 0.19 | 0.19 |
| 33 08 0522 | Geogrid, Biaxial 9.8' x 164' Roll, TM = 18,500 PSF | SF | 0.37 | 0.36 | 0.35 | 0.33 | 0.33 |
| 33 08 0523 | Geogrid, Uniaxial 4.3' x 98' Roll, TM = 50,000 PSF | SF | 0.61 | 0.59 | 0.59 | 0.57 | 0.56 |
| 33 08 0524 | Geogrid, Uniaxial 4.3' x 98' Roll, TM = 130,000 PSF | SF | 1.21 | 1.18 | 1.17 | 1.15 | 1.14 |
| 33 08 0525 | Geogrid, Nylon Geomatrix/PVC-coated Polyester | SF | 1.45 | 1.43 | 1.42 | 1.40 | 1.39 |
| 33 08 0531 | 60 Mil Geotextile, Nonwoven | SY | 1.53 | 1.31 | 1.21 | 1.00 | 0.93 |
| 33 08 0532 | 80 Mil Geotextile, Nonwoven | SY | 1.57 | 1.35 | 1.25 | 1.04 | 0.97 |
| 33 08 0533 | 105 Mil Geotextile, Nonwoven | SY | 1.92 | 1.65 | 1.52 | 1.26 | 1.17 |
| 33 08 0534 | 130 Mil Geotextile, Nonwoven | SY | 2.56 | 2.20 | 2.03 | 1.68 | 1.56 |
| 33 08 0535 | 170 Mil Geotextile, Nonwoven | SY | 2.96 | 2.60 | 2.43 | 2.08 | 1.96 |
| | **Capping - Drainage Layer** | | | | | | |
| 17 03 0425 | Sand, 6" Lifts, On-Site | CY | 13.88 | 11.35 | 9.36 | 6.87 | 6.54 |
| 17 03 0426 | Sand, 6" Lifts, Off-Site | CY | 15.55 | 14.34 | 13.42 | 12.22 | 12.04 |
| 17 03 0430 | Gravel, 6" Lifts | CY | 15.46 | 13.36 | 12.05 | 9.99 | 9.49 |
| 33 08 0511 | Drainage Netting, 1/4" Thick High-density Polyethylene | SF | 0.23 | 0.21 | 0.20 | 0.19 | 0.18 |
| 33 08 0512 | Drainage Netting, Geotextile Fabric Heat-bonded 1 Side | SF | 0.41 | 0.39 | 0.38 | 0.36 | 0.35 |
| 33 08 0513 | Drainage Netting, Geotextile Fabric Heat-bonded 2 Sides | SF | 0.48 | 0.45 | 0.44 | 0.42 | 0.41 |
| 33 08 0521 | Geogrid, Biaxial 9.8' x 164' Roll, TM = 14,000 PSF | SF | 0.23 | 0.21 | 0.21 | 0.19 | 0.19 |
| 33 08 0522 | Geogrid, Biaxial 9.8' x 164' Roll, TM = 18,500 PSF | SF | 0.37 | 0.36 | 0.35 | 0.33 | 0.33 |
| 33 08 0523 | Geogrid, Uniaxial 4.3' x 98' Roll, TM = 50,000 PSF | SF | 0.61 | 0.59 | 0.59 | 0.57 | 0.56 |
| 33 08 0524 | Geogrid, Uniaxial 4.3' x 98' Roll, TM = 130,000 PSF | SF | 1.21 | 1.18 | 1.17 | 1.15 | 1.14 |

**Environmental Remediation: Assemblies Cost Book**

# Cost Data for Remediation

## Capping

| Assembly | Description | Unit | Unit Cost by Safety Level | | | | |
|---|---|---|---|---|---|---|---|
| | | | A | B | C | D | E |
| | **Capping - Drainage Layer** | | | | | | |
| 33 08 0525 | Geogrid, Nylon Geomatrix/PVC-coated Polyester | SF | 1.45 | 1.43 | 1.42 | 1.40 | 1.39 |
| 33 08 0531 | 60 Mil Geotextile, Nonwoven | SY | 1.53 | 1.31 | 1.21 | 1.00 | 0.93 |
| 33 08 0532 | 80 Mil Geotextile, Nonwoven | SY | 1.57 | 1.35 | 1.25 | 1.04 | 0.97 |
| 33 08 0533 | 105 Mil Geotextile, Nonwoven | SY | 1.92 | 1.65 | 1.52 | 1.26 | 1.17 |
| 33 08 0534 | 130 Mil Geotextile, Nonwoven | SY | 2.56 | 2.20 | 2.03 | 1.68 | 1.56 |
| 33 08 0535 | 170 Mil Geotextile, Nonwoven | SY | 2.96 | 2.60 | 2.43 | 2.08 | 1.96 |
| | **Capping - Impervious Layer** | | | | | | |
| 33 08 0503 | Polymeric Liner Anchor Trench, 3' x 1.5' | LF | 1.58 | 1.25 | 1.07 | 0.76 | 0.67 |
| 33 08 0504 | Herbicide Application | ACRE | 487.34 | 400.41 | 343.04 | 257.83 | 239.51 |
| 33 08 0505 | Soil/Bentonite Liner | SF | 0.78 | 0.77 | 0.76 | 0.75 | 0.74 |
| 33 08 0506 | Clay 10E-7, 6" Lifts, On-Site | CY | 29.50 | 23.82 | 19.83 | 14.25 | 13.21 |
| 33 08 0507 | Clay 10E-7, 6" Lifts, Off-Site | CY | 31.11 | 26.22 | 22.84 | 18.04 | 17.11 |
| 33 08 0508 | Fabric Membrane Liner/Bentonite Liner | SF | 0.59 | 0.52 | 0.48 | 0.41 | 0.40 |
| 33 08 0511 | Drainage Netting, 1/4" Thick High-density Polyethylene | SF | 0.23 | 0.21 | 0.20 | 0.19 | 0.18 |
| 33 08 0512 | Drainage Netting, Geotextile Fabric Heat-bonded 1 Side | SF | 0.41 | 0.39 | 0.38 | 0.36 | 0.35 |
| 33 08 0513 | Drainage Netting, Geotextile Fabric Heat-bonded 2 Sides | SF | 0.48 | 0.45 | 0.44 | 0.42 | 0.41 |
| 33 08 0521 | Geogrid, Biaxial 9.8' x 164' Roll, TM = 14,000 PSF | SF | 0.23 | 0.21 | 0.21 | 0.19 | 0.19 |
| 33 08 0522 | Geogrid, Biaxial 9.8' x 164' Roll, TM = 18,500 PSF | SF | 0.37 | 0.36 | 0.35 | 0.33 | 0.33 |
| 33 08 0523 | Geogrid, Uniaxial 4.3' x 98' Roll, TM = 50,000 PSF | SF | 0.61 | 0.59 | 0.59 | 0.57 | 0.56 |
| 33 08 0524 | Geogrid, Uniaxial 4.3' x 98' Roll, TM = 130,000 PSF | SF | 1.21 | 1.18 | 1.17 | 1.15 | 1.14 |
| 33 08 0525 | Geogrid, Nylon Geomatrix/PVC-coated Polyester | SF | 1.45 | 1.43 | 1.42 | 1.40 | 1.39 |
| 33 08 0531 | 60 Mil Geotextile, Nonwoven | SY | 1.53 | 1.31 | 1.21 | 1.00 | 0.93 |
| 33 08 0532 | 80 Mil Geotextile, Nonwoven | SY | 1.57 | 1.35 | 1.25 | 1.04 | 0.97 |
| 33 08 0533 | 105 Mil Geotextile, Nonwoven | SY | 1.92 | 1.65 | 1.52 | 1.26 | 1.17 |
| 33 08 0534 | 130 Mil Geotextile, Nonwoven | SY | 2.56 | 2.20 | 2.03 | 1.68 | 1.56 |
| 33 08 0535 | 170 Mil Geotextile, Nonwoven | SY | 2.96 | 2.60 | 2.43 | 2.08 | 1.96 |
| 33 08 0541 | 20 Mil Polymeric Liner, Very Low Density Polyethylene | SF | 1.25 | 1.02 | 0.91 | 0.69 | 0.61 |
| 33 08 0542 | 30 Mil Polymeric Liner, Very Low Density Polyethylene | SF | 1.98 | 1.59 | 1.41 | 1.03 | 0.90 |
| 33 08 0543 | 40 Mil Polymeric Liner, Very Low Density Polyethylene | SF | 2.38 | 1.91 | 1.70 | 1.24 | 1.09 |
| 33 08 0544 | 60 Mil Polymeric Liner, Very Low Density Polyethylene | SF | 3.54 | 2.83 | 2.51 | 1.83 | 1.60 |
| 33 08 0545 | 80 Mil Polymeric Liner, Very Low Density Polyethylene | SF | 4.35 | 3.49 | 3.10 | 2.27 | 1.98 |
| 33 08 0546 | 100 Mil Polymeric Liner, Very Low Density Polyethylene | SF | 5.50 | 4.41 | 3.91 | 2.85 | 2.49 |
| 33 08 0551 | 36 Mil Polymeric Liner, Chlorosulfanated Polyethylene | SF | 0.75 | 0.73 | 0.72 | 0.70 | 0.70 |

**Environmental Remediation: Assemblies Cost Book**

# Cost Data for Remediation

## Capping

| Assembly | Description | Unit | A | B | C | D | E |
|---|---|---|---|---|---|---|---|
| | **Capping - Impervious Layer** | | | | | | |
| 33 08 0552 | 45 Mil Polymeric Liner, Chlorosulfanated Polyethylene | SF | 0.83 | 0.81 | 0.80 | 0.79 | 0.78 |
| 33 08 0561 | 20 Mil Polymeric Liner, PVC | SF | 0.24 | 0.23 | 0.22 | 0.21 | 0.20 |
| 33 08 0562 | 30 Mil Polymeric Liner, PVC | SF | 0.32 | 0.31 | 0.30 | 0.29 | 0.28 |
| 33 08 0563 | 40 Mil Polymeric Liner, PVC | SF | 0.38 | 0.36 | 0.36 | 0.34 | 0.34 |
| 33 08 0564 | 60 Mil Polymeric Liner, PVC | SF | 0.72 | 0.70 | 0.69 | 0.67 | 0.67 |
| 33 08 0565 | 80 Mil Polymeric Liner, PVC | SF | 0.98 | 0.96 | 0.95 | 0.93 | 0.93 |
| 33 08 0566 | 100 Mil Polymeric Liner, PVC | SF | 1.28 | 1.26 | 1.25 | 1.23 | 1.22 |
| 33 08 0571 | 40 Mil Polymeric Liner, High-density Polyethylene | SF | 2.35 | 1.88 | 1.67 | 1.21 | 1.06 |
| 33 08 0572 | 60 Mil Polymeric Liner, High-density Polyethylene | SF | 3.49 | 2.80 | 2.48 | 1.81 | 1.58 |
| 33 08 0573 | 80 Mil Polymeric Liner, High-density Polyethylene | SF | 4.32 | 3.47 | 3.08 | 2.25 | 1.97 |
| | **Capping - Gas Vent Layer** | | | | | | |
| 17 03 0425 | Sand, 6" Lifts, On-Site | CY | 13.88 | 11.35 | 9.36 | 6.87 | 6.54 |
| 17 03 0426 | Sand, 6" Lifts, Off-Site | CY | 15.55 | 14.34 | 13.42 | 12.22 | 12.04 |
| 17 03 0430 | Gravel, 6" Lifts | CY | 15.46 | 13.36 | 12.05 | 9.99 | 9.49 |
| 33 07 0201 | 6" Inside Diameter (Vertical Pipe Spaced @ 200 LF), Gas Vent Piping System | LF | 28.29 | 24.75 | 22.57 | 19.10 | 18.26 |
| 33 08 0511 | Drainage Netting, 1/4" Thick High-density Polyethylene | SF | 0.23 | 0.21 | 0.20 | 0.19 | 0.18 |
| 33 08 0512 | Drainage Netting, Geotextile Fabric Heat-bonded 1 Side | SF | 0.41 | 0.39 | 0.38 | 0.36 | 0.35 |
| 33 08 0513 | Drainage Netting, Geotextile Fabric Heat-bonded 2 Sides | SF | 0.48 | 0.45 | 0.44 | 0.42 | 0.41 |
| 33 08 0521 | Geogrid, Biaxial 9.8' x 164' Roll, TM = 14,000 PSF | SF | 0.23 | 0.21 | 0.21 | 0.19 | 0.19 |
| 33 08 0522 | Geogrid, Biaxial 9.8' x 164' Roll, TM = 18,500 PSF | SF | 0.37 | 0.36 | 0.35 | 0.33 | 0.33 |
| 33 08 0523 | Geogrid, Uniaxial 4.3' x 98' Roll, TM = 50,000 PSF | SF | 0.61 | 0.59 | 0.59 | 0.57 | 0.56 |
| 33 08 0524 | Geogrid, Uniaxial 4.3' x 98' Roll, TM = 130,000 PSF | SF | 1.21 | 1.18 | 1.17 | 1.15 | 1.14 |
| 33 08 0525 | Geogrid, Nylon Geomatrix/PVC-coated Polyester | SF | 1.45 | 1.43 | 1.42 | 1.40 | 1.39 |
| 33 08 0531 | 60 Mil Geotextile, Nonwoven | SY | 1.53 | 1.31 | 1.21 | 1.00 | 0.93 |
| 33 08 0532 | 80 Mil Geotextile, Nonwoven | SY | 1.57 | 1.35 | 1.25 | 1.04 | 0.97 |
| 33 08 0533 | 105 Mil Geotextile, Nonwoven | SY | 1.92 | 1.65 | 1.52 | 1.26 | 1.17 |
| 33 08 0534 | 130 Mil Geotextile, Nonwoven | SY | 2.56 | 2.20 | 2.03 | 1.68 | 1.56 |
| 33 08 0535 | 170 Mil Geotextile, Nonwoven | SY | 2.96 | 2.60 | 2.43 | 2.08 | 1.96 |
| 33 13 9101 | 1,500 CFM Fluidized Bed Gas Scrubber, Single Stage, Off-Gas | EACH | 40,103 | 40,103 | 40,103 | 40,103 | 40,103 |
| 33 13 9102 | 3,000 CFM Fluidized Bed Gas Scrubber, Single Stage, Off-Gas | EACH | 56,109 | 56,109 | 56,109 | 56,109 | 56,109 |
| 33 13 9103 | 4,400 CFM Fluidized Bed Gas Scrubber, Single Stage, Off-Gas | EACH | 66,868 | 66,868 | 66,868 | 66,868 | 66,868 |

**Unit Cost by Safety Level**

**Environmental Remediation: Assemblies Cost Book**

# Cost Data for Remediation

## Capping

| Assembly | Description | Unit | Unit Cost by Safety Level A | B | C | D | E |
|---|---|---|---|---|---|---|---|
| | **Capping - Gas Vent Layer** | | | | | | |
| 33 13 9104 | 5,800 CFM Fluidized Bed Gas Scrubber, Single Stage, Off-Gas | EACH | 77,910 | 77,910 | 77,910 | 77,910 | 77,910 |
| 33 13 9105 | 7,700 CFM Fluidized Bed Gas Scrubber, Single Stage, Off-Gas | EACH | 98,880 | 98,880 | 98,880 | 98,880 | 98,880 |
| 33 13 9106 | 9,600 CFM Fluidized Bed Gas Scrubber, Single Stage, Off-Gas | EACH | 111,742 | 111,742 | 111,742 | 111,742 | 111,742 |
| 33 13 9107 | 12,000 CFM Fluidized Bed Gas Scrubber, Single Stage, Off-Gas | EACH | 124,815 | 124,815 | 124,815 | 124,815 | 124,815 |
| 33 13 9108 | 17,000 CFM Fluidized Bed Gas Scrubber, Single Stage, Off-Gas | EACH | 149,460 | 149,460 | 149,460 | 149,460 | 149,460 |
| 33 13 9109 | 23,600 CFM Fluidized Bed Gas Scrubber, Single Stage, Off-Gas | EACH | 175,077 | 175,077 | 175,077 | 175,077 | 175,077 |
| 33 13 9110 | 30,200 CFM Fluidized Bed Gas Scrubber, Single Stage, Off-Gas | EACH | 196,630 | 196,630 | 196,630 | 196,630 | 196,630 |
| 33 13 9111 | 38,700 CFM Fluidized Bed Gas Scrubber, Single Stage, Off-Gas | EACH | 235,320 | 235,320 | 235,320 | 235,320 | 235,320 |
| 33 13 9112 | 47,200 CFM Fluidized Bed Gas Scrubber, Single Stage, Off-Gas | EACH | 257,933 | 257,933 | 257,933 | 257,933 | 257,933 |
| 33 13 9113 | 58,000 CFM Fluidized Bed Gas Scrubber, Single Stage, Off-Gas | EACH | 289,203 | 289,203 | 289,203 | 289,203 | 289,203 |
| 33 13 9114 | 79,700 CFM Fluidized Bed Gas Scrubber, Single Stage, Off-Gas | EACH | 338,493 | 338,493 | 338,493 | 338,493 | 338,493 |
| 33 13 9115 | 90,900 CFM Fluidized Bed Gas Scrubber, Single Stage, Off-Gas | EACH | 363,580 | 363,580 | 363,580 | 363,580 | 363,580 |
| 33 13 9116 | 113,300 CFM Fluidized Bed Gas Scrubber, Single Stage, Off-Gas | EACH | 498,023 | 498,023 | 498,023 | 498,023 | 498,023 |
| 33 14 9301 | Flare, 2.25 Million BTU/Hour, Landfill Off-Gas | EACH | 39,827 | 39,521 | 39,374 | 39,077 | 38,978 |
| 33 14 9302 | Flare, 7.5 Million BTU/Hour, Landfill Off-Gas | EACH | 54,031 | 53,673 | 53,503 | 53,156 | 53,040 |
| 33 14 9303 | Flare, 10.5 Million BTU/Hour, Landfill Off-Gas | EACH | 66,608 | 65,933 | 65,599 | 64,943 | 64,731 |
| 33 14 9304 | Flare, 22.5 Million BTU/Hour, Landfill Off-Gas | EACH | 96,121 | 95,404 | 95,049 | 94,353 | 94,128 |
| 33 14 9305 | Flare, 30.0 Million BTU/Hour, Landfill Off-Gas | EACH | 94,517 | 93,780 | 93,415 | 92,700 | 92,469 |
| 33 14 9306 | Flare, 36.0 Million BTU/Hour, Landfill Off-Gas | EACH | 123,668 | 122,577 | 122,039 | 120,981 | 120,639 |
| 33 14 9307 | Flare, 60.0 Million BTU/Hour, Landfill Off-Gas | EACH | 223,873 | 222,758 | 222,208 | 221,126 | 220,777 |
| 33 14 9308 | Flare, 120.0 Million BTU/Hour, Landfill Off-Gas | EACH | 415,427 | 413,818 | 413,022 | 411,461 | 410,956 |
| 33 23 0402 | 4" Filter Sock | LF | 5.57 | 4.65 | 3.93 | 3.02 | 2.90 |
| 33 23 0403 | 6" Filter Sock | LF | 5.70 | 4.78 | 4.06 | 3.15 | 3.03 |
| 33 23 0404 | 8" Filter Sock | LF | 6.35 | 5.43 | 4.71 | 3.80 | 3.68 |
| 33 23 1503 | 3,000 PSI, 6" Thick, Concrete Pad | CY | 85.80 | 82.00 | 80.29 | 76.60 | 75.31 |
| 33 23 2104 | 3/8", 1' Thick, Sodium Bentonite Pellets | CF | 275.35 | 229.42 | 193.32 | 147.97 | 141.88 |

## Environmental Remediation: Assemblies Cost Book

# Cost Data for Remediation

## Capping

| Assembly | Description | Unit | Unit Cost by Safety Level | | | | |
|---|---|---|---|---|---|---|---|
| | | | A | B | C | D | E |
| | **Capping - Gas Vent Layer** | | | | | | |
| 33 26 0512 | 4" High-density Polyethylene, Transfer Pipe | LF | 6.75 | 5.61 | 5.10 | 4.00 | 3.62 |
| 33 26 0513 | 6" High-density Polyethylene, Transfer Pipe | LF | 8.82 | 7.47 | 6.86 | 5.55 | 5.10 |
| 33 26 0514 | 8" High-density Polyethylene, Transfer Pipe | LF | 11.20 | 9.76 | 9.11 | 7.72 | 7.23 |
| 33 26 0516 | 12" High-density Polyethylene, Transfer Pipe | LF | 19.22 | 17.26 | 16.37 | 14.47 | 13.81 |
| 33 26 0518 | 16" High-density Polyethylene, Transfer Pipe | LF | 30.24 | 27.16 | 25.76 | 22.78 | 21.74 |
| 33 26 0520 | 24" High-density Polyethylene, Transfer Pipe | LF | 61.27 | 55.88 | 53.44 | 48.22 | 46.39 |
| 33 26 0802 | 4" Slotted PVC Pipe | LF | 8.67 | 7.65 | 7.22 | 6.23 | 5.87 |
| 33 26 0803 | 6" Slotted PVC Pipe | LF | 14.25 | 12.89 | 12.31 | 10.99 | 10.51 |
| 33 26 0804 | 8" Slotted PVC Pipe | LF | 22.19 | 20.15 | 19.28 | 17.31 | 16.59 |
| 33 26 0805 | 10" Slotted PVC Pipe | LF | 36.76 | 34.04 | 32.88 | 30.26 | 29.29 |
| 33 26 0806 | 12" Slotted PVC Pipe | LF | 26.50 | 22.42 | 20.68 | 16.74 | 15.30 |
| 33 27 0301 | 4" High-density Polyethylene, Tee | LF | 103.49 | 86.91 | 79.39 | 63.34 | 57.72 |
| 33 27 0302 | 6" High-density Polyethylene, Tee | LF | 165.88 | 146.28 | 137.40 | 118.43 | 111.78 |
| 33 27 0303 | 8" High-density Polyethylene, Tee | LF | 336.63 | 305.83 | 291.88 | 262.06 | 251.62 |
| 33 27 0304 | 12" High-density Polyethylene, Tee | LF | 599.95 | 546.05 | 521.64 | 469.46 | 451.19 |
| 33 27 0305 | 18" High-density Polyethylene, Tee | LF | 1,655 | 1,440 | 1,342 | 1,133 | 1,060 |
| 33 27 0306 | 24" High-density Polyethylene, Tee | LF | 2,720 | 2,434 | 2,304 | 2,026 | 1,929 |
| 33 27 0311 | 4", 90 Degree, High-density Polyethylene, Elbow | EACH | 94.59 | 79.19 | 72.22 | 57.31 | 52.09 |
| 33 27 0312 | 6", 90 Degree, High-density Polyethylene, Elbow | EACH | 150.00 | 132.04 | 123.90 | 106.51 | 100.42 |
| 33 27 0313 | 8", 90 Degree, High-density Polyethylene, Elbow | EACH | 253.57 | 229.62 | 218.77 | 195.58 | 187.46 |
| 33 27 0314 | 10", 90 Degree, High-density Polyethylene, Elbow | EACH | 405.00 | 369.07 | 352.79 | 318.01 | 305.83 |
| 33 27 0315 | 12", 90 Degree, High-density Polyethylene, Elbow | EACH | 629.04 | 581.13 | 559.43 | 513.06 | 496.81 |
| 33 31 0131 | 48 CFM, 4.9 HP Maximum, Blower, Positive Displacement, with Motor | EACH | 2,288 | 1,983 | 1,853 | 1,559 | 1,451 |
| 33 31 0132 | 111 CFM, 5.6 HP Maximum, Blower, Positive Displacement, with Motor | EACH | 2,499 | 2,169 | 2,028 | 1,709 | 1,593 |
| 33 31 0133 | 143 CFM, 12.2 HP Maximum, Blower, Positive Displacement, with Motor | EACH | 2,683 | 2,323 | 2,169 | 1,821 | 1,694 |
| 33 31 0134 | 271 CFM, 11.8 HP Maximum, Blower, Positive Displacement, with Motor | EACH | 2,904 | 2,508 | 2,338 | 1,956 | 1,816 |
| 33 31 0135 | 134 CFM, 15 HP Maximum, Blower, Positive Displacement, with Motor | EACH | 3,084 | 2,688 | 2,519 | 2,136 | 1,996 |
| 33 31 0136 | 278 CFM, 19.7 HP Maximum, Blower, Positive Displacement, with Motor | EACH | 3,105 | 2,709 | 2,540 | 2,157 | 2,017 |
| 33 31 0137 | 399 CFM, 18.3 HP Maximum, Blower, Positive Displacement, with Motor | EACH | 3,558 | 3,082 | 2,878 | 2,417 | 2,248 |

**Environmental Remediation: Assemblies Cost Book**

*Page 3-40*

# Cost Data for Remediation

## Capping

| Assembly | Description | Unit | Unit Cost by Safety Level A | B | C | D | E |
|---|---|---|---|---|---|---|---|
| | **Capping - Gas Vent Layer** | | | | | | |
| 33 31 0138 | 247 CFM, 25.4 HP Maximum, Blower, Positive Displacement, with Motor | EACH | 3,993 | 3,516 | 3,312 | 2,852 | 2,683 |
| 33 31 0139 | 427 CFM, 28.3 HP Maximum, Blower, Positive Displacement, with Motor | EACH | 4,094 | 3,617 | 3,413 | 2,952 | 2,784 |
| 33 31 0140 | 665 CFM, 29 HP Maximum, Blower, Positive Displacement, with Motor | EACH | 4,232 | 3,755 | 3,551 | 3,090 | 2,921 |
| 33 31 0141 | 393 CFM, 38.8 HP Maximum, Blower, Positive Displacement, with Motor | EACH | 5,259 | 4,758 | 4,544 | 4,060 | 3,882 |
| 33 31 0142 | 615 CFM, 49.6 HP Maximum, Blower, Positive Displacement, with Motor | EACH | 5,429 | 4,928 | 4,714 | 4,230 | 4,052 |
| 33 31 0143 | 1,248 CFM, 46.5 HP Maximum, Blower, Positive Displacement, with Motor | EACH | 6,951 | 6,313 | 6,040 | 5,423 | 5,196 |
| 33 31 0144 | 574 CFM, 55 HP Maximum, Blower, Positive Displacement, with Motor | EACH | 7,577 | 6,938 | 6,665 | 6,048 | 5,822 |
| 33 31 0145 | 1,071 CFM, 67.1 HP Maximum, Blower, Positive Displacement, with Motor | EACH | 8,061 | 7,391 | 7,104 | 6,456 | 6,218 |
| 33 31 0146 | 1,878 CFM, 65.5 HP Maximum, Blower, Positive Displacement, with Motor | EACH | 8,973 | 8,278 | 7,981 | 7,309 | 7,063 |
| 33 41 0201 | Blower and Motor Maintenance and Repair | EACH | 865.17 | 671.48 | 581.58 | 393.96 | 329.64 |
| 33 42 0101 | Electrical Charge | KWH | 0.06 | 0.06 | 0.06 | 0.06 | 0.06 |
| 33 43 0101 | Equipment Building 10' Ceiling, Built-Up Roof, Concrete Block Exterior | SF | 38.02 | 38.02 | 38.02 | 38.02 | 38.02 |
| 33 43 0102 | 5.58' x 5.5' x 8.33' Hazardous Material Storage Building, 4 Drum Capacity | EACH | 3,202 | 3,202 | 3,202 | 3,202 | 3,202 |
| 33 43 0103 | 10.25' x 5.5' x 8.33' Hazardous Material Storage Building, 4 Drum Capacity | EACH | 9,162 | 9,162 | 9,162 | 9,162 | 9,162 |
| 33 43 0104 | 10.25' x 5.5' x 8.33' Hazardous Material Storage Building, 8 Drum Capacity | EACH | 11,448 | 11,448 | 11,448 | 11,448 | 11,448 |
| | **Capping - Final Cover** | | | | | | |
| 17 03 0422 | Unclassified Fill, 6" Lifts, On-Site | CY | 14.23 | 11.65 | 9.60 | 7.05 | 6.72 |
| 17 03 0423 | Unclassified Fill, 6" Lifts, Off-Site | CY | 11.08 | 9.81 | 8.83 | 7.58 | 7.40 |
| | **Operations & Maintenance** | | | | | | |
| 18 01 0313 | Jointed Concrete, Place & Finish, 4" - 12" | SY | 10.99 | 9.18 | 8.38 | 6.63 | 6.02 |
| 18 01 0314 | Jointed Mesh Reinforced Concrete, Place & Finish, 4" - 12" | SY | 17.21 | 14.26 | 12.96 | 10.11 | 9.09 |
| 18 01 0315 | Jointed Reinforced Concrete, Place & Finish, 4" - 12" | SY | 28.21 | 24.00 | 22.15 | 18.08 | 16.60 |
| 18 01 0316 | Field Mix Concrete | CY | 90.38 | 85.86 | 83.82 | 79.44 | 77.89 |
| 18 02 0301 | Asphalt Pavement - 10" Subgrade, 9" Base, 1 1/2" Topping | SY | 61.11 | 50.94 | 43.58 | 33.57 | 31.83 |
| 18 05 0202 | Rock Cover, Riprap, Light (10 to 100 Lb Pieces) | CY | 26.72 | 24.15 | 22.59 | 20.09 | 19.46 |
| 18 05 0203 | Rock Cover, Riprap, Medium (10 to 200 Lb Pieces) | CY | 27.05 | 24.48 | 22.92 | 20.42 | 19.79 |

**Environmental Remediation: Assemblies Cost Book**

*Page 3-41*

# Cost Data for Remediation

## Capping

| Assembly | Description | Unit | Unit Cost by Safety Level A | B | C | D | E |
|---|---|---|---|---|---|---|---|
| | **Operations & Maintenance** | | | | | | |
| 18 05 0204 | Rock Cover, Riprap, Heavy (25 to 500 Lb Pieces) | CY | 27.39 | 24.82 | 23.26 | 20.76 | 20.13 |
| 18 05 0205 | Rock Cover, Bank-run Gravel | CY | 32.96 | 27.35 | 23.49 | 17.98 | 16.89 |
| 18 05 0301 | Topsoil, 6" Lifts, Off-Site | CY | 32.76 | 29.51 | 27.43 | 24.25 | 23.53 |
| 18 05 0402 | Seeding, Vegetative Cover | ACRE | 1,874 | 1,823 | 1,791 | 1,741 | 1,729 |
| 18 05 0409 | Fertilize, 800 Lbs/Acre, Push Rotary | ACRE | 140.58 | 118.67 | 105.24 | 83.82 | 78.56 |
| 18 05 0413 | Watering with 3,000-Gallon Tank Truck, per Pass | ACRE | 111.81 | 90.08 | 75.73 | 54.43 | 49.85 |
| 18 05 0415 | Mowing | ACRE | 64.67 | 51.13 | 43.38 | 30.18 | 26.60 |
| 33 07 0201 | 6" Inside Diameter (Vertical Pipe Spaced @ 200 LF), Gas Vent Piping System | LF | 28.29 | 24.75 | 22.57 | 19.10 | 18.26 |
| 33 08 0541 | 20 Mil Polymeric Liner, Very Low Density Polyethylene | SF | 1.25 | 1.02 | 0.91 | 0.69 | 0.61 |
| 33 08 0542 | 30 Mil Polymeric Liner, Very Low Density Polyethylene | SF | 1.98 | 1.59 | 1.41 | 1.03 | 0.90 |
| 33 08 0543 | 40 Mil Polymeric Liner, Very Low Density Polyethylene | SF | 2.38 | 1.91 | 1.70 | 1.24 | 1.09 |
| 33 08 0544 | 60 Mil Polymeric Liner, Very Low Density Polyethylene | SF | 3.54 | 2.83 | 2.51 | 1.83 | 1.60 |
| 33 08 0545 | 80 Mil Polymeric Liner, Very Low Density Polyethylene | SF | 4.35 | 3.49 | 3.10 | 2.27 | 1.98 |
| 33 08 0546 | 100 Mil Polymeric Liner, Very Low Density Polyethylene | SF | 5.50 | 4.41 | 3.91 | 2.85 | 2.49 |
| 33 08 0551 | 36 Mil Polymeric Liner, Chlorosulfanated Polyethylene | SF | 0.75 | 0.73 | 0.72 | 0.70 | 0.70 |
| 33 08 0552 | 45 Mil Polymeric Liner, Chlorosulfanated Polyethylene | SF | 0.83 | 0.81 | 0.80 | 0.79 | 0.78 |
| 33 08 0561 | 20 Mil Polymeric Liner, PVC | SF | 0.24 | 0.23 | 0.22 | 0.21 | 0.20 |
| 33 08 0562 | 30 Mil Polymeric Liner, PVC | SF | 0.32 | 0.31 | 0.30 | 0.29 | 0.28 |
| 33 08 0563 | 40 Mil Polymeric Liner, PVC | SF | 0.38 | 0.36 | 0.36 | 0.34 | 0.34 |
| 33 08 0564 | 60 Mil Polymeric Liner, PVC | SF | 0.72 | 0.70 | 0.69 | 0.67 | 0.67 |
| 33 08 0565 | 80 Mil Polymeric Liner, PVC | SF | 0.98 | 0.96 | 0.95 | 0.93 | 0.93 |
| 33 08 0566 | 100 Mil Polymeric Liner, PVC | SF | 1.28 | 1.26 | 1.25 | 1.23 | 1.22 |
| 33 08 0571 | 40 Mil Polymeric Liner, High-density Polyethylene | SF | 2.35 | 1.88 | 1.67 | 1.21 | 1.06 |
| 33 08 0572 | 60 Mil Polymeric Liner, High-density Polyethylene | SF | 3.49 | 2.80 | 2.48 | 1.81 | 1.58 |
| 33 08 0573 | 80 Mil Polymeric Liner, High-density Polyethylene | SF | 4.32 | 3.47 | 3.08 | 2.25 | 1.97 |
| 33 17 0801 | Decontaminate Light Equipment | EACH | 189.11 | 146.65 | 127.13 | 86.01 | 71.80 |
| 33 17 0802 | Decontaminate Medium Equipment | EACH | 378.22 | 293.30 | 254.27 | 172.03 | 143.59 |
| 33 17 0803 | Decontaminate Heavy Equipment | EACH | 560.32 | 434.51 | 376.69 | 254.86 | 212.73 |
| 33 23 0402 | 4" Filter Sock | LF | 5.57 | 4.65 | 3.93 | 3.02 | 2.90 |
| 33 23 0403 | 6" Filter Sock | LF | 5.70 | 4.78 | 4.06 | 3.15 | 3.03 |
| 33 23 0404 | 8" Filter Sock | LF | 6.35 | 5.43 | 4.71 | 3.80 | 3.68 |
| 33 23 1503 | 3,000 PSI, 6" Thick, Concrete Pad | CY | 85.80 | 82.00 | 80.29 | 76.60 | 75.31 |
| 33 23 2104 | 3/8", 1' Thick, Sodium Bentonite Pellets | CF | 275.35 | 229.42 | 193.32 | 147.97 | 141.88 |

**Environmental Remediation: Assemblies Cost Book**

*Page 3-42*

# Cost Data for Remediation

## Capping

| Assembly | Description | Unit | A | B | C | D | E |
|---|---|---|---|---|---|---|---|
| | **Operations & Maintenance** | | | | | | |
| 33 26 0512 | 4" High-density Polyethylene, Transfer Pipe | LF | 6.75 | 5.61 | 5.10 | 4.00 | 3.62 |
| 33 26 0513 | 6" High-density Polyethylene, Transfer Pipe | LF | 8.82 | 7.47 | 6.86 | 5.55 | 5.10 |
| 33 26 0514 | 8" High-density Polyethylene, Transfer Pipe | LF | 11.20 | 9.76 | 9.11 | 7.72 | 7.23 |
| 33 26 0516 | 12" High-density Polyethylene, Transfer Pipe | LF | 19.22 | 17.26 | 16.37 | 14.47 | 13.81 |
| 33 26 0518 | 16" High-density Polyethylene, Transfer Pipe | LF | 30.24 | 27.16 | 25.76 | 22.78 | 21.74 |
| 33 26 0520 | 24" High-density Polyethylene, Transfer Pipe | LF | 61.27 | 55.88 | 53.44 | 48.22 | 46.39 |
| 33 26 0802 | 4" Slotted PVC Pipe | LF | 8.67 | 7.65 | 7.22 | 6.23 | 5.87 |
| 33 26 0803 | 6" Slotted PVC Pipe | LF | 14.25 | 12.89 | 12.31 | 10.99 | 10.51 |
| 33 26 0804 | 8" Slotted PVC Pipe | LF | 22.19 | 20.15 | 19.28 | 17.31 | 16.59 |
| 33 26 0805 | 10" Slotted PVC Pipe | LF | 36.76 | 34.04 | 32.88 | 30.26 | 29.29 |
| 33 26 0806 | 12" Slotted PVC Pipe | LF | 26.50 | 22.42 | 20.68 | 16.74 | 15.30 |
| 33 27 0301 | 4" High-density Polyethylene, Tee | LF | 103.49 | 86.91 | 79.39 | 63.34 | 57.72 |
| 33 27 0302 | 6" High-density Polyethylene, Tee | LF | 165.88 | 146.28 | 137.40 | 118.43 | 111.78 |
| 33 27 0303 | 8" High-density Polyethylene, Tee | LF | 336.63 | 305.83 | 291.88 | 262.06 | 251.62 |
| 33 27 0304 | 12" High-density Polyethylene, Tee | LF | 599.95 | 546.05 | 521.64 | 469.46 | 451.19 |
| 33 27 0305 | 18" High-density Polyethylene, Tee | LF | 1,655 | 1,440 | 1,342 | 1,133 | 1,060 |
| 33 27 0306 | 24" High-density Polyethylene, Tee | LF | 2,720 | 2,434 | 2,304 | 2,026 | 1,929 |
| 33 27 0311 | 4", 90 Degree, High-density Polyethylene, Elbow | EACH | 94.59 | 79.19 | 72.22 | 57.31 | 52.09 |
| 33 27 0312 | 6", 90 Degree, High-density Polyethylene, Elbow | EACH | 150.00 | 132.04 | 123.90 | 106.51 | 100.42 |
| 33 27 0313 | 8", 90 Degree, High-density Polyethylene, Elbow | EACH | 253.57 | 229.62 | 218.77 | 195.58 | 187.46 |
| 33 27 0314 | 10", 90 Degree, High-density Polyethylene, Elbow | EACH | 405.00 | 369.07 | 352.79 | 318.01 | 305.83 |
| 33 27 0315 | 12", 90 Degree, High-density Polyethylene, Elbow | EACH | 629.04 | 581.13 | 559.43 | 513.06 | 496.81 |

**Environmental Remediation: Assemblies Cost Book**

# Cost Data for Remediation
## Carbon Adsorption - Gas

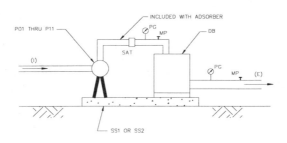

PROFILE

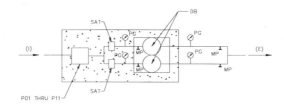

PLAN

VARIABLES

| | | | |
|---|---|---|---|
| SAMPLE PORT | = MP | PUMP SIZE | = P01 THRU P11 |
| CHECK VALVE | = CV | DUAL BED ADSORBER | = DB |
| PRESSURE GAUGE | = PG | 8" STRUCTURAL SLAB | = SS1 |
| THERMOSTAT | = TH | 12" STRUCTURAL SLAB | = SS2 |
| HUMIDITY CONTROL | | 25' X 6" FLEX S.S. HOSE | = FH |
| SATURATION INDICATOR | = SAT | INFLUENT PIPING | = I |
| | | EFFLUENT PIPING | = E |

### General:

*Treating waste streams by adsorption involves transferring and concentrating contaminants from one medium (gas) to another (the adsorbent). The most commonly used adsorbent is granular-activated carbon. In gas phase carbon adsorption, the contaminated gas comes in contact with the carbon by passing through one or more adsorbers which are usually the fixed bed type. When the adsorbent becomes saturated with organics (that is, when "breakthrough" occurs), the carbon must be removed, thermally regenerated or recycled, and replaced.*

### Typical Treatment Train:
*Air stripping, soil vapor extraction, adjustment of relative humidity by heating of gas stream, sampling and analysis.*

### Common Cost Components:

1. Piping (influent and effluent)
2. Adsorbers (dual bed units, modular -- disposable or permanent)
3. Structural slab
4. Blower system
5. Heater
6. Electrical usage
7. Operations and Maintenance

### Other Cost Considerations:
*Electrical distribution.*

# Cost Data for Remediation

## Carbon Adsorption - Gas

| Assembly | Description | Unit | A | B | C | D | E |
|---|---|---|---|---|---|---|---|
| | **Capital Costs** | | | | | | |
| 18 02 0322 | 8" Structural Slab on Grade | SF | 9.75 | 8.38 | 7.71 | 6.38 | 5.95 |
| 18 02 0324 | 12" Structural Slab on Grade | SF | 11.57 | 10.04 | 9.30 | 7.82 | 7.34 |
| 33 02 1501 | Saturation Indicator | EACH | 106.53 | 106.53 | 106.53 | 106.53 | 106.53 |
| 33 02 1502 | Thermostat & Humidity Control Devices | EACH | 329.49 | 279.97 | 258.78 | 210.91 | 193.37 |
| 33 02 1506 | Monitoring Port with Gas Monitor | EACH | 28.81 | 22.42 | 19.69 | 13.52 | 11.26 |
| 33 13 1901 | 50 CFM, 110 Lb Fill, Closed Upflow, 7.0" Pressure Drop | EACH | 618.14 | 583.98 | 569.35 | 536.32 | 524.22 |
| 33 13 1902 | 50 CFM, 110 Lb Fill, Closed Upflow, 7.0" Pressure Drop High-density Polyethylene | EACH | 634.04 | 599.88 | 585.25 | 552.22 | 540.12 |
| 33 13 1903 | 100 CFM, 150 Lb Fill, Closed Upflow, 5" Pressure Drop | EACH | 773.84 | 728.29 | 708.78 | 664.75 | 648.60 |
| 33 13 1904 | 100 CFM, 200 Lb Fill, Closed Upflow, 6.8" Pressure Drop | EACH | 887.79 | 842.24 | 822.73 | 778.70 | 762.55 |
| 33 13 1905 | 100 CFM, 200 Lb Fill, Closed Upflow, 6.8" Pressure Drop High-density Polyethylene | EACH | 887.79 | 842.24 | 822.73 | 778.70 | 762.55 |
| 33 13 1906 | 150 CFM, 300 Lb Fill, Closed Upflow, 7.9" Pressure Drop | EACH | 1,646 | 1,555 | 1,516 | 1,428 | 1,395 |
| 33 13 1907 | 250 CFM, 400 Lb Fill, Closed Upflow, 11.3" Pressure Drop | EACH | 2,091 | 1,954 | 1,896 | 1,764 | 1,715 |
| 33 13 1908 | 250 CFM, 400 Lb Fill, Closed Upflow, 11.3" Pressure Drop High-density Polyethylene | EACH | 2,979 | 2,842 | 2,784 | 2,651 | 2,603 |
| 33 13 1909 | 500 CFM, 1,200 Lb Fill, Closed Upflow, 8.5" Pressure Drop High-density Polyethylene | EACH | 5,260 | 5,099 | 5,022 | 4,866 | 4,814 |
| 33 13 1910 | 500 CFM, 1,400 Lb Fill, Closed Upflow, 11.5" Pressure Drop | EACH | 6,767 | 6,445 | 6,291 | 5,979 | 5,875 |
| 33 13 1911 | 500 CFM, 200 Lb Fill, Radial Flow, 4.8" Pressure Drop | EACH | 1,776 | 1,684 | 1,645 | 1,557 | 1,525 |
| 33 13 1912 | 750 CFM, 3,200 Lb Fill, Closed Upflow, 11.5" Pressure Drop | EACH | 14,894 | 14,573 | 14,419 | 14,106 | 14,002 |
| 33 13 1913 | 1,000 CFM, 400 Lb Fill, Radial Flow, 4.8" Pressure Drop | EACH | 3,233 | 3,096 | 3,038 | 2,906 | 2,857 |
| 33 13 1914 | 1,500 CFM, 5,700 Lb Fill, Closed Upflow, 11.5" Pressure Drop | EACH | 20,352 | 19,708 | 19,400 | 18,776 | 18,568 |
| 33 13 1915 | 1,500 CFM, 300 Lb Fill, Radial Flow, 4.8" Pressure Drop | EACH | 2,769 | 2,678 | 2,639 | 2,551 | 2,519 |
| 33 13 1916 | 3,000 CFM, 1,600 Lb Fill, Radial Flow, 4.8" Pressure Drop | EACH | 9,891 | 9,569 | 9,415 | 9,103 | 8,999 |
| 33 13 1917 | 5,000 CFM, 4,700 Lb Fill, Radial Flow | EACH | 20,187 | 19,865 | 19,711 | 19,398 | 19,294 |
| 33 13 1918 | 8,000 CFM, 6,300 Lb Fill, Radial Flow | EACH | 25,666 | 25,022 | 24,714 | 24,089 | 23,881 |
| 33 13 1919 | 2,200 CFM, 3,000 Lb Fill, 6 x 6 Closed Traverse 8" Pressure Drop | EACH | 26,255 | 25,933 | 25,779 | 25,467 | 25,363 |
| 33 13 1920 | 4,000 CFM, 5,300 Lb Fill, 8 x 8 Closed Traverse 8" Pressure Drop | EACH | 30,462 | 29,818 | 29,510 | 28,886 | 28,678 |
| 33 13 1921 | Dual Bed, 500 CFM Series/1,000 CFM Parallel, 2,000 Lb Fill | EACH | 9,852 | 9,530 | 9,376 | 9,063 | 8,959 |
| 33 13 1922 | Dual Bed, 1,000 CFM Series/2,000 CFM Parallel, 2,000 Lb Fill | EACH | 8,703 | 8,059 | 7,751 | 7,127 | 6,919 |

**Environmental Remediation: Assemblies Cost Book**

# Cost Data for Remediation

## Carbon Adsorption - Gas

| Assembly | Description | Unit | Unit Cost by Safety Level | | | | |
|---|---|---|---|---|---|---|---|
| | | | A | B | C | D | E |
| | **Capital Costs** | | | | | | |
| 33 13 1931 | 500 CFM Radial Unit Accumulator Cabinet | EACH | 1,581 | 1,420 | 1,344 | 1,187 | 1,135 |
| 33 13 1932 | 1,000 CFM & 1,500 CFM Radial Unit Accumulator Cabinet | EACH | 1,971 | 1,810 | 1,733 | 1,577 | 1,525 |
| 33 13 1933 | 3,000 CFM Radial Unit Accumulator Cabinet | EACH | 4,525 | 4,203 | 4,049 | 3,737 | 3,633 |
| 33 13 1941 | Coal-based, 4-mm Pellet, for Solvent Recovery < 2,000 Lb | LB | 1.78 | 1.78 | 1.78 | 1.78 | 1.78 |
| 33 13 1942 | Coal-based, 4 mm Pellet, for Solvent Recovery 2,000 - 10,000 Lb | LB | 1.25 | 1.25 | 1.25 | 1.25 | 1.25 |
| 33 13 1943 | Coal-based, 4 mm Pellet, for Solvent Recovery > 10,000 Lb | LB | 1.04 | 1.04 | 1.04 | 1.04 | 1.04 |
| 33 13 1944 | KOH Impregnated for H2S, Acidic Gas or Mercaptans, < 2,000 Lb | LB | 2.43 | 2.43 | 2.43 | 2.43 | 2.43 |
| 33 13 1945 | KOH Impregnated for H2S, Acidic Gas or Mercaptans, 2,000 - 10,000 Lb | LB | 2.17 | 2.17 | 2.17 | 2.17 | 2.17 |
| 33 13 1946 | KOH Impregnated for H2S, Acidic Gas or Mercaptans, >10,000 Lb | LB | 1.78 | 1.78 | 1.78 | 1.78 | 1.78 |
| 33 13 1947 | Coconut-based, 4 x 8 Sieve, General Purpose, < 2,000 Lb | LB | 1.78 | 1.78 | 1.78 | 1.78 | 1.78 |
| 33 13 1948 | Coconut-based, 4 x 8 Sieve, General Purpose, 2,000 - 10,000 Lb | LB | 1.20 | 1.20 | 1.20 | 1.20 | 1.20 |
| 33 13 1949 | Coconut-based, 4 x 8 Sieve, General Purpose, > 10,000 Lb | LB | 1.04 | 1.04 | 1.04 | 1.04 | 1.04 |
| 33 13 1950 | 25' x 6" Flexible Stainless Steel High-pressure Hose | EACH | 1,288 | 1,239 | 1,217 | 1,170 | 1,153 |
| 33 13 2058 | Reactivation or Thermal Regeneration of Carbon | LB | 0.05 | 0.05 | 0.05 | 0.05 | 0.05 |
| 33 13 2059 | Remove Carbon from Vessels, 10,000 - 20,000 Lb Minimum, Transport & Reactivate | LB | 0.02 | 0.02 | 0.02 | 0.02 | 0.02 |
| 33 19 0107 | Remove/Reinstall Carbon Adsorber Unit | EACH | 441.95 | 340.67 | 297.31 | 199.41 | 163.52 |
| 33 27 0404 | 4" Iron Body Check Valve | EACH | 354.29 | 315.96 | 299.54 | 262.49 | 248.90 |
| 33 31 0101 | 50 CFM, 7" Pressure, 3/4 HP, Blower System | EACH | 714.44 | 655.11 | 629.72 | 572.37 | 551.35 |
| 33 31 0102 | 100 CFM, 5" Pressure, 1/3 HP, Blower System | EACH | 2,481 | 2,417 | 2,390 | 2,329 | 2,306 |
| 33 31 0103 | 150 CFM, 8" Pressure, 3/4 HP, Blower System | EACH | 782.32 | 712.56 | 682.69 | 615.26 | 590.53 |
| 33 31 0104 | 250 CFM, 12" Pressure, 1 1/2 HP, Blower System | EACH | 908.42 | 827.30 | 792.57 | 714.16 | 685.41 |
| 33 31 0105 | 500 CFM, 5" Pressure, 1 HP, Blower System | EACH | 3,812 | 3,731 | 3,696 | 3,618 | 3,589 |
| 33 31 0106 | 500 CFM, 9" Pressure, 2 HP, Blower System | EACH | 1,056 | 964.35 | 925.05 | 836.32 | 803.79 |
| 33 31 0107 | 500 CFM, 12" Pressure, 5 HP, Blower System | EACH | 2,023 | 1,849 | 1,774 | 1,606 | 1,544 |
| 33 31 0108 | 750 CFM, 12" Pressure, 5 HP, Blower System | EACH | 1,509 | 1,335 | 1,260 | 1,091 | 1,029 |
| 33 31 0109 | 1,000 CFM, 5" Pressure, 1 1/2 HP, Blower System | EACH | 1,007 | 926.01 | 891.28 | 812.87 | 784.12 |
| 33 31 0112 | 1,500 CFM, 5" Pressure, 3 HP, Blower System | EACH | 1,220 | 1,104 | 1,054 | 941.45 | 900.24 |
| 33 31 0113 | 1,500 CFM, 12" Pressure, 10 HP, Blower System | EACH | 2,756 | 2,488 | 2,373 | 2,114 | 2,019 |
| 33 31 0114 | 2,200 CFM, 8" Pressure, 7 1/2 HP, Blower System | EACH | 2,690 | 2,422 | 2,307 | 2,047 | 1,952 |
| 33 31 0116 | 3,000 CFM, 5" Pressure, 7 1/2 HP, Blower System | EACH | 2,873 | 2,605 | 2,490 | 2,231 | 2,135 |

**Environmental Remediation: Assemblies Cost Book**

# Cost Data for Remediation

## Carbon Adsorption - Gas

| Assembly | Description | Unit | Unit Cost by Safety Level | | | | |
|---|---|---|---|---|---|---|---|
| | | | A | B | C | D | E |
| | **Capital Costs** | | | | | | |
| 33 31 0117 | 4,000 CFM, 8" Pressure, 15 HP, Blower System | EACH | 3,005 | 2,778 | 2,680 | 2,460 | 2,379 |
| 33 31 0120 | 5,000 CFM, 8" Pressure, 20 HP, Blower System | EACH | 4,245 | 3,965 | 3,845 | 3,574 | 3,475 |
| 33 31 0122 | 8,000 CFM, 8" Pressure, 25 HP, Blower System | EACH | 3,964 | 3,761 | 3,675 | 3,479 | 3,407 |
| 33 31 0209 | Pressure Gauge | EACH | 140.16 | 121.61 | 113.67 | 95.74 | 89.16 |
| 33 31 0301 | 7.5 KW, 25,600 BTU, Hazardous Air Heater | EACH | 6,192 | 5,861 | 5,718 | 5,398 | 5,281 |
| 33 31 0302 | 10 KW, 34,150 BTU, Hazardous Air Heater | EACH | 6,404 | 6,073 | 5,930 | 5,610 | 5,493 |
| 33 31 0303 | 15 KW, 51,200 BTU, Hazardous Air Heater | EACH | 6,713 | 6,360 | 6,207 | 5,865 | 5,741 |
| 33 31 0304 | 20 KW, 68,300 BTU, Hazardous Air Heater | EACH | 6,986 | 6,632 | 6,480 | 6,138 | 6,014 |
| 33 41 0201 | Blower and Motor Maintenance and Repair | EACH | 865.17 | 671.48 | 581.58 | 393.96 | 329.64 |
| 33 42 0101 | Electrical Charge | KWH | 0.06 | 0.06 | 0.06 | 0.06 | 0.06 |
| | **Operations & Maintenance** | | | | | | |
| 33 13 1901 | 50 CFM, 110 Lb Fill, Closed Upflow, 7.0" Pressure Drop | EACH | 618.14 | 583.98 | 569.35 | 536.32 | 524.22 |
| 33 13 1902 | 50 CFM, 110 Lb Fill, Closed Upflow, 7.0" Pressure Drop High-density Polyethylene | EACH | 634.04 | 599.88 | 585.25 | 552.22 | 540.12 |
| 33 13 1903 | 100 CFM, 150 Lb Fill, Closed Upflow, 5" Pressure Drop | EACH | 773.84 | 728.29 | 708.78 | 664.75 | 648.60 |
| 33 13 1904 | 100 CFM, 200 Lb Fill, Closed Upflow, 6.8" Pressure Drop | EACH | 887.79 | 842.24 | 822.73 | 778.70 | 762.55 |
| 33 13 1905 | 100 CFM, 200 Lb Fill, Closed Upflow, 6.8" Pressure Drop High-density Polyethylene | EACH | 887.79 | 842.24 | 822.73 | 778.70 | 762.55 |
| 33 13 1906 | 150 CFM, 300 Lb Fill, Closed Upflow, 7.9" Pressure Drop | EACH | 1,646 | 1,555 | 1,516 | 1,428 | 1,395 |
| 33 13 1907 | 250 CFM, 400 Lb Fill, Closed Upflow, 11.3" Pressure Drop | EACH | 2,091 | 1,954 | 1,896 | 1,764 | 1,715 |
| 33 13 1908 | 250 CFM, 400 Lb Fill, Closed Upflow, 11.3" Pressure Drop High-density Polyethylene | EACH | 2,979 | 2,842 | 2,784 | 2,651 | 2,603 |
| 33 13 1909 | 500 CFM, 1,200 Lb Fill, Closed Upflow, 8.5" Pressure Drop High-density Polyethylene | EACH | 5,260 | 5,099 | 5,022 | 4,866 | 4,814 |
| 33 13 1910 | 500 CFM, 1,400 Lb Fill, Closed Upflow, 11.5" Pressure Drop | EACH | 6,767 | 6,445 | 6,291 | 5,979 | 5,875 |
| 33 13 1911 | 500 CFM, 200 Lb Fill, Radial Flow, 4.8" Pressure Drop | EACH | 1,776 | 1,684 | 1,645 | 1,557 | 1,525 |
| 33 13 1912 | 750 CFM, 3,200 Lb Fill, Closed Upflow, 11.5" Pressure Drop | EACH | 14,894 | 14,573 | 14,419 | 14,106 | 14,002 |
| 33 13 1913 | 1,000 CFM, 400 Lb Fill, Radial Flow, 4.8" Pressure Drop | EACH | 3,233 | 3,096 | 3,038 | 2,906 | 2,857 |
| 33 13 1914 | 1,500 CFM, 5,700 Lb Fill, Closed Upflow, 11.5" Pressure Drop | EACH | 20,352 | 19,708 | 19,400 | 18,776 | 18,568 |
| 33 13 1915 | 1,500 CFM, 300 Lb Fill, Radial Flow, 4.8" Pressure Drop | EACH | 2,769 | 2,678 | 2,639 | 2,551 | 2,519 |
| 33 13 1916 | 3,000 CFM, 1,600 Lb Fill, Radial Flow, 4.8" Pressure Drop | EACH | 9,891 | 9,569 | 9,415 | 9,103 | 8,999 |
| 33 13 1917 | 5,000 CFM, 4,700 Lb Fill, Radial Flow | EACH | 20,187 | 19,865 | 19,711 | 19,398 | 19,294 |
| 33 13 1918 | 8,000 CFM, 6,300 Lb Fill, Radial Flow | EACH | 25,666 | 25,022 | 24,714 | 24,089 | 23,881 |

**Environmental Remediation: Assemblies Cost Book**

*Page 3-47*

# Cost Data for Remediation

## Carbon Adsorption - Gas

| Assembly | Description | Unit | Unit Cost by Safety Level | | | | |
|---|---|---|---|---|---|---|---|
| | | | A | B | C | D | E |
| | **Operations & Maintenance** | | | | | | |
| 33 13 1919 | 2,200 CFM, 3,000 Lb Fill, 6 x 6 Closed Traverse 8" Pressure Drop | EACH | 26,255 | 25,933 | 25,779 | 25,467 | 25,363 |
| 33 13 1920 | 4,000 CFM, 5,300 Lb Fill, 8 x 8 Closed Traverse 8" Pressure Drop | EACH | 30,462 | 29,818 | 29,510 | 28,886 | 28,678 |
| 33 13 1941 | Coal-based, 4-mm Pellet, for Solvent Recovery < 2,000 Lb | LB | 1.78 | 1.78 | 1.78 | 1.78 | 1.78 |
| 33 13 1942 | Coal-based, 4 mm Pellet, for Solvent Recovery 2,000 - 10,000 Lb | LB | 1.25 | 1.25 | 1.25 | 1.25 | 1.25 |
| 33 13 1943 | Coal-based, 4 mm Pellet, for Solvent Recovery > 10,000 Lb | LB | 1.04 | 1.04 | 1.04 | 1.04 | 1.04 |
| 33 13 1944 | KOH Impregnated for H2S, Acidic Gas or Mercaptans, < 2,000 Lb | LB | 2.43 | 2.43 | 2.43 | 2.43 | 2.43 |
| 33 13 1945 | KOH Impregnated for H2S, Acidic Gas or Mercaptans, 2,000 - 10,000 Lb | LB | 2.17 | 2.17 | 2.17 | 2.17 | 2.17 |
| 33 13 1946 | KOH Impregnated for H2S, Acidic Gas or Mercaptans, >10,000 Lb | LB | 1.78 | 1.78 | 1.78 | 1.78 | 1.78 |
| 33 13 1947 | Coconut-based, 4 x 8 Sieve, General Purpose, < 2,000 Lb | LB | 1.78 | 1.78 | 1.78 | 1.78 | 1.78 |
| 33 13 1948 | Coconut-based, 4 x 8 Sieve, General Purpose, 2,000 - 10,000 Lb | LB | 1.20 | 1.20 | 1.20 | 1.20 | 1.20 |
| 33 13 1949 | Coconut-based, 4 x 8 Sieve, General Purpose, > 10,000 Lb | LB | 1.04 | 1.04 | 1.04 | 1.04 | 1.04 |
| 33 13 2058 | Reactivation or Thermal Regeneration of Carbon | LB | 0.05 | 0.05 | 0.05 | 0.05 | 0.05 |
| 33 13 2059 | Remove Carbon from Vessels, 10,000 - 20,000 Lb Minimum, Transport & Reactivate | LB | 0.02 | 0.02 | 0.02 | 0.02 | 0.02 |
| 33 19 0107 | Remove/Reinstall Carbon Adsorber Unit | EACH | 441.95 | 340.67 | 297.31 | 199.41 | 163.52 |
| 33 41 0201 | Blower and Motor Maintenance and Repair | EACH | 865.17 | 671.48 | 581.58 | 393.96 | 329.64 |
| 33 42 0101 | Electrical Charge | KWH | 0.06 | 0.06 | 0.06 | 0.06 | 0.06 |

# Cost Data for Remediation
## Carbon Adsorption - Liquid

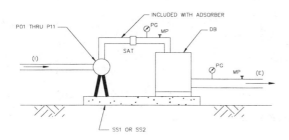

PROFILE

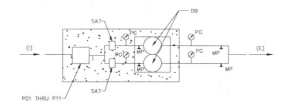

PLAN

### VARIABLES

| | | | |
|---|---|---|---|
| SAMPLE PORT | = MP | PUMP SIZE | = P01 THRU P11 |
| CHECK VALVE | = CV | DUAL BED ADSORBER | = DB |
| PRESSURE GAUGE | = PG | 8" STRUCTURAL SLAB | = SS1 |
| THERMOSTAT | = TH | 12" STRUCTURAL SLAB | = SS2 |
| HUMIDITY CONTROL | | 25' X 6" FLEX S.S. HOSE | = FH |
| SATURATION INDICATOR | = SAT | INFLUENT PIPING | = I |
| | | EFFLUENT PIPING | = E |

### General:

Most liquid-phase adsorption treatment applications involve the use of granular activated carbon adsorbers which operate in a downflow series mode, either by gravity or under pressure. The effluent from a gravity adsorber may require pumping if a pressure water system is designed. The most critical operating parameter is the time it takes for breakthrough (the point when the adsorbent becomes saturated with organics). The carbon must be removed, thermally regenerated or recycled, and replaced. Other considerations include backwashing the adsorber if suspended solids are present in the influent and connecting the adsorbers in parallel to provide increased hydraulic capacity.

### Typical Treatment Train:

Groundwater recovery, sedimentation, filtration, metals removal, oil removal, air stripping, pH adjustment, sampling and analysis.

### Common Cost Components:

1. Piping (influent and effluent)
2. Adsorbers (dual bed units, modular -- disposable or permanent)
3. Pumps
4. Structural slab
5. Saturation indicator
6. Electrical usage
7. Operations and Maintenance

### Other Cost Considerations:

Electrical distribution, fencing.

# Cost Data for Remediation

## Carbon Adsorption - Liquid

| Assembly | Description | Unit | Unit Cost by Safety Level A | B | C | D | E |
|---|---|---|---|---|---|---|---|
| | **Capital Costs** | | | | | | |
| 18 02 0322 | 8" Structural Slab on Grade | SF | 9.75 | 8.38 | 7.71 | 6.38 | 5.95 |
| 18 02 0324 | 12" Structural Slab on Grade | SF | 11.57 | 10.04 | 9.30 | 7.82 | 7.34 |
| 33 02 1501 | Saturation Indicator | EACH | 106.53 | 106.53 | 106.53 | 106.53 | 106.53 |
| 33 13 2001 | 5 GPM, 85 Lb Fill, DOT 5B Drum, Disposable | EACH | 640.61 | 606.19 | 591.46 | 558.19 | 546.00 |
| 33 13 2002 | 5 GPM, 85 Lb Fill, High-density Polyethylene DOT Spec 34, Disposable | EACH | 634.25 | 599.83 | 585.10 | 551.83 | 539.64 |
| 33 13 2003 | 15 GPM, 165 Lb Fill, DOT 5B Drum, Disposable | EACH | 877.39 | 831.51 | 811.86 | 767.51 | 751.25 |
| 33 13 2004 | 15 GPM, 165 Lb Fill, High-density Polyethylene DOT Spec 34, Disposable | EACH | 877.39 | 831.51 | 811.86 | 767.51 | 751.25 |
| 33 13 2005 | 20 GPM, 250 Lb Fill, Disposable | EACH | 1,424 | 1,333 | 1,293 | 1,205 | 1,172 |
| 33 13 2006 | 25 GPM, 330 Lb Fill, Disposable | EACH | 2,198 | 2,060 | 2,002 | 1,868 | 1,820 |
| 33 13 2007 | 25 GPM, 330 Lb Fill, High-density Polyethylene, Disposable | EACH | 3,025 | 2,888 | 2,829 | 2,696 | 2,647 |
| 33 13 2008 | 25 GPM, 330 Lb Fill, High-density Polyethylene, Permanent | EACH | 3,025 | 2,888 | 2,829 | 2,696 | 2,647 |
| 33 13 2009 | 25 GPM, 330 Lb Fill, High-density Polyethylene-lined Steel, Permanent | EACH | 2,922 | 2,863 | 2,832 | 2,775 | 2,758 |
| 33 13 2010 | 25 GPM, 330 Lb Fill, 316L Stainless Steel, Permanent | EACH | 2,922 | 2,863 | 2,832 | 2,775 | 2,758 |
| 33 13 2011 | 35 GPM, 1,050 Lb Fill, Disposable | EACH | 3,846 | 3,699 | 3,623 | 3,480 | 3,436 |
| 33 13 2012 | 35 GPM, 660 Lb Fill, High-density Polyethylene-lined Steel, Permanent | EACH | 5,277 | 5,130 | 5,054 | 4,911 | 4,867 |
| 33 13 2013 | 35 GPM, 660 Lb Fill, 316L Stainless Steel, Permanent | EACH | 5,449 | 5,302 | 5,226 | 5,083 | 5,040 |
| 33 13 2014 | 50 GPM, 1,650 Lb Fill, Disposable | EACH | 5,331 | 5,135 | 5,034 | 4,843 | 4,785 |
| 33 13 2015 | 50 GPM, 1,050 Lb Fill, High-density Polyethylene Disposable | EACH | 5,503 | 5,308 | 5,206 | 5,016 | 4,957 |
| 33 13 2016 | 50 GPM, 1,050 Lb Fill, High-density Polyethylene Permanent | EACH | 5,503 | 5,308 | 5,206 | 5,016 | 4,957 |
| 33 13 2017 | 50 GPM, 880 Lb Fill, High-density Polyethylene-lined Steel, Permanent | EACH | 6,948 | 6,752 | 6,650 | 6,460 | 6,402 |
| 33 13 2018 | 50 GPM, 880 Lb Fill, 316L Stainless Steel, Permanent | EACH | 7,266 | 7,070 | 6,968 | 6,778 | 6,720 |
| 33 13 2019 | 75 GPM, 1,650 Lb Fill, High-density Polyethylene-lined Steel, Permanent | EACH | 10,517 | 10,223 | 10,071 | 9,785 | 9,698 |
| 33 13 2020 | 75 GPM, 1,650 Lb Fill, 316L Stainless Steel, Permanent | EACH | 10,899 | 10,605 | 10,452 | 10,167 | 10,079 |
| 33 13 2021 | 100 GPM, 3,000 Lb Fill, Disposable | EACH | 9,600 | 9,306 | 9,154 | 8,868 | 8,781 |
| 33 13 2022 | 100 GPM, 3,000 Lb Fill, High-density Polyethylene-lined Steel, Permanent | EACH | 13,416 | 13,122 | 12,970 | 12,684 | 12,597 |
| 33 13 2023 | 100 GPM, 3,000 Lb Fill, 316L Stainless Steel, Permanent | EACH | 14,609 | 14,315 | 14,162 | 13,877 | 13,789 |
| 33 13 2024 | 200 GPM, 6,000 Lb Fill, Disposable | EACH | 17,531 | 16,943 | 16,638 | 16,067 | 15,892 |
| 33 13 2025 | 200 GPM, 6,000 Lb Fill, High-density Polyethylene-lined Steel, Permanent | EACH | 21,996 | 21,408 | 21,103 | 20,532 | 20,357 |

**Environmental Remediation: Assemblies Cost Book**

# Cost Data for Remediation

## Carbon Adsorption - Liquid

| Assembly | Description | Unit | Unit Cost by Safety Level | | | | |
|---|---|---|---|---|---|---|---|
| | | | A | B | C | D | E |
| | **Capital Costs** | | | | | | |
| 33 13 2026 | 200 GPM, 6,000 Lb Fill, 316L Stainless Steel, Permanent | EACH | 24,050 | 23,462 | 23,157 | 22,586 | 22,411 |
| 33 13 2027 | Dual Bed, 50 GPM Series, 100 GPM Parallel, 1,760 Lb Fill Each | EACH | 15,437 | 14,850 | 14,544 | 13,973 | 13,799 |
| 33 13 2028 | Dual Bed, 75 GPM Series, 150 GPM Parallel, 3,300 Lb Fill Each | EACH | 21,797 | 21,210 | 20,904 | 20,333 | 20,159 |
| 33 13 2029 | Dual Bed, 2 - 4' Diameter, 65 GPM Series, 130 GPM Parallel, 2,000 Lb Each | EACH | 51,577 | 47,071 | 44,731 | 40,353 | 39,011 |
| 33 13 2030 | Dual Bed, 2 - 7.5' Diameter, 175 GPM Series, 350 GPM Parallel, 10,000 Lb Each | EACH | 117,006 | 110,786 | 107,712 | 101,676 | 99,726 |
| 33 13 2031 | Dual Bed, 2 - 10' Diameter, 350 GPM Series, 700 GPM Parallel, 20,000 Lb Each | EACH | 170,779 | 163,916 | 160,659 | 154,007 | 151,773 |
| 33 13 2032 | Dual Bed, 2 - 12' Diameter, 350 GPM Series, 700 GPM Parallel, 20,000 Lb Each | EACH | 181,026 | 174,163 | 170,906 | 164,254 | 162,019 |
| 33 13 2041 | Prefilter/Postfilter Housing & Cartridge to 20 GPM | EACH | 421.57 | 384.85 | 365.78 | 330.09 | 319.16 |
| 33 13 2042 | Replacement High-capacity Prefilter Cartridges | EACH | 37.10 | 37.10 | 37.10 | 37.10 | 37.10 |
| 33 13 2051 | Coal-based General Purpose, 8 x 30 Sieve, 900 Iodine, < 2,000 Lb | LB | 1.30 | 1.30 | 1.30 | 1.30 | 1.30 |
| 33 13 2052 | Coal-based General Purpose, 8 x 30 Sieve, 900 Iodine, 2,000 - 10,000 Lb | LB | 1.01 | 1.01 | 1.01 | 1.01 | 1.01 |
| 33 13 2053 | Coal-based General Purpose, 8 x 30 Sieve, 900 Iodine, > 10,000 Lb | LB | 0.77 | 0.77 | 0.77 | 0.77 | 0.77 |
| 33 13 2054 | Coconut-based, High Capacity, 12 x 30 Sieve, 1100 Iodine, < 2,000 Lb | LB | 1.30 | 1.30 | 1.30 | 1.30 | 1.30 |
| 33 13 2055 | Coconut-based, High Capacity, 12 x 30 Sieve, 1100 Iodine, 2,000 - 10,000 Lb | LB | 1.01 | 1.01 | 1.01 | 1.01 | 1.01 |
| 33 13 2056 | Coconut-based, High Capacity, 12 x 30 Sieve, 1100 Iodine, > 10,000 Lb | LB | 0.77 | 0.77 | 0.77 | 0.77 | 0.77 |
| 33 13 2057 | Activated Aluminas for Highly Oxidizable Contaminants | LB | 0.67 | 0.67 | 0.67 | 0.67 | 0.67 |
| 33 13 2058 | Reactivation or Thermal Regeneration of Carbon | LB | 0.05 | 0.05 | 0.05 | 0.05 | 0.05 |
| 33 13 2059 | Remove Carbon from Vessels, 10,000 - 20,000 Lb Minimum, Transport & Reactivate | LB | 0.02 | 0.02 | 0.02 | 0.02 | 0.02 |
| 33 19 0107 | Remove/Reinstall Carbon Adsorber Unit | EACH | 441.95 | 340.67 | 297.31 | 199.41 | 163.52 |
| 33 29 0117 | 15 GPM, 1/2 HP, Transfer Pump with Motor, Valves, Piping | EACH | 1,588 | 1,399 | 1,319 | 1,136 | 1,069 |
| 33 29 0118 | 20 GPM, 1/2 HP, Transfer Pump with Motor, Valves, Piping | EACH | 1,773 | 1,556 | 1,464 | 1,255 | 1,178 |
| 33 29 0119 | 25 GPM, 1 HP, Transfer Pump with Motor, Valves, Piping | EACH | 2,125 | 1,873 | 1,765 | 1,522 | 1,433 |
| 33 29 0120 | 35 GPM, 1 HP, Transfer Pump with Motor, Valves, Piping | EACH | 2,381 | 2,096 | 1,974 | 1,698 | 1,596 |
| 33 29 0121 | 50 GPM, 1.5 HP, Transfer Pump with Motor, Valves, Piping | EACH | 3,393 | 2,979 | 2,802 | 2,402 | 2,256 |
| 33 29 0122 | 75 GPM, 3 HP, Transfer Pump with Motor, Valves, Piping | EACH | 5,178 | 4,622 | 4,383 | 3,846 | 3,648 |
| 33 29 0123 | 100 GPM, 5 HP, Transfer Pump with Motor, Valves, Piping | EACH | 4,666 | 4,210 | 4,014 | 3,573 | 3,411 |

**Environmental Remediation: Assemblies Cost Book**

# Cost Data for Remediation

## Carbon Adsorption - Liquid

| Assembly | | | Description | Unit | A | B | C | D | E |
|---|---|---|---|---|---|---|---|---|---|
| | | | **Capital Costs** | | | | | | |
| 33 | 29 | 0124 | 150 GPM, 5 HP, Transfer Pump with Motor, Valves, Piping | EACH | 6,949 | 6,048 | 5,663 | 4,792 | 4,473 |
| 33 | 29 | 0125 | 200 GPM, 10 HP, Transfer Pump with Motor, Valves, Piping | EACH | 5,710 | 5,129 | 4,880 | 4,318 | 4,112 |
| 33 | 29 | 0126 | 350 GPM, 10 HP, Transfer Pump with Motor, Valves, Piping | EACH | 8,286 | 7,280 | 6,849 | 5,876 | 5,519 |
| 33 | 29 | 0127 | 700 GPM, 30 HP, Transfer Pump with Motor, Valves, Piping | EACH | 10,437 | 9,267 | 8,752 | 7,620 | 7,214 |
| 33 | 41 | 0101 | Pump & Motor Maintenance/Repair | EACH | 865.17 | 671.48 | 581.58 | 393.96 | 329.64 |
| 33 | 42 | 0101 | Electrical Charge | KWH | 0.06 | 0.06 | 0.06 | 0.06 | 0.06 |
| | | | **Operations & Maintenance** | | | | | | |
| 33 | 13 | 2001 | 5 GPM, 85 Lb Fill, DOT 5B Drum, Disposable | EACH | 640.61 | 606.19 | 591.46 | 558.19 | 546.00 |
| 33 | 13 | 2002 | 5 GPM, 85 Lb Fill, High-density Polyethylene DOT Spec 34, Disposable | EACH | 634.25 | 599.83 | 585.10 | 551.83 | 539.64 |
| 33 | 13 | 2003 | 15 GPM, 165 Lb Fill, DOT 5B Drum, Disposable | EACH | 877.39 | 831.51 | 811.86 | 767.51 | 751.25 |
| 33 | 13 | 2004 | 15 GPM, 165 Lb Fill, High-density Polyethylene DOT Spec 34, Disposable | EACH | 877.39 | 831.51 | 811.86 | 767.51 | 751.25 |
| 33 | 13 | 2005 | 20 GPM, 250 Lb Fill, Disposable | EACH | 1,424 | 1,333 | 1,293 | 1,205 | 1,172 |
| 33 | 13 | 2006 | 25 GPM, 330 Lb Fill, Disposable | EACH | 2,198 | 2,060 | 2,002 | 1,868 | 1,820 |
| 33 | 13 | 2007 | 25 GPM, 330 Lb Fill, High-density Polyethylene, Disposable | EACH | 3,025 | 2,888 | 2,829 | 2,696 | 2,647 |
| 33 | 13 | 2011 | 35 GPM, 1,050 Lb Fill, Disposable | EACH | 3,846 | 3,699 | 3,623 | 3,480 | 3,436 |
| 33 | 13 | 2014 | 50 GPM, 1,650 Lb Fill, Disposable | EACH | 5,331 | 5,135 | 5,034 | 4,843 | 4,785 |
| 33 | 13 | 2015 | 50 GPM, 1,050 Lb Fill, High-density Polyethylene Disposable | EACH | 5,503 | 5,308 | 5,206 | 5,016 | 4,957 |
| 33 | 13 | 2021 | 100 GPM, 3,000 Lb Fill, Disposable | EACH | 9,600 | 9,306 | 9,154 | 8,868 | 8,781 |
| 33 | 13 | 2024 | 200 GPM, 6,000 Lb Fill, Disposable | EACH | 17,531 | 16,943 | 16,638 | 16,067 | 15,892 |
| 33 | 13 | 2051 | Coal-based General Purpose, 8 x 30 Sieve, 900 Iodine, < 2,000 Lb | LB | 1.30 | 1.30 | 1.30 | 1.30 | 1.30 |
| 33 | 13 | 2052 | Coal-based General Purpose, 8 x 30 Sieve, 900 Iodine, 2,000 - 10,000 Lb | LB | 1.01 | 1.01 | 1.01 | 1.01 | 1.01 |
| 33 | 13 | 2053 | Coal-based General Purpose, 8 x 30 Sieve, 900 Iodine, > 10,000 Lb | LB | 0.77 | 0.77 | 0.77 | 0.77 | 0.77 |
| 33 | 13 | 2054 | Coconut-based, High Capacity, 12 x 30 Sieve, 1100 Iodine, < 2,000 Lb | LB | 1.30 | 1.30 | 1.30 | 1.30 | 1.30 |
| 33 | 13 | 2055 | Coconut-based, High Capacity, 12 x 30 Sieve, 1100 Iodine, 2,000 - 10,000 Lb | LB | 1.01 | 1.01 | 1.01 | 1.01 | 1.01 |
| 33 | 13 | 2056 | Coconut-based, High Capacity, 12 x 30 Sieve, 1100 Iodine, > 10,000 Lb | LB | 0.77 | 0.77 | 0.77 | 0.77 | 0.77 |
| 33 | 13 | 2057 | Activated Aluminas for Highly Oxidizable Contaminants | LB | 0.67 | 0.67 | 0.67 | 0.67 | 0.67 |
| 33 | 13 | 2058 | Reactivation or Thermal Regeneration of Carbon | LB | 0.05 | 0.05 | 0.05 | 0.05 | 0.05 |
| 33 | 13 | 2059 | Remove Carbon from Vessels, 10,000 - 20,000 Lb Minimum, Transport & Reactivate | LB | 0.02 | 0.02 | 0.02 | 0.02 | 0.02 |

**Environmental Remediation: Assemblies Cost Book**

# Cost Data for Remediation

## Carbon Adsorption - Liquid

| Assembly | Description | Unit | Unit Cost by Safety Level | | | | |
|---|---|---|---|---|---|---|---|
| | | | A | B | C | D | E |
| | **Operations & Maintenance** | | | | | | |
| 33 19 0107 | Remove/Reinstall Carbon Adsorber Unit | EACH | 441.95 | 340.67 | 297.31 | 199.41 | 163.52 |
| 33 41 0101 | Pump & Motor Maintenance/Repair | EACH | 865.17 | 671.48 | 581.58 | 393.96 | 329.64 |
| 33 42 0101 | Electrical Charge | KWH | 0.06 | 0.06 | 0.06 | 0.06 | 0.06 |

# Cost Data for Remediation
## Chemical Precipitation

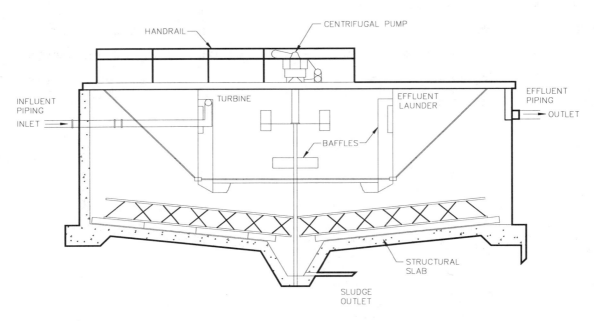

Incline Slope Circular Rake Clarifier

### General:

The precipitation process is typically used to treat aqueous waste streams that have high metals concentrations. Metals are removed from the solution by shifting the chemical equilibrium in a direction which reduces their solubility. The process must have a proper chemical addition to insure that all metal concentrations are reduced to a minimum.

### Typical Treatment Train:

Groundwater recovery, media filtration, and dewatering, transporting, and disposing of treatment residuals. In addition, depending on the required treatment standards, the water effluent stream may require polishing through ion exchange, reverse osmosis, or some other finishing method.

### Common Cost Components:

1. Clarifier
2. Pumps
3. System reagents (primary, secondary
4. pH adjustment (influent, effluent
5. Holding tanks
6. Structural slab
7. Electrical usage
8. Piping
9. Operations and Maintenance

### Other Cost Considerations:

Electrical distribution, fencing.

# Cost Data for Remediation

## Chemical Precipitation

| Assembly | Description | Unit | A | B | C | D | E |
|---|---|---|---|---|---|---|---|
| | **Capital Costs** | | | | | | |
| 18 02 0322 | 8" Structural Slab on Grade | SF | 9.75 | 8.38 | 7.71 | 6.38 | 5.95 |
| 18 02 0324 | 12" Structural Slab on Grade | SF | 11.57 | 10.04 | 9.30 | 7.82 | 7.34 |
| 19 01 0202 | 1", Class 200, PVC Piping | LF | 7.48 | 5.84 | 5.07 | 3.48 | 2.93 |
| 19 01 0204 | 2", Class 200, PVC Piping | LF | 8.86 | 6.96 | 6.07 | 4.23 | 3.60 |
| 19 01 0206 | 3", Class 200, PVC Piping | LF | 11.81 | 9.34 | 8.20 | 5.80 | 4.98 |
| 19 01 0207 | 4", Class 200, PVC Piping | LF | 12.33 | 9.86 | 8.72 | 6.32 | 5.50 |
| 19 01 0208 | 6", Class 200, PVC Piping | LF | 14.95 | 12.20 | 10.93 | 8.27 | 7.36 |
| 19 01 0212 | 8", Class 150, PVC Piping | LF | 18.36 | 15.27 | 13.84 | 10.85 | 9.82 |
| 19 01 0214 | 12", Class 150, PVC Piping | LF | 28.75 | 24.63 | 22.72 | 18.73 | 17.36 |
| 19 04 0401 | 550 Gallon, Stainless Steel Aboveground Wastewater Holding Tank, Rental | MONTH | 320.65 | 320.65 | 320.65 | 320.65 | 320.65 |
| 19 04 0402 | 630 Gallon, Polyethylene Aboveground Wastewater Holding Tank, Rental | MONTH | 320.65 | 320.65 | 320.65 | 320.65 | 320.65 |
| 19 04 0403 | 4,000 Gallon Polyethylene Wastewater Tank, Rental | MONTH | 545.90 | 545.90 | 545.90 | 545.90 | 545.90 |
| 19 04 0404 | 4,000 Gallon, Polyethylene Trailer-mounted Wastewater Holding Tank, Rental | MONTH | 1,124 | 1,124 | 1,124 | 1,124 | 1,124 |
| 19 04 0405 | 6,000 Gallon, Polyethylene Aboveground Wastewater Holding Tank, Rental | MONTH | 641.30 | 641.30 | 641.30 | 641.30 | 641.30 |
| 19 04 0406 | 21,000 Gallon Steel Wastewater Holding Tank, Rental | MONTH | 1,219 | 1,219 | 1,219 | 1,219 | 1,219 |
| 19 04 0408 | 21,000 Gallon Steel, Open Top, Tank Rental | MONTH | 1,124 | 1,124 | 1,124 | 1,124 | 1,124 |
| 19 04 0418 | 55 Gallon Nalgene Horizontal XLPE Tank without legs | EACH | 295.00 | 295.00 | 295.00 | 295.00 | 295.00 |
| 19 04 0419 | 110 Gallon Nalgene Horizontal XLPE Tank without legs | EACH | 410.00 | 410.00 | 410.00 | 410.00 | 410.00 |
| 19 04 0420 | 200 Gallon Nalgene Horizontal XLPE Tank without legs | EACH | 510.00 | 510.00 | 510.00 | 510.00 | 510.00 |
| 19 04 0421 | 300 Gallon Nalgene Horizontal XLPE Tank without legs | EACH | 555.00 | 555.00 | 555.00 | 555.00 | 555.00 |
| 19 04 0422 | 500 Gallon Nalgene Horizontal XLPE Tank without legs | EACH | 695.00 | 695.00 | 695.00 | 695.00 | 695.00 |
| 19 04 0423 | 1,000 Gallon Nalgene Horizontal XLPE Tank without legs | EACH | 1,495 | 1,495 | 1,495 | 1,495 | 1,495 |
| 19 04 0424 | 1,650 Gallon Nalgene Horizontal XLPE Tank without legs | EACH | 1,995 | 1,995 | 1,995 | 1,995 | 1,995 |
| 19 04 0425 | 2,500 Gallon Nalgene Horizontal XLPE Tank without legs | EACH | 3,250 | 3,250 | 3,250 | 3,250 | 3,250 |
| 19 04 0426 | 55 Gallon Tank Fiberglass Saddle | EACH | 295.00 | 295.00 | 295.00 | 295.00 | 295.00 |
| 19 04 0427 | 110 Gallon Tank Fiberglass Saddle | EACH | 410.00 | 410.00 | 410.00 | 410.00 | 410.00 |
| 19 04 0428 | 200 Gallon Tank Fiberglass Saddle | EACH | 585.00 | 585.00 | 585.00 | 585.00 | 585.00 |
| 19 04 0429 | 300 Gallon Tank Fiberglass Saddle | EACH | 995.00 | 995.00 | 995.00 | 995.00 | 995.00 |
| 19 04 0430 | 500 Gallon Tank Fiberglass Saddle | EACH | 1,995 | 1,995 | 1,995 | 1,995 | 1,995 |
| 19 04 0431 | 60 Gallon Nalgene Horizontal XLPE Tank with legs | EACH | 531.94 | 487.69 | 467.40 | 424.55 | 409.70 |
| 19 04 0432 | 125 Gallon Nalgene Horizontal XLPE Tank with legs | EACH | 636.94 | 592.69 | 572.40 | 529.55 | 514.70 |

**Environmental Remediation: Assemblies Cost Book**

*Page 3-55*

1998 by ECHOS. All rights reserved

# Cost Data for Remediation

## Chemical Precipitation

| Assembly | Description | Unit | Unit Cost by Safety Level | | | | |
|---|---|---|---|---|---|---|---|
| | | | A | B | C | D | E |
| | **Capital Costs** | | | | | | |
| 19 04 0433 | 225 Gallon Nalgene Horizontal XLPE Tank with legs | EACH | 791.94 | 747.69 | 727.40 | 684.55 | 669.70 |
| 19 04 0434 | 300 Gallon Nalgene Horizontal XLPE Tank with legs | EACH | 811.94 | 767.69 | 747.40 | 704.55 | 689.70 |
| 19 04 0435 | 500 Gallon Nalgene Horizontal XLPE Tank with legs | EACH | 916.94 | 872.69 | 852.40 | 809.55 | 794.70 |
| 19 04 0436 | 1,575 Gallon Conical Bottom Vertical XLPE Tank | EACH | 2,441 | 2,386 | 2,361 | 2,307 | 2,288 |
| 19 04 0437 | 2,200 Gallon Conical Bottom Vertical XLPE Tank | EACH | 2,886 | 2,831 | 2,806 | 2,752 | 2,733 |
| 19 04 0438 | 2,600 Gallon Conical Bottom Vertical XLPE Tank | EACH | 3,496 | 3,441 | 3,416 | 3,362 | 3,343 |
| 19 04 0439 | 3,000 Gallon Conical Bottom Vertical XLPE Tank | EACH | 3,945 | 3,879 | 3,849 | 3,784 | 3,762 |
| 19 04 0440 | 4,200 Gallon Conical Bottom Vertical XLPE Tank | EACH | 5,595 | 5,529 | 5,499 | 5,434 | 5,412 |
| 19 04 0441 | 6,000 Gallon Conical Bottom Vertical XLPE Tank | EACH | 7,795 | 7,729 | 7,699 | 7,634 | 7,612 |
| 19 04 0442 | 8,000 Gallon Conical Bottom Vertical XLPE Tank | EACH | 10,495 | 10,429 | 10,399 | 10,334 | 10,312 |
| 19 04 0443 | 1,575 Gallon Conical Tank Stand | EACH | 1,495 | 1,495 | 1,495 | 1,495 | 1,495 |
| 19 04 0444 | 2,200 Gallon Conical Tank Stand | EACH | 1,695 | 1,695 | 1,695 | 1,695 | 1,695 |
| 19 04 0445 | 2,600 Gallon Conical Tank Stand | EACH | 1,795 | 1,795 | 1,795 | 1,795 | 1,795 |
| 19 04 0446 | 3,000 Gallon Conical Tank Stand | EACH | 1,795 | 1,795 | 1,795 | 1,795 | 1,795 |
| 33 02 1512 | pH Chemical Feed Controller without Input Card | EACH | 695.00 | 695.00 | 695.00 | 695.00 | 695.00 |
| 33 02 1513 | pH Input Card - Standard | EACH | 217.11 | 205.17 | 200.06 | 188.51 | 184.28 |
| 33 02 1514 | pH/ORP Output Card Dual Propor Alarm | EACH | 133.11 | 121.17 | 116.06 | 104.51 | 100.28 |
| 33 02 1515 | pH/ORP Input Card Isolated 0/4-20 mA | EACH | 180.11 | 168.17 | 163.06 | 151.51 | 147.28 |
| 33 02 1516 | pH/ORP Input Card Isolated 0-5/10 VDC | EACH | 180.11 | 168.17 | 163.06 | 151.51 | 147.28 |
| 33 02 1517 | pH Controller with Non-isolated 4-20 mA Output | EACH | 1,037 | 1,025 | 1,020 | 1,009 | 1,004 |
| 33 02 1518 | pH Controller with Isolated 4-20 mA Output | EACH | 1,127 | 1,115 | 1,110 | 1,099 | 1,094 |
| 33 02 1519 | pH Pump Pulser Board for Controller | EACH | 268.11 | 256.17 | 251.06 | 239.51 | 235.28 |
| 33 02 1520 | Analog pH Analyzer with 2 - 12 pH Dials | EACH | 604.06 | 598.08 | 595.53 | 589.76 | 587.64 |
| 33 02 1521 | pH Coax Cable (4') | EACH | 83.26 | 77.28 | 74.73 | 68.96 | 66.84 |
| 33 02 1522 | pH Coax Cable (8') | EACH | 99.22 | 93.25 | 90.69 | 84.92 | 82.80 |
| 33 02 1523 | pH Coax Cable (12') | EACH | 115.18 | 109.21 | 106.66 | 100.88 | 98.77 |
| 33 02 1524 | Submersion Assembly | EACH | 221.06 | 215.08 | 212.53 | 206.76 | 204.64 |
| 33 12 0601 | 10 Gallon Batch Neutralizer | EACH | 19,837 | 19,603 | 19,503 | 19,276 | 19,193 |
| 33 12 0602 | 25 Gallon Batch Neutralizer | EACH | 23,346 | 23,112 | 23,011 | 22,785 | 22,702 |
| 33 12 0603 | 50 Gallon Batch Neutralizer | EACH | 30,066 | 29,832 | 29,732 | 29,505 | 29,422 |
| 33 12 0604 | 100 Gallon Batch Neutralizer | EACH | 32,955 | 32,720 | 32,620 | 32,394 | 32,311 |
| 33 12 0605 | 200 Gallon Batch Neutralizer | EACH | 37,857 | 37,623 | 37,523 | 37,296 | 37,213 |
| 33 12 0606 | Pumping System | EACH | 2,027 | 2,027 | 2,027 | 2,027 | 2,027 |

## Environmental Remediation: Assemblies Cost Book

# Cost Data for Remediation

## Chemical Precipitation

| Assembly | Description | Unit | A | B | C | D | E |
|---|---|---|---|---|---|---|---|
| | **Capital Costs** | | | | | | |
| 33 13 0401 | 9' Diameter, 25 GPM, Waste Flow, Contact Clarifier | EACH | 125,549 | 109,509 | 101,224 | 85,638 | 80,833 |
| 33 13 0402 | 12' Diameter, 45 GPM, Waste Flow, Contact Clarifier | EACH | 143,480 | 127,441 | 119,155 | 103,569 | 98,764 |
| 33 13 0403 | 15' Diameter, 70 GPM, Waste Flow, Contact Clarifier | EACH | 148,807 | 132,768 | 124,482 | 108,896 | 104,091 |
| 33 13 0404 | 20' Diameter, 130 GPM, Waste Flow, Contact Clarifier | EACH | 158,323 | 137,909 | 127,364 | 107,527 | 101,412 |
| 33 13 0405 | 25' Diameter, 208 GPM, Waste Flow, Contact Clarifier | EACH | 162,595 | 142,181 | 131,636 | 111,799 | 105,683 |
| 33 13 0406 | 30' Diameter, 305 GPM, Waste Flow, Contact Clarifier | EACH | 172,420 | 152,006 | 141,461 | 121,624 | 115,508 |
| 33 13 0407 | 40' Diameter, 546 GPM, Waste Flow, Contact Clarifier | EACH | 300,630 | 272,561 | 258,061 | 230,785 | 222,377 |
| 33 13 0408 | 50' Diameter, 850 GPM, Waste Flow, Contact Clarifier | EACH | 386,586 | 349,161 | 329,828 | 293,460 | 282,249 |
| 33 13 0409 | 60' Diameter, 1,194 GPM, Waste Flow, Contact Clarifier | EACH | 490,608 | 445,698 | 422,498 | 378,857 | 365,403 |
| 33 13 0410 | 70' Diameter, 1,645 GPM, Waste Flow, Contact Clarifier | EACH | 596,008 | 539,871 | 510,871 | 456,320 | 439,503 |
| 33 13 0411 | 2 GPM Vertical Plate Clarifier with Mix Tank | EACH | 7,498 | 7,476 | 7,466 | 7,445 | 7,437 |
| 33 13 0412 | 10 GPM Vertical Plate Clarifier with Mix Tank | EACH | 10,798 | 10,776 | 10,766 | 10,745 | 10,737 |
| 33 13 0413 | 15 GPM Vertical Plate Clarifier with Mix Tank | EACH | 14,198 | 14,176 | 14,166 | 14,145 | 14,137 |
| 33 13 0414 | 30 GPM Vertical Plate Clarifier with Mix Tank | EACH | 19,198 | 19,176 | 19,166 | 19,145 | 19,137 |
| 33 13 0415 | 1/4 HP, Single Propeller 3-1/2" Diameter Mixer | EACH | 1,163 | 1,160 | 1,159 | 1,156 | 1,155 |
| 33 13 0416 | 1/3 HP, Single Propeller 4" Diameter Mixer | EACH | 1,313 | 1,310 | 1,309 | 1,306 | 1,305 |
| 33 13 0417 | 1/2 HP, Single Propeller 4-1/2" Diameter Mixer | EACH | 1,488 | 1,485 | 1,484 | 1,481 | 1,480 |
| 33 13 0418 | 3/4 HP, Single Propeller 5" Diameter Mixer | EACH | 1,653 | 1,650 | 1,649 | 1,646 | 1,645 |
| 33 13 0419 | 1 HP, Single Propeller 5-1/2" Diameter Mixer | EACH | 1,913 | 1,910 | 1,909 | 1,906 | 1,905 |
| 33 13 0420 | 1-1/2 HP, Single Propeller 6" Diameter Mixer | EACH | 2,173 | 2,170 | 2,169 | 2,166 | 2,165 |
| 33 13 0421 | 2 HP, Single Propeller 6-1/2" Diameter Mixer | EACH | 2,438 | 2,435 | 2,434 | 2,431 | 2,430 |
| 33 13 0422 | 1/4 HP, Double Propeller 3" Diameter Mixer | EACH | 1,313 | 1,310 | 1,309 | 1,306 | 1,305 |
| 33 13 0423 | 1/3 HP, Double Propeller 3-1/2" Diameter Mixer | EACH | 1,388 | 1,385 | 1,384 | 1,381 | 1,380 |
| 33 13 0424 | 1/2 HP, Double Propeller 4" Diameter Mixer | EACH | 1,563 | 1,560 | 1,559 | 1,556 | 1,555 |
| 33 13 0425 | 3/4 HP, Double Propeller 4-1/2" Diameter Mixer | EACH | 1,713 | 1,710 | 1,709 | 1,706 | 1,705 |
| 33 13 0426 | 1 HP, Double Propeller 5" Diameter Mixer | EACH | 2,003 | 2,000 | 1,999 | 1,996 | 1,995 |
| 33 13 0427 | 1-1/2 HP, Double Propeller 5-1/2" Diameter Mixer | EACH | 2,263 | 2,260 | 2,259 | 2,256 | 2,255 |
| 33 13 0428 | 2 HP, Double Propeller 6" Diameter Mixer | EACH | 2,563 | 2,560 | 2,559 | 2,556 | 2,555 |
| 33 26 0101 | 1" Carbon Steel Piping | LF | 6.46 | 5.28 | 4.77 | 3.63 | 3.21 |
| 33 26 0102 | 2" Carbon Steel Piping | LF | 10.50 | 8.72 | 7.95 | 6.23 | 5.60 |
| 33 26 0103 | 3" Carbon Steel Piping | LF | 18.02 | 15.20 | 13.99 | 11.26 | 10.26 |
| 33 26 0104 | 4" Carbon Steel Piping | LF | 23.20 | 19.86 | 18.43 | 15.21 | 14.02 |
| 33 26 0105 | 6" Carbon Steel Piping | LF | 36.37 | 30.24 | 27.46 | 21.52 | 19.44 |

**Environmental Remediation: Assemblies Cost Book**

# Cost Data for Remediation

## Chemical Precipitation

| Assembly | Description | Unit | Unit Cost by Safety Level | | | | |
|---|---|---|---|---|---|---|---|
| | | | A | B | C | D | E |
| | **Capital Costs** | | | | | | |
| 33 26 0106 | 8" Carbon Steel Piping | LF | 40.77 | 34.08 | 31.15 | 24.68 | 22.36 |
| 33 26 0108 | 12" Carbon Steel Piping | LF | 75.72 | 65.62 | 61.18 | 51.41 | 47.89 |
| 33 26 0201 | 1" Stainless Steel Piping, Schedule 40, Threaded | LF | 18.67 | 16.12 | 15.03 | 12.56 | 11.65 |
| 33 26 0202 | 2" Stainless Steel Piping, Schedule 40, Threaded | LF | 33.47 | 29.21 | 27.39 | 23.28 | 21.77 |
| 33 26 0203 | 3" Stainless Steel Piping, Schedule 40, Threaded | LF | 57.00 | 50.61 | 47.88 | 41.71 | 39.45 |
| 33 26 0204 | 4" Stainless Steel Piping, Schedule 40, Threaded | LF | 80.15 | 71.72 | 68.12 | 59.98 | 57.00 |
| 33 26 0205 | 6" Stainless Steel Piping, Schedule 10, Type 316 | LF | 78.58 | 70.01 | 66.14 | 57.84 | 54.94 |
| 33 26 0206 | 8" Stainless Steel Piping, Schedule 40, Welded | LF | 197.22 | 183.32 | 176.89 | 163.42 | 158.80 |
| 33 26 0209 | 12" Stainless Steel Piping, Schedule 40, Welded | LF | 312.43 | 295.44 | 287.57 | 271.12 | 265.47 |
| 33 29 0102 | 10 GPM, 1/2 HP, Centrifugal Pump | EACH | 1,063 | 952.32 | 905.07 | 798.39 | 759.27 |
| 33 29 0103 | 50 GPM, 100' Head, 3 HP, Centrifugal Pump | EACH | 2,658 | 2,500 | 2,432 | 2,279 | 2,223 |
| 33 29 0105 | 5 HP, 100 GPM, Centrifugal Pump | EACH | 2,475 | 2,301 | 2,227 | 2,059 | 1,998 |
| 33 29 0108 | 10 HP, 200 GPM, Centrifugal Pump | EACH | 2,941 | 2,714 | 2,616 | 2,396 | 2,316 |
| 33 29 0109 | 10 HP, 250 GPM, Centrifugal Pump | EACH | 3,032 | 2,805 | 2,707 | 2,487 | 2,407 |
| 33 29 0110 | 15 HP, 300 GPM, Centrifugal Pump | EACH | 3,441 | 3,160 | 3,040 | 2,770 | 2,670 |
| 33 29 0111 | 20 HP, 500 GPM, Centrifugal Pump | EACH | 4,544 | 4,241 | 4,111 | 3,818 | 3,710 |
| 33 29 0112 | 30 HP, 750 GPM, Centrifugal Pump | EACH | 5,213 | 4,808 | 4,635 | 4,244 | 4,100 |
| 33 29 0113 | 40 HP, 1050 GPM, Centrifugal Pump | EACH | 5,655 | 5,199 | 5,004 | 4,564 | 4,403 |
| 33 29 0114 | 60 HP, 1500 GPM, Centrifugal Pump | EACH | 9,842 | 8,823 | 8,350 | 7,363 | 7,024 |
| 33 29 0115 | 75 HP, 2000 GPM, Centrifugal Pump | EACH | 12,456 | 11,233 | 10,666 | 9,482 | 9,075 |
| 33 33 0101 | Sodium Hydroxide Solution, 190 Lb Drummed Liquid | EACH | 62.34 | 57.81 | 55.07 | 50.64 | 49.53 |
| 33 33 0103 | Sodium Hydroxide, 500 Lb Container, Beads | EACH | 234.06 | 229.53 | 226.79 | 222.36 | 221.25 |
| 33 33 0104 | Sodium Hydroxide, 100 Lb Container, Flakes | EACH | 89.90 | 85.37 | 82.63 | 78.20 | 77.09 |
| 33 33 0110 | Sulfuric Acid Solution, 220 Lb Drummed Liquid | EACH | 45.38 | 40.85 | 38.11 | 33.68 | 32.57 |
| 33 33 0111 | Sulfuric Acid Solution, 750 Lb Drummed Liquid | EACH | 93.61 | 89.08 | 86.34 | 81.91 | 80.80 |
| 33 33 0112 | Soda Ash, 50 Lb Bag, Powdered | EACH | 28.42 | 23.89 | 21.15 | 16.72 | 15.61 |
| 33 33 0113 | Soda Ash, 100 Lb Bag, Powdered | EACH | 41.14 | 36.61 | 33.87 | 29.44 | 28.33 |
| 33 33 0114 | Quicklime, 1/4" Nominal Granules, Bulk Quantity | TON | 96.49 | 96.49 | 96.48 | 96.48 | 96.47 |
| 33 33 0115 | Quicklime, 3/4" Nominal Granules, Bulk Quantity | TON | 96.49 | 96.49 | 96.48 | 96.48 | 96.47 |
| 33 33 0116 | Quicklime, Combination 1/4" & 3/4" Granules, Bulk Quantity | TON | 96.49 | 96.49 | 96.48 | 96.48 | 96.47 |
| 33 33 0117 | Hydrated Lime, Powdered, Bulk | TON | 97.55 | 97.55 | 97.54 | 97.54 | 97.53 |
| 33 33 0118 | Hydrated Lime, Powdered, 50 Lb Bag | EACH | 25.76 | 21.23 | 18.49 | 14.06 | 12.94 |
| 33 33 0119 | Aluminum Sulfate (Alum), Bulk | TON | 240.07 | 240.06 | 240.06 | 240.05 | 240.05 |

## Environmental Remediation: Assemblies Cost Book

*Page 3-58*

# Cost Data for Remediation

## Chemical Precipitation

| Assembly | Description | Unit | A | B | C | D | E |
|---|---|---|---|---|---|---|---|
| | **Capital Costs** | | | | | | |
| 33 33 0120 | Ferric Sulfate, Bulk | TON | 274.04 | 274.04 | 274.03 | 274.03 | 274.02 |
| 33 33 0121 | Ferric Chloride, Bulk | TON | 241.18 | 241.18 | 241.17 | 241.17 | 241.16 |
| 33 33 0122 | Montmorillonite Clay Flocculant Aid, Bulk | LB | 0.14 | 0.13 | 0.13 | 0.12 | 0.12 |
| 33 33 0123 | Sodium Hectorite Clay Flocculant Aid, Bulk | LB | 0.17 | 0.17 | 0.16 | 0.15 | 0.15 |
| 33 33 0124 | Super Dispersible Sodium Flocculant Aid, Bulk | LB | 0.19 | 0.19 | 0.18 | 0.18 | 0.17 |
| 33 41 0101 | Pump & Motor Maintenance/Repair | EACH | 865.17 | 671.48 | 581.58 | 393.96 | 329.64 |
| 33 42 0101 | Electrical Charge | KWH | 0.06 | 0.06 | 0.06 | 0.06 | 0.06 |
| | **Operations & Maintenance** | | | | | | |
| 19 04 0401 | 550 Gallon, Stainless Steel Aboveground Wastewater Holding Tank, Rental | MONTH | 320.65 | 320.65 | 320.65 | 320.65 | 320.65 |
| 19 04 0402 | 630 Gallon, Polyethylene Aboveground Wastewater Holding Tank, Rental | MONTH | 320.65 | 320.65 | 320.65 | 320.65 | 320.65 |
| 19 04 0403 | 4,000 Gallon Polyethylene Wastewater Tank, Rental | MONTH | 545.90 | 545.90 | 545.90 | 545.90 | 545.90 |
| 19 04 0404 | 4,000 Gallon, Polyethylene Trailer-mounted Wastewater Holding Tank, Rental | MONTH | 1,124 | 1,124 | 1,124 | 1,124 | 1,124 |
| 19 04 0405 | 6,000 Gallon, Polyethylene Aboveground Wastewater Holding Tank, Rental | MONTH | 641.30 | 641.30 | 641.30 | 641.30 | 641.30 |
| 19 04 0406 | 21,000 Gallon Steel Wastewater Holding Tank, Rental | MONTH | 1,219 | 1,219 | 1,219 | 1,219 | 1,219 |
| 19 04 0408 | 21,000 Gallon Steel, Open Top, Tank Rental | MONTH | 1,124 | 1,124 | 1,124 | 1,124 | 1,124 |
| 33 33 0101 | Sodium Hydroxide Solution, 190 Lb Drummed Liquid | EACH | 62.34 | 57.81 | 55.07 | 50.64 | 49.53 |
| 33 33 0103 | Sodium Hydroxide, 500 Lb Container, Beads | EACH | 234.06 | 229.53 | 226.79 | 222.36 | 221.25 |
| 33 33 0104 | Sodium Hydroxide, 100 Lb Container, Flakes | EACH | 89.90 | 85.37 | 82.63 | 78.20 | 77.09 |
| 33 33 0110 | Sulfuric Acid Solution, 220 Lb Drummed Liquid | EACH | 45.38 | 40.85 | 38.11 | 33.68 | 32.57 |
| 33 33 0111 | Sulfuric Acid Solution, 750 Lb Drummed Liquid | EACH | 93.61 | 89.08 | 86.34 | 81.91 | 80.80 |
| 33 33 0112 | Soda Ash, 50 Lb Bag, Powdered | EACH | 28.42 | 23.89 | 21.15 | 16.72 | 15.61 |
| 33 33 0113 | Soda Ash, 100 Lb Bag, Powdered | EACH | 41.14 | 36.61 | 33.87 | 29.44 | 28.33 |
| 33 33 0114 | Quicklime, 1/4" Nominal Granules, Bulk Quantity | TON | 96.49 | 96.49 | 96.48 | 96.48 | 96.47 |
| 33 33 0115 | Quicklime, 3/4" Nominal Granules, Bulk Quantity | TON | 96.49 | 96.49 | 96.48 | 96.48 | 96.47 |
| 33 33 0116 | Quicklime, Combination 1/4" & 3/4" Granules, Bulk Quantity | TON | 96.49 | 96.49 | 96.48 | 96.48 | 96.47 |
| 33 33 0117 | Hydrated Lime, Powdered, Bulk | TON | 97.55 | 97.55 | 97.54 | 97.54 | 97.53 |
| 33 33 0118 | Hydrated Lime, Powdered, 50 Lb Bag | EACH | 25.76 | 21.23 | 18.49 | 14.06 | 12.94 |
| 33 33 0119 | Aluminum Sulfate (Alum), Bulk | TON | 240.07 | 240.06 | 240.06 | 240.05 | 240.05 |
| 33 33 0120 | Ferric Sulfate, Bulk | TON | 274.04 | 274.04 | 274.03 | 274.03 | 274.02 |
| 33 33 0121 | Ferric Chloride, Bulk | TON | 241.18 | 241.18 | 241.17 | 241.17 | 241.16 |
| 33 33 0122 | Montmorillonite Clay Flocculant Aid, Bulk | LB | 0.14 | 0.13 | 0.13 | 0.12 | 0.12 |

**Environmental Remediation: Assemblies Cost Book**

*Page 3-59*

## Cost Data for Remediation

### Chemical Precipitation

| Assembly | Description | Unit | Unit Cost by Safety Level | | | | |
|---|---|---|---|---|---|---|---|
| | | | A | B | C | D | E |
| | **Operations & Maintenance** | | | | | | |
| 33 33 0123 | Sodium Hectorite Clay Flocculant Aid, Bulk | LB | 0.17 | 0.17 | 0.16 | 0.15 | 0.15 |
| 33 33 0124 | Super Dispersible Sodium Flocculant Aid, Bulk | LB | 0.19 | 0.19 | 0.18 | 0.18 | 0.17 |
| 33 41 0101 | Pump & Motor Maintenance/Repair | EACH | 865.17 | 671.48 | 581.58 | 393.96 | 329.64 |
| 33 42 0101 | Electrical Charge | KWH | 0.06 | 0.06 | 0.06 | 0.06 | 0.06 |

**Environmental Remediation: Assemblies Cost Book**

# Cost Data for Remediation
## Coagulation/Flocculation

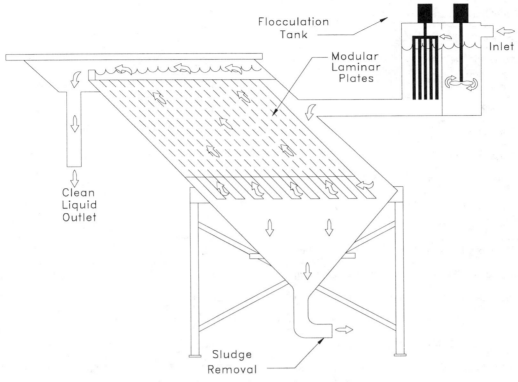

### General:

*Coagulation and flocculation are two processes used to remove extremely fine particles and/or colloids from water and waste water. Coagulation reduces the net electrical repulsive forces at the particle surface by electrolytes in the solution, while flocculation involves three main steps: (1) addition of coagulants to the waste stream, (2) rapid mixing to disperse the coagulant, and (3) slow and gentle mixing to promote agglomeration of the colloids.*

### Typical Treatment Train:

*Groundwater recovery, pretreatment, polishing treatment for aqueous residuals, polishing treatment for solid residuals, sampling and analysis, disposal.*

### Common Cost Components:

1. Clarifier
2. Coagulants (primary, secondary)
3. Structural slab
4. Electrical usage
5. Piping
6. Operations and Maintenance

### Other Cost Considerations:

*Electrical distribution, fencing.*

---

**Environmental Remediation: Assemblies Cost Book**

# Cost Data for Remediation

## Coagulation/Flocculation

| Assembly | Description | Unit | A | B | C | D | E |
|---|---|---|---|---|---|---|---|
| | **Capital Costs** | | | | | | |
| 18 02 0322 | 8" Structural Slab on Grade | SF | 9.75 | 8.38 | 7.71 | 6.38 | 5.95 |
| 18 02 0324 | 12" Structural Slab on Grade | SF | 11.57 | 10.04 | 9.30 | 7.82 | 7.34 |
| 19 01 0202 | 1", Class 200, PVC Piping | LF | 7.48 | 5.84 | 5.07 | 3.48 | 2.93 |
| 19 01 0204 | 2", Class 200, PVC Piping | LF | 8.86 | 6.96 | 6.07 | 4.23 | 3.60 |
| 19 01 0206 | 3", Class 200, PVC Piping | LF | 11.81 | 9.34 | 8.20 | 5.80 | 4.98 |
| 19 01 0207 | 4", Class 200, PVC Piping | LF | 12.33 | 9.86 | 8.72 | 6.32 | 5.50 |
| 19 01 0208 | 6", Class 200, PVC Piping | LF | 14.95 | 12.20 | 10.93 | 8.27 | 7.36 |
| 19 01 0212 | 8", Class 150, PVC Piping | LF | 18.36 | 15.27 | 13.84 | 10.85 | 9.82 |
| 19 01 0214 | 12", Class 150, PVC Piping | LF | 28.75 | 24.63 | 22.72 | 18.73 | 17.36 |
| 33 13 0401 | 9' Diameter, 25 GPM, Waste Flow, Contact Clarifier | EACH | 125,549 | 109,509 | 101,224 | 85,638 | 80,833 |
| 33 13 0402 | 12' Diameter, 45 GPM, Waste Flow, Contact Clarifier | EACH | 143,480 | 127,441 | 119,155 | 103,569 | 98,764 |
| 33 13 0403 | 15' Diameter, 70 GPM, Waste Flow, Contact Clarifier | EACH | 148,807 | 132,768 | 124,482 | 108,896 | 104,091 |
| 33 13 0404 | 20' Diameter, 130 GPM, Waste Flow, Contact Clarifier | EACH | 158,323 | 137,909 | 127,364 | 107,527 | 101,412 |
| 33 13 0405 | 25' Diameter, 208 GPM, Waste Flow, Contact Clarifier | EACH | 162,595 | 142,181 | 131,636 | 111,799 | 105,683 |
| 33 13 0406 | 30' Diameter, 305 GPM, Waste Flow, Contact Clarifier | EACH | 172,420 | 152,006 | 141,461 | 121,624 | 115,508 |
| 33 13 0407 | 40' Diameter, 546 GPM, Waste Flow, Contact Clarifier | EACH | 300,630 | 272,561 | 258,061 | 230,785 | 222,377 |
| 33 13 0408 | 50' Diameter, 850 GPM, Waste Flow, Contact Clarifier | EACH | 386,586 | 349,161 | 329,828 | 293,460 | 282,249 |
| 33 13 0409 | 60' Diameter, 1,194 GPM, Waste Flow, Contact Clarifier | EACH | 490,608 | 445,698 | 422,498 | 378,857 | 365,403 |
| 33 13 0410 | 70' Diameter, 1,645 GPM, Waste Flow, Contact Clarifier | EACH | 596,008 | 539,871 | 510,871 | 456,320 | 439,503 |
| 33 13 0411 | 2 GPM Vertical Plate Clarifier with Mix Tank | EACH | 7,498 | 7,476 | 7,466 | 7,445 | 7,437 |
| 33 13 0412 | 10 GPM Vertical Plate Clarifier with Mix Tank | EACH | 10,798 | 10,776 | 10,766 | 10,745 | 10,737 |
| 33 13 0413 | 15 GPM Vertical Plate Clarifier with Mix Tank | EACH | 14,198 | 14,176 | 14,166 | 14,145 | 14,137 |
| 33 13 0414 | 30 GPM Vertical Plate Clarifier with Mix Tank | EACH | 19,198 | 19,176 | 19,166 | 19,145 | 19,137 |
| 33 26 0101 | 1" Carbon Steel Piping | LF | 6.46 | 5.28 | 4.77 | 3.63 | 3.21 |
| 33 26 0102 | 2" Carbon Steel Piping | LF | 10.50 | 8.72 | 7.95 | 6.23 | 5.60 |
| 33 26 0103 | 3" Carbon Steel Piping | LF | 18.02 | 15.20 | 13.99 | 11.26 | 10.26 |
| 33 26 0104 | 4" Carbon Steel Piping | LF | 23.20 | 19.86 | 18.43 | 15.21 | 14.02 |
| 33 26 0105 | 6" Carbon Steel Piping | LF | 36.37 | 30.24 | 27.46 | 21.52 | 19.44 |
| 33 26 0106 | 8" Carbon Steel Piping | LF | 40.77 | 34.08 | 31.15 | 24.68 | 22.36 |
| 33 26 0108 | 12" Carbon Steel Piping | LF | 75.72 | 65.62 | 61.18 | 51.41 | 47.89 |
| 33 26 0201 | 1" Stainless Steel Piping, Schedule 40, Threaded | LF | 18.67 | 16.12 | 15.03 | 12.56 | 11.65 |
| 33 26 0202 | 2" Stainless Steel Piping, Schedule 40, Threaded | LF | 33.47 | 29.21 | 27.39 | 23.28 | 21.77 |
| 33 26 0203 | 3" Stainless Steel Piping, Schedule 40, Threaded | LF | 57.00 | 50.61 | 47.88 | 41.71 | 39.45 |

**Environmental Remediation: Assemblies Cost Book**

# Cost Data for Remediation

## Coagulation/Flocculation

| Assembly | Description | Unit | Unit Cost by Safety Level | | | | |
|---|---|---|---|---|---|---|---|
| | | | A | B | C | D | E |
| | **Capital Costs** | | | | | | |
| 33 26 0204 | 4" Stainless Steel Piping, Schedule 40, Threaded | LF | 80.15 | 71.72 | 68.12 | 59.98 | 57.00 |
| 33 26 0205 | 6" Stainless Steel Piping, Schedule 10, Type 316 | LF | 78.58 | 70.01 | 66.14 | 57.84 | 54.94 |
| 33 26 0206 | 8" Stainless Steel Piping, Schedule 40, Welded | LF | 197.22 | 183.32 | 176.89 | 163.42 | 158.80 |
| 33 26 0209 | 12" Stainless Steel Piping, Schedule 40, Welded | LF | 312.43 | 295.44 | 287.57 | 271.12 | 265.47 |
| 33 29 0105 | 5 HP, 100 GPM, Centrifugal Pump | EACH | 2,475 | 2,301 | 2,227 | 2,059 | 1,998 |
| 33 29 0108 | 10 HP, 200 GPM, Centrifugal Pump | EACH | 2,941 | 2,714 | 2,616 | 2,396 | 2,316 |
| 33 29 0109 | 10 HP, 250 GPM, Centrifugal Pump | EACH | 3,032 | 2,805 | 2,707 | 2,487 | 2,407 |
| 33 29 0110 | 15 HP, 300 GPM, Centrifugal Pump | EACH | 3,441 | 3,160 | 3,040 | 2,770 | 2,670 |
| 33 29 0111 | 20 HP, 500 GPM, Centrifugal Pump | EACH | 4,544 | 4,241 | 4,111 | 3,818 | 3,710 |
| 33 29 0112 | 30 HP, 750 GPM, Centrifugal Pump | EACH | 5,213 | 4,808 | 4,635 | 4,244 | 4,100 |
| 33 29 0113 | 40 HP, 1050 GPM, Centrifugal Pump | EACH | 5,655 | 5,199 | 5,004 | 4,564 | 4,403 |
| 33 29 0114 | 60 HP, 1500 GPM, Centrifugal Pump | EACH | 9,842 | 8,823 | 8,350 | 7,363 | 7,024 |
| 33 29 0115 | 75 HP, 2000 GPM, Centrifugal Pump | EACH | 12,456 | 11,233 | 10,666 | 9,482 | 9,075 |
| 33 33 0112 | Soda Ash, 50 Lb Bag, Powdered | EACH | 28.42 | 23.89 | 21.15 | 16.72 | 15.61 |
| 33 33 0113 | Soda Ash, 100 Lb Bag, Powdered | EACH | 41.14 | 36.61 | 33.87 | 29.44 | 28.33 |
| 33 33 0114 | Quicklime, 1/4" Nominal Granules, Bulk Quantity | TON | 96.49 | 96.49 | 96.48 | 96.48 | 96.47 |
| 33 33 0115 | Quicklime, 3/4" Nominal Granules, Bulk Quantity | TON | 96.49 | 96.49 | 96.48 | 96.48 | 96.47 |
| 33 33 0116 | Quicklime, Combination 1/4" & 3/4" Granules, Bulk Quantity | TON | 96.49 | 96.49 | 96.48 | 96.48 | 96.47 |
| 33 33 0117 | Hydrated Lime, Powdered, Bulk | TON | 97.55 | 97.55 | 97.54 | 97.54 | 97.53 |
| 33 33 0118 | Hydrated Lime, Powdered, 50 Lb Bag | EACH | 25.76 | 21.23 | 18.49 | 14.06 | 12.94 |
| 33 33 0119 | Aluminum Sulfate (Alum), Bulk | TON | 240.07 | 240.06 | 240.06 | 240.05 | 240.05 |
| 33 33 0120 | Ferric Sulfate, Bulk | TON | 274.04 | 274.04 | 274.03 | 274.03 | 274.02 |
| 33 33 0121 | Ferric Chloride, Bulk | TON | 241.18 | 241.18 | 241.17 | 241.17 | 241.16 |
| 33 33 0122 | Montmorillonite Clay Flocculant Aid, Bulk | LB | 0.14 | 0.13 | 0.13 | 0.12 | 0.12 |
| 33 33 0123 | Sodium Hectorite Clay Flocculant Aid, Bulk | LB | 0.17 | 0.17 | 0.16 | 0.15 | 0.15 |
| 33 33 0124 | Super Dispersible Sodium Flocculant Aid, Bulk | LB | 0.19 | 0.19 | 0.18 | 0.18 | 0.17 |
| 33 41 0101 | Pump & Motor Maintenance/Repair | EACH | 865.17 | 671.48 | 581.58 | 393.96 | 329.64 |
| 33 42 0101 | Electrical Charge | KWH | 0.06 | 0.06 | 0.06 | 0.06 | 0.06 |
| | **Operations & Maintenance** | | | | | | |
| 33 29 0105 | 5 HP, 100 GPM, Centrifugal Pump | EACH | 2,475 | 2,301 | 2,227 | 2,059 | 1,998 |
| 33 29 0108 | 10 HP, 200 GPM, Centrifugal Pump | EACH | 2,941 | 2,714 | 2,616 | 2,396 | 2,316 |
| 33 29 0109 | 10 HP, 250 GPM, Centrifugal Pump | EACH | 3,032 | 2,805 | 2,707 | 2,487 | 2,407 |
| 33 29 0110 | 15 HP, 300 GPM, Centrifugal Pump | EACH | 3,441 | 3,160 | 3,040 | 2,770 | 2,670 |

**Environmental Remediation: Assemblies Cost Book**

# Cost Data for Remediation

## Coagulation/Flocculation

| Assembly | Description | Unit | A | B | C | D | E |
|---|---|---|---|---|---|---|---|
| | **Operations & Maintenance** | | | | | | |
| 33 29 0111 | 20 HP, 500 GPM, Centrifugal Pump | EACH | 4,544 | 4,241 | 4,111 | 3,818 | 3,710 |
| 33 29 0112 | 30 HP, 750 GPM, Centrifugal Pump | EACH | 5,213 | 4,808 | 4,635 | 4,244 | 4,100 |
| 33 29 0113 | 40 HP, 1050 GPM, Centrifugal Pump | EACH | 5,655 | 5,199 | 5,004 | 4,564 | 4,403 |
| 33 29 0114 | 60 HP, 1500 GPM, Centrifugal Pump | EACH | 9,842 | 8,823 | 8,350 | 7,363 | 7,024 |
| 33 29 0115 | 75 HP, 2000 GPM, Centrifugal Pump | EACH | 12,456 | 11,233 | 10,666 | 9,482 | 9,075 |
| 33 33 0112 | Soda Ash, 50 Lb Bag, Powdered | EACH | 28.42 | 23.89 | 21.15 | 16.72 | 15.61 |
| 33 33 0113 | Soda Ash, 100 Lb Bag, Powdered | EACH | 41.14 | 36.61 | 33.87 | 29.44 | 28.33 |
| 33 33 0114 | Quicklime, 1/4" Nominal Granules, Bulk Quantity | TON | 96.49 | 96.49 | 96.48 | 96.48 | 96.47 |
| 33 33 0115 | Quicklime, 3/4" Nominal Granules, Bulk Quantity | TON | 96.49 | 96.49 | 96.48 | 96.48 | 96.47 |
| 33 33 0116 | Quicklime, Combination 1/4" & 3/4" Granules, Bulk Quantity | TON | 96.49 | 96.49 | 96.48 | 96.48 | 96.47 |
| 33 33 0117 | Hydrated Lime, Powdered, Bulk | TON | 97.55 | 97.55 | 97.54 | 97.54 | 97.53 |
| 33 33 0118 | Hydrated Lime, Powdered, 50 Lb Bag | EACH | 25.76 | 21.23 | 18.49 | 14.06 | 12.94 |
| 33 33 0119 | Aluminum Sulfate (Alum), Bulk | TON | 240.07 | 240.06 | 240.06 | 240.05 | 240.05 |
| 33 33 0120 | Ferric Sulfate, Bulk | TON | 274.04 | 274.04 | 274.03 | 274.03 | 274.02 |
| 33 33 0121 | Ferric Chloride, Bulk | TON | 241.18 | 241.18 | 241.17 | 241.17 | 241.16 |
| 33 33 0122 | Montmorillonite Clay Flocculant Aid, Bulk | LB | 0.14 | 0.13 | 0.13 | 0.12 | 0.12 |
| 33 33 0123 | Sodium Hectorite Clay Flocculant Aid, Bulk | LB | 0.17 | 0.17 | 0.16 | 0.15 | 0.15 |
| 33 33 0124 | Super Dispersible Sodium Flocculant Aid, Bulk | LB | 0.19 | 0.19 | 0.18 | 0.18 | 0.17 |
| 33 41 0101 | Pump & Motor Maintenance/Repair | EACH | 865.17 | 671.48 | 581.58 | 393.96 | 329.64 |
| 33 42 0101 | Electrical Charge | KWH | 0.06 | 0.06 | 0.06 | 0.06 | 0.06 |

**Unit Cost by Safety Level**

**Environmental Remediation: Assemblies Cost Book**

# Cost Data for Remediation
## Commercial Disposal (Incineration)

### General:
The last link in the "cradle to grave" hazardous waste management system is disposal. By using an off-site disposal service, the generator may access the resources of a comprehensive waste manager who can assist with the administrative responsibilities connected with hazardous waste generation. Where drums or multiple impoundments are present, it is often more cost effective to consolidate compatible hazardous waste contents in a tank truck.

### Typical Treatment Train:
On-site pretreatment, sampling and analysis (compatibility testing, PCBs, et. al.), transportation, decontamination.

### Common Cost Components:
1. Waste consolidation
2. Drums
3. Overpacks for leading drums
4. Containment berms
5. Sampling
6. Transportation (refer to the transportation section for additional options)
7. Disposal fee/tax

### Other Cost Considerations:
Electrical distribution, access road, clear and grub, fencing.

**Environmental Remediation: Assemblies Cost Book**

# Cost Data for Remediation

## Commercial Disposal (Incineration)

| Assembly | Description | Unit | A | B | C | D | E |
|---|---|---|---|---|---|---|---|
| | **Capital Cost** | | | | | | |
| 33 19 0101 | Liquid Loading Into 5,000 Gallon Bulk Tank Truck | EACH | 997.03 | 798.90 | 667.78 | 473.57 | 432.04 |
| 33 19 0102 | Bulk Solid Hazardous Waste Loading Into Truck | CY | 4.51 | 3.67 | 3.01 | 2.18 | 2.07 |
| 33 19 0103 | Load Drums on Disposal Vehicle | EACH | 6.21 | 4.95 | 4.16 | 2.93 | 2.64 |
| 33 19 0104 | Load Bulk Liquid/Sludge Waste Into 55 Gallon Drums, Drums Separate | EACH | 17.53 | 13.51 | 11.79 | 7.91 | 6.48 |
| 33 19 0105 | Load Bulk Solid Waste Into 55 Gallon Drums, Drums Separate | EACH | 62.31 | 48.03 | 41.92 | 28.12 | 23.06 |
| 33 19 0106 | Recontainerize Drums, Not Including Salvage Drums | EACH | 105.53 | 84.84 | 70.65 | 50.34 | 46.31 |
| 33 19 0201 | Drummed Waste, Minimum Charge for Shipment | EACH | 2,350 | 2,350 | 2,350 | 2,350 | 2,350 |
| 33 19 0202 | Bulk Hazardous Waste, Minimum Charge for Shipment | EACH | 2,350 | 2,350 | 2,350 | 2,350 | 2,350 |
| 33 19 0203 | PCB Hazardous Waste, Minimum Charge for Shipment | EACH | 2,350 | 2,350 | 2,350 | 2,350 | 2,350 |
| 33 19 0204 | Transport 55 Gallon Drums of Hazardous Waste, Maximum 80 (per Mile) | MILE | 1.50 | 1.50 | 1.50 | 1.50 | 1.50 |
| 33 19 0205 | Transport Bulk Solid Hazardous Waste, Maximum 20 CY (per Mile) | MILE | 1.44 | 1.44 | 1.44 | 1.44 | 1.44 |
| 33 19 0206 | Transport Bulk Solid Hazardous Waste, Maximum 18 Ton (per Mile) | MILE | 1.44 | 1.44 | 1.44 | 1.44 | 1.44 |
| 33 19 0207 | Transport Bulk Liquid/Sludge Hazardous Waste, Maximum 5,000 Gallon (per Mile) | MILE | 1.50 | 1.50 | 1.50 | 1.50 | 1.50 |
| 33 19 0324 | State HTW Disposal Tax/Fee (Bulk Solid) | CY | 0.00 | 0.00 | 0.00 | 0.00 | 0.00 |
| 33 19 0325 | State HTW Disposal Tax/Fee (Bulk Liquid) | GAL | 0.00 | 0.00 | 0.00 | 0.00 | 0.00 |
| 33 19 0326 | State HTW Disposal Tax/Fee (Drums) | DRM | 0.00 | 0.00 | 0.00 | 0.00 | 0.00 |
| 33 19 9501 | Energetic Drummed Solid Waste Incineration, 55 Gallon | EACH | 787.50 | 787.50 | 787.50 | 787.50 | 787.50 |
| 33 19 9502 | Nonenergetic Drummed Solid Waste Incineration, 55 Gallon | EACH | 537.50 | 537.50 | 537.50 | 537.50 | 537.50 |
| 33 19 9503 | Energetic, Requiring Repack, Drummed Solid Waste Incineration, 55 Gallon | EACH | 787.50 | 787.50 | 787.50 | 787.50 | 787.50 |
| 33 19 9504 | Nonenergetic, Requiring Repack, Drummed Solid Waste Incineration, 55 Gallon | EACH | 537.50 | 537.50 | 537.50 | 537.50 | 537.50 |
| 33 19 9505 | Reactive or Corrosive Drummed Solids Incineration | LB | 1.88 | 1.88 | 1.88 | 1.88 | 1.88 |
| 33 19 9506 | Amenable-to-Bulking Drummed Liquids Incineration, 55 Gallon | EACH | 515.00 | 515.00 | 515.00 | 515.00 | 515.00 |
| 33 19 9507 | Drummed Liquids Requiring Repack Incineration, 55 Gallon | EACH | 887.50 | 887.50 | 887.50 | 887.50 | 887.50 |
| 33 19 9508 | Drummed Reactive or Corrosive Liquids Incineration, 55 Gallon | EACH | 647.50 | 647.50 | 647.50 | 647.50 | 647.50 |
| 33 19 9509 | Energetic Drummed Sludge Incineration, 55 Gallon | EACH | 477.50 | 477.50 | 477.50 | 477.50 | 477.50 |
| 33 19 9510 | Nonenergetic Drummed Sludge Incineration, 55 Gallon | EACH | 320.00 | 320.00 | 320.00 | 320.00 | 320.00 |
| 33 19 9511 | Drummed Sludge Requiring Repack Incineration, 55 Gallon | EACH | 1,038 | 1,038 | 1,038 | 1,038 | 1,038 |

**Environmental Remediation: Assemblies Cost Book**

# Cost Data for Remediation

## Commercial Disposal (Incineration)

| Assembly | Description | Unit | Unit Cost by Safety Level A | B | C | D | E |
|---|---|---|---|---|---|---|---|
| | **Capital Cost** | | | | | | |
| 33 19 9512 | Lean Water Incineration for Non-PCB 55 Gallon Drummed Waste | EACH | 507.50 | 507.50 | 507.50 | 507.50 | 507.50 |
| 33 19 9513 | Lab Packs Containing Nonreactive Material Incineration | LB | 2.25 | 2.25 | 2.25 | 2.25 | 2.25 |
| 33 19 9514 | Fluorinated Aerosol Cans Incineration | LB | 2.53 | 2.53 | 2.53 | 2.53 | 2.53 |
| 33 19 9515 | Nonfluorinated Aerosol Cans Incineration | LB | 1.84 | 1.84 | 1.84 | 1.84 | 1.84 |
| 33 19 9516 | Drummed Waste Containing Over 5% Halogen, Extra Charges | EACH | 6.34 | 6.34 | 6.34 | 6.34 | 6.34 |
| 33 19 9517 | Drummed Waste Containing Over 10% Ash, Extra Charges | EACH | 3.50 | 3.50 | 3.50 | 3.50 | 3.50 |
| 33 19 9518 | Waste Packed in 85 Gallon Metal Drums, Extra Charges | EACH | 57.88 | 57.88 | 57.88 | 57.88 | 57.88 |
| 33 19 9519 | Waste Packed in 85 Gallon Plastic Drums, Extra Charges | EACH | 135.00 | 135.00 | 135.00 | 135.00 | 135.00 |
| 33 19 9520 | Incineration of Bulk Solid Waste (2,000 Lb/CY) | CY | 1,526 | 1,526 | 1,526 | 1,526 | 1,526 |
| 33 19 9521 | Bulk Liquids, 2,000 to 12,000 BTU, Incineration | LB | 0.93 | 0.93 | 0.93 | 0.93 | 0.93 |
| 33 19 9522 | Reactive or Corrosive Bulk Liquids Incineration | GAL | 11.43 | 11.43 | 11.43 | 11.43 | 11.43 |
| 33 19 9523 | Lean Water Incineration for Non-PCB Bulk Liquids | GAL | 3.88 | 3.88 | 3.88 | 3.88 | 3.88 |
| 33 19 9524 | Aqueous Liquids, Approximately 90% Water Content, Incineration | LB | 0.93 | 0.93 | 0.93 | 0.93 | 0.93 |
| 33 19 9525 | 55 Gallon Drum with Chlorine Content 26 - 50%, Extra Charges | EACH | 148.50 | 148.50 | 148.50 | 148.50 | 148.50 |
| 33 19 9526 | 55 Gallon Drum with Chlorine Content > 50%, Extra Charges | EACH | 225.00 | 225.00 | 225.00 | 225.00 | 225.00 |
| 33 19 9527 | Bulk Waste Containing Over 5% Halogen, Extra Charges | GAL | 0.17 | 0.17 | 0.17 | 0.17 | 0.17 |
| 33 19 9528 | Bulk Waste Containing Over 10% Ash, Extra Charges | GAL | 0.09 | 0.09 | 0.09 | 0.09 | 0.09 |
| 33 19 9529 | Drummed or Boxed PCB Capacitors Incineration | LB | 1.58 | 1.58 | 1.58 | 1.58 | 1.58 |
| 33 19 9530 | PCB Debris (Clothing, Rags, etc.) Incineration | LB | 1.28 | 1.28 | 1.28 | 1.28 | 1.28 |
| 33 19 9531 | PCB Contaminated Soil Incineration | LB | 1.58 | 1.58 | 1.58 | 1.58 | 1.58 |
| 33 19 9532 | PCB Light Ballasts Incineration | LB | 1.59 | 1.59 | 1.59 | 1.59 | 1.59 |
| 33 19 9533 | Drummed PCB Transformer Incineration, 55 Gallon | EACH | 522.50 | 522.50 | 522.50 | 522.50 | 522.50 |
| 33 19 9534 | Incineration of Drummed PCB Solid Waste, 55 Gallon | EACH | 60.95 | 60.95 | 60.95 | 60.95 | 60.95 |
| 33 19 9535 | Drummed PCB Liquids < 100,000 PPM Incineration, 55 Gallon | EACH | 312.50 | 312.50 | 312.50 | 312.50 | 312.50 |
| 33 19 9536 | Drummed PCB Liquids > 100,000 PPM Incineration, 55 Gallon | EACH | 362.50 | 362.50 | 362.50 | 362.50 | 362.50 |
| 33 19 9537 | Lean Water Incineration for 55 Gallon Drummed PCB Waste | EACH | 278.25 | 278.25 | 278.25 | 278.25 | 278.25 |
| 33 19 9538 | Palletized PCB Solid Waste, Single-layer Stack, Incineration | EACH | 567.50 | 567.50 | 567.50 | 567.50 | 567.50 |
| 33 19 9539 | Empty Uncrushed Drums Incineration | EACH | 57.50 | 57.50 | 57.50 | 57.50 | 57.50 |

**Environmental Remediation: Assemblies Cost Book**

# Cost Data for Remediation

## Commercial Disposal (Incineration)

| Assembly | | | Description | Unit | Unit Cost by Safety Level | | | | |
|---|---|---|---|---|---|---|---|---|---|
| | | | | | A | B | C | D | E |
| | | | **Capital Cost** | | | | | | |
| 33 | 19 | 9540 | Incineration of Bulk PCB Solid Waste (2,000 Lb/CY) | CY | 1,802 | 1,802 | 1,802 | 1,802 | 1,802 |
| 33 | 19 | 9541 | Lean Water Incineration of Bulk PCB Waste | GAL | 4.02 | 4.02 | 4.02 | 4.02 | 4.02 |
| 33 | 19 | 9542 | Incineration of Boxes & Nondrummed PCB Solid Wastes | CY | 333.45 | 333.45 | 333.45 | 333.45 | 333.45 |
| 33 | 19 | 9911 | Air Blower to Inflate Portable Containment Berm | EACH | 160.59 | 160.59 | 160.59 | 160.59 | 160.59 |
| 33 | 19 | 9921 | DOT Steel Drum, 55 Gallon | EACH | 65.19 | 65.19 | 65.19 | 65.19 | 65.19 |
| 33 | 19 | 9922 | Polyethylene Closed Head Drum, 55 Gallon | EACH | 43.63 | 43.63 | 43.63 | 43.63 | 43.63 |
| 33 | 19 | 9923 | Polyethylene Open Head Drum, 55 Gallon | EACH | 49.90 | 49.90 | 49.90 | 49.90 | 49.90 |
| 33 | 19 | 9924 | DOT Steel Salvage Drum, 85 Gallon, 16 Gauge | EACH | 57.06 | 57.06 | 57.06 | 57.06 | 57.06 |
| 33 | 19 | 9925 | DOT Steel Salvage Drum Composite Overpack, 85 Gallon, 16 Gauge | EACH | 70.49 | 70.49 | 70.49 | 70.49 | 70.49 |

**Environmental Remediation: Assemblies Cost Book**

# Cost Data for Remediation
## Composting

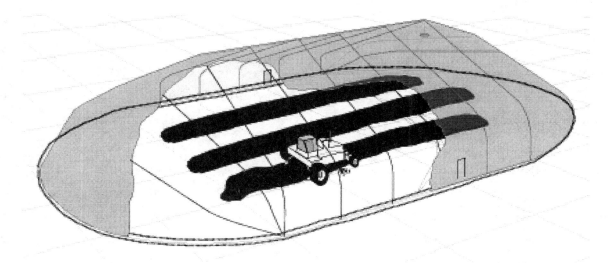

### General:
Composting is a process wherein biodegradable organic wastes such as sewage sludge, grass trimmings, and food scraps are converted into a humus-like material. Petroleum, oils, and lubricants (POL) such as diesel fuel, gasoline, and kerosene, and some explosive-compounds are also biodegradable under certain aerobic conditions. In the composting process, indigenous (naturally occurring) or exogenous (supplemented) microorganisms convert, to an extent, biodegradable organics into carbon dioxide, water, humus, leachate, and additional biomass (cells). There are three common methods of composting biodegradable wastes; static piles, aerated static piles, and windrows. Although the methods differ in whether the soil is periodically overturned upon initial placement, each shares common criteria necessary for aerobic composting.

### Typical Treatment Train:
N/A

### Common Cost Components:
1. Clearing: removal of vegetation such as trees, shrubs, and brush.
2. Grubbing: removal of stumps, roots, and debris from the soil by using dozers or other heavy equipment
3. Soil stripping: removal of topsoil
4. Excavation of contaminated soils

### Other Cost Considerations:
Overhead Electrical Distribution, Access Roads, Clear and Grub, Fencing, Water Distribution, Cleanup and Landscaping.

**Environmental Remediation: Assemblies Cost Book**

# Cost Data for Remediation

## Composting

| Assembly | Description | Unit | A | B | C | D | E |
|---|---|---|---|---|---|---|---|
| | **Capital Costs** | | | | | | |
| 17 03 0102 | Rough Grading, 12G, 1 Pass | SY | 1.28 | 1.04 | 0.85 | 0.62 | 0.59 |
| 17 03 0106 | Fine Grading, 12G, 2 Passes | SY | 0.62 | 0.50 | 0.42 | 0.30 | 0.28 |
| 17 03 0419 | Crushed Stone, 1/2" to 3/4" | CY | 23.45 | 21.67 | 20.70 | 18.97 | 18.47 |
| 17 03 0425 | Sand, 6" Lifts, On-Site | CY | 13.88 | 11.35 | 9.36 | 6.87 | 6.54 |
| 17 03 0426 | Sand, 6" Lifts, Off-Site | CY | 15.55 | 14.34 | 13.42 | 12.22 | 12.04 |
| 17 03 0430 | Gravel, 6" Lifts | CY | 15.46 | 13.36 | 12.05 | 9.99 | 9.49 |
| 17 03 9911 | 2' High Earthen Berm | CY | 4.03 | 4.03 | 4.03 | 4.03 | 4.03 |
| 18 01 0206 | Concrete Curb, 8" x 8" | LF | 7.58 | 6.24 | 5.66 | 4.36 | 3.89 |
| 18 01 0207 | Concrete Curb, 12" High x 8" Deep | LF | 10.27 | 8.45 | 7.67 | 5.92 | 5.27 |
| 18 02 0301 | Asphalt Pavement - 10" Subgrade, 9" Base, 1 1/2" Topping | SY | 61.11 | 50.94 | 43.58 | 33.57 | 31.83 |
| 19 01 0207 | 4", Class 200, PVC Piping | LF | 12.33 | 9.86 | 8.72 | 6.32 | 5.50 |
| 19 01 0208 | 6", Class 200, PVC Piping | LF | 14.95 | 12.20 | 10.93 | 8.27 | 7.36 |
| 19 04 0403 | 4,000 Gallon Polyethylene Wastewater Tank, Rental | MONTH | 545.90 | 545.90 | 545.90 | 545.90 | 545.90 |
| 19 04 0405 | 6,000 Gallon, Polyethylene Aboveground Wastewater Holding Tank, Rental | MONTH | 641.30 | 641.30 | 641.30 | 641.30 | 641.30 |
| 19 04 0406 | 21,000 Gallon Steel Wastewater Holding Tank, Rental | MONTH | 1,219 | 1,219 | 1,219 | 1,219 | 1,219 |
| 33 07 9905 | Purchase & Install Portable Building, 18' Ceiling | SF | 6.09 | 5.63 | 5.41 | 4.95 | 4.80 |
| 33 07 9907 | Disassemble Portable Building, 18' Ceiling, without Slab | SF | 1.26 | 0.98 | 0.85 | 0.57 | 0.48 |
| 33 08 0506 | Clay 10E-7, 6" Lifts, On-Site | CY | 29.50 | 23.82 | 19.83 | 14.25 | 13.21 |
| 33 08 0507 | Clay 10E-7, 6" Lifts, Off-Site | CY | 31.11 | 26.22 | 22.84 | 18.04 | 17.11 |
| 33 08 0511 | Drainage Netting, 1/4" Thick High-density Polyethylene | SF | 0.23 | 0.21 | 0.20 | 0.19 | 0.18 |
| 33 08 0512 | Drainage Netting, Geotextile Fabric Heat-bonded 1 Side | SF | 0.41 | 0.39 | 0.38 | 0.36 | 0.35 |
| 33 08 0513 | Drainage Netting, Geotextile Fabric Heat-bonded 2 Sides | SF | 0.48 | 0.45 | 0.44 | 0.42 | 0.41 |
| 33 08 0531 | 60 Mil Geotextile, Nonwoven | SY | 1.53 | 1.31 | 1.21 | 1.00 | 0.93 |
| 33 08 0532 | 80 Mil Geotextile, Nonwoven | SY | 1.57 | 1.35 | 1.25 | 1.04 | 0.97 |
| 33 08 0533 | 105 Mil Geotextile, Nonwoven | SY | 1.92 | 1.65 | 1.52 | 1.26 | 1.17 |
| 33 08 0534 | 130 Mil Geotextile, Nonwoven | SY | 2.56 | 2.20 | 2.03 | 1.68 | 1.56 |
| 33 08 0535 | 170 Mil Geotextile, Nonwoven | SY | 2.96 | 2.60 | 2.43 | 2.08 | 1.96 |
| 33 08 0541 | 20 Mil Polymeric Liner, Very Low Density Polyethylene | SF | 1.25 | 1.02 | 0.91 | 0.69 | 0.61 |
| 33 08 0542 | 30 Mil Polymeric Liner, Very Low Density Polyethylene | SF | 1.98 | 1.59 | 1.41 | 1.03 | 0.90 |
| 33 08 0543 | 40 Mil Polymeric Liner, Very Low Density Polyethylene | SF | 2.38 | 1.91 | 1.70 | 1.24 | 1.09 |
| 33 08 0544 | 60 Mil Polymeric Liner, Very Low Density Polyethylene | SF | 3.54 | 2.83 | 2.51 | 1.83 | 1.60 |
| 33 08 0545 | 80 Mil Polymeric Liner, Very Low Density Polyethylene | SF | 4.35 | 3.49 | 3.10 | 2.27 | 1.98 |
| 33 08 0546 | 100 Mil Polymeric Liner, Very Low Density Polyethylene | SF | 5.50 | 4.41 | 3.91 | 2.85 | 2.49 |

**Environmental Remediation: Assemblies Cost Book**

*Page 3-70*

# Cost Data for Remediation

## Composting

| Assembly | Description | Unit | Unit Cost by Safety Level A | B | C | D | E |
|---|---|---|---|---|---|---|---|
| | **Capital Costs** | | | | | | |
| 33 08 0551 | 36 Mil Polymeric Liner, Chlorosulfanated Polyethylene | SF | 0.75 | 0.73 | 0.72 | 0.70 | 0.70 |
| 33 08 0552 | 45 Mil Polymeric Liner, Chlorosulfanated Polyethylene | SF | 0.83 | 0.81 | 0.80 | 0.79 | 0.78 |
| 33 08 0561 | 20 Mil Polymeric Liner, PVC | SF | 0.24 | 0.23 | 0.22 | 0.21 | 0.20 |
| 33 08 0562 | 30 Mil Polymeric Liner, PVC | SF | 0.32 | 0.31 | 0.30 | 0.29 | 0.28 |
| 33 08 0563 | 40 Mil Polymeric Liner, PVC | SF | 0.38 | 0.36 | 0.36 | 0.34 | 0.34 |
| 33 08 0564 | 60 Mil Polymeric Liner, PVC | SF | 0.72 | 0.70 | 0.69 | 0.67 | 0.67 |
| 33 08 0565 | 80 Mil Polymeric Liner, PVC | SF | 0.98 | 0.96 | 0.95 | 0.93 | 0.93 |
| 33 08 0566 | 100 Mil Polymeric Liner, PVC | SF | 1.28 | 1.26 | 1.25 | 1.23 | 1.22 |
| 33 08 0571 | 40 Mil Polymeric Liner, High-density Polyethylene | SF | 2.35 | 1.88 | 1.67 | 1.21 | 1.06 |
| 33 08 0572 | 60 Mil Polymeric Liner, High-density Polyethylene | SF | 3.49 | 2.80 | 2.48 | 1.81 | 1.58 |
| 33 08 0573 | 80 Mil Polymeric Liner, High-density Polyethylene | SF | 4.32 | 3.47 | 3.08 | 2.25 | 1.97 |
| 33 08 0584 | Plastic Laminate Waste Pile Cover | SF | 0.16 | 0.15 | 0.14 | 0.13 | 0.13 |
| 33 08 0590 | Waste Pile Cover, 135 Lb Tear, 2 - 2.5 Year Life | SY | 2.03 | 1.93 | 1.89 | 1.80 | 1.77 |
| 33 08 0591 | Waste Pile Cover, 185 Lb Tear, 3 - 4 Year Life | SY | 2.98 | 2.89 | 2.85 | 2.76 | 2.72 |
| 33 08 0592 | Waste Pile Cover, 250 Lb Tear, 4 - 5 Year Life | SY | 4.00 | 3.89 | 3.85 | 3.74 | 3.70 |
| 33 10 9657 | 1,000 Gallon Single-wall Steel Aboveground Tank | EACH | 2,118 | 1,844 | 1,727 | 1,463 | 1,366 |
| 33 10 9658 | 2,000 Gallon Single-wall Steel Aboveground Tank | EACH | 3,457 | 2,997 | 2,777 | 2,332 | 2,183 |
| 33 10 9659 | 3,000 Gallon Single-wall Steel Aboveground Tank | EACH | 4,433 | 3,959 | 3,733 | 3,274 | 3,121 |
| 33 10 9660 | 5,000 Gallon Single-wall Steel Aboveground Tank | EACH | 6,079 | 5,494 | 5,214 | 4,646 | 4,457 |
| 33 10 9661 | 8,000 Gallon Single-wall Steel Aboveground Tank | EACH | 7,876 | 7,205 | 6,884 | 6,234 | 6,017 |
| 33 10 9662 | 10,000 Gallon Single-wall Steel Aboveground Tank | EACH | 9,147 | 8,415 | 8,065 | 7,356 | 7,119 |
| 33 10 9663 | 12,000 Gallon Single-wall Steel Aboveground Tank | EACH | 10,460 | 9,655 | 9,270 | 8,490 | 8,230 |
| 33 10 9664 | 15,000 Gallon Single-wall Steel Aboveground Tank | EACH | 11,591 | 10,697 | 10,269 | 9,402 | 9,113 |
| 33 10 9665 | 20,000 Gallon Single-wall Steel Aboveground Tank | EACH | 14,594 | 13,445 | 12,895 | 11,780 | 11,408 |
| 33 10 9666 | 30,000 Gallon Single-wall Steel Aboveground Tank | EACH | 6,427 | 5,085 | 4,444 | 3,143 | 2,709 |
| 33 11 0701 | Vibrating Grizzly Screen w/3" Screen Openings | EA | 30,047 | 30,037 | 30,032 | 30,023 | 30,020 |
| 33 11 0702 | 75 CY/hr Rotary Auger Mixer/Blender | EA | 23,016 | 22,426 | 22,118 | 21,544 | 21,370 |
| 33 11 0703 | 100 CY/hr Rotary Auger Mixer/Blender | EA | 24,214 | 24,207 | 24,204 | 24,197 | 24,195 |
| 33 11 0704 | 150 CY/hr Rotary Auger Mixer/Blender | EA | 30,984 | 30,975 | 30,970 | 30,962 | 30,959 |
| 33 11 0705 | 25 CY/hr Alfalfa Tub Grinder | EA | 23,949 | 23,943 | 23,940 | 23,935 | 23,933 |
| 33 11 0706 | 50 CY/hr Alfalfa Tub Grinder | EA | 36,443 | 36,435 | 36,430 | 36,422 | 36,419 |
| 33 11 0707 | 75 CY/hr Alfalfa Tub Grinder | EA | 46,855 | 46,844 | 46,839 | 46,828 | 46,825 |
| 33 11 0711 | Startup Compost Supply | CY | 8.56 | 8.55 | 8.54 | 8.53 | 8.53 |

**Environmental Remediation: Assemblies Cost Book**

# Cost Data for Remediation

## Composting

| Assembly | Description | Unit | Unit Cost by Safety Level A | B | C | D | E |
|---|---|---|---|---|---|---|---|
| | **Capital Costs** | | | | | | |
| 33 23 1306 | High Sump Level Switch for Avoiding Overflow | EACH | 214.65 | 214.65 | 214.65 | 214.65 | 214.65 |
| 33 26 0104 | 4" Carbon Steel Piping | LF | 23.20 | 19.86 | 18.43 | 15.21 | 14.02 |
| 33 26 0105 | 6" Carbon Steel Piping | LF | 36.37 | 30.24 | 27.46 | 21.52 | 19.44 |
| 33 26 0901 | 4" Diameter Perforated PVC Pipe | LF | 13.24 | 10.73 | 9.22 | 6.77 | 6.15 |
| 33 26 0902 | 6" Diameter Perforated PVC Pipe | LF | 14.13 | 11.62 | 10.11 | 7.66 | 7.04 |
| 33 26 0903 | 8" Diameter Perforated PVC Pipe | LF | 14.56 | 12.01 | 10.49 | 8.00 | 7.37 |
| 33 26 0904 | 10" Diameter Perforated PVC Pipe | LF | 16.84 | 14.07 | 12.42 | 9.71 | 9.02 |
| 33 27 0104 | 4" PVC, Schedule 40, Tee | EACH | 137.04 | 108.84 | 96.77 | 69.51 | 59.51 |
| 33 27 0105 | 6" PVC, Schedule 40, Tee | EACH | 205.86 | 169.47 | 153.89 | 118.72 | 105.82 |
| 33 27 0114 | 4" PVC, 90 Degree, Elbow | EACH | 101.72 | 80.57 | 71.51 | 51.07 | 43.57 |
| 33 27 0115 | 6" PVC, 90 Degree, Elbow | EACH | 135.44 | 111.27 | 100.92 | 77.56 | 68.99 |
| 33 27 0552 | 4" Carbon Steel 90-degree Elbow, Schedule 80 | EACH | 190.36 | 152.53 | 135.42 | 98.81 | 85.97 |
| 33 27 0554 | 6" Carbon Steel 90-degree Elbow, Schedule 80 | EACH | 297.54 | 245.36 | 221.75 | 171.24 | 153.54 |
| 33 27 0562 | 4" Carbon Steel Tee, Schedule 80 | EACH | 307.17 | 250.43 | 224.76 | 169.84 | 150.59 |
| 33 27 0564 | 6" Carbon Steel Tee, Schedule 80 | EACH | 452.39 | 374.65 | 339.48 | 264.23 | 237.85 |
| 33 29 0401 | 25 GPM, 1 1/2" Discharge, Cast-iron Sump Pump | EACH | 2,432 | 2,274 | 2,206 | 2,053 | 1,997 |
| 33 29 0402 | 75 GPM, 2" Discharge, Cast-iron Sump Pump | EACH | 3,011 | 2,820 | 2,738 | 2,552 | 2,484 |
| 33 29 0403 | 100 GPM, 2 1/2" Discharge, Cast-iron Sump Pump | EACH | 3,243 | 3,016 | 2,918 | 2,698 | 2,618 |
| 33 29 0404 | 150 GPM, 3" Discharge, Cast-iron Sump Pump | EACH | 3,703 | 3,422 | 3,302 | 3,032 | 2,932 |
| 33 29 0405 | 200 GPM, 3" Discharge, Cast-iron Sump Pump | EACH | 3,904 | 3,601 | 3,471 | 3,178 | 3,070 |
| 33 29 0406 | 300 GPM, 4" Discharge, Cast-iron Sump Pump | EACH | 4,817 | 4,453 | 4,297 | 3,945 | 3,816 |
| 33 29 0407 | 500 GPM, 5" Discharge, Cast-iron Sump Pump | EACH | 4,969 | 4,610 | 4,456 | 4,110 | 3,982 |
| 33 29 0408 | 800 GPM, 6" Discharge, Cast-iron Sump Pump | EACH | 12,550 | 12,113 | 11,911 | 11,488 | 11,342 |
| 33 29 0409 | 1,000 GPM, 6" Discharge, Cast-iron Sump Pump | EACH | 15,213 | 14,703 | 14,467 | 13,974 | 13,804 |
| 33 31 0150 | 500 CFM, 6" Pressure, 1 HP, Blower | EACH | 974.03 | 882.23 | 842.93 | 754.20 | 721.67 |
| 33 31 0151 | 750 CFM, 8" Pressure, 1.5 HP, Blower | EACH | 1,207 | 1,100 | 1,055 | 952.48 | 914.88 |
| 33 31 0152 | 1,000 CFM, 8" Pressure, 2.5 HP, Blower | EACH | 1,487 | 1,373 | 1,324 | 1,213 | 1,172 |
| 33 31 0153 | 1,500 CFM, 8" Pressure, 4 HP, Blower | EACH | 1,783 | 1,642 | 1,582 | 1,447 | 1,397 |
| 33 31 0154 | 2,000 CFM, 7" Pressure, 6 HP, Blower | EACH | 2,658 | 2,389 | 2,274 | 2,015 | 1,920 |
| 33 31 0155 | 3,000 CFM, 6" Pressure, 7.5 HP, Blower | EACH | 2,500 | 2,232 | 2,117 | 1,858 | 1,762 |
| 33 31 0156 | 4,000 CFM, 6" Pressure, 10 HP, Blower | EACH | 2,747 | 2,456 | 2,332 | 2,051 | 1,948 |
| 33 31 0157 | 5,000 CFM, 6" Pressure, 10 HP, Blower | EACH | 3,037 | 2,746 | 2,622 | 2,341 | 2,238 |
| | **Operations & Maintenance** | | | | | | |

**Environmental Remediation: Assemblies Cost Book**

*Page 3-72*

1998 by ECHOS. All rights reserved

# Cost Data for Remediation

## Composting

| Assembly | Description | Unit | \[Unit Cost by Safety Level\] A | B | C | D | E |
|---|---|---|---|---|---|---|---|
| | **Operations & Maintenance** | | | | | | |
| 17 03 0220 | 910, 1.25 CY, Wheel Loader | HOUR | 130.20 | 104.06 | 87.23 | 61.63 | 55.86 |
| 17 03 0221 | 916, 1.5 CY, Wheel Loader | HOUR | 114.32 | 90.83 | 76.64 | 53.69 | 47.92 |
| 17 03 0222 | 926, 2.0 CY, Wheel Loader | HOUR | 154.30 | 124.14 | 103.30 | 73.68 | 67.91 |
| 17 03 0223 | 950, 3.0 CY, Wheel Loader | HOUR | 180.10 | 145.64 | 120.50 | 86.58 | 80.81 |
| 17 03 0224 | 966, 4.0 CY, Wheel Loader | HOUR | 211.58 | 171.88 | 141.48 | 102.32 | 96.55 |
| 17 03 0225 | 980, 5.25 CY, Wheel Loader | HOUR | 298.84 | 244.38 | 199.68 | 145.79 | 139.74 |
| 17 03 0284 | 8 CY, Dump Truck | HOUR | 142.67 | 115.22 | 95.47 | 68.47 | 63.71 |
| 17 03 0285 | 12 CY, Dump Truck | HOUR | 119.55 | 95.96 | 80.05 | 56.91 | 52.15 |
| 17 03 0287 | 20 CY, Semi Dump | HOUR | 161.11 | 130.59 | 107.76 | 77.69 | 72.93 |
| 17 03 0288 | 26 CY, Semi Dump | HOUR | 149.99 | 121.33 | 100.35 | 72.13 | 67.37 |
| 17 03 0431 | 580K, 1.0 CY, Backhoe with Front-end Loader | HOUR | 127.72 | 101.78 | 85.60 | 60.23 | 54.18 |
| 18 05 0410 | Fertilize, 800 Lbs/Acre, Spray from Truck | ACRE | 107.45 | 90.70 | 78.22 | 61.72 | 59.08 |
| 18 05 0411 | Crushed Limestone, pH Adjustment, 800 Lbs/Acre | ACRE | 156.71 | 134.79 | 121.37 | 99.94 | 94.68 |
| 18 05 0413 | Watering with 3,000-Gallon Tank Truck, per Pass | ACRE | 111.81 | 90.08 | 75.73 | 54.43 | 49.85 |
| 33 11 0712 | Bulking Agents, Compost Amendments | CY | 11.56 | 11.55 | 11.54 | 11.53 | 11.53 |
| 33 11 0713 | Agricultural Wastes, Compost Amendments | CY | 12.76 | 12.75 | 12.74 | 12.73 | 12.73 |
| 33 11 0714 | Porous Wood Chip Aeration Base | CY | 11.62 | 11.59 | 11.58 | 11.56 | 11.55 |
| 33 11 0721 | Compost Plant Operating Crew | HR | 1.89 | 1.46 | 1.27 | 0.85 | 0.70 |
| 33 11 0722 | 12' Wide Windrow Turner | HR | 3.30 | 2.68 | 2.20 | 1.60 | 1.51 |
| 33 11 9901 | Application of Bioculture to Contaminated Soil | ACRE | 109.25 | 87.52 | 73.18 | 51.87 | 47.29 |
| 33 11 9902 | Light Petroleum Hydrocarbon Degraders, Microorganisms | LB | 19.88 | 19.88 | 19.88 | 19.88 | 19.88 |
| 33 11 9903 | Light Petroleum Hydrocarbon Degraders, 100 Lb Bag, Microorganisms | EACH | 1,948 | 1,948 | 1,948 | 1,948 | 1,948 |
| 33 11 9904 | Light Petroleum Hydrocarbon Degraders, 1,000 Lb Bag, Microorganisms | EACH | 17,490 | 17,490 | 17,490 | 17,490 | 17,490 |
| 33 11 9905 | Heavy Petroleum Hydrocarbon/Creosote Degraders, Microorganisms | LB | 13.25 | 13.25 | 13.25 | 13.25 | 13.25 |
| 33 11 9906 | Heavy Petroleum Hydrocarbon/Creosote Degraders, 100 Lb Bag, Microorganisms | EACH | 1,007 | 1,007 | 1,007 | 1,007 | 1,007 |
| 33 11 9907 | Heavy Petroleum Hydrocarbon/Creosote Degraders, 1,000 Lb Bag, Microorganisms | EACH | 8,745 | 8,745 | 8,745 | 8,745 | 8,745 |
| 33 11 9951 | Bionutrients, 50 Lb Bag | EACH | 58.30 | 58.30 | 58.30 | 58.30 | 58.30 |
| 33 42 0101 | Electrical Charge | KWH | 0.06 | 0.06 | 0.06 | 0.06 | 0.06 |
| 33 42 0120 | Generator Supplied Electricity | kWh | 0.08 | 0.08 | 0.08 | 0.08 | 0.08 |
| 33 42 0301 | Water | KGAL | 8.39 | 8.39 | 8.39 | 8.39 | 8.39 |

**Environmental Remediation: Assemblies Cost Book**

# Cost Data for Remediation

## Composting

| Assembly | Description | Unit | Unit Cost by Safety Level | | | | |
|---|---|---|---|---|---|---|---|
| | | | A | B | C | D | E |
| | **Operations & Maintenance** | | | | | | |
| 33 42 0302 | Process Water, Supplied by Water Line | KGAL | 2.56 | 2.56 | 2.56 | 2.56 | 2.56 |

# Cost Data for Remediation
## Decontamination Facilities

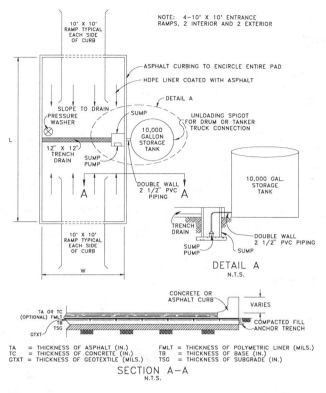

### General:

Decontamination is the process of removing or neutralizing contaminants that have accumulated on personnel or equipment. The costs presented here include the capital cost of constructing an equipment decontamination facility, operation of that facility, and personnel decontamination.

### Typical Treatment Train:

Applicable to any hazardous waste site project.

### Common Cost Components:

1. Decontamination pad
2. Liners
3. Wash water storage tank
4. Sump pumps
5. Decontamination equipment
6. Piping
7. Trailer facilities (change rooms, showers)
8. Storage containers
9. Operations and maintenance

### Other Cost Considerations:

Electrical distribution, access road, parking lot, fencing.

**Environmental Remediation: Assemblies Cost Book**

# Cost Data for Remediation

## Decontamination Facilities

| Assembly | Description | Unit | Unit Cost by Safety Level | | | | |
|---|---|---|---|---|---|---|---|
| | | | A | B | C | D | E |
| | **Capital Costs** | | | | | | |
| 17 01 0106 | Heavy Brush, Light Trees, Clear, Grub, Haul | ACRE | 11,338 | 9,081 | 7,594 | 5,382 | 4,905 |
| 17 01 0107 | Medium Brush, Medium Trees, Clear, Grub, Haul | ACRE | 13,478 | 10,783 | 9,029 | 6,389 | 5,807 |
| 17 01 0108 | Light Brush, Heavy Trees, Clear, Grub, Haul | ACRE | 17,877 | 14,287 | 11,977 | 8,462 | 7,669 |
| 17 03 0109 | Pad Subgrade Preparation | CY | 8.13 | 6.56 | 5.44 | 3.90 | 3.63 |
| 17 03 0257 | Cat 215, 1.0 CY, Soil, Shallow, Trenching | CY | 2.42 | 1.95 | 1.62 | 1.16 | 1.07 |
| 17 03 0501 | Compact Subgrade, 2 Lifts | CY | 0.84 | 0.68 | 0.56 | 0.40 | 0.37 |
| 17 03 0510 | Dry Roll Gravel, Steel Roller | SY | 1.22 | 0.96 | 0.82 | 0.57 | 0.49 |
| 17 03 0511 | Compact Soil with Vibrating Plate | CY | 7.20 | 5.57 | 4.84 | 3.26 | 2.70 |
| 17 03 0515 | Compact With Pogosticks | CY | 5.91 | 4.58 | 3.98 | 2.68 | 2.23 |
| 17 03 9903 | Hand Place Small Earth Fill Berm | CY | 112.16 | 86.46 | 75.45 | 50.61 | 41.50 |
| 18 01 0102 | Gravel, Delivered & Dumped | CY | 24.33 | 22.93 | 22.03 | 20.66 | 20.36 |
| 18 01 0103 | Gravel (90%) & Sand Base (10%), with Calcium Chloride 3/4 - 1 Lb/CY | CY | 25.70 | 24.15 | 23.16 | 21.65 | 21.31 |
| 18 01 0104 | Asphalt, Intermediate Course (Line Item Includes 5% Waste) | TON | 76.50 | 65.72 | 59.09 | 48.56 | 45.99 |
| 18 01 0201 | Concrete Curb, 6" x 6" | LF | 2.78 | 2.38 | 2.21 | 1.83 | 1.70 |
| 18 01 0202 | Concrete Curb & Gutter, 6" x 24", Formed | LF | 20.78 | 16.61 | 14.83 | 10.80 | 9.32 |
| 18 01 0203 | Asphalt Curb 8" W x 6" H | LF | 12.00 | 9.93 | 8.44 | 6.40 | 6.03 |
| 18 01 0204 | Asphalt Curb 8" W x 8" H | LF | 12.37 | 10.29 | 8.80 | 6.76 | 6.40 |
| 18 01 0205 | Curb Inlet Frame Grate & Box, 525 Lb | EACH | 6,004 | 4,987 | 4,534 | 3,550 | 3,200 |
| 18 01 0310 | Prime Coat | SY | 0.55 | 0.52 | 0.51 | 0.49 | 0.48 |
| 18 01 0311 | Tack Coat | SY | 0.70 | 0.64 | 0.60 | 0.53 | 0.52 |
| 18 01 0312 | Asphalt Wearing Course, 1 Pass (Line Item Includes 5% Waste) | TON | 83.61 | 72.83 | 66.20 | 55.67 | 53.10 |
| 18 02 0203 | 26" x 26", 5' Deep Area Drain with Grate | EACH | 3,425 | 2,959 | 2,752 | 2,302 | 2,141 |
| 18 02 0204 | 27" x 20", 5' Deep Area Drain with Grate | EACH | 3,346 | 2,880 | 2,673 | 2,223 | 2,062 |
| 18 02 0205 | 24" Diameter, 5' Deep Area Drain with Grate | EACH | 3,131 | 2,673 | 2,470 | 2,027 | 1,869 |
| 18 02 0320 | 4" Structural Slab on Grade | SF | 6.45 | 5.45 | 4.98 | 4.00 | 3.67 |
| 18 02 0321 | 6" Structural Slab on Grade | SF | 6.98 | 5.95 | 5.47 | 4.48 | 4.14 |
| 18 02 0322 | 8" Structural Slab on Grade | SF | 9.75 | 8.38 | 7.71 | 6.38 | 5.95 |
| 18 02 0323 | 10" Structural Slab on Grade | SF | 10.56 | 9.13 | 8.43 | 7.04 | 6.58 |
| 18 02 0324 | 12" Structural Slab on Grade | SF | 11.57 | 10.04 | 9.30 | 7.82 | 7.34 |
| 18 02 0330 | 4" Mesh Reinforced Slab on Grade | SF | 5.48 | 4.60 | 4.18 | 3.32 | 3.03 |
| 18 02 0331 | 6" Mesh Reinforced Slab on Grade | SF | 6.01 | 5.10 | 4.67 | 3.79 | 3.49 |

**Environmental Remediation: Assemblies Cost Book**

*Page 3-76*

1998 by ECHOS. All rights reserved

# Cost Data for Remediation

## Decontamination Facilities

| Assembly | Description | Unit | Unit Cost by Safety Level A | B | C | D | E |
|---|---|---|---|---|---|---|---|
| | **Capital Costs** | | | | | | |
| 18 02 0332 | 8" Mesh Reinforced Slab on Grade | SF | 7.58 | 6.50 | 5.96 | 4.91 | 4.58 |
| 18 02 0333 | 10" Mesh Reinforced Slab on Grade | SF | 8.10 | 7.00 | 6.44 | 5.37 | 5.03 |
| 18 02 0334 | 12" Mesh Reinforced Slab on Grade | SF | 8.78 | 7.63 | 7.05 | 5.93 | 5.58 |
| 18 02 0340 | 4" Unreinforced Slab on Grade | SF | 4.34 | 3.64 | 3.29 | 2.61 | 2.38 |
| 18 02 0341 | 6" Unreinforced Slab on Grade | SF | 4.87 | 4.14 | 3.79 | 3.08 | 2.85 |
| 18 02 0342 | 8" Unreinforced Slab on Grade | SF | 6.45 | 5.54 | 5.07 | 4.19 | 3.93 |
| 18 02 0343 | 10" Unreinforced Slab on Grade | SF | 6.97 | 6.04 | 5.56 | 4.66 | 4.38 |
| 18 02 0344 | 12" Unreinforced Slab on Grade | SF | 7.64 | 6.67 | 6.16 | 5.22 | 4.93 |
| 18 04 0205 | CIP Walls Form & Strip (4 Uses) | SF | 8.86 | 6.99 | 6.19 | 4.38 | 3.73 |
| 19 02 0311 | 12' x 36" Diameter Reinforced Concrete Pipe Wet Well for Lift Station | EACH | 9,840 | 8,577 | 7,816 | 6,581 | 6,270 |
| 19 02 0312 | 24' x 60" Diameter Reinforced Concrete Pipe Wet Well for Lift Station | EACH | 39,142 | 33,897 | 30,681 | 25,554 | 24,296 |
| 19 02 0313 | 5' x 5' x 5' Reinforced Concrete Sump | EACH | 3,948 | 3,282 | 2,979 | 2,335 | 2,110 |
| 19 02 0604 | 12" x 12" CIP Concrete In-Ground Trench Drain with Metal Grate | LF | 137.91 | 117.00 | 107.92 | 87.71 | 80.38 |
| 19 03 0112 | 8" Corrugated Metal Pipe, Plain | LF | 8.69 | 7.91 | 7.51 | 6.75 | 6.51 |
| 19 03 0113 | 10" Corrugated Metal Pipe, Plain | LF | 9.99 | 9.21 | 8.81 | 8.05 | 7.81 |
| 19 03 0114 | 12" Corrugated Metal Pipe, Plain | LF | 11.23 | 10.38 | 9.96 | 9.13 | 8.87 |
| 19 03 0115 | 15" Corrugated Metal Pipe, Plain | LF | 13.37 | 12.43 | 11.95 | 11.04 | 10.74 |
| 19 03 0116 | 18" Corrugated Metal Pipe, Plain | LF | 16.76 | 15.33 | 14.61 | 13.22 | 12.77 |
| 19 03 0117 | 24" Corrugated Metal Pipe, Plain | LF | 24.55 | 22.64 | 21.68 | 19.82 | 19.23 |
| 19 03 0158 | 8" Extra Strength, Nonreinforced Concrete Pipe | LF | 12.86 | 10.79 | 9.75 | 7.74 | 7.10 |
| 19 03 0159 | 10" Extra Strength, Nonreinforced Concrete Pipe | LF | 16.79 | 14.59 | 13.49 | 11.36 | 10.68 |
| 19 03 0160 | 12" Extra Strength, Nonreinforced Concrete Pipe | LF | 17.97 | 15.63 | 14.46 | 12.20 | 11.48 |
| 19 03 0161 | 15" Extra Strength, Nonreinforced Concrete Pipe | LF | 19.95 | 17.36 | 16.06 | 13.55 | 12.74 |
| 19 03 0162 | 18" Extra Strength, Nonreinforced Concrete Pipe | LF | 24.93 | 22.06 | 20.62 | 17.83 | 16.94 |
| 19 03 0163 | 21" Extra Strength, Nonreinforced Concrete Pipe | LF | 30.32 | 27.45 | 26.01 | 23.22 | 22.33 |
| 19 03 0164 | 24" Extra Strength, Nonreinforced Concrete Pipe | LF | 39.69 | 36.49 | 34.88 | 31.77 | 30.78 |
| 19 03 0174 | 12" Reinforced Concrete Pipe, Class 4, with Gaskets | LF | 14.33 | 13.02 | 12.29 | 11.01 | 10.65 |
| 19 03 0175 | 15" Reinforced Concrete Pipe, Class 4, with Gaskets | LF | 16.47 | 14.84 | 13.94 | 12.36 | 11.91 |
| 19 03 0176 | 18" Reinforced Concrete Pipe, Class 4, with Gaskets | LF | 23.67 | 21.05 | 19.60 | 17.05 | 16.33 |
| 19 03 0178 | 24" Reinforced Concrete Pipe, Class 4, with Gaskets | LF | 34.58 | 31.22 | 29.36 | 26.09 | 25.17 |
| 19 04 0601 | 275 Gallon Steel Sump, Aboveground with Supports & Fittings | EACH | 1,325 | 1,153 | 1,080 | 913.28 | 852.29 |

**Environmental Remediation: Assemblies Cost Book**

# Cost Data for Remediation

## Decontamination Facilities

| Assembly | Description | Unit | Unit Cost by Safety Level | | | | |
|---|---|---|---|---|---|---|---|
| | | | A | B | C | D | E |
| | **Capital Costs** | | | | | | |
| 19 04 0602 | 550 Gallon Steel Sump, Aboveground with Supports & Fittings | EACH | 1,598 | 1,390 | 1,301 | 1,099 | 1,026 |
| 19 04 0603 | 1,000 Gallon Steel Sump, Aboveground with Supports & Fittings | EACH | 2,118 | 1,844 | 1,727 | 1,463 | 1,366 |
| 19 04 0604 | 1,500 Gallon Steel Sump, Aboveground with Supports & Fittings | EACH | 2,697 | 2,356 | 2,209 | 1,879 | 1,758 |
| 19 04 0605 | 2,000 Gallon Steel Sump, Aboveground with Supports & Fittings | EACH | 3,457 | 2,997 | 2,777 | 2,332 | 2,183 |
| 19 04 0606 | 5,000 Gallon Steel Sump, Aboveground with Supports & Fittings | EACH | 6,079 | 5,494 | 5,214 | 4,646 | 4,457 |
| 19 04 0621 | 550 Gallon Horizontal Plastic Sump with 4" NPT Connection | EACH | 1,699 | 1,596 | 1,551 | 1,452 | 1,415 |
| 19 04 0622 | 1,000 Gallon Horizontal Plastic Sump with 4" NPT Connection | EACH | 2,179 | 2,042 | 1,984 | 1,852 | 1,803 |
| 19 04 0623 | 2,000 Gallon Horizontal Plastic Sump with 6" NPT Connection | EACH | 3,089 | 2,918 | 2,845 | 2,680 | 2,619 |
| 19 04 0624 | 4,000 Gallon Horizontal Plastic Sump with 6" NPT Connection | EACH | 4,130 | 3,927 | 3,841 | 3,645 | 3,573 |
| 19 04 0625 | 6,000 Gallon Horizontal Plastic Sump with 6" NPT Connection | EACH | 5,248 | 4,975 | 4,858 | 4,594 | 4,497 |
| 19 04 0626 | 8,000 Gallon Horizontal Plastic Sump with 6" NPT Connection | EACH | 5,857 | 5,516 | 5,369 | 5,039 | 4,918 |
| 19 04 0627 | 10,000 Gallon Horizontal Plastic Sump with 6" NPT Connection | EACH | 7,082 | 6,622 | 6,402 | 5,956 | 5,807 |
| 19 04 0628 | 12,000 Gallon Horizontal Plastic Sump with 6" NPT Connection | EACH | 8,308 | 7,722 | 7,442 | 6,875 | 6,686 |
| 19 04 0629 | 15,000 Gallon Horizontal Plastic Sump with 6" NPT Connection | EACH | 14,184 | 13,540 | 13,232 | 12,608 | 12,400 |
| 19 04 0631 | 20,000 Gallon Horizontal Plastic Sump with 6" NPT Connection | EACH | 15,799 | 15,084 | 14,742 | 14,048 | 13,817 |
| 19 04 0632 | 25,000 Gallon Horizontal Plastic Sump with 6" NPT Connection | EACH | 24,407 | 23,487 | 23,047 | 22,155 | 21,858 |
| 19 04 0633 | 30,000 Gallon Horizontal Plastic Sump with 6" NPT Connection | EACH | 35,112 | 34,039 | 33,525 | 32,485 | 32,138 |
| 19 04 0634 | 40,000 Gallon Horizontal Plastic Sump with 6" NPT Connection | EACH | 40,844 | 39,234 | 38,465 | 36,904 | 36,383 |
| 19 04 0635 | 48,000 Gallon Horizontal Plastic Sump with 6" NPT Connection | EACH | 59,758 | 57,612 | 56,586 | 54,505 | 53,811 |
| 33 08 0503 | Polymeric Liner Anchor Trench, 3' x 1.5' | LF | 1.58 | 1.25 | 1.07 | 0.76 | 0.67 |
| 33 08 0511 | Drainage Netting, 1/4" Thick High-density Polyethylene | SF | 0.23 | 0.21 | 0.20 | 0.19 | 0.18 |
| 33 08 0512 | Drainage Netting, Geotextile Fabric Heat-bonded 1 Side | SF | 0.41 | 0.39 | 0.38 | 0.36 | 0.35 |
| 33 08 0513 | Drainage Netting, Geotextile Fabric Heat-bonded 2 Sides | SF | 0.48 | 0.45 | 0.44 | 0.42 | 0.41 |

**Environmental Remediation: Assemblies Cost Book**

# Cost Data for Remediation

## Decontamination Facilities

| Assembly | Description | Unit | Unit Cost by Safety Level A | B | C | D | E |
|---|---|---|---|---|---|---|---|
| | **Capital Costs** | | | | | | |
| 33 08 0531 | 60 Mil Geotextile, Nonwoven | SY | 1.53 | 1.31 | 1.21 | 1.00 | 0.93 |
| 33 08 0532 | 80 Mil Geotextile, Nonwoven | SY | 1.57 | 1.35 | 1.25 | 1.04 | 0.97 |
| 33 08 0533 | 105 Mil Geotextile, Nonwoven | SY | 1.92 | 1.65 | 1.52 | 1.26 | 1.17 |
| 33 08 0534 | 130 Mil Geotextile, Nonwoven | SY | 2.56 | 2.20 | 2.03 | 1.68 | 1.56 |
| 33 08 0535 | 170 Mil Geotextile, Nonwoven | SY | 2.96 | 2.60 | 2.43 | 2.08 | 1.96 |
| 33 08 0541 | 20 Mil Polymeric Liner, Very Low Density Polyethylene | SF | 1.25 | 1.02 | 0.91 | 0.69 | 0.61 |
| 33 08 0542 | 30 Mil Polymeric Liner, Very Low Density Polyethylene | SF | 1.98 | 1.59 | 1.41 | 1.03 | 0.90 |
| 33 08 0543 | 40 Mil Polymeric Liner, Very Low Density Polyethylene | SF | 2.38 | 1.91 | 1.70 | 1.24 | 1.09 |
| 33 08 0544 | 60 Mil Polymeric Liner, Very Low Density Polyethylene | SF | 3.54 | 2.83 | 2.51 | 1.83 | 1.60 |
| 33 08 0545 | 80 Mil Polymeric Liner, Very Low Density Polyethylene | SF | 4.35 | 3.49 | 3.10 | 2.27 | 1.98 |
| 33 08 0546 | 100 Mil Polymeric Liner, Very Low Density Polyethylene | SF | 5.50 | 4.41 | 3.91 | 2.85 | 2.49 |
| 33 08 0551 | 36 Mil Polymeric Liner, Chlorosulfanated Polyethylene | SF | 0.75 | 0.73 | 0.72 | 0.70 | 0.70 |
| 33 08 0552 | 45 Mil Polymeric Liner, Chlorosulfanated Polyethylene | SF | 0.83 | 0.81 | 0.80 | 0.79 | 0.78 |
| 33 08 0561 | 20 Mil Polymeric Liner, PVC | SF | 0.24 | 0.23 | 0.22 | 0.21 | 0.20 |
| 33 08 0562 | 30 Mil Polymeric Liner, PVC | SF | 0.32 | 0.31 | 0.30 | 0.29 | 0.28 |
| 33 08 0563 | 40 Mil Polymeric Liner, PVC | SF | 0.38 | 0.36 | 0.36 | 0.34 | 0.34 |
| 33 08 0564 | 60 Mil Polymeric Liner, PVC | SF | 0.72 | 0.70 | 0.69 | 0.67 | 0.67 |
| 33 08 0565 | 80 Mil Polymeric Liner, PVC | SF | 0.98 | 0.96 | 0.95 | 0.93 | 0.93 |
| 33 08 0566 | 100 Mil Polymeric Liner, PVC | SF | 1.28 | 1.26 | 1.25 | 1.23 | 1.22 |
| 33 08 0571 | 40 Mil Polymeric Liner, High-density Polyethylene | SF | 2.35 | 1.88 | 1.67 | 1.21 | 1.06 |
| 33 08 0572 | 60 Mil Polymeric Liner, High-density Polyethylene | SF | 3.49 | 2.80 | 2.48 | 1.81 | 1.58 |
| 33 08 0573 | 80 Mil Polymeric Liner, High-density Polyethylene | SF | 4.32 | 3.47 | 3.08 | 2.25 | 1.97 |
| 33 09 0701 | CIP Reinforced Concrete Containment Wall | CY | 279.72 | 243.62 | 224.28 | 189.16 | 178.78 |
| 33 17 0814 | 1,800 PSI Pressure Washer, 6 HP, 4.8 GPM | EACH | 1,683 | 1,683 | 1,683 | 1,683 | 1,683 |
| 33 17 0815 | 1,800 PSI Steam Cleaner, 6 HP | EACH | 3,004 | 3,004 | 3,004 | 3,004 | 3,004 |
| 33 17 0816 | 3,000 PSI Pressure Washer, 4.5 GPM | EACH | 8,868 | 8,868 | 8,868 | 8,868 | 8,868 |
| 33 17 0817 | 55 Gallon Drum of Liquid Soap | EACH | 973.08 | 973.08 | 973.08 | 973.08 | 973.08 |
| 33 17 0818 | 1,800 PSI Pressure Washer Rental | MONTH | 768.50 | 768.50 | 768.50 | 768.50 | 768.50 |
| 33 17 0819 | 1,800 PSI Steam Cleaner Rental | MONTH | 1,179 | 1,179 | 1,179 | 1,179 | 1,179 |
| 33 17 0821 | 8' x 24' Decontamination Trailer with 4 Showers, HVAC, 2 Sinks | MONTH | 2,120 | 2,120 | 2,120 | 2,120 | 2,120 |
| 33 17 0822 | 8' x 36' Decontamination Trailer with 2 Showers, Fans | MONTH | 2,332 | 2,332 | 2,332 | 2,332 | 2,332 |
| 33 17 0823 | Operation of Pressure Washer, Including Water, Soap, Electricity, Labor | HOUR | 79.38 | 63.31 | 56.43 | 40.91 | 35.21 |

## Environmental Remediation: Assemblies Cost Book

*Page 3-79*

1998 by ECHOS. All rights reserved

# Cost Data for Remediation

## Decontamination Facilities

| Assembly | Description | Unit | Unit Cost by Safety Level A | B | C | D | E |
|---|---|---|---|---|---|---|---|
| | **Capital Costs** | | | | | | |
| 33 17 0824 | Operation of Steam Cleaner, Including Water, Soap, Electricity, Labor | HOUR | 81.81 | 65.74 | 58.86 | 43.33 | 37.64 |
| 33 17 0825 | 8' - 6" Railroad Wood Crossties | EACH | 55.33 | 49.51 | 45.88 | 40.19 | 38.83 |
| 33 19 9901 | 750 Gallon, 10' x 10' x 17" Inflatable Containment Berm, Single Lined | EACH | 2,289 | 2,272 | 2,265 | 2,249 | 2,243 |
| 33 19 9921 | DOT Steel Drum, 55 Gallon | EACH | 65.19 | 65.19 | 65.19 | 65.19 | 65.19 |
| 33 19 9922 | Polyethylene Closed Head Drum, 55 Gallon | EACH | 43.63 | 43.63 | 43.63 | 43.63 | 43.63 |
| 33 19 9923 | Polyethylene Open Head Drum, 55 Gallon | EACH | 49.90 | 49.90 | 49.90 | 49.90 | 49.90 |
| 33 19 9926 | 20 Gallon, Fiberglass Drums, DOT 21C-250 | EACH | 20.86 | 20.86 | 20.86 | 20.86 | 20.86 |
| 33 23 1306 | High Sump Level Switch for Avoiding Overflow | EACH | 214.65 | 214.65 | 214.65 | 214.65 | 214.65 |
| 33 26 0512 | 4" High-density Polyethylene, Transfer Pipe | LF | 6.75 | 5.61 | 5.10 | 4.00 | 3.62 |
| 33 26 0513 | 6" High-density Polyethylene, Transfer Pipe | LF | 8.82 | 7.47 | 6.86 | 5.55 | 5.10 |
| 33 26 0514 | 8" High-density Polyethylene, Transfer Pipe | LF | 11.20 | 9.76 | 9.11 | 7.72 | 7.23 |
| 33 26 0516 | 12" High-density Polyethylene, Transfer Pipe | LF | 19.22 | 17.26 | 16.37 | 14.47 | 13.81 |
| 33 26 0518 | 16" High-density Polyethylene, Transfer Pipe | LF | 30.24 | 27.16 | 25.76 | 22.78 | 21.74 |
| 33 26 0520 | 24" High-density Polyethylene, Transfer Pipe | LF | 61.27 | 55.88 | 53.44 | 48.22 | 46.39 |
| 33 26 0601 | (1 1/2", 3") Stainless Steel Double-wall Piping, with Fittings | LF | 134.61 | 117.05 | 109.53 | 92.56 | 86.33 |
| 33 26 0602 | (2", 4") Stainless Steel Double-wall Piping, with Fittings | LF | 185.98 | 163.46 | 153.81 | 132.04 | 124.05 |
| 33 26 0603 | (2 1/2", 4") Stainless Steel Double-wall Piping, with Fittings | LF | 206.86 | 182.17 | 171.60 | 147.74 | 138.99 |
| 33 26 0604 | (4", 6") Stainless Steel Double-wall Piping, with Fittings | LF | 339.64 | 280.79 | 254.18 | 197.21 | 177.23 |
| 33 26 0621 | (1 1/2", 3") PVC Double-wall Piping, with Fittings | LF | 33.15 | 27.05 | 24.43 | 18.53 | 16.36 |
| 33 26 0622 | (2", 4") PVC Double-wall Piping, with Fittings | LF | 46.22 | 37.98 | 34.43 | 26.46 | 23.55 |
| 33 26 0623 | (2 1/2", 4") PVC Double-wall Piping, with Fittings | LF | 49.22 | 40.39 | 36.61 | 28.07 | 24.94 |
| 33 26 0624 | (4", 6") PVC Double-wall Piping, with Fittings | LF | 70.61 | 58.00 | 52.59 | 40.40 | 35.93 |
| 33 26 0704 | 50' Flexible, Product Discharge Hose | EACH | 159.00 | 159.00 | 159.00 | 159.00 | 159.00 |
| 33 29 0102 | 10 GPM, 1/2 HP, Centrifugal Pump | EACH | 1,063 | 952.32 | 905.07 | 798.39 | 759.27 |
| 33 29 0103 | 50 GPM, 100' Head, 3 HP, Centrifugal Pump | EACH | 2,658 | 2,500 | 2,432 | 2,279 | 2,223 |
| 33 29 0104 | 75 GPM, 100' Head, 5 HP, Centrifugal Pump | EACH | 2,946 | 2,754 | 2,672 | 2,487 | 2,419 |
| 33 29 0106 | 100 GPM, 150' Head, 7.5 HP, Centrifugal Pump | EACH | 3,419 | 3,191 | 3,093 | 2,873 | 2,793 |
| 33 29 0107 | 125 GPM, 150' Head, 10 HP, Centrifugal Pump | EACH | 3,815 | 3,512 | 3,382 | 3,088 | 2,981 |
| 33 29 0108 | 10 HP, 200 GPM, Centrifugal Pump | EACH | 2,941 | 2,714 | 2,616 | 2,396 | 2,316 |
| 33 29 0109 | 10 HP, 250 GPM, Centrifugal Pump | EACH | 3,032 | 2,805 | 2,707 | 2,487 | 2,407 |
| 33 29 0110 | 15 HP, 300 GPM, Centrifugal Pump | EACH | 3,441 | 3,160 | 3,040 | 2,770 | 2,670 |
| 33 29 0111 | 20 HP, 500 GPM, Centrifugal Pump | EACH | 4,544 | 4,241 | 4,111 | 3,818 | 3,710 |

**Environmental Remediation: Assemblies Cost Book**

# Cost Data for Remediation

## Decontamination Facilities

| Assembly | Description | Unit | Unit Cost by Safety Level A | B | C | D | E |
|---|---|---|---|---|---|---|---|
| | **Capital Costs** | | | | | | |
| 33 29 0112 | 30 HP, 750 GPM, Centrifugal Pump | EACH | 5,213 | 4,808 | 4,635 | 4,244 | 4,100 |
| 33 29 0113 | 40 HP, 1050 GPM, Centrifugal Pump | EACH | 5,655 | 5,199 | 5,004 | 4,564 | 4,403 |
| 33 29 0114 | 60 HP, 1500 GPM, Centrifugal Pump | EACH | 9,842 | 8,823 | 8,350 | 7,363 | 7,024 |
| 33 29 0115 | 75 HP, 2000 GPM, Centrifugal Pump | EACH | 12,456 | 11,233 | 10,666 | 9,482 | 9,075 |
| 33 29 0116 | 100 HP, 3000 GPM, Centrifugal Pump | EACH | 15,575 | 14,046 | 13,338 | 11,857 | 11,348 |
| 33 29 0401 | 25 GPM, 1 1/2" Discharge, Cast-iron Sump Pump | EACH | 2,432 | 2,274 | 2,206 | 2,053 | 1,997 |
| 33 29 0402 | 75 GPM, 2" Discharge, Cast-iron Sump Pump | EACH | 3,011 | 2,820 | 2,738 | 2,552 | 2,484 |
| 33 29 0403 | 100 GPM, 2 1/2" Discharge, Cast-iron Sump Pump | EACH | 3,243 | 3,016 | 2,918 | 2,698 | 2,618 |
| 33 29 0404 | 150 GPM, 3" Discharge, Cast-iron Sump Pump | EACH | 3,703 | 3,422 | 3,302 | 3,032 | 2,932 |
| 33 29 0405 | 200 GPM, 3" Discharge, Cast-iron Sump Pump | EACH | 3,904 | 3,601 | 3,471 | 3,178 | 3,070 |
| 33 29 0406 | 300 GPM, 4" Discharge, Cast-iron Sump Pump | EACH | 4,817 | 4,453 | 4,297 | 3,945 | 3,816 |
| 33 29 0407 | 500 GPM, 5" Discharge, Cast-iron Sump Pump | EACH | 4,969 | 4,610 | 4,456 | 4,110 | 3,982 |
| 33 29 0408 | 800 GPM, 6" Discharge, Cast-iron Sump Pump | EACH | 12,550 | 12,113 | 11,911 | 11,488 | 11,342 |
| 33 29 0409 | 1,000 GPM, 6" Discharge, Cast-iron Sump Pump | EACH | 15,213 | 14,703 | 14,467 | 13,974 | 13,804 |
| 33 29 0410 | 1,600 GPM, 8" Discharge, Cast-iron Sump Pump | EACH | 17,561 | 16,949 | 16,666 | 16,074 | 15,870 |
| 33 29 0411 | 2,000 GPM, 8" Discharge, Cast-iron Sump Pump | EACH | 19,430 | 18,751 | 18,436 | 17,778 | 17,552 |
| | **Operations & Maintenance** | | | | | | |
| 33 17 0817 | 55 Gallon Drum of Liquid Soap | EACH | 973.08 | 973.08 | 973.08 | 973.08 | 973.08 |
| 33 17 0818 | 1,800 PSI Pressure Washer Rental | MONTH | 768.50 | 768.50 | 768.50 | 768.50 | 768.50 |
| 33 17 0819 | 1,800 PSI Steam Cleaner Rental | MONTH | 1,179 | 1,179 | 1,179 | 1,179 | 1,179 |
| 33 17 0821 | 8' x 24' Decontamination Trailer with 4 Showers, HVAC, 2 Sinks | MONTH | 2,120 | 2,120 | 2,120 | 2,120 | 2,120 |
| 33 17 0822 | 8' x 36' Decontamination Trailer with 2 Showers, Fans | MONTH | 2,332 | 2,332 | 2,332 | 2,332 | 2,332 |
| 33 17 0823 | Operation of Pressure Washer, Including Water, Soap, Electricity, Labor | HOUR | 79.38 | 63.31 | 56.43 | 40.91 | 35.21 |
| 33 17 0824 | Operation of Steam Cleaner, Including Water, Soap, Electricity, Labor | HOUR | 81.81 | 65.74 | 58.86 | 43.33 | 37.64 |
| 33 19 9921 | DOT Steel Drum, 55 Gallon | EACH | 65.19 | 65.19 | 65.19 | 65.19 | 65.19 |
| 33 19 9922 | Polyethylene Closed Head Drum, 55 Gallon | EACH | 43.63 | 43.63 | 43.63 | 43.63 | 43.63 |
| 33 19 9923 | Polyethylene Open Head Drum, 55 Gallon | EACH | 49.90 | 49.90 | 49.90 | 49.90 | 49.90 |
| 33 19 9926 | 20 Gallon, Fiberglass Drums, DOT 21C-250 | EACH | 20.86 | 20.86 | 20.86 | 20.86 | 20.86 |
| 33 23 0512 | 1" Submersible Pump Rental, Month | MONTH | 89.40 | 89.40 | 89.40 | 89.40 | 89.40 |
| 33 41 0101 | Pump & Motor Maintenance/Repair | EACH | 865.17 | 671.48 | 581.58 | 393.96 | 329.64 |
| 33 42 0101 | Electrical Charge | KWH | 0.06 | 0.06 | 0.06 | 0.06 | 0.06 |

**Environmental Remediation: Assemblies Cost Book**

*Page 3-81*

# Cost Data for Remediation
## Dewatering (Sludge)

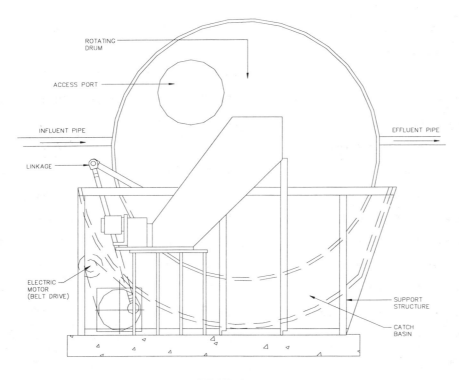

PROFILE

### General:
Dewatering is a physical unit operation which reduces the moisture content of slurries or sludges to enable handling and to prepare the materials for final treatment and/or disposal. Though there are other methods for dewatering, the costs presented here are focused on vacuum filtration.

### Typical Treatment Train:
Pretreatment, sampling and analysis, disposal of dewatered cake.

### Common Cost Components:
1. Vacuum filtration unit
2. Structural slab
3. Chemical conditioners
4. Electrical usage
5. Piping
6. Operations and Maintenance

### Other Cost Considerations:
Electrical distribution, chemical storage building, conditioning tank, conveyor.

**Environmental Remediation: Assemblies Cost Book**

# Cost Data for Remediation

## Dewatering (Sludge)

| Assembly | Description | Unit | Unit Cost by Safety Level | | | | |
|---|---|---|---|---|---|---|---|
| | | | A | B | C | D | E |
| | **Capital Costs** | | | | | | |
| 18 02 0322 | 8" Structural Slab on Grade | SF | 9.75 | 8.38 | 7.71 | 6.38 | 5.95 |
| 18 02 0324 | 12" Structural Slab on Grade | SF | 11.57 | 10.04 | 9.30 | 7.82 | 7.34 |
| 19 01 0202 | 1", Class 200, PVC Piping | LF | 7.48 | 5.84 | 5.07 | 3.48 | 2.93 |
| 19 01 0204 | 2", Class 200, PVC Piping | LF | 8.86 | 6.96 | 6.07 | 4.23 | 3.60 |
| 19 01 0206 | 3", Class 200, PVC Piping | LF | 11.81 | 9.34 | 8.20 | 5.80 | 4.98 |
| 19 01 0207 | 4", Class 200, PVC Piping | LF | 12.33 | 9.86 | 8.72 | 6.32 | 5.50 |
| 19 01 0208 | 6", Class 200, PVC Piping | LF | 14.95 | 12.20 | 10.93 | 8.27 | 7.36 |
| 19 01 0212 | 8", Class 150, PVC Piping | LF | 18.36 | 15.27 | 13.84 | 10.85 | 9.82 |
| 19 01 0214 | 12", Class 150, PVC Piping | LF | 28.75 | 24.63 | 22.72 | 18.73 | 17.36 |
| 33 13 1101 | 3.1 SF of Nominal Filtering Area, 1' Drum Diameter, 4 HP | EACH | 28,381 | 27,914 | 27,715 | 27,264 | 27,098 |
| 33 13 1102 | 6.2 SF of Nominal Filtering Area, 2' Drum Diameter, 6 HP | EACH | 33,387 | 32,920 | 32,721 | 32,270 | 32,104 |
| 33 13 1103 | 12.6 SF of Nominal Filtering Area, 2' Drum Diameter, 9 HP | EACH | 37,763 | 37,141 | 36,875 | 36,274 | 36,054 |
| 33 13 1104 | 18.8 SF of Nominal Filtering Area, 2' Drum Diameter, 9 HP | EACH | 42,115 | 41,494 | 41,227 | 40,626 | 40,406 |
| 33 13 1105 | 25.1 SF of Nominal Filtering Area, 2' Drum Diameter, 11 HP | EACH | 47,148 | 46,370 | 46,037 | 45,286 | 45,010 |
| 33 13 1106 | 28.3 SF of Nominal Filtering Area, 3' Drum Diameter, 11 HP | EACH | 49,402 | 48,624 | 48,291 | 47,540 | 47,264 |
| 33 13 1107 | 37.7 SF of Nominal Filtering Area, 3' Drum Diameter, 13 HP | EACH | 58,106 | 57,329 | 56,996 | 56,244 | 55,969 |
| 33 13 1108 | 56.5 SF of Nominal Filtering Area, 3' Drum Diameter, 24 HP | EACH | 66,402 | 65,469 | 65,069 | 64,167 | 63,837 |
| 33 13 1109 | 75.4 SF of Nominal Filtering Area, 6' Drum Diameter, 33 HP | EACH | 73,082 | 72,149 | 71,750 | 70,848 | 70,517 |
| 33 13 1110 | 94.2 SF of Nominal Filtering Area, 6' Drum Diameter, 38 HP | EACH | 78,479 | 77,313 | 76,813 | 75,686 | 75,273 |
| 33 13 1111 | 113.0 SF of Nominal Filtering Area, 6' Drum Diameter, 49 HP | EACH | 87,159 | 85,993 | 85,494 | 84,366 | 83,953 |
| 33 13 1112 | 170.0 SF of Nominal Filtering Area, 6' Drum Diameter, 60 HP | EACH | 133,036 | 131,613 | 131,004 | 129,630 | 129,125 |
| 33 13 1113 | 226.0 SF of Nominal Filtering Area, 6' Drum Diameter, 79 HP | EACH | 159,377 | 157,954 | 157,345 | 155,971 | 155,466 |
| 33 13 1114 | 302.0 SF of Nominal Filtering Area, 8' Drum Diameter, 99 HP | EACH | 179,876 | 177,995 | 177,189 | 175,371 | 174,704 |
| 33 13 1115 | 402.0 SF of Nominal Filtering Area, 8' Drum Diameter, 120 HP | EACH | 200,153 | 198,071 | 197,179 | 195,166 | 194,428 |
| 33 13 1116 | 503.0 SF of Nominal Filtering Area, 8' Drum Diameter, 140 HP | EACH | 228,539 | 226,206 | 225,208 | 222,953 | 222,126 |
| 33 13 1131 | Operation and Maintenance of Vacuum Filtration Unit | HOUR | 401.10 | 330.31 | 293.30 | 224.48 | 203.55 |
| 33 13 1950 | 25' x 6" Flexible Stainless Steel High-pressure Hose | EACH | 1,288 | 1,239 | 1,217 | 1,170 | 1,153 |
| 33 26 0101 | 1" Carbon Steel Piping | LF | 6.46 | 5.28 | 4.77 | 3.63 | 3.21 |
| 33 26 0102 | 2" Carbon Steel Piping | LF | 10.50 | 8.72 | 7.95 | 6.23 | 5.60 |
| 33 26 0103 | 3" Carbon Steel Piping | LF | 18.02 | 15.20 | 13.99 | 11.26 | 10.26 |

**Environmental Remediation: Assemblies Cost Book**

*Page 3-83*

# Cost Data for Remediation

## Dewatering (Sludge)

| Assembly | Description | Unit | Unit Cost by Safety Level A | B | C | D | E |
|---|---|---|---|---|---|---|---|
| | **Capital Costs** | | | | | | |
| 33 26 0104 | 4" Carbon Steel Piping | LF | 23.20 | 19.86 | 18.43 | 15.21 | 14.02 |
| 33 26 0105 | 6" Carbon Steel Piping | LF | 36.37 | 30.24 | 27.46 | 21.52 | 19.44 |
| 33 26 0106 | 8" Carbon Steel Piping | LF | 40.77 | 34.08 | 31.15 | 24.68 | 22.36 |
| 33 26 0108 | 12" Carbon Steel Piping | LF | 75.72 | 65.62 | 61.18 | 51.41 | 47.89 |
| 33 26 0201 | 1" Stainless Steel Piping, Schedule 40, Threaded | LF | 18.67 | 16.12 | 15.03 | 12.56 | 11.65 |
| 33 26 0202 | 2" Stainless Steel Piping, Schedule 40, Threaded | LF | 33.47 | 29.21 | 27.39 | 23.28 | 21.77 |
| 33 26 0203 | 3" Stainless Steel Piping, Schedule 40, Threaded | LF | 57.00 | 50.61 | 47.88 | 41.71 | 39.45 |
| 33 26 0204 | 4" Stainless Steel Piping, Schedule 40, Threaded | LF | 80.15 | 71.72 | 68.12 | 59.98 | 57.00 |
| 33 26 0205 | 6" Stainless Steel Piping, Schedule 10, Type 316 | LF | 78.58 | 70.01 | 66.14 | 57.84 | 54.94 |
| 33 26 0206 | 8" Stainless Steel Piping, Schedule 40, Welded | LF | 197.22 | 183.32 | 176.89 | 163.42 | 158.80 |
| 33 26 0209 | 12" Stainless Steel Piping, Schedule 40, Welded | LF | 312.43 | 295.44 | 287.57 | 271.12 | 265.47 |
| 33 26 0301 | 1 1/2" Polypropylene Pipe Including Fittings | LF | 7.56 | 6.11 | 5.48 | 4.08 | 3.57 |
| 33 26 0302 | 2" Polypropylene Pipe Including Fittings | LF | 8.79 | 7.16 | 6.46 | 4.88 | 4.30 |
| 33 26 0303 | 3" Polypropylene Pipe Including Fittings | LF | 13.80 | 11.43 | 10.42 | 8.13 | 7.29 |
| 33 26 0304 | 4" Polypropylene Pipe Including Fittings | LF | 18.36 | 15.28 | 13.96 | 10.99 | 9.90 |
| 33 26 0702 | 50' x 6" Brown Gum Rubber, Chemical-resistant, Flexible Hose | EACH | 1,123 | 1,038 | 1,001 | 918.34 | 888.04 |
| 33 26 0703 | 50' X 6" Nitrile Rubber, Low pH Res, Flex Hose | EA | 0.00 | 0.00 | 0.00 | 0.00 | 0.00 |
| 33 33 0114 | Quicklime, 1/4" Nominal Granules, Bulk Quantity | TON | 96.49 | 96.49 | 96.48 | 96.48 | 96.47 |
| 33 33 0115 | Quicklime, 3/4" Nominal Granules, Bulk Quantity | TON | 96.49 | 96.49 | 96.48 | 96.48 | 96.47 |
| 33 33 0116 | Quicklime, Combination 1/4" & 3/4" Granules, Bulk Quantity | TON | 96.49 | 96.49 | 96.48 | 96.48 | 96.47 |
| 33 33 0117 | Hydrated Lime, Powdered, Bulk | TON | 97.55 | 97.55 | 97.54 | 97.54 | 97.53 |
| 33 33 0118 | Hydrated Lime, Powdered, 50 Lb Bag | EACH | 25.76 | 21.23 | 18.49 | 14.06 | 12.94 |
| 33 33 0121 | Ferric Chloride, Bulk | TON | 241.18 | 241.18 | 241.17 | 241.17 | 241.16 |
| 33 42 0101 | Electrical Charge | KWH | 0.06 | 0.06 | 0.06 | 0.06 | 0.06 |
| | **Operations & Maintenance** | | | | | | |
| 33 13 1131 | Operation and Maintenance of Vacuum Filtration Unit | HOUR | 401.10 | 330.31 | 293.30 | 224.48 | 203.55 |
| 33 33 0114 | Quicklime, 1/4" Nominal Granules, Bulk Quantity | TON | 96.49 | 96.49 | 96.48 | 96.48 | 96.47 |
| 33 33 0115 | Quicklime, 3/4" Nominal Granules, Bulk Quantity | TON | 96.49 | 96.49 | 96.48 | 96.48 | 96.47 |
| 33 33 0116 | Quicklime, Combination 1/4" & 3/4" Granules, Bulk Quantity | TON | 96.49 | 96.49 | 96.48 | 96.48 | 96.47 |
| 33 33 0117 | Hydrated Lime, Powdered, Bulk | TON | 97.55 | 97.55 | 97.54 | 97.54 | 97.53 |
| 33 33 0118 | Hydrated Lime, Powdered, 50 Lb Bag | EACH | 25.76 | 21.23 | 18.49 | 14.06 | 12.94 |
| 33 33 0121 | Ferric Chloride, Bulk | TON | 241.18 | 241.18 | 241.17 | 241.17 | 241.16 |
| 33 42 0101 | Electrical Charge | KWH | 0.06 | 0.06 | 0.06 | 0.06 | 0.06 |

**Environmental Remediation: Assemblies Cost Book**

# Cost Data for Remediation
## Discharge to POTW

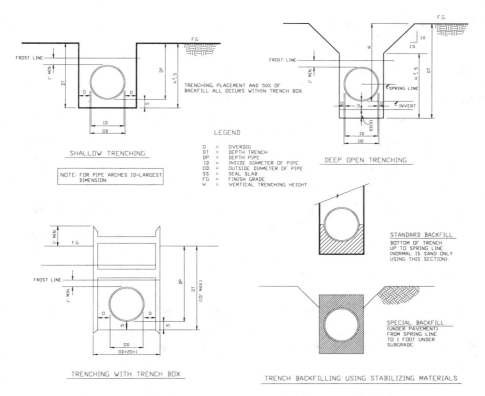

### General:

Discharge to a publicly-owned treatment works (POTW) consists of discharging untreated or pretreated wastes to a POTW for treatment and disposal. The types of aqueous wastes commonly discharged to a POTW include contaminated groundwater, leachate, and surface runoff. When considering discharge to POTW as a disposal alternative, the major technical considerations include the treatment technologies utilized by the POTW, the treatment capacity, the distance of conveyance, the contaminant characteristics of the influent, expected flow rates, and duration of discharge.

### Typical Treatment Train:

Surface or groundwater recovery, pretreatment, sampling and analysis.

### Common Cost Components:

1. Disposal fee
2. Sewer connection (clearing, piping, lift stations, replacement cover, sewer connection fee)
3. Tanker trucks
4. Holding tank
5. Operations and maintenance

### Other Cost Considerations:

Electrical distribution.

**Environmental Remediation: Assemblies Cost Book**

# Cost Data for Remediation

## Discharge to POTW

| Assembly | Description | Unit | Unit Cost by Safety Level | | | | |
|---|---|---|---|---|---|---|---|
| | | | A | B | C | D | E |
| | **Capital Costs** | | | | | | |
| 17 01 0106 | Heavy Brush, Light Trees, Clear, Grub, Haul | ACRE | 11,338 | 9,081 | 7,594 | 5,382 | 4,905 |
| 17 01 0107 | Medium Brush, Medium Trees, Clear, Grub, Haul | ACRE | 13,478 | 10,783 | 9,029 | 6,389 | 5,807 |
| 17 01 0108 | Light Brush, Heavy Trees, Clear, Grub, Haul | ACRE | 17,877 | 14,287 | 11,977 | 8,462 | 7,669 |
| 17 01 0210 | Clear Trees to 6" Diameter with D8 Cat | EACH | 12.03 | 9.85 | 8.04 | 5.88 | 5.66 |
| 17 01 0211 | Clear Trees to 12" Diameter with D8 Cat | EACH | 22.45 | 18.39 | 15.00 | 10.98 | 10.56 |
| 17 01 0212 | Clear Trees to 24" Diameter with D8 Cat | EACH | 33.68 | 27.58 | 22.50 | 16.46 | 15.84 |
| 17 01 0213 | Clear Trees to 36" Diameter with D8 Cat | EACH | 67.36 | 55.17 | 45.00 | 32.93 | 31.67 |
| 17 02 0201 | Demolish Bituminous Road with Power Equipment | CY | 46.17 | 36.60 | 30.96 | 21.62 | 19.19 |
| 17 02 0203 | Demolish Bituminous Pavement with Air Equipment | CY | 80.15 | 63.38 | 53.76 | 37.41 | 32.98 |
| 17 02 0205 | Demolish Unreinforced Concrete to 6" Thick with Air Equipment | CY | 126.90 | 100.35 | 85.12 | 59.24 | 52.23 |
| 17 02 0206 | Demolish Mesh Reinforced Concrete to 6" Thick with Air Equipment | CY | 183.47 | 145.09 | 123.07 | 85.65 | 75.51 |
| 17 02 0207 | Demolish Rod Reinforced Concrete to 6" Thick with Air Equipment | CY | 199.06 | 157.42 | 133.53 | 92.92 | 81.92 |
| 17 02 0208 | Demolish Mesh Reinforced Concrete to 6" Thick with Power Equipment | CY | 96.73 | 76.68 | 64.87 | 45.30 | 40.20 |
| 17 02 0209 | Demolish Rod Reinforced Concrete to 6" Thick with Power Equipment | CY | 106.91 | 84.75 | 71.70 | 50.07 | 44.43 |
| 17 02 0210 | Demolish Unreinforced Concrete 7" to 24" Thick with Power Equipment | CY | 258.44 | 204.87 | 173.31 | 121.03 | 107.40 |
| 17 02 0211 | Demolish Reinforced Concrete 7" to 24" Thick with Power Equipment | CY | 356.37 | 282.51 | 238.98 | 166.90 | 148.10 |
| 17 02 0212 | Demolish Unreinforced Concrete to 6" Thick with Power Equipment | CY | 67.53 | 53.53 | 45.29 | 31.63 | 28.06 |
| 17 02 0215 | Demolish Bituminous Sidewalk | CY | 154.36 | 121.53 | 103.59 | 71.64 | 62.41 |
| 17 02 0220 | Demolish Bituminous Curbs | LF | 2.66 | 2.09 | 1.78 | 1.23 | 1.07 |
| 17 02 0221 | Demolish Unreinforced Concrete Curbs | LF | 7.26 | 5.71 | 4.87 | 3.37 | 2.93 |
| 17 02 0222 | Demolish Reinforced Concrete Curbs | LF | 9.90 | 7.79 | 6.64 | 4.59 | 4.00 |
| 17 03 0105 | Fine Grading, Hand | SY | 4.21 | 3.24 | 2.83 | 1.90 | 1.56 |
| 17 03 0259 | Cat 225, 1.5 CY, Soil/Sand, Trenching | CY | 1.84 | 1.49 | 1.23 | 0.88 | 0.82 |
| 17 03 0260 | Cat 225, 1.5 CY, Soil/Sand, 10' - 20' Deep Trench Box, Trench | CY | 7.22 | 6.04 | 5.20 | 4.04 | 3.82 |
| 17 03 0261 | Cat 225, 1.5 CY, Soil/Sand with Boulders, Trenching | CY | 3.91 | 3.15 | 2.62 | 1.87 | 1.74 |
| 17 03 0272 | Pull Trench Box, Cat 225, 1.5 CY, Soil/Sand, Trenching | CY | 6.67 | 5.49 | 4.65 | 3.48 | 3.27 |
| 17 03 0282 | Soil, 5 Miles, Dump Truck, Load/Haul Spoil From Trench | CY | 6.97 | 5.60 | 4.66 | 3.33 | 3.06 |
| 17 03 0283 | Sand, 5 Mile, Dump Truck, Load/Haul Spoil From Trench | CY | 6.28 | 5.05 | 4.20 | 3.00 | 2.76 |

## Environmental Remediation: Assemblies Cost Book

# Cost Data for Remediation

## Discharge to POTW

| Assembly | Description | Unit | A | B | C | D | E |
|---|---|---|---|---|---|---|---|
| | **Capital Costs** | | | | | | |
| 17 03 0306 | Cat 225, 1.5 CY, Rock, No Haul or Borrow, Trenching | BCY | 149.48 | 122.48 | 106.81 | 80.48 | 73.45 |
| 17 03 0317 | Rock, 5 Miles, Dump Truck, Load/Haul Spoil from Trench | CY | 10.31 | 8.31 | 6.90 | 4.93 | 4.56 |
| 17 03 0401 | 950, 3.00 CY, Backfill with Excavated Material | CY | 2.54 | 2.05 | 1.70 | 1.22 | 1.14 |
| 17 03 0405 | 950, 3.00 CY, Delivered & Dumped, Backfill with Sand | CY | 38.31 | 35.03 | 32.95 | 29.74 | 29.00 |
| 17 03 0410 | 950, Delivered & Dumped, Backfill with Cement Stabilized Base Material | CY | 46.93 | 43.66 | 41.58 | 38.37 | 37.63 |
| 17 03 0420 | Backfill Trench, Borrow Material, Delivered & Dumped Only | CY | 9.86 | 8.74 | 7.93 | 6.83 | 6.63 |
| 17 03 0426 | Sand, 6" Lifts, Off-Site | CY | 15.55 | 14.34 | 13.42 | 12.22 | 12.04 |
| 17 03 0511 | Compact Soil with Vibrating Plate | CY | 7.20 | 5.57 | 4.84 | 3.26 | 2.70 |
| 17 03 0515 | Compact With Pogosticks | CY | 5.91 | 4.58 | 3.98 | 2.68 | 2.23 |
| 17 03 0516 | Compact with 50% Pogosticks, 50% Hand Roller | CY | 4.24 | 3.30 | 2.85 | 1.94 | 1.63 |
| 17 03 1001 | Wellpoint for Trench, Install & Remove <500', 1 Month | LFHD | 55.06 | 55.06 | 55.06 | 55.06 | 55.06 |
| 17 03 1002 | 2" Diameter Contractor's Trash Pump, 75 GPM | DAY | 50.20 | 47.01 | 45.64 | 42.55 | 41.42 |
| 18 01 0102 | Gravel, Delivered & Dumped | CY | 24.33 | 22.93 | 22.03 | 20.66 | 20.36 |
| 18 01 0103 | Gravel (90%) & Sand Base (10%), with Calcium Chloride 3/4 - 1 Lb/CY | CY | 25.70 | 24.15 | 23.16 | 21.65 | 21.31 |
| 18 01 0104 | Asphalt, Intermediate Course (Line Item Includes 5% Waste) | TON | 76.50 | 65.72 | 59.09 | 48.56 | 45.99 |
| 18 01 0310 | Prime Coat | SY | 0.55 | 0.52 | 0.51 | 0.49 | 0.48 |
| 18 01 0311 | Tack Coat | SY | 0.70 | 0.64 | 0.60 | 0.53 | 0.52 |
| 18 01 0312 | Asphalt Wearing Course, 1 Pass (Line Item Includes 5% Waste) | TON | 83.61 | 72.83 | 66.20 | 55.67 | 53.10 |
| 18 02 0320 | 4" Structural Slab on Grade | SF | 6.45 | 5.45 | 4.98 | 4.00 | 3.67 |
| 18 02 0321 | 6" Structural Slab on Grade | SF | 6.98 | 5.95 | 5.47 | 4.48 | 4.14 |
| 18 02 0322 | 8" Structural Slab on Grade | SF | 9.75 | 8.38 | 7.71 | 6.38 | 5.95 |
| 18 02 0323 | 10" Structural Slab on Grade | SF | 10.56 | 9.13 | 8.43 | 7.04 | 6.58 |
| 18 02 0324 | 12" Structural Slab on Grade | SF | 11.57 | 10.04 | 9.30 | 7.82 | 7.34 |
| 18 02 0330 | 4" Mesh Reinforced Slab on Grade | SF | 5.48 | 4.60 | 4.18 | 3.32 | 3.03 |
| 18 02 0331 | 6" Mesh Reinforced Slab on Grade | SF | 6.01 | 5.10 | 4.67 | 3.79 | 3.49 |
| 18 02 0332 | 8" Mesh Reinforced Slab on Grade | SF | 7.58 | 6.50 | 5.96 | 4.91 | 4.58 |
| 18 02 0333 | 10" Mesh Reinforced Slab on Grade | SF | 8.10 | 7.00 | 6.44 | 5.37 | 5.03 |
| 18 02 0334 | 12" Mesh Reinforced Slab on Grade | SF | 8.78 | 7.63 | 7.05 | 5.93 | 5.58 |
| 18 02 0340 | 4" Unreinforced Slab on Grade | SF | 4.34 | 3.64 | 3.29 | 2.61 | 2.38 |
| 18 02 0341 | 6" Unreinforced Slab on Grade | SF | 4.87 | 4.14 | 3.79 | 3.08 | 2.85 |
| 18 02 0342 | 8" Unreinforced Slab on Grade | SF | 6.45 | 5.54 | 5.07 | 4.19 | 3.93 |

**Environmental Remediation: Assemblies Cost Book**

# Cost Data for Remediation

## Discharge to POTW

| Assembly | Description | Unit | A | B | C | D | E |
|---|---|---|---|---|---|---|---|
| | **Capital Costs** | | | | | | |
| 18 02 0343 | 10" Unreinforced Slab on Grade | SF | 6.97 | 6.04 | 5.56 | 4.66 | 4.38 |
| 18 02 0344 | 12" Unreinforced Slab on Grade | SF | 7.64 | 6.67 | 6.16 | 5.22 | 4.93 |
| 18 04 0203 | Pour & Cure Concrete, Continuous Footing | CY | 140.13 | 126.22 | 118.87 | 105.35 | 101.29 |
| 18 05 0402 | Seeding, Vegetative Cover | ACRE | 1,874 | 1,823 | 1,791 | 1,741 | 1,729 |
| 18 05 0405 | Sodding, Vegetative Cover | ACRE | 25,697 | 22,064 | 20,238 | 16,710 | 15,589 |
| 19 01 0202 | 1", Class 200, PVC Piping | LF | 7.48 | 5.84 | 5.07 | 3.48 | 2.93 |
| 19 01 0204 | 2", Class 200, PVC Piping | LF | 8.86 | 6.96 | 6.07 | 4.23 | 3.60 |
| 19 01 0206 | 3", Class 200, PVC Piping | LF | 11.81 | 9.34 | 8.20 | 5.80 | 4.98 |
| 19 01 0207 | 4", Class 200, PVC Piping | LF | 12.33 | 9.86 | 8.72 | 6.32 | 5.50 |
| 19 01 0208 | 6", Class 200, PVC Piping | LF | 14.95 | 12.20 | 10.93 | 8.27 | 7.36 |
| 19 01 0212 | 8", Class 150, PVC Piping | LF | 18.36 | 15.27 | 13.84 | 10.85 | 9.82 |
| 19 01 0213 | 10", Class 150, PVC Piping | LF | 22.90 | 19.37 | 17.73 | 14.32 | 13.14 |
| 19 01 0214 | 12", Class 150, PVC Piping | LF | 28.75 | 24.63 | 22.72 | 18.73 | 17.36 |
| 19 02 0101 | 4", Class 50, Bell & Spigot Sanitary Sewer, Cast-iron Pipe | LF | 8.49 | 7.49 | 7.05 | 6.08 | 5.72 |
| 19 02 0103 | 6", Class 50, Bell & Spigot Sanitary Sewer, Cast-iron Pipe | LF | 13.25 | 11.79 | 11.16 | 9.75 | 9.23 |
| 19 02 0104 | 8", Class 50, Bell & Spigot Sanitary Sewer, Cast-iron Pipe | LF | 18.98 | 17.26 | 16.52 | 14.86 | 14.25 |
| 19 02 0105 | 10", Class 50, Bell & Spigot Sanitary Sewer, Cast-iron Pipe | LF | 27.66 | 25.51 | 24.59 | 22.51 | 21.75 |
| 19 02 0106 | 12", Class 50, Bell & Spigot Sanitary Sewer, Cast-iron Pipe | LF | 36.36 | 34.21 | 33.29 | 31.21 | 30.45 |
| 19 02 0107 | 15", Class 50, Bell & Spigot Sanitary Sewer, Cast-iron Pipe | LF | 51.76 | 49.01 | 47.83 | 45.17 | 44.19 |
| 19 02 0110 | 4" Extra-strength Vitrified Clay Pipe, Class 200, Premium Joints | LF | 17.47 | 14.14 | 12.39 | 9.15 | 8.17 |
| 19 02 0112 | 6" Extra-strength Vitrified Clay Pipe, Class 200, Premium Joints | LF | 20.06 | 16.40 | 14.48 | 10.92 | 9.84 |
| 19 02 0113 | 8" Extra-strength Vitrified Clay Pipe, Class 200, Premium Joints | LF | 22.16 | 18.37 | 16.38 | 12.70 | 11.58 |
| 19 02 0114 | 10" Extra-strength Vitrified Clay Pipe, Class 200, Premium Joints | LF | 27.02 | 22.79 | 20.57 | 16.46 | 15.22 |
| 19 02 0115 | 12" Extra-strength Vitrified Clay Pipe, Class 200, Premium Joints | LF | 28.25 | 23.86 | 21.55 | 17.28 | 15.98 |
| 19 02 0116 | 15" Extra-strength Vitrified Clay Pipe, Class 200, Premium Joints | LF | 39.29 | 34.29 | 31.67 | 26.81 | 25.34 |
| 19 02 0117 | 18" Extra-strength Vitrified Clay Pipe, Class 200, Premium Joints | LF | 50.85 | 45.06 | 42.02 | 36.40 | 34.70 |
| 19 02 0118 | 24" Extra-strength Vitrified Clay Pipe, Class 200, Premium Joints | LF | 85.93 | 78.09 | 73.96 | 66.33 | 64.03 |
| 19 02 0125 | 4" PVC Pipe Sanitary | LF | 13.31 | 10.99 | 9.71 | 7.45 | 6.80 |
| 19 02 0126 | 6" PVC Pipe Sanitary | LF | 15.68 | 13.19 | 11.82 | 9.39 | 8.70 |

**Unit Cost by Safety Level**

---

**Environmental Remediation: Assemblies Cost Book**

*Page 3-88*

1998 by ECHOS. All rights reserved

# Cost Data for Remediation

## Discharge to POTW

| Assembly | Description | Unit | Unit Cost by Safety Level | | | | |
|---|---|---|---|---|---|---|---|
| | | | A | B | C | D | E |
| | **Capital Costs** | | | | | | |
| 19 02 0127 | 8" PVC Pipe Sanitary | LF | 17.03 | 14.43 | 13.00 | 10.47 | 9.74 |
| 19 02 0128 | 10" PVC Pipe Sanitary | LF | 22.78 | 19.45 | 17.70 | 14.46 | 13.48 |
| 19 02 0129 | 12" PVC Pipe Sanitary | LF | 26.69 | 23.26 | 21.45 | 18.11 | 17.11 |
| 19 02 0130 | 15" PVC Pipe Sanitary | LF | 42.35 | 36.56 | 33.52 | 27.90 | 26.20 |
| 19 02 0201 | Precast, CIP Base, 4' Diameter, 6' Deep, Manhole | EACH | 1,459 | 1,276 | 1,179 | 1,001 | 947.36 |
| 19 02 0202 | Precast, CIP Base, 4' Diameter, 8' Deep, Manhole | EACH | 1,957 | 1,697 | 1,557 | 1,304 | 1,229 |
| 19 02 0203 | Precast, CIP Base, 4' Diameter, 12' Deep, Manhole | EACH | 2,835 | 2,459 | 2,257 | 1,891 | 1,783 |
| 19 02 0210 | Precast Manhole, Per Foot, 4' Diameter x Depth | LF | 219.59 | 190.64 | 174.83 | 146.66 | 138.52 |
| 19 02 0301 | 10,000 GPD (7 GPM) Lift Station | EACH | 2,828 | 2,659 | 2,587 | 2,423 | 2,363 |
| 19 02 0302 | 25,000 GPD (18 GPM) Lift Station | EACH | 3,539 | 3,313 | 3,217 | 2,998 | 2,918 |
| 19 02 0303 | 50,000 GPD (35 GPM) Lift Station | EACH | 4,960 | 4,621 | 4,476 | 4,149 | 4,029 |
| 19 02 0304 | 100,000 GPD (70 GPM) Lift Station | EACH | 7,842 | 7,509 | 7,366 | 7,043 | 6,925 |
| 19 02 0305 | 200,000 GPD Packaged Lift Station | EACH | 133,007 | 123,412 | 118,907 | 109,609 | 106,455 |
| 19 02 0306 | 500,000 GPD Packaged Lift Station | EACH | 157,190 | 144,431 | 138,440 | 126,076 | 121,881 |
| 19 02 0307 | 800,000 GPD Packaged Lift Station | EACH | 188,049 | 172,867 | 165,739 | 151,027 | 146,036 |
| 19 02 0311 | 12' x 36" Diameter Reinforced Concrete Pipe Wet Well for Lift Station | EACH | 9,840 | 8,577 | 7,816 | 6,581 | 6,270 |
| 19 02 0312 | 24' x 60" Diameter Reinforced Concrete Pipe Wet Well for Lift Station | EACH | 39,142 | 33,897 | 30,681 | 25,554 | 24,296 |
| 19 03 0157 | 6" Extra Strength, Nonreinforced Concrete Pipe | LF | 10.78 | 8.92 | 7.98 | 6.17 | 5.59 |
| 19 03 0158 | 8" Extra Strength, Nonreinforced Concrete Pipe | LF | 12.86 | 10.79 | 9.75 | 7.74 | 7.10 |
| 19 03 0159 | 10" Extra Strength, Nonreinforced Concrete Pipe | LF | 16.79 | 14.59 | 13.49 | 11.36 | 10.68 |
| 19 03 0160 | 12" Extra Strength, Nonreinforced Concrete Pipe | LF | 17.97 | 15.63 | 14.46 | 12.20 | 11.48 |
| 19 03 0161 | 15" Extra Strength, Nonreinforced Concrete Pipe | LF | 19.95 | 17.36 | 16.06 | 13.55 | 12.74 |
| 19 03 0162 | 18" Extra Strength, Nonreinforced Concrete Pipe | LF | 24.93 | 22.06 | 20.62 | 17.83 | 16.94 |
| 19 03 0163 | 21" Extra Strength, Nonreinforced Concrete Pipe | LF | 30.32 | 27.45 | 26.01 | 23.22 | 22.33 |
| 19 03 0164 | 24" Extra Strength, Nonreinforced Concrete Pipe | LF | 39.69 | 36.49 | 34.88 | 31.77 | 30.78 |
| 19 03 0165 | 12" Reinforced Concrete Pipe, Class 3, with Gaskets | LF | 14.33 | 13.02 | 12.29 | 11.01 | 10.65 |
| 19 03 0166 | 15" Reinforced Concrete Pipe, Class 3, with Gaskets | LF | 16.47 | 14.84 | 13.94 | 12.36 | 11.91 |
| 19 03 0167 | 18" Reinforced Concrete Pipe, Class 3, with Gaskets | LF | 23.67 | 21.05 | 19.60 | 17.05 | 16.33 |
| 19 03 0168 | 24" Reinforced Concrete Pipe, Class 3, with Gaskets | LF | 34.58 | 31.22 | 29.36 | 26.09 | 25.17 |
| 19 03 0174 | 12" Reinforced Concrete Pipe, Class 4, with Gaskets | LF | 14.33 | 13.02 | 12.29 | 11.01 | 10.65 |
| 19 03 0175 | 15" Reinforced Concrete Pipe, Class 4, with Gaskets | LF | 16.47 | 14.84 | 13.94 | 12.36 | 11.91 |
| 19 03 0176 | 18" Reinforced Concrete Pipe, Class 4, with Gaskets | LF | 23.67 | 21.05 | 19.60 | 17.05 | 16.33 |

**Environmental Remediation: Assemblies Cost Book**

*Page 3-89*

# Cost Data for Remediation

## Discharge to POTW

| Assembly | Description | Unit | A | B | C | D | E |
|---|---|---|---|---|---|---|---|
| | **Capital Costs** | | | | | | |
| 19 03 0177 | 21" Reinforced Concrete Pipe, Class 4, with Gaskets | LF | 27.08 | 24.38 | 22.87 | 20.24 | 19.50 |
| 19 03 0178 | 24" Reinforced Concrete Pipe, Class 4, with Gaskets | LF | 34.58 | 31.22 | 29.36 | 26.09 | 25.17 |
| 19 03 0182 | 12" Reinforced Concrete Pipe, Class 5, with Gaskets | LF | 14.33 | 13.02 | 12.29 | 11.01 | 10.65 |
| 19 03 0183 | 15" Reinforced Concrete Pipe, Class 5, with Gaskets | LF | 16.47 | 14.84 | 13.94 | 12.36 | 11.91 |
| 19 03 0184 | 18" Reinforced Concrete Pipe, Class 5, with Gaskets | LF | 23.67 | 21.05 | 19.60 | 17.05 | 16.33 |
| 19 03 0185 | 21" Reinforced Concrete Pipe, Class 5, with Gaskets | LF | 27.08 | 24.38 | 22.87 | 20.24 | 19.50 |
| 19 03 0186 | 24" Reinforced Concrete Pipe, Class 5, with Gaskets | LF | 34.58 | 31.22 | 29.36 | 26.09 | 25.17 |
| 19 04 0401 | 550 Gallon, Stainless Steel Aboveground Wastewater Holding Tank, Rental | MONTH | 320.65 | 320.65 | 320.65 | 320.65 | 320.65 |
| 19 04 0403 | 4,000 Gallon Polyethylene Wastewater Tank, Rental | MONTH | 545.90 | 545.90 | 545.90 | 545.90 | 545.90 |
| 19 04 0405 | 6,000 Gallon, Polyethylene Aboveground Wastewater Holding Tank, Rental | MONTH | 641.30 | 641.30 | 641.30 | 641.30 | 641.30 |
| 19 04 0407 | 21,000 Gallon, Steel Closed Stationary Aboveground Wastewater Holding Tank, Rental | MONTH | 1,219 | 1,219 | 1,219 | 1,219 | 1,219 |
| 19 04 0621 | 550 Gallon Horizontal Plastic Sump with 4" NPT Connection | EACH | 1,699 | 1,596 | 1,551 | 1,452 | 1,415 |
| 19 04 0624 | 4,000 Gallon Horizontal Plastic Sump with 6" NPT Connection | EACH | 4,130 | 3,927 | 3,841 | 3,645 | 3,573 |
| 19 04 0625 | 6,000 Gallon Horizontal Plastic Sump with 6" NPT Connection | EACH | 5,248 | 4,975 | 4,858 | 4,594 | 4,497 |
| 19 04 0631 | 20,000 Gallon Horizontal Plastic Sump with 6" NPT Connection | EACH | 15,799 | 15,084 | 14,742 | 14,048 | 13,817 |
| 33 19 0208 | Haul 6,000 Gallon Wastewater, Includes Loading & Unloading | MI | 33.47 | 28.85 | 25.42 | 20.87 | 20.14 |
| 33 19 7101 | Sewer Connection Fee | EACH | 3,399 | 2,620 | 2,287 | 1,534 | 1,258 |
| 33 19 7102 | Wastewater Disposal Fee | KGAL | 1.50 | 1.50 | 1.50 | 1.50 | 1.50 |
| 33 19 7301 | Haul & Dispose Debris, 16.5 CY Truck, 10 Mile, Landfill | CY | 107.37 | 106.15 | 105.16 | 103.95 | 103.81 |
| 33 26 0101 | 1" Carbon Steel Piping | LF | 6.46 | 5.28 | 4.77 | 3.63 | 3.21 |
| 33 26 0102 | 2" Carbon Steel Piping | LF | 10.50 | 8.72 | 7.95 | 6.23 | 5.60 |
| 33 26 0103 | 3" Carbon Steel Piping | LF | 18.02 | 15.20 | 13.99 | 11.26 | 10.26 |
| 33 26 0104 | 4" Carbon Steel Piping | LF | 23.20 | 19.86 | 18.43 | 15.21 | 14.02 |
| 33 26 0105 | 6" Carbon Steel Piping | LF | 36.37 | 30.24 | 27.46 | 21.52 | 19.44 |
| 33 26 0106 | 8" Carbon Steel Piping | LF | 40.77 | 34.08 | 31.15 | 24.68 | 22.36 |
| 33 26 0107 | 10" Carbon Steel Piping | LF | 54.40 | 46.05 | 42.38 | 34.30 | 31.39 |
| 33 26 0108 | 12" Carbon Steel Piping | LF | 75.72 | 65.62 | 61.18 | 51.41 | 47.89 |
| 33 26 0201 | 1" Stainless Steel Piping, Schedule 40, Threaded | LF | 18.67 | 16.12 | 15.03 | 12.56 | 11.65 |
| 33 26 0202 | 2" Stainless Steel Piping, Schedule 40, Threaded | LF | 33.47 | 29.21 | 27.39 | 23.28 | 21.77 |

**Environmental Remediation: Assemblies Cost Book**

# Cost Data for Remediation

## Discharge to POTW

| Assembly | Description | Unit | Unit Cost by Safety Level | | | | |
|---|---|---|---|---|---|---|---|
| | | | A | B | C | D | E |
| | **Capital Costs** | | | | | | |
| 33 26 0203 | 3" Stainless Steel Piping, Schedule 40, Threaded | LF | 57.00 | 50.61 | 47.88 | 41.71 | 39.45 |
| 33 26 0204 | 4" Stainless Steel Piping, Schedule 40, Threaded | LF | 80.15 | 71.72 | 68.12 | 59.98 | 57.00 |
| 33 26 0205 | 6" Stainless Steel Piping, Schedule 10, Type 316 | LF | 78.58 | 70.01 | 66.14 | 57.84 | 54.94 |
| 33 26 0206 | 8" Stainless Steel Piping, Schedule 40, Welded | LF | 197.22 | 183.32 | 176.89 | 163.42 | 158.80 |
| 33 26 0208 | 10" Stainless Steel Piping, Schedule 40, Welded | LF | 280.77 | 265.48 | 258.40 | 243.59 | 238.50 |
| 33 26 0209 | 12" Stainless Steel Piping, Schedule 40, Welded | LF | 312.43 | 295.44 | 287.57 | 271.12 | 265.47 |
| 33 29 0102 | 10 GPM, 1/2 HP, Centrifugal Pump | EACH | 1,063 | 952.32 | 905.07 | 798.39 | 759.27 |
| 33 29 0103 | 50 GPM, 100' Head, 3 HP, Centrifugal Pump | EACH | 2,658 | 2,500 | 2,432 | 2,279 | 2,223 |
| 33 29 0104 | 75 GPM, 100' Head, 5 HP, Centrifugal Pump | EACH | 2,946 | 2,754 | 2,672 | 2,487 | 2,419 |
| 33 29 0106 | 100 GPM, 150' Head, 7.5 HP, Centrifugal Pump | EACH | 3,419 | 3,191 | 3,093 | 2,873 | 2,793 |
| 33 29 0107 | 125 GPM, 150' Head, 10 HP, Centrifugal Pump | EACH | 3,815 | 3,512 | 3,382 | 3,088 | 2,981 |
| 33 29 0108 | 10 HP, 200 GPM, Centrifugal Pump | EACH | 2,941 | 2,714 | 2,616 | 2,396 | 2,316 |
| 33 29 0109 | 10 HP, 250 GPM, Centrifugal Pump | EACH | 3,032 | 2,805 | 2,707 | 2,487 | 2,407 |
| 33 29 0110 | 15 HP, 300 GPM, Centrifugal Pump | EACH | 3,441 | 3,160 | 3,040 | 2,770 | 2,670 |
| 33 29 0111 | 20 HP, 500 GPM, Centrifugal Pump | EACH | 4,544 | 4,241 | 4,111 | 3,818 | 3,710 |
| 33 29 0112 | 30 HP, 750 GPM, Centrifugal Pump | EACH | 5,213 | 4,808 | 4,635 | 4,244 | 4,100 |
| 33 29 0113 | 40 HP, 1050 GPM, Centrifugal Pump | EACH | 5,655 | 5,199 | 5,004 | 4,564 | 4,403 |
| 33 29 0114 | 60 HP, 1500 GPM, Centrifugal Pump | EACH | 9,842 | 8,823 | 8,350 | 7,363 | 7,024 |
| 33 29 0115 | 75 HP, 2000 GPM, Centrifugal Pump | EACH | 12,456 | 11,233 | 10,666 | 9,482 | 9,075 |
| 33 29 0116 | 100 HP, 3000 GPM, Centrifugal Pump | EACH | 15,575 | 14,046 | 13,338 | 11,857 | 11,348 |
| 33 41 0101 | Pump & Motor Maintenance/Repair | EACH | 865.17 | 671.48 | 581.58 | 393.96 | 329.64 |
| 33 42 0101 | Electrical Charge | KWH | 0.06 | 0.06 | 0.06 | 0.06 | 0.06 |
| | **Operations & Maintenance** | | | | | | |
| 19 04 0401 | 550 Gallon, Stainless Steel Aboveground Wastewater Holding Tank, Rental | MONTH | 320.65 | 320.65 | 320.65 | 320.65 | 320.65 |
| 19 04 0403 | 4,000 Gallon Polyethylene Wastewater Tank, Rental | MONTH | 545.90 | 545.90 | 545.90 | 545.90 | 545.90 |
| 19 04 0405 | 6,000 Gallon, Polyethylene Aboveground Wastewater Holding Tank, Rental | MONTH | 641.30 | 641.30 | 641.30 | 641.30 | 641.30 |
| 19 04 0407 | 21,000 Gallon, Steel Closed Stationary Aboveground Wastewater Holding Tank, Rental | MONTH | 1,219 | 1,219 | 1,219 | 1,219 | 1,219 |
| 33 19 0208 | Haul 6,000 Gallon Wastewater, Includes Loading & Unloading | MI | 33.47 | 28.85 | 25.42 | 20.87 | 20.14 |
| 33 19 7102 | Wastewater Disposal Fee | KGAL | 1.50 | 1.50 | 1.50 | 1.50 | 1.50 |
| 33 41 0101 | Pump & Motor Maintenance/Repair | EACH | 865.17 | 671.48 | 581.58 | 393.96 | 329.64 |
| 33 42 0101 | Electrical Charge | KWH | 0.06 | 0.06 | 0.06 | 0.06 | 0.06 |

**Environmental Remediation: Assemblies Cost Book**

# Cost Data for Remediation
## Drum Removal

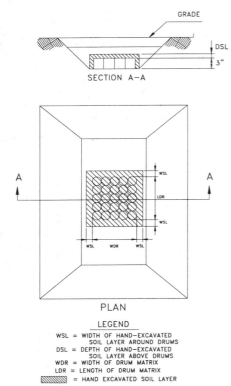

### General:
The assemblies included in this section address the costs associated with removing drums from a hazardous waste site. This includes transferring the drums to a staging area, sampling and analyzing drums, overpacking leaking drums, and consolidating compatible wastes.

### Typical Treatment Train:
Excavation, treatment, transportation, disposal, sampling and analysis.

### Common Cost Components:
1. Construction of staging area
2. Drum transfer to staging area
3. Recontainerizing drums
4. Overpacks
5. Consolidation of drum contents
6. Sampling
7. Rinsing drums

### Other Cost Considerations:
Access road, parking lot.

**Environmental Remediation: Assemblies Cost Book**

# Cost Data for Remediation

## Drum Removal

| Assembly | Description | Unit | Unit Cost by Safety Level A | B | C | D | E |
|---|---|---|---|---|---|---|---|
| | **Capital Costs** | | | | | | |
| 17 03 0276 | 1 CY, Crawler-mounted, Hydraulic Excavator | CY | 7.74 | 6.25 | 5.18 | 3.71 | 3.44 |
| 17 03 0277 | 2 CY, Crawler-mounted, Hydraulic Excavator | CY | 5.16 | 4.19 | 3.45 | 2.49 | 2.35 |
| 17 03 0278 | 3 CY, Crawler-mounted, Hydraulic Excavator | CY | 3.67 | 2.95 | 2.46 | 1.75 | 1.61 |
| 17 03 0279 | 4 CY, Crawler-mounted, Hydraulic Excavator | CY | 3.00 | 2.42 | 2.01 | 1.44 | 1.33 |
| 17 03 1002 | 2" Diameter Contractor's Trash Pump, 75 GPM | DAY | 50.20 | 47.01 | 45.64 | 42.55 | 41.42 |
| 17 03 1003 | 3" Diameter Contractor's Trash Pump, 150 GPM | DAY | 61.72 | 57.05 | 55.05 | 50.54 | 48.88 |
| 17 03 1004 | 4" Diameter Contractor's Trash Pump, 300 GPM | DAY | 80.30 | 74.23 | 71.63 | 65.77 | 63.61 |
| 33 07 9901 | Lease, Portable 40' W x 150' L x 18' High Emissions Containment Building | YEAR | 7,500 | 7,500 | 7,500 | 7,500 | 7,500 |
| 33 07 9902 | Buy, Portable 40' W x 150' L x 18' High Emissions Containment Building | EACH | 23,503 | 23,503 | 23,503 | 23,503 | 23,503 |
| 33 07 9903 | Assemble Emissions Containment Building, 5 Days | EACH | 12,346 | 9,591 | 8,298 | 5,629 | 4,722 |
| 33 07 9904 | Disassemble Emissions Containment Building, 3 Days | EACH | 7,402 | 5,750 | 4,975 | 3,374 | 2,831 |
| 33 10 0101 | Transfer Drums to Staging Area with Grappler | EACH | 63.32 | 50.90 | 42.39 | 30.21 | 27.79 |
| 33 10 0102 | Consolidate Drums into Tanker Truck | GAL | 0.94 | 0.74 | 0.63 | 0.44 | 0.38 |
| 33 10 0103 | Consolidate Drums into Dump Truck | CY | 155.31 | 122.89 | 104.17 | 72.55 | 64.06 |
| 33 10 0104 | Drum Crushing, per Individual Drum | EACH | 2.32 | 1.80 | 1.56 | 1.06 | 0.88 |
| 33 10 0105 | Polyethylene Spill Pallet, 4 55-Gallon Drum Capacity, 85 Gallon Sump | EACH | 323.65 | 323.65 | 323.65 | 323.65 | 323.65 |
| 33 10 0106 | Polyethylene Drum Storage System, 2 Drum Capacity, 135 Gallon Sump, Lockable | EACH | 613.03 | 613.03 | 613.03 | 613.03 | 613.03 |
| 33 10 0107 | 7 Gauge Steel Drum Storage, 4 55-Gallon Drum Capacity, 85 Gallon Sump | EACH | 157.59 | 157.59 | 157.59 | 157.59 | 157.59 |
| 33 10 0108 | Polyethylene Drum Storage System, 1 Drum Capacity, 85 Gallon Sump | EACH | 174.71 | 174.71 | 174.71 | 174.71 | 174.71 |
| 33 10 0109 | 250 Gallon Bulk Container, Polyethylene with Metal Case | EACH | 749.95 | 749.95 | 749.95 | 749.95 | 749.95 |
| 33 10 0110 | 350 Gallon Bulk Container, Polyethylene | EACH | 294.33 | 294.33 | 294.33 | 294.33 | 294.33 |
| 33 10 0111 | 450 Gallon Bulk Container, Polyethylene | EACH | 266.77 | 266.77 | 266.77 | 266.77 | 266.77 |
| 33 10 0112 | 550 Gallon Bulk Container, Polyethylene | EACH | 324.36 | 324.36 | 324.36 | 324.36 | 324.36 |
| 33 10 0113 | 35 CF Dry Flow Hopper, Solids/Granular | EACH | 923.79 | 923.79 | 923.79 | 923.79 | 923.79 |
| 33 10 0114 | 70 CF Dry Flow Hopper, Solids/Granular | EACH | 1,418 | 1,418 | 1,418 | 1,418 | 1,418 |
| 33 10 0115 | 5.58' x 5.58' x 8.33' Hazardous Material Storage Building, 4 Drum Capacity | EACH | 3,202 | 3,202 | 3,202 | 3,202 | 3,202 |
| 33 10 0116 | 10.25' x 10.25' x 8.33' Hazardous Material Storage Building, 10 Drum Capacity | EACH | 9,162 | 9,162 | 9,162 | 9,162 | 9,162 |
| 33 10 0117 | 10.25' x 10.25' x 8.33' Hazardous Material Storage Building, 4-Hour Rated | EACH | 11,448 | 11,448 | 11,448 | 11,448 | 11,448 |

**Environmental Remediation: Assemblies Cost Book**

# Cost Data for Remediation

## Drum Removal

| Assembly | Description | Unit | Unit Cost by Safety Level | | | | |
|---|---|---|---|---|---|---|---|
| | | | A | B | C | D | E |
| | **Capital Costs** | | | | | | |
| 33 17 0801 | Decontaminate Light Equipment | EACH | 189.11 | 146.65 | 127.13 | 86.01 | 71.80 |
| 33 17 0802 | Decontaminate Medium Equipment | EACH | 378.22 | 293.30 | 254.27 | 172.03 | 143.59 |
| 33 17 0803 | Decontaminate Heavy Equipment | EACH | 560.32 | 434.51 | 376.69 | 254.86 | 212.73 |
| 33 17 0809 | Triple-rinse 55 Gallon Drums | EACH | 14.07 | 10.92 | 9.46 | 6.41 | 5.36 |
| 33 19 0106 | Recontainerize Drums, Not Including Salvage Drums | EACH | 105.53 | 84.84 | 70.65 | 50.34 | 46.31 |
| 33 19 7280 | 30 Gallon, 17C, Open | EACH | 57.64 | 57.64 | 57.64 | 57.64 | 57.64 |
| 33 19 7281 | 30 Gallon, 17H, Open | EACH | 57.64 | 57.64 | 57.64 | 57.64 | 57.64 |
| 33 19 7282 | 30 Gallon, 17E, Open | EACH | 59.31 | 59.31 | 59.31 | 59.31 | 59.31 |
| 33 19 9911 | Air Blower to Inflate Portable Containment Berm | EACH | 160.59 | 160.59 | 160.59 | 160.59 | 160.59 |
| 33 19 9921 | DOT Steel Drum, 55 Gallon | EACH | 65.19 | 65.19 | 65.19 | 65.19 | 65.19 |
| 33 19 9922 | Polyethylene Closed Head Drum, 55 Gallon | EACH | 43.63 | 43.63 | 43.63 | 43.63 | 43.63 |
| 33 19 9923 | Polyethylene Open Head Drum, 55 Gallon | EACH | 49.90 | 49.90 | 49.90 | 49.90 | 49.90 |
| 33 19 9924 | DOT Steel Salvage Drum, 85 Gallon, 16 Gauge | EACH | 57.06 | 57.06 | 57.06 | 57.06 | 57.06 |
| 33 19 9925 | DOT Steel Salvage Drum Composite Overpack, 85 Gallon, 16 Gauge | EACH | 70.49 | 70.49 | 70.49 | 70.49 | 70.49 |

**Environmental Remediation: Assemblies Cost Book**

# Cost Data for Remediation
## Ex Situ Vapor Extraction

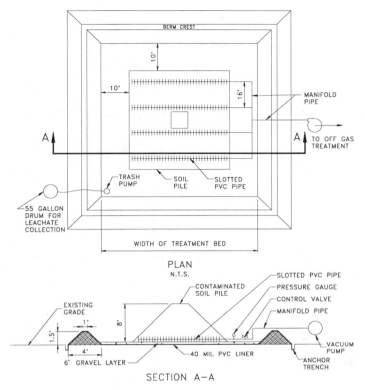

### General:

In ex situ soil vapor extraction (SVE), contaminated soil is excavated from a site and placed in an SVE bed. The treatment bed is usually underlain by an impermeable liner and surrounded by a berm to prevent intrusion of surface runoff. The liner is normally covered by a 6 to 12 inch layer of gravel to protect it from damage. A perforated pipe grid is placed on the gravel surface. The grid is attached to a common header pipe leading to a vacuum pump which pulls air through the soil. Ex situ SVE is a batch process.

### Typical Treatment Train:
Excavation, UST removal, or other processes that result in contaminated spoil, transportation, treatment of air emissions, disposal, sampling and analysis.

### Common Cost Components:

1. Construction of a treatment bed including 10-foot alley on all sides, bed liner, berm, gravel, and cover
2. Leachate collection
3. Vacuum pump(s)
4. Bed loading/unloading
5. Operations and maintenance

### Other Cost Considerations:
Utilities, piping.

**Environmental Remediation: Assemblies Cost Book**

# Cost Data for Remediation

## Ex Situ Vapor Extraction

| | | | | Unit Cost by Safety Level | | | | |
|---|---|---|---|---|---|---|---|---|
| Assembly | Description | Unit | A | B | C | D | E | |

| Assembly | Description | Unit | A | B | C | D | E |
|---|---|---|---|---|---|---|---|
| | **Capital Costs** | | | | | | |
| 17 03 0215 | 931, 1.0 CY, Track Loader | HOUR | 112.08 | 88.96 | 75.15 | 52.57 | 46.80 |
| 17 03 0216 | 943, 1.5 CY, Track Loader | HOUR | 170.50 | 137.43 | 114.12 | 81.62 | 75.57 |
| 17 03 0217 | 953, 2.0 CY, Track Loader | HOUR | 195.08 | 158.13 | 130.48 | 94.07 | 88.30 |
| 17 03 0218 | 963, 2.5 CY, Track Loader | HOUR | 280.58 | 229.38 | 187.48 | 136.82 | 131.05 |
| 17 03 0219 | 973, 3.75 CY, Track Loader | HOUR | 357.58 | 293.33 | 238.84 | 175.16 | 169.11 |
| 17 03 0340 | Standby, 931, 1.0 CY Track Loader | HOUR | 8.58 | 7.15 | 5.72 | 4.29 | 4.29 |
| 17 03 0341 | Standby, 943, 1.5 CY Track Loader | HOUR | 21.48 | 17.90 | 14.32 | 10.74 | 10.74 |
| 17 03 0342 | Standby, 953, 2.0 CY Track Loader | HOUR | 27.35 | 22.79 | 18.23 | 13.67 | 13.67 |
| 17 03 0343 | Standby, 963, 2.5 CY Track Loader | HOUR | 46.00 | 38.33 | 30.67 | 23.00 | 23.00 |
| 17 03 0344 | Standby, 973, 3.75 CY Track Loader | HOUR | 62.65 | 52.21 | 41.77 | 31.33 | 31.33 |
| 17 03 0405 | 950, 3.00 CY, Delivered & Dumped, Backfill with Sand | CY | 38.31 | 35.03 | 32.95 | 29.74 | 29.00 |
| 17 03 0411 | 966, Delivered & Dumped, Backfill with Cement Stabilized Sand | CY | 46.41 | 43.23 | 41.23 | 38.12 | 37.39 |
| 17 03 0419 | Crushed Stone, 1/2" to 3/4" | CY | 23.45 | 21.67 | 20.70 | 18.97 | 18.47 |
| 17 03 0420 | Backfill Trench, Borrow Material, Delivered & Dumped Only | CY | 9.86 | 8.74 | 7.93 | 6.83 | 6.63 |
| 17 03 1002 | 2" Diameter Contractor's Trash Pump, 75 GPM | DAY | 50.20 | 47.01 | 45.64 | 42.55 | 41.42 |
| 17 03 1003 | 3" Diameter Contractor's Trash Pump, 150 GPM | DAY | 61.72 | 57.05 | 55.05 | 50.54 | 48.88 |
| 17 03 1004 | 4" Diameter Contractor's Trash Pump, 300 GPM | DAY | 80.30 | 74.23 | 71.63 | 65.77 | 63.61 |
| 17 03 9903 | Hand Place Small Earth Fill Berm | CY | 112.16 | 86.46 | 75.45 | 50.61 | 41.50 |
| 19 02 0601 | 12" x 12" Underground French Drain | LF | 4.50 | 3.92 | 3.61 | 3.04 | 2.88 |
| 19 02 0602 | 18" x 18" Underground French Drain | LF | 5.93 | 5.29 | 4.94 | 4.31 | 4.13 |
| 19 02 0603 | 24" x 24" Underground French Drain | LF | 7.43 | 6.71 | 6.31 | 5.61 | 5.41 |
| 19 04 0401 | 550 Gallon, Stainless Steel Aboveground Wastewater Holding Tank, Rental | MONTH | 320.65 | 320.65 | 320.65 | 320.65 | 320.65 |
| 19 04 0403 | 4,000 Gallon Polyethylene Wastewater Tank, Rental | MONTH | 545.90 | 545.90 | 545.90 | 545.90 | 545.90 |
| 19 04 0405 | 6,000 Gallon, Polyethylene Aboveground Wastewater Holding Tank, Rental | MONTH | 641.30 | 641.30 | 641.30 | 641.30 | 641.30 |
| 19 04 0407 | 21,000 Gallon, Steel Closed Stationary Aboveground Wastewater Holding Tank, Rental | MONTH | 1,219 | 1,219 | 1,219 | 1,219 | 1,219 |
| 19 04 0621 | 550 Gallon Horizontal Plastic Sump with 4" NPT Connection | EACH | 1,699 | 1,596 | 1,551 | 1,452 | 1,415 |
| 19 04 0624 | 4,000 Gallon Horizontal Plastic Sump with 6" NPT Connection | EACH | 4,130 | 3,927 | 3,841 | 3,645 | 3,573 |
| 19 04 0625 | 6,000 Gallon Horizontal Plastic Sump with 6" NPT Connection | EACH | 5,248 | 4,975 | 4,858 | 4,594 | 4,497 |
| 19 04 0631 | 20,000 Gallon Horizontal Plastic Sump with 6" NPT Connection | EACH | 15,799 | 15,084 | 14,742 | 14,048 | 13,817 |

## Environmental Remediation: Assemblies Cost Book

# Cost Data for Remediation

## Ex Situ Vapor Extraction

| Assembly | Description | Unit | A | B | C | D | E |
|---|---|---|---|---|---|---|---|
| | **Capital Costs** | | | | | | |
| 33 02 1506 | Monitoring Port with Gas Monitor | EACH | 28.81 | 22.42 | 19.69 | 13.52 | 11.26 |
| 33 02 1507 | Continuous Monitoring and Recording of Air Flow | EACH | 9,870 | 9,870 | 9,870 | 9,870 | 9,870 |
| 33 07 9901 | Lease, Portable 40' W x 150' L x 18' High Emissions Containment Building | YEAR | 7,500 | 7,500 | 7,500 | 7,500 | 7,500 |
| 33 07 9902 | Buy, Portable 40' W x 150' L x 18' High Emissions Containment Building | EACH | 23,503 | 23,503 | 23,503 | 23,503 | 23,503 |
| 33 07 9903 | Assemble Emissions Containment Building, 5 Days | EACH | 12,346 | 9,591 | 8,298 | 5,629 | 4,722 |
| 33 07 9904 | Disassemble Emissions Containment Building, 3 Days | EACH | 7,402 | 5,750 | 4,975 | 3,374 | 2,831 |
| 33 08 0503 | Polymeric Liner Anchor Trench, 3' x 1.5' | LF | 1.58 | 1.25 | 1.07 | 0.76 | 0.67 |
| 33 08 0512 | Drainage Netting, Geotextile Fabric Heat-bonded 1 Side | SF | 0.41 | 0.39 | 0.38 | 0.36 | 0.35 |
| 33 08 0513 | Drainage Netting, Geotextile Fabric Heat-bonded 2 Sides | SF | 0.48 | 0.45 | 0.44 | 0.42 | 0.41 |
| 33 08 0541 | 20 Mil Polymeric Liner, Very Low Density Polyethylene | SF | 1.25 | 1.02 | 0.91 | 0.69 | 0.61 |
| 33 08 0542 | 30 Mil Polymeric Liner, Very Low Density Polyethylene | SF | 1.98 | 1.59 | 1.41 | 1.03 | 0.90 |
| 33 08 0543 | 40 Mil Polymeric Liner, Very Low Density Polyethylene | SF | 2.38 | 1.91 | 1.70 | 1.24 | 1.09 |
| 33 08 0544 | 60 Mil Polymeric Liner, Very Low Density Polyethylene | SF | 3.54 | 2.83 | 2.51 | 1.83 | 1.60 |
| 33 08 0545 | 80 Mil Polymeric Liner, Very Low Density Polyethylene | SF | 4.35 | 3.49 | 3.10 | 2.27 | 1.98 |
| 33 08 0546 | 100 Mil Polymeric Liner, Very Low Density Polyethylene | SF | 5.50 | 4.41 | 3.91 | 2.85 | 2.49 |
| 33 08 0551 | 36 Mil Polymeric Liner, Chlorosulfanated Polyethylene | SF | 0.75 | 0.73 | 0.72 | 0.70 | 0.70 |
| 33 08 0552 | 45 Mil Polymeric Liner, Chlorosulfanated Polyethylene | SF | 0.83 | 0.81 | 0.80 | 0.79 | 0.78 |
| 33 08 0561 | 20 Mil Polymeric Liner, PVC | SF | 0.24 | 0.23 | 0.22 | 0.21 | 0.20 |
| 33 08 0562 | 30 Mil Polymeric Liner, PVC | SF | 0.32 | 0.31 | 0.30 | 0.29 | 0.28 |
| 33 08 0563 | 40 Mil Polymeric Liner, PVC | SF | 0.38 | 0.36 | 0.36 | 0.34 | 0.34 |
| 33 08 0564 | 60 Mil Polymeric Liner, PVC | SF | 0.72 | 0.70 | 0.69 | 0.67 | 0.67 |
| 33 08 0565 | 80 Mil Polymeric Liner, PVC | SF | 0.98 | 0.96 | 0.95 | 0.93 | 0.93 |
| 33 08 0566 | 100 Mil Polymeric Liner, PVC | SF | 1.28 | 1.26 | 1.25 | 1.23 | 1.22 |
| 33 08 0571 | 40 Mil Polymeric Liner, High-density Polyethylene | SF | 2.35 | 1.88 | 1.67 | 1.21 | 1.06 |
| 33 08 0572 | 60 Mil Polymeric Liner, High-density Polyethylene | SF | 3.49 | 2.80 | 2.48 | 1.81 | 1.58 |
| 33 08 0573 | 80 Mil Polymeric Liner, High-density Polyethylene | SF | 4.32 | 3.47 | 3.08 | 2.25 | 1.97 |
| 33 08 0584 | Plastic Laminate Waste Pile Cover | SF | 0.16 | 0.15 | 0.14 | 0.13 | 0.13 |
| 33 08 0590 | Waste Pile Cover, 135 Lb Tear, 2 - 2.5 Year Life | SY | 2.03 | 1.93 | 1.89 | 1.80 | 1.77 |
| 33 08 0591 | Waste Pile Cover, 185 Lb Tear, 3 - 4 Year Life | SY | 2.98 | 2.89 | 2.85 | 2.76 | 2.72 |
| 33 08 0592 | Waste Pile Cover, 250 Lb Tear, 4 - 5 Year Life | SY | 4.00 | 3.89 | 3.85 | 3.74 | 3.70 |
| 33 13 2306 | Propane Conversion for Soil Vapor Extractor | EACH | 90.00 | 90.00 | 90.00 | 90.00 | 90.00 |
| 33 13 2307 | Modulating Motor for Automatic Fuel Adjustment | EACH | 1,958 | 1,958 | 1,958 | 1,958 | 1,958 |

**Environmental Remediation: Assemblies Cost Book**

# Cost Data for Remediation

## Ex Situ Vapor Extraction

| Assembly | Description | Unit | Unit Cost by Safety Level | | | | |
|---|---|---|---|---|---|---|---|
| | | | A | B | C | D | E |
| | **Capital Costs** | | | | | | |
| 33 13 2308 | Trailer Mounting for Vapor Extractor Unit | EACH | 945.00 | 945.00 | 945.00 | 945.00 | 945.00 |
| 33 13 2333 | 5 HP, 90 SCFM Vapor Extraction Blower, Monthly Rental | MONTH | 850.00 | 850.00 | 850.00 | 850.00 | 850.00 |
| 33 13 2334 | 7.5 HP, 140 SCFM Vapor Extraction Blower, Monthly Rental | MONTH | 900.00 | 900.00 | 900.00 | 900.00 | 900.00 |
| 33 13 2335 | 10 HP, 190 SCFM Vapor Extraction Blower, Monthly Rental | MONTH | 1,163 | 1,163 | 1,163 | 1,163 | 1,163 |
| 33 13 2336 | 15 HP, 290 SCFM Vapor Extraction Blower, Monthly Rental | MONTH | 1,325 | 1,325 | 1,325 | 1,325 | 1,325 |
| 33 13 2337 | 30 HP, 580 SCFM Vapor Extraction Blower, Monthly Rental | MONTH | 1,625 | 1,625 | 1,625 | 1,625 | 1,625 |
| 33 13 2338 | Purchase, 5.0 HP, 90 SCFM Vapor Extraction Blower | EACH | 1,912 | 1,912 | 1,912 | 1,912 | 1,912 |
| 33 13 2339 | Purchase, 7.5 HP, 140 SCFM Vapor Extraction Blower | EACH | 3,110 | 2,931 | 2,854 | 2,680 | 2,617 |
| 33 13 2340 | Purchase, 10.0 HP, 190 SCFM Vapor Extraction Blower | EACH | 3,359 | 3,179 | 3,102 | 2,929 | 2,865 |
| 33 13 2341 | Purchase, 15.0 HP, 280 SCFM Vapor Extraction Blower | EACH | 5,126 | 4,947 | 4,870 | 4,696 | 4,632 |
| 33 13 2342 | Purchase, 30.0 HP, 580 SCFM Vapor Extraction Blower | EACH | 7,516 | 7,337 | 7,260 | 7,086 | 7,022 |
| 33 13 2343 | Knockout Drum | EACH | 65.19 | 65.19 | 65.19 | 65.19 | 65.19 |
| 33 13 2348 | Install/Assemble Rental Vacuum Extraction Blower | EACH | 783.78 | 604.17 | 527.27 | 353.66 | 290.00 |
| 33 17 0801 | Decontaminate Light Equipment | EACH | 189.11 | 146.65 | 127.13 | 86.01 | 71.80 |
| 33 17 0802 | Decontaminate Medium Equipment | EACH | 378.22 | 293.30 | 254.27 | 172.03 | 143.59 |
| 33 17 0803 | Decontaminate Heavy Equipment | EACH | 560.32 | 434.51 | 376.69 | 254.86 | 212.73 |
| 33 19 9921 | DOT Steel Drum, 55 Gallon | EACH | 65.19 | 65.19 | 65.19 | 65.19 | 65.19 |
| 33 26 0408 | 4" PVC, Schedule 40, Manifold Piping | LF | 13.38 | 10.65 | 9.48 | 6.84 | 5.88 |
| 33 26 0411 | 6" PVC, Schedule 40, Manifold Piping | LF | 18.73 | 15.09 | 13.53 | 10.01 | 8.73 |
| 33 26 0802 | 4" Slotted PVC Pipe | LF | 8.67 | 7.65 | 7.22 | 6.23 | 5.87 |
| 33 26 0803 | 6" Slotted PVC Pipe | LF | 14.25 | 12.89 | 12.31 | 10.99 | 10.51 |
| 33 27 0104 | 4" PVC, Schedule 40, Tee | EACH | 137.04 | 108.84 | 96.77 | 69.51 | 59.51 |
| 33 27 0105 | 6" PVC, Schedule 40, Tee | EACH | 205.86 | 169.47 | 153.89 | 118.72 | 105.82 |
| 33 27 0114 | 4" PVC, 90 Degree, Elbow | EACH | 101.72 | 80.57 | 71.51 | 51.07 | 43.57 |
| 33 27 0115 | 6" PVC, 90 Degree, Elbow | EACH | 135.44 | 111.27 | 100.92 | 77.56 | 68.99 |
| 33 27 0404 | 4" Iron Body Check Valve | EACH | 354.29 | 315.96 | 299.54 | 262.49 | 248.90 |
| 33 27 0405 | 6" Iron Body Check Valve | EACH | 618.60 | 554.57 | 523.95 | 461.86 | 441.15 |
| 33 29 0412 | 15 GPM Submersible Sump Pump | EACH | 492.79 | 460.07 | 446.06 | 414.43 | 402.83 |
| 33 29 0413 | 42 GPM Submersible Sump Pump | EACH | 792.79 | 760.07 | 746.06 | 714.43 | 702.83 |
| 33 29 0414 | 52 GPM Submersible Sump Pump | EACH | 871.35 | 832.08 | 815.27 | 777.32 | 763.40 |
| 33 31 0209 | Pressure Gauge | EACH | 140.16 | 121.61 | 113.67 | 95.74 | 89.16 |
| 33 42 0101 | Electrical Charge | KWH | 0.06 | 0.06 | 0.06 | 0.06 | 0.06 |

**Environmental Remediation: Assemblies Cost Book**

*Page 3-98*

# Cost Data for Remediation
## Excavation, Buried Waste

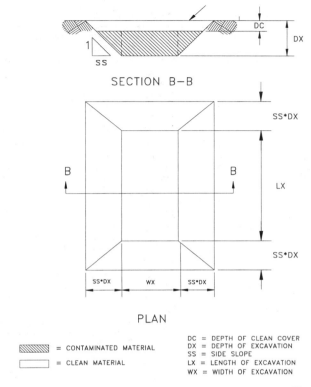

### General:

The assemblies included in this section are associated with the excavation and backfilling of a contaminated site. The area may be bulk excavated or excavated in lifts. Hand excavation is most appropriate for the soil around buried drums.

### Typical Treatment Train:

Excavation, drum removal, removal of rock, concrete, rubble, or tanks, stockpiling, treatment (soil washing, for example), transportation, disposal (incineration, landfilling, etc.), sampling and analysis.

### Common Cost Components:

1. Sidewall protection
2. Staff engineer on-site during excavation
3. Excavation equipment (bulldozers, scrapers, excavators, track loaders, wheel loaders)
4. Dust suppressant

### Other Cost Considerations:

N/A

**Environmental Remediation: Assemblies Cost Book**

# Cost Data for Remediation

## Excavation, Buried Waste

| Assembly | Description | Unit | Unit Cost by Safety Level | | | | |
|---|---|---|---|---|---|---|---|
| | | | A | B | C | D | E |
| | **Capital Costs** | | | | | | |
| 17 03 0206 | D3 with A-blade Bulldozer | HOUR | 151.96 | 122.19 | 101.74 | 72.51 | 66.74 |
| 17 03 0207 | D4 with A-blade Bulldozer | HOUR | 153.56 | 123.53 | 102.80 | 73.31 | 67.54 |
| 17 03 0208 | D5 with A-blade Bulldozer | HOUR | 181.44 | 146.76 | 121.39 | 87.25 | 81.48 |
| 17 03 0209 | D6 with A-blade Bulldozer | HOUR | 208.40 | 169.01 | 139.38 | 100.57 | 94.52 |
| 17 03 0210 | D7 with U-blade Bulldozer | HOUR | 296.12 | 242.11 | 197.86 | 144.43 | 138.38 |
| 17 03 0211 | Hand Excavation, Normal Soil | CY | 80.12 | 61.76 | 53.90 | 36.15 | 29.64 |
| 17 03 0212 | Hand Excavation, Sand/Gravel | CY | 70.10 | 54.04 | 47.16 | 31.63 | 25.94 |
| 17 03 0213 | Hand Excavation, Medium Clay | CY | 93.47 | 72.05 | 62.88 | 42.17 | 34.58 |
| 17 03 0215 | 931, 1.0 CY, Track Loader | HOUR | 112.08 | 88.96 | 75.15 | 52.57 | 46.80 |
| 17 03 0216 | 943, 1.5 CY, Track Loader | HOUR | 170.50 | 137.43 | 114.12 | 81.62 | 75.57 |
| 17 03 0217 | 953, 2.0 CY, Track Loader | HOUR | 195.08 | 158.13 | 130.48 | 94.07 | 88.30 |
| 17 03 0218 | 963, 2.5 CY, Track Loader | HOUR | 280.58 | 229.38 | 187.48 | 136.82 | 131.05 |
| 17 03 0219 | 973, 3.75 CY, Track Loader | HOUR | 357.58 | 293.33 | 238.84 | 175.16 | 169.11 |
| 17 03 0220 | 910, 1.25 CY, Wheel Loader | HOUR | 130.20 | 104.06 | 87.23 | 61.63 | 55.86 |
| 17 03 0221 | 916, 1.5 CY, Wheel Loader | HOUR | 114.32 | 90.83 | 76.64 | 53.69 | 47.92 |
| 17 03 0222 | 926, 2.0 CY, Wheel Loader | HOUR | 154.30 | 124.14 | 103.30 | 73.68 | 67.91 |
| 17 03 0223 | 950, 3.0 CY, Wheel Loader | HOUR | 180.10 | 145.64 | 120.50 | 86.58 | 80.81 |
| 17 03 0224 | 966, 4.0 CY, Wheel Loader | HOUR | 211.58 | 171.88 | 141.48 | 102.32 | 96.55 |
| 17 03 0225 | 980, 5.25 CY, Wheel Loader | HOUR | 298.84 | 244.38 | 199.68 | 145.79 | 139.74 |
| 17 03 0226 | 988, 7.0 CY, Wheel Loader | HOUR | 384.90 | 316.10 | 257.05 | 188.82 | 182.77 |
| 17 03 0227 | 992, 13.5 CY, Wheel Loader | HOUR | 648.92 | 536.11 | 433.06 | 320.83 | 314.78 |
| 17 03 0229 | Hand Excavation, Heavy Clay | CY | 112.16 | 86.46 | 75.45 | 50.61 | 41.50 |
| 17 03 0230 | Crawler-mounted, 1.0 CY, 215 Hydraulic Excavator | HOUR | 210.40 | 170.50 | 140.73 | 101.43 | 95.15 |
| 17 03 0231 | Crawler-mounted, 1.25 CY, 225 Hydraulic Excavator | HOUR | 210.40 | 170.50 | 140.73 | 101.43 | 95.15 |
| 17 03 0232 | Crawler-mounted, 2.0 CY, 235 Hydraulic Excavator | HOUR | 330.90 | 270.92 | 221.07 | 161.68 | 155.40 |
| 17 03 0233 | Crawler-mounted, 3.125 CY, 245 Hydraulic Excavator | HOUR | 363.36 | 297.97 | 242.71 | 177.91 | 171.63 |
| 17 03 0234 | Crawler-mounted, 4.0 CY, Koehring 1166 Hydraulic Excavator | HOUR | 363.36 | 297.97 | 242.71 | 177.91 | 171.63 |
| 17 03 0235 | Crawler-mounted, 5.5 CY, Koehring 1266, Hydraulic Excavator | HOUR | 491.70 | 404.92 | 328.27 | 242.08 | 235.80 |
| 17 03 0236 | Scraper - Standard, 15 CY - 621 with D8L Dozer | HOUR | 569.73 | 468.78 | 380.40 | 280.19 | 272.40 |
| 17 03 0237 | Scraper - Standard, 22 CY - 631 with D9L Dozers | HOUR | 755.52 | 622.44 | 504.37 | 372.17 | 362.87 |
| 17 03 0238 | Scraper - Standard, 34 CY - 651 with D9L Dozers | HOUR | 652.54 | 536.62 | 435.72 | 320.68 | 311.38 |
| 17 03 0240 | Scraper - Elevating, 11 CY - 613 | HOUR | 181.08 | 146.07 | 121.19 | 86.77 | 80.49 |

## Environmental Remediation: Assemblies Cost Book

*Page 3-100*

1998 by ECHOS. All rights reserved

# Cost Data for Remediation

## Excavation, Buried Waste

| Assembly | Description | Unit | A | B | C | D | E |
|---|---|---|---|---|---|---|---|
| | **Capital Costs** | | | | | | |
| 17 03 0241 | Scraper - Elevating, 22 CY - 623 | HOUR | 280.40 | 228.83 | 187.40 | 136.43 | 130.15 |
| 17 03 0243 | Scraper - Tandem, 14 CY - 627 with D8L Dozer | HOUR | 392.35 | 320.97 | 262.15 | 191.50 | 183.71 |
| 17 03 0244 | Scraper - Tandem, 21 CY - 637 with D9L Dozer | HOUR | 555.75 | 457.13 | 371.08 | 273.19 | 265.40 |
| 17 03 0245 | Scraper - Tandem, 32 CY - 657 with D9L Dozers | HOUR | 762.00 | 627.84 | 508.69 | 375.41 | 366.11 |
| 17 03 0310 | D8 with U-blade & Single-shank Ripper, Bulldozer | HOUR | 396.60 | 325.85 | 264.85 | 194.67 | 188.62 |
| 17 03 0311 | D9 with U-blade & Single-shank Ripper, Bulldozer | HOUR | 455.94 | 375.30 | 304.41 | 224.34 | 218.29 |
| 17 03 0312 | D10 with U-blade & Single-shank Ripper, Bulldozer | HOUR | 476.58 | 392.50 | 318.17 | 234.66 | 228.61 |
| 17 03 0318 | Standby, D3 with A-blade Bulldozer | HOUR | 17.90 | 14.92 | 11.93 | 8.95 | 8.95 |
| 17 03 0319 | Standby, D4 with A-blade Bulldozer | HOUR | 18.40 | 15.33 | 12.27 | 9.20 | 9.20 |
| 17 03 0320 | Standby, D5 with A-blade Bulldozer | HOUR | 25.18 | 20.98 | 16.79 | 12.59 | 12.59 |
| 17 03 0321 | Standby, D6 with A-blade Bulldozer | HOUR | 30.75 | 25.63 | 20.50 | 15.38 | 15.38 |
| 17 03 0322 | Standby, D7 with U-blade Bulldozer | HOUR | 51.49 | 42.91 | 34.33 | 25.75 | 25.75 |
| 17 03 0323 | Standby, D8 with U-blade & Single-shank Ripper | HOUR | 82.30 | 68.58 | 54.86 | 41.15 | 41.15 |
| 17 03 0324 | Standby, D9 with U-blade & Single-shank Ripper | HOUR | 98.73 | 82.28 | 65.82 | 49.37 | 49.37 |
| 17 03 0325 | Standby, D10 with U-blade & Single-shank Ripper | HOUR | 102.75 | 85.62 | 68.50 | 51.37 | 51.37 |
| 17 03 0326 | Standby, Scraper - Standard, 15 CY 621 with D8L Dozer | HOUR | 123.67 | 103.06 | 82.44 | 61.83 | 61.83 |
| 17 03 0327 | Standby, Scraper - Standard, 22 CY 631 with D9L Dozers | HOUR | 170.67 | 142.22 | 113.78 | 85.33 | 85.33 |
| 17 03 0328 | Standby, Scraper - Standard, 34 CY 651 with D9L Dozers | HOUR | 145.64 | 121.37 | 97.09 | 72.82 | 72.82 |
| 17 03 0329 | Standby, Scraper - Elevating, 11 CY 613 | HOUR | 26.27 | 21.89 | 17.51 | 13.13 | 13.13 |
| 17 03 0330 | Standby, Scraper - Elevating, 22 CY 623 | HOUR | 53.31 | 44.43 | 35.54 | 26.66 | 26.66 |
| 17 03 0331 | Standby, Scraper - Tandem, 14 CY 627 with D8L Dozer | HOUR | 118.18 | 98.49 | 78.79 | 59.09 | 59.09 |
| 17 03 0332 | Standby, Scraper - Tandem 21 CY 637 with D9L Dozer | HOUR | 143.63 | 119.69 | 95.75 | 71.81 | 71.81 |
| 17 03 0333 | Standby, Scraper - Tandem 32 CY 657 with D9L Dozers | HOUR | 138.87 | 115.73 | 92.58 | 69.44 | 69.44 |
| 17 03 0334 | Standby, Crawler-mounted, 1.0 CY, 215 Hydraulic Excavator | HOUR | 39.85 | 33.21 | 26.57 | 19.93 | 19.93 |
| 17 03 0335 | Standby, Crawler-mounted, 1.25 CY, 225 Hydraulic Excavator | HOUR | 29.90 | 24.91 | 19.93 | 14.95 | 14.95 |
| 17 03 0336 | Standby, Crawler-mounted, 2.0 CY, 235 Hydraulic Excavator | HOUR | 74.45 | 62.04 | 49.63 | 37.22 | 37.22 |
| 17 03 0337 | Standby, Crawler-mounted, 3.125 CY, 245 Hydraulic Excavator | HOUR | 77.03 | 64.19 | 51.35 | 38.51 | 38.51 |
| 17 03 0338 | Standby, Crawler-mounted, 4.0 CY, Koehring 1166 Hydraulic Excavator | HOUR | 73.91 | 61.59 | 49.27 | 36.95 | 36.95 |
| 17 03 0339 | Standby, Crawler-mounted, 5.5 CY, Koehring 1266 Hydraulic Excavator | HOUR | 106.65 | 88.88 | 71.10 | 53.33 | 53.33 |
| 17 03 0340 | Standby, 931, 1.0 CY Track Loader | HOUR | 8.58 | 7.15 | 5.72 | 4.29 | 4.29 |

**Environmental Remediation: Assemblies Cost Book**

# Cost Data for Remediation

## Excavation, Buried Waste

| Assembly | Description | Unit | A | B | C | D | E |
|---|---|---|---|---|---|---|---|
| | **Capital Costs** | | | | | | |
| 17 03 0341 | Standby, 943, 1.5 CY Track Loader | HOUR | 21.48 | 17.90 | 14.32 | 10.74 | 10.74 |
| 17 03 0342 | Standby, 953, 2.0 CY Track Loader | HOUR | 27.35 | 22.79 | 18.23 | 13.67 | 13.67 |
| 17 03 0343 | Standby, 963, 2.5 CY Track Loader | HOUR | 46.00 | 38.33 | 30.67 | 23.00 | 23.00 |
| 17 03 0344 | Standby, 973, 3.75 CY Track Loader | HOUR | 62.65 | 52.21 | 41.77 | 31.33 | 31.33 |
| 17 03 0345 | Standby, 910, 1.25 CY Wheel Loader | HOUR | 15.13 | 12.61 | 10.09 | 7.57 | 7.57 |
| 17 03 0346 | Standby, 916, 1.5 CY Wheel Loader | HOUR | 11.32 | 9.44 | 7.55 | 5.66 | 5.66 |
| 17 03 0347 | Standby, 926, 2.0 CY Wheel Loader | HOUR | 21.40 | 17.83 | 14.27 | 10.70 | 10.70 |
| 17 03 0348 | Standby, 950, 3.0 CY Wheel Loader | HOUR | 27.24 | 22.70 | 18.16 | 13.62 | 13.62 |
| 17 03 0349 | Standby, 966, 4.0 CY Wheel Loader | HOUR | 34.27 | 28.56 | 22.85 | 17.13 | 17.13 |
| 17 03 0350 | Standby, 980, 5.25 CY Wheel Loader | HOUR | 57.52 | 47.94 | 38.35 | 28.76 | 28.76 |
| 17 03 0351 | Standby, 988, 7.0 CY Wheel Loader | HOUR | 81.69 | 68.07 | 54.46 | 40.84 | 40.84 |
| 17 03 0352 | Standby, 992, 13.5 CY Wheel Loader | HOUR | 149.61 | 124.68 | 99.74 | 74.81 | 74.81 |
| 17 03 0415 | Backfill with Excavated Material | CY | 7.50 | 5.97 | 5.12 | 3.63 | 3.21 |
| 17 03 0422 | Unclassified Fill, 6" Lifts, On-Site | CY | 14.23 | 11.65 | 9.60 | 7.05 | 6.72 |
| 17 03 0423 | Unclassified Fill, 6" Lifts, Off-Site | CY | 11.08 | 9.81 | 8.83 | 7.58 | 7.40 |
| 17 03 0425 | Sand, 6" Lifts, On-Site | CY | 13.88 | 11.35 | 9.36 | 6.87 | 6.54 |
| 17 03 0426 | Sand, 6" Lifts, Off-Site | CY | 15.55 | 14.34 | 13.42 | 12.22 | 12.04 |
| 17 03 0428 | Clay, 8" Lifts, Off-Site | CY | 22.97 | 19.74 | 17.38 | 14.21 | 13.67 |
| 17 03 0429 | Clay, 8" Lifts, On-Site | CY | 21.18 | 17.21 | 14.26 | 10.34 | 9.71 |
| 17 03 0430 | Gravel, 6" Lifts | CY | 15.46 | 13.36 | 12.05 | 9.99 | 9.49 |
| 17 03 0433 | Crawler Mounted 0.5 CY Hydraulic Excavator | HR | 0.00 | 0.00 | 0.00 | 0.00 | 0.00 |
| 17 03 0434 | Wheel Mounted 0.5 CY 206 Hydraulic Excavator | HR | 0.00 | 0.00 | 0.00 | 0.00 | 0.00 |
| 17 03 0436 | 0.75 CY Wheel Loader | HR | 0.00 | 0.00 | 0.00 | 0.00 | 0.00 |
| 17 03 0437 | 0.75 CY Backhoe with Front End Loader | HR | 0.00 | 0.00 | 0.00 | 0.00 | 0.00 |
| 17 03 0438 | Standby Crawler Mounted 0.5 CY Hydraulic Excavator | HR | 0.00 | 0.00 | 0.00 | 0.00 | 0.00 |
| 17 03 0439 | Standby Wheel Mounted 0.5 CY Hydraulic Excavator | HR | 0.00 | 0.00 | 0.00 | 0.00 | 0.00 |
| 17 03 0441 | Standby 0.75 CY Wheel Loader | HR | 0.00 | 0.00 | 0.00 | 0.00 | 0.00 |
| 17 03 0442 | Standby 0.75 CY Backhoe with Front End Loader | HR | 0.00 | 0.00 | 0.00 | 0.00 | 0.00 |
| 17 03 0901 | Steel Sheeting, Pull & Salvage, to 15' | SF | 13.65 | 11.21 | 9.59 | 7.20 | 6.70 |
| 17 03 0902 | Steel Sheeting, Pull & Salvage, to 20' | SF | 14.11 | 11.61 | 9.95 | 7.50 | 6.99 |
| 17 03 0903 | Steel Sheeting, Pull & Salvage, to 25' | SF | 12.32 | 10.20 | 8.79 | 6.72 | 6.28 |
| 17 03 0904 | Steel Sheeting, Pull & Salvage, to 40' | SF | 14.52 | 12.05 | 10.40 | 7.98 | 7.46 |
| 33 02 0302 | Organic Vapor Analyzer Rental, per Month | MONTH | 1,920 | 1,920 | 1,920 | 1,920 | 1,920 |

**Environmental Remediation: Assemblies Cost Book**

*Page 3-102*

# Cost Data for Remediation

## Excavation, Buried Waste

| Assembly | Description | Unit | A | B | C | D | E |
|---|---|---|---|---|---|---|---|
| | **Capital Costs** | | | | | | |
| 33 02 0303 | Organic Vapor Analyzer Rental, per Day | DAY | 184.30 | 184.30 | 184.30 | 184.30 | 184.30 |
| 33 07 9901 | Lease, Portable 40' W x 150' L x 18' High Emissions Containment Building | YEAR | 7,500 | 7,500 | 7,500 | 7,500 | 7,500 |
| 33 07 9902 | Buy, Portable 40' W x 150' L x 18' High Emissions Containment Building | EACH | 23,503 | 23,503 | 23,503 | 23,503 | 23,503 |
| 33 07 9903 | Assemble Emissions Containment Building, 5 Days | EACH | 12,346 | 9,591 | 8,298 | 5,629 | 4,722 |
| 33 07 9904 | Disassemble Emissions Containment Building, 3 Days | EACH | 7,402 | 5,750 | 4,975 | 3,374 | 2,831 |
| 33 08 0574 | Treesap-based, Overall Dust Suppressant | SY | 3.73 | 3.04 | 2.57 | 1.89 | 1.74 |
| 33 08 0575 | Chloride-based, Overall Dust Suppressant | SY | 3.73 | 3.04 | 2.57 | 1.89 | 1.74 |
| 33 08 0576 | Oil Emulsion-based, Overall Dust Suppressant | SY | 3.73 | 3.04 | 2.57 | 1.89 | 1.74 |
| 33 08 0577 | Asphalt Emulsion-based, Overall Dust Suppressant | SY | 4.12 | 3.35 | 2.84 | 2.07 | 1.91 |
| 33 08 0578 | Lignosulphonate-based, Overall Dust Suppressant | SY | 3.73 | 3.04 | 2.57 | 1.89 | 1.74 |
| 33 08 0579 | Treesap-based, Stockpile Dust Suppressant | SY | 3.81 | 3.11 | 2.65 | 1.96 | 1.82 |
| 33 08 0580 | Chloride-based, Stockpile Dust Suppressant | SY | 3.81 | 3.11 | 2.65 | 1.96 | 1.82 |
| 33 08 0581 | Oil Emulsion-based, Stockpile Dust Suppressant | SY | 3.81 | 3.11 | 2.65 | 1.96 | 1.82 |
| 33 08 0582 | Asphalt Emulsion-based, Stockpile Dust Suppressant | SY | 4.20 | 3.42 | 2.91 | 2.15 | 1.98 |
| 33 08 0583 | Lignosulphonate-based, Stockpile Dust Suppressant | SY | 3.81 | 3.11 | 2.65 | 1.96 | 1.82 |
| 33 08 0584 | Plastic Laminate Waste Pile Cover | SF | 0.16 | 0.15 | 0.14 | 0.13 | 0.13 |
| 33 08 0585 | Sprayed Water Dust Suppressant | SY | 0.02 | 0.02 | 0.02 | 0.01 | 0.01 |
| 33 17 0801 | Decontaminate Light Equipment | EACH | 189.11 | 146.65 | 127.13 | 86.01 | 71.80 |
| 33 17 0802 | Decontaminate Medium Equipment | EACH | 378.22 | 293.30 | 254.27 | 172.03 | 143.59 |
| 33 17 0803 | Decontaminate Heavy Equipment | EACH | 560.32 | 434.51 | 376.69 | 254.86 | 212.73 |
| 33 17 0807 | Decontaminate Surface By Sandblasting, 137.5 SF/Hour | SF | 2.52 | 1.99 | 1.74 | 1.23 | 1.06 |
| 33 17 0812 | Decontaminate Surface By Steam Cleaning, 137.5 SF/Hour | SF | 0.91 | 0.71 | 0.61 | 0.42 | 0.35 |
| 33 17 0813 | Decontaminate Surface By Pressure Washing, 105 SF/Hour | SF | 1.71 | 1.35 | 1.15 | 0.80 | 0.70 |

*Unit Cost by Safety Level*

**Environmental Remediation: Assemblies Cost Book**

# Cost Data for Remediation
## Extraction Wells

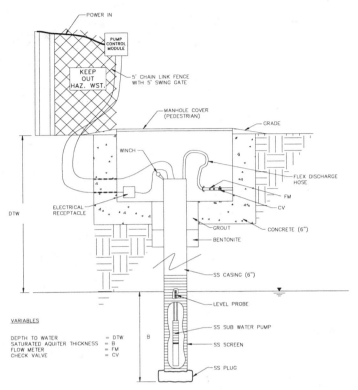

### General:

*Extraction wells are widely used as part of "pump and treat" groundwater remediation. Groundwater extraction as a remedial technology, however, is ineffective due primarily to the presence of continued sources of groundwater contamination, such as contaminated soils. Extraction wells may be installed in both confined and unconfined aquifers.*

### Typical Treatment Train:
*Treatment and disposal of the extracted groundwater, soil treatment to eliminate the source of the groundwater contamination, sampling and analysis.*

### Common Cost Components:

1. Mobilization of the drill rig and crew
2. Drilling, decontamination of equipment
3. Screen, casing, plug, grout, etc.
4. Groundwater and product pumps
5. Well development
6. Piping for effluent
7. Operations and maintenance

### Other Cost Considerations:
*Clearing and grubbing, electrical distribution.*

**Environmental Remediation: Assemblies Cost Book**

# Cost Data for Remediation

## Extraction Wells

| Assembly | Description | Unit | \- Unit Cost by Safety Level \- A | B | C | D | E |
|---|---|---|---|---|---|---|---|
| | **Capital Costs** | | | | | | |
| 17 02 0203 | Demolish Bituminous Pavement with Air Equipment | CY | 80.15 | 63.38 | 53.76 | 37.41 | 32.98 |
| 17 02 0205 | Demolish Unreinforced Concrete to 6" Thick with Air Equipment | CY | 126.90 | 100.35 | 85.12 | 59.24 | 52.23 |
| 17 02 0206 | Demolish Mesh Reinforced Concrete to 6" Thick with Air Equipment | CY | 183.47 | 145.09 | 123.07 | 85.65 | 75.51 |
| 17 02 0207 | Demolish Rod Reinforced Concrete to 6" Thick with Air Equipment | CY | 199.06 | 157.42 | 133.53 | 92.92 | 81.92 |
| 17 02 0208 | Demolish Mesh Reinforced Concrete to 6" Thick with Power Equipment | CY | 96.73 | 76.68 | 64.87 | 45.30 | 40.20 |
| 17 02 0209 | Demolish Rod Reinforced Concrete to 6" Thick with Power Equipment | CY | 106.91 | 84.75 | 71.70 | 50.07 | 44.43 |
| 17 02 0210 | Demolish Unreinforced Concrete 7" to 24" Thick with Power Equipment | CY | 258.44 | 204.87 | 173.31 | 121.03 | 107.40 |
| 17 02 0211 | Demolish Reinforced Concrete 7" to 24" Thick with Power Equipment | CY | 356.37 | 282.51 | 238.98 | 166.90 | 148.10 |
| 17 03 0276 | 1 CY, Crawler-mounted, Hydraulic Excavator | CY | 7.74 | 6.25 | 5.18 | 3.71 | 3.44 |
| 17 03 0277 | 2 CY, Crawler-mounted, Hydraulic Excavator | CY | 5.16 | 4.19 | 3.45 | 2.49 | 2.35 |
| 17 03 0278 | 3 CY, Crawler-mounted, Hydraulic Excavator | CY | 3.67 | 2.95 | 2.46 | 1.75 | 1.61 |
| 17 03 0279 | 4 CY, Crawler-mounted, Hydraulic Excavator | CY | 3.00 | 2.42 | 2.01 | 1.44 | 1.33 |
| 17 03 0282 | Soil, 5 Miles, Dump Truck, Load/Haul Spoil From Trench | CY | 6.97 | 5.60 | 4.66 | 3.33 | 3.06 |
| 17 03 0283 | Sand, 5 Mile, Dump Truck, Load/Haul Spoil From Trench | CY | 6.28 | 5.05 | 4.20 | 3.00 | 2.76 |
| 17 03 0317 | Rock, 5 Miles, Dump Truck, Load/Haul Spoil from Trench | CY | 10.31 | 8.31 | 6.90 | 4.93 | 4.56 |
| 18 01 0313 | Jointed Concrete, Place & Finish, 4" - 12" | SY | 10.99 | 9.18 | 8.38 | 6.63 | 6.02 |
| 18 01 0315 | Jointed Reinforced Concrete, Place & Finish, 4" - 12" | SY | 28.21 | 24.00 | 22.15 | 18.08 | 16.60 |
| 18 01 0316 | Field Mix Concrete | CY | 90.38 | 85.86 | 83.82 | 79.44 | 77.89 |
| 18 02 0301 | Asphalt Pavement - 10" Subgrade, 9" Base, 1 1/2" Topping | SY | 61.11 | 50.94 | 43.58 | 33.57 | 31.83 |
| 18 02 0322 | 8" Structural Slab on Grade | SF | 9.75 | 8.38 | 7.71 | 6.38 | 5.95 |
| 18 04 0106 | 5' Galvanized Chain-link Fence | LF | 19.90 | 19.20 | 18.91 | 18.24 | 18.00 |
| 18 04 0107 | 6' Galvanized Chain-link Fence | LF | 21.07 | 20.35 | 20.04 | 19.35 | 19.09 |
| 18 04 0108 | 7' Galvanized Chain-link Fence | LF | 27.68 | 26.93 | 26.61 | 25.88 | 25.62 |
| 18 04 0111 | Galvanized Barbed Wire, 3-Strand | LF | 1.81 | 1.45 | 1.27 | 0.92 | 0.81 |
| 18 04 0116 | 5' Swing Gate, 12' Double | EACH | 780.94 | 725.61 | 697.51 | 643.78 | 626.90 |
| 18 04 0117 | 6' Swing Gate, 12' Double | EACH | 896.03 | 836.09 | 805.65 | 747.44 | 729.16 |
| 18 04 0118 | 7' Swing Gate, 12' Double | EACH | 329.09 | 329.09 | 329.09 | 329.09 | 329.09 |
| 19 01 0202 | 1", Class 200, PVC Piping | LF | 7.48 | 5.84 | 5.07 | 3.48 | 2.93 |
| 19 01 0204 | 2", Class 200, PVC Piping | LF | 8.86 | 6.96 | 6.07 | 4.23 | 3.60 |

**Environmental Remediation: Assemblies Cost Book**

# Cost Data for Remediation

## Extraction Wells

| Assembly | Description | Unit | Unit Cost by Safety Level A | B | C | D | E |
|---|---|---|---|---|---|---|---|
| | **Capital Costs** | | | | | | |
| 19 01 0206 | 3", Class 200, PVC Piping | LF | 11.81 | 9.34 | 8.20 | 5.80 | 4.98 |
| 19 01 0207 | 4", Class 200, PVC Piping | LF | 12.33 | 9.86 | 8.72 | 6.32 | 5.50 |
| 19 01 0208 | 6", Class 200, PVC Piping | LF | 14.95 | 12.20 | 10.93 | 8.27 | 7.36 |
| 19 01 0212 | 8", Class 150, PVC Piping | LF | 18.36 | 15.27 | 13.84 | 10.85 | 9.82 |
| 19 01 0213 | 10", Class 150, PVC Piping | LF | 22.90 | 19.37 | 17.73 | 14.32 | 13.14 |
| 19 01 0214 | 12", Class 150, PVC Piping | LF | 28.75 | 24.63 | 22.72 | 18.73 | 17.36 |
| 19 07 0120 | 1 1/4", 60 PSI, Polyethylene Pipe | LF | 5.29 | 4.21 | 3.72 | 2.68 | 2.32 |
| 19 07 0121 | 1 1/2", 60 PSI, Polyethylene Pipe | LF | 5.38 | 4.30 | 3.81 | 2.77 | 2.41 |
| 19 07 0122 | 2", 60 PSI, Polyethylene Pipe | LF | 6.05 | 4.85 | 4.31 | 3.15 | 2.75 |
| 33 01 0101 | Mobilize/DeMobilize Drilling Rig & Crew | LS | 4,949 | 4,031 | 3,308 | 2,401 | 2,280 |
| 33 02 0303 | Organic Vapor Analyzer Rental, per Day | DAY | 184.30 | 184.30 | 184.30 | 184.30 | 184.30 |
| 33 10 9656 | 550 Gallon Single-wall Steel Aboveground Tank | EACH | 1,598 | 1,390 | 1,301 | 1,099 | 1,026 |
| 33 17 0808 | Decontaminate Rig, Augers, Screen (Rental Equipment) | DAY | 205.34 | 205.34 | 205.34 | 205.34 | 205.34 |
| 33 23 0101 | 2" PVC, Schedule 40, Well Casing | LF | 17.45 | 14.39 | 11.98 | 8.95 | 8.55 |
| 33 23 0102 | 4" PVC, Schedule 40, Well Casing | LF | 26.98 | 22.38 | 18.77 | 14.24 | 13.63 |
| 33 23 0103 | 6" PVC, Schedule 40, Well Casing | LF | 27.07 | 22.66 | 19.19 | 14.84 | 14.25 |
| 33 23 0104 | 8" PVC, Schedule 40, Well Casing | LF | 34.50 | 29.27 | 25.15 | 19.98 | 19.29 |
| 33 23 0105 | 10" PVC, Schedule 40, Well Casing | LF | 38.28 | 33.04 | 28.92 | 23.75 | 23.06 |
| 33 23 0106 | 12" PVC, Schedule 40, Well Casing | LF | 44.32 | 39.08 | 34.97 | 29.80 | 29.10 |
| 33 23 0111 | 2" PVC, Schedule 80, Well Casing | LF | 17.78 | 14.72 | 12.31 | 9.28 | 8.88 |
| 33 23 0112 | 4" PVC, Schedule 80, Well Casing | LF | 28.55 | 23.95 | 20.34 | 15.81 | 15.20 |
| 33 23 0113 | 6" PVC, Schedule 80, Well Casing | LF | 47.52 | 40.17 | 34.40 | 27.14 | 26.17 |
| 33 23 0115 | 8" PVC, Schedule 80, Well Casing | LF | 61.99 | 52.81 | 45.58 | 36.51 | 35.30 |
| 33 23 0116 | 10" PVC, Schedule 80, Well Casing | LF | 65.25 | 56.06 | 48.84 | 39.77 | 38.56 |
| 33 23 0117 | 12" PVC, Schedule 80, Well Casing | LF | 68.60 | 59.41 | 52.19 | 43.12 | 41.91 |
| 33 23 0121 | 2" Stainless Steel, Well Casing | LF | 31.75 | 28.14 | 25.30 | 21.73 | 21.25 |
| 33 23 0122 | 4" Stainless Steel, Well Casing | LF | 47.66 | 43.06 | 39.45 | 34.92 | 34.31 |
| 33 23 0123 | 6" Stainless Steel Well Casing, 5' Sections, Flush Threaded | LF | 370.73 | 311.96 | 265.77 | 207.74 | 199.94 |
| 33 23 0124 | 6" Stainless Steel Well Casing, 10' Sections, Flush Threaded | LF | 348.53 | 289.76 | 243.57 | 185.54 | 177.74 |
| 33 23 0125 | 8" Stainless Steel Well Casing, 5' Sections, Flush Threaded | LF | 526.42 | 452.94 | 395.18 | 322.61 | 312.87 |
| 33 23 0126 | 8" Stainless Steel Well Casing, 10' Sections, Flush Threaded | LF | 478.02 | 404.54 | 346.78 | 274.21 | 264.47 |

**Environmental Remediation: Assemblies Cost Book**

*Page 3-106*

# Cost Data for Remediation

## Extraction Wells

| Assembly | Description | Unit | A | B | C | D | E |
|---|---|---|---|---|---|---|---|
| | **Capital Costs** | | | | | | |
| 33 23 0127 | 10" Stainless Steel Well Casing, 5' Sections, Flush Threaded | LF | 533.65 | 460.17 | 402.41 | 329.84 | 320.10 |
| 33 23 0128 | 10" Stainless Steel Well Casing, 10' Sections, Flush Threaded | LF | 512.82 | 439.34 | 381.58 | 309.01 | 299.27 |
| 33 23 0129 | 12" Stainless Steel Well Casing, 5' Sections, Flush Threaded | LF | 561.57 | 488.09 | 430.33 | 357.76 | 348.02 |
| 33 23 0130 | 12" Stainless Steel Well Casing, 10' Sections, Flush Threaded | LF | 533.82 | 460.34 | 402.58 | 330.01 | 320.27 |
| 33 23 0131 | 10" Stainless Steel Well Casing, 5' Sections, Welded Ring | LF | 482.01 | 408.53 | 350.77 | 278.20 | 268.46 |
| 33 23 0132 | 10" Stainless Steel Well Casing, 10' Sections, Welded Ring | LF | 459.46 | 385.98 | 328.22 | 255.65 | 245.91 |
| 33 23 0133 | 12" Stainless Steel Well Casing, 5' Sections, Welded Ring | LF | 519.88 | 446.40 | 388.64 | 316.07 | 306.33 |
| 33 23 0134 | 12" Stainless Steel Well Casing, 10' Sections, Welded Ring | LF | 484.79 | 411.31 | 353.55 | 280.98 | 271.24 |
| 33 23 0201 | 2" PVC, Schedule 40, Well Screen | LF | 23.47 | 19.52 | 16.42 | 12.52 | 12.00 |
| 33 23 0202 | 4" PVC, Schedule 40, Well Screen | LF | 38.16 | 32.00 | 27.16 | 21.09 | 20.27 |
| 33 23 0203 | 6" PVC, Schedule 40, Well Screen | LF | 47.13 | 39.78 | 34.01 | 26.75 | 25.78 |
| 33 23 0204 | 8" PVC, Schedule 40, Well Screen | LF | 62.02 | 52.84 | 45.61 | 36.54 | 35.33 |
| 33 23 0205 | 10" PVC, Schedule 40, Well Screen | LF | 67.86 | 58.67 | 51.45 | 42.38 | 41.17 |
| 33 23 0206 | 12" PVC, Schedule 40, Well Screen | LF | 70.91 | 61.73 | 54.50 | 45.43 | 44.22 |
| 33 23 0211 | 2" PVC, Schedule 80, Well Screen | LF | 18.70 | 15.64 | 13.23 | 10.20 | 9.80 |
| 33 23 0212 | 4" PVC, Schedule 80, Well Screen | LF | 30.01 | 25.41 | 21.80 | 17.27 | 16.66 |
| 33 23 0213 | 6" PVC, Schedule 80, Well Screen | LF | 49.72 | 42.37 | 36.60 | 29.34 | 28.37 |
| 33 23 0214 | 8" PVC, Schedule 80, Well Screen | LF | 65.91 | 56.73 | 49.50 | 40.43 | 39.22 |
| 33 23 0215 | 10" PVC, Schedule 80, Well Screen | LF | 73.71 | 64.53 | 57.30 | 48.23 | 47.02 |
| 33 23 0216 | 12" PVC, Schedule 80, Well Screen | LF | 77.41 | 68.22 | 61.00 | 51.93 | 50.72 |
| 33 23 0221 | 2" Stainless Steel, Well Screen | LF | 26.91 | 23.84 | 21.44 | 18.41 | 18.01 |
| 33 23 0222 | 4" Stainless Steel, Well Screen | LF | 47.66 | 43.06 | 39.45 | 34.92 | 34.31 |
| 33 23 0223 | 6" Stainless Steel Well Screen, 5' Sec, Flush Threaded | LF | 366.05 | 307.28 | 261.09 | 203.06 | 195.26 |
| 33 23 0224 | 6" Stainless Steel Well Screen, 10' Sec, Flush Threaded | LF | 355.96 | 297.19 | 251.00 | 192.97 | 185.17 |
| 33 23 0225 | 8" Stainless Steel Well Screen, 5' Sec, Flush Threaded | LF | 462.22 | 388.74 | 330.98 | 258.41 | 248.67 |
| 33 23 0226 | 8" Stainless Steel Well Screen, 10' Sec, Flush Threaded | LF | 449.32 | 375.84 | 318.08 | 245.51 | 235.77 |
| 33 23 0227 | 10" Stainless Steel Well Screen, 5' Sec, Flush Threaded | LF | 496.10 | 422.62 | 364.86 | 292.29 | 282.55 |
| 33 23 0228 | 10" Stainless Steel Well Screen, 10' Sec, Flush Threaded | LF | 475.48 | 402.00 | 344.24 | 271.67 | 261.93 |
| 33 23 0229 | 12" Stainless Steel Well Screen, 5' Sec, Flush Threaded | LF | 519.16 | 445.68 | 387.92 | 315.35 | 305.61 |
| 33 23 0230 | 12" Stainless Steel Well Screen, 10' Sec, Flush Threaded | LF | 490.96 | 417.48 | 359.72 | 287.15 | 277.41 |
| 33 23 0231 | 10" Dia, SS Well Screen, 5' Sec, Welded Ring | LF | 436.05 | 355.88 | 291.43 | 212.17 | 202.43 |

**Environmental Remediation: Assemblies Cost Book**

*Page 3-107*

# Cost Data for Remediation

## Extraction Wells

| Assembly | Description | Unit | Unit Cost by Safety Level | | | | |
|----------|-------------|------|------|------|------|------|------|
| | | | A | B | C | D | E |
| | **Capital Costs** | | | | | | |
| 33 23 0232 | 10" Stainless Steel Well Screen, 10' Sec, Welded Ring | LF | 465.38 | 391.90 | 334.14 | 261.57 | 251.83 |
| 33 23 0233 | 12" Stainless Steel Well Screen, 5' Sec, Welded Ring | LF | 506.81 | 433.33 | 375.57 | 303.00 | 293.26 |
| 33 23 0234 | 12" Stainless Steel Well Screen, 10' Sec, Welded Ring | LF | 484.80 | 411.32 | 353.56 | 280.99 | 271.25 |
| 33 23 0270 | 2" PVC Wire Wrapped Well Screen, 10' Sec | LF | 168.39 | 137.77 | 113.70 | 83.47 | 79.41 |
| 33 23 0271 | 4" PVC Wire Wrapped Well Screen, 10' Sec | LF | 254.40 | 208.47 | 172.37 | 127.02 | 120.93 |
| 33 23 0272 | 6" PVC Wire Wrapped Well Screen, 10' Sec | LF | 409.98 | 336.52 | 278.78 | 206.23 | 196.50 |
| 33 23 0273 | 8" PVC Wire Wrapped Well Screen, 10' Sec | LF | 415.86 | 342.40 | 284.65 | 212.11 | 202.37 |
| 33 23 0274 | 2" Stainless Steel Wire Wrapped Well Screen, 10' Sec | LF | 187.43 | 156.81 | 132.75 | 102.51 | 98.45 |
| 33 23 0275 | 4" Stainless Steel Wire Wrapped Well Screen, 10' Sec | LF | 284.68 | 238.76 | 202.66 | 157.30 | 151.21 |
| 33 23 0276 | 6" Stainless Steel Wire Wrapped Well Screen, 10' Sec | LF | 452.74 | 379.28 | 321.53 | 248.99 | 239.25 |
| 33 23 0277 | 8" Stainless Steel Wire Wrapped Well Screen, 10' Sec | LF | 478.69 | 405.22 | 347.48 | 274.94 | 265.20 |
| 33 23 0280 | 2" Teflon Screen, Flush Threaded, 5' Sec, Slotted | LF | 18.17 | 14.83 | 12.14 | 8.84 | 8.43 |
| 33 23 0281 | 4" Teflon Screen, Flush Threaded, 5' Sec, Slotted | LF | 21.80 | 17.79 | 14.57 | 10.61 | 10.12 |
| 33 23 0301 | 2" PVC, Well Plug | EACH | 29.37 | 24.77 | 21.16 | 16.63 | 16.02 |
| 33 23 0302 | 4" PVC, Well Plug | EACH | 55.65 | 48.91 | 43.62 | 36.97 | 36.07 |
| 33 23 0303 | 6" PVC, Well Plug | EACH | 112.63 | 101.15 | 92.13 | 80.79 | 79.27 |
| 33 23 0304 | 8" PVC, Well Plug | EACH | 132.70 | 119.58 | 109.26 | 96.31 | 94.57 |
| 33 23 0305 | 10" PVC, Well Plug | EACH | 156.31 | 141.01 | 128.97 | 113.85 | 111.82 |
| 33 23 0306 | 12" PVC, Well Plug | EACH | 197.87 | 179.50 | 165.06 | 146.92 | 144.48 |
| 33 23 0311 | 2" Stainless Steel, Well Plug | EACH | 83.24 | 74.06 | 66.83 | 57.76 | 56.55 |
| 33 23 0312 | 4" Stainless Steel, Well Plug | EACH | 112.24 | 102.03 | 94.01 | 83.93 | 82.58 |
| 33 23 0313 | 6" Stainless Steel, Well Plug | EACH | 573.04 | 481.18 | 408.98 | 318.27 | 306.09 |
| 33 23 0314 | 8" Stainless Steel, Well Plug | EACH | 702.53 | 597.56 | 515.04 | 411.38 | 397.46 |
| 33 23 0315 | 10" Stainless Steel, Well Plug | EACH | 891.17 | 768.70 | 672.43 | 551.48 | 535.25 |
| 33 23 0325 | 12" Stainless Steel, Well Plug | EACH | 1,023 | 879.90 | 767.56 | 626.43 | 607.48 |
| 33 23 0401 | 2" Filter Sock | LF | 5.40 | 4.48 | 3.76 | 2.85 | 2.73 |
| 33 23 0402 | 4" Filter Sock | LF | 5.57 | 4.65 | 3.93 | 3.02 | 2.90 |
| 33 23 0403 | 6" Filter Sock | LF | 5.70 | 4.78 | 4.06 | 3.15 | 3.03 |
| 33 23 0404 | 8" Filter Sock | LF | 6.35 | 5.43 | 4.71 | 3.80 | 3.68 |
| 33 23 0501 | 4" Submersible Pump, 4 - 13 GPM, < 180' Head | EACH | 1,976 | 1,721 | 1,588 | 1,340 | 1,264 |
| 33 23 0502 | 4" Submersible Pump, 13 - 26 GPM, < 180' Head | EACH | 3,039 | 2,745 | 2,592 | 2,307 | 2,219 |
| 33 23 0503 | 4" Submersible Pump, 13 - 37 GPM, < 180' Head | EACH | 3,084 | 2,790 | 2,637 | 2,352 | 2,264 |
| 33 23 0504 | 6" Submersible Pump, 50 - 125 GPM, < 150' Head | EACH | 3,041 | 2,621 | 2,403 | 1,995 | 1,870 |

**Environmental Remediation: Assemblies Cost Book**

*Page 3-108*

# Cost Data for Remediation

## Extraction Wells

| Assembly | Description | Unit | \multicolumn{5}{c}{Unit Cost by Safety Level} |
|---|---|---|---|---|---|---|---|
| | | | A | B | C | D | E |
| | **Capital Costs** | | | | | | |
| 33 23 0505 | 6" Submersible Pump, 15 - 135 GPM, 200' < Head < 500' | EACH | 4,472 | 3,885 | 3,579 | 3,008 | 2,834 |
| 33 23 0506 | 2" Submersible Pump Rental, Day | DAY | 63.86 | 63.86 | 63.86 | 63.86 | 63.86 |
| 33 23 0507 | 2" Submersible Pump Rental, Week | WEEK | 191.57 | 191.57 | 191.57 | 191.57 | 191.57 |
| 33 23 0508 | 2" Submersible Pump Rental, Month | MONTH | 478.92 | 478.92 | 478.92 | 478.92 | 478.92 |
| 33 23 0509 | 4" Submersible Pump Rental, Day | DAY | 63.86 | 63.86 | 63.86 | 63.86 | 63.86 |
| 33 23 0510 | 4" Submersible Pump Rental, Week | WEEK | 191.57 | 191.57 | 191.57 | 191.57 | 191.57 |
| 33 23 0511 | 4" Submersible Pump Rental, Month | MONTH | 478.92 | 478.92 | 478.92 | 478.92 | 478.92 |
| 33 23 0601 | Groundwater Pump, 1/3 HP, 230V, Controls, Probe | EACH | 2,583 | 2,379 | 2,292 | 2,095 | 2,023 |
| 33 23 0602 | Groundwater Pump, 3/4 HP, 230V, Controls, Probe | EACH | 2,747 | 2,543 | 2,456 | 2,259 | 2,187 |
| 33 23 0603 | Groundwater Pump, 1 HP, 230V, Controls, Probe | EACH | 2,881 | 2,678 | 2,591 | 2,394 | 2,322 |
| 33 23 0604 | Groundwater Pump, 1 1/2 HP, Controls, Probe | EACH | 3,095 | 2,891 | 2,804 | 2,607 | 2,535 |
| 33 23 0605 | Groundwater Pump, 2 HP, 230V, Controls, Probe | EACH | 3,674 | 3,471 | 3,383 | 3,187 | 3,114 |
| 33 23 0606 | Groundwater Pump, 5 HP, 230V, Controls, Probe | EACH | 4,523 | 4,320 | 4,233 | 4,036 | 3,964 |
| 33 23 0701 | Groundwater Pump, 1 1/2 HP, 230V, without Controls | EACH | 2,270 | 2,067 | 1,980 | 1,783 | 1,711 |
| 33 23 0702 | Groundwater Pump, 5 HP, 230V, without Controls | EACH | 3,487 | 3,284 | 3,196 | 3,000 | 2,927 |
| 33 23 0801 | Product Recovery Pump, 1/3 HP, 230V, Controls, Probe | EACH | 8,731 | 8,527 | 8,440 | 8,243 | 8,171 |
| 33 23 0802 | Product Recovery Pump, 1/3 HP, 115V, Controls, Probe | EACH | 9,109 | 8,905 | 8,818 | 8,621 | 8,549 |
| 33 23 0803 | Product Recovery Pump, 1/3 HP, 230V, Controls, Probe, Highly Corrosive Liquid | EACH | 8,730 | 8,526 | 8,439 | 8,242 | 8,170 |
| 33 23 0811 | Explosionproof Electrical Receptacle for Product Recovery Pump | EACH | 497.01 | 468.06 | 455.66 | 427.68 | 417.42 |
| 33 23 0901 | Product Recovery Pump, 1/3 HP, without Controls, Includes Probe | EACH | 2,237 | 2,033 | 1,946 | 1,749 | 1,677 |
| 33 23 0902 | Product Recovery Pump, 3/4 HP, without Controls | EACH | 1,909 | 1,706 | 1,619 | 1,422 | 1,350 |
| 33 23 1101 | Hollow-stem Auger, 8" Outside Diameter Borehole for 2" Well | LF | 89.98 | 73.28 | 60.15 | 43.66 | 41.45 |
| 33 23 1102 | Hollow-stem Auger, 11" Outside Diameter Borehole for 4" Well | LF | 109.98 | 89.57 | 73.52 | 53.36 | 50.66 |
| 33 23 1103 | Hollow-stem Auger, 13 3/4" Outside Diameter Borehole for 6" Well | LF | 141.40 | 115.16 | 94.53 | 68.61 | 65.13 |
| 33 23 1105 | Hollow-stem Auger, 16" Outside Diameter Borehole for 8" Well | LF | 164.97 | 134.35 | 110.28 | 80.05 | 75.99 |
| 33 23 1106 | Split Spoon Sample, 2" x 24", During Drilling | EACH | 46.67 | 46.67 | 46.67 | 46.67 | 46.67 |
| 33 23 1107 | Coring (Fluid or Air) | LF | 102.14 | 88.16 | 78.67 | 64.96 | 62.17 |
| 33 23 1111 | Well Development Equipment Rental | WEEK | 478.91 | 462.93 | 456.09 | 440.65 | 434.98 |
| 33 23 1121 | Standby for Drilling | EACH | 618.63 | 503.81 | 413.56 | 300.17 | 284.95 |

**Environmental Remediation: Assemblies Cost Book**

# Cost Data for Remediation

## Extraction Wells

| Assembly | Description | Unit | Unit Cost by Safety Level | | | | |
|---|---|---|---|---|---|---|---|
| | | | A | B | C | D | E |
| | **Capital Costs** | | | | | | |
| 33 23 1122 | Move Rig/Equipment Around Site | EACH | 154.66 | 125.95 | 103.39 | 75.04 | 71.24 |
| 33 23 1125 | Concrete Coring (Minimum Charge) | DAY | 2,475 | 2,015 | 1,654 | 1,201 | 1,140 |
| 33 23 1126 | Furnish 55 Gallon Drum for Drill Cuttings & Development Water | EACH | 65.19 | 65.19 | 65.19 | 65.19 | 65.19 |
| 33 23 1131 | Drill & Test Well, Normal Soil, 6" | LF | 131.86 | 107.66 | 88.12 | 64.20 | 61.31 |
| 33 23 1132 | Drill & Test Well, Normal Soil, 8" | LF | 172.28 | 140.67 | 115.13 | 83.88 | 80.11 |
| 33 23 1133 | Drill & Test Well, Normal Soil, 10" | LF | 202.59 | 165.42 | 135.39 | 98.64 | 94.21 |
| 33 23 1134 | Drill & Test Well, Normal Soil, 12" | LF | 224.96 | 183.20 | 150.38 | 109.15 | 103.62 |
| 33 23 1135 | Drill & Test Well, Normal Soil, 14" | LF | 270.44 | 220.25 | 180.79 | 131.22 | 124.57 |
| 33 23 1136 | Drill & Test Well, Normal Soil, 16" | LF | 372.11 | 303.05 | 248.76 | 180.56 | 171.40 |
| 33 23 1141 | Drill & Test Well, Rock, 6" | LF | 152.27 | 124.33 | 101.76 | 74.14 | 70.81 |
| 33 23 1142 | Drill & Test Well, Rock, 8" | LF | 177.46 | 144.90 | 118.60 | 86.40 | 82.52 |
| 33 23 1143 | Drill & Test Well, Rock, 10" | LF | 202.59 | 165.42 | 135.39 | 98.64 | 94.21 |
| 33 23 1144 | Drill & Test Well, Rock, 12" | LF | 224.96 | 183.20 | 150.38 | 109.15 | 103.62 |
| 33 23 1145 | Drill & Test Well, Rock, 14" | LF | 291.12 | 237.09 | 194.62 | 141.26 | 134.09 |
| 33 23 1146 | Drill & Test Well, Rock, 16" | LF | 423.00 | 344.49 | 282.77 | 205.25 | 194.84 |
| 33 23 1151 | Mud Drilling, 4" Diameter Borehole | LF | 73.65 | 59.98 | 49.23 | 35.73 | 33.92 |
| 33 23 1152 | Mud Drilling, 6" Diameter Borehole | LF | 81.01 | 65.98 | 54.16 | 39.31 | 37.31 |
| 33 23 1153 | Mud Drilling, 8" Diameter Borehole | LF | 88.38 | 71.97 | 59.08 | 42.88 | 40.71 |
| 33 23 1154 | Mud Drilling, 10" Diameter Borehole | LF | 100.65 | 81.97 | 67.29 | 48.84 | 46.36 |
| 33 23 1155 | Mud Drilling, 12" Diameter Borehole | LF | 112.92 | 91.97 | 75.49 | 54.79 | 52.01 |
| 33 23 1156 | Mud Drilling, 15" Diameter Borehole | LF | 137.47 | 111.96 | 91.90 | 66.71 | 63.32 |
| 33 23 1161 | Air Rotary 6" Borehole in Unconsolidated | LF | 73.65 | 59.98 | 49.23 | 35.73 | 33.92 |
| 33 23 1162 | Air Rotary 8" Borehole in Unconsolidated | LF | 122.74 | 99.96 | 82.06 | 59.56 | 56.54 |
| 33 23 1163 | Air Rotary 10" Borehole in Unconsolidated | LF | 220.94 | 179.93 | 147.70 | 107.20 | 101.77 |
| 33 23 1164 | Air Rotary 12" Borehole in Unconsolidated | LF | 294.59 | 239.91 | 196.93 | 142.94 | 135.69 |
| 33 23 1165 | Air Rotary 16" Borehole in Unconsolidated | LF | 392.78 | 319.88 | 262.58 | 190.59 | 180.92 |
| 33 23 1168 | Air Rotary 6" Borehole in Consolidated | LF | 58.92 | 47.98 | 39.39 | 28.59 | 27.14 |
| 33 23 1169 | Air Rotary 8" Borehole in Consolidated | LF | 147.29 | 119.96 | 98.47 | 71.47 | 67.85 |
| 33 23 1170 | Air Rotary 10" Borehole in Consolidated | LF | 294.59 | 239.91 | 196.93 | 142.94 | 135.69 |
| 33 23 1171 | Air Rotary 12" Borehole in Consolidated | LF | 613.72 | 499.81 | 410.28 | 297.79 | 282.69 |
| 33 23 1301 | 4 Product Recovery/4 Groundwater Pump Control Panel | EACH | 4,599 | 4,395 | 4,308 | 4,111 | 4,039 |
| 33 23 1302 | 3 Product Recovery/3 Groundwater Pump Control Panel | EACH | 5,839 | 5,635 | 5,548 | 5,351 | 5,279 |
| 33 23 1303 | 1 Product Recovery/4 Groundwater Pump Control Panel | EACH | 5,500 | 5,296 | 5,209 | 5,012 | 4,940 |

## Environmental Remediation: Assemblies Cost Book

*Page 3-110*

1998 by ECHOS. All rights reserved

# Cost Data for Remediation

## Extraction Wells

| Assembly | Description | Unit | A | B | C | D | E |
|---|---|---|---|---|---|---|---|
| | **Capital Costs** | | | | | | |
| 33 23 1304 | 2 Product Recovery/2 Groundwater Pump Control Panel | EACH | 4,842 | 4,639 | 4,552 | 4,355 | 4,283 |
| 33 23 1401 | 2" Screen, Filter Pack | LF | 16.49 | 13.88 | 11.84 | 9.27 | 8.92 |
| 33 23 1402 | 4" Screen, Filter Pack | LF | 29.10 | 24.50 | 20.89 | 16.36 | 15.75 |
| 33 23 1403 | 6" Screen, Filter Pack | LF | 42.19 | 35.53 | 30.29 | 23.72 | 22.83 |
| 33 23 1404 | Gravel Pack, 8" Well | LF | 29.47 | 23.93 | 19.82 | 14.36 | 13.48 |
| 33 23 1405 | Gravel Pack, 10" Well | LF | 43.84 | 35.59 | 29.45 | 21.33 | 20.02 |
| 33 23 1406 | Gravel Pack, 12" Well | LF | 54.65 | 44.40 | 36.79 | 26.70 | 25.07 |
| 33 23 1501 | Surface Seal, Concrete Filled | LS | 5,300 | 4,382 | 3,659 | 2,752 | 2,631 |
| 33 23 1601 | Groundwater or Product Recovery Probe, 100' Cable | EACH | 301.04 | 301.04 | 301.04 | 301.04 | 301.04 |
| 33 23 1602 | Dual Pump Probe, 100' Cable | EACH | 193.98 | 193.98 | 193.98 | 193.98 | 193.98 |
| 33 23 1811 | 2" Well, Portland Cement Grout | LF | 0.92 | 0.92 | 0.92 | 0.92 | 0.92 |
| 33 23 1812 | 4" Well, Portland Cement Grout | LF | 1.38 | 1.38 | 1.38 | 1.38 | 1.38 |
| 33 23 1813 | 6" Well, Portland Cement Grout | LF | 7.80 | 7.80 | 7.80 | 7.80 | 7.80 |
| 33 23 1814 | 8" Well, Portland Cement Grout | LF | 11.02 | 11.02 | 11.02 | 11.02 | 11.02 |
| 33 23 1815 | 10" Well, Portland Cement Grout | LF | 12.85 | 12.85 | 12.85 | 12.85 | 12.85 |
| 33 23 1816 | 12" Well, Portland Cement Grout | LF | 14.69 | 14.69 | 14.69 | 14.69 | 14.69 |
| 33 23 2101 | 2" Well, Bentonite Seal | EACH | 63.00 | 52.67 | 44.55 | 34.34 | 32.97 |
| 33 23 2102 | 4" Well, Bentonite Seal | EACH | 157.54 | 131.70 | 111.39 | 85.87 | 82.44 |
| 33 23 2103 | 6" Well, Bentonite Seal | EACH | 252.01 | 210.68 | 178.18 | 137.37 | 131.89 |
| 33 23 2204 | Hazardous Area, Pedestrian Load, Well Protection | EACH | 6,261 | 5,266 | 4,505 | 3,524 | 3,380 |
| 33 23 2205 | Hazardous Area, Traffic Load, Well Protection | EACH | 6,266 | 5,271 | 4,510 | 3,529 | 3,384 |
| 33 23 2206 | Restricted Area, Well Protection (with 4 Posts & Explosionproof Receptacle) | EACH | 1,264 | 1,133 | 1,077 | 949.88 | 903.62 |
| 33 26 0101 | 1" Carbon Steel Piping | LF | 6.46 | 5.28 | 4.77 | 3.63 | 3.21 |
| 33 26 0102 | 2" Carbon Steel Piping | LF | 10.50 | 8.72 | 7.95 | 6.23 | 5.60 |
| 33 26 0103 | 3" Carbon Steel Piping | LF | 18.02 | 15.20 | 13.99 | 11.26 | 10.26 |
| 33 26 0104 | 4" Carbon Steel Piping | LF | 23.20 | 19.86 | 18.43 | 15.21 | 14.02 |
| 33 26 0105 | 6" Carbon Steel Piping | LF | 36.37 | 30.24 | 27.46 | 21.52 | 19.44 |
| 33 26 0106 | 8" Carbon Steel Piping | LF | 40.77 | 34.08 | 31.15 | 24.68 | 22.36 |
| 33 26 0131 | 1" Carbon Steel Piping Including Fittings & Hangers | LF | 14.60 | 11.56 | 10.25 | 7.30 | 6.22 |
| 33 26 0132 | 2" Carbon Steel Piping Including Fittings & Hangers | LF | 25.80 | 20.51 | 18.25 | 13.14 | 11.26 |
| 33 26 0133 | 3" Carbon Steel Piping Including Fittings & Hangers | LF | 40.05 | 32.18 | 28.81 | 21.21 | 18.42 |
| 33 26 0134 | 4" Carbon Steel Piping Including Fittings & Hangers | LF | 47.11 | 38.30 | 34.52 | 26.00 | 22.87 |
| 33 26 0201 | 1" Stainless Steel Piping, Schedule 40, Threaded | LF | 18.67 | 16.12 | 15.03 | 12.56 | 11.65 |

**Environmental Remediation: Assemblies Cost Book**

# Cost Data for Remediation

## Extraction Wells

| Assembly | Description | Unit | A | B | C | D | E |
|---|---|---|---|---|---|---|---|
| | **Capital Costs** | | | | | | |
| 33 26 0202 | 2" Stainless Steel Piping, Schedule 40, Threaded | LF | 33.47 | 29.21 | 27.39 | 23.28 | 21.77 |
| 33 26 0203 | 3" Stainless Steel Piping, Schedule 40, Threaded | LF | 57.00 | 50.61 | 47.88 | 41.71 | 39.45 |
| 33 26 0204 | 4" Stainless Steel Piping, Schedule 40, Threaded | LF | 80.15 | 71.72 | 68.12 | 59.98 | 57.00 |
| 33 26 0205 | 6" Stainless Steel Piping, Schedule 10, Type 316 | LF | 78.58 | 70.01 | 66.14 | 57.84 | 54.94 |
| 33 26 0206 | 8" Stainless Steel Piping, Schedule 40, Welded | LF | 197.22 | 183.32 | 176.89 | 163.42 | 158.80 |
| 33 26 0207 | 8" Stainless Steel, Schedule 10, Type 316, Manifold Piping | LF | 126.78 | 112.88 | 106.45 | 92.98 | 88.36 |
| 33 26 0210 | 2" Stainless Steel, Schedule 40, Type 304, Connection Piping | LF | 33.47 | 29.21 | 27.39 | 23.28 | 21.77 |
| 33 26 0211 | 4" Stainless Steel, Schedule 40, Type 304, Connection Piping | LF | 80.15 | 71.72 | 68.12 | 59.98 | 57.00 |
| 33 26 0212 | 4" Stainless Steel, Schedule 40, Type 304, Manifold Piping | LF | 80.15 | 71.72 | 68.12 | 59.98 | 57.00 |
| 33 26 0231 | 1" Stainless Steel Piping Including Fittings & Hangers | LF | 25.76 | 21.58 | 19.79 | 15.76 | 14.27 |
| 33 26 0232 | 2" Stainless Steel Piping Including Fittings & Hangers | LF | 42.76 | 36.38 | 33.64 | 27.47 | 25.21 |
| 33 26 0233 | 3" Stainless Steel Piping Including Fittings & Hangers | LF | 62.76 | 53.62 | 49.70 | 40.86 | 37.62 |
| 33 26 0234 | 4" Stainless Steel Piping Including Fittings & Hangers | LF | 123.07 | 106.59 | 99.49 | 83.56 | 77.75 |
| 33 26 0301 | 1 1/2" Polypropylene Pipe Including Fittings | LF | 7.56 | 6.11 | 5.48 | 4.08 | 3.57 |
| 33 26 0302 | 2" Polypropylene Pipe Including Fittings | LF | 8.79 | 7.16 | 6.46 | 4.88 | 4.30 |
| 33 26 0303 | 3" Polypropylene Pipe Including Fittings | LF | 13.80 | 11.43 | 10.42 | 8.13 | 7.29 |
| 33 26 0304 | 4" Polypropylene Pipe Including Fittings | LF | 18.36 | 15.28 | 13.96 | 10.99 | 9.90 |
| 33 26 0405 | 8" PVC, Schedule 40, Manifold Piping | LF | 23.47 | 19.13 | 17.28 | 13.08 | 11.55 |
| 33 26 0406 | 2" PVC, Schedule 40, Connection Piping | LF | 6.14 | 4.87 | 4.33 | 3.11 | 2.66 |
| 33 26 0407 | 4" PVC, Schedule 40, Connection Piping | LF | 13.38 | 10.65 | 9.48 | 6.84 | 5.88 |
| 33 26 0408 | 4" PVC, Schedule 40, Manifold Piping | LF | 13.38 | 10.65 | 9.48 | 6.84 | 5.88 |
| 33 26 0601 | (1 1/2", 3") Stainless Steel Double-wall Piping, with Fittings | LF | 134.61 | 117.05 | 109.53 | 92.56 | 86.33 |
| 33 26 0602 | (2", 4") Stainless Steel Double-wall Piping, with Fittings | LF | 185.98 | 163.46 | 153.81 | 132.04 | 124.05 |
| 33 26 0603 | (2 1/2", 4") Stainless Steel Double-wall Piping, with Fittings | LF | 206.86 | 182.17 | 171.60 | 147.74 | 138.99 |
| 33 26 0604 | (4", 6") Stainless Steel Double-wall Piping, with Fittings | LF | 339.64 | 280.79 | 254.18 | 197.21 | 177.23 |
| 33 26 0621 | (1 1/2", 3") PVC Double-wall Piping, with Fittings | LF | 33.15 | 27.05 | 24.43 | 18.53 | 16.36 |
| 33 26 0622 | (2", 4") PVC Double-wall Piping, with Fittings | LF | 46.22 | 37.98 | 34.43 | 26.46 | 23.55 |
| 33 26 0623 | (2 1/2", 4") PVC Double-wall Piping, with Fittings | LF | 49.22 | 40.39 | 36.61 | 28.07 | 24.94 |
| 33 26 0624 | (4", 6") PVC Double-wall Piping, with Fittings | LF | 70.61 | 58.00 | 52.59 | 40.40 | 35.93 |
| 33 26 0641 | (1 1/2", 3") Carbon Steel Double-wall Piping, with Fittings | LF | 48.16 | 40.64 | 37.42 | 30.15 | 27.48 |
| 33 26 0642 | (2", 4") Carbon Steel Double-wall Piping, with Fittings | LF | 64.71 | 55.69 | 51.83 | 43.11 | 39.91 |
| 33 26 0643 | (2 1/2", 4") Carbon Steel Double-wall Piping, with Fittings | LF | 70.81 | 61.02 | 56.83 | 47.37 | 43.90 |

**Unit Cost by Safety Level**

---

**Environmental Remediation: Assemblies Cost Book**

# Cost Data for Remediation

## Extraction Wells

| Assembly | Description | Unit | Unit Cost by Safety Level | | | | |
|---|---|---|---|---|---|---|---|
| | | | A | B | C | D | E |
| | **Capital Costs** | | | | | | |
| 33 26 0644 | (4", 6") Carbon Steel Double-wall Piping, with Fittings | LF | 114.39 | 100.33 | 94.32 | 80.73 | 75.75 |
| 33 26 0704 | 50' Flexible, Product Discharge Hose | EACH | 159.00 | 159.00 | 159.00 | 159.00 | 159.00 |
| 33 27 0102 | 2" PVC, Schedule 40, Tee | EACH | 68.73 | 53.63 | 46.80 | 32.18 | 27.06 |
| 33 27 0104 | 4" PVC, Schedule 40, Tee | EACH | 137.04 | 108.84 | 96.77 | 69.51 | 59.51 |
| 33 27 0106 | 8" PVC, Schedule 40, Tee | EACH | 336.90 | 291.78 | 272.46 | 228.85 | 212.85 |
| 33 27 0112 | 2" PVC, 90 Degree, Elbow | EACH | 41.36 | 32.22 | 28.30 | 19.46 | 16.22 |
| 33 27 0114 | 4" PVC, 90 Degree, Elbow | EACH | 101.72 | 80.57 | 71.51 | 51.07 | 43.57 |
| 33 27 0116 | 8" PVC, 90 Degree, Elbow | EACH | 200.21 | 172.01 | 159.94 | 132.68 | 122.68 |
| 33 27 0201 | 2" Stainless Steel, Tee | EACH | 173.32 | 140.34 | 126.22 | 94.34 | 82.65 |
| 33 27 0202 | 4" Stainless Steel, Tee | EACH | 583.41 | 504.25 | 470.36 | 393.85 | 365.80 |
| 33 27 0203 | 8" Stainless Steel, Schedule 10, Type 316, Tee | EACH | 2,283 | 1,934 | 1,772 | 1,434 | 1,317 |
| 33 27 0211 | 2" Stainless Steel, 90 Degree, Elbow | EACH | 114.28 | 93.45 | 84.53 | 64.39 | 57.01 |
| 33 27 0212 | 4" Stainless Steel, 90 Degree, Elbow | EACH | 391.27 | 338.50 | 315.91 | 264.90 | 246.20 |
| 33 27 0411 | 1/2" x 1/2" Bronze Pressure Relief Valve | EACH | 65.20 | 57.02 | 53.52 | 45.61 | 42.71 |
| 33 27 0412 | 3/4" x 3/4" Bronze Pressure Relief Valve | EACH | 94.34 | 84.52 | 80.32 | 70.83 | 67.35 |
| 33 27 0413 | 1" x 1" Bronze Pressure Relief Valve | EACH | 136.12 | 123.03 | 117.42 | 104.77 | 100.13 |
| 33 27 0414 | 1 1/4" x 1 1/4" Bronze Pressure Relief Valve | EACH | 242.20 | 228.17 | 222.17 | 208.61 | 203.64 |
| 33 27 0415 | 1 1/2" x 1 1/2" Bronze Pressure Relief Valve | EACH | 388.54 | 372.83 | 366.11 | 350.93 | 345.36 |
| 33 27 0416 | 2" x 2" Bronze Pressure Relief Valve | EACH | 442.30 | 421.15 | 412.09 | 391.65 | 384.15 |
| 33 27 0417 | 2 1/2" x 2 1/2" Bronze Pressure Relief Valve | EACH | 781.08 | 756.90 | 746.56 | 723.19 | 714.62 |
| 33 27 0418 | 1" NPT Brass Vacuum Breaker | EACH | 87.55 | 75.28 | 70.02 | 58.16 | 53.81 |
| 33 27 0421 | 1" Ball Valve, Carbon Steel Trim | EACH | 67.07 | 56.73 | 52.31 | 42.32 | 38.66 |
| 33 27 0422 | 2" Ball Valve, Carbon Steel Trim | EACH | 135.68 | 117.83 | 110.19 | 92.94 | 86.61 |
| 33 27 0423 | 3" Ball Valve, Carbon Steel Trim | EACH | 209.09 | 181.04 | 169.03 | 141.92 | 131.98 |
| 33 27 0424 | 4" Ball Valve, Carbon Steel Trim | EACH | 861.95 | 789.95 | 759.12 | 689.52 | 664.00 |
| 33 28 0102 | Seismic Expansion Joint | EACH | 770.68 | 653.19 | 602.89 | 489.33 | 447.69 |
| 33 28 0103 | Heat Trace for Piping | LF | 7.69 | 6.46 | 5.94 | 4.75 | 4.32 |
| 33 29 0101 | 10 GPM, 10' Head, 1/6 HP, Centrifugal Pump | EACH | 789.18 | 730.44 | 705.29 | 648.51 | 627.69 |
| 33 29 0102 | 10 GPM, 1/2 HP, Centrifugal Pump | EACH | 1,063 | 952.32 | 905.07 | 798.39 | 759.27 |
| 33 29 0103 | 50 GPM, 100' Head, 3 HP, Centrifugal Pump | EACH | 2,658 | 2,500 | 2,432 | 2,279 | 2,223 |
| 33 29 0104 | 75 GPM, 100' Head, 5 HP, Centrifugal Pump | EACH | 2,946 | 2,754 | 2,672 | 2,487 | 2,419 |
| 33 29 0106 | 100 GPM, 150' Head, 7.5 HP, Centrifugal Pump | EACH | 3,419 | 3,191 | 3,093 | 2,873 | 2,793 |
| 33 29 0107 | 125 GPM, 150' Head, 10 HP, Centrifugal Pump | EACH | 3,815 | 3,512 | 3,382 | 3,088 | 2,981 |

**Environmental Remediation: Assemblies Cost Book**

*Page 3-113*

# Cost Data for Remediation

## Extraction Wells

| Assembly | Description | Unit | A | B | C | D | E |
|---|---|---|---|---|---|---|---|
| | **Capital Costs** | | | | | | |
| 33 29 0108 | 10 HP, 200 GPM, Centrifugal Pump | EACH | 2,941 | 2,714 | 2,616 | 2,396 | 2,316 |
| 33 29 0109 | 10 HP, 250 GPM, Centrifugal Pump | EACH | 3,032 | 2,805 | 2,707 | 2,487 | 2,407 |
| 33 29 0110 | 15 HP, 300 GPM, Centrifugal Pump | EACH | 3,441 | 3,160 | 3,040 | 2,770 | 2,670 |
| 33 29 0111 | 20 HP, 500 GPM, Centrifugal Pump | EACH | 4,544 | 4,241 | 4,111 | 3,818 | 3,710 |
| 33 29 0201 | 30 GPM Pneumatic Ejector | EACH | 9,959 | 9,665 | 9,512 | 9,227 | 9,139 |
| 33 29 0202 | 50 GPM Pneumatic Ejector | EACH | 14,115 | 13,789 | 13,620 | 13,302 | 13,205 |
| 33 29 0203 | 100 GPM Pneumatic Ejector | EACH | 23,872 | 23,481 | 23,277 | 22,897 | 22,780 |
| 33 29 0204 | 150 GPM Pneumatic Ejector | EACH | 36,757 | 36,305 | 36,070 | 35,631 | 35,497 |
| 33 29 0205 | 200 GPM Pneumatic Ejector | EACH | 46,839 | 46,252 | 45,946 | 45,375 | 45,201 |
| 33 29 0206 | 250 GPM Pneumatic Ejector | EACH | 60,124 | 59,390 | 59,009 | 58,295 | 58,076 |
| 33 29 0207 | 300 GPM Pneumatic Ejector | EACH | 74,768 | 73,593 | 72,983 | 71,841 | 71,491 |
| 33 29 0301 | 10 GPM, 20 PSI, 1/2 HP, Iron Rotary Pump | EACH | 1,573 | 1,415 | 1,347 | 1,194 | 1,138 |
| 33 29 0302 | 10 GPM, 40 PSI, 3/4 HP, Iron Rotary Pump | EACH | 1,818 | 1,627 | 1,545 | 1,359 | 1,291 |
| 33 29 0303 | 10 GPM, 60 PSI, 1 HP, Iron Rotary Pump | EACH | 1,975 | 1,748 | 1,650 | 1,430 | 1,350 |
| 33 29 0304 | 25 GPM, 20 PSI, 1 HP, Iron Rotary Pump | EACH | 1,690 | 1,532 | 1,464 | 1,311 | 1,255 |
| 33 29 0305 | 25 GPM, 40 PSI, 1 1/2 HP, Iron Rotary Pump | EACH | 1,835 | 1,644 | 1,562 | 1,376 | 1,308 |
| 33 29 0306 | 25 GPM, 60 PSI, 2 HP, Iron Rotary Pump | EACH | 2,193 | 1,966 | 1,868 | 1,648 | 1,568 |
| 33 31 0201 | 6 HP, 80 Gallon, 200 PSI, 13 SCFM, Air Compressor | EACH | 7,971 | 7,612 | 7,459 | 7,112 | 6,985 |
| 33 31 0202 | 6 HP, 80 Gallon, 200 PSI, 28 SCFM, Air Compressor | EACH | 8,102 | 7,713 | 7,547 | 7,171 | 7,033 |
| 33 31 0203 | 10 HP, 80 Gallon, 200 PSI, 35 SCFM, Air Compressor | EACH | 7,697 | 7,273 | 7,091 | 6,681 | 6,531 |
| 33 31 0204 | 15 HP, 120 Gallon, 200 PSI, 50 SCFM, Air Compressor | EACH | 12,032 | 11,535 | 11,323 | 10,843 | 10,667 |
| 33 31 0205 | 20 HP, 120 Gallon, 200 PSI, 70 SCFM, Air Compressor | EACH | 12,892 | 12,330 | 12,090 | 11,547 | 11,347 |
| 33 31 0206 | 30 HP, 240 Gallon, 200 PSI, 143 SCFM, Air Compressor | EACH | 24,885 | 24,263 | 23,997 | 23,396 | 23,176 |
| 33 31 0207 | 2" Steel, 0 - 300 PSI Pressure Gauge | EACH | 63.51 | 51.14 | 45.85 | 33.89 | 29.51 |
| 33 31 0208 | 2" Brass, 0 - 300 PSI Pressure Gauge | EACH | 74.27 | 61.90 | 56.61 | 44.65 | 40.27 |
| 33 41 0101 | Pump & Motor Maintenance/Repair | EACH | 865.17 | 671.48 | 581.58 | 393.96 | 329.64 |
| 33 42 0101 | Electrical Charge | KWH | 0.06 | 0.06 | 0.06 | 0.06 | 0.06 |
| | **Operations & Maintenance** | | | | | | |
| 18 05 0415 | Mowing | ACRE | 64.67 | 51.13 | 43.38 | 30.18 | 26.60 |
| 33 23 0501 | 4" Submersible Pump, 4 - 13 GPM, < 180' Head | EACH | 1,976 | 1,721 | 1,588 | 1,340 | 1,264 |
| 33 23 0502 | 4" Submersible Pump, 13 - 26 GPM, < 180' Head | EACH | 3,039 | 2,745 | 2,592 | 2,307 | 2,219 |
| 33 23 0503 | 4" Submersible Pump, 13 - 37 GPM, < 180' Head | EACH | 3,084 | 2,790 | 2,637 | 2,352 | 2,264 |
| 33 23 0504 | 6" Submersible Pump, 50 - 125 GPM, < 150' Head | EACH | 3,041 | 2,621 | 2,403 | 1,995 | 1,870 |

**Environmental Remediation: Assemblies Cost Book**

# Cost Data for Remediation

## Extraction Wells

| Assembly | Description | Unit | Unit Cost by Safety Level A | B | C | D | E |
|---|---|---|---|---|---|---|---|
| | **Operations & Maintenance** | | | | | | |
| 33 23 0505 | 6" Submersible Pump, 15 - 135 GPM, 200' < Head < 500' | EACH | 4,472 | 3,885 | 3,579 | 3,008 | 2,834 |
| 33 23 0601 | Groundwater Pump, 1/3 HP, 230V, Controls, Probe | EACH | 2,583 | 2,379 | 2,292 | 2,095 | 2,023 |
| 33 23 0602 | Groundwater Pump, 3/4 HP, 230V, Controls, Probe | EACH | 2,747 | 2,543 | 2,456 | 2,259 | 2,187 |
| 33 23 0603 | Groundwater Pump, 1 HP, 230V, Controls, Probe | EACH | 2,881 | 2,678 | 2,591 | 2,394 | 2,322 |
| 33 23 0604 | Groundwater Pump, 1 1/2 HP, Controls, Probe | EACH | 3,095 | 2,891 | 2,804 | 2,607 | 2,535 |
| 33 23 0605 | Groundwater Pump, 2 HP, 230V, Controls, Probe | EACH | 3,674 | 3,471 | 3,383 | 3,187 | 3,114 |
| 33 23 0606 | Groundwater Pump, 5 HP, 230V, Controls, Probe | EACH | 4,523 | 4,320 | 4,233 | 4,036 | 3,964 |
| 33 23 0701 | Groundwater Pump, 1 1/2 HP, 230V, without Controls | EACH | 2,270 | 2,067 | 1,980 | 1,783 | 1,711 |
| 33 23 0702 | Groundwater Pump, 5 HP, 230V, without Controls | EACH | 3,487 | 3,284 | 3,196 | 3,000 | 2,927 |
| 33 23 0801 | Product Recovery Pump, 1/3 HP, 230V, Controls, Probe | EACH | 8,731 | 8,527 | 8,440 | 8,243 | 8,171 |
| 33 23 0802 | Product Recovery Pump, 1/3 HP, 115V, Controls, Probe | EACH | 9,109 | 8,905 | 8,818 | 8,621 | 8,549 |
| 33 23 0803 | Product Recovery Pump, 1/3 HP, 230V, Controls, Probe, Highly Corrosive Liquid | EACH | 8,730 | 8,526 | 8,439 | 8,242 | 8,170 |
| 33 23 0901 | Product Recovery Pump, 1/3 HP, without Controls, Includes Probe | EACH | 2,237 | 2,033 | 1,946 | 1,749 | 1,677 |
| 33 23 0902 | Product Recovery Pump, 3/4 HP, without Controls | EACH | 1,909 | 1,706 | 1,619 | 1,422 | 1,350 |
| 33 29 0101 | 10 GPM, 10' Head, 1/6 HP, Centrifugal Pump | EACH | 789.18 | 730.44 | 705.29 | 648.51 | 627.69 |
| 33 29 0102 | 10 GPM, 1/2 HP, Centrifugal Pump | EACH | 1,063 | 952.32 | 905.07 | 798.39 | 759.27 |
| 33 29 0103 | 50 GPM, 100' Head, 3 HP, Centrifugal Pump | EACH | 2,658 | 2,500 | 2,432 | 2,279 | 2,223 |
| 33 29 0104 | 75 GPM, 100' Head, 5 HP, Centrifugal Pump | EACH | 2,946 | 2,754 | 2,672 | 2,487 | 2,419 |
| 33 29 0106 | 100 GPM, 150' Head, 7.5 HP, Centrifugal Pump | EACH | 3,419 | 3,191 | 3,093 | 2,873 | 2,793 |
| 33 29 0107 | 125 GPM, 150' Head, 10 HP, Centrifugal Pump | EACH | 3,815 | 3,512 | 3,382 | 3,088 | 2,981 |
| 33 29 0108 | 10 HP, 200 GPM, Centrifugal Pump | EACH | 2,941 | 2,714 | 2,616 | 2,396 | 2,316 |
| 33 29 0109 | 10 HP, 250 GPM, Centrifugal Pump | EACH | 3,032 | 2,805 | 2,707 | 2,487 | 2,407 |
| 33 29 0110 | 15 HP, 300 GPM, Centrifugal Pump | EACH | 3,441 | 3,160 | 3,040 | 2,770 | 2,670 |
| 33 29 0111 | 20 HP, 500 GPM, Centrifugal Pump | EACH | 4,544 | 4,241 | 4,111 | 3,818 | 3,710 |
| 33 29 0201 | 30 GPM Pneumatic Ejector | EACH | 9,959 | 9,665 | 9,512 | 9,227 | 9,139 |
| 33 29 0202 | 50 GPM Pneumatic Ejector | EACH | 14,115 | 13,789 | 13,620 | 13,302 | 13,205 |
| 33 29 0203 | 100 GPM Pneumatic Ejector | EACH | 23,872 | 23,481 | 23,277 | 22,897 | 22,780 |
| 33 29 0204 | 150 GPM Pneumatic Ejector | EACH | 36,757 | 36,305 | 36,070 | 35,631 | 35,497 |
| 33 29 0205 | 200 GPM Pneumatic Ejector | EACH | 46,839 | 46,252 | 45,946 | 45,375 | 45,201 |
| 33 29 0206 | 250 GPM Pneumatic Ejector | EACH | 60,124 | 59,390 | 59,009 | 58,295 | 58,076 |
| 33 29 0207 | 300 GPM Pneumatic Ejector | EACH | 74,768 | 73,593 | 72,983 | 71,841 | 71,491 |
| 33 29 0301 | 10 GPM, 20 PSI, 1/2 HP, Iron Rotary Pump | EACH | 1,573 | 1,415 | 1,347 | 1,194 | 1,138 |

## Environmental Remediation: Assemblies Cost Book

# Cost Data for Remediation

## Extraction Wells

| Assembly | Description | Unit | Unit Cost by Safety Level A | B | C | D | E |
|---|---|---|---|---|---|---|---|
| | **Operations & Maintenance** | | | | | | |
| 33 29 0302 | 10 GPM, 40 PSI, 3/4 HP, Iron Rotary Pump | EACH | 1,818 | 1,627 | 1,545 | 1,359 | 1,291 |
| 33 29 0303 | 10 GPM, 60 PSI, 1 HP, Iron Rotary Pump | EACH | 1,975 | 1,748 | 1,650 | 1,430 | 1,350 |
| 33 29 0304 | 25 GPM, 20 PSI, 1 HP, Iron Rotary Pump | EACH | 1,690 | 1,532 | 1,464 | 1,311 | 1,255 |
| 33 29 0305 | 25 GPM, 40 PSI, 1 1/2 HP, Iron Rotary Pump | EACH | 1,835 | 1,644 | 1,562 | 1,376 | 1,308 |
| 33 29 0306 | 25 GPM, 60 PSI, 2 HP, Iron Rotary Pump | EACH | 2,193 | 1,966 | 1,868 | 1,648 | 1,568 |
| 33 41 0101 | Pump & Motor Maintenance/Repair | EACH | 865.17 | 671.48 | 581.58 | 393.96 | 329.64 |
| 33 42 0101 | Electrical Charge | KWH | 0.06 | 0.06 | 0.06 | 0.06 | 0.06 |

**Environmental Remediation: Assemblies Cost Book**

# Cost Data for Remediation
## Field Sampling

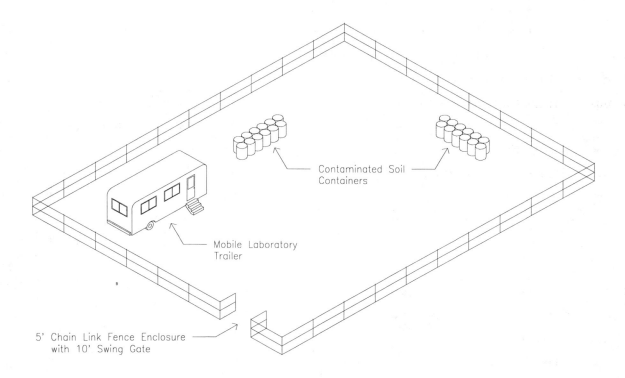

### General:
Some RA applications using mobile laboratories include: excavation and stockpiling of contaminated soil prior to treatment or transport to a disposal facility; and, pre-closure sampling. An on-site laboratory reduces the impact that the analysis time has on excavation and backfill schedules by reducing turn-around-time, packaging, holding times, transportation costs and overhead associated with the use of off-site conventional laboratory facilities. The mobile laboratory also reduces the liability of transporting hazardous waste off-site.

### Typical Treatment Train:
Natural attenuation (monitoring-only plan), on-going remediation (progress measurement), post closure (landfills, storage facilities, etc.), compliance monitoring (release detection).

### Common Cost Components:
1. Mobilize/Demobilize laboratory trailer
2. Mobile laboratory trailer
3. Sampling crew mobilization/per diem
4. Sample preparation, cleanup, containers, disposables, packaging, shipment
5. Professional field and laboratory labor
6. QA/QC samples (replicates/duplicates, trip blanks, equipment blanks, field blanks, spikes, blanks)
7. Analyses (consider turnaround time and QC level)
8. Laboratory utilities allowance

### Other Cost Considerations:
Electrical distribution, gas distribution, water distribution, access road, sanitary sewer, professional labor, fencing.

**Environmental Remediation: Assemblies Cost Book**

# Cost Data for Remediation

## Field Sampling

| Assembly | Description | Unit | \multicolumn{5}{c}{Unit Cost by Safety Level} |
| | | | A | B | C | D | E |
|---|---|---|---|---|---|---|---|
| | **Capital Costs** | | | | | | |
| 17 03 0210 | D7 with U-blade Bulldozer | HOUR | 296.12 | 242.11 | 197.86 | 144.43 | 138.38 |
| 17 03 0230 | Crawler-mounted, 1.0 CY, 215 Hydraulic Excavator | HOUR | 210.40 | 170.50 | 140.73 | 101.43 | 95.15 |
| 33 01 0101 | Mobilize/DeMobilize Drilling Rig & Crew | LS | 4,949 | 4,031 | 3,308 | 2,401 | 2,280 |
| 33 01 0102 | Van or Pickup Rental | DAY | 32.69 | 32.69 | 32.69 | 32.69 | 32.69 |
| 33 01 0103 | Flatbed Truck Rental | DAY | 45.94 | 45.94 | 45.94 | 45.94 | 45.94 |
| 33 01 0104 | Car or Van Mileage Charge | MILE | 0.35 | 0.35 | 0.35 | 0.35 | 0.35 |
| 33 01 0105 | Pickup Truck Mileage Charge | MILE | 0.56 | 0.56 | 0.56 | 0.56 | 0.56 |
| 33 01 0106 | Flatbed Truck Mileage Charge | MILE | 1.01 | 1.01 | 1.01 | 1.01 | 1.01 |
| 33 01 0201 | Mobilize Crew, >= 500 Miles, per Person | EACH | 357.75 | 357.75 | 357.75 | 357.75 | 357.75 |
| 33 01 0202 | Per Diem | DAY | 104.94 | 104.94 | 104.94 | 104.94 | 104.94 |
| 33 01 0203 | Mobilize Crew, 250 Miles, per Person | EACH | 178.88 | 178.88 | 178.88 | 178.88 | 178.88 |
| 33 01 0204 | Mobilize Crew, 100 Miles, per Person | EACH | 71.55 | 71.55 | 71.55 | 71.55 | 71.55 |
| 33 01 0205 | Mobilize Crew, 50 Miles, per Person | EACH | 53.66 | 53.66 | 53.66 | 53.66 | 53.66 |
| 33 01 0206 | Mobilize Crew, Local, per Person | EACH | 17.89 | 17.89 | 17.89 | 17.89 | 17.89 |
| 33 02 0206 | Prefiltering Liquids | EACH | 20.00 | 20.00 | 20.00 | 20.00 | 20.00 |
| 33 02 0207 | Acid Digestion | EACH | 27.50 | 27.50 | 27.50 | 27.50 | 27.50 |
| 33 02 0208 | Drying and Grinding | EACH | 22.50 | 22.50 | 22.50 | 22.50 | 22.50 |
| 33 02 0209 | Dissolution by Microwave | EACH | 40.00 | 40.00 | 40.00 | 40.00 | 40.00 |
| 33 02 0210 | Urine Preparation | EACH | 25.00 | 25.00 | 25.00 | 25.00 | 25.00 |
| 33 02 0211 | Feces Preparation | EACH | 33.33 | 33.33 | 33.33 | 33.33 | 33.33 |
| 33 02 0212 | Blood Preparation | EACH | 33.33 | 33.33 | 33.33 | 33.33 | 33.33 |
| 33 02 0213 | Animal Bone/Tissue Preparation | EACH | 66.67 | 66.67 | 66.67 | 66.67 | 66.67 |
| 33 02 0214 | Gas Flow Proportional Counting | EACH | 27.50 | 27.50 | 27.50 | 27.50 | 27.50 |
| 33 02 0215 | Liquid Scintillation | EACH | 27.50 | 27.50 | 27.50 | 27.50 | 27.50 |
| 33 02 0216 | Alpha Spectroscopy | EACH | 27.50 | 27.50 | 27.50 | 27.50 | 27.50 |
| 33 02 0217 | Gamma Spectroscopy | EACH | 20.00 | 20.00 | 20.00 | 20.00 | 20.00 |
| 33 02 0218 | Fluorometric or Laser | EACH | 27.50 | 27.50 | 27.50 | 27.50 | 27.50 |
| 33 02 0219 | Alpha Scintillation Counting | EACH | 27.50 | 27.50 | 27.50 | 27.50 | 27.50 |
| 33 02 0220 | Radon De-emanation | EACH | 27.50 | 27.50 | 27.50 | 27.50 | 27.50 |
| 33 02 0221 | Beta/Gamma Coincidence | EACH | 27.50 | 27.50 | 27.50 | 27.50 | 27.50 |
| 33 02 0222 | Radioactivity Screening - Tritium Not Present | EACH | 32.50 | 32.50 | 32.50 | 32.50 | 32.50 |
| 33 02 0223 | Radioactivity Screening - Tritium Present | EACH | 10.00 | 10.00 | 10.00 | 10.00 | 10.00 |
| 33 02 0224 | Repackage Sample | EACH | 10.00 | 10.00 | 10.00 | 10.00 | 10.00 |

**Environmental Remediation: Assemblies Cost Book**

# Cost Data for Remediation

## Field Sampling

| Assembly | Description | Unit | Unit Cost by Safety Level | | | | |
|---|---|---|---|---|---|---|---|
| | | | A | B | C | D | E |
| | **Capital Costs** | | | | | | |
| 33 02 0225 | Repackage and Ship Sample | EACH | 23.33 | 23.33 | 23.33 | 23.33 | 23.33 |
| 33 02 0301 | Air Monitoring Station | EACH | 718.68 | 718.68 | 718.68 | 718.68 | 718.68 |
| 33 02 0302 | Organic Vapor Analyzer Rental, per Month | MONTH | 1,920 | 1,920 | 1,920 | 1,920 | 1,920 |
| 33 02 0303 | Organic Vapor Analyzer Rental, per Day | DAY | 184.30 | 184.30 | 184.30 | 184.30 | 184.30 |
| 33 02 0306 | Monitoring Gas Vents | EACH | 20.61 | 15.89 | 13.86 | 9.30 | 7.63 |
| 33 02 0309 | Bacterial Air Sampler, Monthly Rental | MONTH | 436.37 | 436.37 | 436.37 | 436.37 | 436.37 |
| 33 02 0310 | Microbial Air Sampler, Monthly Rental | MONTH | 436.37 | 436.37 | 436.37 | 436.37 | 436.37 |
| 33 02 0311 | Cotton Dust Sampler, Monthly Rental | MONTH | 480.53 | 480.53 | 480.53 | 480.53 | 480.53 |
| 33 02 0312 | Digital Dust Sampler, Monthly Rental | MONTH | 552.97 | 552.97 | 552.97 | 552.97 | 552.97 |
| 33 02 0313 | Flow Calibrator, Monthly Rental | MONTH | 212.00 | 212.00 | 212.00 | 212.00 | 212.00 |
| 33 02 0314 | Personal Low Flow Sampling Pump, Monthly Rental | MONTH | 178.43 | 178.43 | 178.43 | 178.43 | 178.43 |
| 33 02 0315 | Ambient Air Monitor, Monthly Rental | MONTH | 826.80 | 826.80 | 826.80 | 826.80 | 826.80 |
| 33 02 0316 | Ambient CO Analyzer, Monthly Rental | MONTH | 378.95 | 378.95 | 378.95 | 378.95 | 378.95 |
| 33 02 0317 | Ambient Ozone Monitor, Monthly Rental | MONTH | 1,758 | 1,758 | 1,758 | 1,758 | 1,758 |
| 33 02 0318 | Ambient Sulfur Dioxide Analyzer, Monthly Rental | MONTH | 2,690 | 2,690 | 2,690 | 2,690 | 2,690 |
| 33 02 0319 | Calibration Gas Monitor | EACH | 487.60 | 487.60 | 487.60 | 487.60 | 487.60 |
| 33 02 0320 | Chloride Analyzer, Monthly Rental | MONTH | 749.07 | 749.07 | 749.07 | 749.07 | 749.07 |
| 33 02 0321 | Gas Monitors, Monthly Rental | MONTH | 657.20 | 657.20 | 657.20 | 657.20 | 657.20 |
| 33 02 0322 | Hydrogen Sulfide Analyzer, Monthly Rental | MONTH | 249.10 | 249.10 | 249.10 | 249.10 | 249.10 |
| 33 02 0323 | Manual Air Sampling Kit, Monthly Rental | MONTH | 85.68 | 85.68 | 85.68 | 85.68 | 85.68 |
| 33 02 0324 | Manual Air Sampling Kit, Detection Tubes | EACH | 88.69 | 88.69 | 88.69 | 88.69 | 88.69 |
| 33 02 0325 | Mercury Vapor Monitor, Monthly Rental | MONTH | 1,277 | 1,277 | 1,277 | 1,277 | 1,277 |
| 33 02 0326 | Nitrogen Dioxide Analyzer, Monthly Rental | MONTH | 2,385 | 2,385 | 2,385 | 2,385 | 2,385 |
| 33 02 0327 | Personal Alarms, Monthly Rental | MONTH | 318.00 | 318.00 | 318.00 | 318.00 | 318.00 |
| 33 02 0328 | Portable Ambient Air Analyzer, Monthly Rental | MONTH | 1,745 | 1,745 | 1,745 | 1,745 | 1,745 |
| 33 02 0329 | Portable CO2 Monitor, Monthly Rental | MONTH | 485.83 | 485.83 | 485.83 | 485.83 | 485.83 |
| 33 02 0330 | Portable Combustible Gas/Oxygen Indicator, Monthly Rental | MONTH | 355.10 | 355.10 | 355.10 | 355.10 | 355.10 |
| 33 02 0331 | Portable Organic Vapor Analyzer, Monthly Rental | MONTH | 1,920 | 1,920 | 1,920 | 1,920 | 1,920 |
| 33 02 0332 | Portable Oxygen Deficiency Monitor, Monthly Rental | MONTH | 409.87 | 409.87 | 409.87 | 409.87 | 409.87 |
| 33 02 0333 | Portable Toxic Gas Analyzer, Monthly Rental | MONTH | 265.00 | 265.00 | 265.00 | 265.00 | 265.00 |
| 33 02 0334 | Solvent Vapor Indicator, Monthly Rental | MONTH | 1,920 | 1,920 | 1,920 | 1,920 | 1,920 |
| 33 02 0335 | Specific Vapor Analyzer, Monthly Rental | MONTH | 1,290 | 1,290 | 1,290 | 1,290 | 1,290 |
| 33 02 0336 | Stack Sampling Probe, Monthly Rental | MONTH | 648.37 | 648.37 | 648.37 | 648.37 | 648.37 |

**Environmental Remediation: Assemblies Cost Book**

*Page 3-119*

1998 by ECHOS. All rights reserved

# Cost Data for Remediation

## Field Sampling

| Assembly | Description | Unit | A | B | C | D | E |
|---|---|---|---|---|---|---|---|
| | **Capital Costs** | | | | | | |
| 33 02 0337 | Trace-Gas Analyzer, Monthly Rental | MONTH | 1,852 | 1,852 | 1,852 | 1,852 | 1,852 |
| 33 02 0338 | Instrument Shelter, 30.0" x 20.0" x 32.0", Pine with Latex Paint | EACH | 738.96 | 722.90 | 716.02 | 700.49 | 694.80 |
| 33 02 0339 | Instrument Shelter, 18.0" x 13.0" x 20.0", Wood Painted White | EACH | 289.52 | 273.46 | 266.58 | 251.05 | 245.36 |
| 33 02 0340 | Instrument Shelter, 81.0" x 5.5.0" x 16.5", Wood Painted White | EACH | 266.47 | 250.40 | 243.52 | 228.00 | 222.30 |
| 33 02 0341 | Alpha, Beta, Gamma Detector Equipment | EACH | 194.33 | 194.33 | 194.33 | 194.33 | 194.33 |
| 33 02 0401 | Disposable Materials per Sample | EACH | 6.87 | 6.87 | 6.87 | 6.87 | 6.87 |
| 33 02 0402 | Decontamination Materials per Sample | EACH | 6.26 | 6.26 | 6.26 | 6.26 | 6.26 |
| 33 02 0404 | 12' Aluminum Pole with Polyhead Sampler | EACH | 396.66 | 396.66 | 396.66 | 396.66 | 396.66 |
| 33 02 0405 | Monitoring Well Slug Testing Equipment Rental | WEEK | 552.08 | 552.08 | 552.08 | 552.08 | 552.08 |
| 33 02 0521 | Hip Waders Rental | WEEK | 51.44 | 51.44 | 51.44 | 51.44 | 51.44 |
| 33 02 0522 | Boat with Motor, Rental | DAY | 58.83 | 58.83 | 58.83 | 58.83 | 58.83 |
| 33 02 0523 | Boat without Motor, Rental | DAY | 21.73 | 21.73 | 21.73 | 21.73 | 21.73 |
| 33 02 0524 | Glass Coliwasas, Disposable, 7/8" x 42", 200 ml, Case of 12 | EACH | 187.22 | 187.22 | 187.22 | 187.22 | 187.22 |
| 33 02 0525 | Glass Coliwasas, Reusable, 7/8" x 42", 200 ml, Case of 4 | EACH | 67.42 | 67.42 | 67.42 | 67.42 | 67.42 |
| 33 02 0526 | PVC Coliwasas, 1.9" Outside Diameter x 49", 760 ml | EACH | 155.01 | 155.01 | 155.01 | 155.01 | 155.01 |
| 33 02 0527 | Bomb Sampler, 2.5" Diameter x 10", 500 ml | EACH | 524.35 | 524.35 | 524.35 | 524.35 | 524.35 |
| 33 02 0528 | Drum Thief Tubes, 3/4" x 42", 150 ml, Disposable, Case of 25 | EACH | 63.94 | 63.94 | 63.94 | 63.94 | 63.94 |
| 33 02 0529 | Drum Thief Tubes, 1/2" x 42", 75 ml, Disposable, Case of 25 | EACH | 44.75 | 44.75 | 44.75 | 44.75 | 44.75 |
| 33 02 0530 | Bottom Sampler, 3 Lb Brass, 7.5" x 4.5", No Cable Required | EACH | 179.67 | 179.67 | 179.67 | 179.67 | 179.67 |
| 33 02 0531 | Bottom Sampler, 17 Lb Stainless Steel, 6" x 6" x 6", with 100' Cable | EACH | 466.91 | 466.91 | 466.91 | 466.91 | 466.91 |
| 33 02 0532 | Bottom Sampler, 60 Lb Stainless Steel, 9" x 9" | EACH | 915.24 | 915.24 | 915.24 | 915.24 | 915.24 |
| 33 02 0533 | Water Level Indicator, 100' Tape, Electric, Light & Horn | EACH | 461.65 | 461.65 | 461.65 | 461.65 | 461.65 |
| 33 02 0534 | Water Level Indicator, 300' Tape, Electric, Light & Horn | EACH | 688.99 | 688.99 | 688.99 | 688.99 | 688.99 |
| 33 02 0535 | Water Level Indicator, 500' Tape, Electric, Light & Horn | EACH | 996.15 | 996.15 | 996.15 | 996.15 | 996.15 |
| 33 02 0536 | Water Level Indicator, Manual Steel Chalk Tape, with Weight, 100' Cable | EACH | 138.33 | 138.33 | 138.33 | 138.33 | 138.33 |
| 33 02 0538 | DO Meter, Analog, Portable, 6' Cable, Polarographic Probe | EACH | 577.70 | 577.70 | 577.70 | 577.70 | 577.70 |
| 33 02 0539 | DO Meter, Digital, Portable, 10' Cable, Small Diameter Probe | EACH | 1,114 | 1,114 | 1,114 | 1,114 | 1,114 |
| 33 02 0540 | DO Meter, Portable, Probe, 10' Cable, Quick Readings | EACH | 517.60 | 517.60 | 517.60 | 517.60 | 517.60 |

**Environmental Remediation: Assemblies Cost Book**

# Cost Data for Remediation

## Field Sampling

| Assembly | Description | Unit | Unit Cost by Safety Level A | B | C | D | E |
|---|---|---|---|---|---|---|---|
| | **Capital Costs** | | | | | | |
| 33 02 0541 | Conductivity Meter with Dip-type Sensor, Digital, 9V Battery, 7 x 3.2 x 2" | EACH | 977.27 | 977.27 | 977.27 | 977.27 | 977.27 |
| 33 02 0542 | Conductivity Meter, 1" Diameter Probe, 150' Cable, Also Temperature/Level | EACH | 914.25 | 914.25 | 914.25 | 914.25 | 914.25 |
| 33 02 0543 | Conductivity Meter, 10' Cable, Probe, Also Temperature/ Salinity | EACH | 870.79 | 870.79 | 870.79 | 870.79 | 870.79 |
| 33 02 0544 | Thermometer, Stainless Steel, Bi-Metal, 40 - 160 F, 12" Stem | EACH | 62.47 | 62.47 | 62.47 | 62.47 | 62.47 |
| 33 02 0545 | Thermometer, Thermocouple | EACH | 144.48 | 144.48 | 144.48 | 144.48 | 144.48 |
| 33 02 0546 | Thermometer, Digital | EACH | 38.50 | 38.50 | 38.50 | 38.50 | 38.50 |
| 33 02 0547 | pH Meter, Test Paper Kits, 200 Strips, Charts | EACH | 9.89 | 9.89 | 9.89 | 9.89 | 9.89 |
| 33 02 0548 | pH Meter, Digital, LCD, 0 - 14 pH Range | EACH | 361.81 | 361.81 | 361.81 | 361.81 | 361.81 |
| 33 02 0549 | Lysimeter, 1.90" Outside Diameter, PVC Tube Type - PVC Body/Teflon Filter | EACH | 565.44 | 488.90 | 428.73 | 353.14 | 342.99 |
| 33 02 0550 | Lysimeter, 1.90" Outside Diameter, Teflon Tube Type - Teflon Body/Filter | EACH | 636.22 | 559.67 | 499.50 | 423.91 | 413.76 |
| 33 02 0551 | Lysimeter, 2.38" Outside Diameter, PVC Tube Type - PVC Body/Teflon Filter | EACH | 565.44 | 488.90 | 428.73 | 353.14 | 342.99 |
| 33 02 0552 | Lysimeter, 2.38" Outside Diameter, Teflon Tube Type - Teflon Body/Filter | EACH | 544.92 | 468.38 | 408.21 | 332.62 | 322.47 |
| 33 02 0553 | Lysimeter, 1.90" Outside Diameter, PVC Cup Type - PVC Body/Teflon Filter | EACH | 565.44 | 488.90 | 428.73 | 353.14 | 342.99 |
| 33 02 0554 | Lysimeter, 1.90" Outside Diameter, Teflon Cup Type - Teflon Body/Filter | EACH | 630.42 | 553.87 | 493.70 | 418.12 | 407.97 |
| 33 02 0555 | Lysimeter, 1.90" Outside Diameter, PVC Deep Sample Type - Teflon Body/Filter | EACH | 644.64 | 568.09 | 507.92 | 432.33 | 422.18 |
| 33 02 0556 | Lysimeter, 1.90" Outside Diameter, Teflon Deep Sample Type - Teflon Body/Filter | EACH | 741.01 | 664.46 | 604.29 | 528.70 | 518.56 |
| 33 02 0557 | Lysimeter Head Assembly, All PVC Fittings | EACH | 577.72 | 501.17 | 441.00 | 365.41 | 355.26 |
| 33 02 0558 | Lysimeter Head Assembly, Teflon & PVC Fittings | EACH | 684.13 | 607.58 | 547.41 | 471.83 | 461.68 |
| 33 02 0559 | Silica Flour, 50 Lb Pail for Media Continuum | EACH | 174.04 | 151.08 | 133.03 | 110.35 | 107.31 |
| 33 02 0560 | Vacuum Pressure Hand Pump | EACH | 46.40 | 46.40 | 46.40 | 46.40 | 46.40 |
| 33 02 0561 | Nylon Tubing, 1/4" Outside Diameter | LF | 0.47 | 0.47 | 0.47 | 0.47 | 0.47 |
| 33 02 0562 | Teflon Tubing, 1/4" Outside Diameter | LF | 1.21 | 1.21 | 1.21 | 1.21 | 1.21 |
| 33 02 0601 | Drilling 2.5" Diameter Soil Borings, No Sampling | LF | 21.74 | 18.75 | 16.71 | 13.78 | 13.18 |
| 33 02 0602 | Stainless Steel Bottom Sampler | EACH | 378.97 | 378.97 | 378.97 | 378.97 | 378.97 |
| 33 02 0603 | Surface Soil Sampling Equipment | EACH | 280.80 | 280.80 | 280.80 | 280.80 | 280.80 |
| 33 02 0605 | Hand Auger Rental | DAY | 13.25 | 13.25 | 13.25 | 13.25 | 13.25 |
| 33 02 0606 | Power Auger Rental | DAY | 15.96 | 15.96 | 15.96 | 15.96 | 15.96 |

**Environmental Remediation: Assemblies Cost Book**

*Page 3-121*

1998 by ECHOS. All rights reserved

# Cost Data for Remediation

## Field Sampling

| Assembly | Description | Unit | Unit Cost by Safety Level | | | | |
|---|---|---|---|---|---|---|---|
| | | | A | B | C | D | E |
| | **Capital Costs** | | | | | | |
| 33 02 0613 | Soil Recovery Probe, Unslotted, 7/8" x 1' | EACH | 121.64 | 121.64 | 121.64 | 121.64 | 121.64 |
| 33 02 0614 | Soil Recovery Probe, Unslotted, 7/8" x 2' or 9/8" x 1' | EACH | 140.10 | 140.10 | 140.10 | 140.10 | 140.10 |
| 33 02 0615 | Soil Recovery Probe, Unslotted, 9/8" x 2' | EACH | 151.23 | 151.23 | 151.23 | 151.23 | 151.23 |
| 33 02 0616 | Soil Recovery Probe, Liners, 7/8" per LF, Butyrate Plast | LF | 0.86 | 0.86 | 0.86 | 0.86 | 0.86 |
| 33 02 0617 | Soil Recovery Probe, Liners, 9/8" per LF, Butyrate Plast | LF | 1.41 | 1.41 | 1.41 | 1.41 | 1.41 |
| 33 02 0618 | Soil Recovery Probe, Liners, Stainless Steel, 7/8" | EACH | 6.51 | 6.51 | 6.51 | 6.51 | 6.51 |
| 33 02 0619 | Soil Recovery Probe, Liners, Stainless Steel, 9/8" | EACH | 9.54 | 9.54 | 9.54 | 9.54 | 9.54 |
| 33 02 0620 | Soil Gas Vapor Probe, Stainless Steel, Manual, Nonremovable Tip | EACH | 45.93 | 45.93 | 45.93 | 45.93 | 45.93 |
| 33 02 0621 | Soil Gas Vapor Probe, Stainless Steel, Electric, Rotary, Up/Down Hammer | EACH | 226.73 | 226.73 | 226.73 | 226.73 | 226.73 |
| 33 02 0622 | Screw Auger, High-carbon Steel, 1.5" x 6" | EACH | 79.89 | 79.89 | 79.89 | 79.89 | 79.89 |
| 33 02 0623 | Screw Auger, High-carbon Steel, 1.75" x 6" | EACH | 97.33 | 97.33 | 97.33 | 97.33 | 97.33 |
| 33 02 0624 | Soil Auger Buckets, Dutch, 3" Diameter, Carbon/Stainless Steel | EACH | 129.80 | 129.80 | 129.80 | 129.80 | 129.80 |
| 33 02 0625 | Soil Auger Buckets, Mud, 2" Diameter, Carbon/Stainless Steel | EACH | 120.12 | 120.12 | 120.12 | 120.12 | 120.12 |
| 33 02 0626 | Soil Auger Buckets, Reg, 2" Diameter, Carbon/Stainless Steel | EACH | 100.46 | 100.46 | 100.46 | 100.46 | 100.46 |
| 33 02 0627 | Soil Auger Buckets, Sand, 2" Diameter, Carbon/Stainless Steel | EACH | 114.29 | 114.29 | 114.29 | 114.29 | 114.29 |
| 33 02 0628 | Soil Recovery Auger Buckets, Stainless Steel, 2.25" x 8" | EACH | 342.05 | 342.05 | 342.05 | 342.05 | 342.05 |
| 33 02 0629 | Soil Recovery Auger Buckets, Stainless Steel, 3.25" x 10" | EACH | 352.12 | 352.12 | 352.12 | 352.12 | 352.12 |
| 33 02 0630 | Soil Recovery Auger Buckets, Stainless Steel with Carbon Bits, 2.25" x 8" | EACH | 315.75 | 315.75 | 315.75 | 315.75 | 315.75 |
| 33 02 0631 | Soil Recovery Auger Buckets, Stainless Steel with Carbon Bits, 3.25" x 10" | EACH | 325.82 | 325.82 | 325.82 | 325.82 | 325.82 |
| 33 02 0632 | Soil Recovery Auger Buckets, Liners, Butyrate Plast, 3" x 10" or 2" x 8" | EACH | 5.13 | 5.13 | 5.13 | 5.13 | 5.13 |
| 33 02 0633 | Soil Recovery Auger Buckets, Liners, Stainless Steel, 3" x 10" or 2" x 8" | EACH | 17.14 | 17.14 | 17.14 | 17.14 | 17.14 |
| 33 02 0634 | Sludge Sampler, Stainless Steel, 3.25" x 12", Thread On | EACH | 378.97 | 378.97 | 378.97 | 378.97 | 378.97 |
| 33 02 0635 | Tensiometer Soil Moist Probe, Standard, 7/8" Outside Diameter x 18", Fixed | EACH | 83.69 | 83.69 | 83.69 | 83.69 | 83.69 |
| 33 02 0636 | Tensiometer Soil Moist Probe, Jet Fill, 7/8" Outside Diameter x 18", Fixed | EACH | 71.81 | 71.81 | 71.81 | 71.81 | 71.81 |
| 33 02 0637 | Tensiometer Soil Moist Probe, Quick Draw, 18", Quick Read | EACH | 412.34 | 412.34 | 412.34 | 412.34 | 412.34 |
| 33 02 0638 | Soil Test Kit, 50 Tests | EACH | 278.25 | 278.25 | 278.25 | 278.25 | 278.25 |

**Environmental Remediation: Assemblies Cost Book**

# Cost Data for Remediation

## Field Sampling

| Assembly | Description | Unit | A | B | C | D | E |
|---|---|---|---|---|---|---|---|
| | **Capital Costs** | | | | | | |
| 33 02 0639 | CPT Rig, Includes Labor, Sampling, Punching, Decontamination | DAY | 2,844 | 2,844 | 2,844 | 2,844 | 2,844 |
| 33 02 0640 | Mobilize/Demobilize CPT Rig and Crew | EACH | 2,986 | 2,986 | 2,986 | 2,986 | 2,986 |
| 33 02 0641 | Standby Time for CPT Rig and Crew | HOUR | 3,015 | 3,015 | 3,015 | 3,015 | 3,015 |
| 33 02 0642 | Setup Cost per Each Hole, CPT Rig | EACH | 273.83 | 273.83 | 273.83 | 273.83 | 273.83 |
| 33 02 0643 | Decontamination Trailer Rental for CPT Rig | DAY | 92.75 | 92.75 | 92.75 | 92.75 | 92.75 |
| 33 02 0644 | Decontaminate Cone on CPT Rig | DAY | 273.83 | 273.83 | 273.83 | 273.83 | 273.83 |
| 33 02 0645 | Level "D" PPE Rental per 2-Man CPT Crew | DAY | 2,874 | 2,874 | 2,874 | 2,874 | 2,874 |
| 33 02 0646 | Level "C" PPE Rental per 2-Man CPT Crew | DAY | 3,122 | 3,122 | 3,122 | 3,122 | 3,122 |
| 33 02 0647 | In Situ Batch Sampling for Groundwater, CPT Rig (per Sample) | EACH | 92.75 | 92.75 | 92.75 | 92.75 | 92.75 |
| 33 02 0648 | In Situ Soil Sampling, CPT Rig (per Sample) | EACH | 60.95 | 60.95 | 60.95 | 60.95 | 60.95 |
| 33 02 0649 | Hydropunch with CPT Rig, Hard Soil | FT | 10.60 | 10.60 | 10.60 | 10.60 | 10.60 |
| 33 02 0650 | Hydropunch with CPT Rig, Medium Soil | FT | 10.60 | 10.60 | 10.60 | 10.60 | 10.60 |
| 33 02 0651 | Hydropunch with CPT Rig, Soft Soil | FT | 10.60 | 10.60 | 10.60 | 10.60 | 10.60 |
| 33 02 0652 | Grout Hole After Hydropunching | FT | 4.38 | 4.38 | 4.38 | 4.38 | 4.38 |
| 33 02 0653 | CPT Rig with ROST Analytical Equipment for Obtaining VOC in Soil | DAY | 7,100 | 7,100 | 7,100 | 7,100 | 7,100 |
| 33 02 1501 | Saturation Indicator | EACH | 106.53 | 106.53 | 106.53 | 106.53 | 106.53 |
| 33 02 1502 | Thermostat & Humidity Control Devices | EACH | 329.49 | 279.97 | 258.78 | 210.91 | 193.37 |
| 33 02 1506 | Monitoring Port with Gas Monitor | EACH | 28.81 | 22.42 | 19.69 | 13.52 | 11.26 |
| 33 02 1509 | Water Quality Parameter Testing Device | WEEK | 227.90 | 227.90 | 227.90 | 227.90 | 227.90 |
| 33 02 2001 | Sample Preparation (3005/3010/3020/3040/3050) | EACH | 10.00 | 10.00 | 10.00 | 10.00 | 10.00 |
| 33 02 2002 | Sample Preparation (3510/3520/3540/3550/5030) | EACH | 25.00 | 25.00 | 25.00 | 25.00 | 25.00 |
| 33 02 2003 | Sample Preparation, Waste Dilution (3580) | EACH | 20.00 | 20.00 | 20.00 | 20.00 | 20.00 |
| 33 02 2004 | Cleanup (3610/3611/3620/3630/3650/3660) | EACH | 25.00 | 25.00 | 25.00 | 25.00 | 25.00 |
| 33 02 2005 | Cleanup, Gel-Permeation (3640) | EACH | 25.00 | 25.00 | 25.00 | 25.00 | 25.00 |
| 33 02 2019 | 1 Gallon, 128 Oz, Glass Sample Container, Case of 4 | EACH | 12.80 | 12.80 | 12.80 | 12.80 | 12.80 |
| 33 02 2020 | 1 Liter, 32 Oz, Clear Wide Mouth Jar, Case of 12 | EACH | 44.37 | 44.37 | 44.37 | 44.37 | 44.37 |
| 33 02 2021 | 500 ml, 16 Oz, Clear Wide Mouth Jar, Case of 12 | EACH | 61.67 | 61.67 | 61.67 | 61.67 | 61.67 |
| 33 02 2022 | 250 ml, 8 Oz, Clear Wide Mouth Jar, Case of 24 | EACH | 47.09 | 47.09 | 47.09 | 47.09 | 47.09 |
| 33 02 2023 | 120 ml, 4 Oz, Clear Wide Mouth Jar, Case of 24 | EACH | 34.47 | 34.47 | 34.47 | 34.47 | 34.47 |
| 33 02 2024 | 1 Liter, 32 Oz, Boston Round Bottle, Case of 12 | EACH | 28.78 | 28.78 | 28.78 | 28.78 | 28.78 |
| 33 02 2025 | 500 ml, 16 Oz, Boston Round Bottle, Case of 12 | EACH | 19.91 | 19.91 | 19.91 | 19.91 | 19.91 |

**Environmental Remediation: Assemblies Cost Book**

# Cost Data for Remediation

## Field Sampling

| Assembly | Description | Unit | Unit Cost by Safety Level A | B | C | D | E |
|---|---|---|---|---|---|---|---|
| | **Capital Costs** | | | | | | |
| 33 02 2026 | 40 ml, Clear Vial, Case of 72 | EACH | 89.91 | 89.91 | 89.91 | 89.91 | 89.91 |
| 33 02 2027 | 20 ml, Clear Vial, Case of 72 | EACH | 74.92 | 74.92 | 74.92 | 74.92 | 74.92 |
| 33 02 2028 | 250 ml, 8 oz, Clear Wide Mouth Vial with Septa, Case of 24 | EACH | 58.01 | 58.01 | 58.01 | 58.01 | 58.01 |
| 33 02 2029 | 125 ml, 4 oz, Clear Wide Mouth Vial with Septa, Case of 24 | EACH | 53.53 | 53.53 | 53.53 | 53.53 | 53.53 |
| 33 02 2030 | 1 Liter, 32 oz, High-density Polyethylene Bottle, Case of 12 | EACH | 32.64 | 32.64 | 32.64 | 32.64 | 32.64 |
| 33 02 2031 | 500 ml, 16 oz, High-density Polyethylene Bottle, Case of 24 | EACH | 28.06 | 28.06 | 28.06 | 28.06 | 28.06 |
| 33 02 2032 | 250 ml, 8 oz, High-density Polyethylene Bottle, Case of 24 | EACH | 15.58 | 15.58 | 15.58 | 15.58 | 15.58 |
| 33 02 2033 | 125 ml, 4 oz, High-density Polyethylene Bottle, Case of 24 | EACH | 13.71 | 13.71 | 13.71 | 13.71 | 13.71 |
| 33 02 2034 | Custody Seals, Package of 10 | EACH | 15.60 | 15.60 | 15.60 | 15.60 | 15.60 |
| 33 02 2035 | Safe Transport Can Filled with Vermiculite, 1 Gallon, Case of 4 | EACH | 28.94 | 28.94 | 28.94 | 28.94 | 28.94 |
| 33 02 2036 | Documentation Package for QA Verification, Data & Benchwork | EACH | 100.00 | 100.00 | 100.00 | 100.00 | 100.00 |
| 33 02 2037 | Overnight Delivery, 8 oz Letter | EACH | 10.17 | 10.17 | 10.17 | 10.17 | 10.17 |
| 33 02 2038 | Overnight Delivery, 1 Lb Package | LB | 11.08 | 11.08 | 11.08 | 11.08 | 11.08 |
| 33 02 2039 | Overnight Delivery, 2 - 5 Lb Package | LB | 16.08 | 16.08 | 16.08 | 16.08 | 16.08 |
| 33 02 2040 | Overnight Delivery, 6 - 10 Lb Package | LB | 21.58 | 21.58 | 21.58 | 21.58 | 21.58 |
| 33 02 2041 | Overnight Delivery, 11 - 20 Lb Package | LB | 34.25 | 34.25 | 34.25 | 34.25 | 34.25 |
| 33 02 2042 | Overnight Delivery, 21 - 50 Lb Package | LB | 57.27 | 57.27 | 57.27 | 57.27 | 57.27 |
| 33 02 2043 | Overnight Delivery, 51 - 70 Lb Package | LB | 78.27 | 78.27 | 78.27 | 78.27 | 78.27 |
| 33 02 2044 | Overnight Delivery, 71 Lbs or Greater | LB | 171.25 | 171.25 | 171.25 | 171.25 | 171.25 |
| 33 02 2045 | 48 Quart Ice Chest | EACH | 27.62 | 27.62 | 27.62 | 27.62 | 27.62 |
| 33 02 2046 | 60 Quart Ice Chest | EACH | 46.27 | 46.27 | 46.27 | 46.27 | 46.27 |
| 33 02 2047 | 80 Quart Ice Chest | EACH | 79.85 | 79.85 | 79.85 | 79.85 | 79.85 |
| 33 02 2048 | 102 Quart Ice Chest | EACH | 135.64 | 135.64 | 135.64 | 135.64 | 135.64 |
| 33 02 2049 | 120 Quart Ice Chest | EACH | 216.05 | 216.05 | 216.05 | 216.05 | 216.05 |
| 33 02 2050 | Blue Ice Soft Packs (Equivalent to 7 Lbs Ice) | EACH | 1.73 | 1.73 | 1.73 | 1.73 | 1.73 |
| 33 02 2561 | Subcontracted Field Analysis with Portable Gas Chromotograph | DAY | 976.80 | 976.80 | 976.80 | 976.80 | 976.80 |
| 33 02 9906 | Subcontracted Sampling per Hour, One-man Crew | HOUR | 85.25 | 85.25 | 85.25 | 85.25 | 85.25 |
| 33 02 9907 | Subcontracted Sampling per Day, One-man Crew | DAY | 680.00 | 680.00 | 680.00 | 680.00 | 680.00 |
| 33 02 9913 | Mobile Laboratory Trailer, 8' W x 30' L, Rental | MONTH | 7,950 | 7,950 | 7,950 | 7,950 | 7,950 |
| 33 02 9914 | Gas Chromatograph, HP5890A, Rental | MONTH | 1,855 | 1,855 | 1,855 | 1,855 | 1,855 |
| 33 02 9915 | Mass Spectrometer, HP5970B MD, Rental | MONTH | 1,617 | 1,617 | 1,617 | 1,617 | 1,617 |

**Environmental Remediation: Assemblies Cost Book**

*Page 3-124*

1998 by ECHOS. All rights reserved

# Cost Data for Remediation

## Field Sampling

| Assembly | Description | Unit | Unit Cost by Safety Level A | B | C | D | E |
|---|---|---|---|---|---|---|---|
| | **Capital Costs** | | | | | | |
| 33 02 9917 | Multi-purpose Groundwater Well Sampling Vehicle, Rental | DAY | 225.00 | 225.00 | 225.00 | 225.00 | 225.00 |
| 33 02 9918 | Spectrophotometer, Monthly Rental | MONTH | 275.00 | 275.00 | 275.00 | 275.00 | 275.00 |
| 33 02 9919 | Spectrophotometer, Purchase | EACH | 1,900 | 1,900 | 1,900 | 1,900 | 1,900 |
| 33 02 9921 | Gas Chromatograph Weekly Rental | WK | 910.00 | 910.00 | 910.00 | 910.00 | 910.00 |
| 33 02 9922 | Mobilize/Demobilize Laboratory Trailer | EACH | 3,500 | 3,500 | 3,500 | 3,500 | 3,500 |
| 33 02 9923 | Radiological Scintillation Counter | MONTH | 2,850 | 2,850 | 2,850 | 2,850 | 2,850 |
| 33 02 9924 | pH/Temperature/Conductivity Tester Rental | MONTH | 71.00 | 71.00 | 71.00 | 71.00 | 71.00 |
| 33 02 9925 | Infrared Spectrometer Rental | MONTH | 2,650 | 2,650 | 2,650 | 2,650 | 2,650 |
| 33 02 9926 | Computer Rental | MONTH | 200.00 | 200.00 | 200.00 | 200.00 | 200.00 |
| 33 02 9927 | Recalibration/Maintenance per Instrument per Month | MONTH | 235.00 | 235.00 | 235.00 | 235.00 | 235.00 |
| 33 02 9929 | Laboratory Utilities Allowance | MONTH | 200.00 | 200.00 | 200.00 | 200.00 | 200.00 |
| 33 02 9930 | Laboratory Expendables Allowance | EACH | 10.00 | 10.00 | 10.00 | 10.00 | 10.00 |
| 33 02 9931 | Electric Laboratory Scales, Purchase | EACH | 195.00 | 195.00 | 195.00 | 195.00 | 195.00 |
| 33 02 9932 | Fire Extinguisher | EACH | 99.45 | 99.45 | 99.45 | 99.45 | 99.45 |
| 33 07 9901 | Lease, Portable 40' W x 150' L x 18' High Emissions Containment Building | YEAR | 7,500 | 7,500 | 7,500 | 7,500 | 7,500 |
| 33 07 9902 | Buy, Portable 40' W x 150' L x 18' High Emissions Containment Building | EACH | 23,503 | 23,503 | 23,503 | 23,503 | 23,503 |
| 33 07 9903 | Assemble Emissions Containment Building, 5 Days | EACH | 12,346 | 9,591 | 8,298 | 5,629 | 4,722 |
| 33 07 9904 | Disassemble Emissions Containment Building, 3 Days | EACH | 7,402 | 5,750 | 4,975 | 3,374 | 2,831 |
| 33 17 0808 | Decontaminate Rig, Augers, Screen (Rental Equipment) | DAY | 205.34 | 205.34 | 205.34 | 205.34 | 205.34 |
| 33 19 0201 | Drummed Waste, Minimum Charge for Shipment | EACH | 2,350 | 2,350 | 2,350 | 2,350 | 2,350 |
| 33 19 9921 | DOT Steel Drum, 55 Gallon | EACH | 65.19 | 65.19 | 65.19 | 65.19 | 65.19 |
| 33 19 9922 | Polyethylene Closed Head Drum, 55 Gallon | EACH | 43.63 | 43.63 | 43.63 | 43.63 | 43.63 |
| 33 19 9924 | DOT Steel Salvage Drum, 85 Gallon, 16 Gauge | EACH | 57.06 | 57.06 | 57.06 | 57.06 | 57.06 |
| 33 19 9925 | DOT Steel Salvage Drum Composite Overpack, 85 Gallon, 16 Gauge | EACH | 70.49 | 70.49 | 70.49 | 70.49 | 70.49 |
| 33 23 0503 | 4" Submersible Pump, 13 - 37 GPM, < 180' Head | EACH | 3,084 | 2,790 | 2,637 | 2,352 | 2,264 |
| 33 23 0505 | 6" Submersible Pump, 15 - 135 GPM, 200' < Head < 500' | EACH | 4,472 | 3,885 | 3,579 | 3,008 | 2,834 |
| 33 23 0506 | 2" Submersible Pump Rental, Day | DAY | 63.86 | 63.86 | 63.86 | 63.86 | 63.86 |
| 33 23 0507 | 2" Submersible Pump Rental, Week | WEEK | 191.57 | 191.57 | 191.57 | 191.57 | 191.57 |
| 33 23 0508 | 2" Submersible Pump Rental, Month | MONTH | 478.92 | 478.92 | 478.92 | 478.92 | 478.92 |
| 33 23 0509 | 4" Submersible Pump Rental, Day | DAY | 63.86 | 63.86 | 63.86 | 63.86 | 63.86 |
| 33 23 0510 | 4" Submersible Pump Rental, Week | WEEK | 191.57 | 191.57 | 191.57 | 191.57 | 191.57 |

**Environmental Remediation: Assemblies Cost Book**

*Page 3-125*

1998 by ECHOS. All rights reserved

# Cost Data for Remediation

## Field Sampling

| Assembly | Description | Unit | Unit Cost by Safety Level | | | | |
|---|---|---|---|---|---|---|---|
| | | | A | B | C | D | E |
| | **Capital Costs** | | | | | | |
| 33 23 0511 | 4" Submersible Pump Rental, Month | MONTH | 478.92 | 478.92 | 478.92 | 478.92 | 478.92 |
| 33 23 1106 | Split Spoon Sample, 2" x 24", During Drilling | EACH | 46.67 | 46.67 | 46.67 | 46.67 | 46.67 |
| 33 23 1107 | Coring (Fluid or Air) | LF | 102.14 | 88.16 | 78.67 | 64.96 | 62.17 |
| 33 23 1111 | Well Development Equipment Rental | WEEK | 478.91 | 462.93 | 456.09 | 440.65 | 434.98 |
| 33 23 1121 | Standby for Drilling | EACH | 618.63 | 503.81 | 413.56 | 300.17 | 284.95 |
| 33 23 1122 | Move Rig/Equipment Around Site | EACH | 154.66 | 125.95 | 103.39 | 75.04 | 71.24 |
| 33 23 1123 | Load Supplies/Equipment | LS | 1,856 | 1,511 | 1,241 | 900.52 | 854.85 |
| 33 23 1124 | Security Pass/Protocol | LS | 309.32 | 251.91 | 206.78 | 150.09 | 142.47 |
| 33 23 1125 | Concrete Coring (Minimum Charge) | DAY | 2,475 | 2,015 | 1,654 | 1,201 | 1,140 |
| 33 23 1126 | Furnish 55 Gallon Drum for Drill Cuttings & Development Water | EACH | 65.19 | 65.19 | 65.19 | 65.19 | 65.19 |
| 33 23 1127 | Furnish 55 Gallon Drum for Development/Purge Water | EACH | 65.19 | 65.19 | 65.19 | 65.19 | 65.19 |
| 33 23 1128 | Jackhammer Rental, per Day | DAY | 166.03 | 166.03 | 166.03 | 166.03 | 166.03 |
| 33 23 1129 | Concrete Saw Rental, per Day | DAY | 102.00 | 102.00 | 102.00 | 102.00 | 102.00 |
| 33 23 1801 | 2" Well, Grout (Annular Seal) | LF | 83.99 | 68.83 | 56.92 | 41.95 | 39.94 |
| 33 23 1802 | 4" Well, Grout (Annular Seal) | LF | 145.07 | 118.89 | 98.31 | 72.46 | 68.99 |
| 33 23 1803 | 6" Well, Grout (Annular Seal) | LF | 213.78 | 175.20 | 144.88 | 106.78 | 101.67 |
| 33 23 1811 | 2" Well, Portland Cement Grout | LF | 0.92 | 0.92 | 0.92 | 0.92 | 0.92 |
| 33 23 1812 | 4" Well, Portland Cement Grout | LF | 1.38 | 1.38 | 1.38 | 1.38 | 1.38 |
| 33 23 1813 | 6" Well, Portland Cement Grout | LF | 7.80 | 7.80 | 7.80 | 7.80 | 7.80 |
| 33 23 2104 | 3/8", 1' Thick, Sodium Bentonite Pellets | CF | 275.35 | 229.42 | 193.32 | 147.97 | 141.88 |
| 33 23 2401 | Teflon Bailer, 3/4" Outside Diameter x 1', 60 cc | EACH | 179.77 | 179.77 | 179.77 | 179.77 | 179.77 |
| 33 23 2402 | Teflon Bailer, 3/4" Outside Diameter x 3', 180 cc | EACH | 227.74 | 227.74 | 227.74 | 227.74 | 227.74 |
| 33 23 2403 | Teflon Bailer, 1" Outside Diameter x 1', 80 cc | EACH | 191.28 | 191.28 | 191.28 | 191.28 | 191.28 |
| 33 23 2404 | Teflon Bailer, 1 7/8" Outside Diameter x 1', 350 cc | EACH | 181.13 | 181.13 | 181.13 | 181.13 | 181.13 |
| 33 23 2405 | Teflon Bailer, 1 7/8" Outside Diameter x 2', 700 cc | EACH | 198.20 | 198.20 | 198.20 | 198.20 | 198.20 |
| 33 23 2406 | Teflon Bailer, 1 7/8" Outside Diameter x 3', 1,050 cc | EACH | 261.53 | 261.53 | 261.53 | 261.53 | 261.53 |
| 33 23 2407 | Disposable Bailer, Polyethylene, 1.5" Outside Diameter x 36" | EACH | 27.60 | 27.60 | 27.60 | 27.60 | 27.60 |
| 33 23 2411 | Stainless Steel Bailer, 1 3/4" Outside Diameter x 2' | EACH | 124.23 | 124.23 | 124.23 | 124.23 | 124.23 |
| 33 23 2412 | Stainless Steel Bailer, 1 3/4" Outside Diameter x 3' | EACH | 27.15 | 27.15 | 27.15 | 27.15 | 27.15 |
| 33 23 2421 | Emptying Stand | EACH | 302.32 | 302.32 | 302.32 | 302.32 | 302.32 |
| 33 23 2422 | Suspension Cable, Teflon Coated | FT | 1.00 | 1.00 | 1.00 | 1.00 | 1.00 |
| 33 23 2423 | Hand Reel | EACH | 9.89 | 9.89 | 9.89 | 9.89 | 9.89 |

**Environmental Remediation: Assemblies Cost Book**

# Cost Data for Remediation
## Free Product Removal (French Drain)

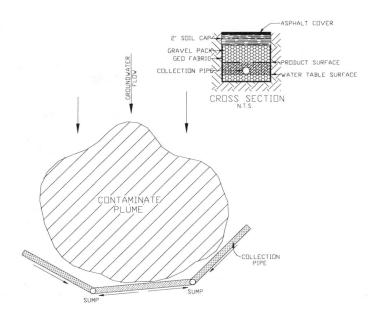

DOWN GRADIENT FREE PRODUCT FRENCH DRAIN
N.T.S.
FIGURE S.7-1

### General:

A free product removal system is used when it is necessary to remove or contain a nonaqueous phase liquid (NAPL) or free product that has accumulated on or below the water table. The basis design of the subsurface free product recovery system closely resembles the design of a French drain system.

### Common Cost Components:

1. Trench excavation
2. Perforated collection pipe
3. Gravel pack
4. Geotextile liner
5. Groundwater pump(s)
6. Free-product pump(s)
7. Concrete sumps
8. Transfer pipe
9. Backfill
10. Storage tank(s)
11. Operations and maintenance

### Typical Treatment Train:

Treatment and disposal of any excavated soils, recovered groundwater, or free product, sampling and analysis.

### Other Cost Considerations:

Slurry wall, decontamination facilities, clearing and grubbing, electrical distribution.

**Environmental Remediation: Assemblies Cost Book**

# Cost Data for Remediation

## Free Product Removal (French Drain)

| Assembly | Description | Unit | Unit Cost by Safety Level | | | | |
|----------|-------------|------|------|------|------|------|------|
| | | | A | B | C | D | E |
| | **Capital Costs** | | | | | | |
| 17 02 0201 | Demolish Bituminous Road with Power Equipment | CY | 46.17 | 36.60 | 30.96 | 21.62 | 19.19 |
| 17 02 0203 | Demolish Bituminous Pavement with Air Equipment | CY | 80.15 | 63.38 | 53.76 | 37.41 | 32.98 |
| 17 02 0205 | Demolish Unreinforced Concrete to 6" Thick with Air Equipment | CY | 126.90 | 100.35 | 85.12 | 59.24 | 52.23 |
| 17 02 0206 | Demolish Mesh Reinforced Concrete to 6" Thick with Air Equipment | CY | 183.47 | 145.09 | 123.07 | 85.65 | 75.51 |
| 17 02 0207 | Demolish Rod Reinforced Concrete to 6" Thick with Air Equipment | CY | 199.06 | 157.42 | 133.53 | 92.92 | 81.92 |
| 17 02 0208 | Demolish Mesh Reinforced Concrete to 6" Thick with Power Equipment | CY | 96.73 | 76.68 | 64.87 | 45.30 | 40.20 |
| 17 02 0209 | Demolish Rod Reinforced Concrete to 6" Thick with Power Equipment | CY | 106.91 | 84.75 | 71.70 | 50.07 | 44.43 |
| 17 02 0210 | Demolish Unreinforced Concrete 7" to 24" Thick with Power Equipment | CY | 258.44 | 204.87 | 173.31 | 121.03 | 107.40 |
| 17 02 0211 | Demolish Reinforced Concrete 7" to 24" Thick with Power Equipment | CY | 356.37 | 282.51 | 238.98 | 166.90 | 148.10 |
| 17 03 0259 | Cat 225, 1.5 CY, Soil/Sand, Trenching | CY | 1.84 | 1.49 | 1.23 | 0.88 | 0.82 |
| 17 03 0261 | Cat 225, 1.5 CY, Soil/Sand with Boulders, Trenching | CY | 3.91 | 3.15 | 2.62 | 1.87 | 1.74 |
| 17 03 0263 | Cat 235, 2.5 CY, Soil/Sand, 10' - 20' Deep Trench Box, Trench | CY | 5.70 | 4.75 | 4.07 | 3.14 | 2.98 |
| 17 03 0273 | Pull Trench Box, Cat 235, 2.5 CY, Soil/Sand, Trenching | CY | 5.31 | 4.37 | 3.68 | 2.75 | 2.59 |
| 17 03 0282 | Soil, 5 Miles, Dump Truck, Load/Haul Spoil From Trench | CY | 6.97 | 5.60 | 4.66 | 3.33 | 3.06 |
| 17 03 0283 | Sand, 5 Mile, Dump Truck, Load/Haul Spoil From Trench | CY | 6.28 | 5.05 | 4.20 | 3.00 | 2.76 |
| 17 03 0307 | Cat 235, 2.0 CY, Rock, No Haul or Borrow, Trenching | BCY | 133.78 | 109.18 | 95.00 | 71.01 | 64.55 |
| 17 03 0317 | Rock, 5 Miles, Dump Truck, Load/Haul Spoil from Trench | CY | 10.31 | 8.31 | 6.90 | 4.93 | 4.56 |
| 17 03 0401 | 950, 3.00 CY, Backfill with Excavated Material | CY | 2.54 | 2.05 | 1.70 | 1.22 | 1.14 |
| 17 03 0419 | Crushed Stone, 1/2" to 3/4" | CY | 23.45 | 21.67 | 20.70 | 18.97 | 18.47 |
| 17 03 0420 | Backfill Trench, Borrow Material, Delivered & Dumped Only | CY | 9.86 | 8.74 | 7.93 | 6.83 | 6.63 |
| 17 03 0511 | Compact Soil with Vibrating Plate | CY | 7.20 | 5.57 | 4.84 | 3.26 | 2.70 |
| 17 03 0514 | Compact Soil by Machine with Roller | CY | 1.18 | 0.93 | 0.79 | 0.55 | 0.47 |
| 17 03 0515 | Compact With Pogosticks | CY | 5.91 | 4.58 | 3.98 | 2.68 | 2.23 |
| 17 03 0516 | Compact with 50% Pogosticks, 50% Hand Roller | CY | 4.24 | 3.30 | 2.85 | 1.94 | 1.63 |
| 17 03 0901 | Steel Sheeting, Pull & Salvage, to 15' | SF | 13.65 | 11.21 | 9.59 | 7.20 | 6.70 |
| 17 03 0902 | Steel Sheeting, Pull & Salvage, to 20' | SF | 14.11 | 11.61 | 9.95 | 7.50 | 6.99 |
| 17 03 0903 | Steel Sheeting, Pull & Salvage, to 25' | SF | 12.32 | 10.20 | 8.79 | 6.72 | 6.28 |
| 17 03 0912 | Cat 235, 2.5 CY, Soil/Sand, < 15' Trench in Sheeting | CY | 4.02 | 3.25 | 2.69 | 1.93 | 1.80 |
| 17 03 0913 | Cat 235, 2.5 CY, Soil/Sand, 15' - 25' Trench in Sheeting | CY | 6.03 | 4.87 | 4.04 | 2.90 | 2.70 |

## Environmental Remediation: Assemblies Cost Book

*Page 3-128*

# Cost Data for Remediation

## Free Product Removal (French Drain)

| Assembly | Description | Unit | \[Unit Cost by Safety Level\] A | B | C | D | E |
|---|---|---|---|---|---|---|---|
| | **Capital Costs** | | | | | | |
| 17 03 1001 | Wellpoint for Trench, Install & Remove <500', 1 Month | LFHD | 55.06 | 55.06 | 55.06 | 55.06 | 55.06 |
| 18 02 0301 | Asphalt Pavement - 10" Subgrade, 9" Base, 1 1/2" Topping | SY | 61.11 | 50.94 | 43.58 | 33.57 | 31.83 |
| 18 02 0321 | 6" Structural Slab on Grade | SF | 6.98 | 5.95 | 5.47 | 4.48 | 4.14 |
| 18 02 0322 | 8" Structural Slab on Grade | SF | 9.75 | 8.38 | 7.71 | 6.38 | 5.95 |
| 18 02 0324 | 12" Structural Slab on Grade | SF | 11.57 | 10.04 | 9.30 | 7.82 | 7.34 |
| 18 02 0331 | 6" Mesh Reinforced Slab on Grade | SF | 6.01 | 5.10 | 4.67 | 3.79 | 3.49 |
| 18 02 0341 | 6" Unreinforced Slab on Grade | SF | 4.87 | 4.14 | 3.79 | 3.08 | 2.85 |
| 18 04 0205 | CIP Walls Form & Strip (4 Uses) | SF | 8.86 | 6.99 | 6.19 | 4.38 | 3.73 |
| 19 01 0202 | 1", Class 200, PVC Piping | LF | 7.48 | 5.84 | 5.07 | 3.48 | 2.93 |
| 19 01 0204 | 2", Class 200, PVC Piping | LF | 8.86 | 6.96 | 6.07 | 4.23 | 3.60 |
| 19 01 0206 | 3", Class 200, PVC Piping | LF | 11.81 | 9.34 | 8.20 | 5.80 | 4.98 |
| 19 01 0207 | 4", Class 200, PVC Piping | LF | 12.33 | 9.86 | 8.72 | 6.32 | 5.50 |
| 19 01 0208 | 6", Class 200, PVC Piping | LF | 14.95 | 12.20 | 10.93 | 8.27 | 7.36 |
| 19 01 0212 | 8", Class 150, PVC Piping | LF | 18.36 | 15.27 | 13.84 | 10.85 | 9.82 |
| 19 02 0310 | 24' x 36" Diameter Reinforced Concrete Pipe Wet Well | EACH | 20,809 | 18,008 | 16,296 | 13,559 | 12,884 |
| 19 02 0311 | 12' x 36" Diameter Reinforced Concrete Pipe Wet Well for Lift Station | EACH | 9,840 | 8,577 | 7,816 | 6,581 | 6,270 |
| 19 02 0312 | 24' x 60" Diameter Reinforced Concrete Pipe Wet Well for Lift Station | EACH | 39,142 | 33,897 | 30,681 | 25,554 | 24,296 |
| 19 04 0621 | 550 Gallon Horizontal Plastic Sump with 4" NPT Connection | EACH | 1,699 | 1,596 | 1,551 | 1,452 | 1,415 |
| 19 04 0622 | 1,000 Gallon Horizontal Plastic Sump with 4" NPT Connection | EACH | 2,179 | 2,042 | 1,984 | 1,852 | 1,803 |
| 19 04 0623 | 2,000 Gallon Horizontal Plastic Sump with 6" NPT Connection | EACH | 3,089 | 2,918 | 2,845 | 2,680 | 2,619 |
| 19 04 0624 | 4,000 Gallon Horizontal Plastic Sump with 6" NPT Connection | EACH | 4,130 | 3,927 | 3,841 | 3,645 | 3,573 |
| 19 04 0625 | 6,000 Gallon Horizontal Plastic Sump with 6" NPT Connection | EACH | 5,248 | 4,975 | 4,858 | 4,594 | 4,497 |
| 19 04 0626 | 8,000 Gallon Horizontal Plastic Sump with 6" NPT Connection | EACH | 5,857 | 5,516 | 5,369 | 5,039 | 4,918 |
| 19 04 0627 | 10,000 Gallon Horizontal Plastic Sump with 6" NPT Connection | EACH | 7,082 | 6,622 | 6,402 | 5,956 | 5,807 |
| 19 04 0628 | 12,000 Gallon Horizontal Plastic Sump with 6" NPT Connection | EACH | 8,308 | 7,722 | 7,442 | 6,875 | 6,686 |
| 19 04 0629 | 15,000 Gallon Horizontal Plastic Sump with 6" NPT Connection | EACH | 14,184 | 13,540 | 13,232 | 12,608 | 12,400 |
| 19 04 0631 | 20,000 Gallon Horizontal Plastic Sump with 6" NPT Connection | EACH | 15,799 | 15,084 | 14,742 | 14,048 | 13,817 |

**Environmental Remediation: Assemblies Cost Book**

*Page 3-129*

# Cost Data for Remediation

## Free Product Removal (French Drain)

### Unit Cost by Safety Level

| Assembly | Description | Unit | A | B | C | D | E |
|---|---|---|---|---|---|---|---|
| | **Capital Costs** | | | | | | |
| 19 04 0632 | 25,000 Gallon Horizontal Plastic Sump with 6" NPT Connection | EACH | 24,407 | 23,487 | 23,047 | 22,155 | 21,858 |
| 19 04 0633 | 30,000 Gallon Horizontal Plastic Sump with 6" NPT Connection | EACH | 35,112 | 34,039 | 33,525 | 32,485 | 32,138 |
| 19 07 0120 | 1 1/4", 60 PSI, Polyethylene Pipe | LF | 5.29 | 4.21 | 3.72 | 2.68 | 2.32 |
| 19 07 0121 | 1 1/2", 60 PSI, Polyethylene Pipe | LF | 5.38 | 4.30 | 3.81 | 2.77 | 2.41 |
| 19 07 0122 | 2", 60 PSI, Polyethylene Pipe | LF | 6.05 | 4.85 | 4.31 | 3.15 | 2.75 |
| 19 07 0123 | 3" Polyethylene 40' Joints, 60 PSI | LF | 7.90 | 6.47 | 5.82 | 4.42 | 3.94 |
| 19 07 0124 | 4" Polyethylene 40' Joints, 60 PSI | LF | 19.10 | 15.75 | 13.90 | 10.64 | 9.71 |
| 19 07 0125 | 6" Polyethylene 40' Joints, 60 PSI | LF | 25.85 | 22.22 | 20.22 | 16.68 | 15.67 |
| 19 07 0126 | 8" Polyethylene 40' Joints, 60 PSI | LF | 35.37 | 31.01 | 28.61 | 24.37 | 23.16 |
| 33 02 0561 | Nylon Tubing, 1/4" Outside Diameter | LF | 0.47 | 0.47 | 0.47 | 0.47 | 0.47 |
| 33 02 0562 | Teflon Tubing, 1/4" Outside Diameter | LF | 1.21 | 1.21 | 1.21 | 1.21 | 1.21 |
| 33 02 1720 | Volatile Organic Analysis (SW 5030/SW 8240) | EACH | 175.00 | 175.00 | 175.00 | 175.00 | 175.00 |
| 33 08 0511 | Drainage Netting, 1/4" Thick High-density Polyethylene | SF | 0.23 | 0.21 | 0.20 | 0.19 | 0.18 |
| 33 08 0512 | Drainage Netting, Geotextile Fabric Heat-bonded 1 Side | SF | 0.41 | 0.39 | 0.38 | 0.36 | 0.35 |
| 33 08 0513 | Drainage Netting, Geotextile Fabric Heat-bonded 2 Sides | SF | 0.48 | 0.45 | 0.44 | 0.42 | 0.41 |
| 33 08 0521 | Geogrid, Biaxial 9.8' x 164' Roll, TM = 14,000 PSF | SF | 0.23 | 0.21 | 0.21 | 0.19 | 0.19 |
| 33 08 0522 | Geogrid, Biaxial 9.8' x 164' Roll, TM = 18,500 PSF | SF | 0.37 | 0.36 | 0.35 | 0.33 | 0.33 |
| 33 08 0523 | Geogrid, Uniaxial 4.3' x 98' Roll, TM = 50,000 PSF | SF | 0.61 | 0.59 | 0.59 | 0.57 | 0.56 |
| 33 08 0524 | Geogrid, Uniaxial 4.3' x 98' Roll, TM = 130,000 PSF | SF | 1.21 | 1.18 | 1.17 | 1.15 | 1.14 |
| 33 08 0532 | 80 Mil Geotextile, Nonwoven | SY | 1.57 | 1.35 | 1.25 | 1.04 | 0.97 |
| 33 08 0534 | 130 Mil Geotextile, Nonwoven | SY | 2.56 | 2.20 | 2.03 | 1.68 | 1.56 |
| 33 08 0535 | 170 Mil Geotextile, Nonwoven | SY | 2.96 | 2.60 | 2.43 | 2.08 | 1.96 |
| 33 09 0701 | CIP Reinforced Concrete Containment Wall | CY | 279.72 | 243.62 | 224.28 | 189.16 | 178.78 |
| 33 10 9656 | 550 Gallon Single-wall Steel Aboveground Tank | EACH | 1,598 | 1,390 | 1,301 | 1,099 | 1,026 |
| 33 10 9657 | 1,000 Gallon Single-wall Steel Aboveground Tank | EACH | 2,118 | 1,844 | 1,727 | 1,463 | 1,366 |
| 33 10 9658 | 2,000 Gallon Single-wall Steel Aboveground Tank | EACH | 3,457 | 2,997 | 2,777 | 2,332 | 2,183 |
| 33 10 9659 | 3,000 Gallon Single-wall Steel Aboveground Tank | EACH | 4,433 | 3,959 | 3,733 | 3,274 | 3,121 |
| 33 10 9660 | 5,000 Gallon Single-wall Steel Aboveground Tank | EACH | 6,079 | 5,494 | 5,214 | 4,646 | 4,457 |
| 33 10 9661 | 8,000 Gallon Single-wall Steel Aboveground Tank | EACH | 7,876 | 7,205 | 6,884 | 6,234 | 6,017 |
| 33 10 9662 | 10,000 Gallon Single-wall Steel Aboveground Tank | EACH | 9,147 | 8,415 | 8,065 | 7,356 | 7,119 |
| 33 10 9663 | 12,000 Gallon Single-wall Steel Aboveground Tank | EACH | 10,460 | 9,655 | 9,270 | 8,490 | 8,230 |
| 33 10 9664 | 15,000 Gallon Single-wall Steel Aboveground Tank | EACH | 11,591 | 10,697 | 10,269 | 9,402 | 9,113 |

## Environmental Remediation: Assemblies Cost Book

*Page 3-130*

# Cost Data for Remediation

## Free Product Removal (French Drain)

| Assembly | Description | Unit | Unit Cost by Safety Level A | B | C | D | E |
|---|---|---|---|---|---|---|---|
| | **Capital Costs** | | | | | | |
| 33 10 9665 | 20,000 Gallon Single-wall Steel Aboveground Tank | EACH | 14,594 | 13,445 | 12,895 | 11,780 | 11,408 |
| 33 10 9666 | 30,000 Gallon Single-wall Steel Aboveground Tank | EACH | 6,427 | 5,085 | 4,444 | 3,143 | 2,709 |
| 33 23 0201 | 2" PVC, Schedule 40, Well Screen | LF | 23.47 | 19.52 | 16.42 | 12.52 | 12.00 |
| 33 23 0202 | 4" PVC, Schedule 40, Well Screen | LF | 38.16 | 32.00 | 27.16 | 21.09 | 20.27 |
| 33 23 0203 | 6" PVC, Schedule 40, Well Screen | LF | 47.13 | 39.78 | 34.01 | 26.75 | 25.78 |
| 33 23 0204 | 8" PVC, Schedule 40, Well Screen | LF | 62.02 | 52.84 | 45.61 | 36.54 | 35.33 |
| 33 23 0205 | 10" PVC, Schedule 40, Well Screen | LF | 67.86 | 58.67 | 51.45 | 42.38 | 41.17 |
| 33 23 0211 | 2" PVC, Schedule 80, Well Screen | LF | 18.70 | 15.64 | 13.23 | 10.20 | 9.80 |
| 33 23 0212 | 4" PVC, Schedule 80, Well Screen | LF | 30.01 | 25.41 | 21.80 | 17.27 | 16.66 |
| 33 23 0213 | 6" PVC, Schedule 80, Well Screen | LF | 49.72 | 42.37 | 36.60 | 29.34 | 28.37 |
| 33 23 0214 | 8" PVC, Schedule 80, Well Screen | LF | 65.91 | 56.73 | 49.50 | 40.43 | 39.22 |
| 33 23 0215 | 10" PVC, Schedule 80, Well Screen | LF | 73.71 | 64.53 | 57.30 | 48.23 | 47.02 |
| 33 23 0221 | 2" Stainless Steel, Well Screen | LF | 26.91 | 23.84 | 21.44 | 18.41 | 18.01 |
| 33 23 0222 | 4" Stainless Steel, Well Screen | LF | 47.66 | 43.06 | 39.45 | 34.92 | 34.31 |
| 33 23 0224 | 6" Stainless Steel Well Screen, 10' Sec, Flush Threaded | LF | 355.96 | 297.19 | 251.00 | 192.97 | 185.17 |
| 33 23 0226 | 8" Stainless Steel Well Screen, 10' Sec, Flush Threaded | LF | 449.32 | 375.84 | 318.08 | 245.51 | 235.77 |
| 33 23 0228 | 10" Stainless Steel Well Screen, 10' Sec, Flush Threaded | LF | 475.48 | 402.00 | 344.24 | 271.67 | 261.93 |
| 33 23 0301 | 2" PVC, Well Plug | EACH | 29.37 | 24.77 | 21.16 | 16.63 | 16.02 |
| 33 23 0302 | 4" PVC, Well Plug | EACH | 55.65 | 48.91 | 43.62 | 36.97 | 36.07 |
| 33 23 0303 | 6" PVC, Well Plug | EACH | 112.63 | 101.15 | 92.13 | 80.79 | 79.27 |
| 33 23 0304 | 8" PVC, Well Plug | EACH | 132.70 | 119.58 | 109.26 | 96.31 | 94.57 |
| 33 23 0305 | 10" PVC, Well Plug | EACH | 156.31 | 141.01 | 128.97 | 113.85 | 111.82 |
| 33 23 0311 | 2" Stainless Steel, Well Plug | EACH | 83.24 | 74.06 | 66.83 | 57.76 | 56.55 |
| 33 23 0312 | 4" Stainless Steel, Well Plug | EACH | 112.24 | 102.03 | 94.01 | 83.93 | 82.58 |
| 33 23 0313 | 6" Stainless Steel, Well Plug | EACH | 573.04 | 481.18 | 408.98 | 318.27 | 306.09 |
| 33 23 0314 | 8" Stainless Steel, Well Plug | EACH | 702.53 | 597.56 | 515.04 | 411.38 | 397.46 |
| 33 23 0315 | 10" Stainless Steel, Well Plug | EACH | 891.17 | 768.70 | 672.43 | 551.48 | 535.25 |
| 33 23 0501 | 4" Submersible Pump, 4 - 13 GPM, < 180' Head | EACH | 1,976 | 1,721 | 1,588 | 1,340 | 1,264 |
| 33 23 0502 | 4" Submersible Pump, 13 - 26 GPM, < 180' Head | EACH | 3,039 | 2,745 | 2,592 | 2,307 | 2,219 |
| 33 23 0503 | 4" Submersible Pump, 13 - 37 GPM, < 180' Head | EACH | 3,084 | 2,790 | 2,637 | 2,352 | 2,264 |
| 33 23 0504 | 6" Submersible Pump, 50 - 125 GPM, < 150' Head | EACH | 3,041 | 2,621 | 2,403 | 1,995 | 1,870 |
| 33 23 0505 | 6" Submersible Pump, 15 - 135 GPM, 200' < Head < 500' | EACH | 4,472 | 3,885 | 3,579 | 3,008 | 2,834 |
| 33 23 0601 | Groundwater Pump, 1/3 HP, 230V, Controls, Probe | EACH | 2,583 | 2,379 | 2,292 | 2,095 | 2,023 |

**Environmental Remediation: Assemblies Cost Book**

*Page 3-131*

1998 by ECHOS. All rights reserved

# Cost Data for Remediation

## Free Product Removal (French Drain)

| Assembly | Description | Unit | Unit Cost by Safety Level | | | | |
|---|---|---|---|---|---|---|---|
| | | | A | B | C | D | E |
| | **Capital Costs** | | | | | | |
| 33 23 0602 | Groundwater Pump, 3/4 HP, 230V, Controls, Probe | EACH | 2,747 | 2,543 | 2,456 | 2,259 | 2,187 |
| 33 23 0603 | Groundwater Pump, 1 HP, 230V, Controls, Probe | EACH | 2,881 | 2,678 | 2,591 | 2,394 | 2,322 |
| 33 23 0604 | Groundwater Pump, 1 1/2 HP, Controls, Probe | EACH | 3,095 | 2,891 | 2,804 | 2,607 | 2,535 |
| 33 23 0605 | Groundwater Pump, 2 HP, 230V, Controls, Probe | EACH | 3,674 | 3,471 | 3,383 | 3,187 | 3,114 |
| 33 23 0606 | Groundwater Pump, 5 HP, 230V, Controls, Probe | EACH | 4,523 | 4,320 | 4,233 | 4,036 | 3,964 |
| 33 23 0701 | Groundwater Pump, 1 1/2 HP, 230V, without Controls | EACH | 2,270 | 2,067 | 1,980 | 1,783 | 1,711 |
| 33 23 0702 | Groundwater Pump, 5 HP, 230V, without Controls | EACH | 3,487 | 3,284 | 3,196 | 3,000 | 2,927 |
| 33 23 0801 | Product Recovery Pump, 1/3 HP, 230V, Controls, Probe | EACH | 8,731 | 8,527 | 8,440 | 8,243 | 8,171 |
| 33 23 0802 | Product Recovery Pump, 1/3 HP, 115V, Controls, Probe | EACH | 9,109 | 8,905 | 8,818 | 8,621 | 8,549 |
| 33 23 0803 | Product Recovery Pump, 1/3 HP, 230V, Controls, Probe, Highly Corrosive Liquid | EACH | 8,730 | 8,526 | 8,439 | 8,242 | 8,170 |
| 33 23 0804 | Product Recovery Pump, 2" Oil Skimmer, 360 GPD, Controls, Pneumatic | EACH | 3,698 | 3,495 | 3,407 | 3,211 | 3,138 |
| 33 23 0805 | Product Recovery Pump, 2" Oil Skimmer, 700 GPD, Controls, Pneumatic | EACH | 3,954 | 3,750 | 3,663 | 3,466 | 3,394 |
| 33 23 0806 | Product Recovery Pump, 4" Oil Skimmer, 2,160 GPD, Controls, Pneumatic | EACH | 4,209 | 4,005 | 3,918 | 3,721 | 3,649 |
| 33 23 0807 | Product Recovery Pump, Shallow Depths, 6 GPM, Controls, Pneumatic | EACH | 3,826 | 3,622 | 3,535 | 3,338 | 3,266 |
| 33 23 0808 | Product Recovery Pump, Deep Depths, 6 GPM, Controls, Pneumatic | EACH | 4,018 | 3,814 | 3,727 | 3,530 | 3,458 |
| 33 23 0809 | Product Recovery Pump, 1/2 HP, Controls, Probe | EACH | 4,241 | 4,037 | 3,950 | 3,753 | 3,681 |
| 33 23 0811 | Explosionproof Electrical Receptacle for Product Recovery Pump | EACH | 497.01 | 468.06 | 455.66 | 427.68 | 417.42 |
| 33 23 0901 | Product Recovery Pump, 1/3 HP, without Controls, Includes Probe | EACH | 2,237 | 2,033 | 1,946 | 1,749 | 1,677 |
| 33 23 0902 | Product Recovery Pump, 3/4 HP, without Controls | EACH | 1,909 | 1,706 | 1,619 | 1,422 | 1,350 |
| 33 23 1301 | 4 Product Recovery/4 Groundwater Pump Control Panel | EACH | 4,599 | 4,395 | 4,308 | 4,111 | 4,039 |
| 33 23 1302 | 3 Product Recovery/3 Groundwater Pump Control Panel | EACH | 5,839 | 5,635 | 5,548 | 5,351 | 5,279 |
| 33 23 1303 | 1 Product Recovery/4 Groundwater Pump Control Panel | EACH | 5,500 | 5,296 | 5,209 | 5,012 | 4,940 |
| 33 23 1304 | 2 Product Recovery/2 Groundwater Pump Control Panel | EACH | 4,842 | 4,639 | 4,552 | 4,355 | 4,283 |
| 33 23 1305 | Water Level Sensor, Float Switch, with 50' Cable | EACH | 549.15 | 549.15 | 549.15 | 549.15 | 549.15 |
| 33 23 1306 | High Sump Level Switch for Avoiding Overflow | EACH | 214.65 | 214.65 | 214.65 | 214.65 | 214.65 |
| 33 23 2204 | Hazardous Area, Pedestrian Load, Well Protection | EACH | 6,261 | 5,266 | 4,505 | 3,524 | 3,380 |
| 33 23 2205 | Hazardous Area, Traffic Load, Well Protection | EACH | 6,266 | 5,271 | 4,510 | 3,529 | 3,384 |
| 33 23 2206 | Restricted Area, Well Protection (with 4 Posts & Explosionproof Receptacle) | EACH | 1,264 | 1,133 | 1,077 | 949.88 | 903.62 |

**Environmental Remediation: Assemblies Cost Book**

# Cost Data for Remediation

## Free Product Removal (French Drain)

| Assembly | Description | Unit | A | B | C | D | E |
|---|---|---|---|---|---|---|---|
| | **Capital Costs** | | | | | | |
| 33 26 0101 | 1" Carbon Steel Piping | LF | 6.46 | 5.28 | 4.77 | 3.63 | 3.21 |
| 33 26 0102 | 2" Carbon Steel Piping | LF | 10.50 | 8.72 | 7.95 | 6.23 | 5.60 |
| 33 26 0103 | 3" Carbon Steel Piping | LF | 18.02 | 15.20 | 13.99 | 11.26 | 10.26 |
| 33 26 0104 | 4" Carbon Steel Piping | LF | 23.20 | 19.86 | 18.43 | 15.21 | 14.02 |
| 33 26 0105 | 6" Carbon Steel Piping | LF | 36.37 | 30.24 | 27.46 | 21.52 | 19.44 |
| 33 26 0106 | 8" Carbon Steel Piping | LF | 40.77 | 34.08 | 31.15 | 24.68 | 22.36 |
| 33 26 0107 | 10" Carbon Steel Piping | LF | 54.40 | 46.05 | 42.38 | 34.30 | 31.39 |
| 33 26 0201 | 1" Stainless Steel Piping, Schedule 40, Threaded | LF | 18.67 | 16.12 | 15.03 | 12.56 | 11.65 |
| 33 26 0202 | 2" Stainless Steel Piping, Schedule 40, Threaded | LF | 33.47 | 29.21 | 27.39 | 23.28 | 21.77 |
| 33 26 0203 | 3" Stainless Steel Piping, Schedule 40, Threaded | LF | 57.00 | 50.61 | 47.88 | 41.71 | 39.45 |
| 33 26 0204 | 4" Stainless Steel Piping, Schedule 40, Threaded | LF | 80.15 | 71.72 | 68.12 | 59.98 | 57.00 |
| 33 26 0205 | 6" Stainless Steel Piping, Schedule 10, Type 316 | LF | 78.58 | 70.01 | 66.14 | 57.84 | 54.94 |
| 33 26 0206 | 8" Stainless Steel Piping, Schedule 40, Welded | LF | 197.22 | 183.32 | 176.89 | 163.42 | 158.80 |
| 33 26 0601 | (1 1/2", 3") Stainless Steel Double-wall Piping, with Fittings | LF | 134.61 | 117.05 | 109.53 | 92.56 | 86.33 |
| 33 26 0602 | (2", 4") Stainless Steel Double-wall Piping, with Fittings | LF | 185.98 | 163.46 | 153.81 | 132.04 | 124.05 |
| 33 26 0603 | (2 1/2", 4") Stainless Steel Double-wall Piping, with Fittings | LF | 206.86 | 182.17 | 171.60 | 147.74 | 138.99 |
| 33 26 0604 | (4", 6") Stainless Steel Double-wall Piping, with Fittings | LF | 339.64 | 280.79 | 254.18 | 197.21 | 177.23 |
| 33 26 0621 | (1 1/2", 3") PVC Double-wall Piping, with Fittings | LF | 33.15 | 27.05 | 24.43 | 18.53 | 16.36 |
| 33 26 0622 | (2", 4") PVC Double-wall Piping, with Fittings | LF | 46.22 | 37.98 | 34.43 | 26.46 | 23.55 |
| 33 26 0623 | (2 1/2", 4") PVC Double-wall Piping, with Fittings | LF | 49.22 | 40.39 | 36.61 | 28.07 | 24.94 |
| 33 26 0624 | (4", 6") PVC Double-wall Piping, with Fittings | LF | 70.61 | 58.00 | 52.59 | 40.40 | 35.93 |
| 33 26 0641 | (1 1/2", 3") Carbon Steel Double-wall Piping, with Fittings | LF | 48.16 | 40.64 | 37.42 | 30.15 | 27.48 |
| 33 26 0642 | (2", 4") Carbon Steel Double-wall Piping, with Fittings | LF | 64.71 | 55.69 | 51.83 | 43.11 | 39.91 |
| 33 26 0643 | (2 1/2", 4") Carbon Steel Double-wall Piping, with Fittings | LF | 70.81 | 61.02 | 56.83 | 47.37 | 43.90 |
| 33 26 0644 | (4", 6") Carbon Steel Double-wall Piping, with Fittings | LF | 114.39 | 100.33 | 94.32 | 80.73 | 75.75 |
| 33 26 0901 | 4" Diameter Perforated PVC Pipe | LF | 13.24 | 10.73 | 9.22 | 6.77 | 6.15 |
| 33 26 0902 | 6" Diameter Perforated PVC Pipe | LF | 14.13 | 11.62 | 10.11 | 7.66 | 7.04 |
| 33 26 0903 | 8" Diameter Perforated PVC Pipe | LF | 14.56 | 12.01 | 10.49 | 8.00 | 7.37 |
| 33 26 0904 | 10" Diameter Perforated PVC Pipe | LF | 16.84 | 14.07 | 12.42 | 9.71 | 9.02 |
| 33 26 0905 | 12" Diameter Perforated PVC Pipe | LF | 18.91 | 15.99 | 14.24 | 11.38 | 10.66 |
| 33 29 0101 | 10 GPM, 10' Head, 1/6 HP, Centrifugal Pump | EACH | 789.18 | 730.44 | 705.29 | 648.51 | 627.69 |
| 33 29 0102 | 10 GPM, 1/2 HP, Centrifugal Pump | EACH | 1,063 | 952.32 | 905.07 | 798.39 | 759.27 |
| 33 29 0103 | 50 GPM, 100' Head, 3 HP, Centrifugal Pump | EACH | 2,658 | 2,500 | 2,432 | 2,279 | 2,223 |

**Environmental Remediation: Assemblies Cost Book**

# Cost Data for Remediation

## Free Product Removal (French Drain)

| Assembly | Description | Unit | Unit Cost by Safety Level A | B | C | D | E |
|---|---|---|---|---|---|---|---|
| | **Capital Costs** | | | | | | |
| 33 29 0104 | 75 GPM, 100' Head, 5 HP, Centrifugal Pump | EACH | 2,946 | 2,754 | 2,672 | 2,487 | 2,419 |
| 33 29 0106 | 100 GPM, 150' Head, 7.5 HP, Centrifugal Pump | EACH | 3,419 | 3,191 | 3,093 | 2,873 | 2,793 |
| 33 29 0107 | 125 GPM, 150' Head, 10 HP, Centrifugal Pump | EACH | 3,815 | 3,512 | 3,382 | 3,088 | 2,981 |
| 33 29 0108 | 10 HP, 200 GPM, Centrifugal Pump | EACH | 2,941 | 2,714 | 2,616 | 2,396 | 2,316 |
| 33 29 0109 | 10 HP, 250 GPM, Centrifugal Pump | EACH | 3,032 | 2,805 | 2,707 | 2,487 | 2,407 |
| 33 29 0110 | 15 HP, 300 GPM, Centrifugal Pump | EACH | 3,441 | 3,160 | 3,040 | 2,770 | 2,670 |
| 33 29 0111 | 20 HP, 500 GPM, Centrifugal Pump | EACH | 4,544 | 4,241 | 4,111 | 3,818 | 3,710 |
| 33 29 0301 | 10 GPM, 20 PSI, 1/2 HP, Iron Rotary Pump | EACH | 1,573 | 1,415 | 1,347 | 1,194 | 1,138 |
| 33 29 0302 | 10 GPM, 40 PSI, 3/4 HP, Iron Rotary Pump | EACH | 1,818 | 1,627 | 1,545 | 1,359 | 1,291 |
| 33 29 0303 | 10 GPM, 60 PSI, 1 HP, Iron Rotary Pump | EACH | 1,975 | 1,748 | 1,650 | 1,430 | 1,350 |
| 33 29 0304 | 25 GPM, 20 PSI, 1 HP, Iron Rotary Pump | EACH | 1,690 | 1,532 | 1,464 | 1,311 | 1,255 |
| 33 29 0305 | 25 GPM, 40 PSI, 1 1/2 HP, Iron Rotary Pump | EACH | 1,835 | 1,644 | 1,562 | 1,376 | 1,308 |
| 33 29 0306 | 25 GPM, 60 PSI, 2 HP, Iron Rotary Pump | EACH | 2,193 | 1,966 | 1,868 | 1,648 | 1,568 |
| 33 31 0202 | 6 HP, 80 Gallon, 200 PSI, 28 SCFM, Air Compressor | EACH | 8,102 | 7,713 | 7,547 | 7,171 | 7,033 |
| 33 41 0101 | Pump & Motor Maintenance/Repair | EACH | 865.17 | 671.48 | 581.58 | 393.96 | 329.64 |
| 33 42 0101 | Electrical Charge | KWH | 0.06 | 0.06 | 0.06 | 0.06 | 0.06 |
| | **Operations & Maintenance** | | | | | | |
| 33 01 0102 | Van or Pickup Rental | DAY | 32.69 | 32.69 | 32.69 | 32.69 | 32.69 |
| 33 01 0201 | Mobilize Crew, >= 500 Miles, per Person | EACH | 357.75 | 357.75 | 357.75 | 357.75 | 357.75 |
| 33 01 0202 | Per Diem | DAY | 104.94 | 104.94 | 104.94 | 104.94 | 104.94 |
| 33 01 0203 | Mobilize Crew, 250 Miles, per Person | EACH | 178.88 | 178.88 | 178.88 | 178.88 | 178.88 |
| 33 01 0204 | Mobilize Crew, 100 Miles, per Person | EACH | 71.55 | 71.55 | 71.55 | 71.55 | 71.55 |
| 33 01 0205 | Mobilize Crew, 50 Miles, per Person | EACH | 53.66 | 53.66 | 53.66 | 53.66 | 53.66 |
| 33 01 0206 | Mobilize Crew, Local, per Person | EACH | 17.89 | 17.89 | 17.89 | 17.89 | 17.89 |
| 33 02 0303 | Organic Vapor Analyzer Rental, per Day | DAY | 184.30 | 184.30 | 184.30 | 184.30 | 184.30 |
| 33 02 0331 | Portable Organic Vapor Analyzer, Monthly Rental | MONTH | 1,920 | 1,920 | 1,920 | 1,920 | 1,920 |
| 33 02 0533 | Water Level Indicator, 100' Tape, Electric, Light & Horn | EACH | 461.65 | 461.65 | 461.65 | 461.65 | 461.65 |
| 33 19 9921 | DOT Steel Drum, 55 Gallon | EACH | 65.19 | 65.19 | 65.19 | 65.19 | 65.19 |
| 33 23 2401 | Teflon Bailer, 3/4" Outside Diameter x 1', 60 cc | EACH | 179.77 | 179.77 | 179.77 | 179.77 | 179.77 |
| 33 23 2402 | Teflon Bailer, 3/4" Outside Diameter x 3', 180 cc | EACH | 227.74 | 227.74 | 227.74 | 227.74 | 227.74 |
| 33 23 2403 | Teflon Bailer, 1" Outside Diameter x 1', 80 cc | EACH | 191.28 | 191.28 | 191.28 | 191.28 | 191.28 |
| 33 23 2404 | Teflon Bailer, 1 7/8" Outside Diameter x 1', 350 cc | EACH | 181.13 | 181.13 | 181.13 | 181.13 | 181.13 |
| 33 23 2405 | Teflon Bailer, 1 7/8" Outside Diameter x 2', 700 cc | EACH | 198.20 | 198.20 | 198.20 | 198.20 | 198.20 |

**Environmental Remediation: Assemblies Cost Book**

# Cost Data for Remediation

## Free Product Removal (French Drain)

| Assembly | Description | Unit | Unit Cost by Safety Level A | B | C | D | E |
|---|---|---|---|---|---|---|---|
| | **Operations & Maintenance** | | | | | | |
| 33 23 2406 | Teflon Bailer, 1 7/8" Outside Diameter x 3', 1,050 cc | EACH | 261.53 | 261.53 | 261.53 | 261.53 | 261.53 |
| 33 23 2407 | Disposable Bailer, Polyethylene, 1.5" Outside Diameter x 36" | EACH | 27.60 | 27.60 | 27.60 | 27.60 | 27.60 |
| 33 23 2411 | Stainless Steel Bailer, 1 3/4" Outside Diameter x 2' | EACH | 124.23 | 124.23 | 124.23 | 124.23 | 124.23 |
| 33 23 2412 | Stainless Steel Bailer, 1 3/4" Outside Diameter x 3' | EACH | 27.15 | 27.15 | 27.15 | 27.15 | 27.15 |
| 33 23 2422 | Suspension Cable, Teflon Coated | FT | 1.00 | 1.00 | 1.00 | 1.00 | 1.00 |
| 33 41 0101 | Pump & Motor Maintenance/Repair | EACH | 865.17 | 671.48 | 581.58 | 393.96 | 329.64 |
| 33 42 0101 | Electrical Charge | KWH | 0.06 | 0.06 | 0.06 | 0.06 | 0.06 |

**Environmental Remediation: Assemblies Cost Book**

# Cost Data for Remediation
## Groundwater Monitoring Wells

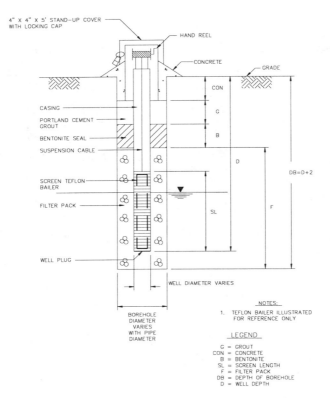

### General:

*The design of a monitoring well depends largely on the hydrogeological environment, chemical nature of the contaminants, and whether the well bore will be used to conduct geologic investigations. Typically, a series of wells will be installed to obtain samples from several different depths within the aquifer.*

### Common Cost Components:

1. Mobilization of Drill Rig and Crew
2. Drilling
3. Well Casing, Screen, Plug
4. Well Cover
5. Slug Testing
6. Well Development
7. Sample Collection During Borehole Advancement
8. Drumming of Drill Cuttings and Development Water
9. Professional Labor Hours (Drilling Supervision, Well Development, Slug Tests)
10. Decontamination

### Typical Treatment Train:

*Sampling and analysis of groundwater, natural attenuation, on-going remediation, compliance monitoring, and post-closure.*

### Other Cost Considerations:

N/A

**Environmental Remediation: Assemblies Cost Book**

# Cost Data for Remediation

## Groundwater Monitoring Wells

| Assembly | Description | Unit | Unit Cost by Safety Level | | | | |
|---|---|---|---|---|---|---|---|
| | | | A | B | C | D | E |
| | **Capital Costs** | | | | | | |
| 33 01 0101 | Mobilize/DeMobilize Drilling Rig & Crew | LS | 4,949 | 4,031 | 3,308 | 2,401 | 2,280 |
| 33 02 0303 | Organic Vapor Analyzer Rental, per Day | DAY | 184.30 | 184.30 | 184.30 | 184.30 | 184.30 |
| 33 02 0405 | Monitoring Well Slug Testing Equipment Rental | WEEK | 552.08 | 552.08 | 552.08 | 552.08 | 552.08 |
| 33 17 0808 | Decontaminate Rig, Augers, Screen (Rental Equipment) | DAY | 205.34 | 205.34 | 205.34 | 205.34 | 205.34 |
| 33 23 0101 | 2" PVC, Schedule 40, Well Casing | LF | 17.45 | 14.39 | 11.98 | 8.95 | 8.55 |
| 33 23 0102 | 4" PVC, Schedule 40, Well Casing | LF | 26.98 | 22.38 | 18.77 | 14.24 | 13.63 |
| 33 23 0103 | 6" PVC, Schedule 40, Well Casing | LF | 27.07 | 22.66 | 19.19 | 14.84 | 14.25 |
| 33 23 0104 | 8" PVC, Schedule 40, Well Casing | LF | 34.50 | 29.27 | 25.15 | 19.98 | 19.29 |
| 33 23 0105 | 10" PVC, Schedule 40, Well Casing | LF | 38.28 | 33.04 | 28.92 | 23.75 | 23.06 |
| 33 23 0106 | 12" PVC, Schedule 40, Well Casing | LF | 44.32 | 39.08 | 34.97 | 29.80 | 29.10 |
| 33 23 0121 | 2" Stainless Steel, Well Casing | LF | 31.75 | 28.14 | 25.30 | 21.73 | 21.25 |
| 33 23 0122 | 4" Stainless Steel, Well Casing | LF | 47.66 | 43.06 | 39.45 | 34.92 | 34.31 |
| 33 23 0123 | 6" Stainless Steel Well Casing, 5' Sections, Flush Threaded | LF | 370.73 | 311.96 | 265.77 | 207.74 | 199.94 |
| 33 23 0125 | 8" Stainless Steel Well Casing, 5' Sections, Flush Threaded | LF | 526.42 | 452.94 | 395.18 | 322.61 | 312.87 |
| 33 23 0127 | 10" Stainless Steel Well Casing, 5' Sections, Flush Threaded | LF | 533.65 | 460.17 | 402.41 | 329.84 | 320.10 |
| 33 23 0129 | 12" Stainless Steel Well Casing, 5' Sections, Flush Threaded | LF | 561.57 | 488.09 | 430.33 | 357.76 | 348.02 |
| 33 23 0201 | 2" PVC, Schedule 40, Well Screen | LF | 23.47 | 19.52 | 16.42 | 12.52 | 12.00 |
| 33 23 0202 | 4" PVC, Schedule 40, Well Screen | LF | 38.16 | 32.00 | 27.16 | 21.09 | 20.27 |
| 33 23 0203 | 6" PVC, Schedule 40, Well Screen | LF | 47.13 | 39.78 | 34.01 | 26.75 | 25.78 |
| 33 23 0204 | 8" PVC, Schedule 40, Well Screen | LF | 62.02 | 52.84 | 45.61 | 36.54 | 35.33 |
| 33 23 0221 | 2" Stainless Steel, Well Screen | LF | 26.91 | 23.84 | 21.44 | 18.41 | 18.01 |
| 33 23 0222 | 4" Stainless Steel, Well Screen | LF | 47.66 | 43.06 | 39.45 | 34.92 | 34.31 |
| 33 23 0223 | 6" Stainless Steel Well Screen, 5' Sec, Flush Threaded | LF | 366.05 | 307.28 | 261.09 | 203.06 | 195.26 |
| 33 23 0225 | 8" Stainless Steel Well Screen, 5' Sec, Flush Threaded | LF | 462.22 | 388.74 | 330.98 | 258.41 | 248.67 |
| 33 23 0301 | 2" PVC, Well Plug | EACH | 29.37 | 24.77 | 21.16 | 16.63 | 16.02 |
| 33 23 0302 | 4" PVC, Well Plug | EACH | 55.65 | 48.91 | 43.62 | 36.97 | 36.07 |
| 33 23 0303 | 6" PVC, Well Plug | EACH | 112.63 | 101.15 | 92.13 | 80.79 | 79.27 |
| 33 23 0304 | 8" PVC, Well Plug | EACH | 132.70 | 119.58 | 109.26 | 96.31 | 94.57 |
| 33 23 0311 | 2" Stainless Steel, Well Plug | EACH | 83.24 | 74.06 | 66.83 | 57.76 | 56.55 |
| 33 23 0312 | 4" Stainless Steel, Well Plug | EACH | 112.24 | 102.03 | 94.01 | 83.93 | 82.58 |
| 33 23 0313 | 6" Stainless Steel, Well Plug | EACH | 573.04 | 481.18 | 408.98 | 318.27 | 306.09 |
| 33 23 0314 | 8" Stainless Steel, Well Plug | EACH | 702.53 | 597.56 | 515.04 | 411.38 | 397.46 |

**Environmental Remediation: Assemblies Cost Book**

*Page 3-137*

# Cost Data for Remediation

## Groundwater Monitoring Wells

| Assembly | Description | Unit | Unit Cost by Safety Level A | B | C | D | E |
|---|---|---|---|---|---|---|---|
| | **Capital Costs** | | | | | | |
| 33 23 0503 | 4" Submersible Pump, 13 - 37 GPM, < 180' Head | EACH | 3,084 | 2,790 | 2,637 | 2,352 | 2,264 |
| 33 23 0506 | 2" Submersible Pump Rental, Day | DAY | 63.86 | 63.86 | 63.86 | 63.86 | 63.86 |
| 33 23 0507 | 2" Submersible Pump Rental, Week | WEEK | 191.57 | 191.57 | 191.57 | 191.57 | 191.57 |
| 33 23 0508 | 2" Submersible Pump Rental, Month | MONTH | 478.92 | 478.92 | 478.92 | 478.92 | 478.92 |
| 33 23 0509 | 4" Submersible Pump Rental, Day | DAY | 63.86 | 63.86 | 63.86 | 63.86 | 63.86 |
| 33 23 0510 | 4" Submersible Pump Rental, Week | WEEK | 191.57 | 191.57 | 191.57 | 191.57 | 191.57 |
| 33 23 0511 | 4" Submersible Pump Rental, Month | MONTH | 478.92 | 478.92 | 478.92 | 478.92 | 478.92 |
| 33 23 1101 | Hollow-stem Auger, 8" Outside Diameter Borehole for 2" Well | LF | 89.98 | 73.28 | 60.15 | 43.66 | 41.45 |
| 33 23 1102 | Hollow-stem Auger, 11" Outside Diameter Borehole for 4" Well | LF | 109.98 | 89.57 | 73.52 | 53.36 | 50.66 |
| 33 23 1103 | Hollow-stem Auger, 13 3/4" Outside Diameter Borehole for 6" Well | LF | 141.40 | 115.16 | 94.53 | 68.61 | 65.13 |
| 33 23 1105 | Hollow-stem Auger, 16" Outside Diameter Borehole for 8" Well | LF | 164.97 | 134.35 | 110.28 | 80.05 | 75.99 |
| 33 23 1106 | Split Spoon Sample, 2" x 24", During Drilling | EACH | 46.67 | 46.67 | 46.67 | 46.67 | 46.67 |
| 33 23 1107 | Coring (Fluid or Air) | LF | 102.14 | 88.16 | 78.67 | 64.96 | 62.17 |
| 33 23 1110 | Continuous Split Spoon Sample | LF | 484.76 | 392.72 | 324.27 | 233.60 | 218.99 |
| 33 23 1111 | Well Development Equipment Rental | WEEK | 478.91 | 462.93 | 456.09 | 440.65 | 434.98 |
| 33 23 1121 | Standby for Drilling | EACH | 618.63 | 503.81 | 413.56 | 300.17 | 284.95 |
| 33 23 1122 | Move Rig/Equipment Around Site | EACH | 154.66 | 125.95 | 103.39 | 75.04 | 71.24 |
| 33 23 1125 | Concrete Coring (Minimum Charge) | DAY | 2,475 | 2,015 | 1,654 | 1,201 | 1,140 |
| 33 23 1126 | Furnish 55 Gallon Drum for Drill Cuttings & Development Water | EACH | 65.19 | 65.19 | 65.19 | 65.19 | 65.19 |
| 33 23 1128 | Jackhammer Rental, per Day | DAY | 166.03 | 166.03 | 166.03 | 166.03 | 166.03 |
| 33 23 1129 | Concrete Saw Rental, per Day | DAY | 102.00 | 102.00 | 102.00 | 102.00 | 102.00 |
| 33 23 1151 | Mud Drilling, 4" Diameter Borehole | LF | 73.65 | 59.98 | 49.23 | 35.73 | 33.92 |
| 33 23 1152 | Mud Drilling, 6" Diameter Borehole | LF | 81.01 | 65.98 | 54.16 | 39.31 | 37.31 |
| 33 23 1153 | Mud Drilling, 8" Diameter Borehole | LF | 88.38 | 71.97 | 59.08 | 42.88 | 40.71 |
| 33 23 1154 | Mud Drilling, 10" Diameter Borehole | LF | 100.65 | 81.97 | 67.29 | 48.84 | 46.36 |
| 33 23 1155 | Mud Drilling, 12" Diameter Borehole | LF | 112.92 | 91.97 | 75.49 | 54.79 | 52.01 |
| 33 23 1156 | Mud Drilling, 15" Diameter Borehole | LF | 137.47 | 111.96 | 91.90 | 66.71 | 63.32 |
| 33 23 1161 | Air Rotary 6" Borehole in Unconsolidated | LF | 73.65 | 59.98 | 49.23 | 35.73 | 33.92 |
| 33 23 1162 | Air Rotary 8" Borehole in Unconsolidated | LF | 122.74 | 99.96 | 82.06 | 59.56 | 56.54 |
| 33 23 1163 | Air Rotary 10" Borehole in Unconsolidated | LF | 220.94 | 179.93 | 147.70 | 107.20 | 101.77 |

## Environmental Remediation: Assemblies Cost Book

*Page 3-138*

1998 by ECHOS. All rights reserved

# Cost Data for Remediation

## Groundwater Monitoring Wells

| Assembly | Description | Unit | Unit Cost by Safety Level | | | | |
|---|---|---|---|---|---|---|---|
| | | | A | B | C | D | E |
| | **Capital Costs** | | | | | | |
| 33 23 1164 | Air Rotary 12" Borehole in Unconsolidated | LF | 294.59 | 239.91 | 196.93 | 142.94 | 135.69 |
| 33 23 1168 | Air Rotary 6" Borehole in Consolidated | LF | 58.92 | 47.98 | 39.39 | 28.59 | 27.14 |
| 33 23 1170 | Air Rotary 10" Borehole in Consolidated | LF | 294.59 | 239.91 | 196.93 | 142.94 | 135.69 |
| 33 23 1171 | Air Rotary 12" Borehole in Consolidated | LF | 613.72 | 499.81 | 410.28 | 297.79 | 282.69 |
| 33 23 1401 | 2" Screen, Filter Pack | LF | 16.49 | 13.88 | 11.84 | 9.27 | 8.92 |
| 33 23 1402 | 4" Screen, Filter Pack | LF | 29.10 | 24.50 | 20.89 | 16.36 | 15.75 |
| 33 23 1403 | 6" Screen, Filter Pack | LF | 42.19 | 35.53 | 30.29 | 23.72 | 22.83 |
| 33 23 1404 | Gravel Pack, 8" Well | LF | 29.47 | 23.93 | 19.82 | 14.36 | 13.48 |
| 33 23 1502 | Surface Pad, Concrete, 4' x 4' x 4" | EACH | 24.21 | 21.82 | 20.74 | 18.43 | 17.61 |
| 33 23 1504 | Surface Pad, Concrete, 2' x 2' x 4" | EACH | 6.05 | 5.45 | 5.18 | 4.61 | 4.40 |
| 33 23 1811 | 2" Well, Portland Cement Grout | LF | 0.92 | 0.92 | 0.92 | 0.92 | 0.92 |
| 33 23 1812 | 4" Well, Portland Cement Grout | LF | 1.38 | 1.38 | 1.38 | 1.38 | 1.38 |
| 33 23 1813 | 6" Well, Portland Cement Grout | LF | 7.80 | 7.80 | 7.80 | 7.80 | 7.80 |
| 33 23 1814 | 8" Well, Portland Cement Grout | LF | 11.02 | 11.02 | 11.02 | 11.02 | 11.02 |
| 33 23 1815 | 10" Well, Portland Cement Grout | LF | 12.85 | 12.85 | 12.85 | 12.85 | 12.85 |
| 33 23 1816 | 12" Well, Portland Cement Grout | LF | 14.69 | 14.69 | 14.69 | 14.69 | 14.69 |
| 33 23 1822 | Well Abandonment, 2" Well | LF | 32.46 | 26.55 | 21.90 | 16.06 | 15.27 |
| 33 23 1823 | Well Abandonment, 4" Well | LF | 46.38 | 37.93 | 31.28 | 22.94 | 21.82 |
| 33 23 1824 | Well Abandonment, 6" Well | LF | 0.00 | 0.00 | 0.00 | 0.00 | 0.00 |
| 33 23 1825 | Well Abandonment, 8" Well | LF | 2.12 | 2.12 | 2.12 | 2.12 | 2.12 |
| 33 23 2101 | 2" Well, Bentonite Seal | EACH | 63.00 | 52.67 | 44.55 | 34.34 | 32.97 |
| 33 23 2102 | 4" Well, Bentonite Seal | EACH | 157.54 | 131.70 | 111.39 | 85.87 | 82.44 |
| 33 23 2103 | 6" Well, Bentonite Seal | EACH | 252.01 | 210.68 | 178.18 | 137.37 | 131.89 |
| 33 23 2105 | 8" Well, Bentonite Seal | EACH | 346.60 | 289.75 | 245.06 | 188.92 | 181.39 |
| 33 23 2213 | 8" x 7.5" Locking Manhole Cover, Watertight | EACH | 345.53 | 288.12 | 242.99 | 186.30 | 178.69 |
| 33 23 2214 | 12" x 7.5" Locking Manhole Cover, Watertight | EACH | 459.57 | 383.02 | 322.86 | 247.27 | 237.12 |
| 33 23 2215 | 4" x 4" x 5' Steel Protective Cover, Lockable | EACH | 539.95 | 448.09 | 375.89 | 285.18 | 273.00 |
| 33 23 2216 | 6" x 6" x 5' Steel Protective Cover, Lockable | EACH | 560.94 | 469.08 | 396.88 | 306.17 | 293.99 |
| 33 23 2217 | 8" x 8" x 5' Steel Protective Cover, Lockable | EACH | 620.41 | 528.55 | 456.35 | 365.64 | 353.46 |
| 33 23 2301 | 5' Guard Posts, Cast Iron, Concrete Fill | EACH | 98.83 | 83.39 | 76.77 | 61.84 | 56.37 |
| 33 23 2401 | Teflon Bailer, 3/4" Outside Diameter x 1', 60 cc | EACH | 179.77 | 179.77 | 179.77 | 179.77 | 179.77 |
| 33 23 2402 | Teflon Bailer, 3/4" Outside Diameter x 3', 180 cc | EACH | 227.74 | 227.74 | 227.74 | 227.74 | 227.74 |
| 33 23 2403 | Teflon Bailer, 1" Outside Diameter x 1', 80 cc | EACH | 191.28 | 191.28 | 191.28 | 191.28 | 191.28 |

**Environmental Remediation: Assemblies Cost Book**

# Cost Data for Remediation

## Groundwater Monitoring Wells

| Assembly | Description | Unit | Unit Cost by Safety Level A | B | C | D | E |
|---|---|---|---|---|---|---|---|
| | **Capital Costs** | | | | | | |
| 33 23 2404 | Teflon Bailer, 1 7/8" Outside Diameter x 1', 350 cc | EACH | 181.13 | 181.13 | 181.13 | 181.13 | 181.13 |
| 33 23 2405 | Teflon Bailer, 1 7/8" Outside Diameter x 2', 700 cc | EACH | 198.20 | 198.20 | 198.20 | 198.20 | 198.20 |
| 33 23 2406 | Teflon Bailer, 1 7/8" Outside Diameter x 3', 1,050 cc | EACH | 261.53 | 261.53 | 261.53 | 261.53 | 261.53 |
| 33 23 2407 | Disposable Bailer, Polyethylene, 1.5" Outside Diameter x 36" | EACH | 27.60 | 27.60 | 27.60 | 27.60 | 27.60 |
| 33 23 2411 | Stainless Steel Bailer, 1 3/4" Outside Diameter x 2' | EACH | 124.23 | 124.23 | 124.23 | 124.23 | 124.23 |
| 33 23 2421 | Emptying Stand | EACH | 302.32 | 302.32 | 302.32 | 302.32 | 302.32 |
| 33 23 2422 | Suspension Cable, Teflon Coated | FT | 1.00 | 1.00 | 1.00 | 1.00 | 1.00 |
| 33 23 2423 | Hand Reel | EACH | 9.89 | 9.89 | 9.89 | 9.89 | 9.89 |

**Environmental Remediation: Assemblies Cost Book**

# Cost Data for Remediation
## Heat Enhanced Vapor Extraction

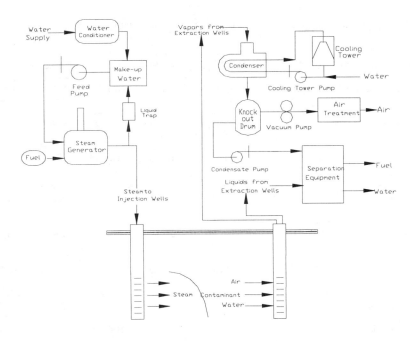

Schematic of Heat Enhanced Vapor Extraction Process
Figure S.4-1
Source: Praxis Environmental Technologies, Inc., San Francisco, Ca.

### General:

Heat-enhanced vapor extraction is an in situ process designed for removing volatile and semivolatile organic compounds from vadose zone soils. In this process, steam is injected into the subsurface through a series of injection wells and contaminants are removed through a series of extraction wells known as vapor extraction points (VEPs).

### Typical Treatment Train:

Carbon adsorption (gas), sampling and analysis.

### Common Cost Components:

1. Steam generation and distribution system (boiler, feedwater treatment and preheating equipment, steam piping, flowmeters, traps)
2. Injection wells
3. Vacuum extraction system, including VEPs, condenser or heat exchanger
4. Condensate holding tanks
5. Operations and maintenance

### Other Cost Considerations:

Clearing and grubbing, fencing, electrical distribution.

# Cost Data for Remediation

## Heat Enhanced Vapor Extraction

| Assembly | Description | Unit | A | B | C | D | E |
|---|---|---|---|---|---|---|---|
| | **Capital Costs** | | | | | | |
| 19 04 0401 | 550 Gallon, Stainless Steel Aboveground Wastewater Holding Tank, Rental | MONTH | 320.65 | 320.65 | 320.65 | 320.65 | 320.65 |
| 19 04 0403 | 4,000 Gallon Polyethylene Wastewater Tank, Rental | MONTH | 545.90 | 545.90 | 545.90 | 545.90 | 545.90 |
| 19 04 0405 | 6,000 Gallon, Polyethylene Aboveground Wastewater Holding Tank, Rental | MONTH | 641.30 | 641.30 | 641.30 | 641.30 | 641.30 |
| 19 04 0407 | 21,000 Gallon, Steel Closed Stationary Aboveground Wastewater Holding Tank, Rental | MONTH | 1,219 | 1,219 | 1,219 | 1,219 | 1,219 |
| 19 04 0621 | 550 Gallon Horizontal Plastic Sump with 4" NPT Connection | EACH | 1,699 | 1,596 | 1,551 | 1,452 | 1,415 |
| 19 04 0624 | 4,000 Gallon Horizontal Plastic Sump with 6" NPT Connection | EACH | 4,130 | 3,927 | 3,841 | 3,645 | 3,573 |
| 19 04 0625 | 6,000 Gallon Horizontal Plastic Sump with 6" NPT Connection | EACH | 5,248 | 4,975 | 4,858 | 4,594 | 4,497 |
| 19 04 0631 | 20,000 Gallon Horizontal Plastic Sump with 6" NPT Connection | EACH | 15,799 | 15,084 | 14,742 | 14,048 | 13,817 |
| 33 01 0101 | Mobilize/DeMobilize Drilling Rig & Crew | LS | 4,949 | 4,031 | 3,308 | 2,401 | 2,280 |
| 33 02 0303 | Organic Vapor Analyzer Rental, per Day | DAY | 184.30 | 184.30 | 184.30 | 184.30 | 184.30 |
| 33 02 0545 | Thermometer, Thermocouple | EACH | 144.48 | 144.48 | 144.48 | 144.48 | 144.48 |
| 33 02 1506 | Monitoring Port with Gas Monitor | EACH | 28.81 | 22.42 | 19.69 | 13.52 | 11.26 |
| 33 02 1507 | Continuous Monitoring and Recording of Air Flow | EACH | 9,870 | 9,870 | 9,870 | 9,870 | 9,870 |
| 33 10 9656 | 550 Gallon Single-wall Steel Aboveground Tank | EACH | 1,598 | 1,390 | 1,301 | 1,099 | 1,026 |
| 33 13 2306 | Propane Conversion for Soil Vapor Extractor | EACH | 90.00 | 90.00 | 90.00 | 90.00 | 90.00 |
| 33 13 2307 | Modulating Motor for Automatic Fuel Adjustment | EACH | 1,958 | 1,958 | 1,958 | 1,958 | 1,958 |
| 33 13 2308 | Trailer Mounting for Vapor Extractor Unit | EACH | 945.00 | 945.00 | 945.00 | 945.00 | 945.00 |
| 33 13 2309 | Security Fence for Vapor Extractor | EACH | 262.00 | 262.00 | 262.00 | 262.00 | 262.00 |
| 33 13 2333 | 5 HP, 90 SCFM Vapor Extraction Blower, Monthly Rental | MONTH | 850.00 | 850.00 | 850.00 | 850.00 | 850.00 |
| 33 13 2334 | 7.5 HP, 140 SCFM Vapor Extraction Blower, Monthly Rental | MONTH | 900.00 | 900.00 | 900.00 | 900.00 | 900.00 |
| 33 13 2335 | 10 HP, 190 SCFM Vapor Extraction Blower, Monthly Rental | MONTH | 1,163 | 1,163 | 1,163 | 1,163 | 1,163 |
| 33 13 2336 | 15 HP, 290 SCFM Vapor Extraction Blower, Monthly Rental | MONTH | 1,325 | 1,325 | 1,325 | 1,325 | 1,325 |
| 33 13 2337 | 30 HP, 580 SCFM Vapor Extraction Blower, Monthly Rental | MONTH | 1,625 | 1,625 | 1,625 | 1,625 | 1,625 |
| 33 13 2338 | Purchase, 5.0 HP, 90 SCFM Vapor Extraction Blower | EACH | 1,912 | 1,912 | 1,912 | 1,912 | 1,912 |
| 33 13 2339 | Purchase, 7.5 HP, 140 SCFM Vapor Extraction Blower | EACH | 3,110 | 2,931 | 2,854 | 2,680 | 2,617 |
| 33 13 2340 | Purchase, 10.0 HP, 190 SCFM Vapor Extraction Blower | EACH | 3,359 | 3,179 | 3,102 | 2,929 | 2,865 |
| 33 13 2341 | Purchase, 15.0 HP, 280 SCFM Vapor Extraction Blower | EACH | 5,126 | 4,947 | 4,870 | 4,696 | 4,632 |
| 33 13 2342 | Purchase, 30.0 HP, 580 SCFM Vapor Extraction Blower | EACH | 7,516 | 7,337 | 7,260 | 7,086 | 7,022 |
| 33 13 2343 | Knockout Drum | EACH | 65.19 | 65.19 | 65.19 | 65.19 | 65.19 |

**Environmental Remediation: Assemblies Cost Book**

# Cost Data for Remediation

## Heat Enhanced Vapor Extraction

| Assembly | Description | Unit | Unit Cost by Safety Level | | | | |
|---|---|---|---|---|---|---|---|
| | | | A | B | C | D | E |
| | **Capital Costs** | | | | | | |
| 33 13 2348 | Install/Assemble Rental Vacuum Extraction Blower | EACH | 783.78 | 604.17 | 527.27 | 353.66 | 290.00 |
| 33 13 2901 | 1,700 pph 100 HP Steam Boiler, Monthly Rental | MONTH | 3,713 | 3,713 | 3,713 | 3,713 | 3,713 |
| 33 13 2902 | 3,450 pph, 100 HP Steam Boiler, Monthly Rental | MONTH | 3,850 | 3,850 | 3,850 | 3,850 | 3,850 |
| 33 13 2903 | 5,000 pph 150 HP Steam Boiler, Monthly Rental | MONTH | 5,963 | 5,963 | 5,963 | 5,963 | 5,963 |
| 33 13 2904 | 12,000 pph 350 HP Steam Boiler, Monthly Rental | MONTH | 10,375 | 10,375 | 10,375 | 10,375 | 10,375 |
| 33 13 2905 | 500 pph Vapor Condenser | EACH | 8,719 | 8,295 | 8,092 | 7,681 | 7,543 |
| 33 13 2906 | 75 Ton Centrifugal Cooling Tower with Pump | EACH | 13,555 | 12,659 | 12,231 | 11,362 | 11,071 |
| 33 13 2907 | 1,500 pph Vapor Condenser | EACH | 11,840 | 11,390 | 11,175 | 10,738 | 10,593 |
| 33 13 2908 | 200 Ton Centrifugal Cooling Tower with Pump | EACH | 25,559 | 24,280 | 23,669 | 22,428 | 22,013 |
| 33 13 2909 | 2,500 pph Vapor Condenser | EACH | 15,054 | 14,558 | 14,321 | 13,840 | 13,679 |
| 33 13 2910 | 300 Ton Centrifugal Cooling Tower with Pump | EACH | 33,685 | 32,291 | 31,627 | 30,276 | 29,824 |
| 33 13 2911 | 6,000 pph Vapor Condenser | EACH | 26,485 | 25,830 | 25,517 | 24,881 | 24,669 |
| 33 13 2912 | 800 Ton Centrifugal Cooling Tower with Pump | EACH | 73,392 | 71,302 | 70,296 | 68,269 | 67,596 |
| 33 13 2913 | 10,000 pph Vapor Condenser | EACH | 38,721 | 37,990 | 37,640 | 36,931 | 36,694 |
| 33 13 2914 | 1,200 Ton Centrifugal Cooling Tower with Pump | EACH | 122,724 | 119,142 | 117,433 | 113,959 | 112,798 |
| 33 13 2915 | Cooling Water Chemical Feeder | EACH | 616.58 | 533.47 | 496.52 | 416.11 | 387.50 |
| 33 13 2916 | Natural Gas Usage, per 1,000 CF | MCF | 5.50 | 5.50 | 5.50 | 5.50 | 5.50 |
| 33 17 0808 | Decontaminate Rig, Augers, Screen (Rental Equipment) | DAY | 205.34 | 205.34 | 205.34 | 205.34 | 205.34 |
| 33 19 9921 | DOT Steel Drum, 55 Gallon | EACH | 65.19 | 65.19 | 65.19 | 65.19 | 65.19 |
| 33 23 0122 | 4" Stainless Steel, Well Casing | LF | 47.66 | 43.06 | 39.45 | 34.92 | 34.31 |
| 33 23 0123 | 6" Stainless Steel Well Casing, 5' Sections, Flush Threaded | LF | 370.73 | 311.96 | 265.77 | 207.74 | 199.94 |
| 33 23 0222 | 4" Stainless Steel, Well Screen | LF | 47.66 | 43.06 | 39.45 | 34.92 | 34.31 |
| 33 23 0223 | 6" Stainless Steel Well Screen, 5' Sec, Flush Threaded | LF | 366.05 | 307.28 | 261.09 | 203.06 | 195.26 |
| 33 23 0312 | 4" Stainless Steel, Well Plug | EACH | 112.24 | 102.03 | 94.01 | 83.93 | 82.58 |
| 33 23 0313 | 6" Stainless Steel, Well Plug | EACH | 573.04 | 481.18 | 408.98 | 318.27 | 306.09 |
| 33 23 0611 | 360 GPD, Pneumatic Groundwater Pump with Controls | EACH | 3,698 | 3,495 | 3,407 | 3,211 | 3,138 |
| 33 23 0612 | 700 GPD, Pneumatic Groundwater Pump with Controls | EACH | 3,954 | 3,750 | 3,663 | 3,466 | 3,394 |
| 33 23 0613 | 2160 GPD, Pneumatic Groundwater Pump with Controls | EACH | 4,209 | 4,005 | 3,918 | 3,721 | 3,649 |
| 33 23 1102 | Hollow-stem Auger, 11" Outside Diameter Borehole for 4" Well | LF | 109.98 | 89.57 | 73.52 | 53.36 | 50.66 |
| 33 23 1103 | Hollow-stem Auger, 13 3/4" Outside Diameter Borehole for 6" Well | LF | 141.40 | 115.16 | 94.53 | 68.61 | 65.13 |
| 33 23 1106 | Split Spoon Sample, 2" x 24", During Drilling | EACH | 46.67 | 46.67 | 46.67 | 46.67 | 46.67 |
| 33 23 1111 | Well Development Equipment Rental | WEEK | 478.91 | 462.93 | 456.09 | 440.65 | 434.98 |

**Environmental Remediation: Assemblies Cost Book**

# Cost Data for Remediation

## Heat Enhanced Vapor Extraction

| Assembly | Description | Unit | Unit Cost by Safety Level | | | | |
|---|---|---|---|---|---|---|---|
| | | | A | B | C | D | E |
| | **Capital Costs** | | | | | | |
| 33 23 1121 | Standby for Drilling | EACH | 618.63 | 503.81 | 413.56 | 300.17 | 284.95 |
| 33 23 1122 | Move Rig/Equipment Around Site | EACH | 154.66 | 125.95 | 103.39 | 75.04 | 71.24 |
| 33 23 1125 | Concrete Coring (Minimum Charge) | DAY | 2,475 | 2,015 | 1,654 | 1,201 | 1,140 |
| 33 23 1153 | Mud Drilling, 8" Diameter Borehole | LF | 88.38 | 71.97 | 59.08 | 42.88 | 40.71 |
| 33 23 1154 | Mud Drilling, 10" Diameter Borehole | LF | 100.65 | 81.97 | 67.29 | 48.84 | 46.36 |
| 33 23 1162 | Air Rotary 8" Borehole in Unconsolidated | LF | 122.74 | 99.96 | 82.06 | 59.56 | 56.54 |
| 33 23 1163 | Air Rotary 10" Borehole in Unconsolidated | LF | 220.94 | 179.93 | 147.70 | 107.20 | 101.77 |
| 33 23 1402 | 4" Screen, Filter Pack | LF | 29.10 | 24.50 | 20.89 | 16.36 | 15.75 |
| 33 23 1403 | 6" Screen, Filter Pack | LF | 42.19 | 35.53 | 30.29 | 23.72 | 22.83 |
| 33 23 1502 | Surface Pad, Concrete, 4' x 4' x 4" | EACH | 24.21 | 21.82 | 20.74 | 18.43 | 17.61 |
| 33 23 1812 | 4" Well, Portland Cement Grout | LF | 1.38 | 1.38 | 1.38 | 1.38 | 1.38 |
| 33 23 1813 | 6" Well, Portland Cement Grout | LF | 7.80 | 7.80 | 7.80 | 7.80 | 7.80 |
| 33 23 2102 | 4" Well, Bentonite Seal | EACH | 157.54 | 131.70 | 111.39 | 85.87 | 82.44 |
| 33 23 2103 | 6" Well, Bentonite Seal | EACH | 252.01 | 210.68 | 178.18 | 137.37 | 131.89 |
| 33 26 0104 | 4" Carbon Steel Piping | LF | 23.20 | 19.86 | 18.43 | 15.21 | 14.02 |
| 33 26 0114 | 4" Carbon Steel Schedule 80 Piping | LF | 37.54 | 34.02 | 32.42 | 29.01 | 27.81 |
| 33 26 0115 | 6" Carbon Steel Schedule 80 Piping | LF | 53.07 | 48.05 | 45.78 | 40.92 | 39.21 |
| 33 26 0204 | 4" Stainless Steel Piping, Schedule 40, Threaded | LF | 80.15 | 71.72 | 68.12 | 59.98 | 57.00 |
| 33 26 0205 | 6" Stainless Steel Piping, Schedule 10, Type 316 | LF | 78.58 | 70.01 | 66.14 | 57.84 | 54.94 |
| 33 26 0702 | 50' x 6" Brown Gum Rubber, Chemical-resistant, Flexible Hose | EACH | 1,123 | 1,038 | 1,001 | 918.34 | 888.04 |
| 33 26 0704 | 50' Flexible, Product Discharge Hose | EACH | 159.00 | 159.00 | 159.00 | 159.00 | 159.00 |
| 33 26 1004 | 4" Galvanized Steel Pipe, Schedule 80 | LF | 40.83 | 37.30 | 35.71 | 32.29 | 31.10 |
| 33 26 1005 | 6" Galvanized Steel Pipe, Schedule 80 | LF | 57.69 | 52.67 | 50.40 | 45.53 | 43.83 |
| 33 27 0202 | 4" Stainless Steel, Tee | EACH | 583.41 | 504.25 | 470.36 | 393.85 | 365.80 |
| 33 27 0204 | 6" Stainless Steel, Tee | EACH | 1,508 | 1,220 | 1,090 | 811.00 | 713.21 |
| 33 27 0212 | 4" Stainless Steel, 90 Degree, Elbow | EACH | 391.27 | 338.50 | 315.91 | 264.90 | 246.20 |
| 33 27 0213 | 6" Stainless Steel, 90 Degree, Elbow | EACH | 425.65 | 376.17 | 355.00 | 307.18 | 289.64 |
| 33 27 0404 | 4" Iron Body Check Valve | EACH | 354.29 | 315.96 | 299.54 | 262.49 | 248.90 |
| 33 27 0405 | 6" Iron Body Check Valve | EACH | 618.60 | 554.57 | 523.95 | 461.86 | 441.15 |
| 33 27 0552 | 4" Carbon Steel 90-degree Elbow, Schedule 80 | EACH | 190.36 | 152.53 | 135.42 | 98.81 | 85.97 |
| 33 27 0554 | 6" Carbon Steel 90-degree Elbow, Schedule 80 | EACH | 297.54 | 245.36 | 221.75 | 171.24 | 153.54 |
| 33 27 0562 | 4" Carbon Steel Tee, Schedule 80 | EACH | 307.17 | 250.43 | 224.76 | 169.84 | 150.59 |
| 33 27 0564 | 6" Carbon Steel Tee, Schedule 80 | EACH | 452.39 | 374.65 | 339.48 | 264.23 | 237.85 |

**Environmental Remediation: Assemblies Cost Book**

*Page 3-144*

1998 by ECHOS. All rights reserved

# Cost Data for Remediation

## Heat Enhanced Vapor Extraction

| Assembly | Description | Unit | A | B | C | D | E |
|---|---|---|---|---|---|---|---|
| | **Capital Costs** | | | | | | |
| 33 27 1002 | 4" Galvanized Steel Tee, Schedule 80 | EACH | 313.86 | 257.12 | 231.45 | 176.53 | 157.27 |
| 33 27 1004 | 6" Galvanized Steel Tee, Schedule 80 | EACH | 465.31 | 387.57 | 352.40 | 277.15 | 250.76 |
| 33 27 1012 | 4" Galvanized Steel 90-degree Elbow, Schedule 80 | EACH | 193.07 | 155.25 | 138.14 | 101.52 | 88.68 |
| 33 27 1013 | 6" Galvanized Steel 90-degree Elbow, Schedule 80 | EACH | 305.50 | 253.32 | 229.71 | 179.20 | 161.49 |
| 33 31 0202 | 6 HP, 80 Gallon, 200 PSI, 28 SCFM, Air Compressor | EACH | 8,102 | 7,713 | 7,547 | 7,171 | 7,033 |
| 33 31 0209 | Pressure Gauge | EACH | 140.16 | 121.61 | 113.67 | 95.74 | 89.16 |
| 33 42 0101 | Electrical Charge | KWH | 0.06 | 0.06 | 0.06 | 0.06 | 0.06 |
| 33 42 0201 | Diesel Fuel | GAL | 1.31 | 1.31 | 1.31 | 1.31 | 1.31 |
| | **Operations & Maintenance** | | | | | | |
| 19 04 0401 | 550 Gallon, Stainless Steel Aboveground Wastewater Holding Tank, Rental | MONTH | 320.65 | 320.65 | 320.65 | 320.65 | 320.65 |
| 19 04 0403 | 4,000 Gallon Polyethylene Wastewater Tank, Rental | MONTH | 545.90 | 545.90 | 545.90 | 545.90 | 545.90 |
| 19 04 0405 | 6,000 Gallon, Polyethylene Aboveground Wastewater Holding Tank, Rental | MONTH | 641.30 | 641.30 | 641.30 | 641.30 | 641.30 |
| 19 04 0407 | 21,000 Gallon, Steel Closed Stationary Aboveground Wastewater Holding Tank, Rental | MONTH | 1,219 | 1,219 | 1,219 | 1,219 | 1,219 |
| 33 02 1507 | Continuous Monitoring and Recording of Air Flow | EACH | 9,870 | 9,870 | 9,870 | 9,870 | 9,870 |
| 33 13 2333 | 5 HP, 90 SCFM Vapor Extraction Blower, Monthly Rental | MONTH | 850.00 | 850.00 | 850.00 | 850.00 | 850.00 |
| 33 13 2334 | 7.5 HP, 140 SCFM Vapor Extraction Blower, Monthly Rental | MONTH | 900.00 | 900.00 | 900.00 | 900.00 | 900.00 |
| 33 13 2335 | 10 HP, 190 SCFM Vapor Extraction Blower, Monthly Rental | MONTH | 1,163 | 1,163 | 1,163 | 1,163 | 1,163 |
| 33 13 2336 | 15 HP, 290 SCFM Vapor Extraction Blower, Monthly Rental | MONTH | 1,325 | 1,325 | 1,325 | 1,325 | 1,325 |
| 33 13 2337 | 30 HP, 580 SCFM Vapor Extraction Blower, Monthly Rental | MONTH | 1,625 | 1,625 | 1,625 | 1,625 | 1,625 |
| 33 13 2901 | 1,700 pph 100 HP Steam Boiler, Monthly Rental | MONTH | 3,713 | 3,713 | 3,713 | 3,713 | 3,713 |
| 33 13 2902 | 3,450 pph, 100 HP Steam Boiler, Monthly Rental | MONTH | 3,850 | 3,850 | 3,850 | 3,850 | 3,850 |
| 33 13 2903 | 5,000 pph 150 HP Steam Boiler, Monthly Rental | MONTH | 5,963 | 5,963 | 5,963 | 5,963 | 5,963 |
| 33 13 2904 | 12,000 pph 350 HP Steam Boiler, Monthly Rental | MONTH | 10,375 | 10,375 | 10,375 | 10,375 | 10,375 |
| 33 13 2916 | Natural Gas Usage, per 1,000 CF | MCF | 5.50 | 5.50 | 5.50 | 5.50 | 5.50 |
| 33 42 0101 | Electrical Charge | KWH | 0.06 | 0.06 | 0.06 | 0.06 | 0.06 |
| 33 42 0201 | Diesel Fuel | GAL | 1.31 | 1.31 | 1.31 | 1.31 | 1.31 |

**Environmental Remediation: Assemblies Cost Book**

# Cost Data for Remediation
## In Situ Biodegradation (Bioventing)

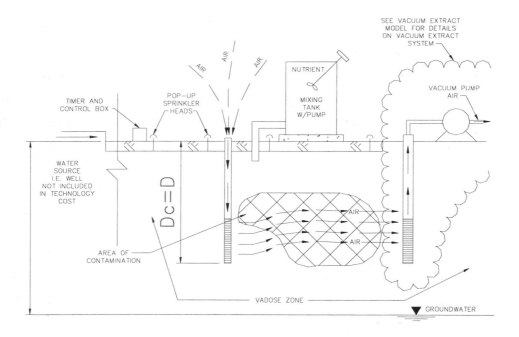

NOTE: DEFAULT SYSTEM HAS NO INJECTION WELLS. THESE CAN BE ADDED AT THE "WELL" SCREEN.

### General:

In situ biodegradation involves microbial transformation of organic contaminants to affect cleanup of groundwater, soils, or other media. Bioventing, one means of performing in situ biodegradation, is similar to soil vapor extraction except that with bioventing, in situ biodegradation is stimulated intentionally. This process utilizes one or more vacuum extraction wells screened outside the contaminated zone to direct oxygen from the surface through the subsurface. Extracted air can be pulled directly through the soil pores from the atmosphere or supplied by one or more injection wells.

### Typical Treatment Train:

Infiltration gallery (for distribution of moisture and/or nutrients), injections wells, sampling and analysis.

### Common Cost Components:

1. VEP installation (vertical or horizontal)
2. VEPs
3. Blowers
4. Connection piping
5. Soil additives (watering, nutrients, microorganisms)
6. Operations and maintenance

### Other Cost Considerations:

Clearing and grubbing, fencing, electrical distribution, monitoring.

# Cost Data for Remediation

## In Situ Biodegradation (Bioventing)

| Assembly | Description | Unit | Unit Cost by Safety Level | | | | |
|---|---|---|---|---|---|---|---|
| | | | A | B | C | D | E |
| | **Capital Costs** | | | | | | |
| 17 03 0206 | D3 with A-blade Bulldozer | HOUR | 151.96 | 122.19 | 101.74 | 72.51 | 66.74 |
| 17 03 0298 | 1/2 CY Crawler-mounted, Hydraulic Excavator | CY | 5.16 | 4.14 | 3.45 | 2.45 | 2.24 |
| 17 03 0421 | 950, 3.0 CY, Backfill with Borrow Material | CY | 10.24 | 8.27 | 6.85 | 4.92 | 4.58 |
| 18 02 0301 | Asphalt Pavement - 10" Subgrade, 9" Base, 1 1/2" Topping | SY | 61.11 | 50.94 | 43.58 | 33.57 | 31.83 |
| 18 05 0409 | Fertilize, 800 Lbs/Acre, Push Rotary | ACRE | 140.58 | 118.67 | 105.24 | 83.82 | 78.56 |
| 18 05 0413 | Watering with 3,000-Gallon Tank Truck, per Pass | ACRE | 111.81 | 90.08 | 75.73 | 54.43 | 49.85 |
| 18 05 0701 | Full Circle Sprinkler Head, 30' Diameter | EACH | 94.41 | 76.34 | 68.61 | 51.14 | 44.74 |
| 18 05 0702 | Semicircle Sprinkler Head, 20' Diameter | EACH | 43.12 | 34.09 | 30.22 | 21.49 | 18.29 |
| 18 05 0703 | Shrub/Tree Sprinkler Head | EACH | 40.45 | 31.42 | 27.55 | 18.82 | 15.62 |
| 18 05 0705 | Control Box | EACH | 1,598 | 1,461 | 1,402 | 1,270 | 1,221 |
| 18 05 0706 | 2 1/2", Cast-iron Body, Gate Valve | EACH | 257.59 | 239.88 | 232.30 | 215.19 | 208.91 |
| 18 05 0707 | 4" Reducer | EACH | 80.38 | 64.15 | 57.20 | 41.51 | 35.76 |
| 18 05 0710 | Testing & Inspection of Sprinkler System | LS | 856.76 | 660.42 | 576.36 | 386.59 | 317.00 |
| 18 05 0711 | Automated Pop-Up Sprinkler System | ACRE | 8,779 | 6,984 | 6,188 | 4,451 | 3,832 |
| 19 01 0201 | 3/4", Class 200, PVC Piping | LF | 7.45 | 5.81 | 5.04 | 3.45 | 2.90 |
| 19 01 0202 | 1", Class 200, PVC Piping | LF | 7.48 | 5.84 | 5.07 | 3.48 | 2.93 |
| 19 01 0205 | 2 1/2", Class 200, PVC Piping | LF | 9.03 | 7.13 | 6.24 | 4.40 | 3.77 |
| 19 04 0403 | 4,000 Gallon Polyethylene Wastewater Tank, Rental | MONTH | 545.90 | 545.90 | 545.90 | 545.90 | 545.90 |
| 19 04 0406 | 21,000 Gallon Steel Wastewater Holding Tank, Rental | MONTH | 1,219 | 1,219 | 1,219 | 1,219 | 1,219 |
| 33 01 0101 | Mobilize/DeMobilize Drilling Rig & Crew | LS | 4,949 | 4,031 | 3,308 | 2,401 | 2,280 |
| 33 02 0303 | Organic Vapor Analyzer Rental, per Day | DAY | 184.30 | 184.30 | 184.30 | 184.30 | 184.30 |
| 33 02 1507 | Continuous Monitoring and Recording of Air Flow | EACH | 9,870 | 9,870 | 9,870 | 9,870 | 9,870 |
| 33 11 9902 | Light Petroleum Hydrocarbon Degraders, Microorganisms | LB | 19.88 | 19.88 | 19.88 | 19.88 | 19.88 |
| 33 11 9903 | Light Petroleum Hydrocarbon Degraders, 100 Lb Bag, Microorganisms | EACH | 1,948 | 1,948 | 1,948 | 1,948 | 1,948 |
| 33 11 9904 | Light Petroleum Hydrocarbon Degraders, 1,000 Lb Bag, Microorganisms | EACH | 17,490 | 17,490 | 17,490 | 17,490 | 17,490 |
| 33 11 9905 | Heavy Petroleum Hydrocarbon/Creosote Degraders, Microorganisms | LB | 13.25 | 13.25 | 13.25 | 13.25 | 13.25 |
| 33 11 9906 | Heavy Petroleum Hydrocarbon/Creosote Degraders, 100 Lb Bag, Microorganisms | EACH | 1,007 | 1,007 | 1,007 | 1,007 | 1,007 |
| 33 11 9907 | Heavy Petroleum Hydrocarbon/Creosote Degraders, 1,000 Lb Bag, Microorganisms | EACH | 8,745 | 8,745 | 8,745 | 8,745 | 8,745 |
| 33 13 2301 | 1 HP, 115V, 98 SCFM, Vapor Recovery System | EACH | 4,298 | 4,118 | 4,042 | 3,868 | 3,804 |
| 33 13 2302 | 1 HP, 230V, 98 SCFM, Vapor Recovery System | EACH | 4,330 | 4,150 | 4,073 | 3,900 | 3,836 |

**Environmental Remediation: Assemblies Cost Book**

# Cost Data for Remediation

## In Situ Biodegradation (Bioventing)

| Assembly | Description | Unit | Unit Cost by Safety Level A | B | C | D | E |
|---|---|---|---|---|---|---|---|
| | **Capital Costs** | | | | | | |
| 33 13 2303 | 1 1/2 HP, 230V, 127 SCFM, Vapor Recovery System | EACH | 4,958 | 4,779 | 4,702 | 4,528 | 4,465 |
| 33 13 2304 | 2 HP, 230V, 160 SCFM, Vapor Recovery System | EACH | 5,527 | 5,347 | 5,270 | 5,096 | 5,033 |
| 33 13 2305 | 5 HP, 230V, 280 SCFM, Vapor Recovery System | EACH | 6,725 | 6,546 | 6,469 | 6,295 | 6,232 |
| 33 13 2306 | Propane Conversion for Soil Vapor Extractor | EACH | 90.00 | 90.00 | 90.00 | 90.00 | 90.00 |
| 33 13 2307 | Modulating Motor for Automatic Fuel Adjustment | EACH | 1,958 | 1,958 | 1,958 | 1,958 | 1,958 |
| 33 13 2308 | Trailer Mounting for Vapor Extractor Unit | EACH | 945.00 | 945.00 | 945.00 | 945.00 | 945.00 |
| 33 13 2309 | Security Fence for Vapor Extractor | EACH | 262.00 | 262.00 | 262.00 | 262.00 | 262.00 |
| 33 13 2320 | Installation Using Chain Trencher, Depth <= 4' | CY | 2.78 | 2.18 | 1.86 | 1.29 | 1.12 |
| 33 13 2322 | 1 HP, 230V, 98 SCFM, Weekly Rental, Vapor Recovery System | WEEK | 266.67 | 266.67 | 266.67 | 266.67 | 266.67 |
| 33 13 2332 | 1 HP, 230V, 98 SCFM, Monthly Rental, Vapor Recovery System | MONTH | 741.67 | 741.67 | 741.67 | 741.67 | 741.67 |
| 33 13 2333 | 5 HP, 90 SCFM Vapor Extraction Blower, Monthly Rental | MONTH | 850.00 | 850.00 | 850.00 | 850.00 | 850.00 |
| 33 13 2334 | 7.5 HP, 140 SCFM Vapor Extraction Blower, Monthly Rental | MONTH | 900.00 | 900.00 | 900.00 | 900.00 | 900.00 |
| 33 13 2335 | 10 HP, 190 SCFM Vapor Extraction Blower, Monthly Rental | MONTH | 1,163 | 1,163 | 1,163 | 1,163 | 1,163 |
| 33 13 2336 | 15 HP, 290 SCFM Vapor Extraction Blower, Monthly Rental | MONTH | 1,325 | 1,325 | 1,325 | 1,325 | 1,325 |
| 33 13 2337 | 30 HP, 580 SCFM Vapor Extraction Blower, Monthly Rental | MONTH | 1,625 | 1,625 | 1,625 | 1,625 | 1,625 |
| 33 13 2338 | Purchase, 5.0 HP, 90 SCFM Vapor Extraction Blower | EACH | 1,912 | 1,912 | 1,912 | 1,912 | 1,912 |
| 33 13 2339 | Purchase, 7.5 HP, 140 SCFM Vapor Extraction Blower | EACH | 3,110 | 2,931 | 2,854 | 2,680 | 2,617 |
| 33 13 2340 | Purchase, 10.0 HP, 190 SCFM Vapor Extraction Blower | EACH | 3,359 | 3,179 | 3,102 | 2,929 | 2,865 |
| 33 13 2341 | Purchase, 15.0 HP, 280 SCFM Vapor Extraction Blower | EACH | 5,126 | 4,947 | 4,870 | 4,696 | 4,632 |
| 33 13 2342 | Purchase, 30.0 HP, 580 SCFM Vapor Extraction Blower | EACH | 7,516 | 7,337 | 7,260 | 7,086 | 7,022 |
| 33 17 0808 | Decontaminate Rig, Augers, Screen (Rental Equipment) | DAY | 205.34 | 205.34 | 205.34 | 205.34 | 205.34 |
| 33 17 0811 | Decontaminate Trenching Equipment | EACH | 3,835 | 3,345 | 3,029 | 2,549 | 2,441 |
| 33 19 9921 | DOT Steel Drum, 55 Gallon | EACH | 65.19 | 65.19 | 65.19 | 65.19 | 65.19 |
| 33 19 9922 | Polyethylene Closed Head Drum, 55 Gallon | EACH | 43.63 | 43.63 | 43.63 | 43.63 | 43.63 |
| 33 23 0101 | 2" PVC, Schedule 40, Well Casing | LF | 17.45 | 14.39 | 11.98 | 8.95 | 8.55 |
| 33 23 0102 | 4" PVC, Schedule 40, Well Casing | LF | 26.98 | 22.38 | 18.77 | 14.24 | 13.63 |
| 33 23 0104 | 8" PVC, Schedule 40, Well Casing | LF | 34.50 | 29.27 | 25.15 | 19.98 | 19.29 |
| 33 23 0111 | 2" PVC, Schedule 80, Well Casing | LF | 17.78 | 14.72 | 12.31 | 9.28 | 8.88 |
| 33 23 0112 | 4" PVC, Schedule 80, Well Casing | LF | 28.55 | 23.95 | 20.34 | 15.81 | 15.20 |
| 33 23 0121 | 2" Stainless Steel, Well Casing | LF | 31.75 | 28.14 | 25.30 | 21.73 | 21.25 |
| 33 23 0122 | 4" Stainless Steel, Well Casing | LF | 47.66 | 43.06 | 39.45 | 34.92 | 34.31 |

**Environmental Remediation: Assemblies Cost Book**

# Cost Data for Remediation

## In Situ Biodegradation (Bioventing)

### Unit Cost by Safety Level

| Assembly | Description | Unit | A | B | C | D | E |
|---|---|---|---|---|---|---|---|
| | **Capital Costs** | | | | | | |
| 33 23 0201 | 2" PVC, Schedule 40, Well Screen | LF | 23.47 | 19.52 | 16.42 | 12.52 | 12.00 |
| 33 23 0202 | 4" PVC, Schedule 40, Well Screen | LF | 38.16 | 32.00 | 27.16 | 21.09 | 20.27 |
| 33 23 0211 | 2" PVC, Schedule 80, Well Screen | LF | 18.70 | 15.64 | 13.23 | 10.20 | 9.80 |
| 33 23 0212 | 4" PVC, Schedule 80, Well Screen | LF | 30.01 | 25.41 | 21.80 | 17.27 | 16.66 |
| 33 23 0221 | 2" Stainless Steel, Well Screen | LF | 26.91 | 23.84 | 21.44 | 18.41 | 18.01 |
| 33 23 0222 | 4" Stainless Steel, Well Screen | LF | 47.66 | 43.06 | 39.45 | 34.92 | 34.31 |
| 33 23 0235 | 4" SDR 17, High-density Polyethylene, Well Screen | LF | 10.00 | 10.00 | 10.00 | 10.00 | 10.00 |
| 33 23 0236 | 6" SDR 17, High-density Polyethylene, Well Screen | LF | 12.00 | 12.00 | 12.00 | 12.00 | 12.00 |
| 33 23 0301 | 2" PVC, Well Plug | EACH | 29.37 | 24.77 | 21.16 | 16.63 | 16.02 |
| 33 23 0302 | 4" PVC, Well Plug | EACH | 55.65 | 48.91 | 43.62 | 36.97 | 36.07 |
| 33 23 0311 | 2" Stainless Steel, Well Plug | EACH | 83.24 | 74.06 | 66.83 | 57.76 | 56.55 |
| 33 23 0312 | 4" Stainless Steel, Well Plug | EACH | 112.24 | 102.03 | 94.01 | 83.93 | 82.58 |
| 33 23 0316 | 4" High-density Polyethylene, Well Plug | EACH | 70.64 | 70.64 | 70.64 | 70.64 | 70.64 |
| 33 23 0317 | 6" High-density Polyethylene, Well Plug | EACH | 86.08 | 86.08 | 86.08 | 86.08 | 86.08 |
| 33 23 1101 | Hollow-stem Auger, 8" Outside Diameter Borehole for 2" Well | LF | 89.98 | 73.28 | 60.15 | 43.66 | 41.45 |
| 33 23 1102 | Hollow-stem Auger, 11" Outside Diameter Borehole for 4" Well | LF | 109.98 | 89.57 | 73.52 | 53.36 | 50.66 |
| 33 23 1106 | Split Spoon Sample, 2" x 24", During Drilling | EACH | 46.67 | 46.67 | 46.67 | 46.67 | 46.67 |
| 33 23 1111 | Well Development Equipment Rental | WEEK | 478.91 | 462.93 | 456.09 | 440.65 | 434.98 |
| 33 23 1121 | Standby for Drilling | EACH | 618.63 | 503.81 | 413.56 | 300.17 | 284.95 |
| 33 23 1122 | Move Rig/Equipment Around Site | EACH | 154.66 | 125.95 | 103.39 | 75.04 | 71.24 |
| 33 23 1125 | Concrete Coring (Minimum Charge) | DAY | 2,475 | 2,015 | 1,654 | 1,201 | 1,140 |
| 33 23 1126 | Furnish 55 Gallon Drum for Drill Cuttings & Development Water | EACH | 65.19 | 65.19 | 65.19 | 65.19 | 65.19 |
| 33 23 1152 | Mud Drilling, 6" Diameter Borehole | LF | 81.01 | 65.98 | 54.16 | 39.31 | 37.31 |
| 33 23 1153 | Mud Drilling, 8" Diameter Borehole | LF | 88.38 | 71.97 | 59.08 | 42.88 | 40.71 |
| 33 23 1161 | Air Rotary 6" Borehole in Unconsolidated | LF | 73.65 | 59.98 | 49.23 | 35.73 | 33.92 |
| 33 23 1162 | Air Rotary 8" Borehole in Unconsolidated | LF | 122.74 | 99.96 | 82.06 | 59.56 | 56.54 |
| 33 23 1181 | Trenching for Horizontal Well Installation | LF | 31.93 | 31.93 | 31.93 | 31.93 | 31.93 |
| 33 23 1185 | Standby for Trenching | EACH | 383.54 | 334.53 | 302.86 | 254.87 | 244.12 |
| 33 23 1187 | Mobilize/DeMobilize Trencher/Crew | EACH | 38,354 | 33,453 | 30,286 | 25,487 | 24,412 |
| 33 23 1401 | 2" Screen, Filter Pack | LF | 16.49 | 13.88 | 11.84 | 9.27 | 8.92 |
| 33 23 1402 | 4" Screen, Filter Pack | LF | 29.10 | 24.50 | 20.89 | 16.36 | 15.75 |

---

**Environmental Remediation: Assemblies Cost Book**

*Page 3-149*

1998 by ECHOS. All rights reserved

# Cost Data for Remediation

## In Situ Biodegradation (Bioventing)

| Assembly | Description | Unit | Unit Cost by Safety Level A | B | C | D | E |
|---|---|---|---|---|---|---|---|
| | **Capital Costs** | | | | | | |
| 33 23 1403 | 6" Screen, Filter Pack | LF | 42.19 | 35.53 | 30.29 | 23.72 | 22.83 |
| 33 23 1407 | Gravel Pack for Horizontal Well Installation | CF | 0.84 | 0.80 | 0.78 | 0.74 | 0.73 |
| 33 23 1502 | Surface Pad, Concrete, 4' x 4' x 4" | EACH | 24.21 | 21.82 | 20.74 | 18.43 | 17.61 |
| 33 23 1811 | 2" Well, Portland Cement Grout | LF | 0.92 | 0.92 | 0.92 | 0.92 | 0.92 |
| 33 23 1812 | 4" Well, Portland Cement Grout | LF | 1.38 | 1.38 | 1.38 | 1.38 | 1.38 |
| 33 23 2101 | 2" Well, Bentonite Seal | EACH | 63.00 | 52.67 | 44.55 | 34.34 | 32.97 |
| 33 23 2102 | 4" Well, Bentonite Seal | EACH | 157.54 | 131.70 | 111.39 | 85.87 | 82.44 |
| 33 26 0210 | 2" Stainless Steel, Schedule 40, Type 304, Connection Piping | LF | 33.47 | 29.21 | 27.39 | 23.28 | 21.77 |
| 33 26 0211 | 4" Stainless Steel, Schedule 40, Type 304, Connection Piping | LF | 80.15 | 71.72 | 68.12 | 59.98 | 57.00 |
| 33 26 0212 | 4" Stainless Steel, Schedule 40, Type 304, Manifold Piping | LF | 80.15 | 71.72 | 68.12 | 59.98 | 57.00 |
| 33 26 0405 | 8" PVC, Schedule 40, Manifold Piping | LF | 23.47 | 19.13 | 17.28 | 13.08 | 11.55 |
| 33 26 0406 | 2" PVC, Schedule 40, Connection Piping | LF | 6.14 | 4.87 | 4.33 | 3.11 | 2.66 |
| 33 26 0407 | 4" PVC, Schedule 40, Connection Piping | LF | 13.38 | 10.65 | 9.48 | 6.84 | 5.88 |
| 33 26 0408 | 4" PVC, Schedule 40, Manifold Piping | LF | 13.38 | 10.65 | 9.48 | 6.84 | 5.88 |
| 33 26 0412 | 8" PVC, Schedule 40, Connection Piping | LF | 23.47 | 19.13 | 17.28 | 13.08 | 11.55 |
| 33 27 0104 | 4" PVC, Schedule 40, Tee | EACH | 137.04 | 108.84 | 96.77 | 69.51 | 59.51 |
| 33 27 0106 | 8" PVC, Schedule 40, Tee | EACH | 336.90 | 291.78 | 272.46 | 228.85 | 212.85 |
| 33 27 0112 | 2" PVC, 90 Degree, Elbow | EACH | 41.36 | 32.22 | 28.30 | 19.46 | 16.22 |
| 33 27 0114 | 4" PVC, 90 Degree, Elbow | EACH | 101.72 | 80.57 | 71.51 | 51.07 | 43.57 |
| 33 27 0116 | 8" PVC, 90 Degree, Elbow | EACH | 200.21 | 172.01 | 159.94 | 132.68 | 122.68 |
| 33 27 0144 | 8" Flange Assemblies, PVC, Schedule 40 | EACH | 31.29 | 31.29 | 31.29 | 31.29 | 31.29 |
| 33 27 0151 | 4" x 2" Reducer, PVC Schedule 40 | EACH | 9.09 | 9.09 | 9.09 | 9.09 | 9.09 |
| 33 27 0153 | 8" x 6" Reducer, PVC Schedule 40 | EACH | 22.98 | 22.98 | 22.98 | 22.98 | 22.98 |
| 33 27 0155 | 8" x 4" Reducer, PVC Schedule 40 | EACH | 67.28 | 67.28 | 67.28 | 67.28 | 67.28 |
| 33 27 0156 | 12" x 8" Reducer, PVC Schedule 40 | EACH | 166.92 | 166.92 | 166.92 | 166.92 | 166.92 |
| 33 27 0202 | 4" Stainless Steel, Tee | EACH | 583.41 | 504.25 | 470.36 | 393.85 | 365.80 |
| 33 27 0211 | 2" Stainless Steel, 90 Degree, Elbow | EACH | 114.28 | 93.45 | 84.53 | 64.39 | 57.01 |
| 33 27 0212 | 4" Stainless Steel, 90 Degree, Elbow | EACH | 391.27 | 338.50 | 315.91 | 264.90 | 246.20 |
| 33 27 0231 | 4" x 2" Stainless Steel Reducer, Schedule 40, Type 304 | EACH | 27.01 | 27.01 | 27.01 | 27.01 | 27.01 |
| 33 27 0242 | 8" x 4" Stainless Steel Reducer, Schedule 10, Type 316 | EACH | 279.00 | 279.00 | 279.00 | 279.00 | 279.00 |
| 33 27 0352 | 4" SDR-17 Flange Assemblies | EACH | 34.09 | 34.09 | 34.09 | 34.09 | 34.09 |
| 33 27 0353 | 6" SDR-17 Flange Assemblies | EACH | 71.20 | 71.20 | 71.20 | 71.20 | 71.20 |

**Environmental Remediation: Assemblies Cost Book**

*Page 3-150*

1998 by ECHOS. All rights reserved

# Cost Data for Remediation

## In Situ Biodegradation (Bioventing)

| Assembly | Description | Unit | A | B | C | D | E |
|---|---|---|---|---|---|---|---|
| | **Capital Costs** | | | | | | |
| 33 27 0404 | 4" Iron Body Check Valve | EACH | 354.29 | 315.96 | 299.54 | 262.49 | 248.90 |
| 33 27 0406 | 8" Iron Body Check Valve | EACH | 1,094 | 980.95 | 926.96 | 817.50 | 780.98 |
| 33 31 0209 | Pressure Gauge | EACH | 140.16 | 121.61 | 113.67 | 95.74 | 89.16 |
| 33 42 0101 | Electrical Charge | KWH | 0.06 | 0.06 | 0.06 | 0.06 | 0.06 |
| | **Operations & Maintenance** | | | | | | |
| 18 05 0409 | Fertilize, 800 Lbs/Acre, Push Rotary | ACRE | 140.58 | 118.67 | 105.24 | 83.82 | 78.56 |
| 33 02 1507 | Continuous Monitoring and Recording of Air Flow | EACH | 9,870 | 9,870 | 9,870 | 9,870 | 9,870 |
| 33 11 9902 | Light Petroleum Hydrocarbon Degraders, Microorganisms | LB | 19.88 | 19.88 | 19.88 | 19.88 | 19.88 |
| 33 11 9903 | Light Petroleum Hydrocarbon Degraders, 100 Lb Bag, Microorganisms | EACH | 1,948 | 1,948 | 1,948 | 1,948 | 1,948 |
| 33 11 9904 | Light Petroleum Hydrocarbon Degraders, 1,000 Lb Bag, Microorganisms | EACH | 17,490 | 17,490 | 17,490 | 17,490 | 17,490 |
| 33 11 9905 | Heavy Petroleum Hydrocarbon/Creosote Degraders, Microorganisms | LB | 13.25 | 13.25 | 13.25 | 13.25 | 13.25 |
| 33 11 9906 | Heavy Petroleum Hydrocarbon/Creosote Degraders, 100 Lb Bag, Microorganisms | EACH | 1,007 | 1,007 | 1,007 | 1,007 | 1,007 |
| 33 11 9907 | Heavy Petroleum Hydrocarbon/Creosote Degraders, 1,000 Lb Bag, Microorganisms | EACH | 8,745 | 8,745 | 8,745 | 8,745 | 8,745 |
| 33 42 0101 | Electrical Charge | KWH | 0.06 | 0.06 | 0.06 | 0.06 | 0.06 |

**Environmental Remediation: Assemblies Cost Book**

# Cost Data for Remediation
## In Situ Biodegradation (Land Treatment)

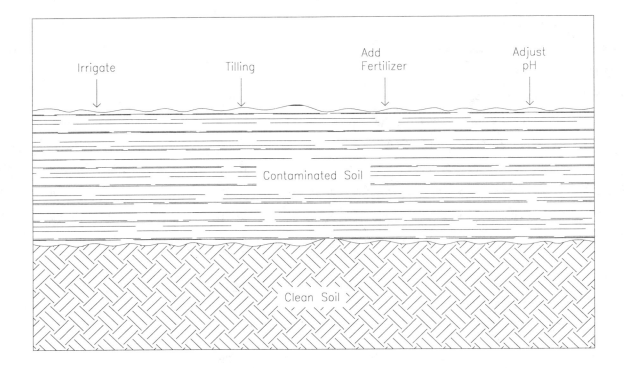

### General:

*In situ biodegradation (land treatment) uses conventional soil management practices to stimulate biodegradation of organic compounds in a layer of contaminated soil up to 24 inches thick. Biodegradation is stimulated by increasing aeration, maintaining moist conditions, providing nutrients, and in some cases, by adding microorganisms.*

### Typical Treatment Train:
*Monitoring.*

### Common Cost Components:

1. Soil preparation (break up, mix, and aerate soil)
2. Soil pH adjustment
3. Tilling
4. Watering
5. Fertilizing
6. Agricultural equipment
7. Equipment standby time

### Other Cost Considerations:
*Fencing, clearing and grubbing.*

**Environmental Remediation: Assemblies Cost Book**

# Cost Data for Remediation

## In Situ Biodegradation (Land Treatment)

| Assembly | Description | Unit | \multicolumn{5}{c}{Unit Cost by Safety Level} |
|---|---|---|---|---|---|---|---|
| | | | A | B | C | D | E |
| | **Capital Costs** | | | | | | |
| 17 03 0206 | D3 with A-blade Bulldozer | HOUR | 151.96 | 122.19 | 101.74 | 72.51 | 66.74 |
| 17 03 0257 | Cat 215, 1.0 CY, Soil, Shallow, Trenching | CY | 2.42 | 1.95 | 1.62 | 1.16 | 1.07 |
| 18 05 0409 | Fertilize, 800 Lbs/Acre, Push Rotary | ACRE | 140.58 | 118.67 | 105.24 | 83.82 | 78.56 |
| 18 05 0410 | Fertilize, 800 Lbs/Acre, Spray from Truck | ACRE | 107.45 | 90.70 | 78.22 | 61.72 | 59.08 |
| 18 05 0411 | Crushed Limestone, pH Adjustment, 800 Lbs/Acre | ACRE | 156.71 | 134.79 | 121.37 | 99.94 | 94.68 |
| 18 05 0412 | Purchase & Spread Dry Granular Limestone for pH Control | SY | 0.05 | 0.04 | 0.04 | 0.03 | 0.03 |
| 18 05 0413 | Watering with 3,000-Gallon Tank Truck, per Pass | ACRE | 111.81 | 90.08 | 75.73 | 54.43 | 49.85 |
| 18 05 0701 | Full Circle Sprinkler Head, 30' Diameter | EACH | 94.41 | 76.34 | 68.61 | 51.14 | 44.74 |
| 18 05 0702 | Semicircle Sprinkler Head, 20' Diameter | EACH | 43.12 | 34.09 | 30.22 | 21.49 | 18.29 |
| 18 05 0703 | Shrub/Tree Sprinkler Head | EACH | 40.45 | 31.42 | 27.55 | 18.82 | 15.62 |
| 18 05 0705 | Control Box | EACH | 1,598 | 1,461 | 1,402 | 1,270 | 1,221 |
| 18 05 0706 | 2 1/2", Cast-iron Body, Gate Valve | EACH | 257.59 | 239.88 | 232.30 | 215.19 | 208.91 |
| 18 05 0707 | 4" Reducer | EACH | 80.38 | 64.15 | 57.20 | 41.51 | 35.76 |
| 18 05 0710 | Testing & Inspection of Sprinkler System | LS | 856.76 | 660.42 | 576.36 | 386.59 | 317.00 |
| 18 05 0711 | Automated Pop-Up Sprinkler System | ACRE | 8,779 | 6,984 | 6,188 | 4,451 | 3,832 |
| 19 01 0201 | 3/4", Class 200, PVC Piping | LF | 7.45 | 5.81 | 5.04 | 3.45 | 2.90 |
| 19 01 0202 | 1", Class 200, PVC Piping | LF | 7.48 | 5.84 | 5.07 | 3.48 | 2.93 |
| 19 01 0204 | 2", Class 200, PVC Piping | LF | 8.86 | 6.96 | 6.07 | 4.23 | 3.60 |
| 19 01 0205 | 2 1/2", Class 200, PVC Piping | LF | 9.03 | 7.13 | 6.24 | 4.40 | 3.77 |
| 33 11 0301 | Soil Tilling, D3 Dozer with Tiller Attachment | HOUR | 157.86 | 126.90 | 105.69 | 75.30 | 69.25 |
| 33 11 0302 | Soil Tilling, D4 Dozer with Tiller Attachment | HOUR | 158.82 | 127.70 | 106.33 | 75.78 | 69.73 |
| 33 11 0303 | Soil Tilling, D7 Dozer with Tiller Attachment | HOUR | 297.42 | 243.20 | 198.73 | 145.08 | 139.03 |
| 33 11 0304 | Soil Tilling, D9 Dozer with Tiller Attachment | HOUR | 431.76 | 355.15 | 288.29 | 212.25 | 206.20 |
| 33 11 9901 | Application of Bioculture to Contaminated Soil | ACRE | 109.25 | 87.52 | 73.18 | 51.87 | 47.29 |
| 33 11 9902 | Light Petroleum Hydrocarbon Degraders, Microorganisms | LB | 19.88 | 19.88 | 19.88 | 19.88 | 19.88 |
| 33 11 9903 | Light Petroleum Hydrocarbon Degraders, 100 Lb Bag, Microorganisms | EACH | 1,948 | 1,948 | 1,948 | 1,948 | 1,948 |
| 33 11 9904 | Light Petroleum Hydrocarbon Degraders, 1,000 Lb Bag, Microorganisms | EACH | 17,490 | 17,490 | 17,490 | 17,490 | 17,490 |
| 33 11 9905 | Heavy Petroleum Hydrocarbon/Creosote Degraders, Microorganisms | LB | 13.25 | 13.25 | 13.25 | 13.25 | 13.25 |
| 33 11 9906 | Heavy Petroleum Hydrocarbon/Creosote Degraders, 100 Lb Bag, Microorganisms | EACH | 1,007 | 1,007 | 1,007 | 1,007 | 1,007 |
| 33 11 9907 | Heavy Petroleum Hydrocarbon/Creosote Degraders, 1,000 Lb Bag, Microorganisms | EACH | 8,745 | 8,745 | 8,745 | 8,745 | 8,745 |

**Environmental Remediation: Assemblies Cost Book**

# Cost Data for Remediation

## In Situ Biodegradation (Land Treatment)

| Assembly | Description | Unit | Unit Cost by Safety Level A | B | C | D | E |
|---|---|---|---|---|---|---|---|
| | **Capital Costs** | | | | | | |
| 33 17 0801 | Decontaminate Light Equipment | EACH | 189.11 | 146.65 | 127.13 | 86.01 | 71.80 |
| 33 17 0802 | Decontaminate Medium Equipment | EACH | 378.22 | 293.30 | 254.27 | 172.03 | 143.59 |
| 33 17 0803 | Decontaminate Heavy Equipment | EACH | 560.32 | 434.51 | 376.69 | 254.86 | 212.73 |
| 33 32 0101 | 7.5 GPH, 860 PSI, 55 Gallon Polyethylene Tank with Steel Overpack | EACH | 4,203 | 3,886 | 3,751 | 3,445 | 3,333 |
| 33 32 0102 | 20.8 GPH, 150 PSI, 55 Gallon Polyethylene Tank with Steel Overpack | EACH | 4,203 | 3,886 | 3,751 | 3,445 | 3,333 |
| 33 32 0103 | 7.5 GPH, 860 PSI, 102 Gallon Polyethylene Tank with Steel Overpack | EACH | 5,601 | 5,206 | 5,036 | 4,654 | 4,513 |
| 33 32 0104 | 20.8 GPH, 150 PSI, 102 Gallon Polyethylene Tank with Steel Overpack | EACH | 5,601 | 5,206 | 5,036 | 4,654 | 4,513 |
| 33 32 0105 | 7.5 GPH, 860 PSI, 167 Gallon Polyethylene Tank with Steel Overpack | EACH | 5,264 | 4,868 | 4,699 | 4,316 | 4,176 |
| 33 32 0106 | 20.8 GPH, 150 PSI, 167 Gallon Polyethylene Tank with Steel Overpack | EACH | 5,264 | 4,868 | 4,699 | 4,316 | 4,176 |
| 33 32 0107 | 7.5 GPH, 860 PSI, 260 Gallon Polyethylene Tank with Steel Overpack | EACH | 6,288 | 5,892 | 5,723 | 5,340 | 5,200 |
| 33 32 0108 | 20.8 GPH, 150 PSI, 260 Gallon Polyethylene Tank with Steel Overpack | EACH | 6,288 | 5,892 | 5,723 | 5,340 | 5,200 |
| 33 33 0117 | Hydrated Lime, Powdered, Bulk | TON | 97.55 | 97.55 | 97.54 | 97.54 | 97.53 |
| 33 33 0118 | Hydrated Lime, Powdered, 50 Lb Bag | EACH | 25.76 | 21.23 | 18.49 | 14.06 | 12.94 |
| 33 34 0101 | Standby D3 Bulldozer with Tiller | HOUR | 17.17 | 14.31 | 11.45 | 8.59 | 8.59 |
| 33 34 0102 | Standby D4 Bulldozer with Tiller | HOUR | 16.89 | 14.08 | 11.26 | 8.45 | 8.45 |
| 33 34 0103 | Standby D7 Bulldozer with Tiller | HOUR | 47.11 | 39.26 | 31.41 | 23.55 | 23.55 |
| 33 34 0104 | Standby D9 Bulldozer with Tiller | HOUR | 82.94 | 69.12 | 55.30 | 41.47 | 41.47 |

---

**Environmental Remediation: Assemblies Cost Book**

# Cost Data for Remediation

## In Situ Biodegradation (Saturated Zone)

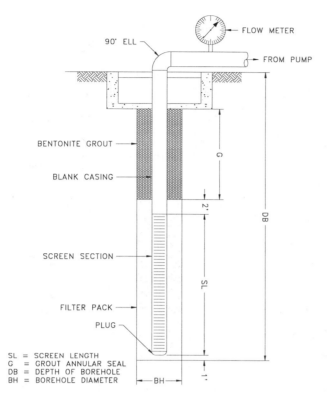

### General:

In situ biodegradation of the saturated zone is accomplished by promoting and accelerating natural biodegradation by introducing oxygen and nutrients into the contaminated aquifer. Injection wells, infiltration galleries, and surface irrigation are all potential methods of distributing oxygen and nutrients to areas of fuel contamination. The costs presented here assume the use of injection wells.

### Typical Treatment Train:

Treatment and disposal of drill cuttings, professional field labor, sampling and analysis.

### Common Cost Components:

1. Mobilization of the drill rig and crew
2. Drilling, decontamination of equipment
3. Well materials and installation
4. Pump(s)
5. Nutrients, microorganisms
6. Holding tank(s)
7. Piping
8. Operations and maintenance

### Other Cost Considerations:

Clearing and grubbing, fencing, electrical distribution.

**Environmental Remediation: Assemblies Cost Book**

# Cost Data for Remediation

## In Situ Biodegradation (Saturated Zone)

| Assembly | Description | Unit | A | B | C | D | E |
|---|---|---|---|---|---|---|---|
| | **Capital Costs** | | | | | | |
| 19 01 0202 | 1", Class 200, PVC Piping | LF | 7.48 | 5.84 | 5.07 | 3.48 | 2.93 |
| 19 01 0204 | 2", Class 200, PVC Piping | LF | 8.86 | 6.96 | 6.07 | 4.23 | 3.60 |
| 19 01 0206 | 3", Class 200, PVC Piping | LF | 11.81 | 9.34 | 8.20 | 5.80 | 4.98 |
| 19 01 0207 | 4", Class 200, PVC Piping | LF | 12.33 | 9.86 | 8.72 | 6.32 | 5.50 |
| 19 01 0208 | 6", Class 200, PVC Piping | LF | 14.95 | 12.20 | 10.93 | 8.27 | 7.36 |
| 19 01 0210 | 4", Class 150, PVC Piping | LF | 12.42 | 9.95 | 8.81 | 6.41 | 5.59 |
| 19 01 0211 | 6", Class 150, PVC Piping | LF | 14.84 | 12.09 | 10.82 | 8.16 | 7.25 |
| 19 01 0212 | 8", Class 150, PVC Piping | LF | 18.36 | 15.27 | 13.84 | 10.85 | 9.82 |
| 19 01 0213 | 10", Class 150, PVC Piping | LF | 22.90 | 19.37 | 17.73 | 14.32 | 13.14 |
| 19 01 0214 | 12", Class 150, PVC Piping | LF | 28.75 | 24.63 | 22.72 | 18.73 | 17.36 |
| 19 01 0310 | 5,000 Gallon Water Tank, Steel, Horizontal @ Grade | EACH | 15,945 | 13,359 | 12,252 | 9,752 | 8,835 |
| 19 01 0311 | 10,000 Gallon Water Tank, Steel, Horizontal @ Grade | EACH | 18,756 | 16,170 | 15,063 | 12,563 | 11,646 |
| 19 01 0312 | 20,000 Gallon Water Tank, Steel, Horizontal @ Grade | EACH | 28,844 | 24,853 | 23,144 | 19,286 | 17,871 |
| 19 01 0313 | 35,000 Gallon Water Tank, Steel, Horizontal @ Grade | EACH | 45,913 | 40,740 | 38,525 | 33,525 | 31,692 |
| 19 04 0602 | 550 Gallon Steel Sump, Aboveground with Supports & Fittings | EACH | 1,598 | 1,390 | 1,301 | 1,099 | 1,026 |
| 19 04 0624 | 4,000 Gallon Horizontal Plastic Sump with 6" NPT Connection | EACH | 4,130 | 3,927 | 3,841 | 3,645 | 3,573 |
| 19 04 0625 | 6,000 Gallon Horizontal Plastic Sump with 6" NPT Connection | EACH | 5,248 | 4,975 | 4,858 | 4,594 | 4,497 |
| 33 01 0101 | Mobilize/DeMobilize Drilling Rig & Crew | LS | 4,949 | 4,031 | 3,308 | 2,401 | 2,280 |
| 33 02 0303 | Organic Vapor Analyzer Rental, per Day | DAY | 184.30 | 184.30 | 184.30 | 184.30 | 184.30 |
| 33 02 0537 | Water Level Chart Recorder for Monitoring Wells on Casings | EACH | 1,221 | 1,221 | 1,221 | 1,221 | 1,221 |
| 33 02 1509 | Water Quality Parameter Testing Device | WEEK | 227.90 | 227.90 | 227.90 | 227.90 | 227.90 |
| 33 02 1511 | Water Flowmeter | EACH | 132.50 | 132.50 | 132.50 | 132.50 | 132.50 |
| 33 02 1913 | Laboratory Bench-scale Studies | EACH | 150.00 | 150.00 | 150.00 | 150.00 | 150.00 |
| 33 11 9902 | Light Petroleum Hydrocarbon Degraders, Microorganisms | LB | 19.88 | 19.88 | 19.88 | 19.88 | 19.88 |
| 33 11 9903 | Light Petroleum Hydrocarbon Degraders, 100 Lb Bag, Microorganisms | EACH | 1,948 | 1,948 | 1,948 | 1,948 | 1,948 |
| 33 11 9904 | Light Petroleum Hydrocarbon Degraders, 1,000 Lb Bag, Microorganisms | EACH | 17,490 | 17,490 | 17,490 | 17,490 | 17,490 |
| 33 11 9905 | Heavy Petroleum Hydrocarbon/Creosote Degraders, Microorganisms | LB | 13.25 | 13.25 | 13.25 | 13.25 | 13.25 |
| 33 11 9906 | Heavy Petroleum Hydrocarbon/Creosote Degraders, 100 Lb Bag, Microorganisms | EACH | 1,007 | 1,007 | 1,007 | 1,007 | 1,007 |

**Environmental Remediation: Assemblies Cost Book**

*Page 3-156*

# Cost Data for Remediation

## In Situ Biodegradation (Saturated Zone)

| Assembly | Description | Unit | A | B | C | D | E |
|---|---|---|---|---|---|---|---|
| | **Capital Costs** | | | | | | |
| 33 11 9907 | Heavy Petroleum Hydrocarbon/Creosote Degraders, 1,000 Lb Bag, Microorganisms | EACH | 8,745 | 8,745 | 8,745 | 8,745 | 8,745 |
| 33 11 9951 | Bionutrients, 50 Lb Bag | EACH | 58.30 | 58.30 | 58.30 | 58.30 | 58.30 |
| 33 17 0808 | Decontaminate Rig, Augers, Screen (Rental Equipment) | DAY | 205.34 | 205.34 | 205.34 | 205.34 | 205.34 |
| 33 23 0101 | 2" PVC, Schedule 40, Well Casing | LF | 17.45 | 14.39 | 11.98 | 8.95 | 8.55 |
| 33 23 0102 | 4" PVC, Schedule 40, Well Casing | LF | 26.98 | 22.38 | 18.77 | 14.24 | 13.63 |
| 33 23 0103 | 6" PVC, Schedule 40, Well Casing | LF | 27.07 | 22.66 | 19.19 | 14.84 | 14.25 |
| 33 23 0104 | 8" PVC, Schedule 40, Well Casing | LF | 34.50 | 29.27 | 25.15 | 19.98 | 19.29 |
| 33 23 0111 | 2" PVC, Schedule 80, Well Casing | LF | 17.78 | 14.72 | 12.31 | 9.28 | 8.88 |
| 33 23 0112 | 4" PVC, Schedule 80, Well Casing | LF | 28.55 | 23.95 | 20.34 | 15.81 | 15.20 |
| 33 23 0113 | 6" PVC, Schedule 80, Well Casing | LF | 47.52 | 40.17 | 34.40 | 27.14 | 26.17 |
| 33 23 0115 | 8" PVC, Schedule 80, Well Casing | LF | 61.99 | 52.81 | 45.58 | 36.51 | 35.30 |
| 33 23 0121 | 2" Stainless Steel, Well Casing | LF | 31.75 | 28.14 | 25.30 | 21.73 | 21.25 |
| 33 23 0122 | 4" Stainless Steel, Well Casing | LF | 47.66 | 43.06 | 39.45 | 34.92 | 34.31 |
| 33 23 0123 | 6" Stainless Steel Well Casing, 5' Sections, Flush Threaded | LF | 370.73 | 311.96 | 265.77 | 207.74 | 199.94 |
| 33 23 0125 | 8" Stainless Steel Well Casing, 5' Sections, Flush Threaded | LF | 526.42 | 452.94 | 395.18 | 322.61 | 312.87 |
| 33 23 0201 | 2" PVC, Schedule 40, Well Screen | LF | 23.47 | 19.52 | 16.42 | 12.52 | 12.00 |
| 33 23 0202 | 4" PVC, Schedule 40, Well Screen | LF | 38.16 | 32.00 | 27.16 | 21.09 | 20.27 |
| 33 23 0203 | 6" PVC, Schedule 40, Well Screen | LF | 47.13 | 39.78 | 34.01 | 26.75 | 25.78 |
| 33 23 0204 | 8" PVC, Schedule 40, Well Screen | LF | 62.02 | 52.84 | 45.61 | 36.54 | 35.33 |
| 33 23 0211 | 2" PVC, Schedule 80, Well Screen | LF | 18.70 | 15.64 | 13.23 | 10.20 | 9.80 |
| 33 23 0212 | 4" PVC, Schedule 80, Well Screen | LF | 30.01 | 25.41 | 21.80 | 17.27 | 16.66 |
| 33 23 0213 | 6" PVC, Schedule 80, Well Screen | LF | 49.72 | 42.37 | 36.60 | 29.34 | 28.37 |
| 33 23 0214 | 8" PVC, Schedule 80, Well Screen | LF | 65.91 | 56.73 | 49.50 | 40.43 | 39.22 |
| 33 23 0221 | 2" Stainless Steel, Well Screen | LF | 26.91 | 23.84 | 21.44 | 18.41 | 18.01 |
| 33 23 0222 | 4" Stainless Steel, Well Screen | LF | 47.66 | 43.06 | 39.45 | 34.92 | 34.31 |
| 33 23 0223 | 6" Stainless Steel Well Screen, 5' Sec, Flush Threaded | LF | 366.05 | 307.28 | 261.09 | 203.06 | 195.26 |
| 33 23 0225 | 8" Stainless Steel Well Screen, 5' Sec, Flush Threaded | LF | 462.22 | 388.74 | 330.98 | 258.41 | 248.67 |
| 33 23 0301 | 2" PVC, Well Plug | EACH | 29.37 | 24.77 | 21.16 | 16.63 | 16.02 |
| 33 23 0302 | 4" PVC, Well Plug | EACH | 55.65 | 48.91 | 43.62 | 36.97 | 36.07 |
| 33 23 0303 | 6" PVC, Well Plug | EACH | 112.63 | 101.15 | 92.13 | 80.79 | 79.27 |
| 33 23 0304 | 8" PVC, Well Plug | EACH | 132.70 | 119.58 | 109.26 | 96.31 | 94.57 |
| 33 23 0311 | 2" Stainless Steel, Well Plug | EACH | 83.24 | 74.06 | 66.83 | 57.76 | 56.55 |
| 33 23 0312 | 4" Stainless Steel, Well Plug | EACH | 112.24 | 102.03 | 94.01 | 83.93 | 82.58 |

**Environmental Remediation: Assemblies Cost Book**

# Cost Data for Remediation

## In Situ Biodegradation (Saturated Zone)

| Assembly | Description | Unit | A | B | C | D | E |
|---|---|---|---|---|---|---|---|
| | **Capital Costs** | | | | | | |
| 33 23 0313 | 6" Stainless Steel, Well Plug | EACH | 573.04 | 481.18 | 408.98 | 318.27 | 306.09 |
| 33 23 0314 | 8" Stainless Steel, Well Plug | EACH | 702.53 | 597.56 | 515.04 | 411.38 | 397.46 |
| 33 23 1101 | Hollow-stem Auger, 8" Outside Diameter Borehole for 2" Well | LF | 89.98 | 73.28 | 60.15 | 43.66 | 41.45 |
| 33 23 1102 | Hollow-stem Auger, 11" Outside Diameter Borehole for 4" Well | LF | 109.98 | 89.57 | 73.52 | 53.36 | 50.66 |
| 33 23 1103 | Hollow-stem Auger, 13 3/4" Outside Diameter Borehole for 6" Well | LF | 141.40 | 115.16 | 94.53 | 68.61 | 65.13 |
| 33 23 1105 | Hollow-stem Auger, 16" Outside Diameter Borehole for 8" Well | LF | 164.97 | 134.35 | 110.28 | 80.05 | 75.99 |
| 33 23 1106 | Split Spoon Sample, 2" x 24", During Drilling | EACH | 46.67 | 46.67 | 46.67 | 46.67 | 46.67 |
| 33 23 1111 | Well Development Equipment Rental | WEEK | 478.91 | 462.93 | 456.09 | 440.65 | 434.98 |
| 33 23 1121 | Standby for Drilling | EACH | 618.63 | 503.81 | 413.56 | 300.17 | 284.95 |
| 33 23 1122 | Move Rig/Equipment Around Site | EACH | 154.66 | 125.95 | 103.39 | 75.04 | 71.24 |
| 33 23 1125 | Concrete Coring (Minimum Charge) | DAY | 2,475 | 2,015 | 1,654 | 1,201 | 1,140 |
| 33 23 1126 | Furnish 55 Gallon Drum for Drill Cuttings & Development Water | EACH | 65.19 | 65.19 | 65.19 | 65.19 | 65.19 |
| 33 23 1152 | Mud Drilling, 6" Diameter Borehole | LF | 81.01 | 65.98 | 54.16 | 39.31 | 37.31 |
| 33 23 1153 | Mud Drilling, 8" Diameter Borehole | LF | 88.38 | 71.97 | 59.08 | 42.88 | 40.71 |
| 33 23 1154 | Mud Drilling, 10" Diameter Borehole | LF | 100.65 | 81.97 | 67.29 | 48.84 | 46.36 |
| 33 23 1155 | Mud Drilling, 12" Diameter Borehole | LF | 112.92 | 91.97 | 75.49 | 54.79 | 52.01 |
| 33 23 1161 | Air Rotary 6" Borehole in Unconsolidated | LF | 73.65 | 59.98 | 49.23 | 35.73 | 33.92 |
| 33 23 1162 | Air Rotary 8" Borehole in Unconsolidated | LF | 122.74 | 99.96 | 82.06 | 59.56 | 56.54 |
| 33 23 1163 | Air Rotary 10" Borehole in Unconsolidated | LF | 220.94 | 179.93 | 147.70 | 107.20 | 101.77 |
| 33 23 1164 | Air Rotary 12" Borehole in Unconsolidated | LF | 294.59 | 239.91 | 196.93 | 142.94 | 135.69 |
| 33 23 1401 | 2" Screen, Filter Pack | LF | 16.49 | 13.88 | 11.84 | 9.27 | 8.92 |
| 33 23 1402 | 4" Screen, Filter Pack | LF | 29.10 | 24.50 | 20.89 | 16.36 | 15.75 |
| 33 23 1403 | 6" Screen, Filter Pack | LF | 42.19 | 35.53 | 30.29 | 23.72 | 22.83 |
| 33 23 1404 | Gravel Pack, 8" Well | LF | 29.47 | 23.93 | 19.82 | 14.36 | 13.48 |
| 33 23 1502 | Surface Pad, Concrete, 4' x 4' x 4" | EACH | 24.21 | 21.82 | 20.74 | 18.43 | 17.61 |
| 33 23 1811 | 2" Well, Portland Cement Grout | LF | 0.92 | 0.92 | 0.92 | 0.92 | 0.92 |
| 33 23 1812 | 4" Well, Portland Cement Grout | LF | 1.38 | 1.38 | 1.38 | 1.38 | 1.38 |
| 33 23 1813 | 6" Well, Portland Cement Grout | LF | 7.80 | 7.80 | 7.80 | 7.80 | 7.80 |
| 33 23 1814 | 8" Well, Portland Cement Grout | LF | 11.02 | 11.02 | 11.02 | 11.02 | 11.02 |
| 33 23 2101 | 2" Well, Bentonite Seal | EACH | 63.00 | 52.67 | 44.55 | 34.34 | 32.97 |

**Environmental Remediation: Assemblies Cost Book**

*Page 3-158*

# Cost Data for Remediation

## In Situ Biodegradation (Saturated Zone)

| | | | Unit Cost by Safety Level | | | | |
|---|---|---|---|---|---|---|---|
| Assembly | Description | Unit | A | B | C | D | E |
| | **Capital Costs** | | | | | | |
| 33 23 2102 | 4" Well, Bentonite Seal | EACH | 157.54 | 131.70 | 111.39 | 85.87 | 82.44 |
| 33 23 2103 | 6" Well, Bentonite Seal | EACH | 252.01 | 210.68 | 178.18 | 137.37 | 131.89 |
| 33 23 2105 | 8" Well, Bentonite Seal | EACH | 346.60 | 289.75 | 245.06 | 188.92 | 181.39 |
| 33 26 0201 | 1" Stainless Steel Piping, Schedule 40, Threaded | LF | 18.67 | 16.12 | 15.03 | 12.56 | 11.65 |
| 33 26 0202 | 2" Stainless Steel Piping, Schedule 40, Threaded | LF | 33.47 | 29.21 | 27.39 | 23.28 | 21.77 |
| 33 26 0203 | 3" Stainless Steel Piping, Schedule 40, Threaded | LF | 57.00 | 50.61 | 47.88 | 41.71 | 39.45 |
| 33 26 0204 | 4" Stainless Steel Piping, Schedule 40, Threaded | LF | 80.15 | 71.72 | 68.12 | 59.98 | 57.00 |
| 33 26 0205 | 6" Stainless Steel Piping, Schedule 10, Type 316 | LF | 78.58 | 70.01 | 66.14 | 57.84 | 54.94 |
| 33 26 0206 | 8" Stainless Steel Piping, Schedule 40, Welded | LF | 197.22 | 183.32 | 176.89 | 163.42 | 158.80 |
| 33 26 0208 | 10" Stainless Steel Piping, Schedule 40, Welded | LF | 280.77 | 265.48 | 258.40 | 243.59 | 238.50 |
| 33 26 0209 | 12" Stainless Steel Piping, Schedule 40, Welded | LF | 312.43 | 295.44 | 287.57 | 271.12 | 265.47 |
| 33 27 0112 | 2" PVC, 90 Degree, Elbow | EACH | 41.36 | 32.22 | 28.30 | 19.46 | 16.22 |
| 33 27 0114 | 4" PVC, 90 Degree, Elbow | EACH | 101.72 | 80.57 | 71.51 | 51.07 | 43.57 |
| 33 27 0115 | 6" PVC, 90 Degree, Elbow | EACH | 135.44 | 111.27 | 100.92 | 77.56 | 68.99 |
| 33 27 0116 | 8" PVC, 90 Degree, Elbow | EACH | 200.21 | 172.01 | 159.94 | 132.68 | 122.68 |
| 33 27 0211 | 2" Stainless Steel, 90 Degree, Elbow | EACH | 114.28 | 93.45 | 84.53 | 64.39 | 57.01 |
| 33 27 0212 | 4" Stainless Steel, 90 Degree, Elbow | EACH | 391.27 | 338.50 | 315.91 | 264.90 | 246.20 |
| 33 27 0213 | 6" Stainless Steel, 90 Degree, Elbow | EACH | 425.65 | 376.17 | 355.00 | 307.18 | 289.64 |
| 33 27 0214 | 8" Stainless Steel, 90 Degree, Elbow | EACH | 566.44 | 516.97 | 495.79 | 447.97 | 430.44 |
| 33 27 0402 | 2" Iron Body Check Valve | EACH | 188.96 | 174.31 | 168.03 | 153.86 | 148.67 |
| 33 27 0404 | 4" Iron Body Check Valve | EACH | 354.29 | 315.96 | 299.54 | 262.49 | 248.90 |
| 33 27 0405 | 6" Iron Body Check Valve | EACH | 618.60 | 554.57 | 523.95 | 461.86 | 441.15 |
| 33 27 0406 | 8" Iron Body Check Valve | EACH | 1,094 | 980.95 | 926.96 | 817.50 | 780.98 |
| 33 29 0101 | 10 GPM, 10' Head, 1/6 HP, Centrifugal Pump | EACH | 789.18 | 730.44 | 705.29 | 648.51 | 627.69 |
| 33 29 0102 | 10 GPM, 1/2 HP, Centrifugal Pump | EACH | 1,063 | 952.32 | 905.07 | 798.39 | 759.27 |
| 33 29 0103 | 50 GPM, 100' Head, 3 HP, Centrifugal Pump | EACH | 2,658 | 2,500 | 2,432 | 2,279 | 2,223 |
| 33 29 0104 | 75 GPM, 100' Head, 5 HP, Centrifugal Pump | EACH | 2,946 | 2,754 | 2,672 | 2,487 | 2,419 |
| 33 29 0105 | 5 HP, 100 GPM, Centrifugal Pump | EACH | 2,475 | 2,301 | 2,227 | 2,059 | 1,998 |
| 33 29 0106 | 100 GPM, 150' Head, 7.5 HP, Centrifugal Pump | EACH | 3,419 | 3,191 | 3,093 | 2,873 | 2,793 |
| 33 29 0107 | 125 GPM, 150' Head, 10 HP, Centrifugal Pump | EACH | 3,815 | 3,512 | 3,382 | 3,088 | 2,981 |
| 33 29 0108 | 10 HP, 200 GPM, Centrifugal Pump | EACH | 2,941 | 2,714 | 2,616 | 2,396 | 2,316 |
| 33 29 0109 | 10 HP, 250 GPM, Centrifugal Pump | EACH | 3,032 | 2,805 | 2,707 | 2,487 | 2,407 |
| 33 29 0110 | 15 HP, 300 GPM, Centrifugal Pump | EACH | 3,441 | 3,160 | 3,040 | 2,770 | 2,670 |

**Environmental Remediation: Assemblies Cost Book**

# Cost Data for Remediation

## In Situ Biodegradation (Saturated Zone)

| Assembly | Description | Unit | Unit Cost by Safety Level | | | | |
|---|---|---|---|---|---|---|---|
| | | | A | B | C | D | E |
| | **Capital Costs** | | | | | | |
| 33 29 0111 | 20 HP, 500 GPM, Centrifugal Pump | EACH | 4,544 | 4,241 | 4,111 | 3,818 | 3,710 |
| 33 29 0112 | 30 HP, 750 GPM, Centrifugal Pump | EACH | 5,213 | 4,808 | 4,635 | 4,244 | 4,100 |
| 33 29 0113 | 40 HP, 1050 GPM, Centrifugal Pump | EACH | 5,655 | 5,199 | 5,004 | 4,564 | 4,403 |
| 33 29 0114 | 60 HP, 1500 GPM, Centrifugal Pump | EACH | 9,842 | 8,823 | 8,350 | 7,363 | 7,024 |
| 33 29 0115 | 75 HP, 2000 GPM, Centrifugal Pump | EACH | 12,456 | 11,233 | 10,666 | 9,482 | 9,075 |
| 33 29 0116 | 100 HP, 3000 GPM, Centrifugal Pump | EACH | 15,575 | 14,046 | 13,338 | 11,857 | 11,348 |
| 33 29 0128 | 100 GPM, Centrifugal Pump, 100' Head, 7.5 HP, Cast Iron | EACH | 3,417 | 3,243 | 3,169 | 3,001 | 2,940 |
| 33 29 0129 | 250 GPM, Centrifugal Pump, 100' Head, 15 HP, Cast Iron | EACH | 4,485 | 4,258 | 4,160 | 3,940 | 3,860 |
| 33 29 0130 | 500 GPM, Centrifugal Pump, 100' Head, 20 HP, Cast Iron | EACH | 5,433 | 5,152 | 5,032 | 4,762 | 4,662 |
| 33 29 0131 | 750 GPM, Centrifugal Pump, 100' Head, 25 HP, Cast Iron | EACH | 5,474 | 5,215 | 5,104 | 4,853 | 4,761 |
| 33 29 0132 | 1,000 GPM, Centrifugal Pump, 100' Head, 40 HP, Cast Iron | EACH | 7,113 | 6,780 | 6,637 | 6,315 | 6,197 |
| 33 29 0133 | 1,500 GPM, Centrifugal Pump, 100' Head, 50 HP, Cast Iron | EACH | 8,869 | 8,461 | 8,272 | 7,878 | 7,742 |
| 33 29 0134 | 2,000 GPM, Centrifugal Pump, 100' Head, 75 HP, Cast Iron | EACH | 11,903 | 11,393 | 11,157 | 10,664 | 10,494 |
| 33 29 0135 | 3,000 GPM, Centrifugal Pump, 100' Head, 100 HP, Cast Iron | EACH | 13,286 | 12,674 | 12,391 | 11,799 | 11,595 |
| 33 29 0136 | 3,500 GPM, Centrifugal Pump, 125' Head, 150 HP, Cast Iron | EACH | 14,254 | 13,575 | 13,260 | 12,602 | 12,376 |
| 33 29 0137 | 4,000 GPM, Centrifugal Pump, 150' Head, 200 HP, Cast Iron | EACH | 18,299 | 17,425 | 17,021 | 16,174 | 15,884 |
| 33 29 0138 | 100 GPM, Centrifugal Pump, 500' Head, 40 HP, Cast Iron | EACH | 5,240 | 4,959 | 4,839 | 4,569 | 4,469 |
| 33 29 0139 | 200 GPM, Centrifugal Pump, 500' Head, 50 HP, Cast Iron | EACH | 5,800 | 5,497 | 5,367 | 5,074 | 4,966 |
| 33 29 0140 | 300 GPM, Centrifugal Pump, 500' Head, 75 HP, Cast Iron | EACH | 6,671 | 6,252 | 6,043 | 5,636 | 5,505 |
| 33 29 0141 | 400 GPM, Centrifugal Pump, 500' Head, 100 HP, Cast Iron | EACH | 7,679 | 7,102 | 6,815 | 6,255 | 6,076 |
| 33 29 0142 | 800 GPM, Centrifugal Pump, 500' Head, 200 HP, Cast Iron | EACH | 16,870 | 16,101 | 15,718 | 14,971 | 14,732 |
| 33 29 0143 | 15 GPM, Centrifugal Pump, 6' Head, 1/8 HP, Bronze | EACH | 602.28 | 557.16 | 537.84 | 494.23 | 478.23 |
| 33 29 0144 | 20 GPM, Centrifugal Pump, 5.5' Head, 1/8 HP, Bronze | EACH | 749.03 | 694.45 | 671.08 | 618.32 | 598.98 |
| 33 29 0145 | 25 GPM, Centrifugal Pump, 4.5' Head, 1/8 HP, Bronze | EACH | 749.03 | 694.45 | 671.08 | 618.32 | 598.98 |
| 33 29 0146 | 30 GPM, Centrifugal Pump, 2' Head, 1/8 HP, Bronze | EACH | 749.03 | 694.45 | 671.08 | 618.32 | 598.98 |
| 33 29 0147 | 30 GPM, Centrifugal Pump, High Velocity, 7' Head, 1/8 HP | EACH | 767.19 | 708.44 | 683.29 | 626.51 | 605.69 |
| 33 29 0148 | 35 GPM, Centrifugal Pump, High Velocity, 4' Head, 1/8 HP | EACH | 767.19 | 708.44 | 683.29 | 626.51 | 605.69 |
| 33 29 0149 | 35 GPM, Centrifugal Pump, High Velocity, 5' Head, 1/8 HP | EACH | 767.19 | 708.44 | 683.29 | 626.51 | 605.69 |
| 33 29 0150 | 50 GPM, Turbine Pump, 100' Head, 2 HP, Iron/Bronze | EACH | 2,766 | 2,361 | 2,188 | 1,797 | 1,653 |
| 33 29 0151 | 100 GPM, Turbine Pump, 100' Head, 3 HP, Iron/Bronze | EACH | 3,487 | 3,031 | 2,836 | 2,396 | 2,235 |
| 33 29 0152 | 250 GPM, Turbine Pump, 150' Head, 15 HP, Iron/Bronze | EACH | 4,820 | 4,412 | 4,223 | 3,829 | 3,693 |

**Environmental Remediation: Assemblies Cost Book**

*Page 3-160*

1998 by ECHOS. All rights reserved

# Cost Data for Remediation

## In Situ Biodegradation (Saturated Zone)

| Assembly | Description | Unit | Unit Cost by Safety Level A | B | C | D | E |
|---|---|---|---|---|---|---|---|
| | **Capital Costs** | | | | | | |
| 33 29 0153 | 500 GPM, Turbine Pump, 150' Head, 25 HP, Iron/Bronze | EACH | 7,100 | 6,629 | 6,412 | 5,956 | 5,799 |
| 33 29 0154 | 1,000 GPM, Turbine Pump, 150' Head, 50 HP, Iron/Bronze | EACH | 10,275 | 9,765 | 9,529 | 9,036 | 8,866 |
| 33 29 0155 | 2,000 GPM, Turbine Pump, 150' Head, 100 HP, Iron/Bronze | EACH | 11,730 | 11,118 | 10,835 | 10,243 | 10,039 |
| 33 29 0156 | 3,000 GPM, Turbine Pump, 150' Head, 150 HP, Iron/Bronze | EACH | 13,412 | 12,648 | 12,294 | 11,553 | 11,299 |
| 33 29 0157 | 4,000 GPM, Turbine Pump, 150' Head, 200 HP, Iron/Bronze | EACH | 18,900 | 18,026 | 17,622 | 16,775 | 16,485 |
| 33 29 0158 | 6,000 GPM, Turbine Pump, 150' Head, 300 HP, Iron/Bronze | EACH | 29,550 | 28,531 | 28,058 | 27,071 | 26,732 |
| 33 29 0159 | 10,000 GPM, Turbine Pump, 100' Head, 300 HP, Iron/Bronze | EACH | 40,460 | 39,237 | 38,670 | 37,486 | 37,079 |
| 33 29 0501 | Adjustable Flow 0 - 100% (25 GPM), 200 PSI, Pump with Motor | EACH | 1,598 | 1,575 | 1,565 | 1,543 | 1,535 |
| 33 29 0601 | Adjustable Flow 0 - 100% (28 GPM), 250 PSI, Pump with Motor | EACH | 1,598 | 1,575 | 1,565 | 1,543 | 1,535 |
| 33 33 0171 | Hydrogen Peroxide, 50% Solution, 500 Lb Drums | EACH | 667.80 | 667.80 | 667.80 | 667.80 | 667.80 |
| 33 33 0172 | Hydrogen Peroxide Feed Tank, Regulator, Injector & Panel Rental | EACH | 3,759 | 3,324 | 3,133 | 2,713 | 2,562 |
| 33 42 0101 | Electrical Charge | KWH | 0.06 | 0.06 | 0.06 | 0.06 | 0.06 |
| | **Operations & Maintenance** | | | | | | |
| 33 11 9902 | Light Petroleum Hydrocarbon Degraders, Microorganisms | LB | 19.88 | 19.88 | 19.88 | 19.88 | 19.88 |
| 33 11 9903 | Light Petroleum Hydrocarbon Degraders, 100 Lb Bag, Microorganisms | EACH | 1,948 | 1,948 | 1,948 | 1,948 | 1,948 |
| 33 11 9904 | Light Petroleum Hydrocarbon Degraders, 1,000 Lb Bag, Microorganisms | EACH | 17,490 | 17,490 | 17,490 | 17,490 | 17,490 |
| 33 11 9905 | Heavy Petroleum Hydrocarbon/Creosote Degraders, Microorganisms | LB | 13.25 | 13.25 | 13.25 | 13.25 | 13.25 |
| 33 11 9906 | Heavy Petroleum Hydrocarbon/Creosote Degraders, 100 Lb Bag, Microorganisms | EACH | 1,007 | 1,007 | 1,007 | 1,007 | 1,007 |
| 33 11 9907 | Heavy Petroleum Hydrocarbon/Creosote Degraders, 1,000 Lb Bag, Microorganisms | EACH | 8,745 | 8,745 | 8,745 | 8,745 | 8,745 |
| 33 11 9951 | Bionutrients, 50 Lb Bag | EACH | 58.30 | 58.30 | 58.30 | 58.30 | 58.30 |
| 33 33 0171 | Hydrogen Peroxide, 50% Solution, 500 Lb Drums | EACH | 667.80 | 667.80 | 667.80 | 667.80 | 667.80 |
| 33 42 0101 | Electrical Charge | KWH | 0.06 | 0.06 | 0.06 | 0.06 | 0.06 |

**Environmental Remediation: Assemblies Cost Book**

# Cost Data for Remediation
## In Situ Solidification

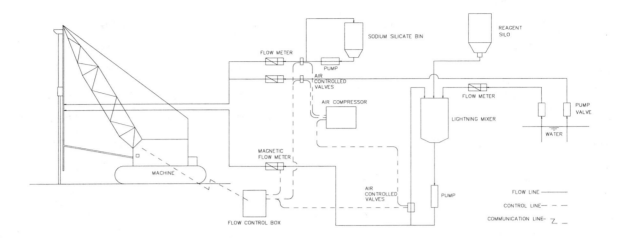

### General:

*In situ solidification is a process in which a reagent is added to transform a liquid, sludge, sediment, or soil into a solid form. Solidification may immobilize the contaminants within the crystalline structure of the solidified material, thus reducing the contaminant leaching potential, although this varies depending upon waste, soil, and reagent characteristics. In situ solidification differs from in situ stabilization in that in situ solidification is a physical treatment, whereas stabilization is a chemical treatment.*

### Typical Treatment Train:

*Site characterization; pre-treated waste sampling; excavation, transportation, and disposal of solidified material; sampling and analysis.*

### Common Cost Components:

1. Mobilization/demobilization of in situ solidification equipment
2. In situ solidification equipment
3. Operational labor
4. Chemical additives
5. Electrical usage
6. Equipment maintenance

### Other Cost Considerations:

*Permits, bench/pilot testing, electrical distribution, clearing and grubbing, access road, fencing, water distribution, gas distribution, sanitary sewer, communications, groundwater monitoring.*

**Environmental Remediation: Assemblies Cost Book**

# Cost Data for Remediation

## In Situ Solidification

| Assembly | Description | Unit | Unit Cost by Safety Level A | B | C | D | E |
|---|---|---|---|---|---|---|---|
| | **Capital Costs** | | | | | | |
| 20 99 9901 | 200 KW Diesel Generator, 3 Phase | EACH | 48,691 | 46,436 | 45,434 | 43,253 | 42,477 |
| 33 15 0401 | Cement Kiln Dust (Bulk) | TON | 5.26 | 5.26 | 5.26 | 5.26 | 5.26 |
| 33 15 0402 | Fly Ash, Class C (Bulk) | TON | 14.84 | 14.84 | 14.84 | 14.84 | 14.84 |
| 33 15 0403 | Fly Ash, Class F (Bulk) | TON | 15.64 | 15.64 | 15.64 | 15.64 | 15.64 |
| 33 15 0404 | Bottom Ash (Bulk) | TON | 2.83 | 2.83 | 2.83 | 2.83 | 2.83 |
| 33 15 0405 | Portland Cement Type I (Bulk) | TON | 77.73 | 77.73 | 77.73 | 77.73 | 77.73 |
| 33 15 0406 | Portland Cement Type K (Bulk) | TON | 103.88 | 103.88 | 103.88 | 103.88 | 103.88 |
| 33 15 0407 | Lime (Crushed Limestone) Bulk | TON | 5.30 | 5.30 | 5.30 | 5.30 | 5.30 |
| 33 15 0408 | Urrichem Proprietary Additive (Bulk) | TON | 980.00 | 980.00 | 980.00 | 980.00 | 980.00 |
| 33 15 0409 | Bitumen (Bulk) | TON | 40.28 | 40.28 | 40.28 | 40.28 | 40.28 |
| 33 15 0410 | Chloranan Proprietary Additive (Bulk) | TON | 636.00 | 636.00 | 636.00 | 636.00 | 636.00 |
| 33 15 0411 | P4 Proprietary Reagent (Bulk) | TON | 725.00 | 725.00 | 725.00 | 725.00 | 725.00 |
| 33 15 0412 | P27 Proprietary Reagent (Bulk) | TON | 275.00 | 275.00 | 275.00 | 275.00 | 275.00 |
| 33 15 0413 | Activated Carbon, Granular or Powdered Form (Bulk) | TON | 3,604 | 3,604 | 3,604 | 3,604 | 3,604 |
| 33 15 0414 | Sodium Silicate (Bulk) | TON | 275.60 | 275.60 | 275.60 | 275.60 | 275.60 |
| 33 15 0421 | Bulk Chemical Transport (40,000 Lb Truckload) | EACH | 2,350 | 2,350 | 2,350 | 2,350 | 2,350 |
| 33 15 0437 | Maintenance of Solidification/Stabilization Unit | YEAR | 14,354 | 11,064 | 9,656 | 6,477 | 5,311 |
| 33 15 0438 | Equipment Cost - In Situ Solidification/Stabilization | MONTH | 169,886 | 141,571 | 113,257 | 84,943 | 84,943 |
| 33 15 0439 | Operational Labor -In Situ Solidification/Stabilization | HR | 827.92 | 638.19 | 556.96 | 373.57 | 306.33 |
| 33 15 0440 | Mobilize/DeMobilize of In Situ Solidification/Stabilization Equipment | EACH | 66,234 | 51,055 | 44,557 | 29,886 | 24,506 |
| 33 17 0816 | 3,000 PSI Pressure Washer, 4.5 GPM | EACH | 8,868 | 8,868 | 8,868 | 8,868 | 8,868 |
| 33 17 0823 | Operation of Pressure Washer, Including Water, Soap, Electricity, Labor | HOUR | 79.38 | 63.31 | 56.43 | 40.91 | 35.21 |
| 33 33 0117 | Hydrated Lime, Powdered, Bulk | TON | 97.55 | 97.55 | 97.54 | 97.54 | 97.53 |
| 33 42 0101 | Electrical Charge | KWH | 0.06 | 0.06 | 0.06 | 0.06 | 0.06 |
| 33 42 0201 | Diesel Fuel | GAL | 1.31 | 1.31 | 1.31 | 1.31 | 1.31 |
| 33 42 0301 | Water | KGAL | 8.39 | 8.39 | 8.39 | 8.39 | 8.39 |

**Environmental Remediation: Assemblies Cost Book**

# Cost Data for Remediation
## In Situ Vitrification

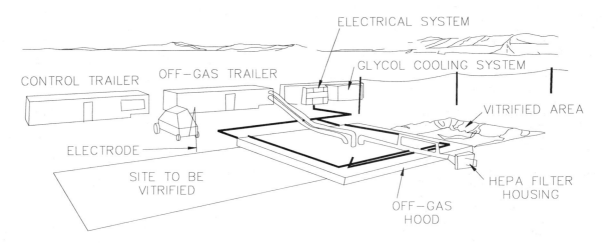

### General:
*In situ vitrification (ISV) involves electrically melting contaminated soil at temperatures well above the soil's initial melting range (between 1,600° C and 2,000° C). Soil vitrification may be conducted in situ or staged, though the costs presented here assume in situ.*

### Typical Treatment Train:
*Dewatering (if the contamination extends below the water table), site preparation other than grading, off-gas treatment, sampling and analysis.*

### Common Cost Components:
1. Site preparation
2. Electrical distribution
3. Mobilization/Demobilization of ISV equipment
4. Operational labor
5. Equipment movement between settings
6. Soil vitrification
7. Equipment maintenance
8. Backfill and replacement cover

### Other Cost Considerations:
Surface water control, on-site generator.

# Cost Data for Remediation

## In Situ Vitrification

| Assembly | Description | Unit | \multicolumn{5}{c}{Unit Cost by Safety Level} |
|---|---|---|---|---|---|---|---|
| | | | A | B | C | D | E |
| | **Capital Costs** | | | | | | |
| 17 03 0101 | Rough Grading, D6 Dozer | SY | 7.02 | 5.72 | 4.70 | 3.41 | 3.24 |
| 17 03 0209 | D6 with A-blade Bulldozer | HOUR | 208.40 | 169.01 | 139.38 | 100.57 | 94.52 |
| 17 03 0259 | Cat 225, 1.5 CY, Soil/Sand, Trenching | CY | 1.84 | 1.49 | 1.23 | 0.88 | 0.82 |
| 17 03 0262 | Cat 235, 2.5 CY, Soil/Sand, Trenching | CY | 2.01 | 1.62 | 1.35 | 0.97 | 0.90 |
| 17 03 0401 | 950, 3.00 CY, Backfill with Excavated Material | CY | 2.54 | 2.05 | 1.70 | 1.22 | 1.14 |
| 17 03 0405 | 950, 3.00 CY, Delivered & Dumped, Backfill with Sand | CY | 38.31 | 35.03 | 32.95 | 29.74 | 29.00 |
| 17 03 0422 | Unclassified Fill, 6" Lifts, On-Site | CY | 14.23 | 11.65 | 9.60 | 7.05 | 6.72 |
| 17 03 0423 | Unclassified Fill, 6" Lifts, Off-Site | CY | 11.08 | 9.81 | 8.83 | 7.58 | 7.40 |
| 17 03 0425 | Sand, 6" Lifts, On-Site | CY | 13.88 | 11.35 | 9.36 | 6.87 | 6.54 |
| 17 03 0426 | Sand, 6" Lifts, Off-Site | CY | 15.55 | 14.34 | 13.42 | 12.22 | 12.04 |
| 17 03 0428 | Clay, 8" Lifts, Off-Site | CY | 22.97 | 19.74 | 17.38 | 14.21 | 13.67 |
| 17 03 0429 | Clay, 8" Lifts, On-Site | CY | 21.18 | 17.21 | 14.26 | 10.34 | 9.71 |
| 17 03 0514 | Compact Soil by Machine with Roller | CY | 1.18 | 0.93 | 0.79 | 0.55 | 0.47 |
| 17 03 0515 | Compact With Pogosticks | CY | 5.91 | 4.58 | 3.98 | 2.68 | 2.23 |
| 17 03 1002 | 2" Diameter Contractor's Trash Pump, 75 GPM | DAY | 50.20 | 47.01 | 45.64 | 42.55 | 41.42 |
| 18 05 0301 | Topsoil, 6" Lifts, Off-Site | CY | 32.76 | 29.51 | 27.43 | 24.25 | 23.53 |
| 18 05 0302 | Topsoil, 6" Lifts, On-Site | CY | 14.14 | 11.54 | 9.45 | 6.88 | 6.57 |
| 18 05 0402 | Seeding, Vegetative Cover | ACRE | 1,874 | 1,823 | 1,791 | 1,741 | 1,729 |
| 18 05 0405 | Sodding, Vegetative Cover | ACRE | 25,697 | 22,064 | 20,238 | 16,710 | 15,589 |
| 20 02 0301 | 1/0 ACSR Conductor | LF | 1.45 | 1.19 | 1.07 | 0.82 | 0.73 |
| 20 02 0302 | 4/0 ACSR Conductor | LF | 2.18 | 1.82 | 1.65 | 1.30 | 1.18 |
| 20 02 0403 | 40' Class 3 Treated Power Pole | EACH | 950.41 | 800.17 | 726.10 | 580.31 | 533.10 |
| 20 02 0421 | Straight-line Structure, 15 KV Pole Top | EACH | 722.43 | 587.04 | 520.30 | 388.92 | 346.37 |
| 20 02 0431 | Terminal Structure, 15 KV Pole Top | EACH | 2,806 | 2,293 | 2,039 | 1,541 | 1,379 |
| 20 02 0522 | 15 KV, 350 MCM, Shielded Cable, Copper | LF | 9.07 | 8.29 | 7.91 | 7.15 | 6.91 |
| 20 02 0546 | 15 KV, 1/0 to 4/0 Conductor, Terminations & Splicing | EACH | 856.33 | 679.41 | 603.67 | 432.66 | 369.96 |
| 20 02 0555 | 15 KV, 3/0 to 4/0 Conductor, Terminations & Splicing | EACH | 1,032 | 819.81 | 728.92 | 523.71 | 448.46 |
| 20 02 0604 | 6" Steel Conduit | LF | 72.02 | 62.75 | 58.79 | 49.83 | 46.55 |
| 20 02 0615 | Concrete Encasement for Ductbank | CY | 231.39 | 194.67 | 178.27 | 142.74 | 130.15 |
| 20 03 9902 | 4" Rigid Steel Conduit | LF | 23.92 | 20.24 | 18.67 | 15.12 | 13.82 |
| 33 15 0203 | Initial Setup ISV Equipment/Decontaminate/Remove Equipment | LS | 2,650 | 2,650 | 2,650 | 2,650 | 2,650 |
| 33 15 0204 | Mobilize ISV Equipment | MILE | 58.30 | 58.30 | 58.30 | 58.30 | 58.30 |
| 33 15 0205 | Demobilize ISV Equipment | MILE | 58.30 | 58.30 | 58.30 | 58.30 | 58.30 |

**Environmental Remediation: Assemblies Cost Book**

# Cost Data for Remediation

## In Situ Vitrification

| Assembly | Description | Unit | \multicolumn{5}{c}{Unit Cost by Safety Level} |
|---|---|---|---|---|---|---|---|
| | | | A | B | C | D | E |
| | **Capital Costs** | | | | | | |
| 33 15 0206 | Vitrify Soil | TON | 318.00 | 318.00 | 318.00 | 318.00 | 318.00 |

**Environmental Remediation: Assemblies Cost Book**

# Cost Data for Remediation
## Incineration (On-Site)

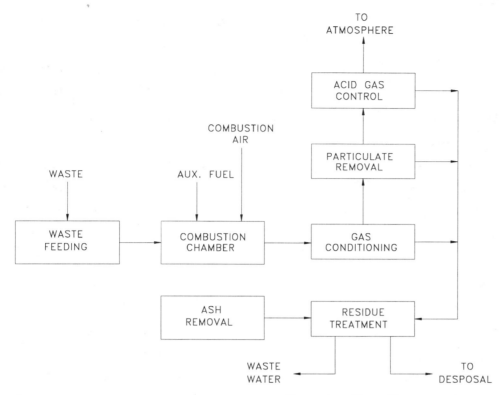

### General:

Incineration is an engineering process that employs thermal decomposition via thermal oxidation at high temperatures. The process detoxifies hazardous wastes, reduces the volume of the wastes, and converts the wastes to ash and gases. Some types of incinerators are not appropriate or economically feasible for certain types of waste: the selection process should consider the physical and chemical characteristics of the waste. The use of on-site mobile incinerators is economically feasible for locations with 5,000+ cubic yards (7,500+ tons) of material to be incinerated. For smaller quantities, refer to Commercial Disposal (Incineration).

### Typical Treatment Train:

Excavation and/or consolidation of waste, off-gas treatment, sampling and analysis.

### Common Cost Components:

1. Mobilization/demobilization of incinerator
2. Solid wastes: particle size reduction, primary/secondary screens, moisture content adjustment, hopper
3. Liquid/sludge wastes: reduction of water content, separation of solids, addition of solvents or dispersants
4. Equipment slab
5. Conveyors
6. Drum shredders
7. Operation of incinerator

### Other Cost Considerations:

Fencing.

**Environmental Remediation: Assemblies Cost Book**

# Cost Data for Remediation

## Incineration (On-Site)

| Assembly | Description | Unit | A | B | C | D | E |
|---|---|---|---|---|---|---|---|
| | **Capital Costs** | | | | | | |
| 17 03 0220 | 910, 1.25 CY, Wheel Loader | HOUR | 130.20 | 104.06 | 87.23 | 61.63 | 55.86 |
| 17 03 0221 | 916, 1.5 CY, Wheel Loader | HOUR | 114.32 | 90.83 | 76.64 | 53.69 | 47.92 |
| 17 03 0222 | 926, 2.0 CY, Wheel Loader | HOUR | 154.30 | 124.14 | 103.30 | 73.68 | 67.91 |
| 17 03 0223 | 950, 3.0 CY, Wheel Loader | HOUR | 180.10 | 145.64 | 120.50 | 86.58 | 80.81 |
| 17 03 0224 | 966, 4.0 CY, Wheel Loader | HOUR | 211.58 | 171.88 | 141.48 | 102.32 | 96.55 |
| 17 03 0225 | 980, 5.25 CY, Wheel Loader | HOUR | 298.84 | 244.38 | 199.68 | 145.79 | 139.74 |
| 17 03 0226 | 988, 7.0 CY, Wheel Loader | HOUR | 384.90 | 316.10 | 257.05 | 188.82 | 182.77 |
| 17 03 0227 | 992, 13.5 CY, Wheel Loader | HOUR | 648.92 | 536.11 | 433.06 | 320.83 | 314.78 |
| 18 02 0320 | 4" Structural Slab on Grade | SF | 6.45 | 5.45 | 4.98 | 4.00 | 3.67 |
| 18 02 0321 | 6" Structural Slab on Grade | SF | 6.98 | 5.95 | 5.47 | 4.48 | 4.14 |
| 18 02 0322 | 8" Structural Slab on Grade | SF | 9.75 | 8.38 | 7.71 | 6.38 | 5.95 |
| 18 02 0323 | 10" Structural Slab on Grade | SF | 10.56 | 9.13 | 8.43 | 7.04 | 6.58 |
| 18 02 0324 | 12" Structural Slab on Grade | SF | 11.57 | 10.04 | 9.30 | 7.82 | 7.34 |
| 18 02 0341 | 6" Unreinforced Slab on Grade | SF | 4.87 | 4.14 | 3.79 | 3.08 | 2.85 |
| 18 02 0342 | 8" Unreinforced Slab on Grade | SF | 6.45 | 5.54 | 5.07 | 4.19 | 3.93 |
| 18 02 0343 | 10" Unreinforced Slab on Grade | SF | 6.97 | 6.04 | 5.56 | 4.66 | 4.38 |
| 18 02 0344 | 12" Unreinforced Slab on Grade | SF | 7.64 | 6.67 | 6.16 | 5.22 | 4.93 |
| 33 13 1117 | Rotary Drum Filter, 5 GPM, Minimum Rental Charge | EACH | 2,176 | 2,176 | 2,176 | 2,176 | 2,176 |
| 33 13 1118 | Rotary Drum Filter, 5 GPM, Additional Rental Charge | EACH | 492.90 | 492.90 | 492.90 | 492.90 | 492.90 |
| 33 13 1119 | Rotary Drum Filter, 5 GPM, Mobilize/DeMobilize | EACH | 4,897 | 4,897 | 4,897 | 4,897 | 4,897 |
| 33 13 1120 | Pretreatment System Operational Labor Cost | DAY | 217.57 | 217.57 | 217.57 | 217.57 | 217.57 |
| 33 13 2401 | Weekly Rental, Portable Drum Shredder, 60 - 100 Drums/Hour | WEEK | 11,408 | 11,408 | 11,408 | 11,408 | 11,408 |
| 33 13 2402 | Daily Rental, Portable Drum Shredder, 60 - 100 Drums/Hour | DAY | 2,282 | 2,282 | 2,282 | 2,282 | 2,282 |
| 33 13 2403 | Monthly Rental, Portable Drum Shredder, 60 - 100 Drums/Hour | MONTH | 45,633 | 45,633 | 45,633 | 45,633 | 45,633 |
| 33 13 2404 | Shredder Mobilize/DeMobilize | EACH | 1,700 | 1,700 | 1,700 | 1,700 | 1,700 |
| 33 13 2405 | Electric Shredder, Minimum Rental Charge, 50 - 75 Tons/Hour | EACH | 41,750 | 41,750 | 41,750 | 41,750 | 41,750 |
| 33 13 2406 | Electric Shredder, Additional Monthly Rental, 50 - 75 Tons/Hour | MONTH | 18,875 | 18,875 | 18,875 | 18,875 | 18,875 |
| 33 13 2407 | Electric Shredder, Purchase, 35 Tons/Hour Output | EACH | 271,283 | 271,283 | 271,283 | 271,283 | 271,283 |
| 33 13 2408 | Portable Drum Shredder, 60 - 100 Drums/Hour Output | EACH | 251,900 | 251,900 | 251,900 | 251,900 | 251,900 |
| 33 14 0103 | Minimum Mobilization/Demobilization Charge | EACH | 457,496 | 457,496 | 457,496 | 457,496 | 457,496 |

**Environmental Remediation: Assemblies Cost Book**

*Page 3-168*

# Cost Data for Remediation

## Incineration (On-Site)

| Assembly | Description | Unit | Unit Cost by Safety Level | | | | |
|---|---|---|---|---|---|---|---|
| | | | A | B | C | D | E |
| | **Capital Costs** | | | | | | |
| 33 14 0104 | Additional Mobilization/Demobilization Charge | MILE | 36.80 | 36.80 | 36.80 | 36.80 | 36.80 |
| 33 14 0105 | Rotary Kiln Incinerator, Operations Cost | TON | 330.18 | 330.18 | 330.18 | 330.18 | 330.18 |
| 33 14 0106 | Rotary Kiln Incinerator, Fixed Cost | EACH | 1,351,500 | 1,351,500 | 1,351,500 | 1,351,500 | 1,351,500 |
| 33 14 0107 | Infrared Furnace, Operations Cost | TON | 318.00 | 318.00 | 318.00 | 318.00 | 318.00 |
| 33 14 0108 | Infrared Furnace, Fixed Cost | EACH | 371,000 | 371,000 | 371,000 | 371,000 | 371,000 |
| 33 14 0109 | Vitrification, Operations Cost | TON | 543.25 | 543.25 | 543.25 | 543.25 | 543.25 |
| 33 14 0110 | Vitrification, Fixed Cost | EACH | 927,500 | 927,500 | 927,500 | 927,500 | 927,500 |
| 33 14 0111 | Liquid Injection Incinerator, Operations Cost | TON | 416.05 | 416.05 | 416.05 | 416.05 | 416.05 |
| 33 14 0112 | Liquid Injection Incinerator, Fixed Cost | EACH | 1,060,000 | 1,060,000 | 1,060,000 | 1,060,000 | 1,060,000 |
| 33 14 0113 | Circulating Bed Combustor, Operations Cost | TON | 492.90 | 492.90 | 492.90 | 492.90 | 492.90 |
| 33 14 0114 | Circulating Bed Combustor, Fixed Cost | EACH | 1,113,000 | 1,113,000 | 1,113,000 | 1,113,000 | 1,113,000 |
| 33 14 0115 | Supercritical Water Oxidation, Operations Cost | TON | 506.15 | 506.15 | 506.15 | 506.15 | 506.15 |
| 33 14 0116 | Supercritical Water Oxidation, Fixed Cost | EACH | 1,325,000 | 1,325,000 | 1,325,000 | 1,325,000 | 1,325,000 |
| 33 14 0117 | Advanced Electric Reactor, Operations Cost | TON | 702.51 | 702.51 | 702.51 | 702.51 | 702.51 |
| 33 14 0118 | Advanced Electric Reactor, Fixed Cost | EACH | 4,001,500 | 4,001,500 | 4,001,500 | 4,001,500 | 4,001,500 |
| 33 15 9501 | Metal Stabilization of Ash | TON | 68.72 | 68.72 | 68.72 | 68.72 | 68.72 |
| 33 18 8401 | 41.5' Automatic Conveyor, 45 FPM, Horizontal 24" Belt, Center Drive | EACH | 8,054 | 7,271 | 6,882 | 6,121 | 5,877 |
| 33 18 8402 | 61.5' Automatic Conveyor, 45 FPM, Horizontal 24" Belt, Center Drive | EACH | 10,682 | 9,637 | 9,119 | 8,105 | 7,779 |
| 33 18 8403 | 34' Automatic Inclined Conveyor, 25 Degree, 24" Belt, Loader/End Idler | EACH | 9,284 | 8,239 | 7,721 | 6,707 | 6,381 |
| 33 18 8501 | Refuse Bottom Hopper, Aluminized Steel, 18" Diameter | EACH | 1,334 | 1,251 | 1,215 | 1,135 | 1,105 |
| 33 18 8601 | 4' x 12' Single-tray Vibrating Screening Unit, with Motor & Accessories | EACH | 16,360 | 15,705 | 15,362 | 14,725 | 14,531 |
| 33 18 8602 | 5' x 16' Double-tray Vibrating Screening Unit, with Motor & Accessories | EACH | 25,858 | 25,105 | 24,711 | 23,979 | 23,757 |
| 33 18 8603 | 6' x 20' Triple-tray Vibrating Screening Unit, with Motor & Accessories | EACH | 32,164 | 30,853 | 30,168 | 28,894 | 28,506 |
| 33 18 8604 | 7' x 24' Triple-tray Vibrating Screening Unit, with Motor & Accessories | EACH | 39,668 | 37,805 | 36,831 | 35,020 | 34,469 |
| 33 31 0401 | Heat Exchanger, 8 GPM, Purchase and Installation | EACH | 10,053 | 9,987 | 9,959 | 9,895 | 9,872 |
| 33 32 0132 | Mixing Tank for Dispersant (3 Month Minimum Rental) | EACH | 26,684 | 26,684 | 26,684 | 26,684 | 26,684 |
| 33 32 0133 | Monthly Rental of Mixing Tank | EACH | 769.67 | 769.67 | 769.67 | 769.67 | 769.67 |
| 33 33 0126 | Dispersant, Viscosity Reduction of Liquids/Sludges | LB | 0.19 | 0.19 | 0.19 | 0.19 | 0.19 |
| 33 33 3001 | Use of Filter Press, Moisture Reduction | TON | 95.40 | 95.40 | 95.40 | 95.40 | 95.40 |

**Environmental Remediation: Assemblies Cost Book**

# Cost Data for Remediation
## Infiltration Gallery

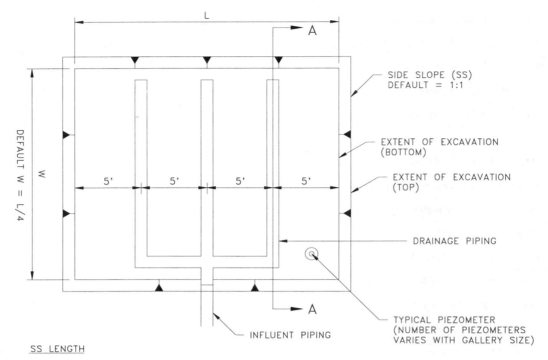

### General:

*An infiltration gallery, as addressed here, is a method for discharging water that has been previously treated in a groundwater treatment system. Infiltration galleries function much like the drainage fields associated with conventional household septic tanks in that perforated, subsurface, drainage piping, surrounded by crushed stone or gravel, is utilized to discharge water into the vadose (unsaturated) cone where it infiltrates downward and ultimately enters the water table (surficial aquifer).*

### Typical Treatment Train:

*Groundwater recovery, treatment of the extracted groundwater, sampling and analysis.*

### Common Cost Components:

1. Demolition of existing cover
2. Excavation of a pit or gallery
3. Installation of a drainage fabric
4. Placement of porous backfill material
5. Installation of perforated drainage piping
6. Placement of additional stone backfill
7. Installation of additional drainage fabric
8. Placement of clean soil backfill
9. Compaction of the soil backfill
10. Installation of a replacement cover
11. Operation and maintenance

### Other Cost Considerations:

*Piping, clearing and grubbing, fencing, electrical distribution.*

**Environmental Remediation: Assemblies Cost Book**

# Cost Data for Remediation

## Infiltration Gallery

| Assembly | Description | Unit | A | B | C | D | E |
|---|---|---|---|---|---|---|---|
| | **Capital Costs** | | | | | | |
| 17 01 0106 | Heavy Brush, Light Trees, Clear, Grub, Haul | ACRE | 11,338 | 9,081 | 7,594 | 5,382 | 4,905 |
| 17 01 0107 | Medium Brush, Medium Trees, Clear, Grub, Haul | ACRE | 13,478 | 10,783 | 9,029 | 6,389 | 5,807 |
| 17 01 0108 | Light Brush, Heavy Trees, Clear, Grub, Haul | ACRE | 17,877 | 14,287 | 11,977 | 8,462 | 7,669 |
| 17 01 0210 | Clear Trees to 6" Diameter with D8 Cat | EACH | 12.03 | 9.85 | 8.04 | 5.88 | 5.66 |
| 17 01 0211 | Clear Trees to 12" Diameter with D8 Cat | EACH | 22.45 | 18.39 | 15.00 | 10.98 | 10.56 |
| 17 01 0212 | Clear Trees to 24" Diameter with D8 Cat | EACH | 33.68 | 27.58 | 22.50 | 16.46 | 15.84 |
| 17 01 0213 | Clear Trees to 36" Diameter with D8 Cat | EACH | 67.36 | 55.17 | 45.00 | 32.93 | 31.67 |
| 17 02 0201 | Demolish Bituminous Road with Power Equipment | CY | 46.17 | 36.60 | 30.96 | 21.62 | 19.19 |
| 17 02 0203 | Demolish Bituminous Pavement with Air Equipment | CY | 80.15 | 63.38 | 53.76 | 37.41 | 32.98 |
| 17 02 0205 | Demolish Unreinforced Concrete to 6" Thick with Air Equipment | CY | 126.90 | 100.35 | 85.12 | 59.24 | 52.23 |
| 17 02 0206 | Demolish Mesh Reinforced Concrete to 6" Thick with Air Equipment | CY | 183.47 | 145.09 | 123.07 | 85.65 | 75.51 |
| 17 02 0207 | Demolish Rod Reinforced Concrete to 6" Thick with Air Equipment | CY | 199.06 | 157.42 | 133.53 | 92.92 | 81.92 |
| 17 02 0208 | Demolish Mesh Reinforced Concrete to 6" Thick with Power Equipment | CY | 96.73 | 76.68 | 64.87 | 45.30 | 40.20 |
| 17 02 0209 | Demolish Rod Reinforced Concrete to 6" Thick with Power Equipment | CY | 106.91 | 84.75 | 71.70 | 50.07 | 44.43 |
| 17 02 0210 | Demolish Unreinforced Concrete 7" to 24" Thick with Power Equipment | CY | 258.44 | 204.87 | 173.31 | 121.03 | 107.40 |
| 17 02 0211 | Demolish Reinforced Concrete 7" to 24" Thick with Power Equipment | CY | 356.37 | 282.51 | 238.98 | 166.90 | 148.10 |
| 17 02 0212 | Demolish Unreinforced Concrete to 6" Thick with Power Equipment | CY | 67.53 | 53.53 | 45.29 | 31.63 | 28.06 |
| 17 02 0215 | Demolish Bituminous Sidewalk | CY | 154.36 | 121.53 | 103.59 | 71.64 | 62.41 |
| 17 02 0220 | Demolish Bituminous Curbs | LF | 2.66 | 2.09 | 1.78 | 1.23 | 1.07 |
| 17 02 0221 | Demolish Unreinforced Concrete Curbs | LF | 7.26 | 5.71 | 4.87 | 3.37 | 2.93 |
| 17 02 0222 | Demolish Reinforced Concrete Curbs | LF | 9.90 | 7.79 | 6.64 | 4.59 | 4.00 |
| 17 03 0105 | Fine Grading, Hand | SY | 4.21 | 3.24 | 2.83 | 1.90 | 1.56 |
| 17 03 0259 | Cat 225, 1.5 CY, Soil/Sand, Trenching | CY | 1.84 | 1.49 | 1.23 | 0.88 | 0.82 |
| 17 03 0261 | Cat 225, 1.5 CY, Soil/Sand with Boulders, Trenching | CY | 3.91 | 3.15 | 2.62 | 1.87 | 1.74 |
| 17 03 0282 | Soil, 5 Miles, Dump Truck, Load/Haul Spoil From Trench | CY | 6.97 | 5.60 | 4.66 | 3.33 | 3.06 |
| 17 03 0283 | Sand, 5 Mile, Dump Truck, Load/Haul Spoil From Trench | CY | 6.28 | 5.05 | 4.20 | 3.00 | 2.76 |
| 17 03 0307 | Cat 235, 2.0 CY, Rock, No Haul or Borrow, Trenching | BCY | 133.78 | 109.18 | 95.00 | 71.01 | 64.55 |
| 17 03 0317 | Rock, 5 Miles, Dump Truck, Load/Haul Spoil from Trench | CY | 10.31 | 8.31 | 6.90 | 4.93 | 4.56 |
| 17 03 0401 | 950, 3.00 CY, Backfill with Excavated Material | CY | 2.54 | 2.05 | 1.70 | 1.22 | 1.14 |

**Environmental Remediation: Assemblies Cost Book**

*Page 3-171*

# Cost Data for Remediation

## Infiltration Gallery

| Assembly | Description | Unit | Unit Cost by Safety Level | | | | |
|---|---|---|---|---|---|---|---|
| | | | A | B | C | D | E |
| | **Capital Costs** | | | | | | |
| 17 03 0419 | Crushed Stone, 1/2" to 3/4" | CY | 23.45 | 21.67 | 20.70 | 18.97 | 18.47 |
| 17 03 0420 | Backfill Trench, Borrow Material, Delivered & Dumped Only | CY | 9.86 | 8.74 | 7.93 | 6.83 | 6.63 |
| 17 03 0426 | Sand, 6" Lifts, Off-Site | CY | 15.55 | 14.34 | 13.42 | 12.22 | 12.04 |
| 17 03 0511 | Compact Soil with Vibrating Plate | CY | 7.20 | 5.57 | 4.84 | 3.26 | 2.70 |
| 18 01 0102 | Gravel, Delivered & Dumped | CY | 24.33 | 22.93 | 22.03 | 20.66 | 20.36 |
| 18 01 0103 | Gravel (90%) & Sand Base (10%), with Calcium Chloride 3/4 - 1 Lb/CY | CY | 25.70 | 24.15 | 23.16 | 21.65 | 21.31 |
| 18 01 0104 | Asphalt, Intermediate Course (Line Item Includes 5% Waste) | TON | 76.50 | 65.72 | 59.09 | 48.56 | 45.99 |
| 18 01 0310 | Prime Coat | SY | 0.55 | 0.52 | 0.51 | 0.49 | 0.48 |
| 18 01 0311 | Tack Coat | SY | 0.70 | 0.64 | 0.60 | 0.53 | 0.52 |
| 18 01 0312 | Asphalt Wearing Course, 1 Pass (Line Item Includes 5% Waste) | TON | 83.61 | 72.83 | 66.20 | 55.67 | 53.10 |
| 18 02 0320 | 4" Structural Slab on Grade | SF | 6.45 | 5.45 | 4.98 | 4.00 | 3.67 |
| 18 02 0321 | 6" Structural Slab on Grade | SF | 6.98 | 5.95 | 5.47 | 4.48 | 4.14 |
| 18 02 0322 | 8" Structural Slab on Grade | SF | 9.75 | 8.38 | 7.71 | 6.38 | 5.95 |
| 18 02 0323 | 10" Structural Slab on Grade | SF | 10.56 | 9.13 | 8.43 | 7.04 | 6.58 |
| 18 02 0324 | 12" Structural Slab on Grade | SF | 11.57 | 10.04 | 9.30 | 7.82 | 7.34 |
| 18 02 0330 | 4" Mesh Reinforced Slab on Grade | SF | 5.48 | 4.60 | 4.18 | 3.32 | 3.03 |
| 18 02 0331 | 6" Mesh Reinforced Slab on Grade | SF | 6.01 | 5.10 | 4.67 | 3.79 | 3.49 |
| 18 02 0332 | 8" Mesh Reinforced Slab on Grade | SF | 7.58 | 6.50 | 5.96 | 4.91 | 4.58 |
| 18 02 0333 | 10" Mesh Reinforced Slab on Grade | SF | 8.10 | 7.00 | 6.44 | 5.37 | 5.03 |
| 18 02 0334 | 12" Mesh Reinforced Slab on Grade | SF | 8.78 | 7.63 | 7.05 | 5.93 | 5.58 |
| 18 02 0340 | 4" Unreinforced Slab on Grade | SF | 4.34 | 3.64 | 3.29 | 2.61 | 2.38 |
| 18 02 0341 | 6" Unreinforced Slab on Grade | SF | 4.87 | 4.14 | 3.79 | 3.08 | 2.85 |
| 18 02 0342 | 8" Unreinforced Slab on Grade | SF | 6.45 | 5.54 | 5.07 | 4.19 | 3.93 |
| 18 02 0343 | 10" Unreinforced Slab on Grade | SF | 6.97 | 6.04 | 5.56 | 4.66 | 4.38 |
| 18 02 0344 | 12" Unreinforced Slab on Grade | SF | 7.64 | 6.67 | 6.16 | 5.22 | 4.93 |
| 18 05 0402 | Seeding, Vegetative Cover | ACRE | 1,874 | 1,823 | 1,791 | 1,741 | 1,729 |
| 18 05 0405 | Sodding, Vegetative Cover | ACRE | 25,697 | 22,064 | 20,238 | 16,710 | 15,589 |
| 19 01 0202 | 1", Class 200, PVC Piping | LF | 7.48 | 5.84 | 5.07 | 3.48 | 2.93 |
| 19 01 0204 | 2", Class 200, PVC Piping | LF | 8.86 | 6.96 | 6.07 | 4.23 | 3.60 |
| 19 01 0206 | 3", Class 200, PVC Piping | LF | 11.81 | 9.34 | 8.20 | 5.80 | 4.98 |
| 19 01 0207 | 4", Class 200, PVC Piping | LF | 12.33 | 9.86 | 8.72 | 6.32 | 5.50 |

**Environmental Remediation: Assemblies Cost Book**

*Page 3-172*

# Cost Data for Remediation

## Infiltration Gallery

| Assembly | Description | Unit | A | B | C | D | E |
|---|---|---|---|---|---|---|---|
| | **Capital Costs** | | | | | | |
| 19 01 0208 | 6", Class 200, PVC Piping | LF | 14.95 | 12.20 | 10.93 | 8.27 | 7.36 |
| 19 01 0212 | 8", Class 150, PVC Piping | LF | 18.36 | 15.27 | 13.84 | 10.85 | 9.82 |
| 19 01 0213 | 10", Class 150, PVC Piping | LF | 22.90 | 19.37 | 17.73 | 14.32 | 13.14 |
| 19 01 0214 | 12", Class 150, PVC Piping | LF | 28.75 | 24.63 | 22.72 | 18.73 | 17.36 |
| 19 02 0101 | 4", Class 50, Bell & Spigot Sanitary Sewer, Cast-iron Pipe | LF | 8.49 | 7.49 | 7.05 | 6.08 | 5.72 |
| 19 02 0103 | 6", Class 50, Bell & Spigot Sanitary Sewer, Cast-iron Pipe | LF | 13.25 | 11.79 | 11.16 | 9.75 | 9.23 |
| 19 02 0104 | 8", Class 50, Bell & Spigot Sanitary Sewer, Cast-iron Pipe | LF | 18.98 | 17.26 | 16.52 | 14.86 | 14.25 |
| 19 02 0105 | 10", Class 50, Bell & Spigot Sanitary Sewer, Cast-iron Pipe | LF | 27.66 | 25.51 | 24.59 | 22.51 | 21.75 |
| 19 02 0106 | 12", Class 50, Bell & Spigot Sanitary Sewer, Cast-iron Pipe | LF | 36.36 | 34.21 | 33.29 | 31.21 | 30.45 |
| 19 02 0107 | 15", Class 50, Bell & Spigot Sanitary Sewer, Cast-iron Pipe | LF | 51.76 | 49.01 | 47.83 | 45.17 | 44.19 |
| 19 02 0110 | 4" Extra-strength Vitrified Clay Pipe, Class 200, Premium Joints | LF | 17.47 | 14.14 | 12.39 | 9.15 | 8.17 |
| 19 02 0112 | 6" Extra-strength Vitrified Clay Pipe, Class 200, Premium Joints | LF | 20.06 | 16.40 | 14.48 | 10.92 | 9.84 |
| 19 02 0113 | 8" Extra-strength Vitrified Clay Pipe, Class 200, Premium Joints | LF | 22.16 | 18.37 | 16.38 | 12.70 | 11.58 |
| 19 02 0114 | 10" Extra-strength Vitrified Clay Pipe, Class 200, Premium Joints | LF | 27.02 | 22.79 | 20.57 | 16.46 | 15.22 |
| 19 02 0115 | 12" Extra-strength Vitrified Clay Pipe, Class 200, Premium Joints | LF | 28.25 | 23.86 | 21.55 | 17.28 | 15.98 |
| 19 02 0116 | 15" Extra-strength Vitrified Clay Pipe, Class 200, Premium Joints | LF | 39.29 | 34.29 | 31.67 | 26.81 | 25.34 |
| 19 02 0117 | 18" Extra-strength Vitrified Clay Pipe, Class 200, Premium Joints | LF | 50.85 | 45.06 | 42.02 | 36.40 | 34.70 |
| 19 02 0118 | 24" Extra-strength Vitrified Clay Pipe, Class 200, Premium Joints | LF | 85.93 | 78.09 | 73.96 | 66.33 | 64.03 |
| 19 02 0125 | 4" PVC Pipe Sanitary | LF | 13.31 | 10.99 | 9.71 | 7.45 | 6.80 |
| 19 02 0126 | 6" PVC Pipe Sanitary | LF | 15.68 | 13.19 | 11.82 | 9.39 | 8.70 |
| 19 02 0127 | 8" PVC Pipe Sanitary | LF | 17.03 | 14.43 | 13.00 | 10.47 | 9.74 |
| 19 02 0128 | 10" PVC Pipe Sanitary | LF | 22.78 | 19.45 | 17.70 | 14.46 | 13.48 |
| 19 02 0129 | 12" PVC Pipe Sanitary | LF | 26.69 | 23.26 | 21.45 | 18.11 | 17.11 |
| 19 02 0130 | 15" PVC Pipe Sanitary | LF | 42.35 | 36.56 | 33.52 | 27.90 | 26.20 |
| 19 03 0157 | 6" Extra Strength, Nonreinforced Concrete Pipe | LF | 10.78 | 8.92 | 7.98 | 6.17 | 5.59 |
| 19 03 0158 | 8" Extra Strength, Nonreinforced Concrete Pipe | LF | 12.86 | 10.79 | 9.75 | 7.74 | 7.10 |
| 19 03 0159 | 10" Extra Strength, Nonreinforced Concrete Pipe | LF | 16.79 | 14.59 | 13.49 | 11.36 | 10.68 |
| 19 03 0160 | 12" Extra Strength, Nonreinforced Concrete Pipe | LF | 17.97 | 15.63 | 14.46 | 12.20 | 11.48 |
| 19 03 0161 | 15" Extra Strength, Nonreinforced Concrete Pipe | LF | 19.95 | 17.36 | 16.06 | 13.55 | 12.74 |

**Environmental Remediation: Assemblies Cost Book**

*Page 3-173*

# Cost Data for Remediation

## Infiltration Gallery

| Assembly | Description | Unit | Unit Cost by Safety Level A | B | C | D | E |
|---|---|---|---|---|---|---|---|
| | **Capital Costs** | | | | | | |
| 19 03 0162 | 18" Extra Strength, Nonreinforced Concrete Pipe | LF | 24.93 | 22.06 | 20.62 | 17.83 | 16.94 |
| 19 03 0163 | 21" Extra Strength, Nonreinforced Concrete Pipe | LF | 30.32 | 27.45 | 26.01 | 23.22 | 22.33 |
| 19 03 0164 | 24" Extra Strength, Nonreinforced Concrete Pipe | LF | 39.69 | 36.49 | 34.88 | 31.77 | 30.78 |
| 19 03 0165 | 12" Reinforced Concrete Pipe, Class 3, with Gaskets | LF | 14.33 | 13.02 | 12.29 | 11.01 | 10.65 |
| 19 03 0166 | 15" Reinforced Concrete Pipe, Class 3, with Gaskets | LF | 16.47 | 14.84 | 13.94 | 12.36 | 11.91 |
| 19 03 0167 | 18" Reinforced Concrete Pipe, Class 3, with Gaskets | LF | 23.67 | 21.05 | 19.60 | 17.05 | 16.33 |
| 19 03 0168 | 24" Reinforced Concrete Pipe, Class 3, with Gaskets | LF | 34.58 | 31.22 | 29.36 | 26.09 | 25.17 |
| 19 03 0174 | 12" Reinforced Concrete Pipe, Class 4, with Gaskets | LF | 14.33 | 13.02 | 12.29 | 11.01 | 10.65 |
| 19 03 0175 | 15" Reinforced Concrete Pipe, Class 4, with Gaskets | LF | 16.47 | 14.84 | 13.94 | 12.36 | 11.91 |
| 19 03 0176 | 18" Reinforced Concrete Pipe, Class 4, with Gaskets | LF | 23.67 | 21.05 | 19.60 | 17.05 | 16.33 |
| 19 03 0177 | 21" Reinforced Concrete Pipe, Class 4, with Gaskets | LF | 27.08 | 24.38 | 22.87 | 20.24 | 19.50 |
| 19 03 0178 | 24" Reinforced Concrete Pipe, Class 4, with Gaskets | LF | 34.58 | 31.22 | 29.36 | 26.09 | 25.17 |
| 19 03 0182 | 12" Reinforced Concrete Pipe, Class 5, with Gaskets | LF | 14.33 | 13.02 | 12.29 | 11.01 | 10.65 |
| 19 03 0183 | 15" Reinforced Concrete Pipe, Class 5, with Gaskets | LF | 16.47 | 14.84 | 13.94 | 12.36 | 11.91 |
| 19 03 0184 | 18" Reinforced Concrete Pipe, Class 5, with Gaskets | LF | 23.67 | 21.05 | 19.60 | 17.05 | 16.33 |
| 19 03 0185 | 21" Reinforced Concrete Pipe, Class 5, with Gaskets | LF | 27.08 | 24.38 | 22.87 | 20.24 | 19.50 |
| 19 03 0186 | 24" Reinforced Concrete Pipe, Class 5, with Gaskets | LF | 34.58 | 31.22 | 29.36 | 26.09 | 25.17 |
| 19 04 0601 | 275 Gallon Steel Sump, Aboveground with Supports & Fittings | EACH | 1,325 | 1,153 | 1,080 | 913.28 | 852.29 |
| 19 04 0602 | 550 Gallon Steel Sump, Aboveground with Supports & Fittings | EACH | 1,598 | 1,390 | 1,301 | 1,099 | 1,026 |
| 19 04 0603 | 1,000 Gallon Steel Sump, Aboveground with Supports & Fittings | EACH | 2,118 | 1,844 | 1,727 | 1,463 | 1,366 |
| 19 04 0604 | 1,500 Gallon Steel Sump, Aboveground with Supports & Fittings | EACH | 2,697 | 2,356 | 2,209 | 1,879 | 1,758 |
| 19 04 0605 | 2,000 Gallon Steel Sump, Aboveground with Supports & Fittings | EACH | 3,457 | 2,997 | 2,777 | 2,332 | 2,183 |
| 19 04 0606 | 5,000 Gallon Steel Sump, Aboveground with Supports & Fittings | EACH | 6,079 | 5,494 | 5,214 | 4,646 | 4,457 |
| 19 04 0621 | 550 Gallon Horizontal Plastic Sump with 4" NPT Connection | EACH | 1,699 | 1,596 | 1,551 | 1,452 | 1,415 |
| 19 04 0622 | 1,000 Gallon Horizontal Plastic Sump with 4" NPT Connection | EACH | 2,179 | 2,042 | 1,984 | 1,852 | 1,803 |
| 19 04 0623 | 2,000 Gallon Horizontal Plastic Sump with 6" NPT Connection | EACH | 3,089 | 2,918 | 2,845 | 2,680 | 2,619 |
| 19 04 0624 | 4,000 Gallon Horizontal Plastic Sump with 6" NPT Connection | EACH | 4,130 | 3,927 | 3,841 | 3,645 | 3,573 |

**Environmental Remediation: Assemblies Cost Book**

# Cost Data for Remediation

## Infiltration Gallery

| Assembly | Description | Unit | A | B | C | D | E |
|---|---|---|---|---|---|---|---|
| | **Capital Costs** | | | | | | |
| 19 04 0625 | 6,000 Gallon Horizontal Plastic Sump with 6" NPT Connection | EACH | 5,248 | 4,975 | 4,858 | 4,594 | 4,497 |
| 19 04 0626 | 8,000 Gallon Horizontal Plastic Sump with 6" NPT Connection | EACH | 5,857 | 5,516 | 5,369 | 5,039 | 4,918 |
| 19 04 0627 | 10,000 Gallon Horizontal Plastic Sump with 6" NPT Connection | EACH | 7,082 | 6,622 | 6,402 | 5,956 | 5,807 |
| 19 04 0628 | 12,000 Gallon Horizontal Plastic Sump with 6" NPT Connection | EACH | 8,308 | 7,722 | 7,442 | 6,875 | 6,686 |
| 19 04 0629 | 15,000 Gallon Horizontal Plastic Sump with 6" NPT Connection | EACH | 14,184 | 13,540 | 13,232 | 12,608 | 12,400 |
| 19 04 0631 | 20,000 Gallon Horizontal Plastic Sump with 6" NPT Connection | EACH | 15,799 | 15,084 | 14,742 | 14,048 | 13,817 |
| 19 04 0632 | 25,000 Gallon Horizontal Plastic Sump with 6" NPT Connection | EACH | 24,407 | 23,487 | 23,047 | 22,155 | 21,858 |
| 19 04 0633 | 30,000 Gallon Horizontal Plastic Sump with 6" NPT Connection | EACH | 35,112 | 34,039 | 33,525 | 32,485 | 32,138 |
| 19 04 0634 | 40,000 Gallon Horizontal Plastic Sump with 6" NPT Connection | EACH | 40,844 | 39,234 | 38,465 | 36,904 | 36,383 |
| 19 04 0635 | 48,000 Gallon Horizontal Plastic Sump with 6" NPT Connection | EACH | 59,758 | 57,612 | 56,586 | 54,505 | 53,811 |
| 33 08 0511 | Drainage Netting, 1/4" Thick High-density Polyethylene | SF | 0.23 | 0.21 | 0.20 | 0.19 | 0.18 |
| 33 08 0512 | Drainage Netting, Geotextile Fabric Heat-bonded 1 Side | SF | 0.41 | 0.39 | 0.38 | 0.36 | 0.35 |
| 33 08 0513 | Drainage Netting, Geotextile Fabric Heat-bonded 2 Sides | SF | 0.48 | 0.45 | 0.44 | 0.42 | 0.41 |
| 33 08 0531 | 60 Mil Geotextile, Nonwoven | SY | 1.53 | 1.31 | 1.21 | 1.00 | 0.93 |
| 33 08 0532 | 80 Mil Geotextile, Nonwoven | SY | 1.57 | 1.35 | 1.25 | 1.04 | 0.97 |
| 33 08 0533 | 105 Mil Geotextile, Nonwoven | SY | 1.92 | 1.65 | 1.52 | 1.26 | 1.17 |
| 33 08 0534 | 130 Mil Geotextile, Nonwoven | SY | 2.56 | 2.20 | 2.03 | 1.68 | 1.56 |
| 33 08 0535 | 170 Mil Geotextile, Nonwoven | SY | 2.96 | 2.60 | 2.43 | 2.08 | 1.96 |
| 33 08 0541 | 20 Mil Polymeric Liner, Very Low Density Polyethylene | SF | 1.25 | 1.02 | 0.91 | 0.69 | 0.61 |
| 33 08 0542 | 30 Mil Polymeric Liner, Very Low Density Polyethylene | SF | 1.98 | 1.59 | 1.41 | 1.03 | 0.90 |
| 33 08 0543 | 40 Mil Polymeric Liner, Very Low Density Polyethylene | SF | 2.38 | 1.91 | 1.70 | 1.24 | 1.09 |
| 33 08 0544 | 60 Mil Polymeric Liner, Very Low Density Polyethylene | SF | 3.54 | 2.83 | 2.51 | 1.83 | 1.60 |
| 33 08 0545 | 80 Mil Polymeric Liner, Very Low Density Polyethylene | SF | 4.35 | 3.49 | 3.10 | 2.27 | 1.98 |
| 33 08 0546 | 100 Mil Polymeric Liner, Very Low Density Polyethylene | SF | 5.50 | 4.41 | 3.91 | 2.85 | 2.49 |
| 33 08 0551 | 36 Mil Polymeric Liner, Chlorosulfanated Polyethylene | SF | 0.75 | 0.73 | 0.72 | 0.70 | 0.70 |
| 33 08 0552 | 45 Mil Polymeric Liner, Chlorosulfanated Polyethylene | SF | 0.83 | 0.81 | 0.80 | 0.79 | 0.78 |
| 33 08 0561 | 20 Mil Polymeric Liner, PVC | SF | 0.24 | 0.23 | 0.22 | 0.21 | 0.20 |
| 33 08 0562 | 30 Mil Polymeric Liner, PVC | SF | 0.32 | 0.31 | 0.30 | 0.29 | 0.28 |

**Environmental Remediation: Assemblies Cost Book**

# Cost Data for Remediation

## Infiltration Gallery

| Assembly | Description | Unit | A | B | C | D | E |
|----------|-------------|------|---|---|---|---|---|
| | **Capital Costs** | | | | | | |
| 33 08 0563 | 40 Mil Polymeric Liner, PVC | SF | 0.38 | 0.36 | 0.36 | 0.34 | 0.34 |
| 33 08 0564 | 60 Mil Polymeric Liner, PVC | SF | 0.72 | 0.70 | 0.69 | 0.67 | 0.67 |
| 33 08 0565 | 80 Mil Polymeric Liner, PVC | SF | 0.98 | 0.96 | 0.95 | 0.93 | 0.93 |
| 33 08 0566 | 100 Mil Polymeric Liner, PVC | SF | 1.28 | 1.26 | 1.25 | 1.23 | 1.22 |
| 33 08 0571 | 40 Mil Polymeric Liner, High-density Polyethylene | SF | 2.35 | 1.88 | 1.67 | 1.21 | 1.06 |
| 33 08 0572 | 60 Mil Polymeric Liner, High-density Polyethylene | SF | 3.49 | 2.80 | 2.48 | 1.81 | 1.58 |
| 33 08 0573 | 80 Mil Polymeric Liner, High-density Polyethylene | SF | 4.32 | 3.47 | 3.08 | 2.25 | 1.97 |
| 33 12 9901 | 1.7 Gallon Bypass Chemical Shot Feeder, In-line Mount, 125 PSIG | EACH | 847.93 | 691.64 | 624.73 | 473.67 | 418.28 |
| 33 12 9902 | 5 Gallon Bypass Chemical Shot Feeder, Floor Mount, 175 PSIG | EACH | 1,236 | 1,001 | 900.54 | 673.47 | 590.22 |
| 33 12 9903 | 12 Gallon Bypass Chemical Shot Feeder, Floor Mount, 175 PSIG | EACH | 1,626 | 1,315 | 1,182 | 881.31 | 771.02 |
| 33 12 9904 | 5 Gallon Bypass Chemical Shot Feeder, Floor Mount, 150 Lb ASME | EACH | 1,236 | 1,001 | 900.54 | 673.47 | 590.22 |
| 33 12 9905 | 10 Gallon Bypass Chemical Shot Feeder, Floor Mount, 150 Lb ASME | EACH | 1,626 | 1,315 | 1,182 | 881.31 | 771.02 |
| 33 12 9906 | 5 Gallon Bypass Chemical Shot Feeder, Floor Mount, 300 Lb ASME | EACH | 1,676 | 1,441 | 1,341 | 1,114 | 1,030 |
| 33 12 9907 | 10 Gallon Bypass Chemical Shot Feeder, Floor Mount, 300 Lb ASME | EACH | 2,084 | 1,772 | 1,639 | 1,338 | 1,228 |
| 33 19 7301 | Haul & Dispose Debris, 16.5 CY Truck, 10 Mile, Landfill | CY | 107.37 | 106.15 | 105.16 | 103.95 | 103.81 |
| 33 23 0101 | 2" PVC, Schedule 40, Well Casing | LF | 17.45 | 14.39 | 11.98 | 8.95 | 8.55 |
| 33 23 0201 | 2" PVC, Schedule 40, Well Screen | LF | 23.47 | 19.52 | 16.42 | 12.52 | 12.00 |
| 33 23 1305 | Water Level Sensor, Float Switch, with 50' Cable | EACH | 549.15 | 549.15 | 549.15 | 549.15 | 549.15 |
| 33 23 1306 | High Sump Level Switch for Avoiding Overflow | EACH | 214.65 | 214.65 | 214.65 | 214.65 | 214.65 |
| 33 23 2209 | 10" x 7.5" Manhole Cover | EACH | 393.44 | 327.83 | 276.26 | 211.47 | 202.77 |
| 33 23 2210 | 12" x 7.5" Manhole Cover | EACH | 476.95 | 400.40 | 340.24 | 264.65 | 254.50 |
| 33 23 2250 | 4" x 3' Protective Enclosure, Schedule 40, Hinged Lid, Lockable | EACH | 370.22 | 312.81 | 267.68 | 210.99 | 203.38 |
| 33 23 2251 | 4" x 4' Protective Enclosure, Schedule 40, Hinged Lid, Lockable | EACH | 471.91 | 395.37 | 335.20 | 259.61 | 249.46 |
| 33 26 0101 | 1" Carbon Steel Piping | LF | 6.46 | 5.28 | 4.77 | 3.63 | 3.21 |
| 33 26 0102 | 2" Carbon Steel Piping | LF | 10.50 | 8.72 | 7.95 | 6.23 | 5.60 |
| 33 26 0103 | 3" Carbon Steel Piping | LF | 18.02 | 15.20 | 13.99 | 11.26 | 10.26 |
| 33 26 0104 | 4" Carbon Steel Piping | LF | 23.20 | 19.86 | 18.43 | 15.21 | 14.02 |
| 33 26 0105 | 6" Carbon Steel Piping | LF | 36.37 | 30.24 | 27.46 | 21.52 | 19.44 |

**Unit Cost by Safety Level**

**Environmental Remediation: Assemblies Cost Book**

# Cost Data for Remediation

## Infiltration Gallery

| Assembly | Description | Unit | Unit Cost by Safety Level A | B | C | D | E |
|---|---|---|---|---|---|---|---|
| | **Capital Costs** | | | | | | |
| 33 26 0106 | 8" Carbon Steel Piping | LF | 40.77 | 34.08 | 31.15 | 24.68 | 22.36 |
| 33 26 0107 | 10" Carbon Steel Piping | LF | 54.40 | 46.05 | 42.38 | 34.30 | 31.39 |
| 33 26 0108 | 12" Carbon Steel Piping | LF | 75.72 | 65.62 | 61.18 | 51.41 | 47.89 |
| 33 26 0201 | 1" Stainless Steel Piping, Schedule 40, Threaded | LF | 18.67 | 16.12 | 15.03 | 12.56 | 11.65 |
| 33 26 0202 | 2" Stainless Steel Piping, Schedule 40, Threaded | LF | 33.47 | 29.21 | 27.39 | 23.28 | 21.77 |
| 33 26 0203 | 3" Stainless Steel Piping, Schedule 40, Threaded | LF | 57.00 | 50.61 | 47.88 | 41.71 | 39.45 |
| 33 26 0204 | 4" Stainless Steel Piping, Schedule 40, Threaded | LF | 80.15 | 71.72 | 68.12 | 59.98 | 57.00 |
| 33 26 0205 | 6" Stainless Steel Piping, Schedule 10, Type 316 | LF | 78.58 | 70.01 | 66.14 | 57.84 | 54.94 |
| 33 26 0206 | 8" Stainless Steel Piping, Schedule 40, Welded | LF | 197.22 | 183.32 | 176.89 | 163.42 | 158.80 |
| 33 26 0208 | 10" Stainless Steel Piping, Schedule 40, Welded | LF | 280.77 | 265.48 | 258.40 | 243.59 | 238.50 |
| 33 26 0209 | 12" Stainless Steel Piping, Schedule 40, Welded | LF | 312.43 | 295.44 | 287.57 | 271.12 | 265.47 |
| 33 26 0901 | 4" Diameter Perforated PVC Pipe | LF | 13.24 | 10.73 | 9.22 | 6.77 | 6.15 |
| 33 26 0902 | 6" Diameter Perforated PVC Pipe | LF | 14.13 | 11.62 | 10.11 | 7.66 | 7.04 |
| 33 26 0903 | 8" Diameter Perforated PVC Pipe | LF | 14.56 | 12.01 | 10.49 | 8.00 | 7.37 |
| 33 26 0904 | 10" Diameter Perforated PVC Pipe | LF | 16.84 | 14.07 | 12.42 | 9.71 | 9.02 |
| 33 26 0905 | 12" Diameter Perforated PVC Pipe | LF | 18.91 | 15.99 | 14.24 | 11.38 | 10.66 |
| 33 26 0951 | 6-hole Distribution Box | EACH | 31.93 | 31.93 | 31.93 | 31.93 | 31.93 |
| 33 29 0102 | 10 GPM, 1/2 HP, Centrifugal Pump | EACH | 1,063 | 952.32 | 905.07 | 798.39 | 759.27 |
| 33 29 0103 | 50 GPM, 100' Head, 3 HP, Centrifugal Pump | EACH | 2,658 | 2,500 | 2,432 | 2,279 | 2,223 |
| 33 29 0104 | 75 GPM, 100' Head, 5 HP, Centrifugal Pump | EACH | 2,946 | 2,754 | 2,672 | 2,487 | 2,419 |
| 33 29 0106 | 100 GPM, 150' Head, 7.5 HP, Centrifugal Pump | EACH | 3,419 | 3,191 | 3,093 | 2,873 | 2,793 |
| 33 29 0107 | 125 GPM, 150' Head, 10 HP, Centrifugal Pump | EACH | 3,815 | 3,512 | 3,382 | 3,088 | 2,981 |
| 33 29 0108 | 10 HP, 200 GPM, Centrifugal Pump | EACH | 2,941 | 2,714 | 2,616 | 2,396 | 2,316 |
| 33 29 0109 | 10 HP, 250 GPM, Centrifugal Pump | EACH | 3,032 | 2,805 | 2,707 | 2,487 | 2,407 |
| 33 29 0110 | 15 HP, 300 GPM, Centrifugal Pump | EACH | 3,441 | 3,160 | 3,040 | 2,770 | 2,670 |
| 33 29 0111 | 20 HP, 500 GPM, Centrifugal Pump | EACH | 4,544 | 4,241 | 4,111 | 3,818 | 3,710 |
| 33 29 0112 | 30 HP, 750 GPM, Centrifugal Pump | EACH | 5,213 | 4,808 | 4,635 | 4,244 | 4,100 |
| 33 29 0113 | 40 HP, 1050 GPM, Centrifugal Pump | EACH | 5,655 | 5,199 | 5,004 | 4,564 | 4,403 |
| 33 29 0114 | 60 HP, 1500 GPM, Centrifugal Pump | EACH | 9,842 | 8,823 | 8,350 | 7,363 | 7,024 |
| 33 29 0115 | 75 HP, 2000 GPM, Centrifugal Pump | EACH | 12,456 | 11,233 | 10,666 | 9,482 | 9,075 |
| 33 29 0116 | 100 HP, 3000 GPM, Centrifugal Pump | EACH | 15,575 | 14,046 | 13,338 | 11,857 | 11,348 |
| 33 29 0401 | 25 GPM, 1 1/2" Discharge, Cast-iron Sump Pump | EACH | 2,432 | 2,274 | 2,206 | 2,053 | 1,997 |
| 33 29 0402 | 75 GPM, 2" Discharge, Cast-iron Sump Pump | EACH | 3,011 | 2,820 | 2,738 | 2,552 | 2,484 |

**Environmental Remediation: Assemblies Cost Book**

*Page 3-177*

1998 by ECHOS. All rights reserved

# Cost Data for Remediation

## Infiltration Gallery

| Assembly | Description | Unit | Unit Cost by Safety Level A | B | C | D | E |
|---|---|---|---|---|---|---|---|
| | **Capital Costs** | | | | | | |
| 33 29 0403 | 100 GPM, 2 1/2" Discharge, Cast-iron Sump Pump | EACH | 3,243 | 3,016 | 2,918 | 2,698 | 2,618 |
| 33 29 0404 | 150 GPM, 3" Discharge, Cast-iron Sump Pump | EACH | 3,703 | 3,422 | 3,302 | 3,032 | 2,932 |
| 33 29 0405 | 200 GPM, 3" Discharge, Cast-iron Sump Pump | EACH | 3,904 | 3,601 | 3,471 | 3,178 | 3,070 |
| 33 29 0406 | 300 GPM, 4" Discharge, Cast-iron Sump Pump | EACH | 4,817 | 4,453 | 4,297 | 3,945 | 3,816 |
| 33 29 0407 | 500 GPM, 5" Discharge, Cast-iron Sump Pump | EACH | 4,969 | 4,610 | 4,456 | 4,110 | 3,982 |
| 33 29 0408 | 800 GPM, 6" Discharge, Cast-iron Sump Pump | EACH | 12,550 | 12,113 | 11,911 | 11,488 | 11,342 |
| 33 29 0409 | 1,000 GPM, 6" Discharge, Cast-iron Sump Pump | EACH | 15,213 | 14,703 | 14,467 | 13,974 | 13,804 |
| 33 29 0410 | 1,600 GPM, 8" Discharge, Cast-iron Sump Pump | EACH | 17,561 | 16,949 | 16,666 | 16,074 | 15,870 |
| 33 29 0411 | 2,000 GPM, 8" Discharge, Cast-iron Sump Pump | EACH | 19,430 | 18,751 | 18,436 | 17,778 | 17,552 |
| 33 41 0101 | Pump & Motor Maintenance/Repair | EACH | 865.17 | 671.48 | 581.58 | 393.96 | 329.64 |
| 33 42 0101 | Electrical Charge | KWH | 0.06 | 0.06 | 0.06 | 0.06 | 0.06 |
| | **Operations & Maintenance** | | | | | | |
| 33 41 0101 | Pump & Motor Maintenance/Repair | EACH | 865.17 | 671.48 | 581.58 | 393.96 | 329.64 |
| 33 42 0101 | Electrical Charge | KWH | 0.06 | 0.06 | 0.06 | 0.06 | 0.06 |

# Cost Data for Remediation
## Injection Wells

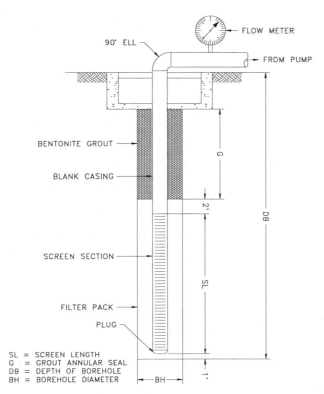

### General:

Injection wells are used for a variety of purposes including aquifer recharge, protection of water supplies, groundwater control, solution mining, waste disposal, and geothermal energy production. Appropriate well construction techniques will vary according to the specific purpose of the well.

### Typical Treatment Train:

Treatment and disposal of the extracted groundwater, soil treatment to eliminate the source of the groundwater contamination, sampling and analysis.

### Common Cost Components:

1. Mobilization of the drill rig and crew
2. Drilling, decontamination of equipment
3. Well development
4. Screen, casing, plug, grout, etc.
5. Pump(s)
6. Feed tank
7. Influent piping
8. Operations and maintenance

### Other Cost Considerations:

Professional field labor, fencing, electrical distribution, clearing and grubbing.

**Environmental Remediation: Assemblies Cost Book**

# Cost Data for Remediation

## Injection Wells

| Assembly | Description | Unit | A | B | C | D | E |
|---|---|---|---|---|---|---|---|
| | **Capital Costs** | | | | | | |
| 19 01 0202 | 1", Class 200, PVC Piping | LF | 7.48 | 5.84 | 5.07 | 3.48 | 2.93 |
| 19 01 0204 | 2", Class 200, PVC Piping | LF | 8.86 | 6.96 | 6.07 | 4.23 | 3.60 |
| 19 01 0206 | 3", Class 200, PVC Piping | LF | 11.81 | 9.34 | 8.20 | 5.80 | 4.98 |
| 19 01 0207 | 4", Class 200, PVC Piping | LF | 12.33 | 9.86 | 8.72 | 6.32 | 5.50 |
| 19 01 0208 | 6", Class 200, PVC Piping | LF | 14.95 | 12.20 | 10.93 | 8.27 | 7.36 |
| 19 01 0210 | 4", Class 150, PVC Piping | LF | 12.42 | 9.95 | 8.81 | 6.41 | 5.59 |
| 19 01 0211 | 6", Class 150, PVC Piping | LF | 14.84 | 12.09 | 10.82 | 8.16 | 7.25 |
| 19 01 0212 | 8", Class 150, PVC Piping | LF | 18.36 | 15.27 | 13.84 | 10.85 | 9.82 |
| 19 01 0213 | 10", Class 150, PVC Piping | LF | 22.90 | 19.37 | 17.73 | 14.32 | 13.14 |
| 19 01 0214 | 12", Class 150, PVC Piping | LF | 28.75 | 24.63 | 22.72 | 18.73 | 17.36 |
| 19 01 0310 | 5,000 Gallon Water Tank, Steel, Horizontal @ Grade | EACH | 15,945 | 13,359 | 12,252 | 9,752 | 8,835 |
| 19 01 0311 | 10,000 Gallon Water Tank, Steel, Horizontal @ Grade | EACH | 18,756 | 16,170 | 15,063 | 12,563 | 11,646 |
| 19 01 0312 | 20,000 Gallon Water Tank, Steel, Horizontal @ Grade | EACH | 28,844 | 24,853 | 23,144 | 19,286 | 17,871 |
| 19 01 0313 | 35,000 Gallon Water Tank, Steel, Horizontal @ Grade | EACH | 45,913 | 40,740 | 38,525 | 33,525 | 31,692 |
| 19 04 0624 | 4,000 Gallon Horizontal Plastic Sump with 6" NPT Connection | EACH | 4,130 | 3,927 | 3,841 | 3,645 | 3,573 |
| 19 04 0625 | 6,000 Gallon Horizontal Plastic Sump with 6" NPT Connection | EACH | 5,248 | 4,975 | 4,858 | 4,594 | 4,497 |
| 33 01 0101 | Mobilize/DeMobilize Drilling Rig & Crew | LS | 4,949 | 4,031 | 3,308 | 2,401 | 2,280 |
| 33 02 0303 | Organic Vapor Analyzer Rental, per Day | DAY | 184.30 | 184.30 | 184.30 | 184.30 | 184.30 |
| 33 02 0537 | Water Level Chart Recorder for Monitoring Wells on Casings | EACH | 1,221 | 1,221 | 1,221 | 1,221 | 1,221 |
| 33 02 1509 | Water Quality Parameter Testing Device | WEEK | 227.90 | 227.90 | 227.90 | 227.90 | 227.90 |
| 33 02 1511 | Water Flowmeter | EACH | 132.50 | 132.50 | 132.50 | 132.50 | 132.50 |
| 33 17 0808 | Decontaminate Rig, Augers, Screen (Rental Equipment) | DAY | 205.34 | 205.34 | 205.34 | 205.34 | 205.34 |
| 33 23 0101 | 2" PVC, Schedule 40, Well Casing | LF | 17.45 | 14.39 | 11.98 | 8.95 | 8.55 |
| 33 23 0102 | 4" PVC, Schedule 40, Well Casing | LF | 26.98 | 22.38 | 18.77 | 14.24 | 13.63 |
| 33 23 0103 | 6" PVC, Schedule 40, Well Casing | LF | 27.07 | 22.66 | 19.19 | 14.84 | 14.25 |
| 33 23 0104 | 8" PVC, Schedule 40, Well Casing | LF | 34.50 | 29.27 | 25.15 | 19.98 | 19.29 |
| 33 23 0111 | 2" PVC, Schedule 80, Well Casing | LF | 17.78 | 14.72 | 12.31 | 9.28 | 8.88 |
| 33 23 0112 | 4" PVC, Schedule 80, Well Casing | LF | 28.55 | 23.95 | 20.34 | 15.81 | 15.20 |
| 33 23 0113 | 6" PVC, Schedule 80, Well Casing | LF | 47.52 | 40.17 | 34.40 | 27.14 | 26.17 |
| 33 23 0115 | 8" PVC, Schedule 80, Well Casing | LF | 61.99 | 52.81 | 45.58 | 36.51 | 35.30 |
| 33 23 0121 | 2" Stainless Steel, Well Casing | LF | 31.75 | 28.14 | 25.30 | 21.73 | 21.25 |

**Environmental Remediation: Assemblies Cost Book**

# Cost Data for Remediation

## Injection Wells

| Assembly | Description | Unit | A | B | C | D | E |
|---|---|---|---|---|---|---|---|
| | **Capital Costs** | | | | | | |
| 33 23 0122 | 4" Stainless Steel, Well Casing | LF | 47.66 | 43.06 | 39.45 | 34.92 | 34.31 |
| 33 23 0123 | 6" Stainless Steel Well Casing, 5' Sections, Flush Threaded | LF | 370.73 | 311.96 | 265.77 | 207.74 | 199.94 |
| 33 23 0125 | 8" Stainless Steel Well Casing, 5' Sections, Flush Threaded | LF | 526.42 | 452.94 | 395.18 | 322.61 | 312.87 |
| 33 23 0201 | 2" PVC, Schedule 40, Well Screen | LF | 23.47 | 19.52 | 16.42 | 12.52 | 12.00 |
| 33 23 0202 | 4" PVC, Schedule 40, Well Screen | LF | 38.16 | 32.00 | 27.16 | 21.09 | 20.27 |
| 33 23 0203 | 6" PVC, Schedule 40, Well Screen | LF | 47.13 | 39.78 | 34.01 | 26.75 | 25.78 |
| 33 23 0204 | 8" PVC, Schedule 40, Well Screen | LF | 62.02 | 52.84 | 45.61 | 36.54 | 35.33 |
| 33 23 0211 | 2" PVC, Schedule 80, Well Screen | LF | 18.70 | 15.64 | 13.23 | 10.20 | 9.80 |
| 33 23 0212 | 4" PVC, Schedule 80, Well Screen | LF | 30.01 | 25.41 | 21.80 | 17.27 | 16.66 |
| 33 23 0213 | 6" PVC, Schedule 80, Well Screen | LF | 49.72 | 42.37 | 36.60 | 29.34 | 28.37 |
| 33 23 0214 | 8" PVC, Schedule 80, Well Screen | LF | 65.91 | 56.73 | 49.50 | 40.43 | 39.22 |
| 33 23 0221 | 2" Stainless Steel, Well Screen | LF | 26.91 | 23.84 | 21.44 | 18.41 | 18.01 |
| 33 23 0222 | 4" Stainless Steel, Well Screen | LF | 47.66 | 43.06 | 39.45 | 34.92 | 34.31 |
| 33 23 0223 | 6" Stainless Steel Well Screen, 5' Sec, Flush Threaded | LF | 366.05 | 307.28 | 261.09 | 203.06 | 195.26 |
| 33 23 0225 | 8" Stainless Steel Well Screen, 5' Sec, Flush Threaded | LF | 462.22 | 388.74 | 330.98 | 258.41 | 248.67 |
| 33 23 0301 | 2" PVC, Well Plug | EACH | 29.37 | 24.77 | 21.16 | 16.63 | 16.02 |
| 33 23 0302 | 4" PVC, Well Plug | EACH | 55.65 | 48.91 | 43.62 | 36.97 | 36.07 |
| 33 23 0303 | 6" PVC, Well Plug | EACH | 112.63 | 101.15 | 92.13 | 80.79 | 79.27 |
| 33 23 0304 | 8" PVC, Well Plug | EACH | 132.70 | 119.58 | 109.26 | 96.31 | 94.57 |
| 33 23 0311 | 2" Stainless Steel, Well Plug | EACH | 83.24 | 74.06 | 66.83 | 57.76 | 56.55 |
| 33 23 0312 | 4" Stainless Steel, Well Plug | EACH | 112.24 | 102.03 | 94.01 | 83.93 | 82.58 |
| 33 23 0313 | 6" Stainless Steel, Well Plug | EACH | 573.04 | 481.18 | 408.98 | 318.27 | 306.09 |
| 33 23 0314 | 8" Stainless Steel, Well Plug | EACH | 702.53 | 597.56 | 515.04 | 411.38 | 397.46 |
| 33 23 1101 | Hollow-stem Auger, 8" Outside Diameter Borehole for 2" Well | LF | 89.98 | 73.28 | 60.15 | 43.66 | 41.45 |
| 33 23 1102 | Hollow-stem Auger, 11" Outside Diameter Borehole for 4" Well | LF | 109.98 | 89.57 | 73.52 | 53.36 | 50.66 |
| 33 23 1103 | Hollow-stem Auger, 13 3/4" Outside Diameter Borehole for 6" Well | LF | 141.40 | 115.16 | 94.53 | 68.61 | 65.13 |
| 33 23 1105 | Hollow-stem Auger, 16" Outside Diameter Borehole for 8" Well | LF | 164.97 | 134.35 | 110.28 | 80.05 | 75.99 |
| 33 23 1106 | Split Spoon Sample, 2" x 24", During Drilling | EACH | 46.67 | 46.67 | 46.67 | 46.67 | 46.67 |
| 33 23 1111 | Well Development Equipment Rental | WEEK | 478.91 | 462.93 | 456.09 | 440.65 | 434.98 |
| 33 23 1121 | Standby for Drilling | EACH | 618.63 | 503.81 | 413.56 | 300.17 | 284.95 |
| 33 23 1122 | Move Rig/Equipment Around Site | EACH | 154.66 | 125.95 | 103.39 | 75.04 | 71.24 |

**Environmental Remediation: Assemblies Cost Book**

# Cost Data for Remediation

## Injection Wells

| Assembly | Description | Unit | Unit Cost by Safety Level A | B | C | D | E |
|---|---|---|---|---|---|---|---|
| | **Capital Costs** | | | | | | |
| 33 23 1125 | Concrete Coring (Minimum Charge) | DAY | 2,475 | 2,015 | 1,654 | 1,201 | 1,140 |
| 33 23 1126 | Furnish 55 Gallon Drum for Drill Cuttings & Development Water | EACH | 65.19 | 65.19 | 65.19 | 65.19 | 65.19 |
| 33 23 1152 | Mud Drilling, 6" Diameter Borehole | LF | 81.01 | 65.98 | 54.16 | 39.31 | 37.31 |
| 33 23 1153 | Mud Drilling, 8" Diameter Borehole | LF | 88.38 | 71.97 | 59.08 | 42.88 | 40.71 |
| 33 23 1154 | Mud Drilling, 10" Diameter Borehole | LF | 100.65 | 81.97 | 67.29 | 48.84 | 46.36 |
| 33 23 1155 | Mud Drilling, 12" Diameter Borehole | LF | 112.92 | 91.97 | 75.49 | 54.79 | 52.01 |
| 33 23 1161 | Air Rotary 6" Borehole in Unconsolidated | LF | 73.65 | 59.98 | 49.23 | 35.73 | 33.92 |
| 33 23 1162 | Air Rotary 8" Borehole in Unconsolidated | LF | 122.74 | 99.96 | 82.06 | 59.56 | 56.54 |
| 33 23 1163 | Air Rotary 10" Borehole in Unconsolidated | LF | 220.94 | 179.93 | 147.70 | 107.20 | 101.77 |
| 33 23 1164 | Air Rotary 12" Borehole in Unconsolidated | LF | 294.59 | 239.91 | 196.93 | 142.94 | 135.69 |
| 33 23 1401 | 2" Screen, Filter Pack | LF | 16.49 | 13.88 | 11.84 | 9.27 | 8.92 |
| 33 23 1402 | 4" Screen, Filter Pack | LF | 29.10 | 24.50 | 20.89 | 16.36 | 15.75 |
| 33 23 1403 | 6" Screen, Filter Pack | LF | 42.19 | 35.53 | 30.29 | 23.72 | 22.83 |
| 33 23 1404 | Gravel Pack, 8" Well | LF | 29.47 | 23.93 | 19.82 | 14.36 | 13.48 |
| 33 23 1502 | Surface Pad, Concrete, 4' x 4' x 4" | EACH | 24.21 | 21.82 | 20.74 | 18.43 | 17.61 |
| 33 23 1811 | 2" Well, Portland Cement Grout | LF | 0.92 | 0.92 | 0.92 | 0.92 | 0.92 |
| 33 23 1812 | 4" Well, Portland Cement Grout | LF | 1.38 | 1.38 | 1.38 | 1.38 | 1.38 |
| 33 23 1813 | 6" Well, Portland Cement Grout | LF | 7.80 | 7.80 | 7.80 | 7.80 | 7.80 |
| 33 23 1814 | 8" Well, Portland Cement Grout | LF | 11.02 | 11.02 | 11.02 | 11.02 | 11.02 |
| 33 23 2101 | 2" Well, Bentonite Seal | EACH | 63.00 | 52.67 | 44.55 | 34.34 | 32.97 |
| 33 23 2102 | 4" Well, Bentonite Seal | EACH | 157.54 | 131.70 | 111.39 | 85.87 | 82.44 |
| 33 23 2103 | 6" Well, Bentonite Seal | EACH | 252.01 | 210.68 | 178.18 | 137.37 | 131.89 |
| 33 23 2105 | 8" Well, Bentonite Seal | EACH | 346.60 | 289.75 | 245.06 | 188.92 | 181.39 |
| 33 26 0201 | 1" Stainless Steel Piping, Schedule 40, Threaded | LF | 18.67 | 16.12 | 15.03 | 12.56 | 11.65 |
| 33 26 0202 | 2" Stainless Steel Piping, Schedule 40, Threaded | LF | 33.47 | 29.21 | 27.39 | 23.28 | 21.77 |
| 33 26 0203 | 3" Stainless Steel Piping, Schedule 40, Threaded | LF | 57.00 | 50.61 | 47.88 | 41.71 | 39.45 |
| 33 26 0204 | 4" Stainless Steel Piping, Schedule 40, Threaded | LF | 80.15 | 71.72 | 68.12 | 59.98 | 57.00 |
| 33 26 0205 | 6" Stainless Steel Piping, Schedule 10, Type 316 | LF | 78.58 | 70.01 | 66.14 | 57.84 | 54.94 |
| 33 26 0206 | 8" Stainless Steel Piping, Schedule 40, Welded | LF | 197.22 | 183.32 | 176.89 | 163.42 | 158.80 |
| 33 26 0208 | 10" Stainless Steel Piping, Schedule 40, Welded | LF | 280.77 | 265.48 | 258.40 | 243.59 | 238.50 |
| 33 26 0209 | 12" Stainless Steel Piping, Schedule 40, Welded | LF | 312.43 | 295.44 | 287.57 | 271.12 | 265.47 |
| 33 27 0112 | 2" PVC, 90 Degree, Elbow | EACH | 41.36 | 32.22 | 28.30 | 19.46 | 16.22 |
| 33 27 0114 | 4" PVC, 90 Degree, Elbow | EACH | 101.72 | 80.57 | 71.51 | 51.07 | 43.57 |

## Environmental Remediation:  Assemblies Cost Book

*Page 3-182*

1998 by ECHOS. All rights reserved

# Cost Data for Remediation

## Injection Wells

| Assembly | Description | Unit | Unit Cost by Safety Level | | | | |
|---|---|---|---|---|---|---|---|
| | | | A | B | C | D | E |
| | **Capital Costs** | | | | | | |
| 33 27 0115 | 6" PVC, 90 Degree, Elbow | EACH | 135.44 | 111.27 | 100.92 | 77.56 | 68.99 |
| 33 27 0116 | 8" PVC, 90 Degree, Elbow | EACH | 200.21 | 172.01 | 159.94 | 132.68 | 122.68 |
| 33 27 0211 | 2" Stainless Steel, 90 Degree, Elbow | EACH | 114.28 | 93.45 | 84.53 | 64.39 | 57.01 |
| 33 27 0212 | 4" Stainless Steel, 90 Degree, Elbow | EACH | 391.27 | 338.50 | 315.91 | 264.90 | 246.20 |
| 33 27 0213 | 6" Stainless Steel, 90 Degree, Elbow | EACH | 425.65 | 376.17 | 355.00 | 307.18 | 289.64 |
| 33 27 0214 | 8" Stainless Steel, 90 Degree, Elbow | EACH | 566.44 | 516.97 | 495.79 | 447.97 | 430.44 |
| 33 27 0402 | 2" Iron Body Check Valve | EACH | 188.96 | 174.31 | 168.03 | 153.86 | 148.67 |
| 33 27 0404 | 4" Iron Body Check Valve | EACH | 354.29 | 315.96 | 299.54 | 262.49 | 248.90 |
| 33 27 0405 | 6" Iron Body Check Valve | EACH | 618.60 | 554.57 | 523.95 | 461.86 | 441.15 |
| 33 27 0406 | 8" Iron Body Check Valve | EACH | 1,094 | 980.95 | 926.96 | 817.50 | 780.98 |
| 33 29 0101 | 10 GPM, 10' Head, 1/6 HP, Centrifugal Pump | EACH | 789.18 | 730.44 | 705.29 | 648.51 | 627.69 |
| 33 29 0102 | 10 GPM, 1/2 HP, Centrifugal Pump | EACH | 1,063 | 952.32 | 905.07 | 798.39 | 759.27 |
| 33 29 0103 | 50 GPM, 100' Head, 3 HP, Centrifugal Pump | EACH | 2,658 | 2,500 | 2,432 | 2,279 | 2,223 |
| 33 29 0104 | 75 GPM, 100' Head, 5 HP, Centrifugal Pump | EACH | 2,946 | 2,754 | 2,672 | 2,487 | 2,419 |
| 33 29 0105 | 5 HP, 100 GPM, Centrifugal Pump | EACH | 2,475 | 2,301 | 2,227 | 2,059 | 1,998 |
| 33 29 0106 | 100 GPM, 150' Head, 7.5 HP, Centrifugal Pump | EACH | 3,419 | 3,191 | 3,093 | 2,873 | 2,793 |
| 33 29 0107 | 125 GPM, 150' Head, 10 HP, Centrifugal Pump | EACH | 3,815 | 3,512 | 3,382 | 3,088 | 2,981 |
| 33 29 0108 | 10 HP, 200 GPM, Centrifugal Pump | EACH | 2,941 | 2,714 | 2,616 | 2,396 | 2,316 |
| 33 29 0109 | 10 HP, 250 GPM, Centrifugal Pump | EACH | 3,032 | 2,805 | 2,707 | 2,487 | 2,407 |
| 33 29 0110 | 15 HP, 300 GPM, Centrifugal Pump | EACH | 3,441 | 3,160 | 3,040 | 2,770 | 2,670 |
| 33 29 0111 | 20 HP, 500 GPM, Centrifugal Pump | EACH | 4,544 | 4,241 | 4,111 | 3,818 | 3,710 |
| 33 29 0112 | 30 HP, 750 GPM, Centrifugal Pump | EACH | 5,213 | 4,808 | 4,635 | 4,244 | 4,100 |
| 33 29 0113 | 40 HP, 1050 GPM, Centrifugal Pump | EACH | 5,655 | 5,199 | 5,004 | 4,564 | 4,403 |
| 33 29 0114 | 60 HP, 1500 GPM, Centrifugal Pump | EACH | 9,842 | 8,823 | 8,350 | 7,363 | 7,024 |
| 33 29 0115 | 75 HP, 2000 GPM, Centrifugal Pump | EACH | 12,456 | 11,233 | 10,666 | 9,482 | 9,075 |
| 33 29 0116 | 100 HP, 3000 GPM, Centrifugal Pump | EACH | 15,575 | 14,046 | 13,338 | 11,857 | 11,348 |
| 33 29 0128 | 100 GPM, Centrifugal Pump, 100' Head, 7.5 HP, Cast Iron | EACH | 3,417 | 3,243 | 3,169 | 3,001 | 2,940 |
| 33 29 0129 | 250 GPM, Centrifugal Pump, 100' Head, 15 HP, Cast Iron | EACH | 4,485 | 4,258 | 4,160 | 3,940 | 3,860 |
| 33 29 0130 | 500 GPM, Centrifugal Pump, 100' Head, 20 HP, Cast Iron | EACH | 5,433 | 5,152 | 5,032 | 4,762 | 4,662 |
| 33 29 0131 | 750 GPM, Centrifugal Pump, 100' Head, 25 HP, Cast Iron | EACH | 5,474 | 5,215 | 5,104 | 4,853 | 4,761 |
| 33 29 0132 | 1,000 GPM, Centrifugal Pump, 100' Head, 40 HP, Cast Iron | EACH | 7,113 | 6,780 | 6,637 | 6,315 | 6,197 |
| 33 29 0133 | 1,500 GPM, Centrifugal Pump, 100' Head, 50 HP, Cast Iron | EACH | 8,869 | 8,461 | 8,272 | 7,878 | 7,742 |
| 33 29 0134 | 2,000 GPM, Centrifugal Pump, 100' Head, 75 HP, Cast Iron | EACH | 11,903 | 11,393 | 11,157 | 10,664 | 10,494 |

**Environmental Remediation: Assemblies Cost Book**

# Cost Data for Remediation

## Injection Wells

| Assembly | Description | Unit | A | B | C | D | E |
|---|---|---|---|---|---|---|---|
| | **Capital Costs** | | | | | | |
| 33 29 0135 | 3,000 GPM, Centrifugal Pump, 100' Head, 100 HP, Cast Iron | EACH | 13,286 | 12,674 | 12,391 | 11,799 | 11,595 |
| 33 29 0136 | 3,500 GPM, Centrifugal Pump, 125' Head, 150 HP, Cast Iron | EACH | 14,254 | 13,575 | 13,260 | 12,602 | 12,376 |
| 33 29 0137 | 4,000 GPM, Centrifugal Pump, 150' Head, 200 HP, Cast Iron | EACH | 18,299 | 17,425 | 17,021 | 16,174 | 15,884 |
| 33 29 0138 | 100 GPM, Centrifugal Pump, 500' Head, 40 HP, Cast Iron | EACH | 5,240 | 4,959 | 4,839 | 4,569 | 4,469 |
| 33 29 0139 | 200 GPM, Centrifugal Pump, 500' Head, 50 HP, Cast Iron | EACH | 5,800 | 5,497 | 5,367 | 5,074 | 4,966 |
| 33 29 0140 | 300 GPM, Centrifugal Pump, 500' Head, 75 HP, Cast Iron | EACH | 6,671 | 6,252 | 6,043 | 5,636 | 5,505 |
| 33 29 0141 | 400 GPM, Centrifugal Pump, 500' Head, 100 HP, Cast Iron | EACH | 7,679 | 7,102 | 6,815 | 6,255 | 6,076 |
| 33 29 0142 | 800 GPM, Centrifugal Pump, 500' Head, 200 HP, Cast Iron | EACH | 16,870 | 16,101 | 15,718 | 14,971 | 14,732 |
| 33 29 0143 | 15 GPM, Centrifugal Pump, 6' Head, 1/8 HP, Bronze | EACH | 602.28 | 557.16 | 537.84 | 494.23 | 478.23 |
| 33 29 0144 | 20 GPM, Centrifugal Pump, 5.5' Head, 1/8 HP, Bronze | EACH | 749.03 | 694.45 | 671.08 | 618.32 | 598.98 |
| 33 29 0145 | 25 GPM, Centrifugal Pump, 4.5' Head, 1/8 HP, Bronze | EACH | 749.03 | 694.45 | 671.08 | 618.32 | 598.98 |
| 33 29 0146 | 30 GPM, Centrifugal Pump, 2' Head, 1/8 HP, Bronze | EACH | 749.03 | 694.45 | 671.08 | 618.32 | 598.98 |
| 33 29 0147 | 30 GPM, Centrifugal Pump, High Velocity, 7' Head, 1/8 HP | EACH | 767.19 | 708.44 | 683.29 | 626.51 | 605.69 |
| 33 29 0148 | 35 GPM, Centrifugal Pump, High Velocity, 4' Head, 1/8 HP | EACH | 767.19 | 708.44 | 683.29 | 626.51 | 605.69 |
| 33 29 0149 | 35 GPM, Centrifugal Pump, High Velocity, 5' Head, 1/8 HP | EACH | 767.19 | 708.44 | 683.29 | 626.51 | 605.69 |
| 33 29 0150 | 50 GPM, Turbine Pump, 100' Head, 2 HP, Iron/Bronze | EACH | 2,766 | 2,361 | 2,188 | 1,797 | 1,653 |
| 33 29 0151 | 100 GPM, Turbine Pump, 100' Head, 3 HP, Iron/Bronze | EACH | 3,487 | 3,031 | 2,836 | 2,396 | 2,235 |
| 33 29 0152 | 250 GPM, Turbine Pump, 150' Head, 15 HP, Iron/Bronze | EACH | 4,820 | 4,412 | 4,223 | 3,829 | 3,693 |
| 33 29 0153 | 500 GPM, Turbine Pump, 150' Head, 25 HP, Iron/Bronze | EACH | 7,100 | 6,629 | 6,412 | 5,956 | 5,799 |
| 33 29 0154 | 1,000 GPM, Turbine Pump, 150' Head, 50 HP, Iron/Bronze | EACH | 10,275 | 9,765 | 9,529 | 9,036 | 8,866 |
| 33 29 0155 | 2,000 GPM, Turbine Pump, 150' Head, 100 HP, Iron/Bronze | EACH | 11,730 | 11,118 | 10,835 | 10,243 | 10,039 |
| 33 29 0156 | 3,000 GPM, Turbine Pump, 150' Head, 150 HP, Iron/Bronze | EACH | 13,412 | 12,648 | 12,294 | 11,553 | 11,299 |
| 33 29 0157 | 4,000 GPM, Turbine Pump, 150' Head, 200 HP, Iron/Bronze | EACH | 18,900 | 18,026 | 17,622 | 16,775 | 16,485 |
| 33 29 0158 | 6,000 GPM, Turbine Pump, 150' Head, 300 HP, Iron/Bronze | EACH | 29,550 | 28,531 | 28,058 | 27,071 | 26,732 |
| 33 29 0159 | 10,000 GPM, Turbine Pump, 100' Head, 300 HP, Iron/Bronze | EACH | 40,460 | 39,237 | 38,670 | 37,486 | 37,079 |
| 33 29 0501 | Adjustable Flow 0 - 100% (25 GPM), 200 PSI, Pump with Motor | EACH | 1,598 | 1,575 | 1,565 | 1,543 | 1,535 |
| 33 29 0601 | Adjustable Flow 0 - 100% (28 GPM), 250 PSI, Pump with Motor | EACH | 1,598 | 1,575 | 1,565 | 1,543 | 1,535 |
| 33 42 0101 | Electrical Charge | KWH | 0.06 | 0.06 | 0.06 | 0.06 | 0.06 |
| | **Operations & Maintenance** | | | | | | |
| 33 42 0101 | Electrical Charge | KWH | 0.06 | 0.06 | 0.06 | 0.06 | 0.06 |

**Environmental Remediation: Assemblies Cost Book**

# Cost Data for Remediation
## Land Farming (Ex Situ)

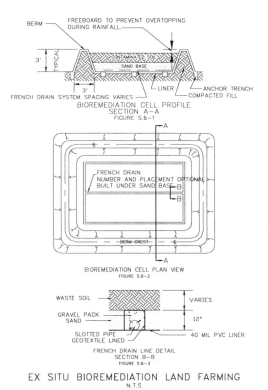

EX SITU BIOREMEDIATION LAND FARMING
N.T.S.

### General:

Ex situ bioremediation (land farming) is a process for treating contaminated soil that requires excavation and movement to a treatment cell. The contaminated soil is spread in a thin layer over an area to enhance volatilization, aeration, biodegradation, and photolysis.

### Typical Treatment Train:
Excavate and move soil to treatment cell, return treated soil to site, sampling and analysis, decontamination, close (demolish) cell.

### Common Cost Components:

1. Construction of treatment cell, including liner, earthen berm, French drain along inner perimeter, sump pump(s)
2. Soil treatment within the cell: soil tilling, application
   of fertilizer, pH adjustment, watering
3. Cell cover

### Other Cost Considerations:
Pilot tests, clearing and grubbing, hauling off debris, temporary holding areas, electrical service, professional (supervisory) labor.

---

**Environmental Remediation: Assemblies Cost Book**

# Cost Data for Remediation

## Land Farming (Ex Situ)

Unit Cost by Safety Level

| Assembly | Description | Unit | A | B | C | D | E |
|---|---|---|---|---|---|---|---|
| | **Capital Costs** | | | | | | |
| 17 01 0501 | Dozer 105 HP D5, Grubbing & Stacking | CY | 13.44 | 10.84 | 8.99 | 6.44 | 5.98 |
| 17 03 0105 | Fine Grading, Hand | SY | 4.21 | 3.24 | 2.83 | 1.90 | 1.56 |
| 17 03 0211 | Hand Excavation, Normal Soil | CY | 80.12 | 61.76 | 53.90 | 36.15 | 29.64 |
| 17 03 0212 | Hand Excavation, Sand/Gravel | CY | 70.10 | 54.04 | 47.16 | 31.63 | 25.94 |
| 17 03 0213 | Hand Excavation, Medium Clay | CY | 93.47 | 72.05 | 62.88 | 42.17 | 34.58 |
| 17 03 0215 | 931, 1.0 CY, Track Loader | HOUR | 112.08 | 88.96 | 75.15 | 52.57 | 46.80 |
| 17 03 0216 | 943, 1.5 CY, Track Loader | HOUR | 170.50 | 137.43 | 114.12 | 81.62 | 75.57 |
| 17 03 0217 | 953, 2.0 CY, Track Loader | HOUR | 195.08 | 158.13 | 130.48 | 94.07 | 88.30 |
| 17 03 0218 | 963, 2.5 CY, Track Loader | HOUR | 280.58 | 229.38 | 187.48 | 136.82 | 131.05 |
| 17 03 0219 | 973, 3.75 CY, Track Loader | HOUR | 357.58 | 293.33 | 238.84 | 175.16 | 169.11 |
| 17 03 0340 | Standby, 931, 1.0 CY Track Loader | HOUR | 8.58 | 7.15 | 5.72 | 4.29 | 4.29 |
| 17 03 0341 | Standby, 943, 1.5 CY Track Loader | HOUR | 21.48 | 17.90 | 14.32 | 10.74 | 10.74 |
| 17 03 0342 | Standby, 953, 2.0 CY Track Loader | HOUR | 27.35 | 22.79 | 18.23 | 13.67 | 13.67 |
| 17 03 0343 | Standby, 963, 2.5 CY Track Loader | HOUR | 46.00 | 38.33 | 30.67 | 23.00 | 23.00 |
| 17 03 0344 | Standby, 973, 3.75 CY Track Loader | HOUR | 62.65 | 52.21 | 41.77 | 31.33 | 31.33 |
| 17 03 0405 | 950, 3.00 CY, Delivered & Dumped, Backfill with Sand | CY | 38.31 | 35.03 | 32.95 | 29.74 | 29.00 |
| 17 03 0420 | Backfill Trench, Borrow Material, Delivered & Dumped Only | CY | 9.86 | 8.74 | 7.93 | 6.83 | 6.63 |
| 17 03 1002 | 2" Diameter Contractor's Trash Pump, 75 GPM | DAY | 50.20 | 47.01 | 45.64 | 42.55 | 41.42 |
| 17 03 1003 | 3" Diameter Contractor's Trash Pump, 150 GPM | DAY | 61.72 | 57.05 | 55.05 | 50.54 | 48.88 |
| 17 03 1004 | 4" Diameter Contractor's Trash Pump, 300 GPM | DAY | 80.30 | 74.23 | 71.63 | 65.77 | 63.61 |
| 17 03 9903 | Hand Place Small Earth Fill Berm | CY | 112.16 | 86.46 | 75.45 | 50.61 | 41.50 |
| 17 03 9904 | Mulch Hay, 50 Lb Average Bale, 36" x 24" x 18" | EACH | 3.83 | 3.83 | 3.83 | 3.83 | 3.83 |
| 18 01 0201 | Concrete Curb, 6" x 6" | LF | 2.78 | 2.38 | 2.21 | 1.83 | 1.70 |
| 18 01 0206 | Concrete Curb, 8" x 8" | LF | 7.58 | 6.24 | 5.66 | 4.36 | 3.89 |
| 18 01 0207 | Concrete Curb, 12" High x 8" Deep | LF | 10.27 | 8.45 | 7.67 | 5.92 | 5.27 |
| 18 02 0320 | 4" Structural Slab on Grade | SF | 6.45 | 5.45 | 4.98 | 4.00 | 3.67 |
| 18 02 0321 | 6" Structural Slab on Grade | SF | 6.98 | 5.95 | 5.47 | 4.48 | 4.14 |
| 18 02 0322 | 8" Structural Slab on Grade | SF | 9.75 | 8.38 | 7.71 | 6.38 | 5.95 |
| 18 02 0323 | 10" Structural Slab on Grade | SF | 10.56 | 9.13 | 8.43 | 7.04 | 6.58 |
| 18 05 0409 | Fertilize, 800 Lbs/Acre, Push Rotary | ACRE | 140.58 | 118.67 | 105.24 | 83.82 | 78.56 |
| 18 05 0410 | Fertilize, 800 Lbs/Acre, Spray from Truck | ACRE | 107.45 | 90.70 | 78.22 | 61.72 | 59.08 |
| 18 05 0411 | Crushed Limestone, pH Adjustment, 800 Lbs/Acre | ACRE | 156.71 | 134.79 | 121.37 | 99.94 | 94.68 |
| 18 05 0412 | Purchase & Spread Dry Granular Limestone for pH Control | SY | 0.05 | 0.04 | 0.04 | 0.03 | 0.03 |

## Environmental Remediation: Assemblies Cost Book

*Page 3-186*

# Cost Data for Remediation

## Land Farming (Ex Situ)

| Assembly | Description | Unit | Unit Cost by Safety Level A | B | C | D | E |
|---|---|---|---|---|---|---|---|
| | **Capital Costs** | | | | | | |
| 18 05 0413 | Watering with 3,000-Gallon Tank Truck, per Pass | ACRE | 111.81 | 90.08 | 75.73 | 54.43 | 49.85 |
| 18 05 0701 | Full Circle Sprinkler Head, 30' Diameter | EACH | 94.41 | 76.34 | 68.61 | 51.14 | 44.74 |
| 18 05 0702 | Semicircle Sprinkler Head, 20' Diameter | EACH | 43.12 | 34.09 | 30.22 | 21.49 | 18.29 |
| 18 05 0703 | Shrub/Tree Sprinkler Head | EACH | 40.45 | 31.42 | 27.55 | 18.82 | 15.62 |
| 18 05 0704 | Full Circle Sprinkler Head, 80' Diameter | EACH | 302.56 | 287.17 | 280.57 | 265.69 | 260.23 |
| 18 05 0705 | Control Box | EACH | 1,598 | 1,461 | 1,402 | 1,270 | 1,221 |
| 18 05 0706 | 2 1/2", Cast-iron Body, Gate Valve | EACH | 257.59 | 239.88 | 232.30 | 215.19 | 208.91 |
| 18 05 0707 | 4" Reducer | EACH | 80.38 | 64.15 | 57.20 | 41.51 | 35.76 |
| 19 01 0201 | 3/4", Class 200, PVC Piping | LF | 7.45 | 5.81 | 5.04 | 3.45 | 2.90 |
| 19 01 0202 | 1", Class 200, PVC Piping | LF | 7.48 | 5.84 | 5.07 | 3.48 | 2.93 |
| 19 01 0204 | 2", Class 200, PVC Piping | LF | 8.86 | 6.96 | 6.07 | 4.23 | 3.60 |
| 19 01 0205 | 2 1/2", Class 200, PVC Piping | LF | 9.03 | 7.13 | 6.24 | 4.40 | 3.77 |
| 19 02 0601 | 12" x 12" Underground French Drain | LF | 4.50 | 3.92 | 3.61 | 3.04 | 2.88 |
| 19 02 0602 | 18" x 18" Underground French Drain | LF | 5.93 | 5.29 | 4.94 | 4.31 | 4.13 |
| 19 02 0603 | 24" x 24" Underground French Drain | LF | 7.43 | 6.71 | 6.31 | 5.61 | 5.41 |
| 19 04 0401 | 550 Gallon, Stainless Steel Aboveground Wastewater Holding Tank, Rental | MONTH | 320.65 | 320.65 | 320.65 | 320.65 | 320.65 |
| 19 04 0403 | 4,000 Gallon Polyethylene Wastewater Tank, Rental | MONTH | 545.90 | 545.90 | 545.90 | 545.90 | 545.90 |
| 19 04 0405 | 6,000 Gallon, Polyethylene Aboveground Wastewater Holding Tank, Rental | MONTH | 641.30 | 641.30 | 641.30 | 641.30 | 641.30 |
| 19 04 0407 | 21,000 Gallon, Steel Closed Stationary Aboveground Wastewater Holding Tank, Rental | MONTH | 1,219 | 1,219 | 1,219 | 1,219 | 1,219 |
| 19 04 0610 | 75 GPM Cast-iron Sump Pump, 2.0" Discharge, Single Stage | EACH | 3,011 | 2,820 | 2,738 | 2,552 | 2,484 |
| 19 04 0611 | 100 GPM Cast-iron Sump Pump, 2.5" Discharge, Single Stage | EACH | 3,243 | 3,016 | 2,918 | 2,698 | 2,618 |
| 19 04 0621 | 550 Gallon Horizontal Plastic Sump with 4" NPT Connection | EACH | 1,699 | 1,596 | 1,551 | 1,452 | 1,415 |
| 19 04 0624 | 4,000 Gallon Horizontal Plastic Sump with 6" NPT Connection | EACH | 4,130 | 3,927 | 3,841 | 3,645 | 3,573 |
| 19 04 0625 | 6,000 Gallon Horizontal Plastic Sump with 6" NPT Connection | EACH | 5,248 | 4,975 | 4,858 | 4,594 | 4,497 |
| 19 04 0631 | 20,000 Gallon Horizontal Plastic Sump with 6" NPT Connection | EACH | 15,799 | 15,084 | 14,742 | 14,048 | 13,817 |
| 33 01 0502 | 4.0 KW, 120/240 VAC, Generator, Daily Rental | EACH | 37.03 | 37.03 | 37.03 | 37.03 | 37.03 |
| 33 01 0503 | 5.0 KW, 120/240 VAC, Generator, Daily Rental | EACH | 11.30 | 11.30 | 11.30 | 11.30 | 11.30 |
| 33 01 0504 | 7.5 KW, 120/240 VAC, Generator, Daily Rental | EACH | 72.80 | 72.80 | 72.80 | 72.80 | 72.80 |

**Environmental Remediation: Assemblies Cost Book**

# Cost Data for Remediation

## Land Farming (Ex Situ)

| Assembly | Description | Unit | \multicolumn{5}{c}{Unit Cost by Safety Level} |||||
|---|---|---|---|---|---|---|---|
| | | | A | B | C | D | E |
| | **Capital Costs** | | | | | | |
| 33 01 0505 | 10 KW, 120/240 VAC, Generator, Daily Rental | EACH | 13.09 | 13.09 | 13.09 | 13.09 | 13.09 |
| 33 01 0507 | Portable Water Pump, 2" 6,000 GPH, Water/Fertilizer Transfer | EACH | 906.12 | 906.12 | 906.12 | 906.12 | 906.12 |
| 33 01 0508 | Water Pump, 2" 8,000 GPH, Portable, Gasoline Powered | EACH | 1,391 | 1,391 | 1,391 | 1,391 | 1,391 |
| 33 01 0509 | Portable Water Pump, 2", 10,000 GPH, Gas Powered, with Wheels | EACH | 1,391 | 1,391 | 1,391 | 1,391 | 1,391 |
| 33 07 9901 | Lease, Portable 40' W x 150' L x 18' High Emissions Containment Building | YEAR | 7,500 | 7,500 | 7,500 | 7,500 | 7,500 |
| 33 07 9902 | Buy, Portable 40' W x 150' L x 18' High Emissions Containment Building | EACH | 23,503 | 23,503 | 23,503 | 23,503 | 23,503 |
| 33 07 9903 | Assemble Emissions Containment Building, 5 Days | EACH | 12,346 | 9,591 | 8,298 | 5,629 | 4,722 |
| 33 07 9904 | Disassemble Emissions Containment Building, 3 Days | EACH | 7,402 | 5,750 | 4,975 | 3,374 | 2,831 |
| 33 08 0503 | Polymeric Liner Anchor Trench, 3' x 1.5' | LF | 1.58 | 1.25 | 1.07 | 0.76 | 0.67 |
| 33 08 0541 | 20 Mil Polymeric Liner, Very Low Density Polyethylene | SF | 1.25 | 1.02 | 0.91 | 0.69 | 0.61 |
| 33 08 0542 | 30 Mil Polymeric Liner, Very Low Density Polyethylene | SF | 1.98 | 1.59 | 1.41 | 1.03 | 0.90 |
| 33 08 0543 | 40 Mil Polymeric Liner, Very Low Density Polyethylene | SF | 2.38 | 1.91 | 1.70 | 1.24 | 1.09 |
| 33 08 0544 | 60 Mil Polymeric Liner, Very Low Density Polyethylene | SF | 3.54 | 2.83 | 2.51 | 1.83 | 1.60 |
| 33 08 0545 | 80 Mil Polymeric Liner, Very Low Density Polyethylene | SF | 4.35 | 3.49 | 3.10 | 2.27 | 1.98 |
| 33 08 0546 | 100 Mil Polymeric Liner, Very Low Density Polyethylene | SF | 5.50 | 4.41 | 3.91 | 2.85 | 2.49 |
| 33 08 0551 | 36 Mil Polymeric Liner, Chlorosulfanated Polyethylene | SF | 0.75 | 0.73 | 0.72 | 0.70 | 0.70 |
| 33 08 0552 | 45 Mil Polymeric Liner, Chlorosulfanated Polyethylene | SF | 0.83 | 0.81 | 0.80 | 0.79 | 0.78 |
| 33 08 0561 | 20 Mil Polymeric Liner, PVC | SF | 0.24 | 0.23 | 0.22 | 0.21 | 0.20 |
| 33 08 0562 | 30 Mil Polymeric Liner, PVC | SF | 0.32 | 0.31 | 0.30 | 0.29 | 0.28 |
| 33 08 0563 | 40 Mil Polymeric Liner, PVC | SF | 0.38 | 0.36 | 0.36 | 0.34 | 0.34 |
| 33 08 0564 | 60 Mil Polymeric Liner, PVC | SF | 0.72 | 0.70 | 0.69 | 0.67 | 0.67 |
| 33 08 0565 | 80 Mil Polymeric Liner, PVC | SF | 0.98 | 0.96 | 0.95 | 0.93 | 0.93 |
| 33 08 0566 | 100 Mil Polymeric Liner, PVC | SF | 1.28 | 1.26 | 1.25 | 1.23 | 1.22 |
| 33 08 0571 | 40 Mil Polymeric Liner, High-density Polyethylene | SF | 2.35 | 1.88 | 1.67 | 1.21 | 1.06 |
| 33 08 0572 | 60 Mil Polymeric Liner, High-density Polyethylene | SF | 3.49 | 2.80 | 2.48 | 1.81 | 1.58 |
| 33 08 0573 | 80 Mil Polymeric Liner, High-density Polyethylene | SF | 4.32 | 3.47 | 3.08 | 2.25 | 1.97 |
| 33 08 0584 | Plastic Laminate Waste Pile Cover | SF | 0.16 | 0.15 | 0.14 | 0.13 | 0.13 |
| 33 08 0590 | Waste Pile Cover, 135 Lb Tear, 2 - 2.5 Year Life | SY | 2.03 | 1.93 | 1.89 | 1.80 | 1.77 |
| 33 08 0591 | Waste Pile Cover, 185 Lb Tear, 3 - 4 Year Life | SY | 2.98 | 2.89 | 2.85 | 2.76 | 2.72 |
| 33 08 0592 | Waste Pile Cover, 250 Lb Tear, 4 - 5 Year Life | SY | 4.00 | 3.89 | 3.85 | 3.74 | 3.70 |
| 33 11 0301 | Soil Tilling, D3 Dozer with Tiller Attachment | HOUR | 157.86 | 126.90 | 105.69 | 75.30 | 69.25 |

**Environmental Remediation: Assemblies Cost Book**

# Cost Data for Remediation

## Land Farming (Ex Situ)

| Assembly | Description | Unit | \\multicolumn{5}{c}{Unit Cost by Safety Level} |
|---|---|---|---|---|---|---|---|

| Assembly | Description | Unit | A | B | C | D | E |
|---|---|---|---|---|---|---|---|
| | **Capital Costs** | | | | | | |
| 33 11 0302 | Soil Tilling, D4 Dozer with Tiller Attachment | HOUR | 158.82 | 127.70 | 106.33 | 75.78 | 69.73 |
| 33 11 0303 | Soil Tilling, D7 Dozer with Tiller Attachment | HOUR | 297.42 | 243.20 | 198.73 | 145.08 | 139.03 |
| 33 11 0304 | Soil Tilling, D9 Dozer with Tiller Attachment | HOUR | 431.76 | 355.15 | 288.29 | 212.25 | 206.20 |
| 33 11 0305 | Soil Tilling, D3 Dozer with Plow | HOUR | 157.86 | 126.90 | 105.69 | 75.30 | 69.25 |
| 33 11 9901 | Application of Bioculture to Contaminated Soil | ACRE | 109.25 | 87.52 | 73.18 | 51.87 | 47.29 |
| 33 11 9902 | Light Petroleum Hydrocarbon Degraders, Microorganisms | LB | 19.88 | 19.88 | 19.88 | 19.88 | 19.88 |
| 33 11 9903 | Light Petroleum Hydrocarbon Degraders, 100 Lb Bag, Microorganisms | EACH | 1,948 | 1,948 | 1,948 | 1,948 | 1,948 |
| 33 11 9904 | Light Petroleum Hydrocarbon Degraders, 1,000 Lb Bag, Microorganisms | EACH | 17,490 | 17,490 | 17,490 | 17,490 | 17,490 |
| 33 11 9905 | Heavy Petroleum Hydrocarbon/Creosote Degraders, Microorganisms | LB | 13.25 | 13.25 | 13.25 | 13.25 | 13.25 |
| 33 11 9906 | Heavy Petroleum Hydrocarbon/Creosote Degraders, 100 Lb Bag, Microorganisms | EACH | 1,007 | 1,007 | 1,007 | 1,007 | 1,007 |
| 33 11 9907 | Heavy Petroleum Hydrocarbon/Creosote Degraders, 1,000 Lb Bag, Microorganisms | EACH | 8,745 | 8,745 | 8,745 | 8,745 | 8,745 |
| 33 17 0801 | Decontaminate Light Equipment | EACH | 189.11 | 146.65 | 127.13 | 86.01 | 71.80 |
| 33 17 0802 | Decontaminate Medium Equipment | EACH | 378.22 | 293.30 | 254.27 | 172.03 | 143.59 |
| 33 17 0803 | Decontaminate Heavy Equipment | EACH | 560.32 | 434.51 | 376.69 | 254.86 | 212.73 |
| 33 19 9921 | DOT Steel Drum, 55 Gallon | EACH | 65.19 | 65.19 | 65.19 | 65.19 | 65.19 |
| 33 29 0412 | 15 GPM Submersible Sump Pump | EACH | 492.79 | 460.07 | 446.06 | 414.43 | 402.83 |
| 33 29 0413 | 42 GPM Submersible Sump Pump | EACH | 792.79 | 760.07 | 746.06 | 714.43 | 702.83 |
| 33 29 0414 | 52 GPM Submersible Sump Pump | EACH | 871.35 | 832.08 | 815.27 | 777.32 | 763.40 |
| 33 32 0101 | 7.5 GPH, 860 PSI, 55 Gallon Polyethylene Tank with Steel Overpack | EACH | 4,203 | 3,886 | 3,751 | 3,445 | 3,333 |
| 33 32 0102 | 20.8 GPH, 150 PSI, 55 Gallon Polyethylene Tank with Steel Overpack | EACH | 4,203 | 3,886 | 3,751 | 3,445 | 3,333 |
| 33 32 0103 | 7.5 GPH, 860 PSI, 102 Gallon Polyethylene Tank with Steel Overpack | EACH | 5,601 | 5,206 | 5,036 | 4,654 | 4,513 |
| 33 32 0104 | 20.8 GPH, 150 PSI, 102 Gallon Polyethylene Tank with Steel Overpack | EACH | 5,601 | 5,206 | 5,036 | 4,654 | 4,513 |
| 33 32 0105 | 7.5 GPH, 860 PSI, 167 Gallon Polyethylene Tank with Steel Overpack | EACH | 5,264 | 4,868 | 4,699 | 4,316 | 4,176 |
| 33 32 0106 | 20.8 GPH, 150 PSI, 167 Gallon Polyethylene Tank with Steel Overpack | EACH | 5,264 | 4,868 | 4,699 | 4,316 | 4,176 |
| 33 32 0107 | 7.5 GPH, 860 PSI, 260 Gallon Polyethylene Tank with Steel Overpack | EACH | 6,288 | 5,892 | 5,723 | 5,340 | 5,200 |
| 33 32 0108 | 20.8 GPH, 150 PSI, 260 Gallon Polyethylene Tank with Steel Overpack | EACH | 6,288 | 5,892 | 5,723 | 5,340 | 5,200 |

**Environmental Remediation: Assemblies Cost Book**

# Cost Data for Remediation

## Land Farming (Ex Situ)

| Assembly | Description | Unit | \multicolumn{5}{c}{Unit Cost by Safety Level} | | | | |
|---|---|---|---|---|---|---|---|
| | | | A | B | C | D | E |
| | **Capital Costs** | | | | | | |
| 33 33 0117 | Hydrated Lime, Powdered, Bulk | TON | 97.55 | 97.55 | 97.54 | 97.54 | 97.53 |
| 33 33 0118 | Hydrated Lime, Powdered, 50 Lb Bag | EACH | 25.76 | 21.23 | 18.49 | 14.06 | 12.94 |
| 33 34 0101 | Standby D3 Bulldozer with Tiller | HOUR | 17.17 | 14.31 | 11.45 | 8.59 | 8.59 |
| 33 34 0102 | Standby D4 Bulldozer with Tiller | HOUR | 16.89 | 14.08 | 11.26 | 8.45 | 8.45 |
| 33 34 0103 | Standby D7 Bulldozer with Tiller | HOUR | 47.11 | 39.26 | 31.41 | 23.55 | 23.55 |
| 33 34 0104 | Standby D9 Bulldozer with Tiller | HOUR | 82.94 | 69.12 | 55.30 | 41.47 | 41.47 |
| 33 34 0105 | Standby D6 Bulldozer with Tiller | HOUR | 17.89 | 14.91 | 11.93 | 8.95 | 8.95 |

**Environmental Remediation: Assemblies Cost Book**

# Cost Data for Remediation
## Landfill Disposal

RCRA SECURE LANDFILLS

### General:

Waste disposal may be achieved through the use of commercial transportation and disposal services: landfill disposal, deep well injection, incineration. Solids, liquids, and sludges, both hazardous and nonhazardous, drummed or in bulk, may be disposed of at regulated landfills. It is often most cost effective to consolidate compatible wastes. It should be noted that compatibility testing is required prior to bulking.

### Typical Treatment Train:

Pretreatment or stabilization of wastes, sampling and analysis.

### Common Cost Components:

1. Loading wastes from a containment berm or waste site
2. Waste containerization
3. Overpacks for leaking drums
4. Consolidation of compatible wastes
5. Compatibility testing (prior to bulking
6. Sampling of the waste
7. Transportation of the waste to the disposal facility
8. Disposal fee (includes decontamination of trucks)
9. State taxes and fees

### Other Cost Considerations:

Containment berms, staging areas, decontamination.

**Environmental Remediation: Assemblies Cost Book**

# Cost Data for Remediation

## Landfill Disposal

| Assembly | Description | Unit | \multicolumn{5}{c}{Unit Cost by Safety Level} |
|---|---|---|---|---|---|---|---|
| | | | A | B | C | D | E |
| | **Capital Costs** | | | | | | |
| 33 19 0101 | Liquid Loading Into 5,000 Gallon Bulk Tank Truck | EACH | 997.03 | 798.90 | 667.78 | 473.57 | 432.04 |
| 33 19 0102 | Bulk Solid Hazardous Waste Loading Into Truck | CY | 4.51 | 3.67 | 3.01 | 2.18 | 2.07 |
| 33 19 0103 | Load Drums on Disposal Vehicle | EACH | 6.21 | 4.95 | 4.16 | 2.93 | 2.64 |
| 33 19 0104 | Load Bulk Liquid/Sludge Waste Into 55 Gallon Drums, Drums Separate | EACH | 17.53 | 13.51 | 11.79 | 7.91 | 6.48 |
| 33 19 0105 | Load Bulk Solid Waste Into 55 Gallon Drums, Drums Separate | EACH | 62.31 | 48.03 | 41.92 | 28.12 | 23.06 |
| 33 19 0106 | Recontainerize Drums, Not Including Salvage Drums | EACH | 105.53 | 84.84 | 70.65 | 50.34 | 46.31 |
| 33 19 0201 | Drummed Waste, Minimum Charge for Shipment | EACH | 2,350 | 2,350 | 2,350 | 2,350 | 2,350 |
| 33 19 0202 | Bulk Hazardous Waste, Minimum Charge for Shipment | EACH | 2,350 | 2,350 | 2,350 | 2,350 | 2,350 |
| 33 19 0204 | Transport 55 Gallon Drums of Hazardous Waste, Maximum 80 (per Mile) | MILE | 1.50 | 1.50 | 1.50 | 1.50 | 1.50 |
| 33 19 0205 | Transport Bulk Solid Hazardous Waste, Maximum 20 CY (per Mile) | MILE | 1.44 | 1.44 | 1.44 | 1.44 | 1.44 |
| 33 19 0206 | Transport Bulk Solid Hazardous Waste, Maximum 18 Ton (per Mile) | MILE | 1.44 | 1.44 | 1.44 | 1.44 | 1.44 |
| 33 19 0207 | Transport Bulk Liquid/Sludge Hazardous Waste, Maximum 5,000 Gallon (per Mile) | MILE | 1.50 | 1.50 | 1.50 | 1.50 | 1.50 |
| 33 19 0301 | Extra Charges for 85 Gallon Overpack | EACH | 40.00 | 40.00 | 40.00 | 40.00 | 40.00 |
| 33 19 0302 | Extra Charges for Leaking Drums | EACH | 175.00 | 175.00 | 175.00 | 175.00 | 175.00 |
| 33 19 0303 | Minimum Charges for Drummed Shipments | EACH | 500.00 | 500.00 | 500.00 | 500.00 | 500.00 |
| 33 19 0304 | Extra Charges for Drums < 90% Full, < 30 Gallon Drum | EACH | 20.00 | 20.00 | 20.00 | 20.00 | 20.00 |
| 33 19 0305 | Extra Charges for Drums < 90% Full, 55 Gallon Drum | EACH | 20.00 | 20.00 | 20.00 | 20.00 | 20.00 |
| 33 19 0306 | Extra Charges for Drums < 90% Full, > 55 Gallon Drum | EACH | 20.00 | 20.00 | 20.00 | 20.00 | 20.00 |
| 33 19 0307 | Minimum Charges for Bulk Shipments | EACH | 1,350 | 1,350 | 1,350 | 1,350 | 1,350 |
| 33 19 0308 | Minimum Charges for Bulk Shipments Requiring Treatment or Stabilization | EACH | 2,267 | 2,267 | 2,267 | 2,267 | 2,267 |
| 33 19 0309 | Minimum Charges on Nonfuel Bulk Liquid Shipments | EACH | 2,400 | 2,400 | 2,400 | 2,400 | 2,400 |
| 33 19 0310 | Extra Charges for Leaking Loads, per Drum | EACH | 200.00 | 200.00 | 200.00 | 200.00 | 200.00 |
| 33 19 0311 | Truck Washout | EACH | 150.00 | 150.00 | 150.00 | 150.00 | 150.00 |
| 33 19 0317 | Waste Stream Evaluation Fee, Not Including 50% Rebate on 1st Shipment | EACH | 385.50 | 385.50 | 385.50 | 385.50 | 385.50 |
| 33 19 0318 | Manifest Discrepancies | EACH | 50.00 | 50.00 | 50.00 | 50.00 | 50.00 |
| 33 19 0319 | Lab Pack Profile Evaluation Fee | EACH | 500.00 | 500.00 | 500.00 | 500.00 | 500.00 |
| 33 19 0320 | Priority Waste Stream Evaluation Fee | EACH | 766.67 | 766.67 | 766.67 | 766.67 | 766.67 |
| 33 19 0321 | Disposal of Empty Drum, Metal 55 Gallon RCRA Empty | EACH | 45.00 | 45.00 | 45.00 | 45.00 | 45.00 |
| 33 19 0322 | Disposal of Empty Drum, Metal 85 Gallon RCRA Empty | EACH | 56.67 | 56.67 | 56.67 | 56.67 | 56.67 |

## Environmental Remediation: Assemblies Cost Book

# Cost Data for Remediation

## Landfill Disposal

| Assembly | Description | Unit | Unit Cost by Safety Level A | B | C | D | E |
|---|---|---|---|---|---|---|---|
| | **Capital Costs** | | | | | | |
| 33 19 0323 | Disposal of Empty Drum, Plastic 55 Gallon RCRA Empty | EACH | 58.33 | 58.33 | 58.33 | 58.33 | 58.33 |
| 33 19 0324 | State HTW Disposal Tax/Fee (Bulk Solid) | CY | 0.00 | 0.00 | 0.00 | 0.00 | 0.00 |
| 33 19 0325 | State HTW Disposal Tax/Fee (Bulk Liquid) | GAL | 0.00 | 0.00 | 0.00 | 0.00 | 0.00 |
| 33 19 0326 | State HTW Disposal Tax/Fee (Drums) | DRM | 0.00 | 0.00 | 0.00 | 0.00 | 0.00 |
| 33 19 7201 | Landfill Hazardous Solid Waste, <= 30 Gallon Drum | EACH | 60.63 | 60.63 | 60.63 | 60.63 | 60.63 |
| 33 19 7202 | Landfill Hazardous Solid Waste, 55 Gallon Drum | EACH | 87.67 | 87.67 | 87.67 | 87.67 | 87.67 |
| 33 19 7203 | Landfill Hazardous Solid Waste, > 55 Gallon Drum | EACH | 148.13 | 148.13 | 148.13 | 148.13 | 148.13 |
| 33 19 7204 | Landfill Nonhazardous Solid Waste, <= 30 Gallon Drum | EACH | 53.33 | 53.33 | 53.33 | 53.33 | 53.33 |
| 33 19 7205 | Landfill Nonhazardous Solid Waste, 55 Gallon Drum | EACH | 76.67 | 76.67 | 76.67 | 76.67 | 76.67 |
| 33 19 7206 | Landfill Nonhazardous Solid Waste, > 55 Gallon Drum | EACH | 107.17 | 107.17 | 107.17 | 107.17 | 107.17 |
| 33 19 7207 | Landfill Drummed Solid Waste Requiring Stabilization, <= 30 Gallon Drum | EACH | 225.00 | 225.00 | 225.00 | 225.00 | 225.00 |
| 33 19 7208 | Landfill Drummed Solid Waste Requiring Stabilization, 55 Gallon Drum | EACH | 236.33 | 236.33 | 236.33 | 236.33 | 236.33 |
| 33 19 7209 | Landfill Drummed Solid Waste Requiring Stabilization, > 55 Gallon Drum | EACH | 373.00 | 373.00 | 373.00 | 373.00 | 373.00 |
| 33 19 7210 | Landfill Drummed Solid/Sludge Requiring Stabilization, <= 30 Gallon Drum | EACH | 227.50 | 227.50 | 227.50 | 227.50 | 227.50 |
| 33 19 7211 | Landfill Drummed Solid/Sludge Requiring Stabilization, 55 Gallon Drum | EACH | 302.50 | 302.50 | 302.50 | 302.50 | 302.50 |
| 33 19 7212 | Landfill Drummed Solid/Sludge Requiring Stabilization, > 55 Gallon Drum | EACH | 470.00 | 470.00 | 470.00 | 470.00 | 470.00 |
| 33 19 7213 | Landfill Drummed Liquid Waste, <= 30 Gallon Drum | EACH | 227.50 | 227.50 | 227.50 | 227.50 | 227.50 |
| 33 19 7214 | Landfill Drummed Liquid Waste, 55 Gallon Drum | EACH | 140.00 | 140.00 | 140.00 | 140.00 | 140.00 |
| 33 19 7215 | Landfill Drummed Liquid Waste, > 55 Gallon Drum | EACH | 237.50 | 237.50 | 237.50 | 237.50 | 237.50 |
| 33 19 7216 | Landfill Drummed Liquid Waste Requiring Stabilization, <= 30 Gallon Drum | EACH | 200.00 | 200.00 | 200.00 | 200.00 | 200.00 |
| 33 19 7217 | Landfill Drummed Liquid Waste Requiring Stabilization, 55 Gallon Drum | EACH | 300.00 | 300.00 | 300.00 | 300.00 | 300.00 |
| 33 19 7218 | Landfill Drummed Liquid Waste Requiring Stabilization, > 55 Gallon Drum | EACH | 154.67 | 154.67 | 154.67 | 154.67 | 154.67 |
| 33 19 7219 | Landfill Drummed Liquid Waste, Solar Evaporation, <= 30 Gallon Drum, pH>2.0 | EACH | 154.67 | 154.67 | 154.67 | 154.67 | 154.67 |
| 33 19 7220 | Landfill Drummed Liquid Waste, Solar Evaporation, 55 Gallon Drum, pH > 2.0 | EACH | 160.00 | 160.00 | 160.00 | 160.00 | 160.00 |
| 33 19 7221 | Landfill Drummed Liquid Waste, Solar Evaporation, > 55 Gallon Drum, pH>2.0 | EACH | 195.33 | 195.33 | 195.33 | 195.33 | 195.33 |
| 33 19 7222 | Landfill Drummed Inorganic/Organic Nonhazardous Liquid/Sludge, <= 30 Gallon Drum | EACH | 110.00 | 110.00 | 110.00 | 110.00 | 110.00 |

## Environmental Remediation: Assemblies Cost Book

# Cost Data for Remediation

## Landfill Disposal

| Assembly | Description | Unit | A | B | C | D | E |
|---|---|---|---|---|---|---|---|
| | **Capital Costs** | | | | | | |
| 33 19 7223 | Landfill Drummed Inorganic/Organic Nonhazardous Liquid/Sludge, 55 Gallon Drum | EACH | 155.00 | 155.00 | 155.00 | 155.00 | 155.00 |
| 33 19 7224 | Landfill Drummed Inorganic/Organic Nonhazardous Liquid/Sludge, >55 Gallon Drum | EACH | 278.00 | 278.00 | 278.00 | 278.00 | 278.00 |
| 33 19 7225 | Landfill Drummed Organic Liquid/Sludge Requiring Stabilization, <= 30 Gallon Drum | EACH | 302.50 | 302.50 | 302.50 | 302.50 | 302.50 |
| 33 19 7226 | Landfill Drummed Organic Liquid/Sludge Requiring Stabilization, 55 Gallon Drum | EACH | 300.00 | 300.00 | 300.00 | 300.00 | 300.00 |
| 33 19 7227 | Landfill Drummed Organic Liquid/Sludge Requiring Stabilization, > 55 Gallon Drum | EACH | 302.50 | 302.50 | 302.50 | 302.50 | 302.50 |
| 33 19 7228 | Landfill Drummed Liquid/Sludge, Solar Evaporation, <= 30 Gallon Drum, pH>2.0 | EACH | 658.33 | 658.33 | 658.33 | 658.33 | 658.33 |
| 33 19 7229 | Landfill Drummed Liquid/Sludge, Solar Evaporation, 55 Gallon Drum, pH > 2.0 | EACH | 180.00 | 180.00 | 180.00 | 180.00 | 180.00 |
| 33 19 7230 | Landfill Drummed Liquid/Sludge, Solar Evaporation, > 55 Gallon Drum, pH > 2.0 | EACH | 210.00 | 210.00 | 210.00 | 210.00 | 210.00 |
| 33 19 7231 | Landfill Mineral Acids, 0% - 40% Concentration, <= 30 Gallon Drum | EACH | 251.25 | 251.25 | 251.25 | 251.25 | 251.25 |
| 33 19 7232 | Landfill Mineral Acids, 0% - 40% Concentration, 55 Gallon Drum | EACH | 332.50 | 332.50 | 332.50 | 332.50 | 332.50 |
| 33 19 7233 | Landfill Mineral Acids, 0% - 40% Concentration, > 55 Gallon Drum | EACH | 516.75 | 516.75 | 516.75 | 516.75 | 516.75 |
| 33 19 7234 | Landfill Mineral Acids, 41% - 60% Concentration, <= 30 Gallon Drum | EACH | 362.50 | 362.50 | 362.50 | 362.50 | 362.50 |
| 33 19 7235 | Landfill Mineral Acids, 41% - 60% Concentration, 55 Gallon Drum | EACH | 485.00 | 485.00 | 485.00 | 485.00 | 485.00 |
| 33 19 7236 | Landfill Mineral Acids, 41% - 60% Concentration, > 55 Gallon Drum | EACH | 751.25 | 751.25 | 751.25 | 751.25 | 751.25 |
| 33 19 7237 | Landfill Mineral Acids, > 61% Concentration, <= 30 Gallon Drum | EACH | 469.50 | 469.50 | 469.50 | 469.50 | 469.50 |
| 33 19 7238 | Landfill Mineral Acids, > 61% Concentration, 55 Gallon Drum | EACH | 626.25 | 626.25 | 626.25 | 626.25 | 626.25 |
| 33 19 7239 | Landfill Mineral Acids, > 61% Concentration, > 55 Gallon Drum | EACH | 970.50 | 970.50 | 970.50 | 970.50 | 970.50 |
| 33 19 7240 | Landfill Fuel Substitutes, 0% - 10% Solids, <= 30 Gallon Drum | EACH | 56.83 | 56.83 | 56.83 | 56.83 | 56.83 |
| 33 19 7241 | Landfill Fuel Substitutes, 0% - 10% Solids, 55 Gallon Drum | EACH | 75.75 | 75.75 | 75.75 | 75.75 | 75.75 |
| 33 19 7242 | Landfill Fuel Substitutes, 0% - 10% Solids, > 55 Gallon Drum | EACH | 117.38 | 117.38 | 117.38 | 117.38 | 117.38 |
| 33 19 7243 | Landfill Fuel Substitutes, 10.1% - 25% Solids, <= 30 Gallon Drum | EACH | 90.90 | 90.90 | 90.90 | 90.90 | 90.90 |
| 33 19 7244 | Landfill Fuel Substitutes, 10.1% - 25% Solids, 55 Gallon Drum | EACH | 121.25 | 121.25 | 121.25 | 121.25 | 121.25 |

**Environmental Remediation: Assemblies Cost Book**

# Cost Data for Remediation

## Landfill Disposal

| Assembly | Description | Unit | A | B | C | D | E |
|---|---|---|---|---|---|---|---|
| | **Capital Costs** | | | | | | |
| 33 19 7245 | Landfill Fuel Substitutes, 10.1% - 25% Solids, > 55 Gallon Drum | EACH | 188.00 | 188.00 | 188.00 | 188.00 | 188.00 |
| 33 19 7246 | Landfill Fuel Substitutes, 25.1% - 50% Solids, <= 30 Gallon Drum | EACH | 185.63 | 185.63 | 185.63 | 185.63 | 185.63 |
| 33 19 7247 | Landfill Fuel Substitutes, 25.1% - 50% Solids, 55 Gallon Drum | EACH | 248.50 | 248.50 | 248.50 | 248.50 | 248.50 |
| 33 19 7248 | Landfill Fuel Substitutes, 25.1% - 50% Solids, > 55 Gallon Drum | EACH | 383.58 | 383.58 | 383.58 | 383.58 | 383.58 |
| 33 19 7249 | Landfill Fuel Substitutes, 50.1% - 75% Solids, <= 30 Gallon Drum | EACH | 230.88 | 230.88 | 230.88 | 230.88 | 230.88 |
| 33 19 7250 | Landfill Fuel Substitutes, 50.1% - 75% Solids, 55 Gallon Drum | EACH | 308.00 | 308.00 | 308.00 | 308.00 | 308.00 |
| 33 19 7251 | Landfill Fuel Substitutes, 50.1% - 75% Solids, > 55 Gallon Drum | EACH | 477.43 | 477.43 | 477.43 | 477.43 | 477.43 |
| 33 19 7252 | Landfill Fuel Substitutes, 75.1% - 100% Solids, <= 30 Gallon Drum | EACH | 375.13 | 375.13 | 375.13 | 375.13 | 375.13 |
| 33 19 7253 | Landfill Fuel Substitutes, 75.1% - 100% Solids, 55 Gallon Drum | EACH | 500.00 | 500.00 | 500.00 | 500.00 | 500.00 |
| 33 19 7254 | Landfill Fuel Substitutes, 75.1% - 100% Solids, > 55 Gallon Drum | EACH | 775.13 | 775.13 | 775.13 | 775.13 | 775.13 |
| 33 19 7255 | Landfill Packaged Lab Chemicals, 5 - 25 Gallon Drum | EACH | 202.50 | 202.50 | 202.50 | 202.50 | 202.50 |
| 33 19 7256 | Landfill Packaged Lab Chemicals, 30 - 40 Gallon Drum | EACH | 202.50 | 202.50 | 202.50 | 202.50 | 202.50 |
| 33 19 7257 | Landfill Packaged Lab Chemicals, > 45 Gallon Drum | EACH | 232.50 | 232.50 | 232.50 | 232.50 | 232.50 |
| 33 19 7258 | Landfill Packaged Lab Chemicals Requiring Stabilization, 5 - 25 Gallon Drum | EACH | 454.50 | 454.50 | 454.50 | 454.50 | 454.50 |
| 33 19 7259 | Landfill Packaged Lab Chemicals Requiring Stabilization, 30 - 40 Gallon Drum | EACH | 454.50 | 454.50 | 454.50 | 454.50 | 454.50 |
| 33 19 7260 | Landfill Packaged Lab Chemicals Requiring Stabilization, > 45 Gallon Drum | EACH | 606.00 | 606.00 | 606.00 | 606.00 | 606.00 |
| 33 19 7261 | Landfill Solid Palletized Waste | EACH | 407.50 | 407.50 | 407.50 | 407.50 | 407.50 |
| 33 19 7262 | Landfill Liquid or Sludge Palletized Waste | EACH | 814.00 | 814.00 | 814.00 | 814.00 | 814.00 |
| 33 19 7263 | Landfill Hazardous Solid Bulk Waste by Ton | TON | 141.67 | 141.67 | 141.67 | 141.67 | 141.67 |
| 33 19 7264 | Landfill Hazardous Solid Bulk Waste by CY | CY | 140.00 | 140.00 | 140.00 | 140.00 | 140.00 |
| 33 19 7265 | Landfill Hazardous Solid Bulk Waste Requiring Stabilization | TON | 241.33 | 241.33 | 241.33 | 241.33 | 241.33 |
| 33 19 7266 | Landfill Hazardous Solid Bulk Waste, Odd-shaped Container, < 10 CF/Container | CF | 7.43 | 7.43 | 7.43 | 7.43 | 7.43 |
| 33 19 7267 | Landfill Hazardous Solid Bulk Waste, Odd-shaped Container, >= 10 CF/Container | CF | 8.18 | 8.18 | 8.18 | 8.18 | 8.18 |
| 33 19 7268 | Landfill Hazardous Solid Bulk Waste, Odd-shaped Container Requiring Stabilization | CF | 43.42 | 43.42 | 43.42 | 43.42 | 43.42 |
| 33 19 7269 | Landfill Nonhazardous Solid Bulk Waste by Ton | TON | 74.67 | 74.67 | 74.67 | 74.67 | 74.67 |

**Environmental Remediation: Assemblies Cost Book**

# Cost Data for Remediation

## Landfill Disposal

| Assembly | Description | Unit | A | B | C | D | E |
|---|---|---|---|---|---|---|---|
| | **Capital Costs** | | | | | | |
| 33 19 7270 | Landfill Nonhazardous Solid Bulk Waste by CY | CY | 101.00 | 101.00 | 101.00 | 101.00 | 101.00 |
| 33 19 7271 | Landfill Jumbo Bags, Direct | EACH | 197.50 | 197.50 | 197.50 | 197.50 | 197.50 |
| 33 19 7272 | Landfill Jumbo Bags Requiring Stabilization | EACH | 335.00 | 335.00 | 335.00 | 335.00 | 335.00 |
| 33 19 7273 | Landfill Hazardous Liquid Bulk Waste | GAL | 1.72 | 1.72 | 1.72 | 1.72 | 1.72 |
| 33 19 7274 | Landfill Nonhazardous Liquid Bulk Waste | GAL | 1.75 | 1.75 | 1.75 | 1.75 | 1.75 |
| 33 19 7275 | Landfill Hazard Liquid Bulk Waste Requiring Stabilization | GAL | 2.96 | 2.96 | 2.96 | 2.96 | 2.96 |
| 33 19 7276 | Landfill Hazard Liquid Bulk Waste, Solar Evaporation, pH > 2.0 | GAL | 2.55 | 2.55 | 2.55 | 2.55 | 2.55 |
| 33 19 7277 | Landfill Hazardous Nonfuel Liquid/Sludge | GAL | 2.50 | 2.50 | 2.50 | 2.50 | 2.50 |
| 33 19 7278 | Landfill Nonhazardous Nonfuel Liquid/Sludge | GAL | 2.55 | 2.55 | 2.55 | 2.55 | 2.55 |
| 33 19 7279 | Landfill Fuel Substitution Liquids | GAL | 1.02 | 1.02 | 1.02 | 1.02 | 1.02 |
| 33 19 7280 | 30 Gallon, 17C, Open | EACH | 57.64 | 57.64 | 57.64 | 57.64 | 57.64 |
| 33 19 7281 | 30 Gallon, 17H, Open | EACH | 57.64 | 57.64 | 57.64 | 57.64 | 57.64 |
| 33 19 7282 | 30 Gallon, 17E, Open | EACH | 59.31 | 59.31 | 59.31 | 59.31 | 59.31 |
| 33 19 9911 | Air Blower to Inflate Portable Containment Berm | EACH | 160.59 | 160.59 | 160.59 | 160.59 | 160.59 |
| 33 19 9921 | DOT Steel Drum, 55 Gallon | EACH | 65.19 | 65.19 | 65.19 | 65.19 | 65.19 |
| 33 19 9922 | Polyethylene Closed Head Drum, 55 Gallon | EACH | 43.63 | 43.63 | 43.63 | 43.63 | 43.63 |
| 33 19 9923 | Polyethylene Open Head Drum, 55 Gallon | EACH | 49.90 | 49.90 | 49.90 | 49.90 | 49.90 |
| 33 19 9924 | DOT Steel Salvage Drum, 85 Gallon, 16 Gauge | EACH | 57.06 | 57.06 | 57.06 | 57.06 | 57.06 |
| 33 19 9925 | DOT Steel Salvage Drum Composite Overpack, 85 Gallon, 16 Gauge | EACH | 70.49 | 70.49 | 70.49 | 70.49 | 70.49 |

# Cost Data for Remediation
## Low Level Radioactive Soil Treatment

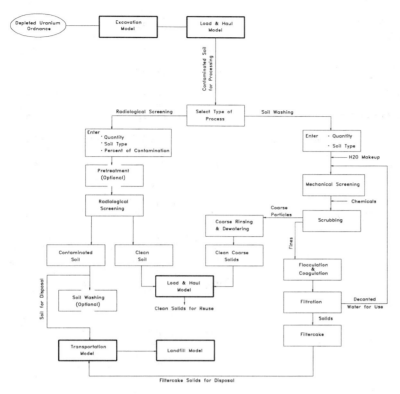

### General:

Treating radioactive waste is a volume reduction activity to separate and consolidate the radioactive component for radioactive waste disposal. The actual reduction in volume for soil washing, based on a U.S. Environmental Protection Agency (USEPA) study, is from 35 to 80 percent. A cost reduction of 25 percent can be obtained by using soil washing and then disposal of contaminated fines as compared to straight forward excavation, transportation, and disposal methods. Radiological screening also has cost advantages in that only the contaminated fractions are separated, potentially allowing the cleaner fractions to be reused at the site depending on if their activity level is below the required standard.

### Typical Treatment Train:

Excavation and/or hauling of contaminated soil to process plant and low level waste repository; landfilling or incinerating contaminated soils or filtercake; loading and hauling of clean material; sampling and analysis.

### Common Cost Components:

1. Mobilization/demobilization of process unit
2. Assemble/shakedown of process unit
3. Rental of process unit
4. Load waste from excavated stockpile to process unit
5. Structural slab
6. Soil conditioning
7. Surfactants
8. Operational labor
9. Electrical usage
10. Maintenance
11. Disposal fees
12. Disassemble/decontamination of process unit

### Other Cost Considerations:

Clearing and grubbing, access road, fencing, cleanup and landscaping, electrical distribution, gas distribution, and water distribution.

**Environmental Remediation: Assemblies Cost Book**

# Cost Data for Remediation

## Low Level Radioactive Soil Treatment

| Assembly | Description | Unit | Unit Cost by Safety Level A | B | C | D | E |
|---|---|---|---|---|---|---|---|
| | **Capital Costs** | | | | | | |
| 17 03 0109 | Pad Subgrade Preparation | CY | 8.13 | 6.56 | 5.44 | 3.90 | 3.63 |
| 17 03 0220 | 910, 1.25 CY, Wheel Loader | HOUR | 130.20 | 104.06 | 87.23 | 61.63 | 55.86 |
| 17 03 0221 | 916, 1.5 CY, Wheel Loader | HOUR | 114.32 | 90.83 | 76.64 | 53.69 | 47.92 |
| 17 03 0222 | 926, 2.0 CY, Wheel Loader | HOUR | 154.30 | 124.14 | 103.30 | 73.68 | 67.91 |
| 17 03 0223 | 950, 3.0 CY, Wheel Loader | HOUR | 180.10 | 145.64 | 120.50 | 86.58 | 80.81 |
| 17 03 0224 | 966, 4.0 CY, Wheel Loader | HOUR | 211.58 | 171.88 | 141.48 | 102.32 | 96.55 |
| 17 03 0225 | 980, 5.25 CY, Wheel Loader | HOUR | 298.84 | 244.38 | 199.68 | 145.79 | 139.74 |
| 17 03 0345 | Standby, 910, 1.25 CY Wheel Loader | HOUR | 15.13 | 12.61 | 10.09 | 7.57 | 7.57 |
| 17 03 0346 | Standby, 916, 1.5 CY Wheel Loader | HOUR | 11.32 | 9.44 | 7.55 | 5.66 | 5.66 |
| 17 03 0347 | Standby, 926, 2.0 CY Wheel Loader | HOUR | 21.40 | 17.83 | 14.27 | 10.70 | 10.70 |
| 17 03 0348 | Standby, 950, 3.0 CY Wheel Loader | HOUR | 27.24 | 22.70 | 18.16 | 13.62 | 13.62 |
| 17 03 0349 | Standby, 966, 4.0 CY Wheel Loader | HOUR | 34.27 | 28.56 | 22.85 | 17.13 | 17.13 |
| 17 03 0350 | Standby, 980, 5.25 CY Wheel Loader | HOUR | 57.52 | 47.94 | 38.35 | 28.76 | 28.76 |
| 17 03 0419 | Crushed Stone, 1/2" to 3/4" | CY | 23.45 | 21.67 | 20.70 | 18.97 | 18.47 |
| 17 03 0420 | Backfill Trench, Borrow Material, Delivered & Dumped Only | CY | 9.86 | 8.74 | 7.93 | 6.83 | 6.63 |
| 17 03 0501 | Compact Subgrade, 2 Lifts | CY | 0.84 | 0.68 | 0.56 | 0.40 | 0.37 |
| 17 03 0510 | Dry Roll Gravel, Steel Roller | SY | 1.22 | 0.96 | 0.82 | 0.57 | 0.49 |
| 17 03 1002 | 2" Diameter Contractor's Trash Pump, 75 GPM | DAY | 50.20 | 47.01 | 45.64 | 42.55 | 41.42 |
| 17 03 1003 | 3" Diameter Contractor's Trash Pump, 150 GPM | DAY | 61.72 | 57.05 | 55.05 | 50.54 | 48.88 |
| 17 03 1004 | 4" Diameter Contractor's Trash Pump, 300 GPM | DAY | 80.30 | 74.23 | 71.63 | 65.77 | 63.61 |
| 17 03 9903 | Hand Place Small Earth Fill Berm | CY | 112.16 | 86.46 | 75.45 | 50.61 | 41.50 |
| 18 01 0102 | Gravel, Delivered & Dumped | CY | 24.33 | 22.93 | 22.03 | 20.66 | 20.36 |
| 18 01 0201 | Concrete Curb, 6" x 6" | LF | 2.78 | 2.38 | 2.21 | 1.83 | 1.70 |
| 18 01 0202 | Concrete Curb & Gutter, 6" x 24", Formed | LF | 20.78 | 16.61 | 14.83 | 10.80 | 9.32 |
| 18 01 0206 | Concrete Curb, 8" x 8" | LF | 7.58 | 6.24 | 5.66 | 4.36 | 3.89 |
| 18 01 0207 | Concrete Curb, 12" High x 8" Deep | LF | 10.27 | 8.45 | 7.67 | 5.92 | 5.27 |
| 18 02 0320 | 4" Structural Slab on Grade | SF | 6.45 | 5.45 | 4.98 | 4.00 | 3.67 |
| 18 02 0321 | 6" Structural Slab on Grade | SF | 6.98 | 5.95 | 5.47 | 4.48 | 4.14 |
| 18 02 0322 | 8" Structural Slab on Grade | SF | 9.75 | 8.38 | 7.71 | 6.38 | 5.95 |
| 18 02 0323 | 10" Structural Slab on Grade | SF | 10.56 | 9.13 | 8.43 | 7.04 | 6.58 |
| 18 02 0324 | 12" Structural Slab on Grade | SF | 11.57 | 10.04 | 9.30 | 7.82 | 7.34 |
| 18 02 0330 | 4" Mesh Reinforced Slab on Grade | SF | 5.48 | 4.60 | 4.18 | 3.32 | 3.03 |
| 18 02 0331 | 6" Mesh Reinforced Slab on Grade | SF | 6.01 | 5.10 | 4.67 | 3.79 | 3.49 |

## Environmental Remediation: Assemblies Cost Book

*Page 3-198*

1998 by ECHOS. All rights reserved

# Cost Data for Remediation

## Low Level Radioactive Soil Treatment

| Assembly | Description | Unit | Unit Cost by Safety Level A | B | C | D | E |
|---|---|---|---|---|---|---|---|
| | **Capital Costs** | | | | | | |
| 18 02 0332 | 8" Mesh Reinforced Slab on Grade | SF | 7.58 | 6.50 | 5.96 | 4.91 | 4.58 |
| 18 02 0333 | 10" Mesh Reinforced Slab on Grade | SF | 8.10 | 7.00 | 6.44 | 5.37 | 5.03 |
| 18 02 0334 | 12" Mesh Reinforced Slab on Grade | SF | 8.78 | 7.63 | 7.05 | 5.93 | 5.58 |
| 18 02 0340 | 4" Unreinforced Slab on Grade | SF | 4.34 | 3.64 | 3.29 | 2.61 | 2.38 |
| 18 02 0341 | 6" Unreinforced Slab on Grade | SF | 4.87 | 4.14 | 3.79 | 3.08 | 2.85 |
| 18 02 0342 | 8" Unreinforced Slab on Grade | SF | 6.45 | 5.54 | 5.07 | 4.19 | 3.93 |
| 18 02 0343 | 10" Unreinforced Slab on Grade | SF | 6.97 | 6.04 | 5.56 | 4.66 | 4.38 |
| 18 02 0344 | 12" Unreinforced Slab on Grade | SF | 7.64 | 6.67 | 6.16 | 5.22 | 4.93 |
| 18 04 0205 | CIP Walls Form & Strip (4 Uses) | SF | 8.86 | 6.99 | 6.19 | 4.38 | 3.73 |
| 19 02 0311 | 12' x 36" Diameter Reinforced Concrete Pipe Wet Well for Lift Station | EACH | 9,840 | 8,577 | 7,816 | 6,581 | 6,270 |
| 19 02 0312 | 24' x 60" Diameter Reinforced Concrete Pipe Wet Well for Lift Station | EACH | 39,142 | 33,897 | 30,681 | 25,554 | 24,296 |
| 19 02 0313 | 5' x 5' x 5' Reinforced Concrete Sump | EACH | 3,948 | 3,282 | 2,979 | 2,335 | 2,110 |
| 19 02 0601 | 12" x 12" Underground French Drain | LF | 4.50 | 3.92 | 3.61 | 3.04 | 2.88 |
| 19 02 0602 | 18" x 18" Underground French Drain | LF | 5.93 | 5.29 | 4.94 | 4.31 | 4.13 |
| 19 02 0603 | 24" x 24" Underground French Drain | LF | 7.43 | 6.71 | 6.31 | 5.61 | 5.41 |
| 19 04 0401 | 550 Gallon, Stainless Steel Aboveground Wastewater Holding Tank, Rental | MONTH | 320.65 | 320.65 | 320.65 | 320.65 | 320.65 |
| 19 04 0403 | 4,000 Gallon Polyethylene Wastewater Tank, Rental | MONTH | 545.90 | 545.90 | 545.90 | 545.90 | 545.90 |
| 19 04 0405 | 6,000 Gallon, Polyethylene Aboveground Wastewater Holding Tank, Rental | MONTH | 641.30 | 641.30 | 641.30 | 641.30 | 641.30 |
| 19 04 0407 | 21,000 Gallon, Steel Closed Stationary Aboveground Wastewater Holding Tank, Rental | MONTH | 1,219 | 1,219 | 1,219 | 1,219 | 1,219 |
| 19 04 0602 | 550 Gallon Steel Sump, Aboveground with Supports & Fittings | EACH | 1,598 | 1,390 | 1,301 | 1,099 | 1,026 |
| 19 04 0603 | 1,000 Gallon Steel Sump, Aboveground with Supports & Fittings | EACH | 2,118 | 1,844 | 1,727 | 1,463 | 1,366 |
| 19 04 0604 | 1,500 Gallon Steel Sump, Aboveground with Supports & Fittings | EACH | 2,697 | 2,356 | 2,209 | 1,879 | 1,758 |
| 19 04 0605 | 2,000 Gallon Steel Sump, Aboveground with Supports & Fittings | EACH | 3,457 | 2,997 | 2,777 | 2,332 | 2,183 |
| 19 04 0606 | 5,000 Gallon Steel Sump, Aboveground with Supports & Fittings | EACH | 6,079 | 5,494 | 5,214 | 4,646 | 4,457 |
| 19 04 0625 | 6,000 Gallon Horizontal Plastic Sump with 6" NPT Connection | EACH | 5,248 | 4,975 | 4,858 | 4,594 | 4,497 |
| 19 04 0626 | 8,000 Gallon Horizontal Plastic Sump with 6" NPT Connection | EACH | 5,857 | 5,516 | 5,369 | 5,039 | 4,918 |

**Environmental Remediation: Assemblies Cost Book**

# Cost Data for Remediation

## Low Level Radioactive Soil Treatment

| Assembly | Description | Unit | \[Unit Cost by Safety Level\] A | B | C | D | E |
|---|---|---|---|---|---|---|---|
| | **Capital Costs** | | | | | | |
| 19 04 0627 | 10,000 Gallon Horizontal Plastic Sump with 6" NPT Connection | EACH | 7,082 | 6,622 | 6,402 | 5,956 | 5,807 |
| 19 04 0628 | 12,000 Gallon Horizontal Plastic Sump with 6" NPT Connection | EACH | 8,308 | 7,722 | 7,442 | 6,875 | 6,686 |
| 19 04 0629 | 15,000 Gallon Horizontal Plastic Sump with 6" NPT Connection | EACH | 14,184 | 13,540 | 13,232 | 12,608 | 12,400 |
| 19 04 0631 | 20,000 Gallon Horizontal Plastic Sump with 6" NPT Connection | EACH | 15,799 | 15,084 | 14,742 | 14,048 | 13,817 |
| 19 04 0632 | 25,000 Gallon Horizontal Plastic Sump with 6" NPT Connection | EACH | 24,407 | 23,487 | 23,047 | 22,155 | 21,858 |
| 19 04 0633 | 30,000 Gallon Horizontal Plastic Sump with 6" NPT Connection | EACH | 35,112 | 34,039 | 33,525 | 32,485 | 32,138 |
| 19 04 0634 | 40,000 Gallon Horizontal Plastic Sump with 6" NPT Connection | EACH | 40,844 | 39,234 | 38,465 | 36,904 | 36,383 |
| 19 04 0635 | 48,000 Gallon Horizontal Plastic Sump with 6" NPT Connection | EACH | 59,758 | 57,612 | 56,586 | 54,505 | 53,811 |
| 33 08 0503 | Polymeric Liner Anchor Trench, 3' x 1.5' | LF | 1.58 | 1.25 | 1.07 | 0.76 | 0.67 |
| 33 08 0563 | 40 Mil Polymeric Liner, PVC | SF | 0.38 | 0.36 | 0.36 | 0.34 | 0.34 |
| 33 09 0701 | CIP Reinforced Concrete Containment Wall | CY | 279.72 | 243.62 | 224.28 | 189.16 | 178.78 |
| 33 13 0915 | Mobilize/Demobilize Soil Washing System | MI | 2.11 | 2.11 | 2.11 | 2.11 | 2.11 |
| 33 13 0916 | Assemble/Disassemble Soil Washing System | EACH | 59,475 | 46,901 | 39,908 | 27,660 | 24,202 |
| 33 13 0917 | Startup/Shakedown - Soil Washing System | EACH | 37,770 | 29,115 | 25,409 | 17,043 | 13,975 |
| 33 13 0918 | Decontaminate Soil Washing System | EACH | 560.32 | 434.51 | 376.69 | 254.86 | 212.73 |
| 33 13 0920 | 25 Tons per Hour Soil Washing System, Monthly Rental | MONTH | 81,000 | 81,000 | 81,000 | 81,000 | 81,000 |
| 33 13 0921 | Operational Labor, 10 - 25 Tons per Hour System | HR | 566.55 | 436.72 | 381.14 | 255.64 | 209.63 |
| 33 13 0927 | Maintenance/Spare Parts, 25 Tons per Hour System | TON | 2.24 | 2.24 | 2.24 | 2.24 | 2.24 |
| 33 18 0401 | Mobilize/Demobilize Radiological Screening System | MI | 2.11 | 2.11 | 2.11 | 2.11 | 2.11 |
| 33 18 0402 | Assemble/Shakedown Radiological Screening System | EACH | 82,472 | 63,572 | 55,481 | 37,213 | 30,515 |
| 33 18 0403 | Disassemble/Decontaminate Radiological Screening System | EACH | 139,073 | 115,894 | 92,715 | 69,536 | 69,536 |
| 33 18 0404 | Rental - 13 Tons per Hour Radiological Screening System | MONTH | 80,000 | 80,000 | 80,000 | 80,000 | 80,000 |
| 33 18 0405 | Operational Labor - 13 Tons per Hour Radiological Screening System | HR | 226.62 | 174.69 | 152.45 | 102.26 | 83.85 |
| 33 18 0406 | Soil Conditioning Charges | TON | 6.50 | 6.50 | 6.50 | 6.50 | 6.50 |
| 33 18 0407 | Maintenance/Spare Parts - 13 Tons per Hour Radiological Screening System | TON | 5.00 | 5.00 | 5.00 | 5.00 | 5.00 |
| 33 18 0901 | Radioactive Waste Disposal Fee | CY | 200.00 | 200.00 | 200.00 | 200.00 | 200.00 |
| 33 19 7103 | Process Water Hauling Fee, Subcontracted | GAL | 0.24 | 0.24 | 0.24 | 0.24 | 0.24 |

## Environmental Remediation: Assemblies Cost Book

# Cost Data for Remediation

## Low Level Radioactive Soil Treatment

| Assembly | Description | Unit | A | B | C | D | E |
|---|---|---|---|---|---|---|---|
| | **Capital Costs** | | | | | | |
| 33 19 7104 | Radioactive Process Water Treatment/Disposal, Subcontracted | GAL | 2.01 | 2.01 | 2.01 | 2.01 | 2.01 |
| 33 23 1306 | High Sump Level Switch for Avoiding Overflow | EACH | 214.65 | 214.65 | 214.65 | 214.65 | 214.65 |
| 33 29 0412 | 15 GPM Submersible Sump Pump | EACH | 492.79 | 460.07 | 446.06 | 414.43 | 402.83 |
| 33 29 0413 | 42 GPM Submersible Sump Pump | EACH | 792.79 | 760.07 | 746.06 | 714.43 | 702.83 |
| 33 29 0414 | 52 GPM Submersible Sump Pump | EACH | 871.35 | 832.08 | 815.27 | 777.32 | 763.40 |
| 33 33 0101 | Sodium Hydroxide Solution, 190 Lb Drummed Liquid | EACH | 62.34 | 57.81 | 55.07 | 50.64 | 49.53 |
| 33 33 0102 | Sodium Hydroxide Solution, 700 Lb Drummed Liquid | EACH | 163.04 | 158.51 | 155.77 | 151.34 | 150.23 |
| 33 33 0103 | Sodium Hydroxide, 500 Lb Container, Beads | EACH | 234.06 | 229.53 | 226.79 | 222.36 | 221.25 |
| 33 33 0104 | Sodium Hydroxide, 100 Lb Container, Flakes | EACH | 89.90 | 85.37 | 82.63 | 78.20 | 77.09 |
| 33 33 0105 | Sodium Hydroxide, 400 Lb Container, Flakes | EACH | 169.60 | 169.60 | 169.60 | 169.60 | 169.60 |
| 33 33 0106 | Magnesium Hydroxide Slurry, 55 Gallon Drums, < 40 Drums | EACH | 249.77 | 249.77 | 249.77 | 249.77 | 249.77 |
| 33 33 0107 | Magnesium Hydroxide Slurry, 55 Gallon Drums, 40 - 71 Drums | EACH | 234.26 | 234.26 | 234.26 | 234.26 | 234.26 |
| 33 33 0108 | Magnesium Hydroxide Slurry, 55 Gallon Drums, 72 Drums | EACH | 107.33 | 107.33 | 107.33 | 107.33 | 107.33 |
| 33 33 0109 | Magnesium Hydroxide, Bulk Quantity, 20 Tons | EACH | 4,452 | 4,452 | 4,452 | 4,452 | 4,452 |
| 33 33 0110 | Sulfuric Acid Solution, 220 Lb Drummed Liquid | EACH | 45.38 | 40.85 | 38.11 | 33.68 | 32.57 |
| 33 33 0111 | Sulfuric Acid Solution, 750 Lb Drummed Liquid | EACH | 93.61 | 89.08 | 86.34 | 81.91 | 80.80 |
| 33 33 0112 | Soda Ash, 50 Lb Bag, Powdered | EACH | 28.42 | 23.89 | 21.15 | 16.72 | 15.61 |
| 33 33 0113 | Soda Ash, 100 Lb Bag, Powdered | EACH | 41.14 | 36.61 | 33.87 | 29.44 | 28.33 |
| 33 33 0114 | Quicklime, 1/4" Nominal Granules, Bulk Quantity | TON | 96.49 | 96.49 | 96.48 | 96.48 | 96.47 |
| 33 33 0115 | Quicklime, 3/4" Nominal Granules, Bulk Quantity | TON | 96.49 | 96.49 | 96.48 | 96.48 | 96.47 |
| 33 33 0116 | Quicklime, Combination 1/4" & 3/4" Granules, Bulk Quantity | TON | 96.49 | 96.49 | 96.48 | 96.48 | 96.47 |
| 33 33 0117 | Hydrated Lime, Powdered, Bulk | TON | 97.55 | 97.55 | 97.54 | 97.54 | 97.53 |
| 33 33 0118 | Hydrated Lime, Powdered, 50 Lb Bag | EACH | 25.76 | 21.23 | 18.49 | 14.06 | 12.94 |
| 33 33 0119 | Aluminum Sulfate (Alum), Bulk | TON | 240.07 | 240.06 | 240.06 | 240.05 | 240.05 |
| 33 33 0120 | Ferric Sulfate, Bulk | TON | 274.04 | 274.04 | 274.03 | 274.03 | 274.02 |
| 33 33 0121 | Ferric Chloride, Bulk | TON | 241.18 | 241.18 | 241.17 | 241.17 | 241.16 |
| 33 33 0122 | Montmorillonite Clay Flocculant Aid, Bulk | LB | 0.14 | 0.13 | 0.13 | 0.12 | 0.12 |
| 33 33 0123 | Sodium Hectorite Clay Flocculant Aid, Bulk | LB | 0.17 | 0.17 | 0.16 | 0.15 | 0.15 |
| 33 33 0124 | Super Dispersible Sodium Flocculant Aid, Bulk | LB | 0.19 | 0.19 | 0.18 | 0.18 | 0.17 |
| 33 33 0125 | Alkylbenzene Sulfonate Surfactant, Bulk | LB | 1.16 | 1.16 | 1.16 | 1.16 | 1.16 |
| 33 33 0127 | Non-ionic Polymeric Flocculant | LB | 3.01 | 3.01 | 3.01 | 3.01 | 3.01 |

**Environmental Remediation: Assemblies Cost Book**

*Page 3-201*

1998 by ECHOS. All rights reserved

# Cost Data for Remediation

## Low Level Radioactive Soil Treatment

| Assembly | Description | Unit | A | B | C | D | E |
|---|---|---|---|---|---|---|---|
| | **Capital Costs** | | | | | | |
| 33 33 0172 | Hydrogen Peroxide Feed Tank, Regulator, Injector & Panel Rental | EACH | 3,759 | 3,324 | 3,133 | 2,713 | 2,562 |
| 33 42 0101 | Electrical Charge | KWH | 0.06 | 0.06 | 0.06 | 0.06 | 0.06 |
| 33 42 0302 | Process Water, Supplied by Water Line | KGAL | 2.56 | 2.56 | 2.56 | 2.56 | 2.56 |

Unit Cost by Safety Level

**Environmental Remediation: Assemblies Cost Book**

# Cost Data for Remediation
## Low Temperature Thermal Desorption

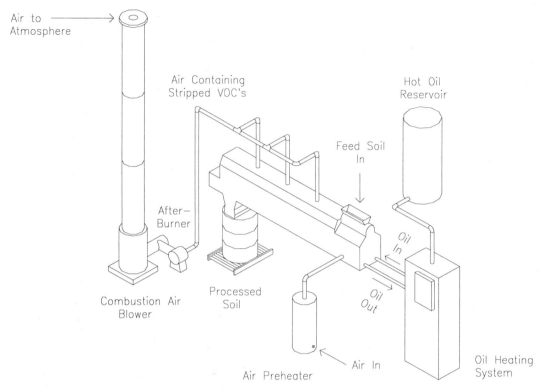

### General:
Thermal desorption is used to remove organic contaminants from soils, sludges, and other solid media. The organic contaminants are removed as vapors or condensed liquids which may then be destroyed in a permitted incinerator or used as a supplemental fuel. Because of low temperatures (100° to 400° C) and gas flow rates, this process is less expensive than incineration. To be economically feasible, however, a location should have a minimum of 5,000 cubic yards (7,500 tons) of material.

### Typical Treatment Train:
Excavation and/or staging of wastes, pretreatment of wastes, disposal of treated material, sampling and analysis.

### Common Cost Components:
1. Mobilization/demobilization of process unit
2. Structural slab
3. Conveyor
4. Hopper
5. Screen
6. Reduction of moisture content
7. Particle size reduction
8. Drum shredder
9. Firing system (direct or indirect)
10. Process operation

### Other Cost Considerations:
Supervisory labor, fencing.

**Environmental Remediation: Assemblies Cost Book**

# Cost Data for Remediation

## Low Temperature Thermal Desorption

| Assembly | Description | Unit | Unit Cost by Safety Level A | B | C | D | E |
|---|---|---|---|---|---|---|---|
| | **Capital Costs** | | | | | | |
| 17 03 0220 | 910, 1.25 CY, Wheel Loader | HOUR | 130.20 | 104.06 | 87.23 | 61.63 | 55.86 |
| 17 03 0221 | 916, 1.5 CY, Wheel Loader | HOUR | 114.32 | 90.83 | 76.64 | 53.69 | 47.92 |
| 17 03 0222 | 926, 2.0 CY, Wheel Loader | HOUR | 154.30 | 124.14 | 103.30 | 73.68 | 67.91 |
| 17 03 0223 | 950, 3.0 CY, Wheel Loader | HOUR | 180.10 | 145.64 | 120.50 | 86.58 | 80.81 |
| 17 03 0224 | 966, 4.0 CY, Wheel Loader | HOUR | 211.58 | 171.88 | 141.48 | 102.32 | 96.55 |
| 17 03 0225 | 980, 5.25 CY, Wheel Loader | HOUR | 298.84 | 244.38 | 199.68 | 145.79 | 139.74 |
| 17 03 0226 | 988, 7.0 CY, Wheel Loader | HOUR | 384.90 | 316.10 | 257.05 | 188.82 | 182.77 |
| 17 03 0227 | 992, 13.5 CY, Wheel Loader | HOUR | 648.92 | 536.11 | 433.06 | 320.83 | 314.78 |
| 18 02 0320 | 4" Structural Slab on Grade | SF | 6.45 | 5.45 | 4.98 | 4.00 | 3.67 |
| 18 02 0321 | 6" Structural Slab on Grade | SF | 6.98 | 5.95 | 5.47 | 4.48 | 4.14 |
| 18 02 0322 | 8" Structural Slab on Grade | SF | 9.75 | 8.38 | 7.71 | 6.38 | 5.95 |
| 18 02 0323 | 10" Structural Slab on Grade | SF | 10.56 | 9.13 | 8.43 | 7.04 | 6.58 |
| 18 02 0324 | 12" Structural Slab on Grade | SF | 11.57 | 10.04 | 9.30 | 7.82 | 7.34 |
| 18 02 0341 | 6" Unreinforced Slab on Grade | SF | 4.87 | 4.14 | 3.79 | 3.08 | 2.85 |
| 18 02 0342 | 8" Unreinforced Slab on Grade | SF | 6.45 | 5.54 | 5.07 | 4.19 | 3.93 |
| 18 02 0343 | 10" Unreinforced Slab on Grade | SF | 6.97 | 6.04 | 5.56 | 4.66 | 4.38 |
| 18 02 0344 | 12" Unreinforced Slab on Grade | SF | 7.64 | 6.67 | 6.16 | 5.22 | 4.93 |
| 33 13 1120 | Pretreatment System Operational Labor Cost | DAY | 217.57 | 217.57 | 217.57 | 217.57 | 217.57 |
| 33 13 2401 | Weekly Rental, Portable Drum Shredder, 60 - 100 Drums/Hour | WEEK | 11,408 | 11,408 | 11,408 | 11,408 | 11,408 |
| 33 13 2402 | Daily Rental, Portable Drum Shredder, 60 - 100 Drums/Hour | DAY | 2,282 | 2,282 | 2,282 | 2,282 | 2,282 |
| 33 13 2403 | Monthly Rental, Portable Drum Shredder, 60 - 100 Drums/Hour | MONTH | 45,633 | 45,633 | 45,633 | 45,633 | 45,633 |
| 33 13 2404 | Shredder Mobilize/DeMobilize | EACH | 1,700 | 1,700 | 1,700 | 1,700 | 1,700 |
| 33 13 2405 | Electric Shredder, Minimum Rental Charge, 50 - 75 Tons/Hour | EACH | 41,750 | 41,750 | 41,750 | 41,750 | 41,750 |
| 33 13 2406 | Electric Shredder, Additional Monthly Rental, 50 - 75 Tons/Hour | MONTH | 18,875 | 18,875 | 18,875 | 18,875 | 18,875 |
| 33 13 2407 | Electric Shredder, Purchase, 35 Tons/Hour Output | EACH | 271,283 | 271,283 | 271,283 | 271,283 | 271,283 |
| 33 13 2408 | Portable Drum Shredder, 60 - 100 Drums/Hour Output | EACH | 251,900 | 251,900 | 251,900 | 251,900 | 251,900 |
| 33 14 0201 | Minimum Mobilize/DeMobilize Charge <=1,000 Mile, Mobile Process Unit | EACH | 84,800 | 84,800 | 84,800 | 84,800 | 84,800 |
| 33 14 0202 | Additional Mobilize/DeMobilize Charge per Mile, Mobile Process Unit | MILE | 68.90 | 68.90 | 68.90 | 68.90 | 68.90 |
| 33 14 0204 | Direct Firing, Rental and Operations Cost | TON | 134.47 | 133.05 | 132.17 | 130.78 | 130.45 |

**Environmental Remediation: Assemblies Cost Book**

*Page 3-204*

# Cost Data for Remediation

## Low Temperature Thermal Desorption

| Assembly | Description | Unit | Unit Cost by Safety Level A | B | C | D | E |
|---|---|---|---|---|---|---|---|
| | **Capital Costs** | | | | | | |
| 33 14 0205 | Indirect Firing, Rental and Operations Cost | TON | 136.97 | 135.55 | 134.67 | 133.28 | 132.95 |
| 33 14 0206 | Min Mob/Demob Chrg for Sm Portable LTTD Unts <= 1000 mi | EA | 5,000 | 5,000 | 5,000 | 5,000 | 5,000 |
| 33 14 0207 | Add. Mob/Demob Chrg for Sm. Port LTTD Units <= 1000 mi | MI | 1.56 | 1.56 | 1.56 | 1.56 | 1.56 |
| 33 18 8401 | 41.5' Automatic Conveyor, 45 FPM, Horizontal 24" Belt, Center Drive | EACH | 8,054 | 7,271 | 6,882 | 6,121 | 5,877 |
| 33 18 8402 | 61.5' Automatic Conveyor, 45 FPM, Horizontal 24" Belt, Center Drive | EACH | 10,682 | 9,637 | 9,119 | 8,105 | 7,779 |
| 33 18 8403 | 34' Automatic Inclined Conveyor, 25 Degree, 24" Belt, Loader/End Idler | EACH | 9,284 | 8,239 | 7,721 | 6,707 | 6,381 |
| 33 18 8501 | Refuse Bottom Hopper, Aluminized Steel, 18" Diameter | EACH | 1,334 | 1,251 | 1,215 | 1,135 | 1,105 |
| 33 18 8601 | 4' x 12' Single-tray Vibrating Screening Unit, with Motor & Accessories | EACH | 16,360 | 15,705 | 15,362 | 14,725 | 14,531 |
| 33 18 8602 | 5' x 16' Double-tray Vibrating Screening Unit, with Motor & Accessories | EACH | 25,858 | 25,105 | 24,711 | 23,979 | 23,757 |
| 33 18 8603 | 6' x 20' Triple-tray Vibrating Screening Unit, with Motor & Accessories | EACH | 32,164 | 30,853 | 30,168 | 28,894 | 28,506 |
| 33 18 8604 | 7' x 24' Triple-tray Vibrating Screening Unit, with Motor & Accessories | EACH | 39,668 | 37,805 | 36,831 | 35,020 | 34,469 |
| 33 33 3001 | Use of Filter Press, Moisture Reduction | TON | 95.40 | 95.40 | 95.40 | 95.40 | 95.40 |

**Environmental Remediation: Assemblies Cost Book**

# Cost Data for Remediation
## Media Filtration

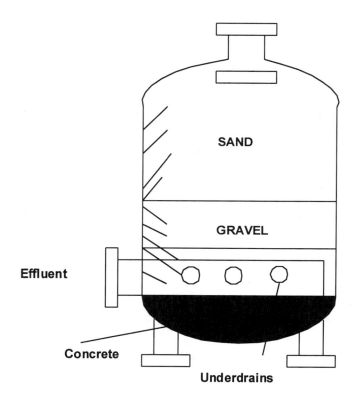

### General:

Granular media filters are used to remove suspended solids from waste water as the liquid is forced through a porous granular medium. The filter bed may be a single material (such as sand), or a dual media, (such as sand with an upper layer of coal) or multimedia (including garnet, sand and coal from top to bottom). The bed is contained within a basin and is supported by an underdrain system that allows the filtered liquid to be drawn off while the filter media is retained in place. The filter is periodically backwashed to prevent plugging.

### Typical Treatment Train:

Removal of solids after precipitation or coagulation reactions, removal of residual biological floc, pretreatment for air strippers and other treatments, disposal of filter material, treatment and disposal of backwash water, sampling and analysis.

### Common Cost Components:

1. Structural slab
2. Piping
3. Filtration unit
4. Operations and maintenance

### Other Cost Considerations:

Pilot studies to determine optimum design (filter configuration, type/size/depth of filter media, filtration rate, etc.)

**Environmental Remediation: Assemblies Cost Book**

# Cost Data for Remediation

## Media Filtration

| Assembly | Description | Unit | Unit Cost by Safety Level | | | | |
|---|---|---|---|---|---|---|---|
| | | | A | B | C | D | E |
| | **Capital Costs** | | | | | | |
| 18 02 0322 | 8" Structural Slab on Grade | SF | 9.75 | 8.38 | 7.71 | 6.38 | 5.95 |
| 18 02 0324 | 12" Structural Slab on Grade | SF | 11.57 | 10.04 | 9.30 | 7.82 | 7.34 |
| 19 01 0202 | 1", Class 200, PVC Piping | LF | 7.48 | 5.84 | 5.07 | 3.48 | 2.93 |
| 19 01 0204 | 2", Class 200, PVC Piping | LF | 8.86 | 6.96 | 6.07 | 4.23 | 3.60 |
| 19 01 0206 | 3", Class 200, PVC Piping | LF | 11.81 | 9.34 | 8.20 | 5.80 | 4.98 |
| 19 01 0207 | 4", Class 200, PVC Piping | LF | 12.33 | 9.86 | 8.72 | 6.32 | 5.50 |
| 19 01 0208 | 6", Class 200, PVC Piping | LF | 14.95 | 12.20 | 10.93 | 8.27 | 7.36 |
| 19 01 0212 | 8", Class 150, PVC Piping | LF | 18.36 | 15.27 | 13.84 | 10.85 | 9.82 |
| 19 01 0214 | 12", Class 150, PVC Piping | LF | 28.75 | 24.63 | 22.72 | 18.73 | 17.36 |
| 33 13 0101 | 3' Diameter Electric Automatic Pressure Filter Unit | EACH | 9,960 | 8,539 | 7,759 | 6,375 | 5,978 |
| 33 13 0102 | 4' Diameter Electric Automatic Pressure Filter Unit | EACH | 14,937 | 13,345 | 12,472 | 10,923 | 10,477 |
| 33 13 0103 | 5' Diameter Electric Automatic Pressure Filter Unit | EACH | 15,593 | 14,001 | 13,128 | 11,579 | 11,133 |
| 33 13 0104 | 6' Diameter Electric Automatic Pressure Filter Unit | EACH | 20,547 | 18,788 | 17,823 | 16,110 | 15,618 |
| 33 13 0105 | 7' Diameter Electric Automatic Pressure Filter Unit | EACH | 23,444 | 21,685 | 20,720 | 19,007 | 18,515 |
| 33 13 0106 | 8' Diameter Electric Automatic Pressure Filter Unit | EACH | 29,455 | 27,301 | 26,120 | 24,023 | 23,421 |
| 33 26 0101 | 1" Carbon Steel Piping | LF | 6.46 | 5.28 | 4.77 | 3.63 | 3.21 |
| 33 26 0102 | 2" Carbon Steel Piping | LF | 10.50 | 8.72 | 7.95 | 6.23 | 5.60 |
| 33 26 0103 | 3" Carbon Steel Piping | LF | 18.02 | 15.20 | 13.99 | 11.26 | 10.26 |
| 33 26 0104 | 4" Carbon Steel Piping | LF | 23.20 | 19.86 | 18.43 | 15.21 | 14.02 |
| 33 26 0105 | 6" Carbon Steel Piping | LF | 36.37 | 30.24 | 27.46 | 21.52 | 19.44 |
| 33 26 0106 | 8" Carbon Steel Piping | LF | 40.77 | 34.08 | 31.15 | 24.68 | 22.36 |
| 33 26 0108 | 12" Carbon Steel Piping | LF | 75.72 | 65.62 | 61.18 | 51.41 | 47.89 |
| 33 26 0201 | 1" Stainless Steel Piping, Schedule 40, Threaded | LF | 18.67 | 16.12 | 15.03 | 12.56 | 11.65 |
| 33 26 0202 | 2" Stainless Steel Piping, Schedule 40, Threaded | LF | 33.47 | 29.21 | 27.39 | 23.28 | 21.77 |
| 33 26 0203 | 3" Stainless Steel Piping, Schedule 40, Threaded | LF | 57.00 | 50.61 | 47.88 | 41.71 | 39.45 |
| 33 26 0204 | 4" Stainless Steel Piping, Schedule 40, Threaded | LF | 80.15 | 71.72 | 68.12 | 59.98 | 57.00 |
| 33 26 0205 | 6" Stainless Steel Piping, Schedule 10, Type 316 | LF | 78.58 | 70.01 | 66.14 | 57.84 | 54.94 |
| 33 26 0206 | 8" Stainless Steel Piping, Schedule 40, Welded | LF | 197.22 | 183.32 | 176.89 | 163.42 | 158.80 |
| 33 26 0209 | 12" Stainless Steel Piping, Schedule 40, Welded | LF | 312.43 | 295.44 | 287.57 | 271.12 | 265.47 |
| 33 41 0101 | Pump & Motor Maintenance/Repair | EACH | 865.17 | 671.48 | 581.58 | 393.96 | 329.64 |
| 33 42 0101 | Electrical Charge | KWH | 0.06 | 0.06 | 0.06 | 0.06 | 0.06 |
| | **Operations & Maintenance** | | | | | | |
| 33 41 0101 | Pump & Motor Maintenance/Repair | EACH | 865.17 | 671.48 | 581.58 | 393.96 | 329.64 |

**Environmental Remediation: Assemblies Cost Book**

# Cost Data for Remediation

## Media Filtration

| Assembly | Description | Unit | Unit Cost by Safety Level | | | | |
|---|---|---|---|---|---|---|---|
| | | | A | B | C | D | E |
| | **Operations & Maintenance** | | | | | | |
| 33 42 0101 | Electrical Charge | KWH | 0.06 | 0.06 | 0.06 | 0.06 | 0.06 |

# Cost Data for Remediation
## Monitoring

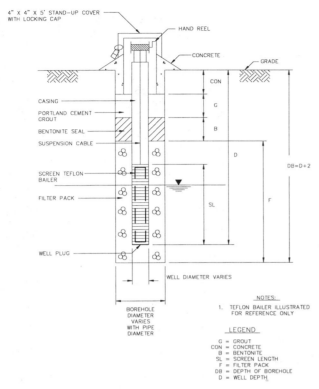

### General:
Environmental monitoring is the process of measuring the physical and/or chemical properties of an environmental medium -- groundwater, surface water, sediment, surface soil, subsurface soil, air.

### Typical Treatment Train:
Natural attenuation (monitoring-only plan), on-going remediation (progress measurement), post closure (landfills, storage facilities, etc.), compliance monitoring (release detection).

### Common Cost Components:
1. Sampling crew mobilization/per diem
2. Sample preparation, cleanup, containers, disposables, packaging, shipment
3. QA/QC samples (replicates/duplicates, trip blanks, equipment blanks, field blanks, spikes, blanks)
4. Analyses (consider turnaround time and QC level)

### Other Cost Considerations:
Groundwater monitoring wells, soil gas monitoring wells, professional labor, fencing.

**Environmental Remediation: Assemblies Cost Book**

# Cost Data for Remediation

## Monitoring

| Assembly | Description | Unit | Unit Cost by Safety Level A | B | C | D | E |
|---|---|---|---|---|---|---|---|
| | **Capital Costs** | | | | | | |
| 33 01 0101 | Mobilize/DeMobilize Drilling Rig & Crew | LS | 4,949 | 4,031 | 3,308 | 2,401 | 2,280 |
| 33 01 0102 | Van or Pickup Rental | DAY | 32.69 | 32.69 | 32.69 | 32.69 | 32.69 |
| 33 01 0103 | Flatbed Truck Rental | DAY | 45.94 | 45.94 | 45.94 | 45.94 | 45.94 |
| 33 01 0104 | Car or Van Mileage Charge | MILE | 0.35 | 0.35 | 0.35 | 0.35 | 0.35 |
| 33 01 0105 | Pickup Truck Mileage Charge | MILE | 0.56 | 0.56 | 0.56 | 0.56 | 0.56 |
| 33 01 0106 | Flatbed Truck Mileage Charge | MILE | 1.01 | 1.01 | 1.01 | 1.01 | 1.01 |
| 33 01 0201 | Mobilize Crew, >= 500 Miles, per Person | EACH | 357.75 | 357.75 | 357.75 | 357.75 | 357.75 |
| 33 01 0202 | Per Diem | DAY | 104.94 | 104.94 | 104.94 | 104.94 | 104.94 |
| 33 01 0203 | Mobilize Crew, 250 Miles, per Person | EACH | 178.88 | 178.88 | 178.88 | 178.88 | 178.88 |
| 33 01 0204 | Mobilize Crew, 100 Miles, per Person | EACH | 71.55 | 71.55 | 71.55 | 71.55 | 71.55 |
| 33 01 0205 | Mobilize Crew, 50 Miles, per Person | EACH | 53.66 | 53.66 | 53.66 | 53.66 | 53.66 |
| 33 01 0206 | Mobilize Crew, Local, per Person | EACH | 17.89 | 17.89 | 17.89 | 17.89 | 17.89 |
| 33 02 0206 | Prefiltering Liquids | EACH | 20.00 | 20.00 | 20.00 | 20.00 | 20.00 |
| 33 02 0207 | Acid Digestion | EACH | 27.50 | 27.50 | 27.50 | 27.50 | 27.50 |
| 33 02 0208 | Drying and Grinding | EACH | 22.50 | 22.50 | 22.50 | 22.50 | 22.50 |
| 33 02 0209 | Dissolution by Microwave | EACH | 40.00 | 40.00 | 40.00 | 40.00 | 40.00 |
| 33 02 0210 | Urine Preparation | EACH | 25.00 | 25.00 | 25.00 | 25.00 | 25.00 |
| 33 02 0211 | Feces Preparation | EACH | 33.33 | 33.33 | 33.33 | 33.33 | 33.33 |
| 33 02 0212 | Blood Preparation | EACH | 33.33 | 33.33 | 33.33 | 33.33 | 33.33 |
| 33 02 0213 | Animal Bone/Tissue Preparation | EACH | 66.67 | 66.67 | 66.67 | 66.67 | 66.67 |
| 33 02 0214 | Gas Flow Proportional Counting | EACH | 27.50 | 27.50 | 27.50 | 27.50 | 27.50 |
| 33 02 0215 | Liquid Scintillation | EACH | 27.50 | 27.50 | 27.50 | 27.50 | 27.50 |
| 33 02 0216 | Alpha Spectroscopy | EACH | 27.50 | 27.50 | 27.50 | 27.50 | 27.50 |
| 33 02 0217 | Gamma Spectroscopy | EACH | 20.00 | 20.00 | 20.00 | 20.00 | 20.00 |
| 33 02 0218 | Fluorometric or Laser | EACH | 27.50 | 27.50 | 27.50 | 27.50 | 27.50 |
| 33 02 0219 | Alpha Scintillation Counting | EACH | 27.50 | 27.50 | 27.50 | 27.50 | 27.50 |
| 33 02 0220 | Radon De-emanation | EACH | 27.50 | 27.50 | 27.50 | 27.50 | 27.50 |
| 33 02 0221 | Beta/Gamma Coincidence | EACH | 27.50 | 27.50 | 27.50 | 27.50 | 27.50 |
| 33 02 0222 | Radioactivity Screening - Tritium Not Present | EACH | 32.50 | 32.50 | 32.50 | 32.50 | 32.50 |
| 33 02 0223 | Radioactivity Screening - Tritium Present | EACH | 10.00 | 10.00 | 10.00 | 10.00 | 10.00 |
| 33 02 0224 | Repackage Sample | EACH | 10.00 | 10.00 | 10.00 | 10.00 | 10.00 |
| 33 02 0225 | Repackage and Ship Sample | EACH | 23.33 | 23.33 | 23.33 | 23.33 | 23.33 |
| 33 02 0301 | Air Monitoring Station | EACH | 718.68 | 718.68 | 718.68 | 718.68 | 718.68 |

**Environmental Remediation: Assemblies Cost Book**

# Cost Data for Remediation

## Monitoring

| Assembly | Description | Unit | \#Unit Cost by Safety Level A | B | C | D | E |
|---|---|---|---|---|---|---|---|
| | **Capital Costs** | | | | | | |
| 33 02 0302 | Organic Vapor Analyzer Rental, per Month | MONTH | 1,920 | 1,920 | 1,920 | 1,920 | 1,920 |
| 33 02 0303 | Organic Vapor Analyzer Rental, per Day | DAY | 184.30 | 184.30 | 184.30 | 184.30 | 184.30 |
| 33 02 0307 | Soil Gas Investigation & Analysis | DAY | 106.00 | 106.00 | 106.00 | 106.00 | 106.00 |
| 33 02 0309 | Bacterial Air Sampler, Monthly Rental | MONTH | 436.37 | 436.37 | 436.37 | 436.37 | 436.37 |
| 33 02 0310 | Microbial Air Sampler, Monthly Rental | MONTH | 436.37 | 436.37 | 436.37 | 436.37 | 436.37 |
| 33 02 0311 | Cotton Dust Sampler, Monthly Rental | MONTH | 480.53 | 480.53 | 480.53 | 480.53 | 480.53 |
| 33 02 0312 | Digital Dust Sampler, Monthly Rental | MONTH | 552.97 | 552.97 | 552.97 | 552.97 | 552.97 |
| 33 02 0313 | Flow Calibrator, Monthly Rental | MONTH | 212.00 | 212.00 | 212.00 | 212.00 | 212.00 |
| 33 02 0314 | Personal Low Flow Sampling Pump, Monthly Rental | MONTH | 178.43 | 178.43 | 178.43 | 178.43 | 178.43 |
| 33 02 0315 | Ambient Air Monitor, Monthly Rental | MONTH | 826.80 | 826.80 | 826.80 | 826.80 | 826.80 |
| 33 02 0316 | Ambient CO Analyzer, Monthly Rental | MONTH | 378.95 | 378.95 | 378.95 | 378.95 | 378.95 |
| 33 02 0317 | Ambient Ozone Monitor, Monthly Rental | MONTH | 1,758 | 1,758 | 1,758 | 1,758 | 1,758 |
| 33 02 0318 | Ambient Sulfur Dioxide Analyzer, Monthly Rental | MONTH | 2,690 | 2,690 | 2,690 | 2,690 | 2,690 |
| 33 02 0319 | Calibration Gas Monitor | EACH | 487.60 | 487.60 | 487.60 | 487.60 | 487.60 |
| 33 02 0320 | Chloride Analyzer, Monthly Rental | MONTH | 749.07 | 749.07 | 749.07 | 749.07 | 749.07 |
| 33 02 0321 | Gas Monitors, Monthly Rental | MONTH | 657.20 | 657.20 | 657.20 | 657.20 | 657.20 |
| 33 02 0322 | Hydrogen Sulfide Analyzer, Monthly Rental | MONTH | 249.10 | 249.10 | 249.10 | 249.10 | 249.10 |
| 33 02 0323 | Manual Air Sampling Kit, Monthly Rental | MONTH | 85.68 | 85.68 | 85.68 | 85.68 | 85.68 |
| 33 02 0324 | Manual Air Sampling Kit, Detection Tubes | EACH | 88.69 | 88.69 | 88.69 | 88.69 | 88.69 |
| 33 02 0325 | Mercury Vapor Monitor, Monthly Rental | MONTH | 1,277 | 1,277 | 1,277 | 1,277 | 1,277 |
| 33 02 0326 | Nitrogen Dioxide Analyzer, Monthly Rental | MONTH | 2,385 | 2,385 | 2,385 | 2,385 | 2,385 |
| 33 02 0327 | Personal Alarms, Monthly Rental | MONTH | 318.00 | 318.00 | 318.00 | 318.00 | 318.00 |
| 33 02 0328 | Portable Ambient Air Analyzer, Monthly Rental | MONTH | 1,745 | 1,745 | 1,745 | 1,745 | 1,745 |
| 33 02 0329 | Portable CO2 Monitor, Monthly Rental | MONTH | 485.83 | 485.83 | 485.83 | 485.83 | 485.83 |
| 33 02 0330 | Portable Combustible Gas/Oxygen Indicator, Monthly Rental | MONTH | 355.10 | 355.10 | 355.10 | 355.10 | 355.10 |
| 33 02 0331 | Portable Organic Vapor Analyzer, Monthly Rental | MONTH | 1,920 | 1,920 | 1,920 | 1,920 | 1,920 |
| 33 02 0332 | Portable Oxygen Deficiency Monitor, Monthly Rental | MONTH | 409.87 | 409.87 | 409.87 | 409.87 | 409.87 |
| 33 02 0333 | Portable Toxic Gas Analyzer, Monthly Rental | MONTH | 265.00 | 265.00 | 265.00 | 265.00 | 265.00 |
| 33 02 0334 | Solvent Vapor Indicator, Monthly Rental | MONTH | 1,920 | 1,920 | 1,920 | 1,920 | 1,920 |
| 33 02 0335 | Specific Vapor Analyzer, Monthly Rental | MONTH | 1,290 | 1,290 | 1,290 | 1,290 | 1,290 |
| 33 02 0336 | Stack Sampling Probe, Monthly Rental | MONTH | 648.37 | 648.37 | 648.37 | 648.37 | 648.37 |
| 33 02 0337 | Trace-Gas Analyzer, Monthly Rental | MONTH | 1,852 | 1,852 | 1,852 | 1,852 | 1,852 |

**Environmental Remediation: Assemblies Cost Book**

# Cost Data for Remediation

## Monitoring

| Assembly | Description | Unit | Unit Cost by Safety Level A | B | C | D | E |
|---|---|---|---|---|---|---|---|
| | **Capital Costs** | | | | | | |
| 33 02 0338 | Instrument Shelter, 30.0" x 20.0" x 32.0", Pine with Latex Paint | EACH | 738.96 | 722.90 | 716.02 | 700.49 | 694.80 |
| 33 02 0339 | Instrument Shelter, 18.0" x 13.0" x 20.0", Wood Painted White | EACH | 289.52 | 273.46 | 266.58 | 251.05 | 245.36 |
| 33 02 0340 | Instrument Shelter, 81.0" x 5.5.0" x 16.5", Wood Painted White | EACH | 266.47 | 250.40 | 243.52 | 228.00 | 222.30 |
| 33 02 0341 | Alpha, Beta, Gamma Detector Equipment | EACH | 194.33 | 194.33 | 194.33 | 194.33 | 194.33 |
| 33 02 0401 | Disposable Materials per Sample | EACH | 6.87 | 6.87 | 6.87 | 6.87 | 6.87 |
| 33 02 0402 | Decontamination Materials per Sample | EACH | 6.26 | 6.26 | 6.26 | 6.26 | 6.26 |
| 33 02 0404 | 12' Aluminum Pole with Polyhead Sampler | EACH | 396.66 | 396.66 | 396.66 | 396.66 | 396.66 |
| 33 02 0405 | Monitoring Well Slug Testing Equipment Rental | WEEK | 552.08 | 552.08 | 552.08 | 552.08 | 552.08 |
| 33 02 0504 | Clarifier Indicator Analyses | EACH | 195.01 | 195.01 | 195.01 | 195.01 | 195.01 |
| 33 02 0505 | Groundwater Contamination Analysis, per Sample | EACH | 126.00 | 126.00 | 126.00 | 126.00 | 126.00 |
| 33 02 0506 | Tank Decontamination Analysis, per Sample | EACH | 115.97 | 115.97 | 115.97 | 115.97 | 115.97 |
| 33 02 0507 | Primary Drinking Water Parameters Analysis, per Sample | EACH | 341.64 | 341.64 | 341.64 | 341.64 | 341.64 |
| 33 02 0508 | Detection/Compliance Monitoring Analysis, per Well | EACH | 990.17 | 990.17 | 990.17 | 990.17 | 990.17 |
| 33 02 0509 | Compliance Monitoring 40 CFR 261, Appendix IX Constituents | EACH | 1,715 | 1,715 | 1,715 | 1,715 | 1,715 |
| 33 02 0510 | Analysis of Wastewater for POTW Permit | EACH | 295.29 | 295.29 | 295.29 | 295.29 | 295.29 |
| 33 02 0511 | Analysis of Lysimeters for Soil-Pore Liquid Monitoring | EACH | 357.33 | 357.33 | 357.33 | 357.33 | 357.33 |
| 33 02 0512 | Rinsate Analysis | EACH | 115.97 | 115.97 | 115.97 | 115.97 | 115.97 |
| 33 02 0521 | Hip Waders Rental | WEEK | 51.44 | 51.44 | 51.44 | 51.44 | 51.44 |
| 33 02 0522 | Boat with Motor, Rental | DAY | 58.83 | 58.83 | 58.83 | 58.83 | 58.83 |
| 33 02 0523 | Boat without Motor, Rental | DAY | 21.73 | 21.73 | 21.73 | 21.73 | 21.73 |
| 33 02 0524 | Glass Coliwasas, Disposable, 7/8" x 42", 200 ml, Case of 12 | EACH | 187.22 | 187.22 | 187.22 | 187.22 | 187.22 |
| 33 02 0525 | Glass Coliwasas, Reusable, 7/8" x 42", 200 ml, Case of 4 | EACH | 67.42 | 67.42 | 67.42 | 67.42 | 67.42 |
| 33 02 0526 | PVC Coliwasas, 1.9" Outside Diameter x 49", 760 ml | EACH | 155.01 | 155.01 | 155.01 | 155.01 | 155.01 |
| 33 02 0527 | Bomb Sampler, 2.5" Diameter x 10", 500 ml | EACH | 524.35 | 524.35 | 524.35 | 524.35 | 524.35 |
| 33 02 0528 | Drum Thief Tubes, 3/4" x 42", 150 ml, Disposable, Case of 25 | EACH | 63.94 | 63.94 | 63.94 | 63.94 | 63.94 |
| 33 02 0529 | Drum Thief Tubes, 1/2" x 42", 75 ml, Disposable, Case of 25 | EACH | 44.75 | 44.75 | 44.75 | 44.75 | 44.75 |
| 33 02 0530 | Bottom Sampler, 3 Lb Brass, 7.5" x 4.5", No Cable Required | EACH | 179.67 | 179.67 | 179.67 | 179.67 | 179.67 |
| 33 02 0531 | Bottom Sampler, 17 Lb Stainless Steel, 6" x 6" x 6", with 100' Cable | EACH | 466.91 | 466.91 | 466.91 | 466.91 | 466.91 |

## Environmental Remediation: Assemblies Cost Book

# Cost Data for Remediation

## Monitoring

| Assembly | Description | Unit | Unit Cost by Safety Level A | B | C | D | E |
|---|---|---|---|---|---|---|---|
| | **Capital Costs** | | | | | | |
| 33 02 0532 | Bottom Sampler, 60 Lb Stainless Steel, 9" x 9" | EACH | 915.24 | 915.24 | 915.24 | 915.24 | 915.24 |
| 33 02 0533 | Water Level Indicator, 100' Tape, Electric, Light & Horn | EACH | 461.65 | 461.65 | 461.65 | 461.65 | 461.65 |
| 33 02 0534 | Water Level Indicator, 300' Tape, Electric, Light & Horn | EACH | 688.99 | 688.99 | 688.99 | 688.99 | 688.99 |
| 33 02 0535 | Water Level Indicator, 500' Tape, Electric, Light & Horn | EACH | 996.15 | 996.15 | 996.15 | 996.15 | 996.15 |
| 33 02 0536 | Water Level Indicator, Manual Steel Chalk Tape, with Weight, 100' Cable | EACH | 138.33 | 138.33 | 138.33 | 138.33 | 138.33 |
| 33 02 0538 | DO Meter, Analog, Portable, 6' Cable, Polarographic Probe | EACH | 577.70 | 577.70 | 577.70 | 577.70 | 577.70 |
| 33 02 0539 | DO Meter, Digital, Portable, 10' Cable, Small Diameter Probe | EACH | 1,114 | 1,114 | 1,114 | 1,114 | 1,114 |
| 33 02 0540 | DO Meter, Portable, Probe, 10' Cable, Quick Readings | EACH | 517.60 | 517.60 | 517.60 | 517.60 | 517.60 |
| 33 02 0541 | Conductivity Meter with Dip-type Sensor, Digital, 9V Battery, 7 x 3.2 x 2" | EACH | 977.27 | 977.27 | 977.27 | 977.27 | 977.27 |
| 33 02 0542 | Conductivity Meter, 1" Diameter Probe, 150' Cable, Also Temperature/Level | EACH | 914.25 | 914.25 | 914.25 | 914.25 | 914.25 |
| 33 02 0543 | Conductivity Meter, 10' Cable, Probe, Also Temperature/ Salinity | EACH | 870.79 | 870.79 | 870.79 | 870.79 | 870.79 |
| 33 02 0544 | Thermometer, Stainless Steel, Bi-Metal, 40 - 160 F, 12" Stem | EACH | 62.47 | 62.47 | 62.47 | 62.47 | 62.47 |
| 33 02 0545 | Thermometer, Thermocouple | EACH | 144.48 | 144.48 | 144.48 | 144.48 | 144.48 |
| 33 02 0546 | Thermometer, Digital | EACH | 38.50 | 38.50 | 38.50 | 38.50 | 38.50 |
| 33 02 0547 | pH Meter, Test Paper Kits, 200 Strips, Charts | EACH | 9.89 | 9.89 | 9.89 | 9.89 | 9.89 |
| 33 02 0548 | pH Meter, Digital, LCD, 0 - 14 pH Range | EACH | 361.81 | 361.81 | 361.81 | 361.81 | 361.81 |
| 33 02 0549 | Lysimeter, 1.90" Outside Diameter, PVC Tube Type - PVC Body/Teflon Filter | EACH | 565.44 | 488.90 | 428.73 | 353.14 | 342.99 |
| 33 02 0550 | Lysimeter, 1.90" Outside Diameter, Teflon Tube Type - Teflon Body/Filter | EACH | 636.22 | 559.67 | 499.50 | 423.91 | 413.76 |
| 33 02 0551 | Lysimeter, 2.38" Outside Diameter, PVC Tube Type - PVC Body/Teflon Filter | EACH | 565.44 | 488.90 | 428.73 | 353.14 | 342.99 |
| 33 02 0552 | Lysimeter, 2.38" Outside Diameter, Teflon Tube Type - Teflon Body/Filter | EACH | 544.92 | 468.38 | 408.21 | 332.62 | 322.47 |
| 33 02 0553 | Lysimeter, 1.90" Outside Diameter, PVC Cup Type - PVC Body/Teflon Filter | EACH | 565.44 | 488.90 | 428.73 | 353.14 | 342.99 |
| 33 02 0554 | Lysimeter, 1.90" Outside Diameter, Teflon Cup Type - Teflon Body/Filter | EACH | 630.42 | 553.87 | 493.70 | 418.12 | 407.97 |
| 33 02 0555 | Lysimeter, 1.90" Outside Diameter, PVC Deep Sample Type - Teflon Body/Filter | EACH | 644.64 | 568.09 | 507.92 | 432.33 | 422.18 |
| 33 02 0556 | Lysimeter, 1.90" Outside Diameter, Teflon Deep Sample Type - Teflon Body/Filter | EACH | 741.01 | 664.46 | 604.29 | 528.70 | 518.56 |
| 33 02 0557 | Lysimeter Head Assembly, All PVC Fittings | EACH | 577.72 | 501.17 | 441.00 | 365.41 | 355.26 |
| 33 02 0558 | Lysimeter Head Assembly, Teflon & PVC Fittings | EACH | 684.13 | 607.58 | 547.41 | 471.83 | 461.68 |

**Environmental Remediation: Assemblies Cost Book**

*Page 3-213*

# Cost Data for Remediation

## Monitoring

| Assembly | Description | Unit | A | B | C | D | E |
|---|---|---|---|---|---|---|---|
| | **Capital Costs** | | | | | | |
| 33 02 0559 | Silica Flour, 50 Lb Pail for Media Continuum | EACH | 174.04 | 151.08 | 133.03 | 110.35 | 107.31 |
| 33 02 0560 | Vacuum Pressure Hand Pump | EACH | 46.40 | 46.40 | 46.40 | 46.40 | 46.40 |
| 33 02 0561 | Nylon Tubing, 1/4" Outside Diameter | LF | 0.47 | 0.47 | 0.47 | 0.47 | 0.47 |
| 33 02 0562 | Teflon Tubing, 1/4" Outside Diameter | LF | 1.21 | 1.21 | 1.21 | 1.21 | 1.21 |
| 33 02 0601 | Drilling 2.5" Diameter Soil Borings, No Sampling | LF | 21.74 | 18.75 | 16.71 | 13.78 | 13.18 |
| 33 02 0602 | Stainless Steel Bottom Sampler | EACH | 378.97 | 378.97 | 378.97 | 378.97 | 378.97 |
| 33 02 0603 | Surface Soil Sampling Equipment | EACH | 280.80 | 280.80 | 280.80 | 280.80 | 280.80 |
| 33 02 0605 | Hand Auger Rental | DAY | 13.25 | 13.25 | 13.25 | 13.25 | 13.25 |
| 33 02 0612 | Soil Analysis | EACH | 506.30 | 506.30 | 506.30 | 506.30 | 506.30 |
| 33 02 0613 | Soil Recovery Probe, Unslotted, 7/8" x 1' | EACH | 121.64 | 121.64 | 121.64 | 121.64 | 121.64 |
| 33 02 0614 | Soil Recovery Probe, Unslotted, 7/8" x 2' or 9/8" x 1' | EACH | 140.10 | 140.10 | 140.10 | 140.10 | 140.10 |
| 33 02 0615 | Soil Recovery Probe, Unslotted, 9/8" x 2' | EACH | 151.23 | 151.23 | 151.23 | 151.23 | 151.23 |
| 33 02 0616 | Soil Recovery Probe, Liners, 7/8" per LF, Butyrate Plast | LF | 0.86 | 0.86 | 0.86 | 0.86 | 0.86 |
| 33 02 0617 | Soil Recovery Probe, Liners, 9/8" per LF, Butyrate Plast | LF | 1.41 | 1.41 | 1.41 | 1.41 | 1.41 |
| 33 02 0618 | Soil Recovery Probe, Liners, Stainless Steel, 7/8" | EACH | 6.51 | 6.51 | 6.51 | 6.51 | 6.51 |
| 33 02 0619 | Soil Recovery Probe, Liners, Stainless Steel, 9/8" | EACH | 9.54 | 9.54 | 9.54 | 9.54 | 9.54 |
| 33 02 0620 | Soil Gas Vapor Probe, Stainless Steel, Manual, Nonremovable Tip | EACH | 45.93 | 45.93 | 45.93 | 45.93 | 45.93 |
| 33 02 0621 | Soil Gas Vapor Probe, Stainless Steel, Electric, Rotary, Up/Down Hammer | EACH | 226.73 | 226.73 | 226.73 | 226.73 | 226.73 |
| 33 02 0622 | Screw Auger, High-carbon Steel, 1.5" x 6" | EACH | 79.89 | 79.89 | 79.89 | 79.89 | 79.89 |
| 33 02 0623 | Screw Auger, High-carbon Steel, 1.75" x 6" | EACH | 97.33 | 97.33 | 97.33 | 97.33 | 97.33 |
| 33 02 0624 | Soil Auger Buckets, Dutch, 3" Diameter, Carbon/Stainless Steel | EACH | 129.80 | 129.80 | 129.80 | 129.80 | 129.80 |
| 33 02 0625 | Soil Auger Buckets, Mud, 2" Diameter, Carbon/Stainless Steel | EACH | 120.12 | 120.12 | 120.12 | 120.12 | 120.12 |
| 33 02 0626 | Soil Auger Buckets, Reg, 2" Diameter, Carbon/Stainless Steel | EACH | 100.46 | 100.46 | 100.46 | 100.46 | 100.46 |
| 33 02 0627 | Soil Auger Buckets, Sand, 2" Diameter, Carbon/Stainless Steel | EACH | 114.29 | 114.29 | 114.29 | 114.29 | 114.29 |
| 33 02 0628 | Soil Recovery Auger Buckets, Stainless Steel, 2.25" x 8" | EACH | 342.05 | 342.05 | 342.05 | 342.05 | 342.05 |
| 33 02 0629 | Soil Recovery Auger Buckets, Stainless Steel, 3.25" x 10" | EACH | 352.12 | 352.12 | 352.12 | 352.12 | 352.12 |
| 33 02 0630 | Soil Recovery Auger Buckets, Stainless Steel with Carbon Bits, 2.25" x 8" | EACH | 315.75 | 315.75 | 315.75 | 315.75 | 315.75 |
| 33 02 0631 | Soil Recovery Auger Buckets, Stainless Steel with Carbon Bits, 3.25" x 10" | EACH | 325.82 | 325.82 | 325.82 | 325.82 | 325.82 |

**Environmental Remediation: Assemblies Cost Book**

# Cost Data for Remediation

## Monitoring

| Assembly | Description | Unit | Unit Cost by Safety Level A | B | C | D | E |
|---|---|---|---|---|---|---|---|
| | **Capital Costs** | | | | | | |
| 33 02 0632 | Soil Recovery Auger Buckets, Liners, Butyrate Plast, 3" x 10" or 2" x 8" | EACH | 5.13 | 5.13 | 5.13 | 5.13 | 5.13 |
| 33 02 0633 | Soil Recovery Auger Buckets, Liners, Stainless Steel, 3" x 10" or 2" x 8" | EACH | 17.14 | 17.14 | 17.14 | 17.14 | 17.14 |
| 33 02 0634 | Sludge Sampler, Stainless Steel, 3.25" x 12", Thread On | EACH | 378.97 | 378.97 | 378.97 | 378.97 | 378.97 |
| 33 02 0635 | Tensiometer Soil Moist Probe, Standard, 7/8" Outside Diameter x 18", Fixed | EACH | 83.69 | 83.69 | 83.69 | 83.69 | 83.69 |
| 33 02 0636 | Tensiometer Soil Moist Probe, Jet Fill, 7/8" Outside Diameter x 18", Fixed | EACH | 71.81 | 71.81 | 71.81 | 71.81 | 71.81 |
| 33 02 0637 | Tensiometer Soil Moist Probe, Quick Draw, 18", Quick Read | EACH | 412.34 | 412.34 | 412.34 | 412.34 | 412.34 |
| 33 02 0638 | Soil Test Kit, 50 Tests | EACH | 278.25 | 278.25 | 278.25 | 278.25 | 278.25 |
| 33 02 0639 | CPT Rig, Includes Labor, Sampling, Punching, Decontamination | DAY | 2,844 | 2,844 | 2,844 | 2,844 | 2,844 |
| 33 02 0640 | Mobilize/Demobilize CPT Rig and Crew | EACH | 2,986 | 2,986 | 2,986 | 2,986 | 2,986 |
| 33 02 0641 | Standby Time for CPT Rig and Crew | HOUR | 3,015 | 3,015 | 3,015 | 3,015 | 3,015 |
| 33 02 0642 | Setup Cost per Each Hole, CPT Rig | EACH | 273.83 | 273.83 | 273.83 | 273.83 | 273.83 |
| 33 02 0643 | Decontamination Trailer Rental for CPT Rig | DAY | 92.75 | 92.75 | 92.75 | 92.75 | 92.75 |
| 33 02 0644 | Decontaminate Cone on CPT Rig | DAY | 273.83 | 273.83 | 273.83 | 273.83 | 273.83 |
| 33 02 0645 | Level "D" PPE Rental per 2-Man CPT Crew | DAY | 2,874 | 2,874 | 2,874 | 2,874 | 2,874 |
| 33 02 0646 | Level "C" PPE Rental per 2-Man CPT Crew | DAY | 3,122 | 3,122 | 3,122 | 3,122 | 3,122 |
| 33 02 0647 | In Situ Batch Sampling for Groundwater, CPT Rig (per Sample) | EACH | 92.75 | 92.75 | 92.75 | 92.75 | 92.75 |
| 33 02 0648 | In Situ Soil Sampling, CPT Rig (per Sample) | EACH | 60.95 | 60.95 | 60.95 | 60.95 | 60.95 |
| 33 02 0649 | Hydropunch with CPT Rig, Hard Soil | FT | 10.60 | 10.60 | 10.60 | 10.60 | 10.60 |
| 33 02 0650 | Hydropunch with CPT Rig, Medium Soil | FT | 10.60 | 10.60 | 10.60 | 10.60 | 10.60 |
| 33 02 0651 | Hydropunch with CPT Rig, Soft Soil | FT | 10.60 | 10.60 | 10.60 | 10.60 | 10.60 |
| 33 02 0652 | Grout Hole After Hydropunching | FT | 4.38 | 4.38 | 4.38 | 4.38 | 4.38 |
| 33 02 1101 | Geotechnical Characteristics Analysis | EACH | 194.42 | 155.02 | 138.16 | 100.07 | 86.11 |
| 33 02 1102 | Soil Moisture Content ASTM D2216 | EACH | 22.50 | 22.50 | 22.50 | 22.50 | 22.50 |
| 33 02 1501 | Saturation Indicator | EACH | 106.53 | 106.53 | 106.53 | 106.53 | 106.53 |
| 33 02 1502 | Thermostat & Humidity Control Devices | EACH | 329.49 | 279.97 | 258.78 | 210.91 | 193.37 |
| 33 02 1506 | Monitoring Port with Gas Monitor | EACH | 28.81 | 22.42 | 19.69 | 13.52 | 11.26 |
| 33 02 1507 | Continuous Monitoring and Recording of Air Flow | EACH | 9,870 | 9,870 | 9,870 | 9,870 | 9,870 |
| 33 02 1509 | Water Quality Parameter Testing Device | WEEK | 227.90 | 227.90 | 227.90 | 227.90 | 227.90 |
| 33 02 1601 | Specific Conductance (EPA 120.1) | EACH | 10.33 | 10.33 | 10.33 | 10.33 | 10.33 |

## Environmental Remediation: Assemblies Cost Book

# Cost Data for Remediation

## Monitoring

| Assembly | Description | Unit | Unit Cost by Safety Level | | | | |
|---|---|---|---|---|---|---|---|
| | | | A | B | C | D | E |
| | **Capital Costs** | | | | | | |
| 33 02 1602 | pH (EPA 150.1) | EACH | 10.00 | 10.00 | 10.00 | 10.00 | 10.00 |
| 33 02 1603 | Total Dissolved Solids (EPA 160.1) | EACH | 15.00 | 15.00 | 15.00 | 15.00 | 15.00 |
| 33 02 1604 | Total Suspended Solids (EPA 160.2) | EACH | 15.00 | 15.00 | 15.00 | 15.00 | 15.00 |
| 33 02 1605 | Total Solids (EPA 160.3) | EACH | 14.33 | 14.33 | 14.33 | 14.33 | 14.33 |
| 33 02 1606 | Total Volatile Solids (EPA 160.4) | EACH | 18.67 | 18.67 | 18.67 | 18.67 | 18.67 |
| 33 02 1607 | Temperature (EPA 170.1) | EACH | 10.00 | 10.00 | 10.00 | 10.00 | 10.00 |
| 33 02 1608 | Nitrogen/Nitrite/Nitrate (EPA 300.1/354.1) | EACH | 19.67 | 19.67 | 19.67 | 19.67 | 19.67 |
| 33 02 1609 | Acidity/Alkalinity (EPA 305.1) | EACH | 15.00 | 15.00 | 15.00 | 15.00 | 15.00 |
| 33 02 1610 | Cyanide (EPA 335.2) | EACH | 45.00 | 45.00 | 45.00 | 45.00 | 45.00 |
| 33 02 1611 | Ammonia Nitrogen (EPA 350.2) | EACH | 30.00 | 30.00 | 30.00 | 30.00 | 30.00 |
| 33 02 1612 | Phosphorus (EPA 365.3) | EACH | 28.00 | 28.00 | 28.00 | 28.00 | 28.00 |
| 33 02 1613 | Oil and Grease (EPA 413.2) | EACH | 55.00 | 55.00 | 55.00 | 55.00 | 55.00 |
| 33 02 1614 | Total Petroleum Hydrocarbons (EPA 418.1) | EACH | 66.67 | 66.67 | 66.67 | 66.67 | 66.67 |
| 33 02 1615 | Phenols (Method 420.1) | EACH | 43.67 | 43.67 | 43.67 | 43.67 | 43.67 |
| 33 02 1617 | Pesticides/PCBs (EPA 608) | EACH | 206.67 | 206.67 | 206.67 | 206.67 | 206.67 |
| 33 02 1618 | Volatile Organic Analysis (EPA 624) | EACH | 175.00 | 175.00 | 175.00 | 175.00 | 175.00 |
| 33 02 1619 | Base Neutral & Acid Extractable Organics (EPA 625) | EACH | 456.67 | 456.67 | 456.67 | 456.67 | 456.67 |
| 33 02 1620 | Target Analyte List Metals (EPA 6010/7000s), Water | EACH | 289.67 | 289.67 | 289.67 | 289.67 | 289.67 |
| 33 02 1621 | Purgeable Halocarbons (EPA 601) | EACH | 113.33 | 113.33 | 113.33 | 113.33 | 113.33 |
| 33 02 1622 | Purgeable Aromatics (EPA 602) | EACH | 96.67 | 96.67 | 96.67 | 96.67 | 96.67 |
| 33 02 1623 | Acrolein & Acrylonitrile (EPA 603) | EACH | 113.33 | 113.33 | 113.33 | 113.33 | 113.33 |
| 33 02 1624 | Phenols (EPA 604) | EACH | 166.67 | 166.67 | 166.67 | 166.67 | 166.67 |
| 33 02 1625 | Benzidines (EPA 605) | EACH | 200.00 | 200.00 | 200.00 | 200.00 | 200.00 |
| 33 02 1626 | Phthalate Esters (EPA 606) | EACH | 180.00 | 180.00 | 180.00 | 180.00 | 180.00 |
| 33 02 1627 | Nitrosamines (EPA 607) | EACH | 166.67 | 166.67 | 166.67 | 166.67 | 166.67 |
| 33 02 1628 | Nitroaromatics & Isophorone (EPA 609) | EACH | 180.00 | 180.00 | 180.00 | 180.00 | 180.00 |
| 33 02 1629 | Polynuclear Aromatic Hydrocarbons, PAH (EPA 610) | EACH | 188.33 | 188.33 | 188.33 | 188.33 | 188.33 |
| 33 02 1630 | Haloethers (EPA 611) | EACH | 150.00 | 150.00 | 150.00 | 150.00 | 150.00 |
| 33 02 1631 | Chlorinated Hydrocarbons (EPA 612) | EACH | 150.00 | 150.00 | 150.00 | 150.00 | 150.00 |
| 33 02 1632 | 2,3,7,8-Tetrachloro-Dibzo-P-Dioxin (EPA 613) | EACH | 182.50 | 182.50 | 182.50 | 182.50 | 182.50 |
| 33 02 1633 | Organophosphorus Pesticides (EPA 614) | EACH | 215.00 | 215.00 | 215.00 | 215.00 | 215.00 |
| 33 02 1634 | Chlorinated Phenoxy Acid Herbicide (EPA 615) | EACH | 225.00 | 225.00 | 225.00 | 225.00 | 225.00 |
| 33 02 1635 | Organochlorine Pesticides & PCBs (EPA 617) | EACH | 156.00 | 156.00 | 156.00 | 156.00 | 156.00 |

## Environmental Remediation: Assemblies Cost Book

*Page 3-216*

# Cost Data for Remediation

## Monitoring

| Assembly | Description | Unit | Unit Cost by Safety Level A | B | C | D | E |
|---|---|---|---|---|---|---|---|
| | **Capital Costs** | | | | | | |
| 33 02 1636 | Triazine Pesticides (EPA 619) | EACH | 183.33 | 183.33 | 183.33 | 183.33 | 183.33 |
| 33 02 1637 | Carbamate & Urea Pesticides (EPA 632) | EACH | 175.00 | 175.00 | 175.00 | 175.00 | 175.00 |
| 33 02 1638 | Color (EPA 110.2/110.3) | EACH | 17.00 | 17.00 | 17.00 | 17.00 | 17.00 |
| 33 02 1639 | Hardness, Total (EPA 130.2) | EACH | 15.50 | 15.50 | 15.50 | 15.50 | 15.50 |
| 33 02 1640 | Odor (EPA 140.1) | EACH | 11.33 | 11.33 | 11.33 | 11.33 | 11.33 |
| 33 02 1641 | Settleable Solids (EPA 160.5) | EACH | 13.67 | 13.67 | 13.67 | 13.67 | 13.67 |
| 33 02 1642 | Total Solids (EPA 166.3) | EACH | 14.33 | 14.33 | 14.33 | 14.33 | 14.33 |
| 33 02 1643 | Turbidity (EPA 180.1) | EACH | 12.67 | 12.67 | 12.67 | 12.67 | 12.67 |
| 33 02 1644 | Metals Screen, Flame, per Each (EPA 200S) | EACH | 11.33 | 11.33 | 11.33 | 11.33 | 11.33 |
| 33 02 1645 | Metals Screen, Furnace, per Each (EPA 200S) | EACH | 32.50 | 32.50 | 32.50 | 32.50 | 32.50 |
| 33 02 1646 | Arsenic, Hydride (EPA 206.3) | EACH | 30.00 | 30.00 | 30.00 | 30.00 | 30.00 |
| 33 02 1647 | Boron, Colorimetric (EPA 212.3) | EACH | 12.00 | 12.00 | 12.00 | 12.00 | 12.00 |
| 33 02 1648 | Chromium, Hexavalent (EPA 218.5) | EACH | 21.67 | 21.67 | 21.67 | 21.67 | 21.67 |
| 33 02 1649 | Mercury, Cold Vapor (EPA 245.1) | EACH | 32.50 | 32.50 | 32.50 | 32.50 | 32.50 |
| 33 02 1650 | Ion Chromatography (EPA 300) | EACH | 18.33 | 18.33 | 18.33 | 18.33 | 18.33 |
| 33 02 1651 | Alkalinity (EPA 310.1/310.2) | EACH | 15.00 | 15.00 | 15.00 | 15.00 | 15.00 |
| 33 02 1652 | Bromide, Titrimetric (EPA 320.1) | EACH | 32.50 | 32.50 | 32.50 | 32.50 | 32.50 |
| 33 02 1653 | Chloride (EPA 325.2/325.3) | EACH | 17.00 | 17.00 | 17.00 | 17.00 | 17.00 |
| 33 02 1654 | Chloride, Total Residual (EPA 330.2/330.5) | EACH | 16.67 | 16.67 | 16.67 | 16.67 | 16.67 |
| 33 02 1655 | Fluoride SPADNS (EPA 340.1) | EACH | 125.00 | 125.00 | 125.00 | 125.00 | 125.00 |
| 33 02 1656 | Fluoride Electrode (EPA 340.2) | EACH | 18.67 | 18.67 | 18.67 | 18.67 | 18.67 |
| 33 02 1657 | Iodide (EPA 345.1) | EACH | 28.33 | 28.33 | 28.33 | 28.33 | 28.33 |
| 33 02 1658 | Nitrogen, Nitrate-Nitrite (EPA 350.3) | EACH | 19.67 | 19.67 | 19.67 | 19.67 | 19.67 |
| 33 02 1659 | Nitrogen, Kjeldahl, Total (EPA 351.2/351.3/351.4) | EACH | 38.33 | 38.33 | 38.33 | 38.33 | 38.33 |
| 33 02 1660 | Nitrogen, Organic (EPA 351.3) | EACH | 58.00 | 58.00 | 58.00 | 58.00 | 58.00 |
| 33 02 1661 | Nitrogen, Nitrate (EPA 352.1) | EACH | 19.67 | 19.67 | 19.67 | 19.67 | 19.67 |
| 33 02 1662 | Nitrogen, Nitrate & Nitrite (EPA 353.2) | EACH | 58.00 | 58.00 | 58.00 | 58.00 | 58.00 |
| 33 02 1663 | Dissolved Oxygen (EPA 360.1) | EACH | 11.50 | 11.50 | 11.50 | 11.50 | 11.50 |
| 33 02 1664 | Phosphorus, Ortho (EPA 365.1) | EACH | 21.33 | 21.33 | 21.33 | 21.33 | 21.33 |
| 33 02 1665 | Phosphorus, Total (EPA 365.2/365.4) | EACH | 27.00 | 27.00 | 27.00 | 27.00 | 27.00 |
| 33 02 1666 | Silica, Dissolved (EPA 370.1) | EACH | 30.00 | 30.00 | 30.00 | 30.00 | 30.00 |
| 33 02 1667 | Sulfate (EPA 375.2/375.3/375.4) | EACH | 16.67 | 16.67 | 16.67 | 16.67 | 16.67 |
| 33 02 1668 | Sulfide (EPA 376.1) | EACH | 16.32 | 16.32 | 16.32 | 16.32 | 16.32 |

**Environmental Remediation: Assemblies Cost Book**

*Page 3-217*

1998 by ECHOS. All rights reserved

# Cost Data for Remediation

## Monitoring

| Assembly | Description | Unit | Unit Cost by Safety Level A | B | C | D | E |
|---|---|---|---|---|---|---|---|
| | **Capital Costs** | | | | | | |
| 33 02 1669 | Sulfite (EPA 377.1) | EACH | 16.32 | 16.32 | 16.32 | 16.32 | 16.32 |
| 33 02 1670 | Metals Screen, 25 Metals Listed in Method EPA 200.7 | EACH | 283.25 | 283.25 | 283.25 | 283.25 | 283.25 |
| 33 02 1671 | Biochemical Oxygen Demand, BOD (EPA 405.1) | EACH | 34.67 | 34.67 | 34.67 | 34.67 | 34.67 |
| 33 02 1672 | Chemical Oxygen Demand, COD (EPA 410.1/410.2/410.3/410.4) | EACH | 28.33 | 28.33 | 28.33 | 28.33 | 28.33 |
| 33 02 1673 | Total Organic Carbon, TOC (EPA 415.1/415.2) | EACH | 30.67 | 30.67 | 30.67 | 30.67 | 30.67 |
| 33 02 1674 | MBAS, Surfactants (EPA 425.1) | EACH | 38.00 | 38.00 | 38.00 | 38.00 | 38.00 |
| 33 02 1675 | NTA (EPA 430.2) | EACH | 25.00 | 25.00 | 25.00 | 25.00 | 25.00 |
| 33 02 1676 | Total Organic Halogens, TOX (EPA 450.1) | EACH | 75.00 | 75.00 | 75.00 | 75.00 | 75.00 |
| 33 02 1677 | Trihalomethanes (EPA 501.1) | EACH | 105.00 | 105.00 | 105.00 | 105.00 | 105.00 |
| 33 02 1688 | 1,2-Dibromo-3-Chloropropane (EPA 504) | EACH | 105.00 | 105.00 | 105.00 | 105.00 | 105.00 |
| 33 02 1689 | Organohalide Pesticides & Aroclors (EPA 505) | EACH | 156.00 | 156.00 | 156.00 | 156.00 | 156.00 |
| 33 02 1693 | Chlorinated Herbicides (EPA 515) | EACH | 225.00 | 225.00 | 225.00 | 225.00 | 225.00 |
| 33 02 1701 | EP Toxicity, Metals (EPA 1310) | EACH | 146.00 | 146.00 | 146.00 | 146.00 | 146.00 |
| 33 02 1702 | TCLP (RCRA) (EPA 1311) | EACH | 1,364 | 1,364 | 1,364 | 1,364 | 1,364 |
| 33 02 1703 | Chromium (SW 7191), with Prep | EACH | 28.33 | 28.33 | 28.33 | 28.33 | 28.33 |
| 33 02 1705 | Targeted TCLP (Metals, Volatiles, Semivolatiles only) | EACH | 669.33 | 669.33 | 669.33 | 669.33 | 669.33 |
| 33 02 1706 | Chromium (EPA 6010) | EACH | 12.00 | 12.00 | 12.00 | 12.00 | 12.00 |
| 33 02 1708 | Purgeable Aromatics (SW 5030/SW 8020) | EACH | 96.67 | 96.67 | 96.67 | 96.67 | 96.67 |
| 33 02 1709 | Target Analyte List Metals (EPA 6010/7000S), Soil | EACH | 289.67 | 289.67 | 289.67 | 289.67 | 289.67 |
| 33 02 1710 | Metals (EPA 6010), per Each Metal | EACH | 11.33 | 11.33 | 11.33 | 11.33 | 11.33 |
| 33 02 1711 | Arsenic (SW 7060), with Prep | EACH | 20.67 | 20.67 | 20.67 | 20.67 | 20.67 |
| 33 02 1712 | Mercury (SW 7470), with Prep | EACH | 29.67 | 29.67 | 29.67 | 29.67 | 29.67 |
| 33 02 1713 | Selenium (SW 7740), with Prep | EACH | 23.33 | 23.33 | 23.33 | 23.33 | 23.33 |
| 33 02 1714 | BTEX Purgeable Aromatics (SW 5030/SW 8020) | EACH | 81.67 | 81.67 | 81.67 | 81.67 | 81.67 |
| 33 02 1715 | BTEX/Gasoline Hydrocarbons (Mod 8020)(PID/FID), with Prep | EACH | 83.33 | 83.33 | 83.33 | 83.33 | 83.33 |
| 33 02 1716 | Phenols (SW 3550/SW 8040) | EACH | 75.33 | 75.33 | 75.33 | 75.33 | 75.33 |
| 33 02 1717 | Pesticides/PCBs (SW 3550/SW 8080) | EACH | 206.67 | 206.67 | 206.67 | 206.67 | 206.67 |
| 33 02 1718 | Chlordane (SW 3550/SW 8080) | EACH | 213.33 | 213.33 | 213.33 | 213.33 | 213.33 |
| 33 02 1719 | Chlorinated Phenoxy Acid Herbicides (SW 3550/SW 8150) | EACH | 233.33 | 233.33 | 233.33 | 233.33 | 233.33 |
| 33 02 1720 | Volatile Organic Analysis (SW 5030/SW 8240) | EACH | 175.00 | 175.00 | 175.00 | 175.00 | 175.00 |
| 33 02 1721 | Base/Neutral & Acid Extractable Organics (SW 3550/SW 8270) | EACH | 456.67 | 456.67 | 456.67 | 456.67 | 456.67 |

## Environmental Remediation: Assemblies Cost Book

*Page 3-218*

1998 by ECHOS. All rights reserved

# Cost Data for Remediation

## Monitoring

| Assembly | Description | Unit | A | B | C | D | E |
|---|---|---|---|---|---|---|---|
| | **Capital Costs** | | | | | | |
| 33 02 1722 | Polynuclear Aromatic Hydrocarbons (PAH) (SW 8310), with Prep | EACH | 188.33 | 188.33 | 188.33 | 188.33 | 188.33 |
| 33 02 1723 | Cyanide (EPA 9010) | EACH | 48.33 | 48.33 | 48.33 | 48.33 | 48.33 |
| 33 02 1724 | Phenols (EPA 9065) | EACH | 43.67 | 43.67 | 43.67 | 43.67 | 43.67 |
| 33 02 1725 | CEC (EPA 9081) | EACH | 142.50 | 142.50 | 142.50 | 142.50 | 142.50 |
| 33 02 1726 | Diesel Hydrocarbon, Mod 8015, GC/FID, with Prep | EACH | 107.50 | 107.50 | 107.50 | 107.50 | 107.50 |
| 33 02 1728 | Gasoline Hydrocarbon, Mod 8015, GC/FID, with Prep | EACH | 86.67 | 86.67 | 86.67 | 86.67 | 86.67 |
| 33 02 1729 | Hydrocarbon Screen GC/FID, Mod 8015, with Prep | EACH | 131.67 | 131.67 | 131.67 | 131.67 | 131.67 |
| 33 02 1730 | Sludge Constituents and Characteristics Analysis Tests | EACH | 143.00 | 143.00 | 143.00 | 143.00 | 143.00 |
| 33 02 1731 | Purgeable Halocarbons (SW 5030/SW 8010) | EACH | 113.33 | 113.33 | 113.33 | 113.33 | 113.33 |
| 33 02 1732 | Total Petroleum Hydrocarbons (SW 5030/SW 8015) | EACH | 66.67 | 66.67 | 66.67 | 66.67 | 66.67 |
| 33 02 1733 | Acrolein, Acrylonitrile, Acetonitrile (SW 5030/SW 8030) | EACH | 113.33 | 113.33 | 113.33 | 113.33 | 113.33 |
| 33 02 1734 | Phthalate Esters (SW 3550/SW 8060/SW 8061) | EACH | 180.00 | 180.00 | 180.00 | 180.00 | 180.00 |
| 33 02 1735 | Nitroaromatics & Cyclic Ketones (SW 3550/SW 8090/SW8091) | EACH | 180.00 | 180.00 | 180.00 | 180.00 | 180.00 |
| 33 02 1736 | Polynuclear Aromatic Hydrocarbons (SW 3550/SW 8100) | EACH | 188.33 | 188.33 | 188.33 | 188.33 | 188.33 |
| 33 02 1737 | Chlorinated Hydrocarbons (SW 3550/SW 8120/SW 8121) | EACH | 150.00 | 150.00 | 150.00 | 150.00 | 150.00 |
| 33 02 1738 | Organophosphorus Pesticides (SW 3550/SW 8140/SW 8141) | EACH | 215.00 | 215.00 | 215.00 | 215.00 | 215.00 |
| 33 02 1739 | Semivolatile Organics, Packed Column (SW 8250), with Prep | EACH | 270.00 | 270.00 | 270.00 | 270.00 | 270.00 |
| 33 02 1740 | Dioxins & Dibenzofurans (SW 3550/SW 8280) | EACH | 190.00 | 190.00 | 190.00 | 190.00 | 190.00 |
| 33 02 1741 | Total Organic Halides, TOX (EPA 9020/9021) | EACH | 75.00 | 75.00 | 75.00 | 75.00 | 75.00 |
| 33 02 1742 | Total Dissolved Sulfide (EPA 9030) | EACH | 16.67 | 16.67 | 16.67 | 16.67 | 16.67 |
| 33 02 1743 | Sulfate (EPA 9035/9036/9038) | EACH | 16.32 | 16.32 | 16.32 | 16.32 | 16.32 |
| 33 02 1744 | pH (EPA 9040/9041/9045) | EACH | 10.00 | 10.00 | 10.00 | 10.00 | 10.00 |
| 33 02 1745 | Specific Conductance, Conductivity (EPA 9050) | EACH | 10.00 | 10.00 | 10.00 | 10.00 | 10.00 |
| 33 02 1746 | Total Organic Carbon, TOC (EPA 9060) | EACH | 35.67 | 35.67 | 35.67 | 35.67 | 35.67 |
| 33 02 1747 | Total Recoverable Oil & Grease (EPA 9070) | EACH | 65.00 | 65.00 | 65.00 | 65.00 | 65.00 |
| 33 02 1748 | Oil & Grease Extraction Method (EPA 9071) | EACH | 48.33 | 48.33 | 48.33 | 48.33 | 48.33 |
| 33 02 1749 | Waste/Membrane Liner Compatibility Test (EPA 9090) | EACH | 107.50 | 107.50 | 107.50 | 107.50 | 107.50 |
| 33 02 1750 | Paint Filter Test (EPA 9095) | EACH | 22.50 | 22.50 | 22.50 | 22.50 | 22.50 |
| 33 02 1751 | Saturated Hydraulic Conductivity/Leachate/Intrinsic Permeability (EPA 9100) | EACH | 261.50 | 261.50 | 261.50 | 261.50 | 261.50 |
| 33 02 1752 | Total Coliform (EPA 9131/9132) | EACH | 26.00 | 26.00 | 26.00 | 26.00 | 26.00 |

**Environmental Remediation: Assemblies Cost Book**

# Cost Data for Remediation

## Monitoring

| Assembly | Description | Unit | Unit Cost by Safety Level A | B | C | D | E |
|---|---|---|---|---|---|---|---|
| | **Capital Costs** | | | | | | |
| 33 02 1753 | Nitrate (EPA 9200) | EACH | 17.33 | 17.33 | 17.33 | 17.33 | 17.33 |
| 33 02 1754 | Chloride (EPA 9250/9251/9252) | EACH | 30.50 | 30.50 | 30.50 | 30.50 | 30.50 |
| 33 02 1756 | Ignitability (EPA 1010/1020) | EACH | 31.67 | 31.67 | 31.67 | 31.67 | 31.67 |
| 33 02 1757 | Corrosivity (EPA 1110) | EACH | 91.67 | 91.67 | 91.67 | 91.67 | 91.67 |
| 33 02 1758 | Reactivity, SW 846 Ch7 P7.4 | EACH | 78.33 | 78.33 | 78.33 | 78.33 | 78.33 |
| 33 02 1759 | Sorbent Cartridges from VOC Sampling (EPA 5040) | EACH | 140.83 | 140.83 | 140.83 | 140.83 | 140.83 |
| 33 02 1760 | Metals, Flame, 7000s, per Each | EACH | 11.33 | 11.33 | 11.33 | 11.33 | 11.33 |
| 33 02 1761 | Metals, Furnace, 7000s, per Each | EACH | 32.50 | 32.50 | 32.50 | 32.50 | 32.50 |
| 33 02 1762 | Chromium, Hexavalent (SW 7195), with Prep | EACH | 17.50 | 17.50 | 17.50 | 17.50 | 17.50 |
| 33 02 1764 | Selenium, Hydride (SW 7741), with Prep | EACH | 38.33 | 38.33 | 38.33 | 38.33 | 38.33 |
| 33 02 1765 | Arsenic, Hydride (SW 7061), with Prep | EACH | 38.33 | 38.33 | 38.33 | 38.33 | 38.33 |
| 33 02 1766 | Total Petroleum Hydrocarbons (EPA 9073) | EACH | 66.67 | 66.67 | 66.67 | 66.67 | 66.67 |
| 33 02 1767 | Chromium, Total (SW 7190) with Prep | EACH | 16.00 | 16.00 | 16.00 | 16.00 | 16.00 |
| 33 02 1768 | Antimony (SW 7040) with Prep | EACH | 34.50 | 34.50 | 34.50 | 34.50 | 34.50 |
| 33 02 1769 | Antimony (SW 7041) with Prep | EACH | 25.00 | 25.00 | 25.00 | 25.00 | 25.00 |
| 33 02 1770 | Thallium (SW 7840) with Prep | EACH | 34.50 | 34.50 | 34.50 | 34.50 | 34.50 |
| 33 02 1771 | Thallium (SW 7841) with Prep | EACH | 25.00 | 25.00 | 25.00 | 25.00 | 25.00 |
| 33 02 1772 | Gasoline Group (Mod 8010/8020, Lead, EDB) | EACH | 90.00 | 90.00 | 90.00 | 90.00 | 90.00 |
| 33 02 1773 | Kerosene Group (Mod 8010/8020, 8100, TPH, Lead, EDB) | EACH | 90.00 | 90.00 | 90.00 | 90.00 | 90.00 |
| 33 02 1774 | BTEX/MTBE (Mod 8020) | EACH | 83.33 | 83.33 | 83.33 | 83.33 | 83.33 |
| 33 02 1775 | Purgeable Organics (SW 3550/SW 8260) | EACH | 175.00 | 175.00 | 175.00 | 175.00 | 175.00 |
| 33 02 1801 | VOST Analysis, Air (30/5040/8240) | EACH | 225.00 | 225.00 | 225.00 | 225.00 | 225.00 |
| 33 02 1802 | Principal Organic Hazardous Constituents, Air (30/5040/8240) | EACH | 192.50 | 192.50 | 192.50 | 192.50 | 192.50 |
| 33 02 1803 | Tentative ID Compounds, GC/MS, Air (30/5040/8240) | EACH | 257.00 | 257.00 | 257.00 | 257.00 | 257.00 |
| 33 02 1804 | Tedlar Bag (Single Component), Air (30/5040/8240) | EACH | 100.00 | 100.00 | 100.00 | 100.00 | 100.00 |
| 33 02 1805 | Sorbent Tube (Single Component), Air (5040/8240) | EACH | 100.00 | 100.00 | 100.00 | 100.00 | 100.00 |
| 33 02 1806 | BTEX, Air (5040/8020) | EACH | 95.00 | 95.00 | 95.00 | 95.00 | 95.00 |
| 33 02 1807 | BTEX/Total Volatile Petroleum Hydrocarbons, Air (5040/8020/8015) | EACH | 111.67 | 111.67 | 111.67 | 111.67 | 111.67 |
| 33 02 1808 | Semivolatiles, Air (10/3550/8270) | EACH | 531.67 | 531.67 | 531.67 | 531.67 | 531.67 |
| 33 02 1809 | Multimetal Train, Air, Various Methods | EACH | 230.50 | 230.50 | 230.50 | 230.50 | 230.50 |
| 33 02 1810 | Pesticides/PCBs, GC, Air, Various Methods | EACH | 150.00 | 150.00 | 150.00 | 150.00 | 150.00 |
| 33 02 1811 | Hydrocarbons, BTEX, Air (NIOSH 1500) | EACH | 102.33 | 102.33 | 102.33 | 102.33 | 102.33 |

## Environmental Remediation: Assemblies Cost Book

*Page 3-220*

# Cost Data for Remediation

## Monitoring

| Assembly | Description | Unit | Unit Cost by Safety Level A | B | C | D | E |
|---|---|---|---|---|---|---|---|
| | **Capital Costs** | | | | | | |
| 33 02 1812 | Polyaromatic Hydrocarbons, Air (NIOSH 5515) | EACH | 462.50 | 462.50 | 462.50 | 462.50 | 462.50 |
| 33 02 1813 | Metals by ICP, Air (NIOSH 7300) | EACH | 249.50 | 249.50 | 249.50 | 249.50 | 249.50 |
| 33 02 1814 | Metals by Graphite Furnace, Air (NIOSH 7300) | EACH | 281.33 | 281.33 | 281.33 | 281.33 | 281.33 |
| 33 02 1815 | Mercury, Air (245.2) | EACH | 35.00 | 35.00 | 35.00 | 35.00 | 35.00 |
| 33 02 1816 | Cyanide, Air (335.2) | EACH | 43.33 | 43.33 | 43.33 | 43.33 | 43.33 |
| 33 02 1817 | Ion Chromatography, Air | EACH | 18.00 | 18.00 | 18.00 | 18.00 | 18.00 |
| 33 02 1818 | Volatile Aromatic Halocarbons, Air, TO-1 | EACH | 95.00 | 95.00 | 95.00 | 95.00 | 95.00 |
| 33 02 1819 | Highly Volatile Nonpolar Organics, TO-2 | EACH | 305.00 | 305.00 | 305.00 | 305.00 | 305.00 |
| 33 02 1820 | Volatile Nonpolar Organics, TO-3 | EACH | 92.50 | 92.50 | 92.50 | 92.50 | 92.50 |
| 33 02 1821 | Organochlorine Pesticides/PCBs, Air, TO-4 | EACH | 242.50 | 242.50 | 242.50 | 242.50 | 242.50 |
| 33 02 1822 | Aldehydes & Ketones, Air, TO-5 | EACH | 112.50 | 112.50 | 112.50 | 112.50 | 112.50 |
| 33 02 1823 | Phosgene in Ambient Air, TO-6 | EACH | 210.00 | 210.00 | 210.00 | 210.00 | 210.00 |
| 33 02 1824 | N-Nitrosodimethylamine, Air, TO-7 | EACH | 182.50 | 182.50 | 182.50 | 182.50 | 182.50 |
| 33 02 1825 | Phenol & Methyl Phenols, Air, TO-8 | EACH | 72.50 | 72.50 | 72.50 | 72.50 | 72.50 |
| 33 02 1826 | Dioxins, Air, TO-9 | EACH | 322.50 | 322.50 | 322.50 | 322.50 | 322.50 |
| 33 02 1827 | Total Nonmethane Organics, Air, TO-12 | EACH | 50.00 | 50.00 | 50.00 | 50.00 | 50.00 |
| 33 02 1828 | Polyaromatic Hydrocarbons, Air, TO-13 | EACH | 100.00 | 100.00 | 100.00 | 100.00 | 100.00 |
| 33 02 1829 | Canister Samples by GC or GC/MS, TO-14 | EACH | 400.00 | 400.00 | 400.00 | 400.00 | 400.00 |
| 33 02 1830 | Hydrocarbon Speciation, C1-C8, GC/FID, Air (TO-12/14) | EACH | 75.00 | 75.00 | 75.00 | 75.00 | 75.00 |
| 33 02 1831 | Hydrocarbon Speciation, C8-C22, GC/FID, Air (TO-12/14) | EACH | 75.00 | 75.00 | 75.00 | 75.00 | 75.00 |
| 33 02 1832 | Hydrocarbon Speciation, C1-C22, GC/FID, Air (TO-12/14) | EACH | 75.00 | 75.00 | 75.00 | 75.00 | 75.00 |
| 33 02 1833 | Hydrocarbon Speciation, C4-C22, GC/FID, Air (TO-12/14) | EACH | 75.00 | 75.00 | 75.00 | 75.00 | 75.00 |
| 33 02 1901 | Freshwater Discharge, LC50, Bioassay Analysis | EACH | 566.67 | 566.67 | 566.67 | 566.67 | 566.67 |
| 33 02 1902 | Saltwater Discharge, Lc50, Bioassay Analysis | EACH | 550.00 | 550.00 | 550.00 | 550.00 | 550.00 |
| 33 02 1903 | Drilling Mud, Lc50, Bioassay Analysis | EACH | 666.67 | 666.67 | 666.67 | 666.67 | 666.67 |
| 33 02 1904 | Freshwater Chronic Toxicity Bioassay Analysis | EACH | 1,150 | 1,150 | 1,150 | 1,150 | 1,150 |
| 33 02 1905 | Saltwater Chronic Toxicity Bioassay Analysis | EACH | 1,483 | 1,483 | 1,483 | 1,483 | 1,483 |
| 33 02 1906 | Ames (Full) Microbiological Analysis | EACH | 1,783 | 1,783 | 1,783 | 1,783 | 1,783 |
| 33 02 1907 | Ames (Screen) Microbiological Analysis | EACH | 175.00 | 175.00 | 175.00 | 175.00 | 175.00 |
| 33 02 1908 | Daphnia IQ Toxicity Analysis | EACH | 316.67 | 316.67 | 316.67 | 316.67 | 316.67 |
| 33 02 1909 | Microtox Microbiological Analysis | EACH | 325.00 | 325.00 | 325.00 | 325.00 | 325.00 |
| 33 02 1910 | Total Heterotrophic Plate Count | EACH | 42.50 | 42.50 | 42.50 | 42.50 | 42.50 |
| 33 02 1911 | Hydrocarbon Degraders | EACH | 77.50 | 77.50 | 77.50 | 77.50 | 77.50 |

**Environmental Remediation: Assemblies Cost Book**

# Cost Data for Remediation

## Monitoring

| Assembly | Description | Unit | Unit Cost by Safety Level | | | | |
|---|---|---|---|---|---|---|---|
| | | | A | B | C | D | E |
| | **Capital Costs** | | | | | | |
| 33 02 1912 | Fluorescent Pseudomonas | EACH | 55.00 | 55.00 | 55.00 | 55.00 | 55.00 |
| 33 02 2001 | Sample Preparation (3005/3010/3020/3040/3050) | EACH | 10.00 | 10.00 | 10.00 | 10.00 | 10.00 |
| 33 02 2002 | Sample Preparation (3510/3520/3540/3550/5030) | EACH | 25.00 | 25.00 | 25.00 | 25.00 | 25.00 |
| 33 02 2003 | Sample Preparation, Waste Dilution (3580) | EACH | 20.00 | 20.00 | 20.00 | 20.00 | 20.00 |
| 33 02 2004 | Cleanup (3610/3611/3620/3630/3650/3660) | EACH | 25.00 | 25.00 | 25.00 | 25.00 | 25.00 |
| 33 02 2005 | Cleanup, Gel-Permeation (3640) | EACH | 25.00 | 25.00 | 25.00 | 25.00 | 25.00 |
| 33 02 2019 | 1 Gallon, 128 Oz, Glass Sample Container, Case of 4 | EACH | 12.80 | 12.80 | 12.80 | 12.80 | 12.80 |
| 33 02 2020 | 1 Liter, 32 Oz, Clear Wide Mouth Jar, Case of 12 | EACH | 44.37 | 44.37 | 44.37 | 44.37 | 44.37 |
| 33 02 2021 | 500 ml, 16 Oz, Clear Wide Mouth Jar, Case of 12 | EACH | 61.67 | 61.67 | 61.67 | 61.67 | 61.67 |
| 33 02 2022 | 250 ml, 8 Oz, Clear Wide Mouth Jar, Case of 24 | EACH | 47.09 | 47.09 | 47.09 | 47.09 | 47.09 |
| 33 02 2023 | 120 ml, 4 Oz, Clear Wide Mouth Jar, Case of 24 | EACH | 34.47 | 34.47 | 34.47 | 34.47 | 34.47 |
| 33 02 2024 | 1 Liter, 32 Oz, Boston Round Bottle, Case of 12 | EACH | 28.78 | 28.78 | 28.78 | 28.78 | 28.78 |
| 33 02 2025 | 500 ml, 16 Oz, Boston Round Bottle, Case of 12 | EACH | 19.91 | 19.91 | 19.91 | 19.91 | 19.91 |
| 33 02 2026 | 40 ml, Clear Vial, Case of 72 | EACH | 89.91 | 89.91 | 89.91 | 89.91 | 89.91 |
| 33 02 2027 | 20 ml, Clear Vial, Case of 72 | EACH | 74.92 | 74.92 | 74.92 | 74.92 | 74.92 |
| 33 02 2028 | 250 ml, 8 oz, Clear Wide Mouth Vial with Septa, Case of 24 | EACH | 58.01 | 58.01 | 58.01 | 58.01 | 58.01 |
| 33 02 2029 | 125 ml, 4 oz, Clear Wide Mouth Vial with Septa, Case of 24 | EACH | 53.53 | 53.53 | 53.53 | 53.53 | 53.53 |
| 33 02 2030 | 1 Liter, 32 oz, High-density Polyethylene Bottle, Case of 12 | EACH | 32.64 | 32.64 | 32.64 | 32.64 | 32.64 |
| 33 02 2031 | 500 ml, 16 oz, High-density Polyethylene Bottle, Case of 24 | EACH | 28.06 | 28.06 | 28.06 | 28.06 | 28.06 |
| 33 02 2032 | 250 ml, 8 oz, High-density Polyethylene Bottle, Case of 24 | EACH | 15.58 | 15.58 | 15.58 | 15.58 | 15.58 |
| 33 02 2033 | 125 ml, 4 oz, High-density Polyethylene Bottle, Case of 24 | EACH | 13.71 | 13.71 | 13.71 | 13.71 | 13.71 |
| 33 02 2034 | Custody Seals, Package of 10 | EACH | 15.60 | 15.60 | 15.60 | 15.60 | 15.60 |
| 33 02 2035 | Safe Transport Can Filled with Vermiculite, 1 Gallon, Case of 4 | EACH | 28.94 | 28.94 | 28.94 | 28.94 | 28.94 |
| 33 02 2036 | Documentation Package for QA Verification, Data & Benchwork | EACH | 100.00 | 100.00 | 100.00 | 100.00 | 100.00 |
| 33 02 2037 | Overnight Delivery, 8 oz Letter | EACH | 10.17 | 10.17 | 10.17 | 10.17 | 10.17 |
| 33 02 2038 | Overnight Delivery, 1 Lb Package | LB | 11.08 | 11.08 | 11.08 | 11.08 | 11.08 |
| 33 02 2039 | Overnight Delivery, 2 - 5 Lb Package | LB | 16.08 | 16.08 | 16.08 | 16.08 | 16.08 |
| 33 02 2040 | Overnight Delivery, 6 - 10 Lb Package | LB | 21.58 | 21.58 | 21.58 | 21.58 | 21.58 |
| 33 02 2041 | Overnight Delivery, 11 - 20 Lb Package | LB | 34.25 | 34.25 | 34.25 | 34.25 | 34.25 |
| 33 02 2042 | Overnight Delivery, 21 - 50 Lb Package | LB | 57.27 | 57.27 | 57.27 | 57.27 | 57.27 |
| 33 02 2043 | Overnight Delivery, 51 - 70 Lb Package | LB | 78.27 | 78.27 | 78.27 | 78.27 | 78.27 |
| 33 02 2044 | Overnight Delivery, 71 Lbs or Greater | LB | 171.25 | 171.25 | 171.25 | 171.25 | 171.25 |

**Environmental Remediation: Assemblies Cost Book**

# Cost Data for Remediation

## Monitoring

| Assembly | Description | Unit | Unit Cost by Safety Level | | | | |
|---|---|---|---|---|---|---|---|
| | | | A | B | C | D | E |
| | **Capital Costs** | | | | | | |
| 33 02 2045 | 48 Quart Ice Chest | EACH | 27.62 | 27.62 | 27.62 | 27.62 | 27.62 |
| 33 02 2046 | 60 Quart Ice Chest | EACH | 46.27 | 46.27 | 46.27 | 46.27 | 46.27 |
| 33 02 2047 | 80 Quart Ice Chest | EACH | 79.85 | 79.85 | 79.85 | 79.85 | 79.85 |
| 33 02 2048 | 102 Quart Ice Chest | EACH | 135.64 | 135.64 | 135.64 | 135.64 | 135.64 |
| 33 02 2049 | 120 Quart Ice Chest | EACH | 216.05 | 216.05 | 216.05 | 216.05 | 216.05 |
| 33 02 2050 | Blue Ice Soft Packs (Equivalent to 7 Lbs Ice) | EACH | 1.73 | 1.73 | 1.73 | 1.73 | 1.73 |
| 33 02 2101 | Color (SM 204B) | EACH | 17.00 | 17.00 | 17.00 | 17.00 | 17.00 |
| 33 02 2102 | Total Solids (SM 209A) | EACH | 14.33 | 14.33 | 14.33 | 14.33 | 14.33 |
| 33 02 2103 | Total Dissolved Solids (SM 209B) | EACH | 15.00 | 15.00 | 15.00 | 15.00 | 15.00 |
| 33 02 2104 | Total Suspended Solids (SM 209C) | EACH | 15.00 | 15.00 | 15.00 | 15.00 | 15.00 |
| 33 02 2105 | Total Volatile Solids (SM 209D) | EACH | 18.67 | 18.67 | 18.67 | 18.67 | 18.67 |
| 33 02 2106 | Turbidity (SM 214A) | EACH | 12.67 | 12.67 | 12.67 | 12.67 | 12.67 |
| 33 02 2107 | Hardness, Total (SM 314B) | EACH | 15.50 | 15.50 | 15.50 | 15.50 | 15.50 |
| 33 02 2108 | Acidity (SM 402) | EACH | 15.00 | 15.00 | 15.00 | 15.00 | 15.00 |
| 33 02 2109 | Alkalinity (SM 403) | EACH | 15.00 | 15.00 | 15.00 | 15.00 | 15.00 |
| 33 02 2110 | Bromide (SM 405) | EACH | 32.50 | 32.50 | 32.50 | 32.50 | 32.50 |
| 33 02 2111 | Chlorine, Residual (SM 408D) | EACH | 16.67 | 16.67 | 16.67 | 16.67 | 16.67 |
| 33 02 2112 | Cyanide, Total (SM 412B/412D) | EACH | 48.33 | 48.33 | 48.33 | 48.33 | 48.33 |
| 33 02 2113 | Fluoride, Electrode (SM 413B) | EACH | 18.67 | 18.67 | 18.67 | 18.67 | 18.67 |
| 33 02 2114 | Fluoride, SPADNS (SM 413C) | EACH | 125.00 | 125.00 | 125.00 | 125.00 | 125.00 |
| 33 02 2115 | Nitrogen, Total (SM 420) | EACH | 38.33 | 38.33 | 38.33 | 38.33 | 38.33 |
| 33 02 2116 | Sulfite (SM 428A) | EACH | 16.32 | 16.32 | 16.32 | 16.32 | 16.32 |
| 33 02 2117 | Oil & Grease (SM 503A) | EACH | 55.00 | 55.00 | 55.00 | 55.00 | 55.00 |
| 33 02 2118 | Total Organic Carbon, TOC (SM 505B) | EACH | 30.67 | 30.67 | 30.67 | 30.67 | 30.67 |
| 33 02 2119 | Biochemical Oxygen Demand BOD (SM 507) | EACH | 34.67 | 34.67 | 34.67 | 34.67 | 34.67 |
| 33 02 2120 | Chemical Oxygen Demand COD (SM 508C) | EACH | 28.33 | 28.33 | 28.33 | 28.33 | 28.33 |
| 33 02 2121 | Surfactants (SM 512B) | EACH | 38.00 | 38.00 | 38.00 | 38.00 | 38.00 |
| 33 02 2122 | Fecal Coliform (SM 909C) | EACH | 26.00 | 26.00 | 26.00 | 26.00 | 26.00 |
| 33 02 2124 | Metal Analysis, Priority 17 Metals | EACH | 131.00 | 131.00 | 131.00 | 131.00 | 131.00 |
| 33 02 2125 | Arsenic, Hydride (SW 7061) with Prep | EACH | 181.33 | 181.33 | 181.33 | 181.33 | 181.33 |
| 33 02 2126 | Arsenic, (SW 7060) with Prep | EACH | 29.00 | 29.00 | 29.00 | 29.00 | 29.00 |
| 33 02 2127 | Metals Screen, 25 Listed in SW 3005/SW 6010 | EACH | 283.33 | 283.33 | 283.33 | 283.33 | 283.33 |
| 33 02 2128 | Chromium, Hexavalent (SW 7195) with Prep | EACH | 21.67 | 21.67 | 21.67 | 21.67 | 21.67 |

**Environmental Remediation: Assemblies Cost Book**

*Page 3-223*

1998 by ECHOS. All rights reserved

# Cost Data for Remediation

## Monitoring

| Assembly | Description | Unit | A | B | C | D | E |
|---|---|---|---|---|---|---|---|
| | **Capital Costs** | | | | | | |
| 33 02 2129 | Mercury, Cold Vapor (SW 7470) with Prep | EACH | 41.67 | 41.67 | 41.67 | 41.67 | 41.67 |
| 33 02 2130 | Cyanide (SW 9010) with Prep | EACH | 43.33 | 43.33 | 43.33 | 43.33 | 43.33 |
| 33 02 2131 | Purgeable Halocarbons (SW 5030/SW 8010) | EACH | 121.67 | 121.67 | 121.67 | 121.67 | 121.67 |
| 33 02 2132 | Purgeable Aromatics (SW 5030/SW 8020) | EACH | 96.67 | 96.67 | 96.67 | 96.67 | 96.67 |
| 33 02 2133 | Pesticides/PCBs (SW 3510/SW 8080) | EACH | 206.67 | 206.67 | 206.67 | 206.67 | 206.67 |
| 33 02 2134 | Polynuclear Aromatic Hydrocarbons, PAH (SW 3510/SW 8310) | EACH | 188.33 | 188.33 | 188.33 | 188.33 | 188.33 |
| 33 02 2135 | Base Neutral & Acid Extractable Organics (SW 3510/SW 8270) | EACH | 434.67 | 434.67 | 434.67 | 434.67 | 434.67 |
| 33 02 2136 | Carbamate & Urea Pesticides (SW 3510/SW 8318) | EACH | 205.00 | 205.00 | 205.00 | 205.00 | 205.00 |
| 33 02 2137 | Gasoline Group (Mod EPA 601/602, Lead, EDB) | EACH | 186.67 | 186.67 | 186.67 | 186.67 | 186.67 |
| 33 02 2138 | Kerosene Group (Mod EPA 601/602, 610, TPH, Lead, EDB) | EACH | 186.67 | 186.67 | 186.67 | 186.67 | 186.67 |
| 33 02 2139 | BTEX/MTBE (Mod EPA 602) | EACH | 80.00 | 80.00 | 80.00 | 80.00 | 80.00 |
| 33 02 2140 | Ethylene Dibromide (EDB) (EPA 501.4) | EACH | 78.33 | 78.33 | 78.33 | 78.33 | 78.33 |
| 33 02 2141 | Chromium, Total (EPA 218.1) | EACH | 19.00 | 19.00 | 19.00 | 19.00 | 19.00 |
| 33 02 2142 | Titanium (EPA 283.2) | EACH | 18.00 | 18.00 | 18.00 | 18.00 | 18.00 |
| 33 02 2143 | Lead (SW 3005/SW 7421) | EACH | 41.67 | 41.67 | 41.67 | 41.67 | 41.67 |
| 33 02 2144 | Ethylene Dibromide (SW 8011) with Prep | EACH | 130.00 | 130.00 | 130.00 | 130.00 | 130.00 |
| 33 02 2145 | Dioxins (SW 8280) with Prep | EACH | 737.00 | 737.00 | 737.00 | 737.00 | 737.00 |
| 33 02 2146 | Selenium (SW 7740) with Prep | EACH | 36.67 | 36.67 | 36.67 | 36.67 | 36.67 |
| 33 02 2147 | Nonhalogenated Volatile Organics (SW 5030/SW 8015) | EACH | 110.00 | 110.00 | 110.00 | 110.00 | 110.00 |
| 33 02 2148 | Organophosphorous Pesticides (SW 3510/SW 8140) | EACH | 215.00 | 215.00 | 215.00 | 215.00 | 215.00 |
| 33 02 2149 | Chlorinated Phenoxy Herbicides (SW 3510/SW 8150) | EACH | 233.33 | 233.33 | 233.33 | 233.33 | 233.33 |
| 33 02 2201 | Tissue/Bone, Americium Isotopic, Alpha Spectroscopy | EACH | 146.67 | 146.67 | 146.67 | 146.67 | 146.67 |
| 33 02 2202 | Tissue/Bone, Curium Isotopic, Alpha Spectroscopy | EACH | 145.00 | 145.00 | 145.00 | 145.00 | 145.00 |
| 33 02 2203 | Tissue/Bone, Neptunium Isotopic, Alpha Spectroscopy | EACH | 145.00 | 145.00 | 145.00 | 145.00 | 145.00 |
| 33 02 2204 | Tissue/Bone, Plutonium Isotopic, Alpha Spectroscopy | EACH | 150.00 | 150.00 | 150.00 | 150.00 | 150.00 |
| 33 02 2205 | Tissue/Bone, Thorium Isotopic, Alpha Spectroscopy | EACH | 146.67 | 146.67 | 146.67 | 146.67 | 146.67 |
| 33 02 2206 | Tissue/Bone, Uranium Isotopic, Alpha Spectroscopy | EACH | 135.00 | 135.00 | 135.00 | 135.00 | 135.00 |
| 33 02 2207 | Tissue/Bone, Strontium - 89, 90, Gas Flow Proportional Counting | EACH | 166.67 | 166.67 | 166.67 | 166.67 | 166.67 |
| 33 02 2208 | Tissue/Bone, Strontium - 90, Gas Flow Proportional Counting | EACH | 155.67 | 155.67 | 155.67 | 155.67 | 155.67 |
| 33 02 2209 | Tissue/Bone, Technetium - 99, Gas Flow Proportional Counting | EACH | 128.33 | 128.33 | 128.33 | 128.33 | 128.33 |

**Environmental Remediation: Assemblies Cost Book**

# Cost Data for Remediation

## Monitoring

| Assembly | Description | Unit | Unit Cost by Safety Level A | B | C | D | E |
|---|---|---|---|---|---|---|---|
| | **Capital Costs** | | | | | | |
| 33 02 2210 | Tissue/Bone, Americium Isotopic, Gamma Spectroscopy | EACH | 131.67 | 131.67 | 131.67 | 131.67 | 131.67 |
| 33 02 2211 | Tissue/Bone, Gamma Isotopic, Gamma Spectroscopy | EACH | 140.00 | 140.00 | 140.00 | 140.00 | 140.00 |
| 33 02 2212 | Tissue/Bone, Iodine - 25 (Filter or Charcoal), Gamma Spectroscopy | EACH | 150.00 | 150.00 | 150.00 | 150.00 | 150.00 |
| 33 02 2213 | Tissue/Bone, Iodine - 129 (Filter or Charcoal), Gamma Spectroscopy | EACH | 150.00 | 150.00 | 150.00 | 150.00 | 150.00 |
| 33 02 2214 | Tissue/Bone, Iodine - 131 (Filter or Charcoal), Gamma Spectroscopy | EACH | 140.00 | 140.00 | 140.00 | 140.00 | 140.00 |
| 33 02 2215 | Tissue/Bone, Krypton - 85, Gamma Spectroscopy | EACH | 102.50 | 102.50 | 102.50 | 102.50 | 102.50 |
| 33 02 2216 | Tissue/Bone, Lead - 210, Gamma Spectroscopy | EACH | 131.67 | 131.67 | 131.67 | 131.67 | 131.67 |
| 33 02 2217 | Tissue/Bone, Radium - 226, 228, Gamma Spectroscopy | EACH | 145.00 | 145.00 | 145.00 | 145.00 | 145.00 |
| 33 02 2218 | Tissue/Bone, Uranium - Total, Gamma Spectroscopy | EACH | 147.50 | 147.50 | 147.50 | 147.50 | 147.50 |
| 33 02 2219 | Tissue/Bone, Carbon - 14, Liquid Scintillation | EACH | 132.67 | 132.67 | 132.67 | 132.67 | 132.67 |
| 33 02 2220 | Tissue/Bone, Uranium Isotopic, Fluorometric | EACH | 175.00 | 175.00 | 175.00 | 175.00 | 175.00 |
| 33 02 2221 | Tissue/Bone, Uranium Isotopic, Laser Phosphorometer | EACH | 75.00 | 75.00 | 75.00 | 75.00 | 75.00 |
| 33 02 2222 | Air, Americium Isotopic, Alpha Spectroscopy | EACH | 146.67 | 146.67 | 146.67 | 146.67 | 146.67 |
| 33 02 2223 | Air, Curium Isotopic, Alpha Spectroscopy | EACH | 145.00 | 145.00 | 145.00 | 145.00 | 145.00 |
| 33 02 2224 | Air, Neptunium - 237, Alpha Spectroscopy | EACH | 145.00 | 145.00 | 145.00 | 145.00 | 145.00 |
| 33 02 2225 | Air, Polonium - 210, Alpha Spectroscopy | EACH | 146.67 | 146.67 | 146.67 | 146.67 | 146.67 |
| 33 02 2226 | Air, Plutonium Isotopic, Alpha Spectroscopy | EACH | 150.00 | 150.00 | 150.00 | 150.00 | 150.00 |
| 33 02 2227 | Air, Radium - 226, Alpha Spectroscopy | EACH | 143.33 | 143.33 | 143.33 | 143.33 | 143.33 |
| 33 02 2228 | Air, Thorium Isotopic, Alpha Spectroscopy | EACH | 146.67 | 146.67 | 146.67 | 146.67 | 146.67 |
| 33 02 2229 | Air, Uranium Isotopic, Alpha Spectroscopy | EACH | 146.67 | 146.67 | 146.67 | 146.67 | 146.67 |
| 33 02 2230 | Air, Gross Beta - Total, Gas Flow Proportional Counting | EACH | 45.00 | 45.00 | 45.00 | 45.00 | 45.00 |
| 33 02 2231 | Air, Lead - 210, Gas Flow Proportional Counting | EACH | 146.67 | 146.67 | 146.67 | 146.67 | 146.67 |
| 33 02 2232 | Air, Strontium - 89, 90, Gas Flow Proportional Counting | EACH | 172.50 | 172.50 | 172.50 | 172.50 | 172.50 |
| 33 02 2233 | Air, Strontium - 90, Gas Flow Proportional Counting | EACH | 135.00 | 135.00 | 135.00 | 135.00 | 135.00 |
| 33 02 2234 | Air, Technetium - 99, Gas Flow Proportional Counting | EACH | 141.67 | 141.67 | 141.67 | 141.67 | 141.67 |
| 33 02 2235 | Air, Gamma Isotopic, Gamma Spectroscopy | EACH | 110.00 | 110.00 | 110.00 | 110.00 | 110.00 |
| 33 02 2236 | Air, Radium - 226, 228, Gamma Spectroscopy | EACH | 176.67 | 176.67 | 176.67 | 176.67 | 176.67 |
| 33 02 2237 | Air, Carbon - 14, Liquid Scintillation | EACH | 157.50 | 157.50 | 157.50 | 157.50 | 157.50 |
| 33 02 2238 | Air, Iron - 55, Liquid Scintillation | EACH | 145.00 | 145.00 | 145.00 | 145.00 | 145.00 |
| 33 02 2239 | Air, Gross Alpha - Total, Gas Flow Proportional Counting or Alpha Scintillation | EACH | 43.33 | 43.33 | 43.33 | 43.33 | 43.33 |
| 33 02 2240 | Air, Gross Alpha & Gross Beta - Total, Gas Flow | EACH | 43.33 | 43.33 | 43.33 | 43.33 | 43.33 |

**Environmental Remediation: Assemblies Cost Book**

*Page 3-225*

1998 by ECHOS. All rights reserved

# Cost Data for Remediation

## Monitoring

| Assembly | Description | Unit | Unit Cost by Safety Level A | B | C | D | E |
|---|---|---|---|---|---|---|---|
| | **Capital Costs** | | | | | | |
| 33 02 2241 | Air, Plutonium Isotopic, Plutonium - 241, Alpha Spectroscopy | EACH | 141.67 | 141.67 | 141.67 | 141.67 | 141.67 |
| 33 02 2242 | Air, Radium - 226, Radon De-emanation | EACH | 78.33 | 78.33 | 78.33 | 78.33 | 78.33 |
| 33 02 2243 | Air, Uranium - Total, Laser Phosphorometer | EACH | 60.00 | 60.00 | 60.00 | 60.00 | 60.00 |
| 33 02 2244 | Air, Uranium - Total, Fluorometric | EACH | 60.00 | 60.00 | 60.00 | 60.00 | 60.00 |
| 33 02 2245 | Liquid, Americium Isotopic, Alpha Spectroscopy | EACH | 131.67 | 131.67 | 131.67 | 131.67 | 131.67 |
| 33 02 2246 | Liquid, Curium Isotopic, Alpha Spectroscopy | EACH | 138.33 | 138.33 | 138.33 | 138.33 | 138.33 |
| 33 02 2247 | Liquid, Neptunium - 237, Alpha Spectroscopy | EACH | 138.33 | 138.33 | 138.33 | 138.33 | 138.33 |
| 33 02 2248 | Liquid, Polonium - 210, Alpha Spectroscopy | EACH | 131.67 | 131.67 | 131.67 | 131.67 | 131.67 |
| 33 02 2249 | Liquid, Plutonium Isotopic, Alpha Spectroscopy | EACH | 150.00 | 150.00 | 150.00 | 150.00 | 150.00 |
| 33 02 2250 | Liquid, Radium - 226, Alpha Spectroscopy | EACH | 126.67 | 126.67 | 126.67 | 126.67 | 126.67 |
| 33 02 2251 | Liquid, Radium - 226, 228, Alpha Spectroscopy | EACH | 166.67 | 166.67 | 166.67 | 166.67 | 166.67 |
| 33 02 2252 | Liquid, Thorium Isotopic, Alpha Spectroscopy | EACH | 131.67 | 131.67 | 131.67 | 131.67 | 131.67 |
| 33 02 2253 | Liquid, Uranium Isotopic, Alpha Spectroscopy | EACH | 131.67 | 131.67 | 131.67 | 131.67 | 131.67 |
| 33 02 2254 | Liquid, Iodine - 131, Beta/Gamma Coincidence | EACH | 125.00 | 125.00 | 125.00 | 125.00 | 125.00 |
| 33 02 2255 | Liquid, Radium - 226, 228, Beta/Gamma Coincidence | EACH | 158.33 | 158.33 | 158.33 | 158.33 | 158.33 |
| 33 02 2256 | Liquid, Radium - 228, Beta/Gamma Coincidence | EACH | 98.33 | 98.33 | 98.33 | 98.33 | 98.33 |
| 33 02 2257 | Liquid, Gross Beta - Total, Gas Flow Proportional Count | EACH | 43.33 | 43.33 | 43.33 | 43.33 | 43.33 |
| 33 02 2258 | Liquid, Calcium - 45, Gas Flow Proportional Counting | EACH | 145.00 | 145.00 | 145.00 | 145.00 | 145.00 |
| 33 02 2259 | Liquid, Cerium - 144, Gas Flow Proportional Counting | EACH | 135.00 | 135.00 | 135.00 | 135.00 | 135.00 |
| 33 02 2260 | Liquid, Chlorine - 36, Gas Flow Proportional Counting | EACH | 135.00 | 135.00 | 135.00 | 135.00 | 135.00 |
| 33 02 2261 | Liquid, Iodine - 131, Gas Flow Proportional Counting | EACH | 135.00 | 135.00 | 135.00 | 135.00 | 135.00 |
| 33 02 2262 | Liquid, Lead - 210, Gas Flow Proportional Counting | EACH | 135.00 | 135.00 | 135.00 | 135.00 | 135.00 |
| 33 02 2263 | Liquid, Radium - 226, 228, Gas Flow Proportional Count | EACH | 166.67 | 166.67 | 166.67 | 166.67 | 166.67 |
| 33 02 2264 | Liquid, Radium - 228, Gas Flow Proportional Counting | EACH | 126.67 | 126.67 | 126.67 | 126.67 | 126.67 |
| 33 02 2265 | Liquid, Ruthenium - 106, Gas Flow Proportional Counting | EACH | 145.00 | 145.00 | 145.00 | 145.00 | 145.00 |
| 33 02 2266 | Liquid, Strontium - 89, 90, Gas Flow Proportional Count | EACH | 148.33 | 148.33 | 148.33 | 148.33 | 148.33 |
| 33 02 2267 | Liquid, Strontium - 90, Gas Flow Proportional Counting | EACH | 138.33 | 138.33 | 138.33 | 138.33 | 138.33 |
| 33 02 2268 | Liquid, Sulfur - 35, Gas Flow Proportional Counting | EACH | 177.50 | 177.50 | 177.50 | 177.50 | 177.50 |
| 33 02 2269 | Liquid, Technetium - 99, Gas Flow Proportional Counting | EACH | 136.67 | 136.67 | 136.67 | 136.67 | 136.67 |
| 33 02 2270 | Liquid, Cerium - 144, Gamma Spectroscopy | EACH | 89.50 | 89.50 | 89.50 | 89.50 | 89.50 |
| 33 02 2271 | Liquid, Gamma Isotopic, Gamma Spectroscopy | EACH | 89.50 | 89.50 | 89.50 | 89.50 | 89.50 |
| 33 02 2272 | Liquid, Iodine - 125, Gamma Spectroscopy | EACH | 89.50 | 89.50 | 89.50 | 89.50 | 89.50 |
| 33 02 2273 | Liquid Iodine - 129, Gamma Spectroscopy | EACH | 89.50 | 89.50 | 89.50 | 89.50 | 89.50 |

## Environmental Remediation: Assemblies Cost Book

# Cost Data for Remediation

## Monitoring

| Assembly | Description | Unit | Unit Cost by Safety Level ||||| 
| | | | A | B | C | D | E |
|---|---|---|---|---|---|---|---|
| | **Capital Costs** | | | | | | |
| 33 02 2274 | Liquid, Ruthenium - 106, Gamma Spectroscopy | EACH | 89.50 | 89.50 | 89.50 | 89.50 | 89.50 |
| 33 02 2275 | Liquid, Sulfur - 35, Gamma Spectroscopy | EACH | 89.50 | 89.50 | 89.50 | 89.50 | 89.50 |
| 33 02 2276 | Liquid, Chlorine - 36, Liquid Scintillation | EACH | 135.00 | 135.00 | 135.00 | 135.00 | 135.00 |
| 33 02 2277 | Liquid, Carbon - 14, Liquid Scintillation | EACH | 142.50 | 142.50 | 142.50 | 142.50 | 142.50 |
| 33 02 2278 | Liquid, Iron - 55, Liquid Scintillation | EACH | 145.00 | 145.00 | 145.00 | 145.00 | 145.00 |
| 33 02 2279 | Liquid, Iron - 125, Liquid Scintillation | EACH | 100.00 | 100.00 | 100.00 | 100.00 | 100.00 |
| 33 02 2280 | Liquid, Iron - 129, Liquid Scintillation | EACH | 100.00 | 100.00 | 100.00 | 100.00 | 100.00 |
| 33 02 2281 | Liquid, Nickel - 63, Liquid Scintillation | EACH | 145.00 | 145.00 | 145.00 | 145.00 | 145.00 |
| 33 02 2282 | Liquid, Promethium - 147, Liquid Scintillation | EACH | 135.00 | 135.00 | 135.00 | 135.00 | 135.00 |
| 33 02 2283 | Liquid, Sulfur - 35, Liquid Scintillation | EACH | 135.00 | 135.00 | 135.00 | 135.00 | 135.00 |
| 33 02 2284 | Liquid, Tritium, (Direct Counting), Liquid Scintillation | EACH | 67.50 | 67.50 | 67.50 | 67.50 | 67.50 |
| 33 02 2285 | Liquid, Tritium, (Distillation), Liquid Scintillation | EACH | 71.67 | 71.67 | 71.67 | 71.67 | 71.67 |
| 33 02 2286 | Liquid, Tritium, (Electrolytic Enrichment), Liquid Scintillation | EACH | 72.50 | 72.50 | 72.50 | 72.50 | 72.50 |
| 33 02 2287 | Liquid, Gross Alpha - Total, Gas Flow &/or Alpha Spectroscopy | EACH | 55.00 | 55.00 | 55.00 | 55.00 | 55.00 |
| 33 02 2288 | Liquid, Suspended or Dissolved, Gross Alpha/Gross Beta | EACH | 80.00 | 80.00 | 80.00 | 80.00 | 80.00 |
| 33 02 2289 | Liquid, Suspended & Dissolved, Gross Alpha/Gross Beta | EACH | 80.00 | 80.00 | 80.00 | 80.00 | 80.00 |
| 33 02 2290 | Liquid, Gross Alpha & Gross Beta - Total | EACH | 55.00 | 55.00 | 55.00 | 55.00 | 55.00 |
| 33 02 2291 | Liquid, Plutonium Isotopic/Plutonium - 241, Alpha Spectroscopy | EACH | 178.33 | 178.33 | 178.33 | 178.33 | 178.33 |
| 33 02 2292 | Liquid, Total Radium, Alpha Scintillation | EACH | 70.00 | 70.00 | 70.00 | 70.00 | 70.00 |
| 33 02 2293 | Liquid, Radium - 226, Radon De-emanation | EACH | 78.33 | 78.33 | 78.33 | 78.33 | 78.33 |
| 33 02 2294 | Liquid, Radium - 226, 228, Radon De-emanation | EACH | 172.50 | 172.50 | 172.50 | 172.50 | 172.50 |
| 33 02 2295 | Liquid, Radon 222 | EACH | 70.00 | 70.00 | 70.00 | 70.00 | 70.00 |
| 33 02 2296 | Liquid, Uranium - Total, Laser Phosphorometer | EACH | 70.00 | 70.00 | 70.00 | 70.00 | 70.00 |
| 33 02 2297 | Liquid, Uranium - Total, Fluorometric | EACH | 50.00 | 50.00 | 50.00 | 50.00 | 50.00 |
| 33 02 2301 | Urine & Feces, Americium Isotopic, Alpha Spectroscopy | EACH | 142.50 | 142.50 | 142.50 | 142.50 | 142.50 |
| 33 02 2302 | Urine & Feces, Curium Isotopic, Alpha Spectroscopy | EACH | 152.50 | 152.50 | 152.50 | 152.50 | 152.50 |
| 33 02 2303 | Urine & Feces, Neptunium - 237, Alpha Spectroscopy | EACH | 152.50 | 152.50 | 152.50 | 152.50 | 152.50 |
| 33 02 2304 | Urine & Feces, Polonium - 210, Alpha Spectroscopy | EACH | 127.50 | 127.50 | 127.50 | 127.50 | 127.50 |
| 33 02 2305 | Urine & Feces, Plutonium Isotopic, Alpha Spectroscopy | EACH | 150.00 | 150.00 | 150.00 | 150.00 | 150.00 |
| 33 02 2306 | Urine & Feces, Thorium Isotopic, Alpha Spectroscopy | EACH | 137.50 | 137.50 | 137.50 | 137.50 | 137.50 |
| 33 02 2307 | Urine & Feces, Uranium Isotopic, Alpha Spectroscopy | EACH | 137.50 | 137.50 | 137.50 | 137.50 | 137.50 |

**Environmental Remediation: Assemblies Cost Book**

*Page 3-227*

# Cost Data for Remediation

## Monitoring

| Assembly | Description | Unit | Unit Cost by Safety Level | | | | |
|---|---|---|---|---|---|---|---|
| | | | A | B | C | D | E |
| | **Capital Costs** | | | | | | |
| 33 02 2308 | Urine & Feces, Gross Beta - Total, Gas Flow Proportional Counting | EACH | 80.00 | 80.00 | 80.00 | 80.00 | 80.00 |
| 33 02 2309 | Urine & Feces, Ruthenium - 106, Gas Flow Proportional Counting | EACH | 122.50 | 122.50 | 122.50 | 122.50 | 122.50 |
| 33 02 2310 | Urine & Feces, Strontium - 89, 90, Gas Flow Proportional Counting | EACH | 180.00 | 180.00 | 180.00 | 180.00 | 180.00 |
| 33 02 2311 | Urine & Feces, Strontium - 90, Gas Flow Proportional Counting | EACH | 180.00 | 180.00 | 180.00 | 180.00 | 180.00 |
| 33 02 2312 | Urine & Feces, Cerium - 144, Gamma Spectroscopy | EACH | 122.50 | 122.50 | 122.50 | 122.50 | 122.50 |
| 33 02 2313 | Urine & Feces, Gamma Isotopic, Gamma Spectroscopy | EACH | 135.00 | 135.00 | 135.00 | 135.00 | 135.00 |
| 33 02 2314 | Urine & Feces, Iodine - 131, Gamma Spectroscopy | EACH | 135.00 | 135.00 | 135.00 | 135.00 | 135.00 |
| 33 02 2315 | Urine & Feces, Radium - 228, Gamma Spectroscopy | EACH | 135.00 | 135.00 | 135.00 | 135.00 | 135.00 |
| 33 02 2316 | Urine & Feces, Ruthenium - 106, Gamma Spectroscopy | EACH | 122.50 | 122.50 | 122.50 | 122.50 | 122.50 |
| 33 02 2317 | Urine & Feces, Carbon - 14, Liquid Scintillation | EACH | 165.00 | 165.00 | 165.00 | 165.00 | 165.00 |
| 33 02 2318 | Urine & Feces, Promethium - 147, Liquid Scintillation | EACH | 102.50 | 102.50 | 102.50 | 102.50 | 102.50 |
| 33 02 2319 | Urine & Feces, Phosphorus - 32, Liquid Scintillation | EACH | 395.00 | 395.00 | 395.00 | 395.00 | 395.00 |
| 33 02 2320 | Urine & Feces, Tritium (Direct Counting), Liquid Scintillation | EACH | 100.00 | 100.00 | 100.00 | 100.00 | 100.00 |
| 33 02 2321 | Urine & Feces, Tritium (Distillation), Liquid Scintillation | EACH | 125.00 | 125.00 | 125.00 | 125.00 | 125.00 |
| 33 02 2322 | Urine & Feces, Mixed Fission Products, Gas Flow | EACH | 95.00 | 95.00 | 95.00 | 95.00 | 95.00 |
| 33 02 2323 | Urine & Feces, Plutonium Isotopic, Plutonium - 241, Alpha Spectroscopy | EACH | 152.50 | 152.50 | 152.50 | 152.50 | 152.50 |
| 33 02 2324 | Urine & Feces, Radium - 226, Radon De-emanation | EACH | 87.50 | 87.50 | 87.50 | 87.50 | 87.50 |
| 33 02 2325 | Urine & Feces, Radium - 226, 228, Radon De-emanation | EACH | 145.00 | 145.00 | 145.00 | 145.00 | 145.00 |
| 33 02 2326 | Urine & Feces, Uranium - Total, Laser Phosphorometer | EACH | 75.00 | 75.00 | 75.00 | 75.00 | 75.00 |
| 33 02 2327 | Urine & Feces, Uranium - Total, Fluorometric | EACH | 70.00 | 70.00 | 70.00 | 70.00 | 70.00 |
| 33 02 2328 | Urine & Feces, Uranium - Total, Radiometric, Alpha Scintillation | EACH | 82.50 | 82.50 | 82.50 | 82.50 | 82.50 |
| 33 02 2329 | Urine & Feces, Uranium - Total, Laser Phosphorometer | EACH | 82.50 | 82.50 | 82.50 | 82.50 | 82.50 |
| 33 02 2330 | Vegetation/Soil/Sediment, Americium Isotopic, Alpha Spectroscopy | EACH | 140.00 | 140.00 | 140.00 | 140.00 | 140.00 |
| 33 02 2331 | Vegetation/Soil/Sediment, Curium Isotopic, Alpha Spectroscopy | EACH | 138.33 | 138.33 | 138.33 | 138.33 | 138.33 |
| 33 02 2332 | Vegetation/Soil/Sediment, Neptunium - 237, Alpha Spectroscopy | EACH | 138.33 | 138.33 | 138.33 | 138.33 | 138.33 |
| 33 02 2333 | Vegetation/Soil/Sediment, Plutonium Isotopic, Alpha Spectroscopy | EACH | 143.33 | 143.33 | 143.33 | 143.33 | 143.33 |
| 33 02 2334 | Vegetation/Soil/Sediment, Thorium Isotopic, Alpha Spectroscopy | EACH | 140.00 | 140.00 | 140.00 | 140.00 | 140.00 |

## Environmental Remediation: Assemblies Cost Book

# Cost Data for Remediation

## Monitoring

| Assembly | Description | Unit | Unit Cost by Safety Level A | B | C | D | E |
|---|---|---|---|---|---|---|---|
| | **Capital Costs** | | | | | | |
| 33 02 2335 | Vegetation/Soil/Sediment, Uranium Isotopic, Alpha Spectroscopy | EACH | 140.00 | 140.00 | 140.00 | 140.00 | 140.00 |
| 33 02 2336 | Vegetation/Soil/Sediment, Gross Beta - Total, Gas Flow Proportional Counting | EACH | 50.00 | 50.00 | 50.00 | 50.00 | 50.00 |
| 33 02 2337 | Vegetation/Soil/Sediment, Ruthenium - 106, Gas Flow Proportional Counting | EACH | 145.00 | 145.00 | 145.00 | 145.00 | 145.00 |
| 33 02 2338 | Vegetation/Soil/Sediment, Strontium - 89, 90, Gas Flow Proportional Counting | EACH | 170.00 | 170.00 | 170.00 | 170.00 | 170.00 |
| 33 02 2339 | Vegetation/Soil/Sediment, Strontium - 90, Gas Flow Proportional Counting | EACH | 135.00 | 135.00 | 135.00 | 135.00 | 135.00 |
| 33 02 2340 | Vegetation/Soil/Sediment, Technetium - 99, Gas Flow Proportional Counting | EACH | 125.00 | 125.00 | 125.00 | 125.00 | 125.00 |
| 33 02 2341 | Vegetation/Soil/Sediment, Americium Isotopic, Gamma Spectroscopy | EACH | 123.33 | 123.33 | 123.33 | 123.33 | 123.33 |
| 33 02 2342 | Vegetation/Soil/Sediment, Gamma Isotopic, Gamma Spectroscopy | EACH | 123.33 | 123.33 | 123.33 | 123.33 | 123.33 |
| 33 02 2343 | Vegetation/Soil/Sediment, Iodine - 129, Gamma Spectroscopy | EACH | 123.33 | 123.33 | 123.33 | 123.33 | 123.33 |
| 33 02 2344 | Vegetation/Soil/Sediment, Lead - 210, Gamma Spectroscopy | EACH | 131.67 | 131.67 | 131.67 | 131.67 | 131.67 |
| 33 02 2345 | Vegetation/Soil/Sediment, Polonium - 210, Gamma Spectroscopy | EACH | 131.67 | 131.67 | 131.67 | 131.67 | 131.67 |
| 33 02 2346 | Vegetation/Soil/Sediment, Radium - 226, 228, Gamma Spectroscopy | EACH | 123.33 | 123.33 | 123.33 | 123.33 | 123.33 |
| 33 02 2347 | Vegetation/Soil/Sediment, Ruthenium - 106, Gamma Spectroscopy | EACH | 90.00 | 90.00 | 90.00 | 90.00 | 90.00 |
| 33 02 2348 | Vegetation/Soil/Sediment, Uranium - Total, Gamma Spectroscopy | EACH | 67.50 | 67.50 | 67.50 | 67.50 | 67.50 |
| 33 02 2349 | Vegetation/Soil/Sediment, Carbon - 14, Liquid Scintillation | EACH | 150.00 | 150.00 | 150.00 | 150.00 | 150.00 |
| 33 02 2350 | Vegetation/Soil/Sediment, Nickel - 63, Liquid Scintillation | EACH | 145.00 | 145.00 | 145.00 | 145.00 | 145.00 |
| 33 02 2351 | Vegetation/Soil/Sediment, Tritium, Liquid Scintillation | EACH | 60.00 | 60.00 | 60.00 | 60.00 | 60.00 |
| 33 02 2352 | Vegetation/Soil/Sediment, Gross Alpha - Total, Gas Flow Proportional Counting | EACH | 43.33 | 43.33 | 43.33 | 43.33 | 43.33 |
| 33 02 2353 | Vegetation/Soil/Sediment, Gross Alpha/Gross Beta - Total, Gas Flow | EACH | 50.00 | 50.00 | 50.00 | 50.00 | 50.00 |
| 33 02 2354 | Vegetation/Soil/Sediment, Plutonium Isotopic/Plutonium 241, Alpha Spectroscopy | EACH | 148.33 | 148.33 | 148.33 | 148.33 | 148.33 |
| 33 02 2355 | Vegetation/Soil/Sediment, Uranium - Total, Laser Phosphorometer | EACH | 60.00 | 60.00 | 60.00 | 60.00 | 60.00 |
| 33 02 2356 | Vegetation/Soil/Sediment, Uranium - Total, Fluorometric | EACH | 50.00 | 50.00 | 50.00 | 50.00 | 50.00 |
| 33 02 2357 | Characterization of Any Unknown (Low Level) Radioactive Sample | EACH | 750.00 | 750.00 | 750.00 | 750.00 | 750.00 |

**Environmental Remediation: Assemblies Cost Book**

# Cost Data for Remediation

## Monitoring

| Assembly | Description | Unit | Unit Cost by Safety Level A | B | C | D | E |
|---|---|---|---|---|---|---|---|
| | **Capital Costs** | | | | | | |
| 33 02 2401 | EPA Method 8330 (11 Compounds) Nitroaromatics/Nitramines | EACH | 287.35 | 287.35 | 287.35 | 287.35 | 287.35 |
| 33 02 2402 | Nitroglycerine | EACH | 236.27 | 236.27 | 236.27 | 236.27 | 236.27 |
| 33 02 2403 | Nitrocellulose | EACH | 191.57 | 191.57 | 191.57 | 191.57 | 191.57 |
| 33 02 2404 | Nitroquanadine | EACH | 191.57 | 191.57 | 191.57 | 191.57 | 191.57 |
| 33 02 2405 | Ammonium Perchlorate | EACH | 108.56 | 108.56 | 108.56 | 108.56 | 108.56 |
| 33 02 9901 | Magnetometer | DAY | 53.00 | 53.00 | 53.00 | 53.00 | 53.00 |
| 33 02 9902 | Electromagnetics | DAY | 53.00 | 53.00 | 53.00 | 53.00 | 53.00 |
| 33 02 9903 | Ground Penetrating Radar | DAY | 50.35 | 50.35 | 50.35 | 50.35 | 50.35 |
| 33 02 9904 | Seismic Refraction | DAY | 2,783 | 2,783 | 2,783 | 2,783 | 2,783 |
| 33 02 9905 | Resistivity | DAY | 2,783 | 2,783 | 2,783 | 2,783 | 2,783 |
| 33 02 9906 | Subcontracted Sampling per Hour, One-man Crew | HOUR | 85.25 | 85.25 | 85.25 | 85.25 | 85.25 |
| 33 02 9907 | Subcontracted Sampling per Day, One-man Crew | DAY | 680.00 | 680.00 | 680.00 | 680.00 | 680.00 |
| 33 02 9908 | Double-ring Infiltrometer Test (DRIT Test) | EACH | 550.00 | 550.00 | 550.00 | 550.00 | 550.00 |
| 33 02 9909 | Falling Head Rigid Wall Consolidometer Tech CAE 1110-2-1906 | EACH | 225.00 | 225.00 | 225.00 | 225.00 | 225.00 |
| 33 02 9911 | Falling Head Permeability Test EPA-9100 | EACH | 215.00 | 215.00 | 215.00 | 215.00 | 215.00 |
| 33 02 9912 | Constant Head Permeability Test ASTM D2432 | EACH | 193.33 | 193.33 | 193.33 | 193.33 | 193.33 |
| 33 02 9913 | Mobile Laboratory Trailer, 8' W x 30' L, Rental | MONTH | 7,950 | 7,950 | 7,950 | 7,950 | 7,950 |
| 33 02 9914 | Gas Chromatograph, HP5890A, Rental | MONTH | 1,855 | 1,855 | 1,855 | 1,855 | 1,855 |
| 33 02 9915 | Mass Spectrometer, HP5970B MD, Rental | MONTH | 1,617 | 1,617 | 1,617 | 1,617 | 1,617 |
| 33 02 9916 | Unconfined Compressive Strength (Pocket Penetrometer) | EACH | 4.00 | 4.00 | 4.00 | 4.00 | 4.00 |
| 33 07 9901 | Lease, Portable 40' W x 150' L x 18' High Emissions Containment Building | YEAR | 7,500 | 7,500 | 7,500 | 7,500 | 7,500 |
| 33 07 9902 | Buy, Portable 40' W x 150' L x 18' High Emissions Containment Building | EACH | 23,503 | 23,503 | 23,503 | 23,503 | 23,503 |
| 33 07 9903 | Assemble Emissions Containment Building, 5 Days | EACH | 12,346 | 9,591 | 8,298 | 5,629 | 4,722 |
| 33 07 9904 | Disassemble Emissions Containment Building, 3 Days | EACH | 7,402 | 5,750 | 4,975 | 3,374 | 2,831 |
| 33 17 0808 | Decontaminate Rig, Augers, Screen (Rental Equipment) | DAY | 205.34 | 205.34 | 205.34 | 205.34 | 205.34 |
| 33 19 0201 | Drummed Waste, Minimum Charge for Shipment | EACH | 2,350 | 2,350 | 2,350 | 2,350 | 2,350 |
| 33 19 9543 | Initial Waste Stream Evaluation, Non-PCB | EACH | 1,375 | 1,375 | 1,375 | 1,375 | 1,375 |
| 33 19 9544 | Initial PCB Waste Stream Evaluation | EACH | 631.25 | 631.25 | 631.25 | 631.25 | 631.25 |
| 33 19 9921 | DOT Steel Drum, 55 Gallon | EACH | 65.19 | 65.19 | 65.19 | 65.19 | 65.19 |
| 33 19 9922 | Polyethylene Closed Head Drum, 55 Gallon | EACH | 43.63 | 43.63 | 43.63 | 43.63 | 43.63 |
| 33 19 9924 | DOT Steel Salvage Drum, 85 Gallon, 16 Gauge | EACH | 57.06 | 57.06 | 57.06 | 57.06 | 57.06 |

**Environmental Remediation: Assemblies Cost Book**

*Page 3-230*

# Cost Data for Remediation

## Monitoring

| Assembly | Description | Unit | Unit Cost by Safety Level | | | | |
|---|---|---|---|---|---|---|---|
| | | | A | B | C | D | E |
| | **Capital Costs** | | | | | | |
| 33 19 9925 | DOT Steel Salvage Drum Composite Overpack, 85 Gallon, 16 Gauge | EACH | 70.49 | 70.49 | 70.49 | 70.49 | 70.49 |
| 33 23 0503 | 4" Submersible Pump, 13 - 37 GPM, < 180' Head | EACH | 3,084 | 2,790 | 2,637 | 2,352 | 2,264 |
| 33 23 0505 | 6" Submersible Pump, 15 - 135 GPM, 200' < Head < 500' | EACH | 4,472 | 3,885 | 3,579 | 3,008 | 2,834 |
| 33 23 1106 | Split Spoon Sample, 2" x 24", During Drilling | EACH | 46.67 | 46.67 | 46.67 | 46.67 | 46.67 |
| 33 23 1111 | Well Development Equipment Rental | WEEK | 478.91 | 462.93 | 456.09 | 440.65 | 434.98 |
| 33 23 1121 | Standby for Drilling | EACH | 618.63 | 503.81 | 413.56 | 300.17 | 284.95 |
| 33 23 1122 | Move Rig/Equipment Around Site | EACH | 154.66 | 125.95 | 103.39 | 75.04 | 71.24 |
| 33 23 1123 | Load Supplies/Equipment | LS | 1,856 | 1,511 | 1,241 | 900.52 | 854.85 |
| 33 23 1124 | Security Pass/Protocol | LS | 309.32 | 251.91 | 206.78 | 150.09 | 142.47 |
| 33 23 1125 | Concrete Coring (Minimum Charge) | DAY | 2,475 | 2,015 | 1,654 | 1,201 | 1,140 |
| 33 23 1126 | Furnish 55 Gallon Drum for Drill Cuttings & Development Water | EACH | 65.19 | 65.19 | 65.19 | 65.19 | 65.19 |
| 33 23 1127 | Furnish 55 Gallon Drum for Development/Purge Water | EACH | 65.19 | 65.19 | 65.19 | 65.19 | 65.19 |
| 33 23 1801 | 2" Well, Grout (Annular Seal) | LF | 83.99 | 68.83 | 56.92 | 41.95 | 39.94 |
| 33 23 1802 | 4" Well, Grout (Annular Seal) | LF | 145.07 | 118.89 | 98.31 | 72.46 | 68.99 |
| 33 23 1803 | 6" Well, Grout (Annular Seal) | LF | 213.78 | 175.20 | 144.88 | 106.78 | 101.67 |
| 33 23 1811 | 2" Well, Portland Cement Grout | LF | 0.92 | 0.92 | 0.92 | 0.92 | 0.92 |
| 33 23 1812 | 4" Well, Portland Cement Grout | LF | 1.38 | 1.38 | 1.38 | 1.38 | 1.38 |
| 33 23 1813 | 6" Well, Portland Cement Grout | LF | 7.80 | 7.80 | 7.80 | 7.80 | 7.80 |
| 33 23 2104 | 3/8", 1' Thick, Sodium Bentonite Pellets | CF | 275.35 | 229.42 | 193.32 | 147.97 | 141.88 |
| 33 23 2401 | Teflon Bailer, 3/4" Outside Diameter x 1', 60 cc | EACH | 179.77 | 179.77 | 179.77 | 179.77 | 179.77 |
| 33 23 2402 | Teflon Bailer, 3/4" Outside Diameter x 3', 180 cc | EACH | 227.74 | 227.74 | 227.74 | 227.74 | 227.74 |
| 33 23 2403 | Teflon Bailer, 1" Outside Diameter x 1', 80 cc | EACH | 191.28 | 191.28 | 191.28 | 191.28 | 191.28 |
| 33 23 2404 | Teflon Bailer, 1 7/8" Outside Diameter x 1', 350 cc | EACH | 181.13 | 181.13 | 181.13 | 181.13 | 181.13 |
| 33 23 2405 | Teflon Bailer, 1 7/8" Outside Diameter x 2', 700 cc | EACH | 198.20 | 198.20 | 198.20 | 198.20 | 198.20 |
| 33 23 2406 | Teflon Bailer, 1 7/8" Outside Diameter x 3', 1,050 cc | EACH | 261.53 | 261.53 | 261.53 | 261.53 | 261.53 |
| 33 23 2407 | Disposable Bailer, Polyethylene, 1.5" Outside Diameter x 36" | EACH | 27.60 | 27.60 | 27.60 | 27.60 | 27.60 |
| 33 23 2421 | Emptying Stand | EACH | 302.32 | 302.32 | 302.32 | 302.32 | 302.32 |
| 33 23 2422 | Suspension Cable, Teflon Coated | FT | 1.00 | 1.00 | 1.00 | 1.00 | 1.00 |
| 33 23 2423 | Hand Reel | EACH | 9.89 | 9.89 | 9.89 | 9.89 | 9.89 |

**Environmental Remediation: Assemblies Cost Book**

# Cost Data for Remediation
## Neutralization

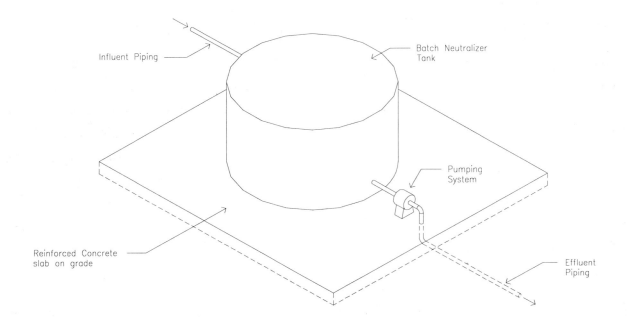

### General:
Neutralization can be used to treat any wastewater requiring pH control. Neutralization renders acidic (pH 0 - 6) or alkaline (pH 9 - 14) wastes noncorrosive by pH adjustment. The resulting residuals include insoluble salts, metal hydroxide sludge, and neutral effluent containing dissolved salts. A final pH of 6 to 9 is usually desired.

### Typical Treatment Train:
Treatment of wastewater prior to biological treatment, discharge to a body of water, disposal, carbon adsorption, air stripping, etc., sampling and analysis.

### Common Cost Components:
1. Structural slab
2. Batch neutralizer
3. Pumping system
4. Influent and effluent piping
5. Neutralizing chemicals for acidic wastes: calcium or
   sodium carbonate, caustic soda (sodium hydroxide), lime, limestone
6. Neutralizing chemicals for alkaline wastes: sulfuric acid, hydrochloric acid, nitric acid
7. Operations and maintenance

### Other Cost Considerations:
Preparation of titration curves to determine the nature of the ions causing the acidity or alkalinity.

**Environmental Remediation: Assemblies Cost Book**

# Cost Data for Remediation

## Neutralization

| Assembly | Description | Unit | Unit Cost by Safety Level A | B | C | D | E |
|---|---|---|---|---|---|---|---|
| | **Capital Costs** | | | | | | |
| 18 02 0322 | 8" Structural Slab on Grade | SF | 9.75 | 8.38 | 7.71 | 6.38 | 5.95 |
| 18 02 0324 | 12" Structural Slab on Grade | SF | 11.57 | 10.04 | 9.30 | 7.82 | 7.34 |
| 33 12 0601 | 10 Gallon Batch Neutralizer | EACH | 19,837 | 19,603 | 19,503 | 19,276 | 19,193 |
| 33 12 0602 | 25 Gallon Batch Neutralizer | EACH | 23,346 | 23,112 | 23,011 | 22,785 | 22,702 |
| 33 12 0603 | 50 Gallon Batch Neutralizer | EACH | 30,066 | 29,832 | 29,732 | 29,505 | 29,422 |
| 33 12 0604 | 100 Gallon Batch Neutralizer | EACH | 32,955 | 32,720 | 32,620 | 32,394 | 32,311 |
| 33 12 0605 | 200 Gallon Batch Neutralizer | EACH | 37,857 | 37,623 | 37,523 | 37,296 | 37,213 |
| 33 12 0606 | Pumping System | EACH | 2,027 | 2,027 | 2,027 | 2,027 | 2,027 |
| 33 26 0231 | 1" Stainless Steel Piping Including Fittings & Hangers | LF | 25.76 | 21.58 | 19.79 | 15.76 | 14.27 |
| 33 26 0232 | 2" Stainless Steel Piping Including Fittings & Hangers | LF | 42.76 | 36.38 | 33.64 | 27.47 | 25.21 |
| 33 26 0233 | 3" Stainless Steel Piping Including Fittings & Hangers | LF | 62.76 | 53.62 | 49.70 | 40.86 | 37.62 |
| 33 26 0234 | 4" Stainless Steel Piping Including Fittings & Hangers | LF | 123.07 | 106.59 | 99.49 | 83.56 | 77.75 |
| 33 26 0301 | 1 1/2" Polypropylene Pipe Including Fittings | LF | 7.56 | 6.11 | 5.48 | 4.08 | 3.57 |
| 33 26 0302 | 2" Polypropylene Pipe Including Fittings | LF | 8.79 | 7.16 | 6.46 | 4.88 | 4.30 |
| 33 26 0303 | 3" Polypropylene Pipe Including Fittings | LF | 13.80 | 11.43 | 10.42 | 8.13 | 7.29 |
| 33 26 0304 | 4" Polypropylene Pipe Including Fittings | LF | 18.36 | 15.28 | 13.96 | 10.99 | 9.90 |
| 33 26 0401 | 1 1/4" PVC Piping Including Fittings & Hangers | LF | 16.63 | 13.07 | 11.49 | 8.05 | 6.82 |
| 33 26 0402 | 2" PVC Piping Including Fittings & Hangers | LF | 17.85 | 14.06 | 12.38 | 8.72 | 7.42 |
| 33 26 0403 | 3" PVC Piping Including Fittings & Hangers | LF | 19.43 | 15.41 | 13.65 | 9.77 | 8.37 |
| 33 26 0404 | 4" PVC Piping Including Fittings & Hangers | LF | 26.66 | 21.15 | 18.74 | 13.41 | 11.49 |
| 33 33 0101 | Sodium Hydroxide Solution, 190 Lb Drummed Liquid | EACH | 62.34 | 57.81 | 55.07 | 50.64 | 49.53 |
| 33 33 0102 | Sodium Hydroxide Solution, 700 Lb Drummed Liquid | EACH | 163.04 | 158.51 | 155.77 | 151.34 | 150.23 |
| 33 33 0103 | Sodium Hydroxide, 500 Lb Container, Beads | EACH | 234.06 | 229.53 | 226.79 | 222.36 | 221.25 |
| 33 33 0104 | Sodium Hydroxide, 100 Lb Container, Flakes | EACH | 89.90 | 85.37 | 82.63 | 78.20 | 77.09 |
| 33 33 0105 | Sodium Hydroxide, 400 Lb Container, Flakes | EACH | 169.60 | 169.60 | 169.60 | 169.60 | 169.60 |
| 33 33 0106 | Magnesium Hydroxide Slurry, 55 Gallon Drums, < 40 Drums | EACH | 249.77 | 249.77 | 249.77 | 249.77 | 249.77 |
| 33 33 0107 | Magnesium Hydroxide Slurry, 55 Gallon Drums, 40 - 71 Drums | EACH | 234.26 | 234.26 | 234.26 | 234.26 | 234.26 |
| 33 33 0108 | Magnesium Hydroxide Slurry, 55 Gallon Drums, 72 Drums | EACH | 107.33 | 107.33 | 107.33 | 107.33 | 107.33 |
| 33 33 0109 | Magnesium Hydroxide, Bulk Quantity, 20 Tons | EACH | 4,452 | 4,452 | 4,452 | 4,452 | 4,452 |
| 33 33 0110 | Sulfuric Acid Solution, 220 Lb Drummed Liquid | EACH | 45.38 | 40.85 | 38.11 | 33.68 | 32.57 |
| 33 33 0111 | Sulfuric Acid Solution, 750 Lb Drummed Liquid | EACH | 93.61 | 89.08 | 86.34 | 81.91 | 80.80 |
| 33 33 0112 | Soda Ash, 50 Lb Bag, Powdered | EACH | 28.42 | 23.89 | 21.15 | 16.72 | 15.61 |

**Environmental Remediation: Assemblies Cost Book**

*Page 3-233*

# Cost Data for Remediation

## Neutralization

| Assembly | Description | Unit | Unit Cost by Safety Level A | B | C | D | E |
|---|---|---|---|---|---|---|---|
| | **Capital Costs** | | | | | | |
| 33 33 0113 | Soda Ash, 100 Lb Bag, Powdered | EACH | 41.14 | 36.61 | 33.87 | 29.44 | 28.33 |
| 33 33 0114 | Quicklime, 1/4" Nominal Granules, Bulk Quantity | TON | 96.49 | 96.49 | 96.48 | 96.48 | 96.47 |
| 33 33 0115 | Quicklime, 3/4" Nominal Granules, Bulk Quantity | TON | 96.49 | 96.49 | 96.48 | 96.48 | 96.47 |
| 33 33 0116 | Quicklime, Combination 1/4" & 3/4" Granules, Bulk Quantity | TON | 96.49 | 96.49 | 96.48 | 96.48 | 96.47 |
| 33 33 0117 | Hydrated Lime, Powdered, Bulk | TON | 97.55 | 97.55 | 97.54 | 97.54 | 97.53 |
| 33 33 0118 | Hydrated Lime, Powdered, 50 Lb Bag | EACH | 25.76 | 21.23 | 18.49 | 14.06 | 12.94 |
| 33 41 0101 | Pump & Motor Maintenance/Repair | EACH | 865.17 | 671.48 | 581.58 | 393.96 | 329.64 |
| 33 42 0101 | Electrical Charge | KWH | 0.06 | 0.06 | 0.06 | 0.06 | 0.06 |
| | **Operations & Maintenance** | | | | | | |
| 33 33 0101 | Sodium Hydroxide Solution, 190 Lb Drummed Liquid | EACH | 62.34 | 57.81 | 55.07 | 50.64 | 49.53 |
| 33 33 0102 | Sodium Hydroxide Solution, 700 Lb Drummed Liquid | EACH | 163.04 | 158.51 | 155.77 | 151.34 | 150.23 |
| 33 33 0103 | Sodium Hydroxide, 500 Lb Container, Beads | EACH | 234.06 | 229.53 | 226.79 | 222.36 | 221.25 |
| 33 33 0104 | Sodium Hydroxide, 100 Lb Container, Flakes | EACH | 89.90 | 85.37 | 82.63 | 78.20 | 77.09 |
| 33 33 0105 | Sodium Hydroxide, 400 Lb Container, Flakes | EACH | 169.60 | 169.60 | 169.60 | 169.60 | 169.60 |
| 33 33 0106 | Magnesium Hydroxide Slurry, 55 Gallon Drums, < 40 Drums | EACH | 249.77 | 249.77 | 249.77 | 249.77 | 249.77 |
| 33 33 0107 | Magnesium Hydroxide Slurry, 55 Gallon Drums, 40 - 71 Drums | EACH | 234.26 | 234.26 | 234.26 | 234.26 | 234.26 |
| 33 33 0108 | Magnesium Hydroxide Slurry, 55 Gallon Drums, 72 Drums | EACH | 107.33 | 107.33 | 107.33 | 107.33 | 107.33 |
| 33 33 0109 | Magnesium Hydroxide, Bulk Quantity, 20 Tons | EACH | 4,452 | 4,452 | 4,452 | 4,452 | 4,452 |
| 33 33 0110 | Sulfuric Acid Solution, 220 Lb Drummed Liquid | EACH | 45.38 | 40.85 | 38.11 | 33.68 | 32.57 |
| 33 33 0111 | Sulfuric Acid Solution, 750 Lb Drummed Liquid | EACH | 93.61 | 89.08 | 86.34 | 81.91 | 80.80 |
| 33 33 0112 | Soda Ash, 50 Lb Bag, Powdered | EACH | 28.42 | 23.89 | 21.15 | 16.72 | 15.61 |
| 33 33 0113 | Soda Ash, 100 Lb Bag, Powdered | EACH | 41.14 | 36.61 | 33.87 | 29.44 | 28.33 |
| 33 33 0114 | Quicklime, 1/4" Nominal Granules, Bulk Quantity | TON | 96.49 | 96.49 | 96.48 | 96.48 | 96.47 |
| 33 33 0115 | Quicklime, 3/4" Nominal Granules, Bulk Quantity | TON | 96.49 | 96.49 | 96.48 | 96.48 | 96.47 |
| 33 33 0116 | Quicklime, Combination 1/4" & 3/4" Granules, Bulk Quantity | TON | 96.49 | 96.49 | 96.48 | 96.48 | 96.47 |
| 33 33 0117 | Hydrated Lime, Powdered, Bulk | TON | 97.55 | 97.55 | 97.54 | 97.54 | 97.53 |
| 33 33 0118 | Hydrated Lime, Powdered, 50 Lb Bag | EACH | 25.76 | 21.23 | 18.49 | 14.06 | 12.94 |
| 33 41 0101 | Pump & Motor Maintenance/Repair | EACH | 865.17 | 671.48 | 581.58 | 393.96 | 329.64 |
| 33 42 0101 | Electrical Charge | KWH | 0.06 | 0.06 | 0.06 | 0.06 | 0.06 |

**Environmental Remediation: Assemblies Cost Book**

# Cost Data for Remediation
## Oil/Water Separation

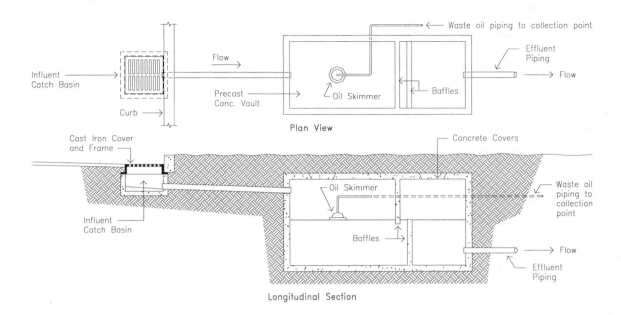

### General:
There are two primary applications of oil/water separation: stormwater runoff in areas such as vehicle or aircraft maintenance facilities and separation of free product from extracted groundwater.

### Typical Treatment Train:
Extraction wells, pipelines, surface runoff collection, treatment and disposal of the extracted or collected water, recycling or disposal of separated oil, sampling and analysis.

### Common Cost Components:
1. Oil/water separator
2. Product storage tank
3. Pump-out unit
4. Influent and effluent piping
5. Product discharge piping
6. Installation
7. Operations and maintenance

### Other Cost Considerations:
Subsurface piping.

# Cost Data for Remediation

## Oil/Water Separation

| Assembly | Description | Unit | Unit Cost by Safety Level ||||| 
|---|---|---|---|---|---|---|---|
| | | | A | B | C | D | E |
| | **Capital Costs** | | | | | | |
| 17 03 0259 | Cat 225, 1.5 CY, Soil/Sand, Trenching | CY | 1.84 | 1.49 | 1.23 | 0.88 | 0.82 |
| 17 03 0282 | Soil, 5 Miles, Dump Truck, Load/Haul Spoil From Trench | CY | 6.97 | 5.60 | 4.66 | 3.33 | 3.06 |
| 17 03 0283 | Sand, 5 Mile, Dump Truck, Load/Haul Spoil From Trench | CY | 6.28 | 5.05 | 4.20 | 3.00 | 2.76 |
| 17 03 0307 | Cat 235, 2.0 CY, Rock, No Haul or Borrow, Trenching | BCY | 133.78 | 109.18 | 95.00 | 71.01 | 64.55 |
| 17 03 0317 | Rock, 5 Miles, Dump Truck, Load/Haul Spoil from Trench | CY | 10.31 | 8.31 | 6.90 | 4.93 | 4.56 |
| 17 03 0401 | 950, 3.00 CY, Backfill with Excavated Material | CY | 2.54 | 2.05 | 1.70 | 1.22 | 1.14 |
| 17 03 0405 | 950, 3.00 CY, Delivered & Dumped, Backfill with Sand | CY | 38.31 | 35.03 | 32.95 | 29.74 | 29.00 |
| 17 03 0420 | Backfill Trench, Borrow Material, Delivered & Dumped Only | CY | 9.86 | 8.74 | 7.93 | 6.83 | 6.63 |
| 17 03 0514 | Compact Soil by Machine with Roller | CY | 1.18 | 0.93 | 0.79 | 0.55 | 0.47 |
| 17 03 0515 | Compact With Pogosticks | CY | 5.91 | 4.58 | 3.98 | 2.68 | 2.23 |
| 17 03 1001 | Wellpoint for Trench, Install & Remove <500', 1 Month | LFHD | 55.06 | 55.06 | 55.06 | 55.06 | 55.06 |
| 17 03 1002 | 2" Diameter Contractor's Trash Pump, 75 GPM | DAY | 50.20 | 47.01 | 45.64 | 42.55 | 41.42 |
| 18 02 0321 | 6" Structural Slab on Grade | SF | 6.98 | 5.95 | 5.47 | 4.48 | 4.14 |
| 18 02 0324 | 12" Structural Slab on Grade | SF | 11.57 | 10.04 | 9.30 | 7.82 | 7.34 |
| 19 01 0202 | 1", Class 200, PVC Piping | LF | 7.48 | 5.84 | 5.07 | 3.48 | 2.93 |
| 19 01 0204 | 2", Class 200, PVC Piping | LF | 8.86 | 6.96 | 6.07 | 4.23 | 3.60 |
| 19 01 0206 | 3", Class 200, PVC Piping | LF | 11.81 | 9.34 | 8.20 | 5.80 | 4.98 |
| 19 01 0207 | 4", Class 200, PVC Piping | LF | 12.33 | 9.86 | 8.72 | 6.32 | 5.50 |
| 19 01 0208 | 6", Class 200, PVC Piping | LF | 14.95 | 12.20 | 10.93 | 8.27 | 7.36 |
| 19 01 0212 | 8", Class 150, PVC Piping | LF | 18.36 | 15.27 | 13.84 | 10.85 | 9.82 |
| 19 02 0101 | 4", Class 50, Bell & Spigot Sanitary Sewer, Cast-iron Pipe | LF | 8.49 | 7.49 | 7.05 | 6.08 | 5.72 |
| 19 02 0103 | 6", Class 50, Bell & Spigot Sanitary Sewer, Cast-iron Pipe | LF | 13.25 | 11.79 | 11.16 | 9.75 | 9.23 |
| 19 02 0104 | 8", Class 50, Bell & Spigot Sanitary Sewer, Cast-iron Pipe | LF | 18.98 | 17.26 | 16.52 | 14.86 | 14.25 |
| 19 02 0105 | 10", Class 50, Bell & Spigot Sanitary Sewer, Cast-iron Pipe | LF | 27.66 | 25.51 | 24.59 | 22.51 | 21.75 |
| 19 02 0106 | 12", Class 50, Bell & Spigot Sanitary Sewer, Cast-iron Pipe | LF | 36.36 | 34.21 | 33.29 | 31.21 | 30.45 |
| 19 02 0107 | 15", Class 50, Bell & Spigot Sanitary Sewer, Cast-iron Pipe | LF | 51.76 | 49.01 | 47.83 | 45.17 | 44.19 |
| 19 02 0201 | Precast, CIP Base, 4' Diameter, 6' Deep, Manhole | EACH | 1,459 | 1,276 | 1,179 | 1,001 | 947.36 |
| 19 02 0202 | Precast, CIP Base, 4' Diameter, 8' Deep, Manhole | EACH | 1,957 | 1,697 | 1,557 | 1,304 | 1,229 |
| 19 02 0203 | Precast, CIP Base, 4' Diameter, 12' Deep, Manhole | EACH | 2,835 | 2,459 | 2,257 | 1,891 | 1,783 |
| 19 04 0411 | Packaged Coalescing 20 GPM Oil/Water Separator | EACH | 10,594 | 9,898 | 9,584 | 8,911 | 8,675 |
| 19 04 0412 | Packaged Coalescing 50 GPM Oil/Water Separator | EACH | 13,803 | 13,087 | 12,764 | 12,070 | 11,826 |
| 19 04 0413 | Packaged Coalescing 100 GPM Oil/Water Separator | EACH | 15,683 | 14,924 | 14,583 | 13,848 | 13,589 |
| 19 04 0414 | Packaged Coalescing 200 GPM Oil/Water Separator | EACH | 21,242 | 20,152 | 19,663 | 18,608 | 18,235 |

**Environmental Remediation: Assemblies Cost Book**

# Cost Data for Remediation

## Oil/Water Separation

| Assembly | Description | Unit | Unit Cost by Safety Level | | | | |
|---|---|---|---|---|---|---|---|
| | | | A | B | C | D | E |
| | **Capital Costs** | | | | | | |
| 19 04 0415 | Packaged Coalescing 300 GPM Oil/Water Separator | EACH | 24,229 | 23,052 | 22,526 | 21,387 | 20,983 |
| 19 04 0416 | Packaged Coalescing 400 GPM Oil/Water Separator | EACH | 29,635 | 27,949 | 27,195 | 25,563 | 24,985 |
| 19 04 0417 | Packaged Coalescing 500 GPM Oil/Water Separator | EACH | 39,821 | 38,227 | 37,518 | 35,976 | 35,428 |
| 19 04 0601 | 275 Gallon Steel Sump, Aboveground with Supports & Fittings | EACH | 1,325 | 1,153 | 1,080 | 913.28 | 852.29 |
| 19 04 0602 | 550 Gallon Steel Sump, Aboveground with Supports & Fittings | EACH | 1,598 | 1,390 | 1,301 | 1,099 | 1,026 |
| 19 04 0603 | 1,000 Gallon Steel Sump, Aboveground with Supports & Fittings | EACH | 2,118 | 1,844 | 1,727 | 1,463 | 1,366 |
| 19 04 0604 | 1,500 Gallon Steel Sump, Aboveground with Supports & Fittings | EACH | 2,697 | 2,356 | 2,209 | 1,879 | 1,758 |
| 19 04 0605 | 2,000 Gallon Steel Sump, Aboveground with Supports & Fittings | EACH | 3,457 | 2,997 | 2,777 | 2,332 | 2,183 |
| 19 04 0606 | 5,000 Gallon Steel Sump, Aboveground with Supports & Fittings | EACH | 6,079 | 5,494 | 5,214 | 4,646 | 4,457 |
| 19 07 0120 | 1 1/4", 60 PSI, Polyethylene Pipe | LF | 5.29 | 4.21 | 3.72 | 2.68 | 2.32 |
| 19 07 0121 | 1 1/2", 60 PSI, Polyethylene Pipe | LF | 5.38 | 4.30 | 3.81 | 2.77 | 2.41 |
| 19 07 0122 | 2", 60 PSI, Polyethylene Pipe | LF | 6.05 | 4.85 | 4.31 | 3.15 | 2.75 |
| 19 07 0123 | 3" Polyethylene 40' Joints, 60 PSI | LF | 7.90 | 6.47 | 5.82 | 4.42 | 3.94 |
| 19 07 0124 | 4" Polyethylene 40' Joints, 60 PSI | LF | 19.10 | 15.75 | 13.90 | 10.64 | 9.71 |
| 19 07 0125 | 6" Polyethylene 40' Joints, 60 PSI | LF | 25.85 | 22.22 | 20.22 | 16.68 | 15.67 |
| 19 07 0126 | 8" Polyethylene 40' Joints, 60 PSI | LF | 35.37 | 31.01 | 28.61 | 24.37 | 23.16 |
| 33 13 1201 | Install Underground 20 GPM Oil/Water Separator | EACH | 1,585 | 1,342 | 1,220 | 983.54 | 908.14 |
| 33 13 1202 | Install Underground 50 GPM Oil/Water Separator | EACH | 1,663 | 1,413 | 1,288 | 1,046 | 969.42 |
| 33 13 1203 | Install Underground 100 GPM Oil/Water Separator | EACH | 1,825 | 1,548 | 1,406 | 1,137 | 1,053 |
| 33 13 1204 | Install Underground 200 GPM Oil/Water Separator | EACH | 2,494 | 2,106 | 1,901 | 1,523 | 1,409 |
| 33 13 1205 | Install Underground 300 GPM Oil/Water Separator | EACH | 2,846 | 2,395 | 2,155 | 1,716 | 1,585 |
| 33 13 1206 | Install Underground 400 GPM Oil/Water Separator | EACH | 3,146 | 2,643 | 2,373 | 1,884 | 1,739 |
| 33 13 1207 | Install Underground 500 GPM Oil/Water Separator | EACH | 3,594 | 3,025 | 2,723 | 2,169 | 2,003 |
| 33 13 1211 | 20 GPM Oil Pump-out Unit with Controls | EACH | 6,533 | 6,166 | 6,000 | 5,646 | 5,522 |
| 33 13 1212 | 50 GPM Oil Pump-out Unit with Controls | EACH | 6,533 | 6,166 | 6,000 | 5,646 | 5,522 |
| 33 13 1213 | 100 GPM Oil Pump-out Unit with Controls | EACH | 6,533 | 6,166 | 6,000 | 5,646 | 5,522 |
| 33 13 1214 | 200 GPM Oil Pump-out Unit with Controls | EACH | 8,773 | 8,289 | 8,070 | 7,601 | 7,437 |
| 33 13 1215 | 300 GPM Oil Pump-out Unit with Controls | EACH | 8,773 | 8,289 | 8,070 | 7,601 | 7,437 |
| 33 13 1216 | 400 GPM Oil Pump-out Unit with Controls | EACH | 9,513 | 8,862 | 8,567 | 7,937 | 7,717 |
| 33 13 1217 | 500 GPM Oil Pump-out Unit with Controls | EACH | 9,232 | 8,580 | 8,285 | 7,654 | 7,433 |

**Environmental Remediation: Assemblies Cost Book**

# Cost Data for Remediation

## Oil/Water Separation

| Assembly | Description | Unit | Unit Cost by Safety Level A | B | C | D | E |
|---|---|---|---|---|---|---|---|
| | **Capital Costs** | | | | | | |
| 33 19 7302 | Disposal of Nonhazardous Liquid Bulk Waste | GAL | 1.75 | 1.75 | 1.75 | 1.75 | 1.75 |
| 33 19 7303 | Disposal of Hazardous Liquid Bulk Waste | GAL | 2.96 | 2.96 | 2.96 | 2.96 | 2.96 |
| 33 19 9921 | DOT Steel Drum, 55 Gallon | EACH | 65.19 | 65.19 | 65.19 | 65.19 | 65.19 |
| 33 26 0101 | 1" Carbon Steel Piping | LF | 6.46 | 5.28 | 4.77 | 3.63 | 3.21 |
| 33 26 0102 | 2" Carbon Steel Piping | LF | 10.50 | 8.72 | 7.95 | 6.23 | 5.60 |
| 33 26 0103 | 3" Carbon Steel Piping | LF | 18.02 | 15.20 | 13.99 | 11.26 | 10.26 |
| 33 26 0104 | 4" Carbon Steel Piping | LF | 23.20 | 19.86 | 18.43 | 15.21 | 14.02 |
| 33 26 0105 | 6" Carbon Steel Piping | LF | 36.37 | 30.24 | 27.46 | 21.52 | 19.44 |
| 33 26 0106 | 8" Carbon Steel Piping | LF | 40.77 | 34.08 | 31.15 | 24.68 | 22.36 |
| 33 26 0107 | 10" Carbon Steel Piping | LF | 54.40 | 46.05 | 42.38 | 34.30 | 31.39 |
| 33 26 0201 | 1" Stainless Steel Piping, Schedule 40, Threaded | LF | 18.67 | 16.12 | 15.03 | 12.56 | 11.65 |
| 33 26 0202 | 2" Stainless Steel Piping, Schedule 40, Threaded | LF | 33.47 | 29.21 | 27.39 | 23.28 | 21.77 |
| 33 26 0203 | 3" Stainless Steel Piping, Schedule 40, Threaded | LF | 57.00 | 50.61 | 47.88 | 41.71 | 39.45 |
| 33 26 0204 | 4" Stainless Steel Piping, Schedule 40, Threaded | LF | 80.15 | 71.72 | 68.12 | 59.98 | 57.00 |
| 33 26 0205 | 6" Stainless Steel Piping, Schedule 10, Type 316 | LF | 78.58 | 70.01 | 66.14 | 57.84 | 54.94 |
| 33 26 0206 | 8" Stainless Steel Piping, Schedule 40, Welded | LF | 197.22 | 183.32 | 176.89 | 163.42 | 158.80 |
| 33 26 0601 | (1 1/2", 3") Stainless Steel Double-wall Piping, with Fittings | LF | 134.61 | 117.05 | 109.53 | 92.56 | 86.33 |
| 33 26 0602 | (2", 4") Stainless Steel Double-wall Piping, with Fittings | LF | 185.98 | 163.46 | 153.81 | 132.04 | 124.05 |
| 33 26 0603 | (2 1/2", 4") Stainless Steel Double-wall Piping, with Fittings | LF | 206.86 | 182.17 | 171.60 | 147.74 | 138.99 |
| 33 26 0604 | (4", 6") Stainless Steel Double-wall Piping, with Fittings | LF | 339.64 | 280.79 | 254.18 | 197.21 | 177.23 |
| 33 26 0621 | (1 1/2", 3") PVC Double-wall Piping, with Fittings | LF | 33.15 | 27.05 | 24.43 | 18.53 | 16.36 |
| 33 26 0622 | (2", 4") PVC Double-wall Piping, with Fittings | LF | 46.22 | 37.98 | 34.43 | 26.46 | 23.55 |
| 33 26 0623 | (2 1/2", 4") PVC Double-wall Piping, with Fittings | LF | 49.22 | 40.39 | 36.61 | 28.07 | 24.94 |
| 33 26 0624 | (4", 6") PVC Double-wall Piping, with Fittings | LF | 70.61 | 58.00 | 52.59 | 40.40 | 35.93 |
| 33 26 0641 | (1 1/2", 3") Carbon Steel Double-wall Piping, with Fittings | LF | 48.16 | 40.64 | 37.42 | 30.15 | 27.48 |
| 33 26 0642 | (2", 4") Carbon Steel Double-wall Piping, with Fittings | LF | 64.71 | 55.69 | 51.83 | 43.11 | 39.91 |
| 33 26 0643 | (2 1/2", 4") Carbon Steel Double-wall Piping, with Fittings | LF | 70.81 | 61.02 | 56.83 | 47.37 | 43.90 |
| 33 26 0644 | (4", 6") Carbon Steel Double-wall Piping, with Fittings | LF | 114.39 | 100.33 | 94.32 | 80.73 | 75.75 |
| 33 41 0101 | Pump & Motor Maintenance/Repair | EACH | 865.17 | 671.48 | 581.58 | 393.96 | 329.64 |
| 33 42 0110 | Electrical Charge for Oil Pump-Out Pump Operation | KWH | 0.06 | 0.06 | 0.06 | 0.06 | 0.06 |
| | **Operations & Maintenance** | | | | | | |
| 33 19 7302 | Disposal of Nonhazardous Liquid Bulk Waste | GAL | 1.75 | 1.75 | 1.75 | 1.75 | 1.75 |
| 33 19 7303 | Disposal of Hazardous Liquid Bulk Waste | GAL | 2.96 | 2.96 | 2.96 | 2.96 | 2.96 |

**Environmental Remediation: Assemblies Cost Book**

*Page 3-238*

1998 by ECHOS. All rights reserved

# Cost Data for Remediation

## Oil/Water Separation

| Assembly | Description | Unit | Unit Cost by Safety Level | | | | |
|---|---|---|---|---|---|---|---|
| | | | A | B | C | D | E |
| | **Operations & Maintenance** | | | | | | |
| 33 41 0101 | Pump & Motor Maintenance/Repair | EACH | 865.17 | 671.48 | 581.58 | 393.96 | 329.64 |
| 33 42 0110 | Electrical Charge for Oil Pump-Out Pump Operation | KWH | 0.06 | 0.06 | 0.06 | 0.06 | 0.06 |

# Cost Data for Remediation
## Ordnance & Explosive Waste Remediation

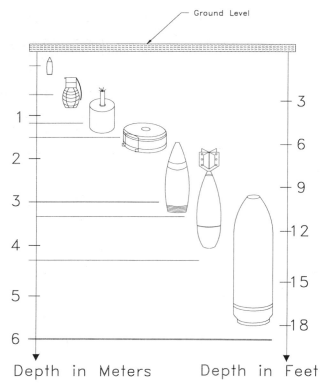

### General:

*This technology provides for the cost of searching for, marking, and removing explosive ordnance from munitions contaminated property. The definition of ordnance and explosive waste is any chemical substance or physical item related to munitions that is designed to cause damage to personnel or material through explosive force, incendiary action or toxic effects. Included in this definition is the residue remaining after the manufacture, destruction or demilling of munitions devices.*

### Typical Treatment Train:
*Incineration, bioremediation, enhanced ultraviolet oxidation, filtration, sampling and analysis*

### Common Cost Components:

1. Operational labor for EOD and Master EOD technicians
2. Magnetometer
3. GPS surveying equipment
4. Ordnance recovery equipment and vehicles
5. Hoist (for heavy bombs or artillery)
6. EOD personnel safety and convenience items

### Other Cost Considerations:
*Military costs associated with ordnance disposal (ordnance pickup costs, explosives costs, etc.); chemical weapons material (CWM) costs; decontamination facilities; loading and hauling; clearing and grubbing; contractor overhead and profit; general conditions (communications and video equipment); carbon adsorption (liquid); fencing; field sampling/mobile laboratory.*

**Environmental Remediation: Assemblies Cost Book**

# Cost Data for Remediation

## Ordnance & Explosive Waste Remediation

| Assembly | Description | Unit | Unit Cost by Safety Level A | B | C | D | E |
|---|---|---|---|---|---|---|---|
| | **Capital Costs** | | | | | | |
| 17 01 0101 | Light Brush without Grub, Clearing | ACRE | 139.11 | 110.49 | 93.27 | 65.31 | 58.24 |
| 17 01 0102 | Medium Brush without Grub, Clearing | ACRE | 330.00 | 264.79 | 220.99 | 157.03 | 143.77 |
| 17 01 0103 | Medium Brush with Average Grub & Some Trees, Clearing | ACRE | 1,219 | 986.58 | 815.67 | 586.64 | 548.39 |
| 17 01 0105 | Heavy Brush with Average Grub, Medium - Heavy Trees, Clearing | ACRE | 1,738 | 1,408 | 1,162 | 837.39 | 785.18 |
| 17 01 0205 | Brush Removal Equipment | EACH | 95.00 | 95.00 | 95.00 | 95.00 | 95.00 |
| 17 03 0310 | D8 with U-blade & Single-shank Ripper, Bulldozer | HOUR | 396.60 | 325.85 | 264.85 | 194.67 | 188.62 |
| 17 03 0406 | 966, 4.00 CY, Delivered & Dumped, Backfill with Sand | CY | 37.79 | 34.61 | 32.60 | 29.49 | 28.76 |
| 17 03 0427 | Sand Bags | EACH | 0.38 | 0.38 | 0.38 | 0.38 | 0.38 |
| 17 03 0431 | 580K, 1.0 CY, Backhoe with Front-end Loader | HOUR | 127.72 | 101.78 | 85.60 | 60.23 | 54.18 |
| 17 03 0432 | Bobcat with Backhoe | MONTH | 1,500 | 1,500 | 1,500 | 1,500 | 1,500 |
| 33 01 0102 | Van or Pickup Rental | DAY | 32.69 | 32.69 | 32.69 | 32.69 | 32.69 |
| 33 01 0103 | Flatbed Truck Rental | DAY | 45.94 | 45.94 | 45.94 | 45.94 | 45.94 |
| 33 01 0108 | Sedan, Automobile, Rental | DAY | 40.87 | 40.87 | 40.87 | 40.87 | 40.87 |
| 33 01 0109 | Truck, 2 Axle, Highway, 21,700 GVW, 4 x 2, 2 Axle | DAY | 110.00 | 110.00 | 110.00 | 110.00 | 110.00 |
| 33 01 0111 | Truck, 2 Axle, Highway, 33,000 GVW, 6 x 2, 2 Axle | DAY | 195.00 | 195.00 | 195.00 | 195.00 | 195.00 |
| 33 01 0407 | Basic Level "A" Encapsulation Suit | EACH | 500.13 | 500.13 | 500.13 | 500.13 | 500.13 |
| 33 01 0412 | Basic Level "B" Suit | EACH | 162.77 | 162.77 | 162.77 | 162.77 | 162.77 |
| 33 01 0463 | EOD Team Work Clothing Ensemble | EACH | 93.79 | 93.79 | 93.79 | 93.79 | 93.79 |
| 33 01 0464 | EOD Personal Safety Equipment | EACH | 19.46 | 19.46 | 19.46 | 19.46 | 19.46 |
| 33 01 0465 | Supervisor's Hazardous Training Course, 8-hours | EACH | 150.00 | 150.00 | 150.00 | 150.00 | 150.00 |
| 33 01 0466 | Hazardous Training, per Worker, 40-hours | EACH | 371.67 | 371.67 | 371.67 | 371.67 | 371.67 |
| 33 01 0467 | Emergency Medical Kit | EACH | 79.18 | 79.18 | 79.18 | 79.18 | 79.18 |
| 33 01 0503 | 5.0 KW, 120/240 VAC, Generator, Daily Rental | EACH | 11.30 | 11.30 | 11.30 | 11.30 | 11.30 |
| 33 01 0505 | 10 KW, 120/240 VAC, Generator, Daily Rental | EACH | 13.09 | 13.09 | 13.09 | 13.09 | 13.09 |
| 33 01 0508 | Water Pump, 2" 8,000 GPH, Portable, Gasoline Powered | EACH | 1,391 | 1,391 | 1,391 | 1,391 | 1,391 |
| 33 01 0509 | Portable Water Pump, 2", 10,000 GPH, Gas Powered, with Wheels | EACH | 1,391 | 1,391 | 1,391 | 1,391 | 1,391 |
| 33 02 0341 | Alpha, Beta, Gamma Detector Equipment | EACH | 194.33 | 194.33 | 194.33 | 194.33 | 194.33 |
| 33 04 0101 | UXO Disposal Technician | HR | 59.81 | 46.10 | 40.24 | 26.99 | 22.13 |
| 33 04 0102 | UXO Supervisor | HR | 79.54 | 61.31 | 53.51 | 35.89 | 29.43 |
| 33 04 0103 | UXO Site Setup | HR | 139.35 | 107.42 | 93.75 | 62.88 | 51.56 |
| 33 04 0104 | Laborer | HR | 55.81 | 43.02 | 37.55 | 25.18 | 20.65 |
| 33 04 0111 | Surface Towed Ordnance Locator, Minimum Charge | EA | 150,695 | 124,666 | 100,683 | 74,773 | 73,504 |

## Environmental Remediation: Assemblies Cost Book

# Cost Data for Remediation

## Ordnance & Explosive Waste Remediation

| Assembly | Description | Unit | A | B | C | D | E |
|---|---|---|---|---|---|---|---|
| | **Capital Costs** | | | | | | |
| 33 04 0112 | Surface Towed Ordnance Locator, Per Each Acre | AC | 6,281 | 5,197 | 4,197 | 3,118 | 3,065 |
| 33 04 0113 | Ordnance Locator, Foerster Ferex K (M26) | WK | 122.46 | 122.46 | 122.46 | 122.46 | 122.46 |
| 33 04 0114 | Ordnance Locator, Foerster Metex | EA | 5,808 | 5,808 | 5,808 | 5,808 | 5,808 |
| 33 04 0115 | Ordnance Locator, Foerster Minex 2 FD | EA | 7,800 | 7,800 | 7,800 | 7,800 | 7,800 |
| 33 04 0116 | Underwater Ordnance Locator, Foerster Model W | EA | 14,030 | 14,030 | 14,030 | 14,030 | 14,030 |
| 33 04 0117 | Ordnance Locator, Schoenstedt, Model GA-72CV | EA | 884.00 | 884.00 | 884.00 | 884.00 | 884.00 |
| 33 04 0118 | Ordnance Locator, Schoenstedt, Model GA-52CV | EA | 847.60 | 847.60 | 847.60 | 847.60 | 847.60 |
| 33 04 0119 | Ordnance Locator, Schoenstedt, MG-220 w/Downhole Attach | EA | 6,240 | 6,240 | 6,240 | 6,240 | 6,240 |
| 33 04 0120 | Portable Cesium Magnetometer, Geometrics, Model G-858 | WK | 109.57 | 109.57 | 109.57 | 109.57 | 109.57 |
| 33 04 0121 | D-Cell Batteries, Long Life Disposable | EA | 0.38 | 0.38 | 0.38 | 0.38 | 0.38 |
| 33 04 0131 | ICAD, Miniature Chemical Agent Detector | EA | 3,767 | 3,767 | 3,767 | 3,767 | 3,767 |
| 33 04 0132 | (MINICAM) Miniature Chemical Agent Monitor | EA | 31,200 | 31,200 | 31,200 | 31,200 | 31,200 |
| 33 04 0140 | 4' X 4' X 4' Transportable Explosives Locker | EA | 2,371 | 2,371 | 2,371 | 2,371 | 2,371 |
| 33 04 0141 | Explosives Storage Locker/Shelter, 16 x 9 x 8 ft. | EA | 16,640 | 16,640 | 16,640 | 16,640 | 16,640 |
| 33 04 0142 | Rope, 5/8", Polyester, Kevlar Center | FT | 2.00 | 2.00 | 2.00 | 2.00 | 2.00 |
| 33 04 0143 | Wood Box, Ordnance Transport Container | EA | 75.16 | 65.10 | 60.73 | 51.00 | 47.48 |
| 33 04 0144 | UXO Field Support Tools and Supplies, Common | EA | 223.50 | 223.50 | 223.50 | 223.50 | 223.50 |
| 33 04 0145 | Crane,22Tn Slf Cntnd,Hyd,Rough Trrn,4WD,for Ordnnc Rmvl | HR | 326.84 | 260.74 | 219.02 | 154.34 | 139.23 |
| 33 04 0146 | Hoist, Heavy Duty, 5000 lb. Lift | MO | 177.36 | 177.36 | 177.36 | 177.36 | 177.36 |
| 33 04 0147 | UXO Tool Kit | EA | 319.28 | 319.28 | 319.28 | 319.28 | 319.28 |
| 33 04 0148 | Alloy Steel Chain, 3/8", Cut Length 100 ft. | EA | 315.44 | 315.44 | 315.44 | 315.44 | 315.44 |
| 33 04 0149 | Nonsparking UXO Shovels | EA | 83.01 | 83.01 | 83.01 | 83.01 | 83.01 |
| 33 04 0150 | Standby Ambulance with EMT and Paramedic | HR | 277.12 | 234.54 | 212.04 | 170.63 | 158.18 |
| 33 04 0152 | Firing Wire | LF | 0.04 | 0.04 | 0.04 | 0.04 | 0.04 |
| 33 04 0153 | Reel for Firing Wire | EA | 31.41 | 31.41 | 31.41 | 31.41 | 31.41 |
| 33 04 0154 | Elec Blstng Set: Blasting Mach, Galvanometer, Test Set | EA | 716.48 | 716.48 | 716.48 | 716.48 | 716.48 |
| 33 04 0155 | Galvanometer battery | EA | 45.05 | 45.05 | 45.05 | 45.05 | 45.05 |
| 33 04 0156 | Electrical Blasting Caps | EA | 241.28 | 241.28 | 241.28 | 241.28 | 241.28 |
| 33 04 0157 | Non-electrical Blasting Caps | EA | 0.51 | 0.51 | 0.51 | 0.51 | 0.51 |
| 33 04 0158 | Igniter | EA | 3.20 | 3.20 | 3.20 | 3.20 | 3.20 |
| 33 04 0159 | Timed fuse | LF | 0.14 | 0.14 | 0.14 | 0.14 | 0.14 |
| 33 04 0160 | C4 Explosive | LB | 12.88 | 12.88 | 12.88 | 12.88 | 12.88 |

**Environmental Remediation: Assemblies Cost Book**

# Cost Data for Remediation

## Ordnance & Explosive Waste Remediation

| Assembly | Description | Unit | Unit Cost by Safety Level A | B | C | D | E |
|---|---|---|---|---|---|---|---|
| | **Capital Costs** | | | | | | |
| 33 04 0161 | Detonation Cord | LF | 0.14 | 0.14 | 0.14 | 0.14 | 0.14 |
| 33 04 0162 | Lightning Detector | EA | 1,171 | 1,171 | 1,171 | 1,171 | 1,171 |
| 33 04 0163 | Storm Detector | EA | 146.29 | 146.29 | 146.29 | 146.29 | 146.29 |
| 33 04 0164 | Polypropylene Shovel | EA | 42.44 | 42.44 | 42.44 | 42.44 | 42.44 |
| 33 10 0118 | R60 Rough Terrain Forklift, 6,000 Lb @ 24" LC | HOUR | 114.98 | 91.38 | 77.08 | 54.02 | 48.25 |
| 33 23 0811 | Explosionproof Electrical Receptacle for Product Recovery Pump | EACH | 497.01 | 468.06 | 455.66 | 427.68 | 417.42 |
| 99 04 1202 | Surveying - 3-man Crew | DAY | 1,922 | 1,505 | 1,291 | 885.68 | 760.00 |
| 99 04 1204 | Hand Held GPS Unit, Battery Powered | EACH | 475.00 | 475.00 | 475.00 | 475.00 | 475.00 |
| 99 04 1205 | Portable GPS Set with Mapping, 5 cm Accuracy | MONTH | 550.00 | 550.00 | 550.00 | 550.00 | 550.00 |
| 99 04 1206 | Data Automation System (Computer) | MONTH | 200.00 | 200.00 | 200.00 | 200.00 | 200.00 |

**Environmental Remediation: Assemblies Cost Book**

# Cost Data for Remediation
## Passive Water Treatment

### General:
*Passive water treatment has applications in the remediation of storm water runoff containing organic constituents, metallic ions, and acidic mine drainage contaminated with heavy metals. Compared to active water treatment methods such as chemical precipitation and neutralization, passive treatment methods generally have lower operations and maintenance (O&M) costs, but require more land area. This section of the report discusses five technologies for treating contaminated surface water and the criteria for selecting one; storm water filtration, runoff detention ponds, anoxic limestone drains, anaerobic compost wetlands, and aerobic wetlands.*

### Typical Treatment Train:
*N/A*

### Common Cost Components:
1. Construct Detention Pond
2. Grading
3. Storm Water Collection Systems
4. Composting Vault

### Other Cost Considerations:
*Land Acquisition, Storm Drainage Piping, Backup Treatment, Sludge Removal, Sludge Hauling & Disposal, Treatment of Compost, Groundwater Monitoring, Sampling and Analysis*

# Cost Data for Remediation

## Passive Water Treatment

| Assembly | Description | Unit | A | B | C | D | E |
|---|---|---|---|---|---|---|---|
| | **Capital Costs** | | | | | | |
| 17 01 0705 | 5 Miles, Dump Truck, Load & Haul Debris | CY | 5.83 | 4.71 | 3.90 | 2.80 | 2.60 |
| 17 01 0706 | 5 Miles, Semi Dump, Load & Haul Debris | CY | 6.67 | 5.45 | 4.46 | 3.25 | 3.11 |
| 17 03 0206 | D3 with A-blade Bulldozer | HOUR | 151.96 | 122.19 | 101.74 | 72.51 | 66.74 |
| 17 03 0207 | D4 with A-blade Bulldozer | HOUR | 153.56 | 123.53 | 102.80 | 73.31 | 67.54 |
| 17 03 0208 | D5 with A-blade Bulldozer | HOUR | 181.44 | 146.76 | 121.39 | 87.25 | 81.48 |
| 17 03 0209 | D6 with A-blade Bulldozer | HOUR | 208.40 | 169.01 | 139.38 | 100.57 | 94.52 |
| 17 03 0210 | D7 with U-blade Bulldozer | HOUR | 296.12 | 242.11 | 197.86 | 144.43 | 138.38 |
| 17 03 0230 | Crawler-mounted, 1.0 CY, 215 Hydraulic Excavator | HOUR | 210.40 | 170.50 | 140.73 | 101.43 | 95.15 |
| 17 03 0231 | Crawler-mounted, 1.25 CY, 225 Hydraulic Excavator | HOUR | 210.40 | 170.50 | 140.73 | 101.43 | 95.15 |
| 17 03 0232 | Crawler-mounted, 2.0 CY, 235 Hydraulic Excavator | HOUR | 330.90 | 270.92 | 221.07 | 161.68 | 155.40 |
| 17 03 0233 | Crawler-mounted, 3.125 CY, 245 Hydraulic Excavator | HOUR | 363.36 | 297.97 | 242.71 | 177.91 | 171.63 |
| 17 03 0234 | Crawler-mounted, 4.0 CY, Koehring 1166 Hydraulic Excavator | HOUR | 363.36 | 297.97 | 242.71 | 177.91 | 171.63 |
| 17 03 0235 | Crawler-mounted, 5.5 CY, Koehring 1266, Hydraulic Excavator | HOUR | 491.70 | 404.92 | 328.27 | 242.08 | 235.80 |
| 17 03 0310 | D8 with U-blade & Single-shank Ripper, Bulldozer | HOUR | 396.60 | 325.85 | 264.85 | 194.67 | 188.62 |
| 17 03 0311 | D9 with U-blade & Single-shank Ripper, Bulldozer | HOUR | 455.94 | 375.30 | 304.41 | 224.34 | 218.29 |
| 17 03 0312 | D10 with U-blade & Single-shank Ripper, Bulldozer | HOUR | 476.58 | 392.50 | 318.17 | 234.66 | 228.61 |
| 17 03 0419 | Crushed Stone, 1/2" to 3/4" | CY | 23.45 | 21.67 | 20.70 | 18.97 | 18.47 |
| 17 03 0425 | Sand, 6" Lifts, On-Site | CY | 13.88 | 11.35 | 9.36 | 6.87 | 6.54 |
| 17 03 0426 | Sand, 6" Lifts, Off-Site | CY | 15.55 | 14.34 | 13.42 | 12.22 | 12.04 |
| 18 02 0301 | Asphalt Pavement - 10" Subgrade, 9" Base, 1 1/2" Topping | SY | 61.11 | 50.94 | 43.58 | 33.57 | 31.83 |
| 18 02 0333 | 10" Mesh Reinforced Slab on Grade | SF | 8.10 | 7.00 | 6.44 | 5.37 | 5.03 |
| 18 05 0202 | Rock Cover, Riprap, Light (10 to 100 Lb Pieces) | CY | 26.72 | 24.15 | 22.59 | 20.09 | 19.46 |
| 18 05 0203 | Rock Cover, Riprap, Medium (10 to 200 Lb Pieces) | CY | 27.05 | 24.48 | 22.92 | 20.42 | 19.79 |
| 18 05 0204 | Rock Cover, Riprap, Heavy (25 to 500 Lb Pieces) | CY | 27.39 | 24.82 | 23.26 | 20.76 | 20.13 |
| 18 05 0402 | Seeding, Vegetative Cover | ACRE | 1,874 | 1,823 | 1,791 | 1,741 | 1,729 |
| 18 05 0409 | Fertilize, 800 Lbs/Acre, Push Rotary | ACRE | 140.58 | 118.67 | 105.24 | 83.82 | 78.56 |
| 18 05 0413 | Watering with 3,000-Gallon Tank Truck, per Pass | ACRE | 111.81 | 90.08 | 75.73 | 54.43 | 49.85 |
| 19 03 0167 | 18" Reinforced Concrete Pipe, Class 3, with Gaskets | LF | 23.67 | 21.05 | 19.60 | 17.05 | 16.33 |
| 19 03 0168 | 24" Reinforced Concrete Pipe, Class 3, with Gaskets | LF | 34.58 | 31.22 | 29.36 | 26.09 | 25.17 |
| 19 03 0170 | 36" Reinforced Concrete Pipe, Class 3, with Gaskets | LF | 61.34 | 56.10 | 53.19 | 48.09 | 46.65 |
| 19 03 0171 | 48" Reinforced Concrete Pipe, Class 3, with Gaskets | LF | 88.98 | 82.77 | 79.32 | 73.27 | 71.56 |
| 19 03 0172 | 60" Reinforced Concrete Pipe, Class 3, with Gaskets | LF | 127.64 | 119.58 | 115.09 | 107.24 | 105.03 |

## Environmental Remediation: Assemblies Cost Book

# Cost Data for Remediation

## Passive Water Treatment

| Assembly | Description | Unit | A | B | C | D | E |
|---|---|---|---|---|---|---|---|
| | **Capital Costs** | | | | | | |
| 33 01 0102 | Van or Pickup Rental | DAY | 32.69 | 32.69 | 32.69 | 32.69 | 32.69 |
| 33 05 0801 | 10' Wide Grass Drainage Swale | LF | 6.61 | 5.80 | 5.24 | 4.43 | 4.27 |
| 33 05 0802 | Grass Ditching, 3' Bottom, 3' Deep, 2:1 Side Slopes | LF | 18.76 | 15.91 | 13.91 | 11.11 | 10.58 |
| 33 05 0803 | Grass Ditching, 5' Bottom, 5' Deep, 4:1 Side Slopes | LF | 57.12 | 48.23 | 42.00 | 33.26 | 31.62 |
| 33 05 0804 | Riprap Ditching, 3' Bottom, 3' Deep, 2:1 Side Slopes | LF | 23.84 | 20.88 | 18.81 | 15.90 | 15.34 |
| 33 05 0805 | Concrete Ditching, 3' Bottom, 3' Deep, 2:1 Side Slopes | LF | 68.58 | 58.45 | 53.68 | 43.86 | 40.53 |
| 33 05 1501 | Install Filter Compost in Confined Space | LF | 262.61 | 240.14 | 228.72 | 206.89 | 200.04 |
| 33 05 1502 | Install Filter Compost in Open Space | LF | 185.54 | 179.55 | 176.98 | 171.19 | 169.07 |
| 33 05 1511 | 2' Wide Iron Sewer Grating | LF | 94.26 | 85.51 | 81.76 | 73.30 | 70.20 |
| 33 05 1512 | Galv.Debris Baffle | SF | 42.28 | 38.39 | 36.72 | 32.96 | 31.58 |
| 33 05 1513 | 4' High Concrete Flow Spreader Pre-Cast | LF | 54.18 | 45.45 | 40.59 | 32.08 | 29.68 |
| 33 05 1514 | 8' High Weir Posts | EA | 343.56 | 308.54 | 293.55 | 259.71 | 247.30 |
| 33 05 1515 | Galv. Overflow Weir | SF | 52.19 | 46.31 | 43.74 | 38.06 | 36.00 |
| 33 05 1516 | 6' x 12' x 6' I.D. Pre-Cast Vault | EA | 13,580 | 11,903 | 10,971 | 9,338 | 8,877 |
| 33 05 1517 | 8' x 18' x 7' I.D. Pre-Cast Vault | EA | 20,285 | 17,773 | 16,375 | 13,928 | 13,236 |
| 33 05 1518 | Vault Access Doors | EA | 553.72 | 507.03 | 487.05 | 441.92 | 425.38 |
| 33 05 1519 | 6' High Reinforced CIP Retaining Wall | LF | 259.22 | 214.12 | 192.85 | 149.14 | 134.38 |
| 33 05 1521 | Tipping Fee for Excavation Spoil | CY | 5.20 | 5.20 | 5.20 | 5.20 | 5.20 |
| 33 05 1531 | No. 3/4 Limestone - Bulk | TN | 46.63 | 44.84 | 43.87 | 42.14 | 41.64 |
| 33 05 1532 | Mushroom Compost - Bulk | CY | 20.45 | 18.81 | 17.92 | 16.32 | 15.86 |
| 33 05 1533 | Wetland Vegetation Planting | CSF | 36.41 | 31.51 | 29.41 | 24.66 | 22.92 |
| 33 05 1534 | Planting Mix - Hand Spread | CY | 47.56 | 41.57 | 39.01 | 33.22 | 31.09 |
| 33 05 1535 | Seed Wetland w/Bionutrients | LB | 1.51 | 1.46 | 1.43 | 1.38 | 1.37 |
| 33 05 1540 | 6 Mil Polyethylene Vapor Barrier | CSF | 22.29 | 17.85 | 15.95 | 11.65 | 10.08 |
| 33 05 1551 | 36 inch Diameter Riser Outlet | EA | 8,322 | 7,202 | 6,641 | 5,553 | 5,207 |
| 33 05 1552 | 48 inch Diameter Outlet Riser | EA | 10,377 | 8,884 | 8,136 | 6,686 | 6,225 |
| 33 05 1553 | 60" Dia. Low Flow Outlet Riser | EA | 3,886 | 3,867 | 3,858 | 3,840 | 3,834 |
| 33 08 0505 | Soil/Bentonite Liner | SF | 0.78 | 0.77 | 0.76 | 0.75 | 0.74 |
| 33 08 0506 | Clay 10E-7, 6" Lifts, On-Site | CY | 29.50 | 23.82 | 19.83 | 14.25 | 13.21 |
| 33 08 0507 | Clay 10E-7, 6" Lifts, Off-Site | CY | 31.11 | 26.22 | 22.84 | 18.04 | 17.11 |
| 33 08 0511 | Drainage Netting, 1/4" Thick High-density Polyethylene | SF | 0.23 | 0.21 | 0.20 | 0.19 | 0.18 |
| 33 08 0531 | 60 Mil Geotextile, Nonwoven | SY | 1.53 | 1.31 | 1.21 | 1.00 | 0.93 |
| 33 08 0532 | 80 Mil Geotextile, Nonwoven | SY | 1.57 | 1.35 | 1.25 | 1.04 | 0.97 |

**Environmental Remediation: Assemblies Cost Book**

*Page 3-246*

1998 by ECHOS. All rights reserved

# Cost Data for Remediation

## Passive Water Treatment

| Assembly | Description | Unit | A | B | C | D | E |
|---|---|---|---|---|---|---|---|
| | **Capital Costs** | | | | | | |
| 33 08 0533 | 105 Mil Geotextile, Nonwoven | SY | 1.92 | 1.65 | 1.52 | 1.26 | 1.17 |
| 33 08 0534 | 130 Mil Geotextile, Nonwoven | SY | 2.56 | 2.20 | 2.03 | 1.68 | 1.56 |
| 33 08 0535 | 170 Mil Geotextile, Nonwoven | SY | 2.96 | 2.60 | 2.43 | 2.08 | 1.96 |
| 33 08 0541 | 20 Mil Polymeric Liner, Very Low Density Polyethylene | SF | 1.25 | 1.02 | 0.91 | 0.69 | 0.61 |
| 33 08 0542 | 30 Mil Polymeric Liner, Very Low Density Polyethylene | SF | 1.98 | 1.59 | 1.41 | 1.03 | 0.90 |
| 33 08 0543 | 40 Mil Polymeric Liner, Very Low Density Polyethylene | SF | 2.38 | 1.91 | 1.70 | 1.24 | 1.09 |
| 33 08 0544 | 60 Mil Polymeric Liner, Very Low Density Polyethylene | SF | 3.54 | 2.83 | 2.51 | 1.83 | 1.60 |
| 33 08 0545 | 80 Mil Polymeric Liner, Very Low Density Polyethylene | SF | 4.35 | 3.49 | 3.10 | 2.27 | 1.98 |
| 33 08 0546 | 100 Mil Polymeric Liner, Very Low Density Polyethylene | SF | 5.50 | 4.41 | 3.91 | 2.85 | 2.49 |
| 33 08 0551 | 36 Mil Polymeric Liner, Chlorosulfanated Polyethylene | SF | 0.75 | 0.73 | 0.72 | 0.70 | 0.70 |
| 33 08 0552 | 45 Mil Polymeric Liner, Chlorosulfanated Polyethylene | SF | 0.83 | 0.81 | 0.80 | 0.79 | 0.78 |
| 33 08 0561 | 20 Mil Polymeric Liner, PVC | SF | 0.24 | 0.23 | 0.22 | 0.21 | 0.20 |
| 33 08 0562 | 30 Mil Polymeric Liner, PVC | SF | 0.32 | 0.31 | 0.30 | 0.29 | 0.28 |
| 33 08 0563 | 40 Mil Polymeric Liner, PVC | SF | 0.38 | 0.36 | 0.36 | 0.34 | 0.34 |
| 33 08 0564 | 60 Mil Polymeric Liner, PVC | SF | 0.72 | 0.70 | 0.69 | 0.67 | 0.67 |
| 33 08 0565 | 80 Mil Polymeric Liner, PVC | SF | 0.98 | 0.96 | 0.95 | 0.93 | 0.93 |
| 33 08 0566 | 100 Mil Polymeric Liner, PVC | SF | 1.28 | 1.26 | 1.25 | 1.23 | 1.22 |
| 33 08 0571 | 40 Mil Polymeric Liner, High-density Polyethylene | SF | 2.35 | 1.88 | 1.67 | 1.21 | 1.06 |
| 33 08 0572 | 60 Mil Polymeric Liner, High-density Polyethylene | SF | 3.49 | 2.80 | 2.48 | 1.81 | 1.58 |
| 33 08 0573 | 80 Mil Polymeric Liner, High-density Polyethylene | SF | 4.32 | 3.47 | 3.08 | 2.25 | 1.97 |
| 33 26 0902 | 6" Diameter Perforated PVC Pipe | LF | 14.13 | 11.62 | 10.11 | 7.66 | 7.04 |
| 33 26 0903 | 8" Diameter Perforated PVC Pipe | LF | 14.56 | 12.01 | 10.49 | 8.00 | 7.37 |
| 33 26 0904 | 10" Diameter Perforated PVC Pipe | LF | 16.84 | 14.07 | 12.42 | 9.71 | 9.02 |
| | **Operations & Maintenance** | | | | | | |
| 17 01 0601 | Selective Thinning | ACRE | 864.95 | 683.08 | 580.29 | 403.04 | 354.04 |
| 17 04 0103 | Load & Haul Debris, 5 Miles, Dumptruck | CY | 8.09 | 6.53 | 5.41 | 3.88 | 3.61 |
| 18 05 0415 | Mowing | ACRE | 64.67 | 51.13 | 43.38 | 30.18 | 26.60 |
| 33 01 0102 | Van or Pickup Rental | DAY | 32.69 | 32.69 | 32.69 | 32.69 | 32.69 |
| 33 05 1501 | Install Filter Compost in Confined Space | LF | 262.61 | 240.14 | 228.72 | 206.89 | 200.04 |
| 33 05 1502 | Install Filter Compost in Open Space | LF | 185.54 | 179.55 | 176.98 | 171.19 | 169.07 |
| 33 05 1503 | Confined Space Filter Compost Removal | LF | 117.96 | 92.27 | 79.22 | 54.27 | 46.44 |
| 33 05 1504 | Open Space Filter Compost Removal | LF | 39.20 | 30.22 | 26.37 | 17.69 | 14.50 |
| 33 05 1505 | Till Filter Compost | LF | 0.87 | 0.67 | 0.58 | 0.39 | 0.32 |

**Environmental Remediation: Assemblies Cost Book**

# Cost Data for Remediation

## Passive Water Treatment

| Assembly | Description | Unit | Unit Cost by Safety Level A | B | C | D | E |
|---|---|---|---|---|---|---|---|
| | **Operations & Maintenance** | | | | | | |
| 33 05 1522 | Tipping Fee for Brush | CY | 6.00 | 6.00 | 6.00 | 6.00 | 6.00 |
| 33 05 1523 | Tipping Fee for Spent Filter Compost | CY | 4.80 | 4.80 | 4.80 | 4.80 | 4.80 |

# Cost Data for Remediation
## Permeable Barriers

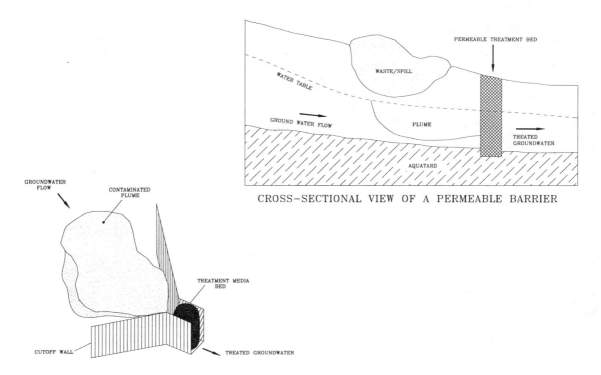

### General:
A permeable barrier is a passive groundwater remediation technique in which contaminants are treated/removed as groundwater flows through an in-situ treatment bed

### Typical Treatment Train:
Combined with cutoff walls such as sheet piling or slurry walls, the design is termed a funnel and gate system. A low permeability cutoff wall diverts and funnels groundwater through a higher permeability gate (treatment bed). To protect the finished treatment area, a vegetative cap or other, more durable cap, such as asphalt is generally added.

### Common Cost Components:
1. Excavation 2. Capping 3. Hauling/Disposal of Solids 4. Groundwater Monitoring Wells 5. Sampling and Analysis

### Other Cost Considerations:
Pumping Groundwater.

**Environmental Remediation: Assemblies Cost Book**

# Cost Data for Remediation

## Permeable Barriers

| Assembly | Description | Unit | Unit Cost by Safety Level | | | | |
|---|---|---|---|---|---|---|---|
| | | | A | B | C | D | E |
| | **Capital Costs** | | | | | | |
| 17 03 0307 | Cat 235, 2.0 CY, Rock, No Haul or Borrow, Trenching | BCY | 133.78 | 109.18 | 95.00 | 71.01 | 64.55 |
| 17 03 0415 | Backfill with Excavated Material | CY | 7.50 | 5.97 | 5.12 | 3.63 | 3.21 |
| 17 03 0420 | Backfill Trench, Borrow Material, Delivered & Dumped Only | CY | 9.86 | 8.74 | 7.93 | 6.83 | 6.63 |
| 17 03 0422 | Unclassified Fill, 6" Lifts, On-Site | CY | 14.23 | 11.65 | 9.60 | 7.05 | 6.72 |
| 17 03 0423 | Unclassified Fill, 6" Lifts, Off-Site | CY | 11.08 | 9.81 | 8.83 | 7.58 | 7.40 |
| 18 02 0301 | Asphalt Pavement - 10" Subgrade, 9" Base, 1 1/2" Topping | SY | 61.11 | 50.94 | 43.58 | 33.57 | 31.83 |
| 18 05 0203 | Rock Cover, Riprap, Medium (10 to 200 Lb Pieces) | CY | 27.05 | 24.48 | 22.92 | 20.42 | 19.79 |
| 18 05 0205 | Rock Cover, Bank-run Gravel | CY | 32.96 | 27.35 | 23.49 | 17.98 | 16.89 |
| 18 05 0301 | Topsoil, 6" Lifts, Off-Site | CY | 32.76 | 29.51 | 27.43 | 24.25 | 23.53 |
| 18 05 0302 | Topsoil, 6" Lifts, On-Site | CY | 14.14 | 11.54 | 9.45 | 6.88 | 6.57 |
| 18 05 0402 | Seeding, Vegetative Cover | ACRE | 1,874 | 1,823 | 1,791 | 1,741 | 1,729 |
| 18 05 0413 | Watering with 3,000-Gallon Tank Truck, per Pass | ACRE | 111.81 | 90.08 | 75.73 | 54.43 | 49.85 |
| 33 06 0301 | Level and Compact Working Surface | CY | 12.15 | 9.89 | 8.12 | 5.89 | 5.58 |
| 33 06 0302 | Construct Dike for Mixing Basin | CY | 12.15 | 9.89 | 8.12 | 5.89 | 5.58 |
| 33 06 0303 | Normal Soil, to 25', Slurry Wall Excavation | CY | 4.28 | 3.45 | 2.86 | 2.05 | 1.91 |
| 33 06 0304 | Clay/Sand with Boulders, to 25', Slurry Wall Excavation | CY | 4.28 | 3.45 | 2.86 | 2.05 | 1.91 |
| 33 06 0305 | Normal Soil, 26' - 75', Slurry Wall Excavation | CY | 9.28 | 7.50 | 6.21 | 4.46 | 4.15 |
| 33 06 0306 | Clay/Sand with Boulders, 26' - 75', Slurry Wall Excavation | CY | 13.79 | 11.14 | 9.23 | 6.62 | 6.16 |
| 33 06 0307 | Normal Soil, 76' - 120', Slurry Wall Excavation | CY | 20.15 | 16.29 | 13.49 | 9.68 | 9.02 |
| 33 06 0308 | Clay/Sand with Boulders, 76' - 120', Slurry Wall Excavation | CY | 30.08 | 24.30 | 20.12 | 14.45 | 13.46 |
| 33 06 0309 | Bentonite, Material Purchase Price per Ton | TON | 54.06 | 54.06 | 54.06 | 54.06 | 54.06 |
| 33 06 0310 | Slurry Mixing, Hydration, and Placement, per Gallon | GAL | 0.11 | 0.09 | 0.08 | 0.05 | 0.04 |
| 33 06 0311 | Soil-Bentonite Backfill Mixing, per Cubic Yard | CY | 5.73 | 4.67 | 3.83 | 2.78 | 2.64 |
| 33 06 0312 | Backfill Slurry Wall Trench, 1,000' Average Haul Distance | CY | 5.09 | 4.17 | 3.40 | 2.48 | 2.38 |
| 33 06 0313 | Demolish Mixing Basins and Regrade Working Surface | SF | 0.15 | 0.12 | 0.10 | 0.07 | 0.06 |
| 33 06 1011 | Temporary Medium Wall Sheet Piling | SF | 25.65 | 22.03 | 19.62 | 16.07 | 15.32 |
| 33 06 1012 | Temporary Heavy Wall Sheet Piling | SF | 35.15 | 30.03 | 26.63 | 21.61 | 20.55 |
| 33 06 1013 | Permanent Medium Wall Sheet Piling | SF | 20.65 | 17.03 | 14.62 | 11.07 | 10.32 |
| 33 06 1014 | Permanent Heavy Wall Sheet Piling | SF | 35.15 | 30.03 | 26.63 | 21.61 | 20.55 |
| 33 06 1021 | Trench - 2-1/2 CY Hydraulic Excavator -113 CY/Hr (86M3)/Hr | CY | 4.28 | 3.45 | 2.86 | 2.05 | 1.91 |
| 33 06 1022 | Excavate Trench in Cobble to 25' Deep | CY | 4.28 | 3.45 | 2.86 | 2.05 | 1.91 |
| 33 06 1023 | Normal Soil - 26' - 75' Slurry Wall Excavation | CY | 9.28 | 7.50 | 6.21 | 4.46 | 4.15 |
| 33 06 1024 | Clay/Sand w/Boulders - 26' - 75' Slurry Wall Excavation | CY | 13.79 | 11.14 | 9.23 | 6.62 | 6.16 |

## Environmental Remediation: Assemblies Cost Book

# Cost Data for Remediation

## Permeable Barriers

| Assembly | Description | Unit | Unit Cost by Safety Level | | | | |
|---|---|---|---|---|---|---|---|
| | | | A | B | C | D | E |
| | **Capital Costs** | | | | | | |
| 33 06 1025 | Normal Soil - 76' - 120' Slurry Wall Excavation | CY | 20.15 | 16.29 | 13.49 | 9.68 | 9.02 |
| 33 06 1026 | Clay/Sand w/Boulders - 76' - 120' Slurry Wall Excavation | CY | 30.08 | 24.30 | 20.12 | 14.45 | 13.46 |
| 33 06 1027 | Key-in Treatment Wall | CY | 133.78 | 109.18 | 95.00 | 71.01 | 64.55 |
| 33 06 1028 | Level and Compact Working Surface | CY | 12.15 | 9.89 | 8.12 | 5.89 | 5.58 |
| 33 06 1031 | Iron Filings | CY | 1,839 | 1,821 | 1,810 | 1,792 | 1,788 |
| 33 06 1032 | Activated Carbon - Coal Derived | CY | 879.29 | 861.29 | 849.83 | 832.21 | 828.15 |
| 33 06 1033 | Crushed Limestone | CY | 63.25 | 60.34 | 58.57 | 55.73 | 55.03 |
| 33 06 1034 | Clinoptilolite | CY | 306.20 | 288.19 | 276.73 | 259.11 | 255.05 |
| 33 06 1035 | Montmorillonite Clay | CY | 311.61 | 293.61 | 282.15 | 264.53 | 260.47 |
| 33 06 1036 | Glauconitic Greensands | CY | 1,099 | 1,081 | 1,070 | 1,052 | 1,048 |
| 33 06 1037 | Fuller's (Diatomaceous) Earth | CY | 399.16 | 381.16 | 369.70 | 352.08 | 348.02 |
| 33 06 1038 | Peat Moss | CY | 82.35 | 79.45 | 77.68 | 74.84 | 74.14 |
| 33 06 1039 | Proprietary Metal Oxidizing Powder | CY | 4,389 | 4,371 | 4,360 | 4,342 | 4,338 |
| 33 06 1040 | Proprietary Iron-Foam Aggregate | CY | 1,989 | 1,971 | 1,960 | 1,942 | 1,938 |
| 33 06 1041 | Proprietary Humic-Acid Adsorbent | CY | 664.29 | 646.29 | 634.83 | 617.21 | 613.15 |
| 33 06 1042 | Pea Gravel | CY | 33.18 | 30.27 | 28.51 | 25.67 | 24.96 |
| 33 06 1043 | User Supplied Treatment Media | CY | 0.00 | 0.00 | 0.00 | 0.00 | 0.00 |
| 33 08 0532 | 80 Mil Geotextile, Nonwoven | SY | 1.57 | 1.35 | 1.25 | 1.04 | 0.97 |
| 33 23 0101 | 2" PVC, Schedule 40, Well Casing | LF | 17.45 | 14.39 | 11.98 | 8.95 | 8.55 |
| 33 23 0102 | 4" PVC, Schedule 40, Well Casing | LF | 26.98 | 22.38 | 18.77 | 14.24 | 13.63 |
| 33 23 0103 | 6" PVC, Schedule 40, Well Casing | LF | 27.07 | 22.66 | 19.19 | 14.84 | 14.25 |
| 33 23 0121 | 2" Stainless Steel, Well Casing | LF | 31.75 | 28.14 | 25.30 | 21.73 | 21.25 |
| 33 23 0122 | 4" Stainless Steel, Well Casing | LF | 47.66 | 43.06 | 39.45 | 34.92 | 34.31 |
| 33 23 0123 | 6" Stainless Steel Well Casing, 5' Sections, Flush Threaded | LF | 370.73 | 311.96 | 265.77 | 207.74 | 199.94 |
| 33 23 0201 | 2" PVC, Schedule 40, Well Screen | LF | 23.47 | 19.52 | 16.42 | 12.52 | 12.00 |
| 33 23 0202 | 4" PVC, Schedule 40, Well Screen | LF | 38.16 | 32.00 | 27.16 | 21.09 | 20.27 |
| 33 23 0203 | 6" PVC, Schedule 40, Well Screen | LF | 47.13 | 39.78 | 34.01 | 26.75 | 25.78 |
| 33 23 0221 | 2" Stainless Steel, Well Screen | LF | 26.91 | 23.84 | 21.44 | 18.41 | 18.01 |
| 33 23 0222 | 4" Stainless Steel, Well Screen | LF | 47.66 | 43.06 | 39.45 | 34.92 | 34.31 |
| 33 23 0223 | 6" Stainless Steel Well Screen, 5' Sec, Flush Threaded | LF | 366.05 | 307.28 | 261.09 | 203.06 | 195.26 |
| 33 23 0301 | 2" PVC, Well Plug | EACH | 29.37 | 24.77 | 21.16 | 16.63 | 16.02 |
| 33 23 0302 | 4" PVC, Well Plug | EACH | 55.65 | 48.91 | 43.62 | 36.97 | 36.07 |
| 33 23 0303 | 6" PVC, Well Plug | EACH | 112.63 | 101.15 | 92.13 | 80.79 | 79.27 |

**Environmental Remediation: Assemblies Cost Book**

*Page 3-251*

1998 by ECHOS. All rights reserved

# Cost Data for Remediation

## Permeable Barriers

| Assembly | Description | Unit | Unit Cost by Safety Level | | | | |
|---|---|---|---|---|---|---|---|
| | | | A | B | C | D | E |
| | **Capital Costs** | | | | | | |
| 33 23 0311 | 2" Stainless Steel, Well Plug | EACH | 83.24 | 74.06 | 66.83 | 57.76 | 56.55 |
| 33 23 0312 | 4" Stainless Steel, Well Plug | EACH | 112.24 | 102.03 | 94.01 | 83.93 | 82.58 |
| 33 23 0313 | 6" Stainless Steel, Well Plug | EACH | 573.04 | 481.18 | 408.98 | 318.27 | 306.09 |

**Environmental Remediation: Assemblies Cost Book**

# Cost Data for Remediation
## Piping

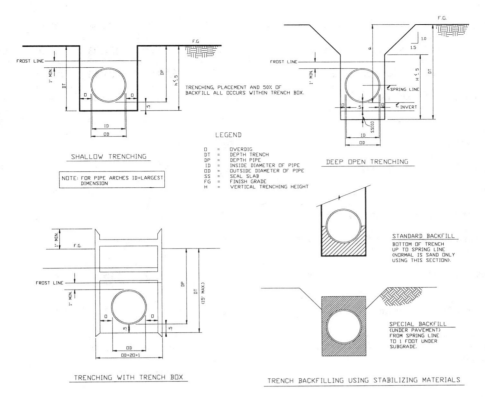

### General:
The piping assemblies included with most of the technologies in these softbooks are for aboveground installation. This section provides costs for subsurface piping.

### Common Cost Components:
1. Trenching
2. Pipe
3. Seal slab for trench
4. Trench box
5. Backfill
6. Dewatering pump
7. Wellpoints
8. Manholes
9. Lift stations
10. Removal of existing cover
11. Replacement of cover

### Typical Treatment Train:
Any technology or process that requires piping.

### Other Cost Considerations:
Clearing and grubbing, electrical distribution.

**Environmental Remediation: Assemblies Cost Book**

# Cost Data for Remediation

## Piping

| Assembly | Description | Unit | Unit Cost by Safety Level A | B | C | D | E |
|---|---|---|---|---|---|---|---|
| | **Capital Costs** | | | | | | |
| 17 01 0106 | Heavy Brush, Light Trees, Clear, Grub, Haul | ACRE | 11,338 | 9,081 | 7,594 | 5,382 | 4,905 |
| 17 01 0107 | Medium Brush, Medium Trees, Clear, Grub, Haul | ACRE | 13,478 | 10,783 | 9,029 | 6,389 | 5,807 |
| 17 01 0108 | Light Brush, Heavy Trees, Clear, Grub, Haul | ACRE | 17,877 | 14,287 | 11,977 | 8,462 | 7,669 |
| 17 01 0210 | Clear Trees to 6" Diameter with D8 Cat | EACH | 12.03 | 9.85 | 8.04 | 5.88 | 5.66 |
| 17 01 0211 | Clear Trees to 12" Diameter with D8 Cat | EACH | 22.45 | 18.39 | 15.00 | 10.98 | 10.56 |
| 17 01 0212 | Clear Trees to 24" Diameter with D8 Cat | EACH | 33.68 | 27.58 | 22.50 | 16.46 | 15.84 |
| 17 01 0213 | Clear Trees to 36" Diameter with D8 Cat | EACH | 67.36 | 55.17 | 45.00 | 32.93 | 31.67 |
| 17 02 0201 | Demolish Bituminous Road with Power Equipment | CY | 46.17 | 36.60 | 30.96 | 21.62 | 19.19 |
| 17 02 0203 | Demolish Bituminous Pavement with Air Equipment | CY | 80.15 | 63.38 | 53.76 | 37.41 | 32.98 |
| 17 02 0205 | Demolish Unreinforced Concrete to 6" Thick with Air Equipment | CY | 126.90 | 100.35 | 85.12 | 59.24 | 52.23 |
| 17 02 0206 | Demolish Mesh Reinforced Concrete to 6" Thick with Air Equipment | CY | 183.47 | 145.09 | 123.07 | 85.65 | 75.51 |
| 17 02 0207 | Demolish Rod Reinforced Concrete to 6" Thick with Air Equipment | CY | 199.06 | 157.42 | 133.53 | 92.92 | 81.92 |
| 17 02 0208 | Demolish Mesh Reinforced Concrete to 6" Thick with Power Equipment | CY | 96.73 | 76.68 | 64.87 | 45.30 | 40.20 |
| 17 02 0209 | Demolish Rod Reinforced Concrete to 6" Thick with Power Equipment | CY | 106.91 | 84.75 | 71.70 | 50.07 | 44.43 |
| 17 02 0210 | Demolish Unreinforced Concrete 7" to 24" Thick with Power Equipment | CY | 258.44 | 204.87 | 173.31 | 121.03 | 107.40 |
| 17 02 0211 | Demolish Reinforced Concrete 7" to 24" Thick with Power Equipment | CY | 356.37 | 282.51 | 238.98 | 166.90 | 148.10 |
| 17 02 0212 | Demolish Unreinforced Concrete to 6" Thick with Power Equipment | CY | 67.53 | 53.53 | 45.29 | 31.63 | 28.06 |
| 17 02 0215 | Demolish Bituminous Sidewalk | CY | 154.36 | 121.53 | 103.59 | 71.64 | 62.41 |
| 17 02 0220 | Demolish Bituminous Curbs | LF | 2.66 | 2.09 | 1.78 | 1.23 | 1.07 |
| 17 02 0221 | Demolish Unreinforced Concrete Curbs | LF | 7.26 | 5.71 | 4.87 | 3.37 | 2.93 |
| 17 02 0222 | Demolish Reinforced Concrete Curbs | LF | 9.90 | 7.79 | 6.64 | 4.59 | 4.00 |
| 17 03 0105 | Fine Grading, Hand | SY | 4.21 | 3.24 | 2.83 | 1.90 | 1.56 |
| 17 03 0259 | Cat 225, 1.5 CY, Soil/Sand, Trenching | CY | 1.84 | 1.49 | 1.23 | 0.88 | 0.82 |
| 17 03 0260 | Cat 225, 1.5 CY, Soil/Sand, 10' - 20' Deep Trench Box, Trench | CY | 7.22 | 6.04 | 5.20 | 4.04 | 3.82 |
| 17 03 0261 | Cat 225, 1.5 CY, Soil/Sand with Boulders, Trenching | CY | 3.91 | 3.15 | 2.62 | 1.87 | 1.74 |
| 17 03 0272 | Pull Trench Box, Cat 225, 1.5 CY, Soil/Sand, Trenching | CY | 6.67 | 5.49 | 4.65 | 3.48 | 3.27 |
| 17 03 0282 | Soil, 5 Miles, Dump Truck, Load/Haul Spoil From Trench | CY | 6.97 | 5.60 | 4.66 | 3.33 | 3.06 |
| 17 03 0283 | Sand, 5 Mile, Dump Truck, Load/Haul Spoil From Trench | CY | 6.28 | 5.05 | 4.20 | 3.00 | 2.76 |

**Environmental Remediation: Assemblies Cost Book**

# Cost Data for Remediation

## Piping

| Assembly | Description | Unit | Unit Cost by Safety Level A | B | C | D | E |
|---|---|---|---|---|---|---|---|
| | **Capital Costs** | | | | | | |
| 17 03 0306 | Cat 225, 1.5 CY, Rock, No Haul or Borrow, Trenching | BCY | 149.48 | 122.48 | 106.81 | 80.48 | 73.45 |
| 17 03 0317 | Rock, 5 Miles, Dump Truck, Load/Haul Spoil from Trench | CY | 10.31 | 8.31 | 6.90 | 4.93 | 4.56 |
| 17 03 0401 | 950, 3.00 CY, Backfill with Excavated Material | CY | 2.54 | 2.05 | 1.70 | 1.22 | 1.14 |
| 17 03 0405 | 950, 3.00 CY, Delivered & Dumped, Backfill with Sand | CY | 38.31 | 35.03 | 32.95 | 29.74 | 29.00 |
| 17 03 0410 | 950, Delivered & Dumped, Backfill with Cement Stabilized Base Material | CY | 46.93 | 43.66 | 41.58 | 38.37 | 37.63 |
| 17 03 0420 | Backfill Trench, Borrow Material, Delivered & Dumped Only | CY | 9.86 | 8.74 | 7.93 | 6.83 | 6.63 |
| 17 03 0426 | Sand, 6" Lifts, Off-Site | CY | 15.55 | 14.34 | 13.42 | 12.22 | 12.04 |
| 17 03 0511 | Compact Soil with Vibrating Plate | CY | 7.20 | 5.57 | 4.84 | 3.26 | 2.70 |
| 17 03 0515 | Compact With Pogosticks | CY | 5.91 | 4.58 | 3.98 | 2.68 | 2.23 |
| 17 03 0516 | Compact with 50% Pogosticks, 50% Hand Roller | CY | 4.24 | 3.30 | 2.85 | 1.94 | 1.63 |
| 17 03 1001 | Wellpoint for Trench, Install & Remove <500', 1 Month | LFHD | 55.06 | 55.06 | 55.06 | 55.06 | 55.06 |
| 17 03 1002 | 2" Diameter Contractor's Trash Pump, 75 GPM | DAY | 50.20 | 47.01 | 45.64 | 42.55 | 41.42 |
| 18 01 0102 | Gravel, Delivered & Dumped | CY | 24.33 | 22.93 | 22.03 | 20.66 | 20.36 |
| 18 01 0103 | Gravel (90%) & Sand Base (10%), with Calcium Chloride 3/4 - 1 Lb/CY | CY | 25.70 | 24.15 | 23.16 | 21.65 | 21.31 |
| 18 01 0104 | Asphalt, Intermediate Course (Line Item Includes 5% Waste) | TON | 76.50 | 65.72 | 59.09 | 48.56 | 45.99 |
| 18 01 0310 | Prime Coat | SY | 0.55 | 0.52 | 0.51 | 0.49 | 0.48 |
| 18 01 0311 | Tack Coat | SY | 0.70 | 0.64 | 0.60 | 0.53 | 0.52 |
| 18 01 0312 | Asphalt Wearing Course, 1 Pass (Line Item Includes 5% Waste) | TON | 83.61 | 72.83 | 66.20 | 55.67 | 53.10 |
| 18 02 0320 | 4" Structural Slab on Grade | SF | 6.45 | 5.45 | 4.98 | 4.00 | 3.67 |
| 18 02 0321 | 6" Structural Slab on Grade | SF | 6.98 | 5.95 | 5.47 | 4.48 | 4.14 |
| 18 02 0322 | 8" Structural Slab on Grade | SF | 9.75 | 8.38 | 7.71 | 6.38 | 5.95 |
| 18 02 0323 | 10" Structural Slab on Grade | SF | 10.56 | 9.13 | 8.43 | 7.04 | 6.58 |
| 18 02 0324 | 12" Structural Slab on Grade | SF | 11.57 | 10.04 | 9.30 | 7.82 | 7.34 |
| 18 02 0330 | 4" Mesh Reinforced Slab on Grade | SF | 5.48 | 4.60 | 4.18 | 3.32 | 3.03 |
| 18 02 0331 | 6" Mesh Reinforced Slab on Grade | SF | 6.01 | 5.10 | 4.67 | 3.79 | 3.49 |
| 18 02 0332 | 8" Mesh Reinforced Slab on Grade | SF | 7.58 | 6.50 | 5.96 | 4.91 | 4.58 |
| 18 02 0333 | 10" Mesh Reinforced Slab on Grade | SF | 8.10 | 7.00 | 6.44 | 5.37 | 5.03 |
| 18 02 0334 | 12" Mesh Reinforced Slab on Grade | SF | 8.78 | 7.63 | 7.05 | 5.93 | 5.58 |
| 18 02 0340 | 4" Unreinforced Slab on Grade | SF | 4.34 | 3.64 | 3.29 | 2.61 | 2.38 |
| 18 02 0341 | 6" Unreinforced Slab on Grade | SF | 4.87 | 4.14 | 3.79 | 3.08 | 2.85 |
| 18 02 0342 | 8" Unreinforced Slab on Grade | SF | 6.45 | 5.54 | 5.07 | 4.19 | 3.93 |

**Environmental Remediation: Assemblies Cost Book**

*Page 3-255*

1998 by ECHOS. All rights reserved

# Cost Data for Remediation

## Piping

| Assembly | Description | Unit | A | B | C | D | E |
|---|---|---|---|---|---|---|---|
| | **Capital Costs** | | | | | | |
| 18 02 0343 | 10" Unreinforced Slab on Grade | SF | 6.97 | 6.04 | 5.56 | 4.66 | 4.38 |
| 18 02 0344 | 12" Unreinforced Slab on Grade | SF | 7.64 | 6.67 | 6.16 | 5.22 | 4.93 |
| 18 04 0203 | Pour & Cure Concrete, Continuous Footing | CY | 140.13 | 126.22 | 118.87 | 105.35 | 101.29 |
| 18 05 0402 | Seeding, Vegetative Cover | ACRE | 1,874 | 1,823 | 1,791 | 1,741 | 1,729 |
| 18 05 0405 | Sodding, Vegetative Cover | ACRE | 25,697 | 22,064 | 20,238 | 16,710 | 15,589 |
| 19 01 0202 | 1", Class 200, PVC Piping | LF | 7.48 | 5.84 | 5.07 | 3.48 | 2.93 |
| 19 01 0204 | 2", Class 200, PVC Piping | LF | 8.86 | 6.96 | 6.07 | 4.23 | 3.60 |
| 19 01 0206 | 3", Class 200, PVC Piping | LF | 11.81 | 9.34 | 8.20 | 5.80 | 4.98 |
| 19 01 0207 | 4", Class 200, PVC Piping | LF | 12.33 | 9.86 | 8.72 | 6.32 | 5.50 |
| 19 01 0208 | 6", Class 200, PVC Piping | LF | 14.95 | 12.20 | 10.93 | 8.27 | 7.36 |
| 19 01 0212 | 8", Class 150, PVC Piping | LF | 18.36 | 15.27 | 13.84 | 10.85 | 9.82 |
| 19 01 0213 | 10", Class 150, PVC Piping | LF | 22.90 | 19.37 | 17.73 | 14.32 | 13.14 |
| 19 01 0214 | 12", Class 150, PVC Piping | LF | 28.75 | 24.63 | 22.72 | 18.73 | 17.36 |
| 19 02 0101 | 4", Class 50, Bell & Spigot Sanitary Sewer, Cast-iron Pipe | LF | 8.49 | 7.49 | 7.05 | 6.08 | 5.72 |
| 19 02 0103 | 6", Class 50, Bell & Spigot Sanitary Sewer, Cast-iron Pipe | LF | 13.25 | 11.79 | 11.16 | 9.75 | 9.23 |
| 19 02 0104 | 8", Class 50, Bell & Spigot Sanitary Sewer, Cast-iron Pipe | LF | 18.98 | 17.26 | 16.52 | 14.86 | 14.25 |
| 19 02 0105 | 10", Class 50, Bell & Spigot Sanitary Sewer, Cast-iron Pipe | LF | 27.66 | 25.51 | 24.59 | 22.51 | 21.75 |
| 19 02 0106 | 12", Class 50, Bell & Spigot Sanitary Sewer, Cast-iron Pipe | LF | 36.36 | 34.21 | 33.29 | 31.21 | 30.45 |
| 19 02 0107 | 15", Class 50, Bell & Spigot Sanitary Sewer, Cast-iron Pipe | LF | 51.76 | 49.01 | 47.83 | 45.17 | 44.19 |
| 19 02 0110 | 4" Extra-strength Vitrified Clay Pipe, Class 200, Premium Joints | LF | 17.47 | 14.14 | 12.39 | 9.15 | 8.17 |
| 19 02 0112 | 6" Extra-strength Vitrified Clay Pipe, Class 200, Premium Joints | LF | 20.06 | 16.40 | 14.48 | 10.92 | 9.84 |
| 19 02 0113 | 8" Extra-strength Vitrified Clay Pipe, Class 200, Premium Joints | LF | 22.16 | 18.37 | 16.38 | 12.70 | 11.58 |
| 19 02 0114 | 10" Extra-strength Vitrified Clay Pipe, Class 200, Premium Joints | LF | 27.02 | 22.79 | 20.57 | 16.46 | 15.22 |
| 19 02 0115 | 12" Extra-strength Vitrified Clay Pipe, Class 200, Premium Joints | LF | 28.25 | 23.86 | 21.55 | 17.28 | 15.98 |
| 19 02 0116 | 15" Extra-strength Vitrified Clay Pipe, Class 200, Premium Joints | LF | 39.29 | 34.29 | 31.67 | 26.81 | 25.34 |
| 19 02 0117 | 18" Extra-strength Vitrified Clay Pipe, Class 200, Premium Joints | LF | 50.85 | 45.06 | 42.02 | 36.40 | 34.70 |
| 19 02 0118 | 24" Extra-strength Vitrified Clay Pipe, Class 200, Premium Joints | LF | 85.93 | 78.09 | 73.96 | 66.33 | 64.03 |
| 19 02 0125 | 4" PVC Pipe Sanitary | LF | 13.31 | 10.99 | 9.71 | 7.45 | 6.80 |
| 19 02 0126 | 6" PVC Pipe Sanitary | LF | 15.68 | 13.19 | 11.82 | 9.39 | 8.70 |

## Environmental Remediation: Assemblies Cost Book

*Page 3-256*

1998 by ECHOS. All rights reserved

# Cost Data for Remediation

## Piping

| Assembly | Description | Unit | Unit Cost by Safety Level A | B | C | D | E |
|---|---|---|---|---|---|---|---|
| | **Capital Costs** | | | | | | |
| 19 02 0127 | 8" PVC Pipe Sanitary | LF | 17.03 | 14.43 | 13.00 | 10.47 | 9.74 |
| 19 02 0128 | 10" PVC Pipe Sanitary | LF | 22.78 | 19.45 | 17.70 | 14.46 | 13.48 |
| 19 02 0129 | 12" PVC Pipe Sanitary | LF | 26.69 | 23.26 | 21.45 | 18.11 | 17.11 |
| 19 02 0130 | 15" PVC Pipe Sanitary | LF | 42.35 | 36.56 | 33.52 | 27.90 | 26.20 |
| 19 02 0201 | Precast, CIP Base, 4' Diameter, 6' Deep, Manhole | EACH | 1,459 | 1,276 | 1,179 | 1,001 | 947.36 |
| 19 02 0202 | Precast, CIP Base, 4' Diameter, 8' Deep, Manhole | EACH | 1,957 | 1,697 | 1,557 | 1,304 | 1,229 |
| 19 02 0203 | Precast, CIP Base, 4' Diameter, 12' Deep, Manhole | EACH | 2,835 | 2,459 | 2,257 | 1,891 | 1,783 |
| 19 02 0210 | Precast Manhole, Per Foot, 4' Diameter x Depth | LF | 219.59 | 190.64 | 174.83 | 146.66 | 138.52 |
| 19 02 0301 | 10,000 GPD (7 GPM) Lift Station | EACH | 2,828 | 2,659 | 2,587 | 2,423 | 2,363 |
| 19 02 0302 | 25,000 GPD (18 GPM) Lift Station | EACH | 3,539 | 3,313 | 3,217 | 2,998 | 2,918 |
| 19 02 0303 | 50,000 GPD (35 GPM) Lift Station | EACH | 4,960 | 4,621 | 4,476 | 4,149 | 4,029 |
| 19 02 0304 | 100,000 GPD (70 GPM) Lift Station | EACH | 7,842 | 7,509 | 7,366 | 7,043 | 6,925 |
| 19 02 0305 | 200,000 GPD Packaged Lift Station | EACH | 133,007 | 123,412 | 118,907 | 109,609 | 106,455 |
| 19 02 0306 | 500,000 GPD Packaged Lift Station | EACH | 157,190 | 144,431 | 138,440 | 126,076 | 121,881 |
| 19 02 0307 | 800,000 GPD Packaged Lift Station | EACH | 188,049 | 172,867 | 165,739 | 151,027 | 146,036 |
| 19 02 0311 | 12' x 36" Diameter Reinforced Concrete Pipe Wet Well for Lift Station | EACH | 9,840 | 8,577 | 7,816 | 6,581 | 6,270 |
| 19 02 0312 | 24' x 60" Diameter Reinforced Concrete Pipe Wet Well for Lift Station | EACH | 39,142 | 33,897 | 30,681 | 25,554 | 24,296 |
| 19 03 0157 | 6" Extra Strength, Nonreinforced Concrete Pipe | LF | 10.78 | 8.92 | 7.98 | 6.17 | 5.59 |
| 19 03 0158 | 8" Extra Strength, Nonreinforced Concrete Pipe | LF | 12.86 | 10.79 | 9.75 | 7.74 | 7.10 |
| 19 03 0159 | 10" Extra Strength, Nonreinforced Concrete Pipe | LF | 16.79 | 14.59 | 13.49 | 11.36 | 10.68 |
| 19 03 0160 | 12" Extra Strength, Nonreinforced Concrete Pipe | LF | 17.97 | 15.63 | 14.46 | 12.20 | 11.48 |
| 19 03 0161 | 15" Extra Strength, Nonreinforced Concrete Pipe | LF | 19.95 | 17.36 | 16.06 | 13.55 | 12.74 |
| 19 03 0162 | 18" Extra Strength, Nonreinforced Concrete Pipe | LF | 24.93 | 22.06 | 20.62 | 17.83 | 16.94 |
| 19 03 0163 | 21" Extra Strength, Nonreinforced Concrete Pipe | LF | 30.32 | 27.45 | 26.01 | 23.22 | 22.33 |
| 19 03 0164 | 24" Extra Strength, Nonreinforced Concrete Pipe | LF | 39.69 | 36.49 | 34.88 | 31.77 | 30.78 |
| 19 03 0165 | 12" Reinforced Concrete Pipe, Class 3, with Gaskets | LF | 14.33 | 13.02 | 12.29 | 11.01 | 10.65 |
| 19 03 0166 | 15" Reinforced Concrete Pipe, Class 3, with Gaskets | LF | 16.47 | 14.84 | 13.94 | 12.36 | 11.91 |
| 19 03 0167 | 18" Reinforced Concrete Pipe, Class 3, with Gaskets | LF | 23.67 | 21.05 | 19.60 | 17.05 | 16.33 |
| 19 03 0168 | 24" Reinforced Concrete Pipe, Class 3, with Gaskets | LF | 34.58 | 31.22 | 29.36 | 26.09 | 25.17 |
| 19 03 0174 | 12" Reinforced Concrete Pipe, Class 4, with Gaskets | LF | 14.33 | 13.02 | 12.29 | 11.01 | 10.65 |
| 19 03 0175 | 15" Reinforced Concrete Pipe, Class 4, with Gaskets | LF | 16.47 | 14.84 | 13.94 | 12.36 | 11.91 |
| 19 03 0176 | 18" Reinforced Concrete Pipe, Class 4, with Gaskets | LF | 23.67 | 21.05 | 19.60 | 17.05 | 16.33 |

**Environmental Remediation: Assemblies Cost Book**

# Cost Data for Remediation

## Piping

| Assembly | Description | Unit | Unit Cost by Safety Level | | | | |
|----------|-------------|------|------|------|------|------|------|
| | | | A | B | C | D | E |
| | **Capital Costs** | | | | | | |
| 19 03 0177 | 21" Reinforced Concrete Pipe, Class 4, with Gaskets | LF | 27.08 | 24.38 | 22.87 | 20.24 | 19.50 |
| 19 03 0178 | 24" Reinforced Concrete Pipe, Class 4, with Gaskets | LF | 34.58 | 31.22 | 29.36 | 26.09 | 25.17 |
| 19 03 0182 | 12" Reinforced Concrete Pipe, Class 5, with Gaskets | LF | 14.33 | 13.02 | 12.29 | 11.01 | 10.65 |
| 19 03 0183 | 15" Reinforced Concrete Pipe, Class 5, with Gaskets | LF | 16.47 | 14.84 | 13.94 | 12.36 | 11.91 |
| 19 03 0184 | 18" Reinforced Concrete Pipe, Class 5, with Gaskets | LF | 23.67 | 21.05 | 19.60 | 17.05 | 16.33 |
| 19 03 0185 | 21" Reinforced Concrete Pipe, Class 5, with Gaskets | LF | 27.08 | 24.38 | 22.87 | 20.24 | 19.50 |
| 19 03 0186 | 24" Reinforced Concrete Pipe, Class 5, with Gaskets | LF | 34.58 | 31.22 | 29.36 | 26.09 | 25.17 |
| 19 04 0601 | 275 Gallon Steel Sump, Aboveground with Supports & Fittings | EACH | 1,325 | 1,153 | 1,080 | 913.28 | 852.29 |
| 19 04 0602 | 550 Gallon Steel Sump, Aboveground with Supports & Fittings | EACH | 1,598 | 1,390 | 1,301 | 1,099 | 1,026 |
| 19 04 0603 | 1,000 Gallon Steel Sump, Aboveground with Supports & Fittings | EACH | 2,118 | 1,844 | 1,727 | 1,463 | 1,366 |
| 19 04 0604 | 1,500 Gallon Steel Sump, Aboveground with Supports & Fittings | EACH | 2,697 | 2,356 | 2,209 | 1,879 | 1,758 |
| 19 04 0605 | 2,000 Gallon Steel Sump, Aboveground with Supports & Fittings | EACH | 3,457 | 2,997 | 2,777 | 2,332 | 2,183 |
| 19 04 0606 | 5,000 Gallon Steel Sump, Aboveground with Supports & Fittings | EACH | 6,079 | 5,494 | 5,214 | 4,646 | 4,457 |
| 19 04 0621 | 550 Gallon Horizontal Plastic Sump with 4" NPT Connection | EACH | 1,699 | 1,596 | 1,551 | 1,452 | 1,415 |
| 19 04 0622 | 1,000 Gallon Horizontal Plastic Sump with 4" NPT Connection | EACH | 2,179 | 2,042 | 1,984 | 1,852 | 1,803 |
| 19 04 0623 | 2,000 Gallon Horizontal Plastic Sump with 6" NPT Connection | EACH | 3,089 | 2,918 | 2,845 | 2,680 | 2,619 |
| 19 04 0624 | 4,000 Gallon Horizontal Plastic Sump with 6" NPT Connection | EACH | 4,130 | 3,927 | 3,841 | 3,645 | 3,573 |
| 19 04 0625 | 6,000 Gallon Horizontal Plastic Sump with 6" NPT Connection | EACH | 5,248 | 4,975 | 4,858 | 4,594 | 4,497 |
| 19 04 0626 | 8,000 Gallon Horizontal Plastic Sump with 6" NPT Connection | EACH | 5,857 | 5,516 | 5,369 | 5,039 | 4,918 |
| 19 04 0627 | 10,000 Gallon Horizontal Plastic Sump with 6" NPT Connection | EACH | 7,082 | 6,622 | 6,402 | 5,956 | 5,807 |
| 19 04 0628 | 12,000 Gallon Horizontal Plastic Sump with 6" NPT Connection | EACH | 8,308 | 7,722 | 7,442 | 6,875 | 6,686 |
| 19 04 0629 | 15,000 Gallon Horizontal Plastic Sump with 6" NPT Connection | EACH | 14,184 | 13,540 | 13,232 | 12,608 | 12,400 |
| 19 04 0631 | 20,000 Gallon Horizontal Plastic Sump with 6" NPT Connection | EACH | 15,799 | 15,084 | 14,742 | 14,048 | 13,817 |
| 19 04 0632 | 25,000 Gallon Horizontal Plastic Sump with 6" NPT Connection | EACH | 24,407 | 23,487 | 23,047 | 22,155 | 21,858 |

**Environmental Remediation: Assemblies Cost Book**

# Cost Data for Remediation

## Piping

| Assembly | Description | Unit | A | B | C | D | E |
|---|---|---|---|---|---|---|---|
| | **Capital Costs** | | | | | | |
| 19 04 0633 | 30,000 Gallon Horizontal Plastic Sump with 6" NPT Connection | EACH | 35,112 | 34,039 | 33,525 | 32,485 | 32,138 |
| 19 04 0634 | 40,000 Gallon Horizontal Plastic Sump with 6" NPT Connection | EACH | 40,844 | 39,234 | 38,465 | 36,904 | 36,383 |
| 19 04 0635 | 48,000 Gallon Horizontal Plastic Sump with 6" NPT Connection | EACH | 59,758 | 57,612 | 56,586 | 54,505 | 53,811 |
| 19 07 0120 | 1 1/4", 60 PSI, Polyethylene Pipe | LF | 5.29 | 4.21 | 3.72 | 2.68 | 2.32 |
| 19 07 0121 | 1 1/2", 60 PSI, Polyethylene Pipe | LF | 5.38 | 4.30 | 3.81 | 2.77 | 2.41 |
| 19 07 0122 | 2", 60 PSI, Polyethylene Pipe | LF | 6.05 | 4.85 | 4.31 | 3.15 | 2.75 |
| 19 07 0123 | 3" Polyethylene 40' Joints, 60 PSI | LF | 7.90 | 6.47 | 5.82 | 4.42 | 3.94 |
| 19 07 0124 | 4" Polyethylene 40' Joints, 60 PSI | LF | 19.10 | 15.75 | 13.90 | 10.64 | 9.71 |
| 19 07 0125 | 6" Polyethylene 40' Joints, 60 PSI | LF | 25.85 | 22.22 | 20.22 | 16.68 | 15.67 |
| 19 07 0126 | 8" Polyethylene 40' Joints, 60 PSI | LF | 35.37 | 31.01 | 28.61 | 24.37 | 23.16 |
| 20 06 0101 | 3 - 9 Lb Magnesium Anodes, Cathodic Protection Point | EACH | 953.82 | 778.29 | 693.26 | 523.02 | 466.94 |
| 20 06 0102 | 3 - 17 Lb Magnesium Anodes, Cathodic Protection Point | EACH | 1,364 | 1,121 | 1,002 | 766.32 | 688.94 |
| 20 06 0103 | 3 - 32 Lb Magnesium Anodes, Cathodic Protection Point | EACH | 1,845 | 1,539 | 1,390 | 1,092 | 995.23 |
| 20 06 0104 | 3 - 48 Lb Magnesium Anodes, Cathodic Protection Point | EACH | 2,585 | 2,166 | 1,960 | 1,553 | 1,420 |
| 33 19 7301 | Haul & Dispose Debris, 16.5 CY Truck, 10 Mile, Landfill | CY | 107.37 | 106.15 | 105.16 | 103.95 | 103.81 |
| 33 26 0101 | 1" Carbon Steel Piping | LF | 6.46 | 5.28 | 4.77 | 3.63 | 3.21 |
| 33 26 0102 | 2" Carbon Steel Piping | LF | 10.50 | 8.72 | 7.95 | 6.23 | 5.60 |
| 33 26 0103 | 3" Carbon Steel Piping | LF | 18.02 | 15.20 | 13.99 | 11.26 | 10.26 |
| 33 26 0104 | 4" Carbon Steel Piping | LF | 23.20 | 19.86 | 18.43 | 15.21 | 14.02 |
| 33 26 0105 | 6" Carbon Steel Piping | LF | 36.37 | 30.24 | 27.46 | 21.52 | 19.44 |
| 33 26 0106 | 8" Carbon Steel Piping | LF | 40.77 | 34.08 | 31.15 | 24.68 | 22.36 |
| 33 26 0107 | 10" Carbon Steel Piping | LF | 54.40 | 46.05 | 42.38 | 34.30 | 31.39 |
| 33 26 0108 | 12" Carbon Steel Piping | LF | 75.72 | 65.62 | 61.18 | 51.41 | 47.89 |
| 33 26 0201 | 1" Stainless Steel Piping, Schedule 40, Threaded | LF | 18.67 | 16.12 | 15.03 | 12.56 | 11.65 |
| 33 26 0202 | 2" Stainless Steel Piping, Schedule 40, Threaded | LF | 33.47 | 29.21 | 27.39 | 23.28 | 21.77 |
| 33 26 0203 | 3" Stainless Steel Piping, Schedule 40, Threaded | LF | 57.00 | 50.61 | 47.88 | 41.71 | 39.45 |
| 33 26 0204 | 4" Stainless Steel Piping, Schedule 40, Threaded | LF | 80.15 | 71.72 | 68.12 | 59.98 | 57.00 |
| 33 26 0205 | 6" Stainless Steel Piping, Schedule 10, Type 316 | LF | 78.58 | 70.01 | 66.14 | 57.84 | 54.94 |
| 33 26 0206 | 8" Stainless Steel Piping, Schedule 40, Welded | LF | 197.22 | 183.32 | 176.89 | 163.42 | 158.80 |
| 33 26 0208 | 10" Stainless Steel Piping, Schedule 40, Welded | LF | 280.77 | 265.48 | 258.40 | 243.59 | 238.50 |
| 33 26 0209 | 12" Stainless Steel Piping, Schedule 40, Welded | LF | 312.43 | 295.44 | 287.57 | 271.12 | 265.47 |

Unit Cost by Safety Level

**Environmental Remediation: Assemblies Cost Book**

*Page 3-259*

# Cost Data for Remediation

## Piping

| Assembly | Description | Unit | Unit Cost by Safety Level A | B | C | D | E |
|---|---|---|---|---|---|---|---|
| | **Capital Costs** | | | | | | |
| 33 26 0512 | 4" High-density Polyethylene, Transfer Pipe | LF | 6.75 | 5.61 | 5.10 | 4.00 | 3.62 |
| 33 26 0513 | 6" High-density Polyethylene, Transfer Pipe | LF | 8.82 | 7.47 | 6.86 | 5.55 | 5.10 |
| 33 26 0514 | 8" High-density Polyethylene, Transfer Pipe | LF | 11.20 | 9.76 | 9.11 | 7.72 | 7.23 |
| 33 26 0516 | 12" High-density Polyethylene, Transfer Pipe | LF | 19.22 | 17.26 | 16.37 | 14.47 | 13.81 |
| 33 26 0518 | 16" High-density Polyethylene, Transfer Pipe | LF | 30.24 | 27.16 | 25.76 | 22.78 | 21.74 |
| 33 26 0520 | 24" High-density Polyethylene, Transfer Pipe | LF | 61.27 | 55.88 | 53.44 | 48.22 | 46.39 |
| 33 26 0601 | (1 1/2", 3") Stainless Steel Double-wall Piping, with Fittings | LF | 134.61 | 117.05 | 109.53 | 92.56 | 86.33 |
| 33 26 0602 | (2", 4") Stainless Steel Double-wall Piping, with Fittings | LF | 185.98 | 163.46 | 153.81 | 132.04 | 124.05 |
| 33 26 0603 | (2 1/2", 4") Stainless Steel Double-wall Piping, with Fittings | LF | 206.86 | 182.17 | 171.60 | 147.74 | 138.99 |
| 33 26 0604 | (4", 6") Stainless Steel Double-wall Piping, with Fittings | LF | 339.64 | 280.79 | 254.18 | 197.21 | 177.23 |
| 33 26 0621 | (1 1/2", 3") PVC Double-wall Piping, with Fittings | LF | 33.15 | 27.05 | 24.43 | 18.53 | 16.36 |
| 33 26 0622 | (2", 4") PVC Double-wall Piping, with Fittings | LF | 46.22 | 37.98 | 34.43 | 26.46 | 23.55 |
| 33 26 0623 | (2 1/2", 4") PVC Double-wall Piping, with Fittings | LF | 49.22 | 40.39 | 36.61 | 28.07 | 24.94 |
| 33 26 0624 | (4", 6") PVC Double-wall Piping, with Fittings | LF | 70.61 | 58.00 | 52.59 | 40.40 | 35.93 |
| 33 26 0641 | (1 1/2", 3") Carbon Steel Double-wall Piping, with Fittings | LF | 48.16 | 40.64 | 37.42 | 30.15 | 27.48 |
| 33 26 0642 | (2", 4") Carbon Steel Double-wall Piping, with Fittings | LF | 64.71 | 55.69 | 51.83 | 43.11 | 39.91 |
| 33 26 0643 | (2 1/2", 4") Carbon Steel Double-wall Piping, with Fittings | LF | 70.81 | 61.02 | 56.83 | 47.37 | 43.90 |
| 33 26 0644 | (4", 6") Carbon Steel Double-wall Piping, with Fittings | LF | 114.39 | 100.33 | 94.32 | 80.73 | 75.75 |
| 33 29 0102 | 10 GPM, 1/2 HP, Centrifugal Pump | EACH | 1,063 | 952.32 | 905.07 | 798.39 | 759.27 |
| 33 29 0103 | 50 GPM, 100' Head, 3 HP, Centrifugal Pump | EACH | 2,658 | 2,500 | 2,432 | 2,279 | 2,223 |
| 33 29 0104 | 75 GPM, 100' Head, 5 HP, Centrifugal Pump | EACH | 2,946 | 2,754 | 2,672 | 2,487 | 2,419 |
| 33 29 0106 | 100 GPM, 150' Head, 7.5 HP, Centrifugal Pump | EACH | 3,419 | 3,191 | 3,093 | 2,873 | 2,793 |
| 33 29 0107 | 125 GPM, 150' Head, 10 HP, Centrifugal Pump | EACH | 3,815 | 3,512 | 3,382 | 3,088 | 2,981 |
| 33 29 0108 | 10 HP, 200 GPM, Centrifugal Pump | EACH | 2,941 | 2,714 | 2,616 | 2,396 | 2,316 |
| 33 29 0109 | 10 HP, 250 GPM, Centrifugal Pump | EACH | 3,032 | 2,805 | 2,707 | 2,487 | 2,407 |
| 33 29 0110 | 15 HP, 300 GPM, Centrifugal Pump | EACH | 3,441 | 3,160 | 3,040 | 2,770 | 2,670 |
| 33 29 0111 | 20 HP, 500 GPM, Centrifugal Pump | EACH | 4,544 | 4,241 | 4,111 | 3,818 | 3,710 |
| 33 29 0112 | 30 HP, 750 GPM, Centrifugal Pump | EACH | 5,213 | 4,808 | 4,635 | 4,244 | 4,100 |
| 33 29 0113 | 40 HP, 1050 GPM, Centrifugal Pump | EACH | 5,655 | 5,199 | 5,004 | 4,564 | 4,403 |
| 33 29 0114 | 60 HP, 1500 GPM, Centrifugal Pump | EACH | 9,842 | 8,823 | 8,350 | 7,363 | 7,024 |
| 33 29 0115 | 75 HP, 2000 GPM, Centrifugal Pump | EACH | 12,456 | 11,233 | 10,666 | 9,482 | 9,075 |
| 33 29 0116 | 100 HP, 3000 GPM, Centrifugal Pump | EACH | 15,575 | 14,046 | 13,338 | 11,857 | 11,348 |
| 33 41 0101 | Pump & Motor Maintenance/Repair | EACH | 865.17 | 671.48 | 581.58 | 393.96 | 329.64 |

**Environmental Remediation: Assemblies Cost Book**

*Page 3-260*

1998 by ECHOS. All rights reserved

# Cost Data for Remediation

## Piping

| Assembly | Description | Unit | Unit Cost by Safety Level | | | | |
|---|---|---|---|---|---|---|---|
| | | | A | B | C | D | E |
| | **Capital Costs** | | | | | | |
| 33 42 0101 | Electrical Charge | KWH | 0.06 | 0.06 | 0.06 | 0.06 | 0.06 |
| | **Operations & Maintenance** | | | | | | |
| 33 41 0101 | Pump & Motor Maintenance/Repair | EACH | 865.17 | 671.48 | 581.58 | 393.96 | 329.64 |
| 33 42 0101 | Electrical Charge | KWH | 0.06 | 0.06 | 0.06 | 0.06 | 0.06 |

**Environmental Remediation: Assemblies Cost Book**

# Cost Data for Remediation
## Sampling and Analysis

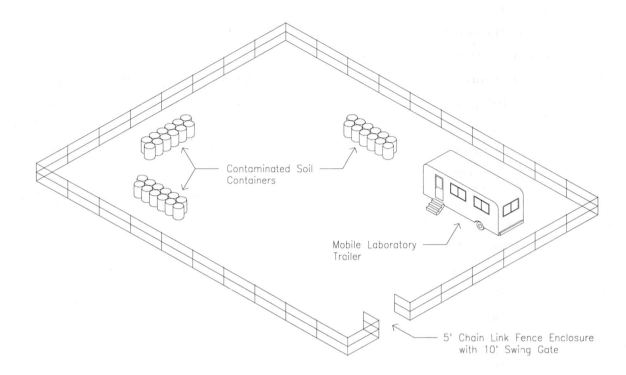

### General:

*The sampling assemblies included in this section include the labor, equipment, and materials required to obtain a sample and transport it to a laboratory. There are assemblies for off-site laboratories, on-site mobile laboratories, conventional drill rigs, and innovative cone penetrometer (CPT) rigs. Analyses costs include sample preparation at the laboratory and the cost to run the analysis and report the findings.*

### Typical Treatment Train:
*Applicable to all treatment trains.*

### Common Cost Components:

1. Sampling crew mobilization/per diem
2. Sample preparation, cleanup, containers, disposables, packaging, shipment
3. QA/QC samples (replicates/duplicates, trip blanks, equipment blanks, field blanks, spikes, blanks)
4. Analyses (consider turnaround time and QC levels)

### Other Cost Considerations:
*Ground-water monitoring wells, soil gas monitoring wells, professional labor.*

# Cost Data for Remediation

## Sampling and Analysis

| Assembly | Description | Unit | Unit Cost by Safety Level A | B | C | D | E |
|---|---|---|---|---|---|---|---|
| | **Capital Costs** | | | | | | |
| 33 01 0101 | Mobilize/DeMobilize Drilling Rig & Crew | LS | 4,949 | 4,031 | 3,308 | 2,401 | 2,280 |
| 33 01 0102 | Van or Pickup Rental | DAY | 32.69 | 32.69 | 32.69 | 32.69 | 32.69 |
| 33 01 0103 | Flatbed Truck Rental | DAY | 45.94 | 45.94 | 45.94 | 45.94 | 45.94 |
| 33 01 0104 | Car or Van Mileage Charge | MILE | 0.35 | 0.35 | 0.35 | 0.35 | 0.35 |
| 33 01 0105 | Pickup Truck Mileage Charge | MILE | 0.56 | 0.56 | 0.56 | 0.56 | 0.56 |
| 33 01 0106 | Flatbed Truck Mileage Charge | MILE | 1.01 | 1.01 | 1.01 | 1.01 | 1.01 |
| 33 01 0201 | Mobilize Crew, >= 500 Miles, per Person | EACH | 357.75 | 357.75 | 357.75 | 357.75 | 357.75 |
| 33 01 0202 | Per Diem | DAY | 104.94 | 104.94 | 104.94 | 104.94 | 104.94 |
| 33 01 0203 | Mobilize Crew, 250 Miles, per Person | EACH | 178.88 | 178.88 | 178.88 | 178.88 | 178.88 |
| 33 01 0204 | Mobilize Crew, 100 Miles, per Person | EACH | 71.55 | 71.55 | 71.55 | 71.55 | 71.55 |
| 33 01 0205 | Mobilize Crew, 50 Miles, per Person | EACH | 53.66 | 53.66 | 53.66 | 53.66 | 53.66 |
| 33 01 0206 | Mobilize Crew, Local, per Person | EACH | 17.89 | 17.89 | 17.89 | 17.89 | 17.89 |
| 33 02 0206 | Prefiltering Liquids | EACH | 20.00 | 20.00 | 20.00 | 20.00 | 20.00 |
| 33 02 0207 | Acid Digestion | EACH | 27.50 | 27.50 | 27.50 | 27.50 | 27.50 |
| 33 02 0208 | Drying and Grinding | EACH | 22.50 | 22.50 | 22.50 | 22.50 | 22.50 |
| 33 02 0209 | Dissolution by Microwave | EACH | 40.00 | 40.00 | 40.00 | 40.00 | 40.00 |
| 33 02 0210 | Urine Preparation | EACH | 25.00 | 25.00 | 25.00 | 25.00 | 25.00 |
| 33 02 0211 | Feces Preparation | EACH | 33.33 | 33.33 | 33.33 | 33.33 | 33.33 |
| 33 02 0212 | Blood Preparation | EACH | 33.33 | 33.33 | 33.33 | 33.33 | 33.33 |
| 33 02 0213 | Animal Bone/Tissue Preparation | EACH | 66.67 | 66.67 | 66.67 | 66.67 | 66.67 |
| 33 02 0214 | Gas Flow Proportional Counting | EACH | 27.50 | 27.50 | 27.50 | 27.50 | 27.50 |
| 33 02 0215 | Liquid Scintillation | EACH | 27.50 | 27.50 | 27.50 | 27.50 | 27.50 |
| 33 02 0216 | Alpha Spectroscopy | EACH | 27.50 | 27.50 | 27.50 | 27.50 | 27.50 |
| 33 02 0217 | Gamma Spectroscopy | EACH | 20.00 | 20.00 | 20.00 | 20.00 | 20.00 |
| 33 02 0218 | Fluorometric or Laser | EACH | 27.50 | 27.50 | 27.50 | 27.50 | 27.50 |
| 33 02 0219 | Alpha Scintillation Counting | EACH | 27.50 | 27.50 | 27.50 | 27.50 | 27.50 |
| 33 02 0220 | Radon De-emanation | EACH | 27.50 | 27.50 | 27.50 | 27.50 | 27.50 |
| 33 02 0221 | Beta/Gamma Coincidence | EACH | 27.50 | 27.50 | 27.50 | 27.50 | 27.50 |
| 33 02 0222 | Radioactivity Screening - Tritium Not Present | EACH | 32.50 | 32.50 | 32.50 | 32.50 | 32.50 |
| 33 02 0223 | Radioactivity Screening - Tritium Present | EACH | 10.00 | 10.00 | 10.00 | 10.00 | 10.00 |
| 33 02 0224 | Repackage Sample | EACH | 10.00 | 10.00 | 10.00 | 10.00 | 10.00 |
| 33 02 0225 | Repackage and Ship Sample | EACH | 23.33 | 23.33 | 23.33 | 23.33 | 23.33 |
| 33 02 0301 | Air Monitoring Station | EACH | 718.68 | 718.68 | 718.68 | 718.68 | 718.68 |

**Environmental Remediation: Assemblies Cost Book**

# Cost Data for Remediation

## Sampling and Analysis

| Assembly | Description | Unit | Unit Cost by Safety Level | | | | |
|---|---|---|---|---|---|---|---|
| | | | A | B | C | D | E |
| | **Capital Costs** | | | | | | |
| 33 02 0302 | Organic Vapor Analyzer Rental, per Month | MONTH | 1,920 | 1,920 | 1,920 | 1,920 | 1,920 |
| 33 02 0303 | Organic Vapor Analyzer Rental, per Day | DAY | 184.30 | 184.30 | 184.30 | 184.30 | 184.30 |
| 33 02 0307 | Soil Gas Investigation & Analysis | DAY | 106.00 | 106.00 | 106.00 | 106.00 | 106.00 |
| 33 02 0309 | Bacterial Air Sampler, Monthly Rental | MONTH | 436.37 | 436.37 | 436.37 | 436.37 | 436.37 |
| 33 02 0310 | Microbial Air Sampler, Monthly Rental | MONTH | 436.37 | 436.37 | 436.37 | 436.37 | 436.37 |
| 33 02 0311 | Cotton Dust Sampler, Monthly Rental | MONTH | 480.53 | 480.53 | 480.53 | 480.53 | 480.53 |
| 33 02 0312 | Digital Dust Sampler, Monthly Rental | MONTH | 552.97 | 552.97 | 552.97 | 552.97 | 552.97 |
| 33 02 0313 | Flow Calibrator, Monthly Rental | MONTH | 212.00 | 212.00 | 212.00 | 212.00 | 212.00 |
| 33 02 0314 | Personal Low Flow Sampling Pump, Monthly Rental | MONTH | 178.43 | 178.43 | 178.43 | 178.43 | 178.43 |
| 33 02 0315 | Ambient Air Monitor, Monthly Rental | MONTH | 826.80 | 826.80 | 826.80 | 826.80 | 826.80 |
| 33 02 0316 | Ambient CO Analyzer, Monthly Rental | MONTH | 378.95 | 378.95 | 378.95 | 378.95 | 378.95 |
| 33 02 0317 | Ambient Ozone Monitor, Monthly Rental | MONTH | 1,758 | 1,758 | 1,758 | 1,758 | 1,758 |
| 33 02 0318 | Ambient Sulfur Dioxide Analyzer, Monthly Rental | MONTH | 2,690 | 2,690 | 2,690 | 2,690 | 2,690 |
| 33 02 0319 | Calibration Gas Monitor | EACH | 487.60 | 487.60 | 487.60 | 487.60 | 487.60 |
| 33 02 0320 | Chloride Analyzer, Monthly Rental | MONTH | 749.07 | 749.07 | 749.07 | 749.07 | 749.07 |
| 33 02 0321 | Gas Monitors, Monthly Rental | MONTH | 657.20 | 657.20 | 657.20 | 657.20 | 657.20 |
| 33 02 0322 | Hydrogen Sulfide Analyzer, Monthly Rental | MONTH | 249.10 | 249.10 | 249.10 | 249.10 | 249.10 |
| 33 02 0323 | Manual Air Sampling Kit, Monthly Rental | MONTH | 85.68 | 85.68 | 85.68 | 85.68 | 85.68 |
| 33 02 0324 | Manual Air Sampling Kit, Detection Tubes | EACH | 88.69 | 88.69 | 88.69 | 88.69 | 88.69 |
| 33 02 0325 | Mercury Vapor Monitor, Monthly Rental | MONTH | 1,277 | 1,277 | 1,277 | 1,277 | 1,277 |
| 33 02 0326 | Nitrogen Dioxide Analyzer, Monthly Rental | MONTH | 2,385 | 2,385 | 2,385 | 2,385 | 2,385 |
| 33 02 0327 | Personal Alarms, Monthly Rental | MONTH | 318.00 | 318.00 | 318.00 | 318.00 | 318.00 |
| 33 02 0328 | Portable Ambient Air Analyzer, Monthly Rental | MONTH | 1,745 | 1,745 | 1,745 | 1,745 | 1,745 |
| 33 02 0329 | Portable CO2 Monitor, Monthly Rental | MONTH | 485.83 | 485.83 | 485.83 | 485.83 | 485.83 |
| 33 02 0330 | Portable Combustible Gas/Oxygen Indicator, Monthly Rental | MONTH | 355.10 | 355.10 | 355.10 | 355.10 | 355.10 |
| 33 02 0331 | Portable Organic Vapor Analyzer, Monthly Rental | MONTH | 1,920 | 1,920 | 1,920 | 1,920 | 1,920 |
| 33 02 0332 | Portable Oxygen Deficiency Monitor, Monthly Rental | MONTH | 409.87 | 409.87 | 409.87 | 409.87 | 409.87 |
| 33 02 0333 | Portable Toxic Gas Analyzer, Monthly Rental | MONTH | 265.00 | 265.00 | 265.00 | 265.00 | 265.00 |
| 33 02 0334 | Solvent Vapor Indicator, Monthly Rental | MONTH | 1,920 | 1,920 | 1,920 | 1,920 | 1,920 |
| 33 02 0335 | Specific Vapor Analyzer, Monthly Rental | MONTH | 1,290 | 1,290 | 1,290 | 1,290 | 1,290 |
| 33 02 0336 | Stack Sampling Probe, Monthly Rental | MONTH | 648.37 | 648.37 | 648.37 | 648.37 | 648.37 |
| 33 02 0337 | Trace-Gas Analyzer, Monthly Rental | MONTH | 1,852 | 1,852 | 1,852 | 1,852 | 1,852 |

**Environmental Remediation: Assemblies Cost Book**

*Page 3-264*

1998 by ECHOS. All rights reserved

# Cost Data for Remediation

## Sampling and Analysis

| Assembly | Description | Unit | A | B | C | D | E |
|---|---|---|---|---|---|---|---|
| | **Capital Costs** | | | | | | |
| 33 02 0338 | Instrument Shelter, 30.0" x 20.0" x 32.0", Pine with Latex Paint | EACH | 738.96 | 722.90 | 716.02 | 700.49 | 694.80 |
| 33 02 0339 | Instrument Shelter, 18.0" x 13.0" x 20.0", Wood Painted White | EACH | 289.52 | 273.46 | 266.58 | 251.05 | 245.36 |
| 33 02 0340 | Instrument Shelter, 81.0" x 5.5.0" x 16.5", Wood Painted White | EACH | 266.47 | 250.40 | 243.52 | 228.00 | 222.30 |
| 33 02 0341 | Alpha, Beta, Gamma Detector Equipment | EACH | 194.33 | 194.33 | 194.33 | 194.33 | 194.33 |
| 33 02 0401 | Disposable Materials per Sample | EACH | 6.87 | 6.87 | 6.87 | 6.87 | 6.87 |
| 33 02 0402 | Decontamination Materials per Sample | EACH | 6.26 | 6.26 | 6.26 | 6.26 | 6.26 |
| 33 02 0404 | 12' Aluminum Pole with Polyhead Sampler | EACH | 396.66 | 396.66 | 396.66 | 396.66 | 396.66 |
| 33 02 0405 | Monitoring Well Slug Testing Equipment Rental | WEEK | 552.08 | 552.08 | 552.08 | 552.08 | 552.08 |
| 33 02 0504 | Clarifier Indicator Analyses | EACH | 195.01 | 195.01 | 195.01 | 195.01 | 195.01 |
| 33 02 0505 | Groundwater Contamination Analysis, per Sample | EACH | 126.00 | 126.00 | 126.00 | 126.00 | 126.00 |
| 33 02 0506 | Tank Decontamination Analysis, per Sample | EACH | 115.97 | 115.97 | 115.97 | 115.97 | 115.97 |
| 33 02 0507 | Primary Drinking Water Parameters Analysis, per Sample | EACH | 341.64 | 341.64 | 341.64 | 341.64 | 341.64 |
| 33 02 0508 | Detection/Compliance Monitoring Analysis, per Well | EACH | 990.17 | 990.17 | 990.17 | 990.17 | 990.17 |
| 33 02 0509 | Compliance Monitoring 40 CFR 261, Appendix IX Constituents | EACH | 1,715 | 1,715 | 1,715 | 1,715 | 1,715 |
| 33 02 0510 | Analysis of Wastewater for POTW Permit | EACH | 295.29 | 295.29 | 295.29 | 295.29 | 295.29 |
| 33 02 0511 | Analysis of Lysimeters for Soil-Pore Liquid Monitoring | EACH | 357.33 | 357.33 | 357.33 | 357.33 | 357.33 |
| 33 02 0512 | Rinsate Analysis | EACH | 115.97 | 115.97 | 115.97 | 115.97 | 115.97 |
| 33 02 0521 | Hip Waders Rental | WEEK | 51.44 | 51.44 | 51.44 | 51.44 | 51.44 |
| 33 02 0522 | Boat with Motor, Rental | DAY | 58.83 | 58.83 | 58.83 | 58.83 | 58.83 |
| 33 02 0523 | Boat without Motor, Rental | DAY | 21.73 | 21.73 | 21.73 | 21.73 | 21.73 |
| 33 02 0524 | Glass Coliwasas, Disposable, 7/8" x 42", 200 ml, Case of 12 | EACH | 187.22 | 187.22 | 187.22 | 187.22 | 187.22 |
| 33 02 0525 | Glass Coliwasas, Reusable, 7/8" x 42", 200 ml, Case of 4 | EACH | 67.42 | 67.42 | 67.42 | 67.42 | 67.42 |
| 33 02 0526 | PVC Coliwasas, 1.9" Outside Diameter x 49", 760 ml | EACH | 155.01 | 155.01 | 155.01 | 155.01 | 155.01 |
| 33 02 0527 | Bomb Sampler, 2.5" Diameter x 10", 500 ml | EACH | 524.35 | 524.35 | 524.35 | 524.35 | 524.35 |
| 33 02 0528 | Drum Thief Tubes, 3/4" x 42", 150 ml, Disposable, Case of 25 | EACH | 63.94 | 63.94 | 63.94 | 63.94 | 63.94 |
| 33 02 0529 | Drum Thief Tubes, 1/2" x 42", 75 ml, Disposable, Case of 25 | EACH | 44.75 | 44.75 | 44.75 | 44.75 | 44.75 |
| 33 02 0530 | Bottom Sampler, 3 Lb Brass, 7.5" x 4.5", No Cable Required | EACH | 179.67 | 179.67 | 179.67 | 179.67 | 179.67 |
| 33 02 0531 | Bottom Sampler, 17 Lb Stainless Steel, 6" x 6" x 6", with 100' Cable | EACH | 466.91 | 466.91 | 466.91 | 466.91 | 466.91 |

**Environmental Remediation: Assemblies Cost Book**

# Cost Data for Remediation

## Sampling and Analysis

| Assembly | Description | Unit | Unit Cost by Safety Level |
|---|---|---|---|
| | | | **A** | **B** | **C** | **D** | **E** |

| Assembly | Description | Unit | A | B | C | D | E |
|---|---|---|---|---|---|---|---|
| | **Capital Costs** | | | | | | |
| 33 02 0532 | Bottom Sampler, 60 Lb Stainless Steel, 9" x 9" | EACH | 915.24 | 915.24 | 915.24 | 915.24 | 915.24 |
| 33 02 0533 | Water Level Indicator, 100' Tape, Electric, Light & Horn | EACH | 461.65 | 461.65 | 461.65 | 461.65 | 461.65 |
| 33 02 0534 | Water Level Indicator, 300' Tape, Electric, Light & Horn | EACH | 688.99 | 688.99 | 688.99 | 688.99 | 688.99 |
| 33 02 0535 | Water Level Indicator, 500' Tape, Electric, Light & Horn | EACH | 996.15 | 996.15 | 996.15 | 996.15 | 996.15 |
| 33 02 0536 | Water Level Indicator, Manual Steel Chalk Tape, with Weight, 100' Cable | EACH | 138.33 | 138.33 | 138.33 | 138.33 | 138.33 |
| 33 02 0538 | DO Meter, Analog, Portable, 6' Cable, Polarographic Probe | EACH | 577.70 | 577.70 | 577.70 | 577.70 | 577.70 |
| 33 02 0539 | DO Meter, Digital, Portable, 10' Cable, Small Diameter Probe | EACH | 1,114 | 1,114 | 1,114 | 1,114 | 1,114 |
| 33 02 0540 | DO Meter, Portable, Probe, 10' Cable, Quick Readings | EACH | 517.60 | 517.60 | 517.60 | 517.60 | 517.60 |
| 33 02 0541 | Conductivity Meter with Dip-type Sensor, Digital, 9V Battery, 7 x 3.2 x 2" | EACH | 977.27 | 977.27 | 977.27 | 977.27 | 977.27 |
| 33 02 0542 | Conductivity Meter, 1" Diameter Probe, 150' Cable, Also Temperature/Level | EACH | 914.25 | 914.25 | 914.25 | 914.25 | 914.25 |
| 33 02 0543 | Conductivity Meter, 10' Cable, Probe, Also Temperature/Salinity | EACH | 870.79 | 870.79 | 870.79 | 870.79 | 870.79 |
| 33 02 0544 | Thermometer, Stainless Steel, Bi-Metal, 40 - 160 F, 12" Stem | EACH | 62.47 | 62.47 | 62.47 | 62.47 | 62.47 |
| 33 02 0545 | Thermometer, Thermocouple | EACH | 144.48 | 144.48 | 144.48 | 144.48 | 144.48 |
| 33 02 0546 | Thermometer, Digital | EACH | 38.50 | 38.50 | 38.50 | 38.50 | 38.50 |
| 33 02 0547 | pH Meter, Test Paper Kits, 200 Strips, Charts | EACH | 9.89 | 9.89 | 9.89 | 9.89 | 9.89 |
| 33 02 0548 | pH Meter, Digital, LCD, 0 - 14 pH Range | EACH | 361.81 | 361.81 | 361.81 | 361.81 | 361.81 |
| 33 02 0549 | Lysimeter, 1.90" Outside Diameter, PVC Tube Type - PVC Body/Teflon Filter | EACH | 565.44 | 488.90 | 428.73 | 353.14 | 342.99 |
| 33 02 0550 | Lysimeter, 1.90" Outside Diameter, Teflon Tube Type - Teflon Body/Filter | EACH | 636.22 | 559.67 | 499.50 | 423.91 | 413.76 |
| 33 02 0551 | Lysimeter, 2.38" Outside Diameter, PVC Tube Type - PVC Body/Teflon Filter | EACH | 565.44 | 488.90 | 428.73 | 353.14 | 342.99 |
| 33 02 0552 | Lysimeter, 2.38" Outside Diameter, Teflon Tube Type - Teflon Body/Filter | EACH | 544.92 | 468.38 | 408.21 | 332.62 | 322.47 |
| 33 02 0553 | Lysimeter, 1.90" Outside Diameter, PVC Cup Type - PVC Body/Teflon Filter | EACH | 565.44 | 488.90 | 428.73 | 353.14 | 342.99 |
| 33 02 0554 | Lysimeter, 1.90" Outside Diameter, Teflon Cup Type - Teflon Body/Filter | EACH | 630.42 | 553.87 | 493.70 | 418.12 | 407.97 |
| 33 02 0555 | Lysimeter, 1.90" Outside Diameter, PVC Deep Sample Type - Teflon Body/Filter | EACH | 644.64 | 568.09 | 507.92 | 432.33 | 422.18 |
| 33 02 0556 | Lysimeter, 1.90" Outside Diameter, Teflon Deep Sample Type - Teflon Body/Filter | EACH | 741.01 | 664.46 | 604.29 | 528.70 | 518.56 |
| 33 02 0557 | Lysimeter Head Assembly, All PVC Fittings | EACH | 577.72 | 501.17 | 441.00 | 365.41 | 355.26 |
| 33 02 0558 | Lysimeter Head Assembly, Teflon & PVC Fittings | EACH | 684.13 | 607.58 | 547.41 | 471.83 | 461.68 |

## Environmental Remediation: Assemblies Cost Book

*Page 3-266*

1998 by ECHOS. All rights reserved

# Cost Data for Remediation

## Sampling and Analysis

| Assembly | Description | Unit | A | B | C | D | E |
|---|---|---|---|---|---|---|---|
| | **Capital Costs** | | | | | | |
| 33 02 0559 | Silica Flour, 50 Lb Pail for Media Continuum | EACH | 174.04 | 151.08 | 133.03 | 110.35 | 107.31 |
| 33 02 0560 | Vacuum Pressure Hand Pump | EACH | 46.40 | 46.40 | 46.40 | 46.40 | 46.40 |
| 33 02 0561 | Nylon Tubing, 1/4" Outside Diameter | LF | 0.47 | 0.47 | 0.47 | 0.47 | 0.47 |
| 33 02 0562 | Teflon Tubing, 1/4" Outside Diameter | LF | 1.21 | 1.21 | 1.21 | 1.21 | 1.21 |
| 33 02 0601 | Drilling 2.5" Diameter Soil Borings, No Sampling | LF | 21.74 | 18.75 | 16.71 | 13.78 | 13.18 |
| 33 02 0602 | Stainless Steel Bottom Sampler | EACH | 378.97 | 378.97 | 378.97 | 378.97 | 378.97 |
| 33 02 0603 | Surface Soil Sampling Equipment | EACH | 280.80 | 280.80 | 280.80 | 280.80 | 280.80 |
| 33 02 0605 | Hand Auger Rental | DAY | 13.25 | 13.25 | 13.25 | 13.25 | 13.25 |
| 33 02 0612 | Soil Analysis | EACH | 506.30 | 506.30 | 506.30 | 506.30 | 506.30 |
| 33 02 0613 | Soil Recovery Probe, Unslotted, 7/8" x 1' | EACH | 121.64 | 121.64 | 121.64 | 121.64 | 121.64 |
| 33 02 0614 | Soil Recovery Probe, Unslotted, 7/8" x 2' or 9/8" x 1' | EACH | 140.10 | 140.10 | 140.10 | 140.10 | 140.10 |
| 33 02 0615 | Soil Recovery Probe, Unslotted, 9/8" x 2' | EACH | 151.23 | 151.23 | 151.23 | 151.23 | 151.23 |
| 33 02 0616 | Soil Recovery Probe, Liners, 7/8" per LF, Butyrate Plast | LF | 0.86 | 0.86 | 0.86 | 0.86 | 0.86 |
| 33 02 0617 | Soil Recovery Probe, Liners, 9/8" per LF, Butyrate Plast | LF | 1.41 | 1.41 | 1.41 | 1.41 | 1.41 |
| 33 02 0618 | Soil Recovery Probe, Liners, Stainless Steel, 7/8" | EACH | 6.51 | 6.51 | 6.51 | 6.51 | 6.51 |
| 33 02 0619 | Soil Recovery Probe, Liners, Stainless Steel, 9/8" | EACH | 9.54 | 9.54 | 9.54 | 9.54 | 9.54 |
| 33 02 0620 | Soil Gas Vapor Probe, Stainless Steel, Manual, Nonremovable Tip | EACH | 45.93 | 45.93 | 45.93 | 45.93 | 45.93 |
| 33 02 0621 | Soil Gas Vapor Probe, Stainless Steel, Electric, Rotary, Up/Down Hammer | EACH | 226.73 | 226.73 | 226.73 | 226.73 | 226.73 |
| 33 02 0622 | Screw Auger, High-carbon Steel, 1.5" x 6" | EACH | 79.89 | 79.89 | 79.89 | 79.89 | 79.89 |
| 33 02 0623 | Screw Auger, High-carbon Steel, 1.75" x 6" | EACH | 97.33 | 97.33 | 97.33 | 97.33 | 97.33 |
| 33 02 0624 | Soil Auger Buckets, Dutch, 3" Diameter, Carbon/Stainless Steel | EACH | 129.80 | 129.80 | 129.80 | 129.80 | 129.80 |
| 33 02 0625 | Soil Auger Buckets, Mud, 2" Diameter, Carbon/Stainless Steel | EACH | 120.12 | 120.12 | 120.12 | 120.12 | 120.12 |
| 33 02 0626 | Soil Auger Buckets, Reg, 2" Diameter, Carbon/Stainless Steel | EACH | 100.46 | 100.46 | 100.46 | 100.46 | 100.46 |
| 33 02 0627 | Soil Auger Buckets, Sand, 2" Diameter, Carbon/Stainless Steel | EACH | 114.29 | 114.29 | 114.29 | 114.29 | 114.29 |
| 33 02 0628 | Soil Recovery Auger Buckets, Stainless Steel, 2.25" x 8" | EACH | 342.05 | 342.05 | 342.05 | 342.05 | 342.05 |
| 33 02 0629 | Soil Recovery Auger Buckets, Stainless Steel, 3.25" x 10" | EACH | 352.12 | 352.12 | 352.12 | 352.12 | 352.12 |
| 33 02 0630 | Soil Recovery Auger Buckets, Stainless Steel with Carbon Bits, 2.25" x 8" | EACH | 315.75 | 315.75 | 315.75 | 315.75 | 315.75 |
| 33 02 0631 | Soil Recovery Auger Buckets, Stainless Steel with Carbon Bits, 3.25" x 10" | EACH | 325.82 | 325.82 | 325.82 | 325.82 | 325.82 |

**Environmental Remediation: Assemblies Cost Book**

# Cost Data for Remediation

## Sampling and Analysis

| Assembly | Description | Unit | Unit Cost by Safety Level | | | | |
|---|---|---|---|---|---|---|---|
| | | | A | B | C | D | E |
| | **Capital Costs** | | | | | | |
| 33 02 0632 | Soil Recovery Auger Buckets, Liners, Butyrate Plast, 3" x 10" or 2" x 8" | EACH | 5.13 | 5.13 | 5.13 | 5.13 | 5.13 |
| 33 02 0633 | Soil Recovery Auger Buckets, Liners, Stainless Steel, 3" x 10" or 2" x 8" | EACH | 17.14 | 17.14 | 17.14 | 17.14 | 17.14 |
| 33 02 0634 | Sludge Sampler, Stainless Steel, 3.25" x 12", Thread On | EACH | 378.97 | 378.97 | 378.97 | 378.97 | 378.97 |
| 33 02 0635 | Tensiometer Soil Moist Probe, Standard, 7/8" Outside Diameter x 18", Fixed | EACH | 83.69 | 83.69 | 83.69 | 83.69 | 83.69 |
| 33 02 0636 | Tensiometer Soil Moist Probe, Jet Fill, 7/8" Outside Diameter x 18", Fixed | EACH | 71.81 | 71.81 | 71.81 | 71.81 | 71.81 |
| 33 02 0637 | Tensiometer Soil Moist Probe, Quick Draw, 18", Quick Read | EACH | 412.34 | 412.34 | 412.34 | 412.34 | 412.34 |
| 33 02 0638 | Soil Test Kit, 50 Tests | EACH | 278.25 | 278.25 | 278.25 | 278.25 | 278.25 |
| 33 02 0639 | CPT Rig, Includes Labor, Sampling, Punching, Decontamination | DAY | 2,844 | 2,844 | 2,844 | 2,844 | 2,844 |
| 33 02 0640 | Mobilize/Demobilize CPT Rig and Crew | EACH | 2,986 | 2,986 | 2,986 | 2,986 | 2,986 |
| 33 02 0641 | Standby Time for CPT Rig and Crew | HOUR | 3,015 | 3,015 | 3,015 | 3,015 | 3,015 |
| 33 02 0642 | Setup Cost per Each Hole, CPT Rig | EACH | 273.83 | 273.83 | 273.83 | 273.83 | 273.83 |
| 33 02 0643 | Decontamination Trailer Rental for CPT Rig | DAY | 92.75 | 92.75 | 92.75 | 92.75 | 92.75 |
| 33 02 0644 | Decontaminate Cone on CPT Rig | DAY | 273.83 | 273.83 | 273.83 | 273.83 | 273.83 |
| 33 02 0645 | Level "D" PPE Rental per 2-Man CPT Crew | DAY | 2,874 | 2,874 | 2,874 | 2,874 | 2,874 |
| 33 02 0646 | Level "C" PPE Rental per 2-Man CPT Crew | DAY | 3,122 | 3,122 | 3,122 | 3,122 | 3,122 |
| 33 02 0647 | In Situ Batch Sampling for Groundwater, CPT Rig (per Sample) | EACH | 92.75 | 92.75 | 92.75 | 92.75 | 92.75 |
| 33 02 0648 | In Situ Soil Sampling, CPT Rig (per Sample) | EACH | 60.95 | 60.95 | 60.95 | 60.95 | 60.95 |
| 33 02 0649 | Hydropunch with CPT Rig, Hard Soil | FT | 10.60 | 10.60 | 10.60 | 10.60 | 10.60 |
| 33 02 0650 | Hydropunch with CPT Rig, Medium Soil | FT | 10.60 | 10.60 | 10.60 | 10.60 | 10.60 |
| 33 02 0651 | Hydropunch with CPT Rig, Soft Soil | FT | 10.60 | 10.60 | 10.60 | 10.60 | 10.60 |
| 33 02 0652 | Grout Hole After Hydropunching | FT | 4.38 | 4.38 | 4.38 | 4.38 | 4.38 |
| 33 02 1101 | Geotechnical Characteristics Analysis | EACH | 194.42 | 155.02 | 138.16 | 100.07 | 86.11 |
| 33 02 1102 | Soil Moisture Content ASTM D2216 | EACH | 22.50 | 22.50 | 22.50 | 22.50 | 22.50 |
| 33 02 1501 | Saturation Indicator | EACH | 106.53 | 106.53 | 106.53 | 106.53 | 106.53 |
| 33 02 1502 | Thermostat & Humidity Control Devices | EACH | 329.49 | 279.97 | 258.78 | 210.91 | 193.37 |
| 33 02 1506 | Monitoring Port with Gas Monitor | EACH | 28.81 | 22.42 | 19.69 | 13.52 | 11.26 |
| 33 02 1507 | Continuous Monitoring and Recording of Air Flow | EACH | 9,870 | 9,870 | 9,870 | 9,870 | 9,870 |
| 33 02 1509 | Water Quality Parameter Testing Device | WEEK | 227.90 | 227.90 | 227.90 | 227.90 | 227.90 |
| 33 02 1601 | Specific Conductance (EPA 120.1) | EACH | 10.33 | 10.33 | 10.33 | 10.33 | 10.33 |

**Environmental Remediation: Assemblies Cost Book**

# Cost Data for Remediation

## Sampling and Analysis

| Assembly | Description | Unit | Unit Cost by Safety Level A | B | C | D | E |
|---|---|---|---|---|---|---|---|
| | **Capital Costs** | | | | | | |
| 33 02 1602 | pH (EPA 150.1) | EACH | 10.00 | 10.00 | 10.00 | 10.00 | 10.00 |
| 33 02 1603 | Total Dissolved Solids (EPA 160.1) | EACH | 15.00 | 15.00 | 15.00 | 15.00 | 15.00 |
| 33 02 1604 | Total Suspended Solids (EPA 160.2) | EACH | 15.00 | 15.00 | 15.00 | 15.00 | 15.00 |
| 33 02 1605 | Total Solids (EPA 160.3) | EACH | 14.33 | 14.33 | 14.33 | 14.33 | 14.33 |
| 33 02 1606 | Total Volatile Solids (EPA 160.4) | EACH | 18.67 | 18.67 | 18.67 | 18.67 | 18.67 |
| 33 02 1607 | Temperature (EPA 170.1) | EACH | 10.00 | 10.00 | 10.00 | 10.00 | 10.00 |
| 33 02 1608 | Nitrogen/Nitrite/Nitrate (EPA 300.1/354.1) | EACH | 19.67 | 19.67 | 19.67 | 19.67 | 19.67 |
| 33 02 1609 | Acidity/Alkalinity (EPA 305.1) | EACH | 15.00 | 15.00 | 15.00 | 15.00 | 15.00 |
| 33 02 1610 | Cyanide (EPA 335.2) | EACH | 45.00 | 45.00 | 45.00 | 45.00 | 45.00 |
| 33 02 1611 | Ammonia Nitrogen (EPA 350.2) | EACH | 30.00 | 30.00 | 30.00 | 30.00 | 30.00 |
| 33 02 1612 | Phosphorus (EPA 365.3) | EACH | 28.00 | 28.00 | 28.00 | 28.00 | 28.00 |
| 33 02 1613 | Oil and Grease (EPA 413.2) | EACH | 55.00 | 55.00 | 55.00 | 55.00 | 55.00 |
| 33 02 1614 | Total Petroleum Hydrocarbons (EPA 418.1) | EACH | 66.67 | 66.67 | 66.67 | 66.67 | 66.67 |
| 33 02 1615 | Phenols (Method 420.1) | EACH | 43.67 | 43.67 | 43.67 | 43.67 | 43.67 |
| 33 02 1617 | Pesticides/PCBs (EPA 608) | EACH | 206.67 | 206.67 | 206.67 | 206.67 | 206.67 |
| 33 02 1618 | Volatile Organic Analysis (EPA 624) | EACH | 175.00 | 175.00 | 175.00 | 175.00 | 175.00 |
| 33 02 1619 | Base Neutral & Acid Extractable Organics (EPA 625) | EACH | 456.67 | 456.67 | 456.67 | 456.67 | 456.67 |
| 33 02 1620 | Target Analyte List Metals (EPA 6010/7000s), Water | EACH | 289.67 | 289.67 | 289.67 | 289.67 | 289.67 |
| 33 02 1621 | Purgeable Halocarbons (EPA 601) | EACH | 113.33 | 113.33 | 113.33 | 113.33 | 113.33 |
| 33 02 1622 | Purgeable Aromatics (EPA 602) | EACH | 96.67 | 96.67 | 96.67 | 96.67 | 96.67 |
| 33 02 1623 | Acrolein & Acrylonitrile (EPA 603) | EACH | 113.33 | 113.33 | 113.33 | 113.33 | 113.33 |
| 33 02 1624 | Phenols (EPA 604) | EACH | 166.67 | 166.67 | 166.67 | 166.67 | 166.67 |
| 33 02 1625 | Benzidines (EPA 605) | EACH | 200.00 | 200.00 | 200.00 | 200.00 | 200.00 |
| 33 02 1626 | Phthalate Esters (EPA 606) | EACH | 180.00 | 180.00 | 180.00 | 180.00 | 180.00 |
| 33 02 1627 | Nitrosamines (EPA 607) | EACH | 166.67 | 166.67 | 166.67 | 166.67 | 166.67 |
| 33 02 1628 | Nitroaromatics & Isophorone (EPA 609) | EACH | 180.00 | 180.00 | 180.00 | 180.00 | 180.00 |
| 33 02 1629 | Polynuclear Aromatic Hydrocarbons, PAH (EPA 610) | EACH | 188.33 | 188.33 | 188.33 | 188.33 | 188.33 |
| 33 02 1630 | Haloethers (EPA 611) | EACH | 150.00 | 150.00 | 150.00 | 150.00 | 150.00 |
| 33 02 1631 | Chlorinated Hydrocarbons (EPA 612) | EACH | 150.00 | 150.00 | 150.00 | 150.00 | 150.00 |
| 33 02 1632 | 2,3,7,8-Tetrachloro-Dibzo-P-Dioxin (EPA 613) | EACH | 182.50 | 182.50 | 182.50 | 182.50 | 182.50 |
| 33 02 1633 | Organophosphorus Pesticides (EPA 614) | EACH | 215.00 | 215.00 | 215.00 | 215.00 | 215.00 |
| 33 02 1634 | Chlorinated Phenoxy Acid Herbicide (EPA 615) | EACH | 225.00 | 225.00 | 225.00 | 225.00 | 225.00 |
| 33 02 1635 | Organochlorine Pesticides & PCBs (EPA 617) | EACH | 156.00 | 156.00 | 156.00 | 156.00 | 156.00 |

**Environmental Remediation: Assemblies Cost Book**

*Page 3-269*

1998 by ECHOS. All rights reserved

# Cost Data for Remediation

## Sampling and Analysis

| Assembly | Description | Unit | Unit Cost by Safety Level A | B | C | D | E |
|---|---|---|---|---|---|---|---|
| | **Capital Costs** | | | | | | |
| 33 02 1636 | Triazine Pesticides (EPA 619) | EACH | 183.33 | 183.33 | 183.33 | 183.33 | 183.33 |
| 33 02 1637 | Carbamate & Urea Pesticides (EPA 632) | EACH | 175.00 | 175.00 | 175.00 | 175.00 | 175.00 |
| 33 02 1638 | Color (EPA 110.2/110.3) | EACH | 17.00 | 17.00 | 17.00 | 17.00 | 17.00 |
| 33 02 1639 | Hardness, Total (EPA 130.2) | EACH | 15.50 | 15.50 | 15.50 | 15.50 | 15.50 |
| 33 02 1640 | Odor (EPA 140.1) | EACH | 11.33 | 11.33 | 11.33 | 11.33 | 11.33 |
| 33 02 1641 | Settleable Solids (EPA 160.5) | EACH | 13.67 | 13.67 | 13.67 | 13.67 | 13.67 |
| 33 02 1642 | Total Solids (EPA 166.3) | EACH | 14.33 | 14.33 | 14.33 | 14.33 | 14.33 |
| 33 02 1643 | Turbidity (EPA 180.1) | EACH | 12.67 | 12.67 | 12.67 | 12.67 | 12.67 |
| 33 02 1644 | Metals Screen, Flame, per Each (EPA 200S) | EACH | 11.33 | 11.33 | 11.33 | 11.33 | 11.33 |
| 33 02 1645 | Metals Screen, Furnace, per Each (EPA 200S) | EACH | 32.50 | 32.50 | 32.50 | 32.50 | 32.50 |
| 33 02 1646 | Arsenic, Hydride (EPA 206.3) | EACH | 30.00 | 30.00 | 30.00 | 30.00 | 30.00 |
| 33 02 1647 | Boron, Colorimetric (EPA 212.3) | EACH | 12.00 | 12.00 | 12.00 | 12.00 | 12.00 |
| 33 02 1648 | Chromium, Hexavalent (EPA 218.5) | EACH | 21.67 | 21.67 | 21.67 | 21.67 | 21.67 |
| 33 02 1649 | Mercury, Cold Vapor (EPA 245.1) | EACH | 32.50 | 32.50 | 32.50 | 32.50 | 32.50 |
| 33 02 1650 | Ion Chromatography (EPA 300) | EACH | 18.33 | 18.33 | 18.33 | 18.33 | 18.33 |
| 33 02 1651 | Alkalinity (EPA 310.1/310.2) | EACH | 15.00 | 15.00 | 15.00 | 15.00 | 15.00 |
| 33 02 1652 | Bromide, Titrimetric (EPA 320.1) | EACH | 32.50 | 32.50 | 32.50 | 32.50 | 32.50 |
| 33 02 1653 | Chloride (EPA 325.2/325.3) | EACH | 17.00 | 17.00 | 17.00 | 17.00 | 17.00 |
| 33 02 1654 | Chloride, Total Residual (EPA 330.2/330.5) | EACH | 16.67 | 16.67 | 16.67 | 16.67 | 16.67 |
| 33 02 1655 | Fluoride SPADNS (EPA 340.1) | EACH | 125.00 | 125.00 | 125.00 | 125.00 | 125.00 |
| 33 02 1656 | Fluoride Electrode (EPA 340.2) | EACH | 18.67 | 18.67 | 18.67 | 18.67 | 18.67 |
| 33 02 1657 | Iodide (EPA 345.1) | EACH | 28.33 | 28.33 | 28.33 | 28.33 | 28.33 |
| 33 02 1658 | Nitrogen, Nitrate-Nitrite (EPA 350.3) | EACH | 19.67 | 19.67 | 19.67 | 19.67 | 19.67 |
| 33 02 1659 | Nitrogen, Kjeldahl, Total (EPA 351.2/351.3/351.4) | EACH | 38.33 | 38.33 | 38.33 | 38.33 | 38.33 |
| 33 02 1660 | Nitrogen, Organic (EPA 351.3) | EACH | 58.00 | 58.00 | 58.00 | 58.00 | 58.00 |
| 33 02 1661 | Nitrogen, Nitrate (EPA 352.1) | EACH | 19.67 | 19.67 | 19.67 | 19.67 | 19.67 |
| 33 02 1662 | Nitrogen, Nitrate & Nitrite (EPA 353.2) | EACH | 58.00 | 58.00 | 58.00 | 58.00 | 58.00 |
| 33 02 1663 | Dissolved Oxygen (EPA 360.1) | EACH | 11.50 | 11.50 | 11.50 | 11.50 | 11.50 |
| 33 02 1664 | Phosphorus, Ortho (EPA 365.1) | EACH | 21.33 | 21.33 | 21.33 | 21.33 | 21.33 |
| 33 02 1665 | Phosphorus, Total (EPA 365.2/365.4) | EACH | 27.00 | 27.00 | 27.00 | 27.00 | 27.00 |
| 33 02 1666 | Silica, Dissolved (EPA 370.1) | EACH | 30.00 | 30.00 | 30.00 | 30.00 | 30.00 |
| 33 02 1667 | Sulfate (EPA 375.2/375.3/375.4) | EACH | 16.67 | 16.67 | 16.67 | 16.67 | 16.67 |
| 33 02 1668 | Sulfide (EPA 376.1) | EACH | 16.32 | 16.32 | 16.32 | 16.32 | 16.32 |

**Environmental Remediation: Assemblies Cost Book**

# Cost Data for Remediation

## Sampling and Analysis

| Assembly | Description | Unit | Unit Cost by Safety Level A | B | C | D | E |
|---|---|---|---|---|---|---|---|
| | **Capital Costs** | | | | | | |
| 33 02 1669 | Sulfite (EPA 377.1) | EACH | 16.32 | 16.32 | 16.32 | 16.32 | 16.32 |
| 33 02 1670 | Metals Screen, 25 Metals Listed in Method EPA 200.7 | EACH | 283.25 | 283.25 | 283.25 | 283.25 | 283.25 |
| 33 02 1671 | Biochemical Oxygen Demand, BOD (EPA 405.1) | EACH | 34.67 | 34.67 | 34.67 | 34.67 | 34.67 |
| 33 02 1672 | Chemical Oxygen Demand, COD (EPA 410.1/410.2/410.3/410.4) | EACH | 28.33 | 28.33 | 28.33 | 28.33 | 28.33 |
| 33 02 1673 | Total Organic Carbon, TOC (EPA 415.1/415.2) | EACH | 30.67 | 30.67 | 30.67 | 30.67 | 30.67 |
| 33 02 1674 | MBAS, Surfactants (EPA 425.1) | EACH | 38.00 | 38.00 | 38.00 | 38.00 | 38.00 |
| 33 02 1675 | NTA (EPA 430.2) | EACH | 25.00 | 25.00 | 25.00 | 25.00 | 25.00 |
| 33 02 1676 | Total Organic Halogens, TOX (EPA 450.1) | EACH | 75.00 | 75.00 | 75.00 | 75.00 | 75.00 |
| 33 02 1677 | Trihalomethanes (EPA 501.1) | EACH | 105.00 | 105.00 | 105.00 | 105.00 | 105.00 |
| 33 02 1688 | 1,2-Dibromo-3-Chloropropane (EPA 504) | EACH | 105.00 | 105.00 | 105.00 | 105.00 | 105.00 |
| 33 02 1689 | Organohalide Pesticides & Aroclors (EPA 505) | EACH | 156.00 | 156.00 | 156.00 | 156.00 | 156.00 |
| 33 02 1693 | Chlorinated Herbicides (EPA 515) | EACH | 225.00 | 225.00 | 225.00 | 225.00 | 225.00 |
| 33 02 1701 | EP Toxicity, Metals (EPA 1310) | EACH | 146.00 | 146.00 | 146.00 | 146.00 | 146.00 |
| 33 02 1702 | TCLP (RCRA) (EPA 1311) | EACH | 1,364 | 1,364 | 1,364 | 1,364 | 1,364 |
| 33 02 1703 | Chromium (SW 7191), with Prep | EACH | 28.33 | 28.33 | 28.33 | 28.33 | 28.33 |
| 33 02 1705 | Targeted TCLP (Metals, Volatiles, Semivolatiles only) | EACH | 669.33 | 669.33 | 669.33 | 669.33 | 669.33 |
| 33 02 1706 | Chromium (EPA 6010) | EACH | 12.00 | 12.00 | 12.00 | 12.00 | 12.00 |
| 33 02 1708 | Purgeable Aromatics (SW 5030/SW 8020) | EACH | 96.67 | 96.67 | 96.67 | 96.67 | 96.67 |
| 33 02 1709 | Target Analyte List Metals (EPA 6010/7000S), Soil | EACH | 289.67 | 289.67 | 289.67 | 289.67 | 289.67 |
| 33 02 1710 | Metals (EPA 6010), per Each Metal | EACH | 11.33 | 11.33 | 11.33 | 11.33 | 11.33 |
| 33 02 1711 | Arsenic (SW 7060), with Prep | EACH | 20.67 | 20.67 | 20.67 | 20.67 | 20.67 |
| 33 02 1712 | Mercury (SW 7470), with Prep | EACH | 29.67 | 29.67 | 29.67 | 29.67 | 29.67 |
| 33 02 1713 | Selenium (SW 7740), with Prep | EACH | 23.33 | 23.33 | 23.33 | 23.33 | 23.33 |
| 33 02 1714 | BTEX Purgeable Aromatics (SW 5030/SW 8020) | EACH | 81.67 | 81.67 | 81.67 | 81.67 | 81.67 |
| 33 02 1715 | BTEX/Gasoline Hydrocarbons (Mod 8020)(PID/FID), with Prep | EACH | 83.33 | 83.33 | 83.33 | 83.33 | 83.33 |
| 33 02 1716 | Phenols (SW 3550/SW 8040) | EACH | 75.33 | 75.33 | 75.33 | 75.33 | 75.33 |
| 33 02 1717 | Pesticides/PCBs (SW 3550/SW 8080) | EACH | 206.67 | 206.67 | 206.67 | 206.67 | 206.67 |
| 33 02 1718 | Chlordane (SW 3550/SW 8080) | EACH | 213.33 | 213.33 | 213.33 | 213.33 | 213.33 |
| 33 02 1719 | Chlorinated Phenoxy Acid Herbicides (SW 3550/SW 8150) | EACH | 233.33 | 233.33 | 233.33 | 233.33 | 233.33 |
| 33 02 1720 | Volatile Organic Analysis (SW 5030/SW 8240) | EACH | 175.00 | 175.00 | 175.00 | 175.00 | 175.00 |
| 33 02 1721 | Base/Neutral & Acid Extractable Organics (SW 3550/SW 8270) | EACH | 456.67 | 456.67 | 456.67 | 456.67 | 456.67 |

---

**Environmental Remediation: Assemblies Cost Book**

# Cost Data for Remediation

## Sampling and Analysis

| Assembly | Description | Unit | Unit Cost by Safety Level A | B | C | D | E |
|---|---|---|---|---|---|---|---|
| | **Capital Costs** | | | | | | |
| 33 02 1722 | Polynuclear Aromatic Hydrocarbons (PAH) (SW 8310), with Prep | EACH | 188.33 | 188.33 | 188.33 | 188.33 | 188.33 |
| 33 02 1723 | Cyanide (EPA 9010) | EACH | 48.33 | 48.33 | 48.33 | 48.33 | 48.33 |
| 33 02 1724 | Phenols (EPA 9065) | EACH | 43.67 | 43.67 | 43.67 | 43.67 | 43.67 |
| 33 02 1725 | CEC (EPA 9081) | EACH | 142.50 | 142.50 | 142.50 | 142.50 | 142.50 |
| 33 02 1726 | Diesel Hydrocarbon, Mod 8015, GC/FID, with Prep | EACH | 107.50 | 107.50 | 107.50 | 107.50 | 107.50 |
| 33 02 1728 | Gasoline Hydrocarbon, Mod 8015, GC/FID, with Prep | EACH | 86.67 | 86.67 | 86.67 | 86.67 | 86.67 |
| 33 02 1729 | Hydrocarbon Screen GC/FID, Mod 8015, with Prep | EACH | 131.67 | 131.67 | 131.67 | 131.67 | 131.67 |
| 33 02 1730 | Sludge Constituents and Characteristics Analysis Tests | EACH | 143.00 | 143.00 | 143.00 | 143.00 | 143.00 |
| 33 02 1731 | Purgeable Halocarbons (SW 5030/SW 8010) | EACH | 113.33 | 113.33 | 113.33 | 113.33 | 113.33 |
| 33 02 1732 | Total Petroleum Hydrocarbons (SW 5030/SW 8015) | EACH | 66.67 | 66.67 | 66.67 | 66.67 | 66.67 |
| 33 02 1733 | Acrolein, Acrylonitrile, Acetonitrile (SW 5030/SW 8030) | EACH | 113.33 | 113.33 | 113.33 | 113.33 | 113.33 |
| 33 02 1734 | Phthalate Esters (SW 3550/SW 8060/SW 8061) | EACH | 180.00 | 180.00 | 180.00 | 180.00 | 180.00 |
| 33 02 1735 | Nitroaromatics & Cyclic Ketones (SW 3550/SW 8090/SW8091) | EACH | 180.00 | 180.00 | 180.00 | 180.00 | 180.00 |
| 33 02 1736 | Polynuclear Aromatic Hydrocarbons (SW 3550/SW 8100) | EACH | 188.33 | 188.33 | 188.33 | 188.33 | 188.33 |
| 33 02 1737 | Chlorinated Hydrocarbons (SW 3550/SW 8120/SW 8121) | EACH | 150.00 | 150.00 | 150.00 | 150.00 | 150.00 |
| 33 02 1738 | Organophosphorus Pesticides (SW 3550/SW 8140/SW 8141) | EACH | 215.00 | 215.00 | 215.00 | 215.00 | 215.00 |
| 33 02 1739 | Semivolatile Organics, Packed Column (SW 8250), with Prep | EACH | 270.00 | 270.00 | 270.00 | 270.00 | 270.00 |
| 33 02 1740 | Dioxins & Dibenzofurans (SW 3550/SW 8280) | EACH | 190.00 | 190.00 | 190.00 | 190.00 | 190.00 |
| 33 02 1741 | Total Organic Halides, TOX (EPA 9020/9021) | EACH | 75.00 | 75.00 | 75.00 | 75.00 | 75.00 |
| 33 02 1742 | Total Dissolved Sulfide (EPA 9030) | EACH | 16.67 | 16.67 | 16.67 | 16.67 | 16.67 |
| 33 02 1743 | Sulfate (EPA 9035/9036/9038) | EACH | 16.32 | 16.32 | 16.32 | 16.32 | 16.32 |
| 33 02 1744 | pH (EPA 9040/9041/9045) | EACH | 10.00 | 10.00 | 10.00 | 10.00 | 10.00 |
| 33 02 1745 | Specific Conductance, Conductivity (EPA 9050) | EACH | 10.00 | 10.00 | 10.00 | 10.00 | 10.00 |
| 33 02 1746 | Total Organic Carbon, TOC (EPA 9060) | EACH | 35.67 | 35.67 | 35.67 | 35.67 | 35.67 |
| 33 02 1747 | Total Recoverable Oil & Grease (EPA 9070) | EACH | 65.00 | 65.00 | 65.00 | 65.00 | 65.00 |
| 33 02 1748 | Oil & Grease Extraction Method (EPA 9071) | EACH | 48.33 | 48.33 | 48.33 | 48.33 | 48.33 |
| 33 02 1749 | Waste/Membrane Liner Compatibility Test (EPA 9090) | EACH | 107.50 | 107.50 | 107.50 | 107.50 | 107.50 |
| 33 02 1750 | Paint Filter Test (EPA 9095) | EACH | 22.50 | 22.50 | 22.50 | 22.50 | 22.50 |
| 33 02 1751 | Saturated Hydraulic Conductivity/Leachate/Intrinsic Permeability (EPA 9100) | EACH | 261.50 | 261.50 | 261.50 | 261.50 | 261.50 |
| 33 02 1752 | Total Coliform (EPA 9131/9132) | EACH | 26.00 | 26.00 | 26.00 | 26.00 | 26.00 |

**Environmental Remediation: Assemblies Cost Book**

# Cost Data for Remediation

## Sampling and Analysis

| Assembly | Description | Unit | Unit Cost by Safety Level |||||
|---|---|---|---|---|---|---|---|
| | | | A | B | C | D | E |
| | **Capital Costs** | | | | | | |
| 33 02 1753 | Nitrate (EPA 9200) | EACH | 17.33 | 17.33 | 17.33 | 17.33 | 17.33 |
| 33 02 1754 | Chloride (EPA 9250/9251/9252) | EACH | 30.50 | 30.50 | 30.50 | 30.50 | 30.50 |
| 33 02 1756 | Ignitability (EPA 1010/1020) | EACH | 31.67 | 31.67 | 31.67 | 31.67 | 31.67 |
| 33 02 1757 | Corrosivity (EPA 1110) | EACH | 91.67 | 91.67 | 91.67 | 91.67 | 91.67 |
| 33 02 1758 | Reactivity, SW 846 Ch7 P7.4 | EACH | 78.33 | 78.33 | 78.33 | 78.33 | 78.33 |
| 33 02 1759 | Sorbent Cartridges from VOC Sampling (EPA 5040) | EACH | 140.83 | 140.83 | 140.83 | 140.83 | 140.83 |
| 33 02 1760 | Metals, Flame, 7000s, per Each | EACH | 11.33 | 11.33 | 11.33 | 11.33 | 11.33 |
| 33 02 1761 | Metals, Furnace, 7000s, per Each | EACH | 32.50 | 32.50 | 32.50 | 32.50 | 32.50 |
| 33 02 1762 | Chromium, Hexavalent (SW 7195), with Prep | EACH | 17.50 | 17.50 | 17.50 | 17.50 | 17.50 |
| 33 02 1764 | Selenium, Hydride (SW 7741), with Prep | EACH | 38.33 | 38.33 | 38.33 | 38.33 | 38.33 |
| 33 02 1765 | Arsenic, Hydride (SW 7061), with Prep | EACH | 38.33 | 38.33 | 38.33 | 38.33 | 38.33 |
| 33 02 1766 | Total Petroleum Hydrocarbons (EPA 9073) | EACH | 66.67 | 66.67 | 66.67 | 66.67 | 66.67 |
| 33 02 1767 | Chromium, Total (SW 7190) with Prep | EACH | 16.00 | 16.00 | 16.00 | 16.00 | 16.00 |
| 33 02 1768 | Antimony (SW 7040) with Prep | EACH | 34.50 | 34.50 | 34.50 | 34.50 | 34.50 |
| 33 02 1769 | Antimony (SW 7041) with Prep | EACH | 25.00 | 25.00 | 25.00 | 25.00 | 25.00 |
| 33 02 1770 | Thallium (SW 7840) with Prep | EACH | 34.50 | 34.50 | 34.50 | 34.50 | 34.50 |
| 33 02 1771 | Thallium (SW 7841) with Prep | EACH | 25.00 | 25.00 | 25.00 | 25.00 | 25.00 |
| 33 02 1772 | Gasoline Group (Mod 8010/8020, Lead, EDB) | EACH | 90.00 | 90.00 | 90.00 | 90.00 | 90.00 |
| 33 02 1773 | Kerosene Group (Mod 8010/8020, 8100, TPH, Lead, EDB) | EACH | 90.00 | 90.00 | 90.00 | 90.00 | 90.00 |
| 33 02 1774 | BTEX/MTBE (Mod 8020) | EACH | 83.33 | 83.33 | 83.33 | 83.33 | 83.33 |
| 33 02 1775 | Purgeable Organics (SW 3550/SW 8260) | EACH | 175.00 | 175.00 | 175.00 | 175.00 | 175.00 |
| 33 02 1801 | VOST Analysis, Air (30/5040/8240) | EACH | 225.00 | 225.00 | 225.00 | 225.00 | 225.00 |
| 33 02 1802 | Principal Organic Hazardous Constituents, Air (30/5040/8240) | EACH | 192.50 | 192.50 | 192.50 | 192.50 | 192.50 |
| 33 02 1803 | Tentative ID Compounds, GC/MS, Air (30/5040/8240) | EACH | 257.00 | 257.00 | 257.00 | 257.00 | 257.00 |
| 33 02 1804 | Tedlar Bag (Single Component), Air (30/5040/8240) | EACH | 100.00 | 100.00 | 100.00 | 100.00 | 100.00 |
| 33 02 1805 | Sorbent Tube (Single Component), Air (5040/8240) | EACH | 100.00 | 100.00 | 100.00 | 100.00 | 100.00 |
| 33 02 1806 | BTEX, Air (5040/8020) | EACH | 95.00 | 95.00 | 95.00 | 95.00 | 95.00 |
| 33 02 1807 | BTEX/Total Volatile Petroleum Hydrocarbons, Air (5040/8020/8015) | EACH | 111.67 | 111.67 | 111.67 | 111.67 | 111.67 |
| 33 02 1808 | Semivolatiles, Air (10/3550/8270) | EACH | 531.67 | 531.67 | 531.67 | 531.67 | 531.67 |
| 33 02 1809 | Multimetal Train, Air, Various Methods | EACH | 230.50 | 230.50 | 230.50 | 230.50 | 230.50 |
| 33 02 1810 | Pesticides/PCBs, GC, Air, Various Methods | EACH | 150.00 | 150.00 | 150.00 | 150.00 | 150.00 |
| 33 02 1811 | Hydrocarbons, BTEX, Air (NIOSH 1500) | EACH | 102.33 | 102.33 | 102.33 | 102.33 | 102.33 |

**Environmental Remediation: Assemblies Cost Book**

*Page 3-273*

1998 by ECHOS. All rights reserved

# Cost Data for Remediation

## Sampling and Analysis

| Assembly | Description | Unit | A | B | C | D | E |
|---|---|---|---|---|---|---|---|
| | **Capital Costs** | | | | | | |
| 33 02 1812 | Polyaromatic Hydrocarbons, Air (NIOSH 5515) | EACH | 462.50 | 462.50 | 462.50 | 462.50 | 462.50 |
| 33 02 1813 | Metals by ICP, Air (NIOSH 7300) | EACH | 249.50 | 249.50 | 249.50 | 249.50 | 249.50 |
| 33 02 1814 | Metals by Graphite Furnace, Air (NIOSH 7300) | EACH | 281.33 | 281.33 | 281.33 | 281.33 | 281.33 |
| 33 02 1815 | Mercury, Air (245.2) | EACH | 35.00 | 35.00 | 35.00 | 35.00 | 35.00 |
| 33 02 1816 | Cyanide, Air (335.2) | EACH | 43.33 | 43.33 | 43.33 | 43.33 | 43.33 |
| 33 02 1817 | Ion Chromatography, Air | EACH | 18.00 | 18.00 | 18.00 | 18.00 | 18.00 |
| 33 02 1818 | Volatile Aromatic Halocarbons, Air, TO-1 | EACH | 95.00 | 95.00 | 95.00 | 95.00 | 95.00 |
| 33 02 1819 | Highly Volatile Nonpolar Organics, TO-2 | EACH | 305.00 | 305.00 | 305.00 | 305.00 | 305.00 |
| 33 02 1820 | Volatile Nonpolar Organics, TO-3 | EACH | 92.50 | 92.50 | 92.50 | 92.50 | 92.50 |
| 33 02 1821 | Organochlorine Pesticides/PCBs, Air, TO-4 | EACH | 242.50 | 242.50 | 242.50 | 242.50 | 242.50 |
| 33 02 1822 | Aldehydes & Ketones, Air, TO-5 | EACH | 112.50 | 112.50 | 112.50 | 112.50 | 112.50 |
| 33 02 1823 | Phosgene in Ambient Air, TO-6 | EACH | 210.00 | 210.00 | 210.00 | 210.00 | 210.00 |
| 33 02 1824 | N-Nitrosodimethylamine, Air, TO-7 | EACH | 182.50 | 182.50 | 182.50 | 182.50 | 182.50 |
| 33 02 1825 | Phenol & Methyl Phenols, Air, TO-8 | EACH | 72.50 | 72.50 | 72.50 | 72.50 | 72.50 |
| 33 02 1826 | Dioxins, Air, TO-9 | EACH | 322.50 | 322.50 | 322.50 | 322.50 | 322.50 |
| 33 02 1827 | Total Nonmethane Organics, Air, TO-12 | EACH | 50.00 | 50.00 | 50.00 | 50.00 | 50.00 |
| 33 02 1828 | Polyaromatic Hydrocarbons, Air, TO-13 | EACH | 100.00 | 100.00 | 100.00 | 100.00 | 100.00 |
| 33 02 1829 | Canister Samples by GC or GC/MS, TO-14 | EACH | 400.00 | 400.00 | 400.00 | 400.00 | 400.00 |
| 33 02 1830 | Hydrocarbon Speciation, C1-C8, GC/FID, Air (TO-12/14) | EACH | 75.00 | 75.00 | 75.00 | 75.00 | 75.00 |
| 33 02 1831 | Hydrocarbon Speciation, C8-C22, GC/FID, Air (TO-12/14) | EACH | 75.00 | 75.00 | 75.00 | 75.00 | 75.00 |
| 33 02 1832 | Hydrocarbon Speciation, C1-C22, GC/FID, Air (TO-12/14) | EACH | 75.00 | 75.00 | 75.00 | 75.00 | 75.00 |
| 33 02 1833 | Hydrocarbon Speciation, C4-C22, GC/FID, Air (TO-12/14) | EACH | 75.00 | 75.00 | 75.00 | 75.00 | 75.00 |
| 33 02 1901 | Freshwater Discharge, LC50, Bioassay Analysis | EACH | 566.67 | 566.67 | 566.67 | 566.67 | 566.67 |
| 33 02 1902 | Saltwater Discharge, Lc50, Bioassay Analysis | EACH | 550.00 | 550.00 | 550.00 | 550.00 | 550.00 |
| 33 02 1903 | Drilling Mud, Lc50, Bioassay Analysis | EACH | 666.67 | 666.67 | 666.67 | 666.67 | 666.67 |
| 33 02 1904 | Freshwater Chronic Toxicity Bioassay Analysis | EACH | 1,150 | 1,150 | 1,150 | 1,150 | 1,150 |
| 33 02 1905 | Saltwater Chronic Toxicity Bioassay Analysis | EACH | 1,483 | 1,483 | 1,483 | 1,483 | 1,483 |
| 33 02 1906 | Ames (Full) Microbiological Analysis | EACH | 1,783 | 1,783 | 1,783 | 1,783 | 1,783 |
| 33 02 1907 | Ames (Screen) Microbiological Analysis | EACH | 175.00 | 175.00 | 175.00 | 175.00 | 175.00 |
| 33 02 1908 | Daphnia IQ Toxicity Analysis | EACH | 316.67 | 316.67 | 316.67 | 316.67 | 316.67 |
| 33 02 1909 | Microtox Microbiological Analysis | EACH | 325.00 | 325.00 | 325.00 | 325.00 | 325.00 |
| 33 02 1910 | Total Heterotrophic Plate Count | EACH | 42.50 | 42.50 | 42.50 | 42.50 | 42.50 |
| 33 02 1911 | Hydrocarbon Degraders | EACH | 77.50 | 77.50 | 77.50 | 77.50 | 77.50 |

**Environmental Remediation: Assemblies Cost Book**

# Cost Data for Remediation

## Sampling and Analysis

| Assembly | Description | Unit | Unit Cost by Safety Level A | B | C | D | E |
|---|---|---|---|---|---|---|---|
| | **Capital Costs** | | | | | | |
| 33 02 1912 | Fluorescent Pseudomonas | EACH | 55.00 | 55.00 | 55.00 | 55.00 | 55.00 |
| 33 02 2001 | Sample Preparation (3005/3010/3020/3040/3050) | EACH | 10.00 | 10.00 | 10.00 | 10.00 | 10.00 |
| 33 02 2002 | Sample Preparation (3510/3520/3540/3550/5030) | EACH | 25.00 | 25.00 | 25.00 | 25.00 | 25.00 |
| 33 02 2003 | Sample Preparation, Waste Dilution (3580) | EACH | 20.00 | 20.00 | 20.00 | 20.00 | 20.00 |
| 33 02 2004 | Cleanup (3610/3611/3620/3630/3650/3660) | EACH | 25.00 | 25.00 | 25.00 | 25.00 | 25.00 |
| 33 02 2005 | Cleanup, Gel-Permeation (3640) | EACH | 25.00 | 25.00 | 25.00 | 25.00 | 25.00 |
| 33 02 2019 | 1 Gallon, 128 Oz, Glass Sample Container, Case of 4 | EACH | 12.80 | 12.80 | 12.80 | 12.80 | 12.80 |
| 33 02 2020 | 1 Liter, 32 Oz, Clear Wide Mouth Jar, Case of 12 | EACH | 44.37 | 44.37 | 44.37 | 44.37 | 44.37 |
| 33 02 2021 | 500 ml, 16 Oz, Clear Wide Mouth Jar, Case of 12 | EACH | 61.67 | 61.67 | 61.67 | 61.67 | 61.67 |
| 33 02 2022 | 250 ml, 8 Oz, Clear Wide Mouth Jar, Case of 24 | EACH | 47.09 | 47.09 | 47.09 | 47.09 | 47.09 |
| 33 02 2023 | 120 ml, 4 Oz, Clear Wide Mouth Jar, Case of 24 | EACH | 34.47 | 34.47 | 34.47 | 34.47 | 34.47 |
| 33 02 2024 | 1 Liter, 32 Oz, Boston Round Bottle, Case of 12 | EACH | 28.78 | 28.78 | 28.78 | 28.78 | 28.78 |
| 33 02 2025 | 500 ml, 16 Oz, Boston Round Bottle, Case of 12 | EACH | 19.91 | 19.91 | 19.91 | 19.91 | 19.91 |
| 33 02 2026 | 40 ml, Clear Vial, Case of 72 | EACH | 89.91 | 89.91 | 89.91 | 89.91 | 89.91 |
| 33 02 2027 | 20 ml, Clear Vial, Case of 72 | EACH | 74.92 | 74.92 | 74.92 | 74.92 | 74.92 |
| 33 02 2028 | 250 ml, 8 oz, Clear Wide Mouth Vial with Septa, Case of 24 | EACH | 58.01 | 58.01 | 58.01 | 58.01 | 58.01 |
| 33 02 2029 | 125 ml, 4 oz, Clear Wide Mouth Vial with Septa, Case of 24 | EACH | 53.53 | 53.53 | 53.53 | 53.53 | 53.53 |
| 33 02 2030 | 1 Liter, 32 oz, High-density Polyethylene Bottle, Case of 12 | EACH | 32.64 | 32.64 | 32.64 | 32.64 | 32.64 |
| 33 02 2031 | 500 ml, 16 oz, High-density Polyethylene Bottle, Case of 24 | EACH | 28.06 | 28.06 | 28.06 | 28.06 | 28.06 |
| 33 02 2032 | 250 ml, 8 oz, High-density Polyethylene Bottle, Case of 24 | EACH | 15.58 | 15.58 | 15.58 | 15.58 | 15.58 |
| 33 02 2033 | 125 ml, 4 oz, High-density Polyethylene Bottle, Case of 24 | EACH | 13.71 | 13.71 | 13.71 | 13.71 | 13.71 |
| 33 02 2034 | Custody Seals, Package of 10 | EACH | 15.60 | 15.60 | 15.60 | 15.60 | 15.60 |
| 33 02 2035 | Safe Transport Can Filled with Vermiculite, 1 Gallon, Case of 4 | EACH | 28.94 | 28.94 | 28.94 | 28.94 | 28.94 |
| 33 02 2036 | Documentation Package for QA Verification, Data & Benchwork | EACH | 100.00 | 100.00 | 100.00 | 100.00 | 100.00 |
| 33 02 2037 | Overnight Delivery, 8 oz Letter | EACH | 10.17 | 10.17 | 10.17 | 10.17 | 10.17 |
| 33 02 2038 | Overnight Delivery, 1 Lb Package | LB | 11.08 | 11.08 | 11.08 | 11.08 | 11.08 |
| 33 02 2039 | Overnight Delivery, 2 - 5 Lb Package | LB | 16.08 | 16.08 | 16.08 | 16.08 | 16.08 |
| 33 02 2040 | Overnight Delivery, 6 - 10 Lb Package | LB | 21.58 | 21.58 | 21.58 | 21.58 | 21.58 |
| 33 02 2041 | Overnight Delivery, 11 - 20 Lb Package | LB | 34.25 | 34.25 | 34.25 | 34.25 | 34.25 |
| 33 02 2042 | Overnight Delivery, 21 - 50 Lb Package | LB | 57.27 | 57.27 | 57.27 | 57.27 | 57.27 |
| 33 02 2043 | Overnight Delivery, 51 - 70 Lb Package | LB | 78.27 | 78.27 | 78.27 | 78.27 | 78.27 |
| 33 02 2044 | Overnight Delivery, 71 Lbs or Greater | LB | 171.25 | 171.25 | 171.25 | 171.25 | 171.25 |

**Environmental Remediation: Assemblies Cost Book**

# Cost Data for Remediation

## Sampling and Analysis

| Assembly | Description | Unit | Unit Cost by Safety Level A | B | C | D | E |
|---|---|---|---|---|---|---|---|
| | **Capital Costs** | | | | | | |
| 33 02 2045 | 48 Quart Ice Chest | EACH | 27.62 | 27.62 | 27.62 | 27.62 | 27.62 |
| 33 02 2046 | 60 Quart Ice Chest | EACH | 46.27 | 46.27 | 46.27 | 46.27 | 46.27 |
| 33 02 2047 | 80 Quart Ice Chest | EACH | 79.85 | 79.85 | 79.85 | 79.85 | 79.85 |
| 33 02 2048 | 102 Quart Ice Chest | EACH | 135.64 | 135.64 | 135.64 | 135.64 | 135.64 |
| 33 02 2049 | 120 Quart Ice Chest | EACH | 216.05 | 216.05 | 216.05 | 216.05 | 216.05 |
| 33 02 2050 | Blue Ice Soft Packs (Equivalent to 7 Lbs Ice) | EACH | 1.73 | 1.73 | 1.73 | 1.73 | 1.73 |
| 33 02 2101 | Color (SM 204B) | EACH | 17.00 | 17.00 | 17.00 | 17.00 | 17.00 |
| 33 02 2102 | Total Solids (SM 209A) | EACH | 14.33 | 14.33 | 14.33 | 14.33 | 14.33 |
| 33 02 2103 | Total Dissolved Solids (SM 209B) | EACH | 15.00 | 15.00 | 15.00 | 15.00 | 15.00 |
| 33 02 2104 | Total Suspended Solids (SM 209C) | EACH | 15.00 | 15.00 | 15.00 | 15.00 | 15.00 |
| 33 02 2105 | Total Volatile Solids (SM 209D) | EACH | 18.67 | 18.67 | 18.67 | 18.67 | 18.67 |
| 33 02 2106 | Turbidity (SM 214A) | EACH | 12.67 | 12.67 | 12.67 | 12.67 | 12.67 |
| 33 02 2107 | Hardness, Total (SM 314B) | EACH | 15.50 | 15.50 | 15.50 | 15.50 | 15.50 |
| 33 02 2108 | Acidity (SM 402) | EACH | 15.00 | 15.00 | 15.00 | 15.00 | 15.00 |
| 33 02 2109 | Alkalinity (SM 403) | EACH | 15.00 | 15.00 | 15.00 | 15.00 | 15.00 |
| 33 02 2110 | Bromide (SM 405) | EACH | 32.50 | 32.50 | 32.50 | 32.50 | 32.50 |
| 33 02 2111 | Chlorine, Residual (SM 408D) | EACH | 16.67 | 16.67 | 16.67 | 16.67 | 16.67 |
| 33 02 2112 | Cyanide, Total (SM 412B/412D) | EACH | 48.33 | 48.33 | 48.33 | 48.33 | 48.33 |
| 33 02 2113 | Fluoride, Electrode (SM 413B) | EACH | 18.67 | 18.67 | 18.67 | 18.67 | 18.67 |
| 33 02 2114 | Fluoride, SPADNS (SM 413C) | EACH | 125.00 | 125.00 | 125.00 | 125.00 | 125.00 |
| 33 02 2115 | Nitrogen, Total (SM 420) | EACH | 38.33 | 38.33 | 38.33 | 38.33 | 38.33 |
| 33 02 2116 | Sulfite (SM 428A) | EACH | 16.32 | 16.32 | 16.32 | 16.32 | 16.32 |
| 33 02 2117 | Oil & Grease (SM 503A) | EACH | 55.00 | 55.00 | 55.00 | 55.00 | 55.00 |
| 33 02 2118 | Total Organic Carbon, TOC (SM 505B) | EACH | 30.67 | 30.67 | 30.67 | 30.67 | 30.67 |
| 33 02 2119 | Biochemical Oxygen Demand BOD (SM 507) | EACH | 34.67 | 34.67 | 34.67 | 34.67 | 34.67 |
| 33 02 2120 | Chemical Oxygen Demand COD (SM 508C) | EACH | 28.33 | 28.33 | 28.33 | 28.33 | 28.33 |
| 33 02 2121 | Surfactants (SM 512B) | EACH | 38.00 | 38.00 | 38.00 | 38.00 | 38.00 |
| 33 02 2122 | Fecal Coliform (SM 909C) | EACH | 26.00 | 26.00 | 26.00 | 26.00 | 26.00 |
| 33 02 2124 | Metal Analysis, Priority 17 Metals | EACH | 131.00 | 131.00 | 131.00 | 131.00 | 131.00 |
| 33 02 2125 | Arsenic, Hydride (SW 7061) with Prep | EACH | 181.33 | 181.33 | 181.33 | 181.33 | 181.33 |
| 33 02 2126 | Arsenic, (SW 7060) with Prep | EACH | 29.00 | 29.00 | 29.00 | 29.00 | 29.00 |
| 33 02 2127 | Metals Screen, 25 Listed in SW 3005/SW 6010 | EACH | 283.33 | 283.33 | 283.33 | 283.33 | 283.33 |
| 33 02 2128 | Chromium, Hexavalent (SW 7195) with Prep | EACH | 21.67 | 21.67 | 21.67 | 21.67 | 21.67 |

**Environmental Remediation: Assemblies Cost Book**

*Page 3-276*

1998 by ECHOS. All rights reserved

# Cost Data for Remediation

## Sampling and Analysis

| Assembly | Description | Unit | Unit Cost by Safety Level | | | | |
|---|---|---|---|---|---|---|---|
| | | | A | B | C | D | E |
| | **Capital Costs** | | | | | | |
| 33 02 2129 | Mercury, Cold Vapor (SW 7470) with Prep | EACH | 41.67 | 41.67 | 41.67 | 41.67 | 41.67 |
| 33 02 2130 | Cyanide (SW 9010) with Prep | EACH | 43.33 | 43.33 | 43.33 | 43.33 | 43.33 |
| 33 02 2131 | Purgeable Halocarbons (SW 5030/SW 8010) | EACH | 121.67 | 121.67 | 121.67 | 121.67 | 121.67 |
| 33 02 2132 | Purgeable Aromatics (SW 5030/SW 8020) | EACH | 96.67 | 96.67 | 96.67 | 96.67 | 96.67 |
| 33 02 2133 | Pesticides/PCBs (SW 3510/SW 8080) | EACH | 206.67 | 206.67 | 206.67 | 206.67 | 206.67 |
| 33 02 2134 | Polynuclear Aromatic Hydrocarbons, PAH (SW 3510/SW 8310) | EACH | 188.33 | 188.33 | 188.33 | 188.33 | 188.33 |
| 33 02 2135 | Base Neutral & Acid Extractable Organics (SW 3510/SW 8270) | EACH | 434.67 | 434.67 | 434.67 | 434.67 | 434.67 |
| 33 02 2136 | Carbamate & Urea Pesticides (SW 3510/SW 8318) | EACH | 205.00 | 205.00 | 205.00 | 205.00 | 205.00 |
| 33 02 2137 | Gasoline Group (Mod EPA 601/602, Lead, EDB) | EACH | 186.67 | 186.67 | 186.67 | 186.67 | 186.67 |
| 33 02 2138 | Kerosene Group (Mod EPA 601/602, 610, TPH, Lead, EDB) | EACH | 186.67 | 186.67 | 186.67 | 186.67 | 186.67 |
| 33 02 2139 | BTEX/MTBE (Mod EPA 602) | EACH | 80.00 | 80.00 | 80.00 | 80.00 | 80.00 |
| 33 02 2140 | Ethylene Dibromide (EDB) (EPA 501.4) | EACH | 78.33 | 78.33 | 78.33 | 78.33 | 78.33 |
| 33 02 2141 | Chromium, Total (EPA 218.1) | EACH | 19.00 | 19.00 | 19.00 | 19.00 | 19.00 |
| 33 02 2142 | Titanium (EPA 283.2) | EACH | 18.00 | 18.00 | 18.00 | 18.00 | 18.00 |
| 33 02 2143 | Lead (SW 3005/SW 7421) | EACH | 41.67 | 41.67 | 41.67 | 41.67 | 41.67 |
| 33 02 2144 | Ethylene Dibromide (SW 8011) with Prep | EACH | 130.00 | 130.00 | 130.00 | 130.00 | 130.00 |
| 33 02 2145 | Dioxins (SW 8280) with Prep | EACH | 737.00 | 737.00 | 737.00 | 737.00 | 737.00 |
| 33 02 2146 | Selenium (SW 7740) with Prep | EACH | 36.67 | 36.67 | 36.67 | 36.67 | 36.67 |
| 33 02 2147 | Nonhalogenated Volatile Organics (SW 5030/SW 8015) | EACH | 110.00 | 110.00 | 110.00 | 110.00 | 110.00 |
| 33 02 2148 | Organophosphorous Pesticides (SW 3510/SW 8140) | EACH | 215.00 | 215.00 | 215.00 | 215.00 | 215.00 |
| 33 02 2149 | Chlorinated Phenoxy Herbicides (SW 3510/SW 8150) | EACH | 233.33 | 233.33 | 233.33 | 233.33 | 233.33 |
| 33 02 2201 | Tissue/Bone, Americium Isotopic, Alpha Spectroscopy | EACH | 146.67 | 146.67 | 146.67 | 146.67 | 146.67 |
| 33 02 2202 | Tissue/Bone, Curium Isotopic, Alpha Spectroscopy | EACH | 145.00 | 145.00 | 145.00 | 145.00 | 145.00 |
| 33 02 2203 | Tissue/Bone, Neptunium Isotopic, Alpha Spectroscopy | EACH | 145.00 | 145.00 | 145.00 | 145.00 | 145.00 |
| 33 02 2204 | Tissue/Bone, Plutonium Isotopic, Alpha Spectroscopy | EACH | 150.00 | 150.00 | 150.00 | 150.00 | 150.00 |
| 33 02 2205 | Tissue/Bone, Thorium Isotopic, Alpha Spectroscopy | EACH | 146.67 | 146.67 | 146.67 | 146.67 | 146.67 |
| 33 02 2206 | Tissue/Bone, Uranium Isotopic, Alpha Spectroscopy | EACH | 135.00 | 135.00 | 135.00 | 135.00 | 135.00 |
| 33 02 2207 | Tissue/Bone, Strontium - 89, 90, Gas Flow Proportional Counting | EACH | 166.67 | 166.67 | 166.67 | 166.67 | 166.67 |
| 33 02 2208 | Tissue/Bone, Strontium - 90, Gas Flow Proportional Counting | EACH | 155.67 | 155.67 | 155.67 | 155.67 | 155.67 |
| 33 02 2209 | Tissue/Bone, Technetium - 99, Gas Flow Proportional Counting | EACH | 128.33 | 128.33 | 128.33 | 128.33 | 128.33 |

**Environmental Remediation: Assemblies Cost Book**

# Cost Data for Remediation

## Sampling and Analysis

| Assembly | Description | Unit | A | B | C | D | E |
|---|---|---|---|---|---|---|---|
| | **Capital Costs** | | | | | | |
| 33 02 2210 | Tissue/Bone, Americium Isotopic, Gamma Spectroscopy | EACH | 131.67 | 131.67 | 131.67 | 131.67 | 131.67 |
| 33 02 2211 | Tissue/Bone, Gamma Isotopic, Gamma Spectroscopy | EACH | 140.00 | 140.00 | 140.00 | 140.00 | 140.00 |
| 33 02 2212 | Tissue/Bone, Iodine - 25 (Filter or Charcoal), Gamma Spectroscopy | EACH | 150.00 | 150.00 | 150.00 | 150.00 | 150.00 |
| 33 02 2213 | Tissue/Bone, Iodine - 129 (Filter or Charcoal), Gamma Spectroscopy | EACH | 150.00 | 150.00 | 150.00 | 150.00 | 150.00 |
| 33 02 2214 | Tissue/Bone, Iodine - 131 (Filter or Charcoal), Gamma Spectroscopy | EACH | 140.00 | 140.00 | 140.00 | 140.00 | 140.00 |
| 33 02 2215 | Tissue/Bone, Krypton - 85, Gamma Spectroscopy | EACH | 102.50 | 102.50 | 102.50 | 102.50 | 102.50 |
| 33 02 2216 | Tissue/Bone, Lead - 210, Gamma Spectroscopy | EACH | 131.67 | 131.67 | 131.67 | 131.67 | 131.67 |
| 33 02 2217 | Tissue/Bone, Radium - 226, 228, Gamma Spectroscopy | EACH | 145.00 | 145.00 | 145.00 | 145.00 | 145.00 |
| 33 02 2218 | Tissue/Bone, Uranium - Total, Gamma Spectroscopy | EACH | 147.50 | 147.50 | 147.50 | 147.50 | 147.50 |
| 33 02 2219 | Tissue/Bone, Carbon - 14, Liquid Scintillation | EACH | 132.67 | 132.67 | 132.67 | 132.67 | 132.67 |
| 33 02 2220 | Tissue/Bone, Uranium Isotopic, Fluorometric | EACH | 175.00 | 175.00 | 175.00 | 175.00 | 175.00 |
| 33 02 2221 | Tissue/Bone, Uranium Isotopic, Laser Phosphorometer | EACH | 75.00 | 75.00 | 75.00 | 75.00 | 75.00 |
| 33 02 2222 | Air, Americium Isotopic, Alpha Spectroscopy | EACH | 146.67 | 146.67 | 146.67 | 146.67 | 146.67 |
| 33 02 2223 | Air, Curium Isotopic, Alpha Spectroscopy | EACH | 145.00 | 145.00 | 145.00 | 145.00 | 145.00 |
| 33 02 2224 | Air, Neptunium - 237, Alpha Spectroscopy | EACH | 145.00 | 145.00 | 145.00 | 145.00 | 145.00 |
| 33 02 2225 | Air, Polonium - 210, Alpha Spectroscopy | EACH | 146.67 | 146.67 | 146.67 | 146.67 | 146.67 |
| 33 02 2226 | Air, Plutonium Isotopic, Alpha Spectroscopy | EACH | 150.00 | 150.00 | 150.00 | 150.00 | 150.00 |
| 33 02 2227 | Air, Radium - 226, Alpha Spectroscopy | EACH | 143.33 | 143.33 | 143.33 | 143.33 | 143.33 |
| 33 02 2228 | Air, Thorium Isotopic, Alpha Spectroscopy | EACH | 146.67 | 146.67 | 146.67 | 146.67 | 146.67 |
| 33 02 2229 | Air, Uranium Isotopic, Alpha Spectroscopy | EACH | 146.67 | 146.67 | 146.67 | 146.67 | 146.67 |
| 33 02 2230 | Air, Gross Beta - Total, Gas Flow Proportional Counting | EACH | 45.00 | 45.00 | 45.00 | 45.00 | 45.00 |
| 33 02 2231 | Air, Lead - 210, Gas Flow Proportional Counting | EACH | 146.67 | 146.67 | 146.67 | 146.67 | 146.67 |
| 33 02 2232 | Air, Strontium - 89, 90, Gas Flow Proportional Counting | EACH | 172.50 | 172.50 | 172.50 | 172.50 | 172.50 |
| 33 02 2233 | Air, Strontium - 90, Gas Flow Proportional Counting | EACH | 135.00 | 135.00 | 135.00 | 135.00 | 135.00 |
| 33 02 2234 | Air, Technetium - 99, Gas Flow Proportional Counting | EACH | 141.67 | 141.67 | 141.67 | 141.67 | 141.67 |
| 33 02 2235 | Air, Gamma Isotopic, Gamma Spectroscopy | EACH | 110.00 | 110.00 | 110.00 | 110.00 | 110.00 |
| 33 02 2236 | Air, Radium - 226, 228, Gamma Spectroscopy | EACH | 176.67 | 176.67 | 176.67 | 176.67 | 176.67 |
| 33 02 2237 | Air, Carbon - 14, Liquid Scintillation | EACH | 157.50 | 157.50 | 157.50 | 157.50 | 157.50 |
| 33 02 2238 | Air, Iron - 55, Liquid Scintillation | EACH | 145.00 | 145.00 | 145.00 | 145.00 | 145.00 |
| 33 02 2239 | Air, Gross Alpha - Total, Gas Flow Proportional Counting or Alpha Scintillation | EACH | 43.33 | 43.33 | 43.33 | 43.33 | 43.33 |
| 33 02 2240 | Air, Gross Alpha & Gross Beta - Total, Gas Flow | EACH | 43.33 | 43.33 | 43.33 | 43.33 | 43.33 |

## Environmental Remediation: Assemblies Cost Book

# Cost Data for Remediation

## Sampling and Analysis

| Assembly | Description | Unit | Unit Cost by Safety Level | | | | |
|---|---|---|---|---|---|---|---|
| | | | A | B | C | D | E |
| | **Capital Costs** | | | | | | |
| 33 02 2241 | Air, Plutonium Isotopic, Plutonium - 241, Alpha Spectroscopy | EACH | 141.67 | 141.67 | 141.67 | 141.67 | 141.67 |
| 33 02 2242 | Air, Radium - 226, Radon De-emanation | EACH | 78.33 | 78.33 | 78.33 | 78.33 | 78.33 |
| 33 02 2243 | Air, Uranium - Total, Laser Phosphorometer | EACH | 60.00 | 60.00 | 60.00 | 60.00 | 60.00 |
| 33 02 2244 | Air, Uranium - Total, Fluorometric | EACH | 60.00 | 60.00 | 60.00 | 60.00 | 60.00 |
| 33 02 2245 | Liquid, Americium Isotopic, Alpha Spectroscopy | EACH | 131.67 | 131.67 | 131.67 | 131.67 | 131.67 |
| 33 02 2246 | Liquid, Curium Isotopic, Alpha Spectroscopy | EACH | 138.33 | 138.33 | 138.33 | 138.33 | 138.33 |
| 33 02 2247 | Liquid, Neptunium - 237, Alpha Spectroscopy | EACH | 138.33 | 138.33 | 138.33 | 138.33 | 138.33 |
| 33 02 2248 | Liquid, Polonium - 210, Alpha Spectroscopy | EACH | 131.67 | 131.67 | 131.67 | 131.67 | 131.67 |
| 33 02 2249 | Liquid, Plutonium Isotopic, Alpha Spectroscopy | EACH | 150.00 | 150.00 | 150.00 | 150.00 | 150.00 |
| 33 02 2250 | Liquid, Radium - 226, Alpha Spectroscopy | EACH | 126.67 | 126.67 | 126.67 | 126.67 | 126.67 |
| 33 02 2251 | Liquid, Radium - 226, 228, Alpha Spectroscopy | EACH | 166.67 | 166.67 | 166.67 | 166.67 | 166.67 |
| 33 02 2252 | Liquid, Thorium Isotopic, Alpha Spectroscopy | EACH | 131.67 | 131.67 | 131.67 | 131.67 | 131.67 |
| 33 02 2253 | Liquid, Uranium Isotopic, Alpha Spectroscopy | EACH | 131.67 | 131.67 | 131.67 | 131.67 | 131.67 |
| 33 02 2254 | Liquid, Iodine - 131, Beta/Gamma Coincidence | EACH | 125.00 | 125.00 | 125.00 | 125.00 | 125.00 |
| 33 02 2255 | Liquid, Radium - 226, 228, Beta/Gamma Coincidence | EACH | 158.33 | 158.33 | 158.33 | 158.33 | 158.33 |
| 33 02 2256 | Liquid, Radium - 228, Beta/Gamma Coincidence | EACH | 98.33 | 98.33 | 98.33 | 98.33 | 98.33 |
| 33 02 2257 | Liquid, Gross Beta - Total, Gas Flow Proportional Count | EACH | 43.33 | 43.33 | 43.33 | 43.33 | 43.33 |
| 33 02 2258 | Liquid, Calcium - 45, Gas Flow Proportional Counting | EACH | 145.00 | 145.00 | 145.00 | 145.00 | 145.00 |
| 33 02 2259 | Liquid, Cerium - 144, Gas Flow Proportional Counting | EACH | 135.00 | 135.00 | 135.00 | 135.00 | 135.00 |
| 33 02 2260 | Liquid, Chlorine - 36, Gas Flow Proportional Counting | EACH | 135.00 | 135.00 | 135.00 | 135.00 | 135.00 |
| 33 02 2261 | Liquid, Iodine - 131, Gas Flow Proportional Counting | EACH | 135.00 | 135.00 | 135.00 | 135.00 | 135.00 |
| 33 02 2262 | Liquid, Lead - 210, Gas Flow Proportional Counting | EACH | 135.00 | 135.00 | 135.00 | 135.00 | 135.00 |
| 33 02 2263 | Liquid, Radium - 226, 228, Gas Flow Proportional Count | EACH | 166.67 | 166.67 | 166.67 | 166.67 | 166.67 |
| 33 02 2264 | Liquid, Radium - 228, Gas Flow Proportional Counting | EACH | 126.67 | 126.67 | 126.67 | 126.67 | 126.67 |
| 33 02 2265 | Liquid, Ruthenium - 106, Gas Flow Proportional Counting | EACH | 145.00 | 145.00 | 145.00 | 145.00 | 145.00 |
| 33 02 2266 | Liquid, Strontium - 89, 90, Gas Flow Proportional Count | EACH | 148.33 | 148.33 | 148.33 | 148.33 | 148.33 |
| 33 02 2267 | Liquid, Strontium - 90, Gas Flow Proportional Counting | EACH | 138.33 | 138.33 | 138.33 | 138.33 | 138.33 |
| 33 02 2268 | Liquid, Sulfur - 35, Gas Flow Proportional Counting | EACH | 177.50 | 177.50 | 177.50 | 177.50 | 177.50 |
| 33 02 2269 | Liquid, Technetium - 99, Gas Flow Proportional Counting | EACH | 136.67 | 136.67 | 136.67 | 136.67 | 136.67 |
| 33 02 2270 | Liquid, Cerium - 144, Gamma Spectroscopy | EACH | 89.50 | 89.50 | 89.50 | 89.50 | 89.50 |
| 33 02 2271 | Liquid, Gamma Isotopic, Gamma Spectroscopy | EACH | 89.50 | 89.50 | 89.50 | 89.50 | 89.50 |
| 33 02 2272 | Liquid, Iodine - 125, Gamma Spectroscopy | EACH | 89.50 | 89.50 | 89.50 | 89.50 | 89.50 |
| 33 02 2273 | Liquid Iodine - 129, Gamma Spectroscopy | EACH | 89.50 | 89.50 | 89.50 | 89.50 | 89.50 |

**Environmental Remediation: Assemblies Cost Book**

# Cost Data for Remediation

## Sampling and Analysis

| Assembly | Description | Unit | Unit Cost by Safety Level | | | | |
|---|---|---|---|---|---|---|---|
| | | | A | B | C | D | E |
| | **Capital Costs** | | | | | | |
| 33 02 2274 | Liquid, Ruthenium - 106, Gamma Spectroscopy | EACH | 89.50 | 89.50 | 89.50 | 89.50 | 89.50 |
| 33 02 2275 | Liquid, Sulfur - 35, Gamma Spectroscopy | EACH | 89.50 | 89.50 | 89.50 | 89.50 | 89.50 |
| 33 02 2276 | Liquid, Chlorine - 36, Liquid Scintillation | EACH | 135.00 | 135.00 | 135.00 | 135.00 | 135.00 |
| 33 02 2277 | Liquid, Carbon - 14, Liquid Scintillation | EACH | 142.50 | 142.50 | 142.50 | 142.50 | 142.50 |
| 33 02 2278 | Liquid, Iron - 55, Liquid Scintillation | EACH | 145.00 | 145.00 | 145.00 | 145.00 | 145.00 |
| 33 02 2279 | Liquid, Iron - 125, Liquid Scintillation | EACH | 100.00 | 100.00 | 100.00 | 100.00 | 100.00 |
| 33 02 2280 | Liquid, Iron - 129, Liquid Scintillation | EACH | 100.00 | 100.00 | 100.00 | 100.00 | 100.00 |
| 33 02 2281 | Liquid, Nickel - 63, Liquid Scintillation | EACH | 145.00 | 145.00 | 145.00 | 145.00 | 145.00 |
| 33 02 2282 | Liquid, Promethium - 147, Liquid Scintillation | EACH | 135.00 | 135.00 | 135.00 | 135.00 | 135.00 |
| 33 02 2283 | Liquid, Sulfur - 35, Liquid Scintillation | EACH | 135.00 | 135.00 | 135.00 | 135.00 | 135.00 |
| 33 02 2284 | Liquid, Tritium, (Direct Counting), Liquid Scintillation | EACH | 67.50 | 67.50 | 67.50 | 67.50 | 67.50 |
| 33 02 2285 | Liquid, Tritium, (Distillation), Liquid Scintillation | EACH | 71.67 | 71.67 | 71.67 | 71.67 | 71.67 |
| 33 02 2286 | Liquid, Tritium, (Electrolytic Enrichment), Liquid Scintillation | EACH | 72.50 | 72.50 | 72.50 | 72.50 | 72.50 |
| 33 02 2287 | Liquid, Gross Alpha - Total, Gas Flow &/or Alpha Spectroscopy | EACH | 55.00 | 55.00 | 55.00 | 55.00 | 55.00 |
| 33 02 2288 | Liquid, Suspended or Dissolved, Gross Alpha/Gross Beta | EACH | 80.00 | 80.00 | 80.00 | 80.00 | 80.00 |
| 33 02 2289 | Liquid, Suspended & Dissolved, Gross Alpha/Gross Beta | EACH | 80.00 | 80.00 | 80.00 | 80.00 | 80.00 |
| 33 02 2290 | Liquid, Gross Alpha & Gross Beta - Total | EACH | 55.00 | 55.00 | 55.00 | 55.00 | 55.00 |
| 33 02 2291 | Liquid, Plutonium Isotopic/Plutonium - 241, Alpha Spectroscopy | EACH | 178.33 | 178.33 | 178.33 | 178.33 | 178.33 |
| 33 02 2292 | Liquid, Total Radium, Alpha Scintillation | EACH | 70.00 | 70.00 | 70.00 | 70.00 | 70.00 |
| 33 02 2293 | Liquid, Radium - 226, Radon De-emanation | EACH | 78.33 | 78.33 | 78.33 | 78.33 | 78.33 |
| 33 02 2294 | Liquid, Radium - 226, 228, Radon De-emanation | EACH | 172.50 | 172.50 | 172.50 | 172.50 | 172.50 |
| 33 02 2295 | Liquid, Radon 222 | EACH | 70.00 | 70.00 | 70.00 | 70.00 | 70.00 |
| 33 02 2296 | Liquid, Uranium - Total, Laser Phosphorometer | EACH | 70.00 | 70.00 | 70.00 | 70.00 | 70.00 |
| 33 02 2297 | Liquid, Uranium - Total, Fluorometric | EACH | 50.00 | 50.00 | 50.00 | 50.00 | 50.00 |
| 33 02 2301 | Urine & Feces, Americium Isotopic, Alpha Spectroscopy | EACH | 142.50 | 142.50 | 142.50 | 142.50 | 142.50 |
| 33 02 2302 | Urine & Feces, Curium Isotopic, Alpha Spectroscopy | EACH | 152.50 | 152.50 | 152.50 | 152.50 | 152.50 |
| 33 02 2303 | Urine & Feces, Neptunium - 237, Alpha Spectroscopy | EACH | 152.50 | 152.50 | 152.50 | 152.50 | 152.50 |
| 33 02 2304 | Urine & Feces, Polonium - 210, Alpha Spectroscopy | EACH | 127.50 | 127.50 | 127.50 | 127.50 | 127.50 |
| 33 02 2305 | Urine & Feces, Plutonium Isotopic, Alpha Spectroscopy | EACH | 150.00 | 150.00 | 150.00 | 150.00 | 150.00 |
| 33 02 2306 | Urine & Feces, Thorium Isotopic, Alpha Spectroscopy | EACH | 137.50 | 137.50 | 137.50 | 137.50 | 137.50 |
| 33 02 2307 | Urine & Feces, Uranium Isotopic, Alpha Spectroscopy | EACH | 137.50 | 137.50 | 137.50 | 137.50 | 137.50 |

**Environmental Remediation: Assemblies Cost Book**

# Cost Data for Remediation

## Sampling and Analysis

| Assembly | Description | Unit | Unit Cost by Safety Level | | | | |
|---|---|---|---|---|---|---|---|
| | | | A | B | C | D | E |
| | **Capital Costs** | | | | | | |
| 33 02 2308 | Urine & Feces, Gross Beta - Total, Gas Flow Proportional Counting | EACH | 80.00 | 80.00 | 80.00 | 80.00 | 80.00 |
| 33 02 2309 | Urine & Feces, Ruthenium - 106, Gas Flow Proportional Counting | EACH | 122.50 | 122.50 | 122.50 | 122.50 | 122.50 |
| 33 02 2310 | Urine & Feces, Strontium - 89, 90, Gas Flow Proportional Counting | EACH | 180.00 | 180.00 | 180.00 | 180.00 | 180.00 |
| 33 02 2311 | Urine & Feces, Strontium - 90, Gas Flow Proportional Counting | EACH | 180.00 | 180.00 | 180.00 | 180.00 | 180.00 |
| 33 02 2312 | Urine & Feces, Cerium - 144, Gamma Spectroscopy | EACH | 122.50 | 122.50 | 122.50 | 122.50 | 122.50 |
| 33 02 2313 | Urine & Feces, Gamma Isotopic, Gamma Spectroscopy | EACH | 135.00 | 135.00 | 135.00 | 135.00 | 135.00 |
| 33 02 2314 | Urine & Feces, Iodine - 131, Gamma Spectroscopy | EACH | 135.00 | 135.00 | 135.00 | 135.00 | 135.00 |
| 33 02 2315 | Urine & Feces, Radium - 228, Gamma Spectroscopy | EACH | 135.00 | 135.00 | 135.00 | 135.00 | 135.00 |
| 33 02 2316 | Urine & Feces, Ruthenium - 106, Gamma Spectroscopy | EACH | 122.50 | 122.50 | 122.50 | 122.50 | 122.50 |
| 33 02 2317 | Urine & Feces, Carbon - 14, Liquid Scintillation | EACH | 165.00 | 165.00 | 165.00 | 165.00 | 165.00 |
| 33 02 2318 | Urine & Feces, Promethium - 147, Liquid Scintillation | EACH | 102.50 | 102.50 | 102.50 | 102.50 | 102.50 |
| 33 02 2319 | Urine & Feces, Phosphorus - 32, Liquid Scintillation | EACH | 395.00 | 395.00 | 395.00 | 395.00 | 395.00 |
| 33 02 2320 | Urine & Feces, Tritium (Direct Counting), Liquid Scintillation | EACH | 100.00 | 100.00 | 100.00 | 100.00 | 100.00 |
| 33 02 2321 | Urine & Feces, Tritium (Distillation), Liquid Scintillation | EACH | 125.00 | 125.00 | 125.00 | 125.00 | 125.00 |
| 33 02 2322 | Urine & Feces, Mixed Fission Products, Gas Flow | EACH | 95.00 | 95.00 | 95.00 | 95.00 | 95.00 |
| 33 02 2323 | Urine & Feces, Plutonium Isotopic, Plutonium - 241, Alpha Spectroscopy | EACH | 152.50 | 152.50 | 152.50 | 152.50 | 152.50 |
| 33 02 2324 | Urine & Feces, Radium - 226, Radon De-emanation | EACH | 87.50 | 87.50 | 87.50 | 87.50 | 87.50 |
| 33 02 2325 | Urine & Feces, Radium - 226, 228, Radon De-emanation | EACH | 145.00 | 145.00 | 145.00 | 145.00 | 145.00 |
| 33 02 2326 | Urine & Feces, Uranium - Total, Laser Phosphorometer | EACH | 75.00 | 75.00 | 75.00 | 75.00 | 75.00 |
| 33 02 2327 | Urine & Feces, Uranium - Total, Fluorometric | EACH | 70.00 | 70.00 | 70.00 | 70.00 | 70.00 |
| 33 02 2328 | Urine & Feces, Uranium - Total, Radiometric, Alpha Scintillation | EACH | 82.50 | 82.50 | 82.50 | 82.50 | 82.50 |
| 33 02 2329 | Urine & Feces, Uranium - Total, Laser Phosphorometer | EACH | 82.50 | 82.50 | 82.50 | 82.50 | 82.50 |
| 33 02 2330 | Vegetation/Soil/Sediment, Americium Isotopic, Alpha Spectroscopy | EACH | 140.00 | 140.00 | 140.00 | 140.00 | 140.00 |
| 33 02 2331 | Vegetation/Soil/Sediment, Curium Isotopic, Alpha Spectroscopy | EACH | 138.33 | 138.33 | 138.33 | 138.33 | 138.33 |
| 33 02 2332 | Vegetation/Soil/Sediment, Neptunium - 237, Alpha Spectroscopy | EACH | 138.33 | 138.33 | 138.33 | 138.33 | 138.33 |
| 33 02 2333 | Vegetation/Soil/Sediment, Plutonium Isotopic, Alpha Spectroscopy | EACH | 143.33 | 143.33 | 143.33 | 143.33 | 143.33 |
| 33 02 2334 | Vegetation/Soil/Sediment, Thorium Isotopic, Alpha Spectroscopy | EACH | 140.00 | 140.00 | 140.00 | 140.00 | 140.00 |

**Environmental Remediation: Assemblies Cost Book**

# Cost Data for Remediation

## Sampling and Analysis

| Assembly | Description | Unit | Unit Cost by Safety Level A | B | C | D | E |
|---|---|---|---|---|---|---|---|
| | **Capital Costs** | | | | | | |
| 33 02 2335 | Vegetation/Soil/Sediment, Uranium Isotopic, Alpha Spectroscopy | EACH | 140.00 | 140.00 | 140.00 | 140.00 | 140.00 |
| 33 02 2336 | Vegetation/Soil/Sediment, Gross Beta - Total, Gas Flow Proportional Counting | EACH | 50.00 | 50.00 | 50.00 | 50.00 | 50.00 |
| 33 02 2337 | Vegetation/Soil/Sediment, Ruthenium - 106, Gas Flow Proportional Counting | EACH | 145.00 | 145.00 | 145.00 | 145.00 | 145.00 |
| 33 02 2338 | Vegetation/Soil/Sediment, Strontium - 89, 90, Gas Flow Proportional Counting | EACH | 170.00 | 170.00 | 170.00 | 170.00 | 170.00 |
| 33 02 2339 | Vegetation/Soil/Sediment, Strontium - 90, Gas Flow Proportional Counting | EACH | 135.00 | 135.00 | 135.00 | 135.00 | 135.00 |
| 33 02 2340 | Vegetation/Soil/Sediment, Technetium - 99, Gas Flow Proportional Counting | EACH | 125.00 | 125.00 | 125.00 | 125.00 | 125.00 |
| 33 02 2341 | Vegetation/Soil/Sediment, Americium Isotopic, Gamma Spectroscopy | EACH | 123.33 | 123.33 | 123.33 | 123.33 | 123.33 |
| 33 02 2342 | Vegetation/Soil/Sediment, Gamma Isotopic, Gamma Spectroscopy | EACH | 123.33 | 123.33 | 123.33 | 123.33 | 123.33 |
| 33 02 2343 | Vegetation/Soil/Sediment, Iodine - 129, Gamma Spectroscopy | EACH | 123.33 | 123.33 | 123.33 | 123.33 | 123.33 |
| 33 02 2344 | Vegetation/Soil/Sediment, Lead - 210, Gamma Spectroscopy | EACH | 131.67 | 131.67 | 131.67 | 131.67 | 131.67 |
| 33 02 2345 | Vegetation/Soil/Sediment, Polonium - 210, Gamma Spectroscopy | EACH | 131.67 | 131.67 | 131.67 | 131.67 | 131.67 |
| 33 02 2346 | Vegetation/Soil/Sediment, Radium - 226, 228, Gamma Spectroscopy | EACH | 123.33 | 123.33 | 123.33 | 123.33 | 123.33 |
| 33 02 2347 | Vegetation/Soil/Sediment, Ruthenium - 106, Gamma Spectroscopy | EACH | 90.00 | 90.00 | 90.00 | 90.00 | 90.00 |
| 33 02 2348 | Vegetation/Soil/Sediment, Uranium - Total, Gamma Spectroscopy | EACH | 67.50 | 67.50 | 67.50 | 67.50 | 67.50 |
| 33 02 2349 | Vegetation/Soil/Sediment, Carbon - 14, Liquid Scintillation | EACH | 150.00 | 150.00 | 150.00 | 150.00 | 150.00 |
| 33 02 2350 | Vegetation/Soil/Sediment, Nickel - 63, Liquid Scintillation | EACH | 145.00 | 145.00 | 145.00 | 145.00 | 145.00 |
| 33 02 2351 | Vegetation/Soil/Sediment, Tritium, Liquid Scintillation | EACH | 60.00 | 60.00 | 60.00 | 60.00 | 60.00 |
| 33 02 2352 | Vegetation/Soil/Sediment, Gross Alpha - Total, Gas Flow Proportional Counting | EACH | 43.33 | 43.33 | 43.33 | 43.33 | 43.33 |
| 33 02 2353 | Vegetation/Soil/Sediment, Gross Alpha/Gross Beta - Total, Gas Flow | EACH | 50.00 | 50.00 | 50.00 | 50.00 | 50.00 |
| 33 02 2354 | Vegetation/Soil/Sediment, Plutonium Isotopic/Plutonium 241, Alpha Spectroscopy | EACH | 148.33 | 148.33 | 148.33 | 148.33 | 148.33 |
| 33 02 2355 | Vegetation/Soil/Sediment, Uranium - Total, Laser Phosphorometer | EACH | 60.00 | 60.00 | 60.00 | 60.00 | 60.00 |
| 33 02 2356 | Vegetation/Soil/Sediment, Uranium - Total, Fluorometric | EACH | 50.00 | 50.00 | 50.00 | 50.00 | 50.00 |
| 33 02 2357 | Characterization of Any Unknown (Low Level) Radioactive Sample | EACH | 750.00 | 750.00 | 750.00 | 750.00 | 750.00 |

## Environmental Remediation: Assemblies Cost Book

# Cost Data for Remediation

## Sampling and Analysis

| Assembly | Description | Unit | A | B | C | D | E |
|---|---|---|---|---|---|---|---|
| | **Capital Costs** | | | | | | |
| 33 02 2401 | EPA Method 8330 (11 Compounds) Nitroaromatics/Nitramines | EACH | 287.35 | 287.35 | 287.35 | 287.35 | 287.35 |
| 33 02 2402 | Nitroglycerine | EACH | 236.27 | 236.27 | 236.27 | 236.27 | 236.27 |
| 33 02 2403 | Nitrocellulose | EACH | 191.57 | 191.57 | 191.57 | 191.57 | 191.57 |
| 33 02 2404 | Nitroquanadine | EACH | 191.57 | 191.57 | 191.57 | 191.57 | 191.57 |
| 33 02 2405 | Ammonium Perchlorate | EACH | 108.56 | 108.56 | 108.56 | 108.56 | 108.56 |
| 33 02 9901 | Magnetometer | DAY | 53.00 | 53.00 | 53.00 | 53.00 | 53.00 |
| 33 02 9902 | Electromagnetics | DAY | 53.00 | 53.00 | 53.00 | 53.00 | 53.00 |
| 33 02 9903 | Ground Penetrating Radar | DAY | 50.35 | 50.35 | 50.35 | 50.35 | 50.35 |
| 33 02 9904 | Seismic Refraction | DAY | 2,783 | 2,783 | 2,783 | 2,783 | 2,783 |
| 33 02 9905 | Resistivity | DAY | 2,783 | 2,783 | 2,783 | 2,783 | 2,783 |
| 33 02 9906 | Subcontracted Sampling per Hour, One-man Crew | HOUR | 85.25 | 85.25 | 85.25 | 85.25 | 85.25 |
| 33 02 9907 | Subcontracted Sampling per Day, One-man Crew | DAY | 680.00 | 680.00 | 680.00 | 680.00 | 680.00 |
| 33 02 9908 | Double-ring Infiltrometer Test (DRIT Test) | EACH | 550.00 | 550.00 | 550.00 | 550.00 | 550.00 |
| 33 02 9909 | Falling Head Rigid Wall Consolidometer Tech CAE 1110-2-1906 | EACH | 225.00 | 225.00 | 225.00 | 225.00 | 225.00 |
| 33 02 9911 | Falling Head Permeability Test EPA-9100 | EACH | 215.00 | 215.00 | 215.00 | 215.00 | 215.00 |
| 33 02 9912 | Constant Head Permeability Test ASTM D2432 | EACH | 193.33 | 193.33 | 193.33 | 193.33 | 193.33 |
| 33 02 9913 | Mobile Laboratory Trailer, 8' W x 30' L, Rental | MONTH | 7,950 | 7,950 | 7,950 | 7,950 | 7,950 |
| 33 02 9914 | Gas Chromatograph, HP5890A, Rental | MONTH | 1,855 | 1,855 | 1,855 | 1,855 | 1,855 |
| 33 02 9915 | Mass Spectrometer, HP5970B MD, Rental | MONTH | 1,617 | 1,617 | 1,617 | 1,617 | 1,617 |
| 33 02 9916 | Unconfined Compressive Strength (Pocket Penetrometer) | EACH | 4.00 | 4.00 | 4.00 | 4.00 | 4.00 |
| 33 07 9901 | Lease, Portable 40' W x 150' L x 18' High Emissions Containment Building | YEAR | 7,500 | 7,500 | 7,500 | 7,500 | 7,500 |
| 33 07 9902 | Buy, Portable 40' W x 150' L x 18' High Emissions Containment Building | EACH | 23,503 | 23,503 | 23,503 | 23,503 | 23,503 |
| 33 07 9903 | Assemble Emissions Containment Building, 5 Days | EACH | 12,346 | 9,591 | 8,298 | 5,629 | 4,722 |
| 33 07 9904 | Disassemble Emissions Containment Building, 3 Days | EACH | 7,402 | 5,750 | 4,975 | 3,374 | 2,831 |
| 33 17 0808 | Decontaminate Rig, Augers, Screen (Rental Equipment) | DAY | 205.34 | 205.34 | 205.34 | 205.34 | 205.34 |
| 33 19 0201 | Drummed Waste, Minimum Charge for Shipment | EACH | 2,350 | 2,350 | 2,350 | 2,350 | 2,350 |
| 33 19 9543 | Initial Waste Stream Evaluation, Non-PCB | EACH | 1,375 | 1,375 | 1,375 | 1,375 | 1,375 |
| 33 19 9544 | Initial PCB Waste Stream Evaluation | EACH | 631.25 | 631.25 | 631.25 | 631.25 | 631.25 |
| 33 19 9921 | DOT Steel Drum, 55 Gallon | EACH | 65.19 | 65.19 | 65.19 | 65.19 | 65.19 |
| 33 19 9922 | Polyethylene Closed Head Drum, 55 Gallon | EACH | 43.63 | 43.63 | 43.63 | 43.63 | 43.63 |
| 33 19 9924 | DOT Steel Salvage Drum, 85 Gallon, 16 Gauge | EACH | 57.06 | 57.06 | 57.06 | 57.06 | 57.06 |

**Environmental Remediation: Assemblies Cost Book**

# Cost Data for Remediation

## Sampling and Analysis

| Assembly | | | Description | Unit | Unit Cost by Safety Level | | | | |
|---|---|---|---|---|---|---|---|---|---|
| | | | | | A | B | C | D | E |
| | | | **Capital Costs** | | | | | | |
| 33 | 19 | 9925 | DOT Steel Salvage Drum Composite Overpack, 85 Gallon, 16 Gauge | EACH | 70.49 | 70.49 | 70.49 | 70.49 | 70.49 |
| 33 | 23 | 0503 | 4" Submersible Pump, 13 - 37 GPM, < 180' Head | EACH | 3,084 | 2,790 | 2,637 | 2,352 | 2,264 |
| 33 | 23 | 0505 | 6" Submersible Pump, 15 - 135 GPM, 200' < Head < 500' | EACH | 4,472 | 3,885 | 3,579 | 3,008 | 2,834 |
| 33 | 23 | 1106 | Split Spoon Sample, 2" x 24", During Drilling | EACH | 46.67 | 46.67 | 46.67 | 46.67 | 46.67 |
| 33 | 23 | 1111 | Well Development Equipment Rental | WEEK | 478.91 | 462.93 | 456.09 | 440.65 | 434.98 |
| 33 | 23 | 1121 | Standby for Drilling | EACH | 618.63 | 503.81 | 413.56 | 300.17 | 284.95 |
| 33 | 23 | 1122 | Move Rig/Equipment Around Site | EACH | 154.66 | 125.95 | 103.39 | 75.04 | 71.24 |
| 33 | 23 | 1123 | Load Supplies/Equipment | LS | 1,856 | 1,511 | 1,241 | 900.52 | 854.85 |
| 33 | 23 | 1124 | Security Pass/Protocol | LS | 309.32 | 251.91 | 206.78 | 150.09 | 142.47 |
| 33 | 23 | 1125 | Concrete Coring (Minimum Charge) | DAY | 2,475 | 2,015 | 1,654 | 1,201 | 1,140 |
| 33 | 23 | 1126 | Furnish 55 Gallon Drum for Drill Cuttings & Development Water | EACH | 65.19 | 65.19 | 65.19 | 65.19 | 65.19 |
| 33 | 23 | 1127 | Furnish 55 Gallon Drum for Development/Purge Water | EACH | 65.19 | 65.19 | 65.19 | 65.19 | 65.19 |
| 33 | 23 | 1801 | 2" Well, Grout (Annular Seal) | LF | 83.99 | 68.83 | 56.92 | 41.95 | 39.94 |
| 33 | 23 | 1802 | 4" Well, Grout (Annular Seal) | LF | 145.07 | 118.89 | 98.31 | 72.46 | 68.99 |
| 33 | 23 | 1803 | 6" Well, Grout (Annular Seal) | LF | 213.78 | 175.20 | 144.88 | 106.78 | 101.67 |
| 33 | 23 | 1811 | 2" Well, Portland Cement Grout | LF | 0.92 | 0.92 | 0.92 | 0.92 | 0.92 |
| 33 | 23 | 1812 | 4" Well, Portland Cement Grout | LF | 1.38 | 1.38 | 1.38 | 1.38 | 1.38 |
| 33 | 23 | 1813 | 6" Well, Portland Cement Grout | LF | 7.80 | 7.80 | 7.80 | 7.80 | 7.80 |
| 33 | 23 | 2104 | 3/8", 1' Thick, Sodium Bentonite Pellets | CF | 275.35 | 229.42 | 193.32 | 147.97 | 141.88 |
| 33 | 23 | 2401 | Teflon Bailer, 3/4" Outside Diameter x 1', 60 cc | EACH | 179.77 | 179.77 | 179.77 | 179.77 | 179.77 |
| 33 | 23 | 2402 | Teflon Bailer, 3/4" Outside Diameter x 3', 180 cc | EACH | 227.74 | 227.74 | 227.74 | 227.74 | 227.74 |
| 33 | 23 | 2403 | Teflon Bailer, 1" Outside Diameter x 1', 80 cc | EACH | 191.28 | 191.28 | 191.28 | 191.28 | 191.28 |
| 33 | 23 | 2404 | Teflon Bailer, 1 7/8" Outside Diameter x 1', 350 cc | EACH | 181.13 | 181.13 | 181.13 | 181.13 | 181.13 |
| 33 | 23 | 2405 | Teflon Bailer, 1 7/8" Outside Diameter x 2', 700 cc | EACH | 198.20 | 198.20 | 198.20 | 198.20 | 198.20 |
| 33 | 23 | 2406 | Teflon Bailer, 1 7/8" Outside Diameter x 3', 1,050 cc | EACH | 261.53 | 261.53 | 261.53 | 261.53 | 261.53 |
| 33 | 23 | 2407 | Disposable Bailer, Polyethylene, 1.5" Outside Diameter x 36" | EACH | 27.60 | 27.60 | 27.60 | 27.60 | 27.60 |
| 33 | 23 | 2421 | Emptying Stand | EACH | 302.32 | 302.32 | 302.32 | 302.32 | 302.32 |
| 33 | 23 | 2422 | Suspension Cable, Teflon Coated | FT | 1.00 | 1.00 | 1.00 | 1.00 | 1.00 |
| 33 | 23 | 2423 | Hand Reel | EACH | 9.89 | 9.89 | 9.89 | 9.89 | 9.89 |

# Cost Data for Remediation
## Slurry Walls

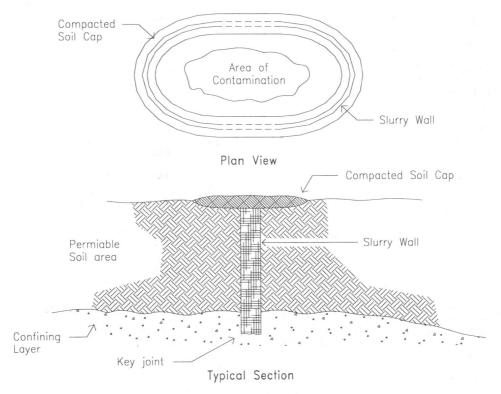

### General:
A slurry wall is a vertical subsurface barrier used to contain, capture, or redirect groundwater flow in the vicinity of a contaminated site. It is a vertical trench which is excavated under a slurry and backfilled with a material that forms a low permeability barrier. The slurry, which is usually a mixture of bentonite and water, hydraulically shores the trench to prevent collapse. In addition, the slurry forms a filter cake on the trench walls to prevent high fluid losses into the surrounding soil.

### Typical Treatment Train:
Treatment and disposal of contaminated spoil.

### Common Cost Components:
1. Trenching
2. Slurry preparation and placement
3. Surface grading and compaction
4. Backfill, may include borrow material
5. Soil/vegetative cap
6. Containment berms for slurry mixing or temporary storage of contaminated material

### Other Cost Considerations:
Clearing and grubbing, transportation from off-site mixing facilities.

**Environmental Remediation: Assemblies Cost Book**

# Cost Data for Remediation

## Slurry Walls

| Assembly | Description | Unit | A | B | C | D | E |
|---|---|---|---|---|---|---|---|
| | **Capital Costs** | | | | | | |
| 17 03 0307 | Cat 235, 2.0 CY, Rock, No Haul or Borrow, Trenching | BCY | 133.78 | 109.18 | 95.00 | 71.01 | 64.55 |
| 17 03 0420 | Backfill Trench, Borrow Material, Delivered & Dumped Only | CY | 9.86 | 8.74 | 7.93 | 6.83 | 6.63 |
| 17 03 0422 | Unclassified Fill, 6" Lifts, On-Site | CY | 14.23 | 11.65 | 9.60 | 7.05 | 6.72 |
| 17 03 0423 | Unclassified Fill, 6" Lifts, Off-Site | CY | 11.08 | 9.81 | 8.83 | 7.58 | 7.40 |
| 17 03 0425 | Sand, 6" Lifts, On-Site | CY | 13.88 | 11.35 | 9.36 | 6.87 | 6.54 |
| 17 03 0426 | Sand, 6" Lifts, Off-Site | CY | 15.55 | 14.34 | 13.42 | 12.22 | 12.04 |
| 17 03 0428 | Clay, 8" Lifts, Off-Site | CY | 22.97 | 19.74 | 17.38 | 14.21 | 13.67 |
| 17 03 0429 | Clay, 8" Lifts, On-Site | CY | 21.18 | 17.21 | 14.26 | 10.34 | 9.71 |
| 18 02 0301 | Asphalt Pavement - 10" Subgrade, 9" Base, 1 1/2" Topping | SY | 61.11 | 50.94 | 43.58 | 33.57 | 31.83 |
| 18 05 0202 | Rock Cover, Riprap, Light (10 to 100 Lb Pieces) | CY | 26.72 | 24.15 | 22.59 | 20.09 | 19.46 |
| 18 05 0203 | Rock Cover, Riprap, Medium (10 to 200 Lb Pieces) | CY | 27.05 | 24.48 | 22.92 | 20.42 | 19.79 |
| 18 05 0204 | Rock Cover, Riprap, Heavy (25 to 500 Lb Pieces) | CY | 27.39 | 24.82 | 23.26 | 20.76 | 20.13 |
| 18 05 0205 | Rock Cover, Bank-run Gravel | CY | 32.96 | 27.35 | 23.49 | 17.98 | 16.89 |
| 18 05 0301 | Topsoil, 6" Lifts, Off-Site | CY | 32.76 | 29.51 | 27.43 | 24.25 | 23.53 |
| 18 05 0302 | Topsoil, 6" Lifts, On-Site | CY | 14.14 | 11.54 | 9.45 | 6.88 | 6.57 |
| 18 05 0402 | Seeding, Vegetative Cover | ACRE | 1,874 | 1,823 | 1,791 | 1,741 | 1,729 |
| 18 05 0405 | Sodding, Vegetative Cover | ACRE | 25,697 | 22,064 | 20,238 | 16,710 | 15,589 |
| 18 05 0413 | Watering with 3,000-Gallon Tank Truck, per Pass | ACRE | 111.81 | 90.08 | 75.73 | 54.43 | 49.85 |
| 33 06 0301 | Level and Compact Working Surface | CY | 12.15 | 9.89 | 8.12 | 5.89 | 5.58 |
| 33 06 0302 | Construct Dike for Mixing Basin | CY | 12.15 | 9.89 | 8.12 | 5.89 | 5.58 |
| 33 06 0303 | Normal Soil, to 25', Slurry Wall Excavation | CY | 4.28 | 3.45 | 2.86 | 2.05 | 1.91 |
| 33 06 0304 | Clay/Sand with Boulders, to 25', Slurry Wall Excavation | CY | 4.28 | 3.45 | 2.86 | 2.05 | 1.91 |
| 33 06 0305 | Normal Soil, 26' - 75', Slurry Wall Excavation | CY | 9.28 | 7.50 | 6.21 | 4.46 | 4.15 |
| 33 06 0306 | Clay/Sand with Boulders, 26' - 75', Slurry Wall Excavation | CY | 13.79 | 11.14 | 9.23 | 6.62 | 6.16 |
| 33 06 0307 | Normal Soil, 76' - 120', Slurry Wall Excavation | CY | 20.15 | 16.29 | 13.49 | 9.68 | 9.02 |
| 33 06 0308 | Clay/Sand with Boulders, 76' - 120', Slurry Wall Excavation | CY | 30.08 | 24.30 | 20.12 | 14.45 | 13.46 |
| 33 06 0309 | Bentonite, Material Purchase Price per Ton | TON | 54.06 | 54.06 | 54.06 | 54.06 | 54.06 |
| 33 06 0310 | Slurry Mixing, Hydration, and Placement, per Gallon | GAL | 0.11 | 0.09 | 0.08 | 0.05 | 0.04 |
| 33 06 0311 | Soil-Bentonite Backfill Mixing, per Cubic Yard | CY | 5.73 | 4.67 | 3.83 | 2.78 | 2.64 |
| 33 06 0312 | Backfill Slurry Wall Trench, 1,000' Average Haul Distance | CY | 5.09 | 4.17 | 3.40 | 2.48 | 2.38 |
| 33 06 0313 | Demolish Mixing Basins and Regrade Working Surface | SF | 0.15 | 0.12 | 0.10 | 0.07 | 0.06 |
| 33 08 0505 | Soil/Bentonite Liner | SF | 0.78 | 0.77 | 0.76 | 0.75 | 0.74 |
| 33 08 0506 | Clay 10E-7, 6" Lifts, On-Site | CY | 29.50 | 23.82 | 19.83 | 14.25 | 13.21 |

**Unit Cost by Safety Level**

## Environmental Remediation: Assemblies Cost Book

# Cost Data for Remediation

## Slurry Walls

| Assembly | Description | Unit | A | B | C | D | E |
|---|---|---|---|---|---|---|---|
| | **Capital Costs** | | | | | | |
| 33 08 0507 | Clay 10E-7, 6" Lifts, Off-Site | CY | 31.11 | 26.22 | 22.84 | 18.04 | 17.11 |
| 33 08 0508 | Fabric Membrane Liner/Bentonite Liner | SF | 0.59 | 0.52 | 0.48 | 0.41 | 0.40 |
| 33 08 0531 | 60 Mil Geotextile, Nonwoven | SY | 1.53 | 1.31 | 1.21 | 1.00 | 0.93 |
| 33 08 0532 | 80 Mil Geotextile, Nonwoven | SY | 1.57 | 1.35 | 1.25 | 1.04 | 0.97 |
| 33 08 0533 | 105 Mil Geotextile, Nonwoven | SY | 1.92 | 1.65 | 1.52 | 1.26 | 1.17 |
| 33 08 0534 | 130 Mil Geotextile, Nonwoven | SY | 2.56 | 2.20 | 2.03 | 1.68 | 1.56 |
| 33 08 0535 | 170 Mil Geotextile, Nonwoven | SY | 2.96 | 2.60 | 2.43 | 2.08 | 1.96 |
| 33 19 9911 | Air Blower to Inflate Portable Containment Berm | EACH | 160.59 | 160.59 | 160.59 | 160.59 | 160.59 |

**Unit Cost by Safety Level**

**Environmental Remediation: Assemblies Cost Book**

# Cost Data for Remediation
## Soil Flushing

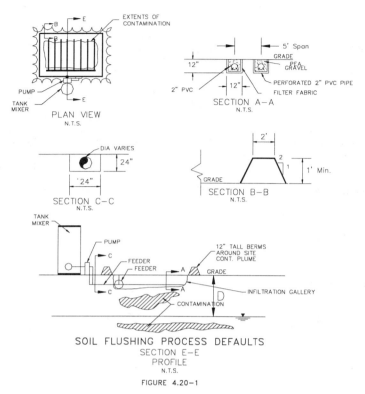

FIGURE 4.20-1

### General:

The use of soil flushing to remove soil contaminants involves the flushing of organic and/or inorganic constituents from in situ soils for recovery and treatment. The site or treatment area is flooded with an appropriate flushing solution, and the contaminated flushing solution is collected in a series of shallow wellpoints or subsurface drains. The flushing solution is then treated and/or recycled back into the site. The treatment area is flushed until it is completely saturated. This is known as pore volume flush. To effectively perform soil flushing, numerous pore volume flushes (5 to 50) may be required.

### Typical Treatment Train:

Extraction wells, water treatment, biodegradation, sampling and analysis.

### Common Cost Components:

1. Piping
2. Pumping system
3. Mixing tank
4. Flushing solution: water, surfactants and water, acidic aqueous solutions, basic aqueous solutions
5. Electricity
6. Operational labor
7. Earth berm around treatment area
8. Demolition and disposal of pipe at completion

### Other Cost Considerations:

Treatability testing, clearing and grubbing, electrical distribution.

**Environmental Remediation: Assemblies Cost Book**

# Cost Data for Remediation

## Soil Flushing

| Assembly | Description | Unit | A | B | C | D | E |
|---|---|---|---|---|---|---|---|
| | **Capital Costs** | | | | | | |
| 17 01 0511 | Dozer 200 HP D7 LGP, Wet Grubbing & Stacking | CY | 5.84 | 4.77 | 3.91 | 2.84 | 2.70 |
| 17 02 0610 | Piping Demolition | LF | 15.80 | 12.45 | 10.60 | 7.34 | 6.41 |
| 17 03 0256 | Cat 215, 1.0 CY, Sand, Shallow, Trenching | CY | 1.99 | 1.60 | 1.33 | 0.95 | 0.88 |
| 17 03 0257 | Cat 215, 1.0 CY, Soil, Shallow, Trenching | CY | 2.42 | 1.95 | 1.62 | 1.16 | 1.07 |
| 17 03 0259 | Cat 225, 1.5 CY, Soil/Sand, Trenching | CY | 1.84 | 1.49 | 1.23 | 0.88 | 0.82 |
| 17 03 0261 | Cat 225, 1.5 CY, Soil/Sand with Boulders, Trenching | CY | 3.91 | 3.15 | 2.62 | 1.87 | 1.74 |
| 17 03 0262 | Cat 235, 2.5 CY, Soil/Sand, Trenching | CY | 2.01 | 1.62 | 1.35 | 0.97 | 0.90 |
| 17 03 0264 | Cat 235, 2.5 CY, Soil/Sand with Boulders, Trenching | CY | 4.28 | 3.45 | 2.86 | 2.05 | 1.91 |
| 17 03 0265 | Cat 245, 3.0 CY, Soil/Sand, Trenching | CY | 2.78 | 2.23 | 1.86 | 1.33 | 1.22 |
| 17 03 0267 | Cat 245, 3.0 CY, Soil/Sand with Boulders, Trenching | CY | 5.18 | 4.17 | 3.47 | 2.47 | 2.28 |
| 17 03 0268 | Koehring 1166, 4.0 CY, Soil/Sand, Trenching | CY | 2.27 | 1.83 | 1.52 | 1.09 | 1.01 |
| 17 03 0270 | Koehring 1166, 4.0 CY, Soil/Sand with Boulders, Trenching | CY | 3.00 | 2.42 | 2.01 | 1.44 | 1.33 |
| 17 03 0419 | Crushed Stone, 1/2" to 3/4" | CY | 23.45 | 21.67 | 20.70 | 18.97 | 18.47 |
| 18 02 0322 | 8" Structural Slab on Grade | SF | 9.75 | 8.38 | 7.71 | 6.38 | 5.95 |
| 18 02 0324 | 12" Structural Slab on Grade | SF | 11.57 | 10.04 | 9.30 | 7.82 | 7.34 |
| 19 01 0202 | 1", Class 200, PVC Piping | LF | 7.48 | 5.84 | 5.07 | 3.48 | 2.93 |
| 19 01 0204 | 2", Class 200, PVC Piping | LF | 8.86 | 6.96 | 6.07 | 4.23 | 3.60 |
| 19 01 0206 | 3", Class 200, PVC Piping | LF | 11.81 | 9.34 | 8.20 | 5.80 | 4.98 |
| 19 01 0207 | 4", Class 200, PVC Piping | LF | 12.33 | 9.86 | 8.72 | 6.32 | 5.50 |
| 19 01 0208 | 6", Class 200, PVC Piping | LF | 14.95 | 12.20 | 10.93 | 8.27 | 7.36 |
| 19 01 0212 | 8", Class 150, PVC Piping | LF | 18.36 | 15.27 | 13.84 | 10.85 | 9.82 |
| 19 01 0214 | 12", Class 150, PVC Piping | LF | 28.75 | 24.63 | 22.72 | 18.73 | 17.36 |
| 19 02 0601 | 12" x 12" Underground French Drain | LF | 4.50 | 3.92 | 3.61 | 3.04 | 2.88 |
| 19 02 0602 | 18" x 18" Underground French Drain | LF | 5.93 | 5.29 | 4.94 | 4.31 | 4.13 |
| 19 02 0603 | 24" x 24" Underground French Drain | LF | 7.43 | 6.71 | 6.31 | 5.61 | 5.41 |
| 19 04 0401 | 550 Gallon, Stainless Steel Aboveground Wastewater Holding Tank, Rental | MONTH | 320.65 | 320.65 | 320.65 | 320.65 | 320.65 |
| 19 04 0405 | 6,000 Gallon, Polyethylene Aboveground Wastewater Holding Tank, Rental | MONTH | 641.30 | 641.30 | 641.30 | 641.30 | 641.30 |
| 19 04 0407 | 21,000 Gallon, Steel Closed Stationary Aboveground Wastewater Holding Tank, Rental | MONTH | 1,219 | 1,219 | 1,219 | 1,219 | 1,219 |
| 33 08 0531 | 60 Mil Geotextile, Nonwoven | SY | 1.53 | 1.31 | 1.21 | 1.00 | 0.93 |
| 33 23 0301 | 2" PVC, Well Plug | EACH | 29.37 | 24.77 | 21.16 | 16.63 | 16.02 |
| 33 26 0801 | 2" Slotted PVC Pipe | LF | 5.64 | 4.83 | 4.48 | 3.69 | 3.40 |

**Environmental Remediation: Assemblies Cost Book**

# Cost Data for Remediation

## Soil Flushing

| Assembly | Description | Unit | Unit Cost by Safety Level A | B | C | D | E |
|---|---|---|---|---|---|---|---|
| | **Capital Costs** | | | | | | |
| 33 27 0101 | 1" PVC, Schedule 40, Tee | EACH | 32.07 | 24.87 | 21.79 | 14.83 | 12.28 |
| 33 27 0102 | 2" PVC, Schedule 40, Tee | EACH | 68.73 | 53.63 | 46.80 | 32.18 | 27.06 |
| 33 27 0103 | 3" PVC, Schedule 40, Tee | EACH | 100.20 | 79.05 | 69.99 | 49.55 | 42.05 |
| 33 27 0104 | 4" PVC, Schedule 40, Tee | EACH | 137.04 | 108.84 | 96.77 | 69.51 | 59.51 |
| 33 27 0105 | 6" PVC, Schedule 40, Tee | EACH | 205.86 | 169.47 | 153.89 | 118.72 | 105.82 |
| 33 27 0106 | 8" PVC, Schedule 40, Tee | EACH | 336.90 | 291.78 | 272.46 | 228.85 | 212.85 |
| 33 27 0111 | 1" PVC, 90 Degree, Elbow | EACH | 20.18 | 15.67 | 13.74 | 9.37 | 7.78 |
| 33 27 0112 | 2" PVC, 90 Degree, Elbow | EACH | 41.36 | 32.22 | 28.30 | 19.46 | 16.22 |
| 33 27 0113 | 3" PVC, 90 Degree, Elbow | EACH | 83.17 | 65.36 | 57.74 | 40.52 | 34.21 |
| 33 27 0114 | 4" PVC, 90 Degree, Elbow | EACH | 101.72 | 80.57 | 71.51 | 51.07 | 43.57 |
| 33 27 0115 | 6" PVC, 90 Degree, Elbow | EACH | 135.44 | 111.27 | 100.92 | 77.56 | 68.99 |
| 33 27 0116 | 8" PVC, 90 Degree, Elbow | EACH | 200.21 | 172.01 | 159.94 | 132.68 | 122.68 |
| 33 27 0322 | 12" Polyethylene SDR-21, Tee | EACH | 734.48 | 639.07 | 591.08 | 498.44 | 469.06 |
| 33 27 0332 | 12" Polyethylene SDR-21, 90-degree Elbow | EACH | 490.43 | 442.40 | 418.24 | 371.60 | 356.81 |
| 33 27 0401 | 1" Iron Body Check Valve | EACH | 169.57 | 148.68 | 139.74 | 119.55 | 112.15 |
| 33 27 0402 | 2" Iron Body Check Valve | EACH | 188.96 | 174.31 | 168.03 | 153.86 | 148.67 |
| 33 27 0403 | 3" Iron Body Check Valve | EACH | 231.32 | 214.06 | 206.67 | 189.99 | 183.87 |
| 33 27 0404 | 4" Iron Body Check Valve | EACH | 354.29 | 315.96 | 299.54 | 262.49 | 248.90 |
| 33 27 0405 | 6" Iron Body Check Valve | EACH | 618.60 | 554.57 | 523.95 | 461.86 | 441.15 |
| 33 27 0406 | 8" Iron Body Check Valve | EACH | 1,094 | 980.95 | 926.96 | 817.50 | 780.98 |
| 33 27 0408 | 12" Iron Body Check Valve | EACH | 1,847 | 1,661 | 1,571 | 1,391 | 1,330 |
| 33 28 0201 | 1" Hanger | EACH | 115.89 | 90.73 | 79.33 | 54.97 | 46.43 |
| 33 28 0202 | 2" Hanger | EACH | 117.08 | 91.91 | 80.52 | 56.16 | 47.62 |
| 33 28 0203 | 3" Hanger | EACH | 104.41 | 82.43 | 72.60 | 51.33 | 43.80 |
| 33 28 0204 | 4" Hanger | EACH | 132.79 | 105.00 | 92.57 | 65.68 | 56.16 |
| 33 28 0205 | 6" Hanger | EACH | 146.70 | 117.21 | 104.03 | 75.49 | 65.39 |
| 33 28 0206 | 8" Hanger | EACH | 175.90 | 142.08 | 126.96 | 94.24 | 82.65 |
| 33 28 0207 | 12" Hanger | EACH | 190.12 | 156.31 | 141.19 | 108.46 | 96.88 |
| 33 29 0105 | 5 HP, 100 GPM, Centrifugal Pump | EACH | 2,475 | 2,301 | 2,227 | 2,059 | 1,998 |
| 33 29 0108 | 10 HP, 200 GPM, Centrifugal Pump | EACH | 2,941 | 2,714 | 2,616 | 2,396 | 2,316 |
| 33 29 0109 | 10 HP, 250 GPM, Centrifugal Pump | EACH | 3,032 | 2,805 | 2,707 | 2,487 | 2,407 |
| 33 29 0110 | 15 HP, 300 GPM, Centrifugal Pump | EACH | 3,441 | 3,160 | 3,040 | 2,770 | 2,670 |
| 33 29 0111 | 20 HP, 500 GPM, Centrifugal Pump | EACH | 4,544 | 4,241 | 4,111 | 3,818 | 3,710 |

## Environmental Remediation: Assemblies Cost Book

# Cost Data for Remediation

## Soil Flushing

| Assembly | Description | Unit | A | B | C | D | E |
|---|---|---|---|---|---|---|---|
| | **Capital Costs** | | | | | | |
| 33 29 0112 | 30 HP, 750 GPM, Centrifugal Pump | EACH | 5,213 | 4,808 | 4,635 | 4,244 | 4,100 |
| 33 29 0113 | 40 HP, 1050 GPM, Centrifugal Pump | EACH | 5,655 | 5,199 | 5,004 | 4,564 | 4,403 |
| 33 29 0114 | 60 HP, 1500 GPM, Centrifugal Pump | EACH | 9,842 | 8,823 | 8,350 | 7,363 | 7,024 |
| 33 29 0115 | 75 HP, 2000 GPM, Centrifugal Pump | EACH | 12,456 | 11,233 | 10,666 | 9,482 | 9,075 |
| 33 32 0101 | 7.5 GPH, 860 PSI, 55 Gallon Polyethylene Tank with Steel Overpack | EACH | 4,203 | 3,886 | 3,751 | 3,445 | 3,333 |
| 33 32 0102 | 20.8 GPH, 150 PSI, 55 Gallon Polyethylene Tank with Steel Overpack | EACH | 4,203 | 3,886 | 3,751 | 3,445 | 3,333 |
| 33 32 0103 | 7.5 GPH, 860 PSI, 102 Gallon Polyethylene Tank with Steel Overpack | EACH | 5,601 | 5,206 | 5,036 | 4,654 | 4,513 |
| 33 32 0104 | 20.8 GPH, 150 PSI, 102 Gallon Polyethylene Tank with Steel Overpack | EACH | 5,601 | 5,206 | 5,036 | 4,654 | 4,513 |
| 33 32 0105 | 7.5 GPH, 860 PSI, 167 Gallon Polyethylene Tank with Steel Overpack | EACH | 5,264 | 4,868 | 4,699 | 4,316 | 4,176 |
| 33 32 0106 | 20.8 GPH, 150 PSI, 167 Gallon Polyethylene Tank with Steel Overpack | EACH | 5,264 | 4,868 | 4,699 | 4,316 | 4,176 |
| 33 32 0107 | 7.5 GPH, 860 PSI, 260 Gallon Polyethylene Tank with Steel Overpack | EACH | 6,288 | 5,892 | 5,723 | 5,340 | 5,200 |
| 33 32 0108 | 20.8 GPH, 150 PSI, 260 Gallon Polyethylene Tank with Steel Overpack | EACH | 6,288 | 5,892 | 5,723 | 5,340 | 5,200 |
| 33 32 0109 | 7.5 GPH, 860 PSI, 50 Gallon 304 Stainless Steel Tank | EACH | 4,882 | 4,566 | 4,430 | 4,124 | 4,012 |
| 33 32 0110 | 20.8 GPH, 150 PSI, 50 Gallon 304 Stainless Steel Tank | EACH | 4,882 | 4,566 | 4,430 | 4,124 | 4,012 |
| 33 32 0111 | 7.5 GPH, 860 PSI, 100 Gallon 304 Stainless Steel Tank | EACH | 5,513 | 5,118 | 4,948 | 4,566 | 4,425 |
| 33 32 0112 | 20.8 GPH, 150 PSI, 100 Gallon 304 Stainless Steel Tank | EACH | 5,513 | 5,118 | 4,948 | 4,566 | 4,425 |
| 33 32 0113 | 7.5 GPH, 860 PSI, 150 Gallon 304 Stainless Steel Tank | EACH | 6,251 | 5,855 | 5,686 | 5,303 | 5,163 |
| 33 32 0114 | 20.8 GPH, 150 PSI, 150 Gallon 304 Stainless Steel Tank | EACH | 6,251 | 5,855 | 5,686 | 5,303 | 5,163 |
| 33 32 0115 | 7.5 GPH, 860 PSI, 250 Gallon 304 Stainless Steel Tank | EACH | 7,927 | 7,532 | 7,362 | 6,980 | 6,839 |
| 33 32 0116 | 20.8 GPH, 150 PSI, 250 Gallon 304 Stainless Steel Tank | EACH | 7,927 | 7,532 | 7,362 | 6,980 | 6,839 |
| 33 32 0117 | 7.5 GPH, 860 PSI, 500 Gallon 304 Stainless Steel Tank | EACH | 9,610 | 9,346 | 9,233 | 8,978 | 8,885 |
| 33 32 0118 | 20.8 GPH, 150 PSI, 500 Gallon 304 Stainless Steel Tank | EACH | 9,610 | 9,346 | 9,233 | 8,978 | 8,885 |
| 33 32 0119 | 7.5 GPH, 860 PSI, Simplex, Packed Plunger | EACH | 3,420 | 3,156 | 3,043 | 2,788 | 2,695 |
| 33 32 0120 | 1.0 GPH, 2,000 PSI, Simplex, Packed Plunger | EACH | 3,611 | 3,347 | 3,234 | 2,979 | 2,886 |
| 33 32 0121 | 20.8 GPH, 150 PSI, Simplex, Diaphragm | EACH | 2,890 | 2,626 | 2,513 | 2,258 | 2,164 |
| 33 32 0122 | 105 GPH, 50 PSI, Simplex, Diaphragm | EACH | 4,797 | 4,534 | 4,421 | 4,166 | 4,072 |
| 33 32 0123 | 0.86 GPH, 700 PSI, Simplex, 316 Stainless Steel, Liquid End | EACH | 2,988 | 2,724 | 2,611 | 2,356 | 2,262 |
| 33 32 0124 | 2.70 GPH, 1,100 PSI, Simplex, 316 Stainless Steel, Liquid End | EACH | 2,796 | 2,532 | 2,419 | 2,164 | 2,070 |

**Environmental Remediation: Assemblies Cost Book**

# Cost Data for Remediation

## Soil Flushing

| Assembly | Description | Unit | Unit Cost by Safety Level A | B | C | D | E |
|---|---|---|---|---|---|---|---|
| | **Capital Costs** | | | | | | |
| 33 32 0125 | 7.0 GPH, 925 PSI, Simplex, 316 Stainless Steel, Liquid End | EACH | 3,457 | 3,193 | 3,080 | 2,825 | 2,731 |
| 33 32 0126 | 18.0 GPH, 350 PSI, Simplex, 316 Stainless Steel, Liquid End | EACH | 2,976 | 2,712 | 2,599 | 2,344 | 2,250 |
| 33 32 0127 | 41.6 GPH, 150 PSI, Duplex, Diaphragm | EACH | 3,721 | 3,457 | 3,344 | 3,089 | 2,996 |
| 33 32 0128 | 0.86 GPH, 700 PSI, Duplex, 316 Stainless Steel, Liquid End | EACH | 3,184 | 2,921 | 2,808 | 2,553 | 2,459 |
| 33 32 0129 | 2.70 GPH, 1,100 PSI, Duplex, 316 Stainless Steel, Liquid End | EACH | 2,993 | 2,729 | 2,616 | 2,361 | 2,267 |
| 33 32 0130 | 7.0 GPH, 925 PSI, Duplex, 316 Stainless Steel, Liquid End | EACH | 3,654 | 3,390 | 3,277 | 3,022 | 2,928 |
| 33 32 0131 | 18 GPH, 350 PSI, Duplex, 316 Stainless Steel, Liquid End | EACH | 4,111 | 3,848 | 3,735 | 3,480 | 3,386 |
| 33 33 0103 | Sodium Hydroxide, 500 Lb Container, Beads | EACH | 234.06 | 229.53 | 226.79 | 222.36 | 221.25 |
| 33 33 0104 | Sodium Hydroxide, 100 Lb Container, Flakes | EACH | 89.90 | 85.37 | 82.63 | 78.20 | 77.09 |
| 33 33 0105 | Sodium Hydroxide, 400 Lb Container, Flakes | EACH | 169.60 | 169.60 | 169.60 | 169.60 | 169.60 |
| 33 33 0110 | Sulfuric Acid Solution, 220 Lb Drummed Liquid | EACH | 45.38 | 40.85 | 38.11 | 33.68 | 32.57 |
| 33 33 0111 | Sulfuric Acid Solution, 750 Lb Drummed Liquid | EACH | 93.61 | 89.08 | 86.34 | 81.91 | 80.80 |
| 33 33 0125 | Alkylbenzene Sulfonate Surfactant, Bulk | LB | 1.16 | 1.16 | 1.16 | 1.16 | 1.16 |
| 33 41 0101 | Pump & Motor Maintenance/Repair | EACH | 865.17 | 671.48 | 581.58 | 393.96 | 329.64 |
| 33 42 0101 | Electrical Charge | KWH | 0.06 | 0.06 | 0.06 | 0.06 | 0.06 |

**Environmental Remediation: Assemblies Cost Book**

# Cost Data for Remediation
## Soil Vapor Extraction

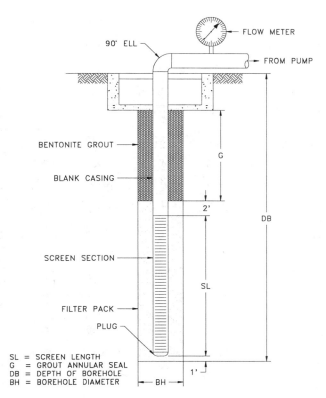

### General:

Soil vapor extraction (SVE) is an in situ process for the removal of volatile organic compounds (VOCs) from the vadose (unsaturated) zone soils. The system consists of a series of vapor extraction wells, commonly referred to as vapor extraction points (VEPs), monitoring wells, and air blowers to draw air through the soil and into the VEPs. It also includes piping to collect the extracted air and systems to remove contaminants from the extracted air. SVE is well suited for the treatment of soil located under structures where soil excavation would be impractical.

### Typical Treatment Train:

Treatment and disposal of contaminated water and/or air, monitoring well installation, treatment and disposal of spoil, sampling and analysis.

### Common Cost Components:

1. VEP installation (vertical or horizontal)
2. Piping
3. Blower(s)
4. Vapor recovery system
5. Condensate collection
6. Organic Vapor Analyzer (OVA)
7. Operations and maintenance

### Other Cost Considerations:

Pilot scale tests, air/water separator if soil moisture content is high, fencing, clearing and grubbing, electrical distribution.

**Environmental Remediation: Assemblies Cost Book**

# Cost Data for Remediation

## Soil Vapor Extraction

| Assembly | Description | Unit | A | B | C | D | E |
|---|---|---|---|---|---|---|---|
| | **Capital Costs** | | | | | | |
| 17 03 0206 | D3 with A-blade Bulldozer | HOUR | 151.96 | 122.19 | 101.74 | 72.51 | 66.74 |
| 17 03 0298 | 1/2 CY Crawler-mounted, Hydraulic Excavator | CY | 5.16 | 4.14 | 3.45 | 2.45 | 2.24 |
| 17 03 0421 | 950, 3.0 CY, Backfill with Borrow Material | CY | 10.24 | 8.27 | 6.85 | 4.92 | 4.58 |
| 18 02 0301 | Asphalt Pavement - 10" Subgrade, 9" Base, 1 1/2" Topping | SY | 61.11 | 50.94 | 43.58 | 33.57 | 31.83 |
| 19 04 0403 | 4,000 Gallon Polyethylene Wastewater Tank, Rental | MONTH | 545.90 | 545.90 | 545.90 | 545.90 | 545.90 |
| 19 04 0406 | 21,000 Gallon Steel Wastewater Holding Tank, Rental | MONTH | 1,219 | 1,219 | 1,219 | 1,219 | 1,219 |
| 33 01 0101 | Mobilize/DeMobilize Drilling Rig & Crew | LS | 4,949 | 4,031 | 3,308 | 2,401 | 2,280 |
| 33 02 0303 | Organic Vapor Analyzer Rental, per Day | DAY | 184.30 | 184.30 | 184.30 | 184.30 | 184.30 |
| 33 02 1507 | Continuous Monitoring and Recording of Air Flow | EACH | 9,870 | 9,870 | 9,870 | 9,870 | 9,870 |
| 33 13 2301 | 1 HP, 115V, 98 SCFM, Vapor Recovery System | EACH | 4,298 | 4,118 | 4,042 | 3,868 | 3,804 |
| 33 13 2302 | 1 HP, 230V, 98 SCFM, Vapor Recovery System | EACH | 4,330 | 4,150 | 4,073 | 3,900 | 3,836 |
| 33 13 2303 | 1 1/2 HP, 230V, 127 SCFM, Vapor Recovery System | EACH | 4,958 | 4,779 | 4,702 | 4,528 | 4,465 |
| 33 13 2304 | 2 HP, 230V, 160 SCFM, Vapor Recovery System | EACH | 5,527 | 5,347 | 5,270 | 5,096 | 5,033 |
| 33 13 2305 | 5 HP, 230V, 280 SCFM, Vapor Recovery System | EACH | 6,725 | 6,546 | 6,469 | 6,295 | 6,232 |
| 33 13 2306 | Propane Conversion for Soil Vapor Extractor | EACH | 90.00 | 90.00 | 90.00 | 90.00 | 90.00 |
| 33 13 2307 | Modulating Motor for Automatic Fuel Adjustment | EACH | 1,958 | 1,958 | 1,958 | 1,958 | 1,958 |
| 33 13 2308 | Trailer Mounting for Vapor Extractor Unit | EACH | 945.00 | 945.00 | 945.00 | 945.00 | 945.00 |
| 33 13 2309 | Security Fence for Vapor Extractor | EACH | 262.00 | 262.00 | 262.00 | 262.00 | 262.00 |
| 33 13 2320 | Installation Using Chain Trencher, Depth <= 4' | CY | 2.78 | 2.18 | 1.86 | 1.29 | 1.12 |
| 33 13 2322 | 1 HP, 230V, 98 SCFM, Weekly Rental, Vapor Recovery System | WEEK | 266.67 | 266.67 | 266.67 | 266.67 | 266.67 |
| 33 13 2332 | 1 HP, 230V, 98 SCFM, Monthly Rental, Vapor Recovery System | MONTH | 741.67 | 741.67 | 741.67 | 741.67 | 741.67 |
| 33 13 2333 | 5 HP, 90 SCFM Vapor Extraction Blower, Monthly Rental | MONTH | 850.00 | 850.00 | 850.00 | 850.00 | 850.00 |
| 33 13 2334 | 7.5 HP, 140 SCFM Vapor Extraction Blower, Monthly Rental | MONTH | 900.00 | 900.00 | 900.00 | 900.00 | 900.00 |
| 33 13 2335 | 10 HP, 190 SCFM Vapor Extraction Blower, Monthly Rental | MONTH | 1,163 | 1,163 | 1,163 | 1,163 | 1,163 |
| 33 13 2336 | 15 HP, 290 SCFM Vapor Extraction Blower, Monthly Rental | MONTH | 1,325 | 1,325 | 1,325 | 1,325 | 1,325 |
| 33 13 2337 | 30 HP, 580 SCFM Vapor Extraction Blower, Monthly Rental | MONTH | 1,625 | 1,625 | 1,625 | 1,625 | 1,625 |
| 33 13 2338 | Purchase, 5.0 HP, 90 SCFM Vapor Extraction Blower | EACH | 1,912 | 1,912 | 1,912 | 1,912 | 1,912 |
| 33 13 2339 | Purchase, 7.5 HP, 140 SCFM Vapor Extraction Blower | EACH | 3,110 | 2,931 | 2,854 | 2,680 | 2,617 |
| 33 13 2340 | Purchase, 10.0 HP, 190 SCFM Vapor Extraction Blower | EACH | 3,359 | 3,179 | 3,102 | 2,929 | 2,865 |
| 33 13 2341 | Purchase, 15.0 HP, 280 SCFM Vapor Extraction Blower | EACH | 5,126 | 4,947 | 4,870 | 4,696 | 4,632 |
| 33 13 2342 | Purchase, 30.0 HP, 580 SCFM Vapor Extraction Blower | EACH | 7,516 | 7,337 | 7,260 | 7,086 | 7,022 |

**Environmental Remediation: Assemblies Cost Book**

# Cost Data for Remediation

## Soil Vapor Extraction

| Assembly | Description | Unit | Unit Cost by Safety Level | | | | |
|---|---|---|---|---|---|---|---|
| | | | A | B | C | D | E |
| | **Capital Costs** | | | | | | |
| 33 17 0808 | Decontaminate Rig, Augers, Screen (Rental Equipment) | DAY | 205.34 | 205.34 | 205.34 | 205.34 | 205.34 |
| 33 17 0811 | Decontaminate Trenching Equipment | EACH | 3,835 | 3,345 | 3,029 | 2,549 | 2,441 |
| 33 19 9921 | DOT Steel Drum, 55 Gallon | EACH | 65.19 | 65.19 | 65.19 | 65.19 | 65.19 |
| 33 19 9922 | Polyethylene Closed Head Drum, 55 Gallon | EACH | 43.63 | 43.63 | 43.63 | 43.63 | 43.63 |
| 33 23 0101 | 2" PVC, Schedule 40, Well Casing | LF | 17.45 | 14.39 | 11.98 | 8.95 | 8.55 |
| 33 23 0102 | 4" PVC, Schedule 40, Well Casing | LF | 26.98 | 22.38 | 18.77 | 14.24 | 13.63 |
| 33 23 0104 | 8" PVC, Schedule 40, Well Casing | LF | 34.50 | 29.27 | 25.15 | 19.98 | 19.29 |
| 33 23 0111 | 2" PVC, Schedule 80, Well Casing | LF | 17.78 | 14.72 | 12.31 | 9.28 | 8.88 |
| 33 23 0112 | 4" PVC, Schedule 80, Well Casing | LF | 28.55 | 23.95 | 20.34 | 15.81 | 15.20 |
| 33 23 0121 | 2" Stainless Steel, Well Casing | LF | 31.75 | 28.14 | 25.30 | 21.73 | 21.25 |
| 33 23 0122 | 4" Stainless Steel, Well Casing | LF | 47.66 | 43.06 | 39.45 | 34.92 | 34.31 |
| 33 23 0201 | 2" PVC, Schedule 40, Well Screen | LF | 23.47 | 19.52 | 16.42 | 12.52 | 12.00 |
| 33 23 0202 | 4" PVC, Schedule 40, Well Screen | LF | 38.16 | 32.00 | 27.16 | 21.09 | 20.27 |
| 33 23 0211 | 2" PVC, Schedule 80, Well Screen | LF | 18.70 | 15.64 | 13.23 | 10.20 | 9.80 |
| 33 23 0212 | 4" PVC, Schedule 80, Well Screen | LF | 30.01 | 25.41 | 21.80 | 17.27 | 16.66 |
| 33 23 0221 | 2" Stainless Steel, Well Screen | LF | 26.91 | 23.84 | 21.44 | 18.41 | 18.01 |
| 33 23 0222 | 4" Stainless Steel, Well Screen | LF | 47.66 | 43.06 | 39.45 | 34.92 | 34.31 |
| 33 23 0235 | 4" SDR 17, High-density Polyethylene, Well Screen | LF | 10.00 | 10.00 | 10.00 | 10.00 | 10.00 |
| 33 23 0236 | 6" SDR 17, High-density Polyethylene, Well Screen | LF | 12.00 | 12.00 | 12.00 | 12.00 | 12.00 |
| 33 23 0301 | 2" PVC, Well Plug | EACH | 29.37 | 24.77 | 21.16 | 16.63 | 16.02 |
| 33 23 0302 | 4" PVC, Well Plug | EACH | 55.65 | 48.91 | 43.62 | 36.97 | 36.07 |
| 33 23 0311 | 2" Stainless Steel, Well Plug | EACH | 83.24 | 74.06 | 66.83 | 57.76 | 56.55 |
| 33 23 0312 | 4" Stainless Steel, Well Plug | EACH | 112.24 | 102.03 | 94.01 | 83.93 | 82.58 |
| 33 23 0316 | 4" High-density Polyethylene, Well Plug | EACH | 70.64 | 70.64 | 70.64 | 70.64 | 70.64 |
| 33 23 0317 | 6" High-density Polyethylene, Well Plug | EACH | 86.08 | 86.08 | 86.08 | 86.08 | 86.08 |
| 33 23 1101 | Hollow-stem Auger, 8" Outside Diameter Borehole for 2" Well | LF | 89.98 | 73.28 | 60.15 | 43.66 | 41.45 |
| 33 23 1102 | Hollow-stem Auger, 11" Outside Diameter Borehole for 4" Well | LF | 109.98 | 89.57 | 73.52 | 53.36 | 50.66 |
| 33 23 1106 | Split Spoon Sample, 2" x 24", During Drilling | EACH | 46.67 | 46.67 | 46.67 | 46.67 | 46.67 |
| 33 23 1111 | Well Development Equipment Rental | WEEK | 478.91 | 462.93 | 456.09 | 440.65 | 434.98 |
| 33 23 1121 | Standby for Drilling | EACH | 618.63 | 503.81 | 413.56 | 300.17 | 284.95 |
| 33 23 1122 | Move Rig/Equipment Around Site | EACH | 154.66 | 125.95 | 103.39 | 75.04 | 71.24 |
| 33 23 1125 | Concrete Coring (Minimum Charge) | DAY | 2,475 | 2,015 | 1,654 | 1,201 | 1,140 |

## Environmental Remediation: Assemblies Cost Book

*Page 3-295*

1998 by ECHOS. All rights reserved

# Cost Data for Remediation

## Soil Vapor Extraction

| Assembly | Description | Unit | Unit Cost by Safety Level A | B | C | D | E |
|---|---|---|---|---|---|---|---|
| | **Capital Costs** | | | | | | |
| 33 23 1126 | Furnish 55 Gallon Drum for Drill Cuttings & Development Water | EACH | 65.19 | 65.19 | 65.19 | 65.19 | 65.19 |
| 33 23 1152 | Mud Drilling, 6" Diameter Borehole | LF | 81.01 | 65.98 | 54.16 | 39.31 | 37.31 |
| 33 23 1153 | Mud Drilling, 8" Diameter Borehole | LF | 88.38 | 71.97 | 59.08 | 42.88 | 40.71 |
| 33 23 1161 | Air Rotary 6" Borehole in Unconsolidated | LF | 73.65 | 59.98 | 49.23 | 35.73 | 33.92 |
| 33 23 1162 | Air Rotary 8" Borehole in Unconsolidated | LF | 122.74 | 99.96 | 82.06 | 59.56 | 56.54 |
| 33 23 1181 | Trenching for Horizontal Well Installation | LF | 31.93 | 31.93 | 31.93 | 31.93 | 31.93 |
| 33 23 1185 | Standby for Trenching | EACH | 383.54 | 334.53 | 302.86 | 254.87 | 244.12 |
| 33 23 1187 | Mobilize/DeMobilize Trencher/Crew | EACH | 38,354 | 33,453 | 30,286 | 25,487 | 24,412 |
| 33 23 1401 | 2" Screen, Filter Pack | LF | 16.49 | 13.88 | 11.84 | 9.27 | 8.92 |
| 33 23 1402 | 4" Screen, Filter Pack | LF | 29.10 | 24.50 | 20.89 | 16.36 | 15.75 |
| 33 23 1403 | 6" Screen, Filter Pack | LF | 42.19 | 35.53 | 30.29 | 23.72 | 22.83 |
| 33 23 1407 | Gravel Pack for Horizontal Well Installation | CF | 0.84 | 0.80 | 0.78 | 0.74 | 0.73 |
| 33 23 1502 | Surface Pad, Concrete, 4' x 4' x 4" | EACH | 24.21 | 21.82 | 20.74 | 18.43 | 17.61 |
| 33 23 1811 | 2" Well, Portland Cement Grout | LF | 0.92 | 0.92 | 0.92 | 0.92 | 0.92 |
| 33 23 1812 | 4" Well, Portland Cement Grout | LF | 1.38 | 1.38 | 1.38 | 1.38 | 1.38 |
| 33 23 2101 | 2" Well, Bentonite Seal | EACH | 63.00 | 52.67 | 44.55 | 34.34 | 32.97 |
| 33 23 2102 | 4" Well, Bentonite Seal | EACH | 157.54 | 131.70 | 111.39 | 85.87 | 82.44 |
| 33 26 0210 | 2" Stainless Steel, Schedule 40, Type 304, Connection Piping | LF | 33.47 | 29.21 | 27.39 | 23.28 | 21.77 |
| 33 26 0211 | 4" Stainless Steel, Schedule 40, Type 304, Connection Piping | LF | 80.15 | 71.72 | 68.12 | 59.98 | 57.00 |
| 33 26 0212 | 4" Stainless Steel, Schedule 40, Type 304, Manifold Piping | LF | 80.15 | 71.72 | 68.12 | 59.98 | 57.00 |
| 33 26 0405 | 8" PVC, Schedule 40, Manifold Piping | LF | 23.47 | 19.13 | 17.28 | 13.08 | 11.55 |
| 33 26 0406 | 2" PVC, Schedule 40, Connection Piping | LF | 6.14 | 4.87 | 4.33 | 3.11 | 2.66 |
| 33 26 0407 | 4" PVC, Schedule 40, Connection Piping | LF | 13.38 | 10.65 | 9.48 | 6.84 | 5.88 |
| 33 26 0408 | 4" PVC, Schedule 40, Manifold Piping | LF | 13.38 | 10.65 | 9.48 | 6.84 | 5.88 |
| 33 26 0412 | 8" PVC, Schedule 40, Connection Piping | LF | 23.47 | 19.13 | 17.28 | 13.08 | 11.55 |
| 33 27 0104 | 4" PVC, Schedule 40, Tee | EACH | 137.04 | 108.84 | 96.77 | 69.51 | 59.51 |
| 33 27 0106 | 8" PVC, Schedule 40, Tee | EACH | 336.90 | 291.78 | 272.46 | 228.85 | 212.85 |
| 33 27 0112 | 2" PVC, 90 Degree, Elbow | EACH | 41.36 | 32.22 | 28.30 | 19.46 | 16.22 |
| 33 27 0114 | 4" PVC, 90 Degree, Elbow | EACH | 101.72 | 80.57 | 71.51 | 51.07 | 43.57 |
| 33 27 0116 | 8" PVC, 90 Degree, Elbow | EACH | 200.21 | 172.01 | 159.94 | 132.68 | 122.68 |
| 33 27 0144 | 8" Flange Assemblies, PVC, Schedule 40 | EACH | 31.29 | 31.29 | 31.29 | 31.29 | 31.29 |

**Environmental Remediation: Assemblies Cost Book**

*Page 3-296*

# Cost Data for Remediation

## Soil Vapor Extraction

| Assembly | Description | Unit | Unit Cost by Safety Level | | | | |
|---|---|---|---|---|---|---|---|
| | | | A | B | C | D | E |
| | **Capital Costs** | | | | | | |
| 33 27 0151 | 4" x 2" Reducer, PVC Schedule 40 | EACH | 9.09 | 9.09 | 9.09 | 9.09 | 9.09 |
| 33 27 0153 | 8" x 6" Reducer, PVC Schedule 40 | EACH | 22.98 | 22.98 | 22.98 | 22.98 | 22.98 |
| 33 27 0155 | 8" x 4" Reducer, PVC Schedule 40 | EACH | 67.28 | 67.28 | 67.28 | 67.28 | 67.28 |
| 33 27 0156 | 12" x 8" Reducer, PVC Schedule 40 | EACH | 166.92 | 166.92 | 166.92 | 166.92 | 166.92 |
| 33 27 0202 | 4" Stainless Steel, Tee | EACH | 583.41 | 504.25 | 470.36 | 393.85 | 365.80 |
| 33 27 0211 | 2" Stainless Steel, 90 Degree, Elbow | EACH | 114.28 | 93.45 | 84.53 | 64.39 | 57.01 |
| 33 27 0212 | 4" Stainless Steel, 90 Degree, Elbow | EACH | 391.27 | 338.50 | 315.91 | 264.90 | 246.20 |
| 33 27 0231 | 4" x 2" Stainless Steel Reducer, Schedule 40, Type 304 | EACH | 27.01 | 27.01 | 27.01 | 27.01 | 27.01 |
| 33 27 0242 | 8" x 4" Stainless Steel Reducer, Schedule 10, Type 316 | EACH | 279.00 | 279.00 | 279.00 | 279.00 | 279.00 |
| 33 27 0352 | 4" SDR-17 Flange Assemblies | EACH | 34.09 | 34.09 | 34.09 | 34.09 | 34.09 |
| 33 27 0353 | 6" SDR-17 Flange Assemblies | EACH | 71.20 | 71.20 | 71.20 | 71.20 | 71.20 |
| 33 27 0404 | 4" Iron Body Check Valve | EACH | 354.29 | 315.96 | 299.54 | 262.49 | 248.90 |
| 33 27 0406 | 8" Iron Body Check Valve | EACH | 1,094 | 980.95 | 926.96 | 817.50 | 780.98 |
| 33 31 0209 | Pressure Gauge | EACH | 140.16 | 121.61 | 113.67 | 95.74 | 89.16 |
| 33 42 0101 | Electrical Charge | KWH | 0.06 | 0.06 | 0.06 | 0.06 | 0.06 |
| | **Operations & Maintenance** | | | | | | |
| 33 02 1507 | Continuous Monitoring and Recording of Air Flow | EACH | 9,870 | 9,870 | 9,870 | 9,870 | 9,870 |
| 33 42 0101 | Electrical Charge | KWH | 0.06 | 0.06 | 0.06 | 0.06 | 0.06 |

**Environmental Remediation: Assemblies Cost Book**

# Cost Data for Remediation
## Soil Washing

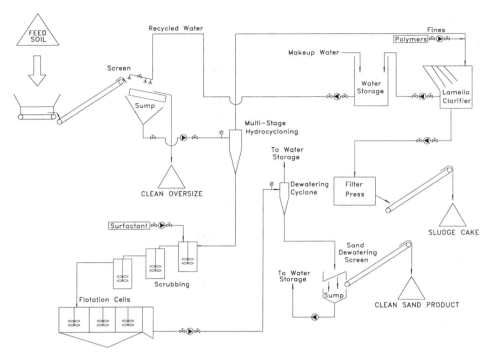

FLOW DIAGRAM OF SOIL WASHING PROCESS
SOURCE: ALTERNATIVE REMEDIAL TECHNOLOGIES, INC., TAMPA, FL.

### General:

Soil washing is a water-based process for mechanically scrubbing and leaching waste constituents from a contaminated soil for recovery and treatment. It is an ex situ toxicity reduction technology. The process removes contaminants from soils in one of two ways: by dissolving or suspending contaminants in the wash solution (which is later treated by conventional wastewater treatment methods) or by concentrating contaminants into a smaller volume of soil through simple particle size separation techniques.

### Typical Treatment Train:

Soil excavation, debris removal, treatment and disposal of the wash solution and/or filtered soil, sampling and analysis.

### Common Cost Components:

1. Soil treatment using a packaged system

### Other Cost Considerations:

Pilot tests.

**Environmental Remediation: Assemblies Cost Book**

# Cost Data for Remediation

## Soil Washing

| Assembly | Description | Unit | Unit Cost by Safety Level A | B | C | D | E |
|---|---|---|---|---|---|---|---|
| | **Capital Costs** | | | | | | |
| 17 03 0109 | Pad Subgrade Preparation | CY | 8.13 | 6.56 | 5.44 | 3.90 | 3.63 |
| 17 03 0220 | 910, 1.25 CY, Wheel Loader | HOUR | 130.20 | 104.06 | 87.23 | 61.63 | 55.86 |
| 17 03 0221 | 916, 1.5 CY, Wheel Loader | HOUR | 114.32 | 90.83 | 76.64 | 53.69 | 47.92 |
| 17 03 0222 | 926, 2.0 CY, Wheel Loader | HOUR | 154.30 | 124.14 | 103.30 | 73.68 | 67.91 |
| 17 03 0223 | 950, 3.0 CY, Wheel Loader | HOUR | 180.10 | 145.64 | 120.50 | 86.58 | 80.81 |
| 17 03 0224 | 966, 4.0 CY, Wheel Loader | HOUR | 211.58 | 171.88 | 141.48 | 102.32 | 96.55 |
| 17 03 0225 | 980, 5.25 CY, Wheel Loader | HOUR | 298.84 | 244.38 | 199.68 | 145.79 | 139.74 |
| 17 03 0345 | Standby, 910, 1.25 CY Wheel Loader | HOUR | 15.13 | 12.61 | 10.09 | 7.57 | 7.57 |
| 17 03 0346 | Standby, 916, 1.5 CY Wheel Loader | HOUR | 11.32 | 9.44 | 7.55 | 5.66 | 5.66 |
| 17 03 0347 | Standby, 926, 2.0 CY Wheel Loader | HOUR | 21.40 | 17.83 | 14.27 | 10.70 | 10.70 |
| 17 03 0348 | Standby, 950, 3.0 CY Wheel Loader | HOUR | 27.24 | 22.70 | 18.16 | 13.62 | 13.62 |
| 17 03 0349 | Standby, 966, 4.0 CY Wheel Loader | HOUR | 34.27 | 28.56 | 22.85 | 17.13 | 17.13 |
| 17 03 0350 | Standby, 980, 5.25 CY Wheel Loader | HOUR | 57.52 | 47.94 | 38.35 | 28.76 | 28.76 |
| 17 03 0419 | Crushed Stone, 1/2" to 3/4" | CY | 23.45 | 21.67 | 20.70 | 18.97 | 18.47 |
| 17 03 0420 | Backfill Trench, Borrow Material, Delivered & Dumped Only | CY | 9.86 | 8.74 | 7.93 | 6.83 | 6.63 |
| 17 03 0510 | Dry Roll Gravel, Steel Roller | SY | 1.22 | 0.96 | 0.82 | 0.57 | 0.49 |
| 17 03 1002 | 2" Diameter Contractor's Trash Pump, 75 GPM | DAY | 50.20 | 47.01 | 45.64 | 42.55 | 41.42 |
| 17 03 1003 | 3" Diameter Contractor's Trash Pump, 150 GPM | DAY | 61.72 | 57.05 | 55.05 | 50.54 | 48.88 |
| 17 03 1004 | 4" Diameter Contractor's Trash Pump, 300 GPM | DAY | 80.30 | 74.23 | 71.63 | 65.77 | 63.61 |
| 17 03 9903 | Hand Place Small Earth Fill Berm | CY | 112.16 | 86.46 | 75.45 | 50.61 | 41.50 |
| 18 01 0102 | Gravel, Delivered & Dumped | CY | 24.33 | 22.93 | 22.03 | 20.66 | 20.36 |
| 18 01 0201 | Concrete Curb, 6" x 6" | LF | 2.78 | 2.38 | 2.21 | 1.83 | 1.70 |
| 18 01 0202 | Concrete Curb & Gutter, 6" x 24", Formed | LF | 20.78 | 16.61 | 14.83 | 10.80 | 9.32 |
| 18 01 0206 | Concrete Curb, 8" x 8" | LF | 7.58 | 6.24 | 5.66 | 4.36 | 3.89 |
| 18 01 0207 | Concrete Curb, 12" High x 8" Deep | LF | 10.27 | 8.45 | 7.67 | 5.92 | 5.27 |
| 18 02 0320 | 4" Structural Slab on Grade | SF | 6.45 | 5.45 | 4.98 | 4.00 | 3.67 |
| 18 02 0321 | 6" Structural Slab on Grade | SF | 6.98 | 5.95 | 5.47 | 4.48 | 4.14 |
| 18 02 0322 | 8" Structural Slab on Grade | SF | 9.75 | 8.38 | 7.71 | 6.38 | 5.95 |
| 18 02 0323 | 10" Structural Slab on Grade | SF | 10.56 | 9.13 | 8.43 | 7.04 | 6.58 |
| 18 02 0324 | 12" Structural Slab on Grade | SF | 11.57 | 10.04 | 9.30 | 7.82 | 7.34 |
| 18 02 0330 | 4" Mesh Reinforced Slab on Grade | SF | 5.48 | 4.60 | 4.18 | 3.32 | 3.03 |
| 18 02 0331 | 6" Mesh Reinforced Slab on Grade | SF | 6.01 | 5.10 | 4.67 | 3.79 | 3.49 |
| 18 02 0332 | 8" Mesh Reinforced Slab on Grade | SF | 7.58 | 6.50 | 5.96 | 4.91 | 4.58 |

**Environmental Remediation: Assemblies Cost Book**

# Cost Data for Remediation

## Soil Washing

| Assembly | Description | Unit | Unit Cost by Safety Level A | B | C | D | E |
|---|---|---|---|---|---|---|---|
| | **Capital Costs** | | | | | | |
| 18 02 0333 | 10" Mesh Reinforced Slab on Grade | SF | 8.10 | 7.00 | 6.44 | 5.37 | 5.03 |
| 18 02 0334 | 12" Mesh Reinforced Slab on Grade | SF | 8.78 | 7.63 | 7.05 | 5.93 | 5.58 |
| 18 02 0340 | 4" Unreinforced Slab on Grade | SF | 4.34 | 3.64 | 3.29 | 2.61 | 2.38 |
| 18 02 0341 | 6" Unreinforced Slab on Grade | SF | 4.87 | 4.14 | 3.79 | 3.08 | 2.85 |
| 18 02 0342 | 8" Unreinforced Slab on Grade | SF | 6.45 | 5.54 | 5.07 | 4.19 | 3.93 |
| 18 02 0343 | 10" Unreinforced Slab on Grade | SF | 6.97 | 6.04 | 5.56 | 4.66 | 4.38 |
| 18 02 0344 | 12" Unreinforced Slab on Grade | SF | 7.64 | 6.67 | 6.16 | 5.22 | 4.93 |
| 19 02 0311 | 12' x 36" Diameter Reinforced Concrete Pipe Wet Well for Lift Station | EACH | 9,840 | 8,577 | 7,816 | 6,581 | 6,270 |
| 19 02 0312 | 24' x 60" Diameter Reinforced Concrete Pipe Wet Well for Lift Station | EACH | 39,142 | 33,897 | 30,681 | 25,554 | 24,296 |
| 19 02 0313 | 5' x 5' x 5' Reinforced Concrete Sump | EACH | 3,948 | 3,282 | 2,979 | 2,335 | 2,110 |
| 19 02 0601 | 12" x 12" Underground French Drain | LF | 4.50 | 3.92 | 3.61 | 3.04 | 2.88 |
| 19 02 0602 | 18" x 18" Underground French Drain | LF | 5.93 | 5.29 | 4.94 | 4.31 | 4.13 |
| 19 02 0603 | 24" x 24" Underground French Drain | LF | 7.43 | 6.71 | 6.31 | 5.61 | 5.41 |
| 19 04 0401 | 550 Gallon, Stainless Steel Aboveground Wastewater Holding Tank, Rental | MONTH | 320.65 | 320.65 | 320.65 | 320.65 | 320.65 |
| 19 04 0403 | 4,000 Gallon Polyethylene Wastewater Tank, Rental | MONTH | 545.90 | 545.90 | 545.90 | 545.90 | 545.90 |
| 19 04 0405 | 6,000 Gallon, Polyethylene Aboveground Wastewater Holding Tank, Rental | MONTH | 641.30 | 641.30 | 641.30 | 641.30 | 641.30 |
| 19 04 0407 | 21,000 Gallon, Steel Closed Stationary Aboveground Wastewater Holding Tank, Rental | MONTH | 1,219 | 1,219 | 1,219 | 1,219 | 1,219 |
| 19 04 0602 | 550 Gallon Steel Sump, Aboveground with Supports & Fittings | EACH | 1,598 | 1,390 | 1,301 | 1,099 | 1,026 |
| 19 04 0603 | 1,000 Gallon Steel Sump, Aboveground with Supports & Fittings | EACH | 2,118 | 1,844 | 1,727 | 1,463 | 1,366 |
| 19 04 0604 | 1,500 Gallon Steel Sump, Aboveground with Supports & Fittings | EACH | 2,697 | 2,356 | 2,209 | 1,879 | 1,758 |
| 19 04 0605 | 2,000 Gallon Steel Sump, Aboveground with Supports & Fittings | EACH | 3,457 | 2,997 | 2,777 | 2,332 | 2,183 |
| 19 04 0606 | 5,000 Gallon Steel Sump, Aboveground with Supports & Fittings | EACH | 6,079 | 5,494 | 5,214 | 4,646 | 4,457 |
| 19 04 0625 | 6,000 Gallon Horizontal Plastic Sump with 6" NPT Connection | EACH | 5,248 | 4,975 | 4,858 | 4,594 | 4,497 |
| 19 04 0626 | 8,000 Gallon Horizontal Plastic Sump with 6" NPT Connection | EACH | 5,857 | 5,516 | 5,369 | 5,039 | 4,918 |
| 19 04 0627 | 10,000 Gallon Horizontal Plastic Sump with 6" NPT Connection | EACH | 7,082 | 6,622 | 6,402 | 5,956 | 5,807 |
| 19 04 0628 | 12,000 Gallon Horizontal Plastic Sump with 6" NPT Connection | EACH | 8,308 | 7,722 | 7,442 | 6,875 | 6,686 |

**Environmental Remediation: Assemblies Cost Book**

# Cost Data for Remediation

## Soil Washing

| Assembly | Description | Unit | A | B | C | D | E |
|---|---|---|---|---|---|---|---|
| | **Capital Costs** | | | | | | |
| 19 04 0629 | 15,000 Gallon Horizontal Plastic Sump with 6" NPT Connection | EACH | 14,184 | 13,540 | 13,232 | 12,608 | 12,400 |
| 19 04 0631 | 20,000 Gallon Horizontal Plastic Sump with 6" NPT Connection | EACH | 15,799 | 15,084 | 14,742 | 14,048 | 13,817 |
| 19 04 0632 | 25,000 Gallon Horizontal Plastic Sump with 6" NPT Connection | EACH | 24,407 | 23,487 | 23,047 | 22,155 | 21,858 |
| 19 04 0633 | 30,000 Gallon Horizontal Plastic Sump with 6" NPT Connection | EACH | 35,112 | 34,039 | 33,525 | 32,485 | 32,138 |
| 19 04 0634 | 40,000 Gallon Horizontal Plastic Sump with 6" NPT Connection | EACH | 40,844 | 39,234 | 38,465 | 36,904 | 36,383 |
| 19 04 0635 | 48,000 Gallon Horizontal Plastic Sump with 6" NPT Connection | EACH | 59,758 | 57,612 | 56,586 | 54,505 | 53,811 |
| 33 08 0503 | Polymeric Liner Anchor Trench, 3' x 1.5' | LF | 1.58 | 1.25 | 1.07 | 0.76 | 0.67 |
| 33 08 0563 | 40 Mil Polymeric Liner, PVC | SF | 0.38 | 0.36 | 0.36 | 0.34 | 0.34 |
| 33 09 0702 | CIP, Reinforced Concrete Containment Wall, Including Forms | CY | 913.17 | 739.71 | 665.37 | 497.70 | 436.28 |
| 33 11 9301 | 200 MBH Gas-fired Water Boiler | EACH | 6,218 | 5,682 | 5,425 | 4,905 | 4,731 |
| 33 11 9302 | 275 MBH Gas-fired Water Boiler | EACH | 8,169 | 7,249 | 6,809 | 5,917 | 5,620 |
| 33 11 9303 | 360 MBH Gas-fired Water Boiler | EACH | 8,247 | 7,327 | 6,887 | 5,995 | 5,698 |
| 33 11 9304 | 520 MBH Gas-fired Water Boiler | EACH | 9,546 | 8,473 | 7,959 | 6,919 | 6,572 |
| 33 11 9305 | 600 MBH Gas-fired Water Boiler | EACH | 10,154 | 9,081 | 8,567 | 7,527 | 7,180 |
| 33 11 9306 | 720 MBH Gas-fired Water Boiler | EACH | 11,730 | 10,442 | 9,826 | 8,578 | 8,161 |
| 33 11 9307 | 960 MBH Gas-fired Water Boiler | EACH | 13,789 | 12,179 | 11,410 | 9,849 | 9,328 |
| 33 11 9308 | 1,220 MBH Gas-fired Water Boiler | EACH | 14,398 | 12,788 | 12,019 | 10,458 | 9,937 |
| 33 11 9309 | 1,440 MBH Gas-fired Water Boiler | EACH | 23,432 | 21,594 | 20,716 | 18,934 | 18,340 |
| 33 11 9310 | 1,680 MBH Gas-fired Water Boiler | EACH | 25,423 | 23,277 | 22,251 | 20,170 | 19,476 |
| 33 11 9311 | 1,920 MBH Gas-fired Water Boiler | EACH | 27,362 | 24,885 | 23,701 | 21,300 | 20,499 |
| 33 11 9312 | 2,160 MBH Gas-fired Water Boiler | EACH | 31,110 | 27,891 | 26,351 | 23,230 | 22,188 |
| 33 11 9313 | 2,400 MBH Gas-fired Water Boiler | EACH | 31,410 | 28,191 | 26,651 | 23,530 | 22,488 |
| 33 11 9314 | 7 GPM Water Heat Exchanger | EACH | 1,265 | 1,199 | 1,171 | 1,107 | 1,084 |
| 33 11 9315 | 16 GPM Water Heat Exchanger | EACH | 1,728 | 1,649 | 1,615 | 1,538 | 1,510 |
| 33 11 9316 | 34 GPM Water Heat Exchanger | EACH | 2,532 | 2,434 | 2,391 | 2,296 | 2,260 |
| 33 11 9317 | 55 GPM Water Heat Exchanger | EACH | 3,609 | 3,477 | 3,420 | 3,293 | 3,246 |
| 33 11 9318 | 74 GPM Water Heat Exchanger | EACH | 4,503 | 4,331 | 4,257 | 4,091 | 4,030 |
| 33 11 9319 | 86 GPM Water Heat Exchanger | EACH | 5,935 | 5,727 | 5,637 | 5,436 | 5,362 |
| 33 11 9320 | 112 GPM Water Heat Exchanger | EACH | 7,395 | 7,142 | 7,022 | 6,777 | 6,695 |

**Environmental Remediation: Assemblies Cost Book**

# Cost Data for Remediation

## Soil Washing

| Assembly | Description | Unit | Unit Cost by Safety Level A | B | C | D | E |
|---|---|---|---|---|---|---|---|
| | **Capital Costs** | | | | | | |
| 33 11 9331 | Expansion Tank, Pipe, & Fittings, 1 - 1,220 MBH Boiler | EACH | 4,657 | 3,871 | 3,535 | 2,776 | 2,497 |
| 33 11 9332 | Expansion Tank, Pipe, & Fittings, 1,440 - 2,400 MBH Boiler | EACH | 20,553 | 17,748 | 16,543 | 13,832 | 12,841 |
| 33 13 0915 | Mobilize/Demobilize Soil Washing System | MI | 2.11 | 2.11 | 2.11 | 2.11 | 2.11 |
| 33 13 0916 | Assemble/Disassemble Soil Washing System | EACH | 59,475 | 46,901 | 39,908 | 27,660 | 24,202 |
| 33 13 0917 | Startup/Shakedown - Soil Washing System | EACH | 37,770 | 29,115 | 25,409 | 17,043 | 13,975 |
| 33 13 0918 | Decontaminate Soil Washing System | EACH | 560.32 | 434.51 | 376.69 | 254.86 | 212.73 |
| 33 13 0920 | 25 Tons per Hour Soil Washing System, Monthly Rental | MONTH | 81,000 | 81,000 | 81,000 | 81,000 | 81,000 |
| 33 13 0921 | Operational Labor, 10 - 25 Tons per Hour System | HR | 566.55 | 436.72 | 381.14 | 255.64 | 209.63 |
| 33 13 0922 | 50 Tons per Hour Soil Washing System, Monthly Rental | MONTH | 95,000 | 95,000 | 95,000 | 95,000 | 95,000 |
| 33 13 0923 | Operational Labor, 50 Tons per Hour System | HR | 755.41 | 582.29 | 508.18 | 340.85 | 279.50 |
| 33 13 0924 | 100 Tons per Hour Soil Washing System, Monthly Rental | MONTH | 120,000 | 120,000 | 120,000 | 120,000 | 120,000 |
| 33 13 0925 | Operational Labor, 100 Tons per Hour System | HR | 944.26 | 727.86 | 635.23 | 426.07 | 349.38 |
| 33 13 0927 | Maintenance/Spare Parts, 25 Tons per Hour System | TON | 2.24 | 2.24 | 2.24 | 2.24 | 2.24 |
| 33 13 0928 | Maintenance/Spare Parts, 50 Tons per Hour System | TON | 1.31 | 1.31 | 1.31 | 1.31 | 1.31 |
| 33 13 0929 | Maintenance/Spare Parts, 100 Tons per Hour System | TON | 0.84 | 0.84 | 0.84 | 0.84 | 0.84 |
| 33 13 2916 | Natural Gas Usage, per 1,000 CF | MCF | 5.50 | 5.50 | 5.50 | 5.50 | 5.50 |
| 33 19 7103 | Process Water Hauling Fee, Subcontracted | GAL | 0.24 | 0.24 | 0.24 | 0.24 | 0.24 |
| 33 19 7104 | Radioactive Process Water Treatment/Disposal, Subcontracted | GAL | 2.01 | 2.01 | 2.01 | 2.01 | 2.01 |
| 33 19 7105 | Non-Radioactive Process Water Treatment/Disposal, Subcontracted | GAL | 1.01 | 1.01 | 1.01 | 1.01 | 1.01 |
| 33 23 1306 | High Sump Level Switch for Avoiding Overflow | EACH | 214.65 | 214.65 | 214.65 | 214.65 | 214.65 |
| 33 29 0103 | 50 GPM, 100' Head, 3 HP, Centrifugal Pump | EACH | 2,658 | 2,500 | 2,432 | 2,279 | 2,223 |
| 33 29 0108 | 10 HP, 200 GPM, Centrifugal Pump | EACH | 2,941 | 2,714 | 2,616 | 2,396 | 2,316 |
| 33 29 0412 | 15 GPM Submersible Sump Pump | EACH | 492.79 | 460.07 | 446.06 | 414.43 | 402.83 |
| 33 29 0413 | 42 GPM Submersible Sump Pump | EACH | 792.79 | 760.07 | 746.06 | 714.43 | 702.83 |
| 33 29 0414 | 52 GPM Submersible Sump Pump | EACH | 871.35 | 832.08 | 815.27 | 777.32 | 763.40 |
| 33 33 0101 | Sodium Hydroxide Solution, 190 Lb Drummed Liquid | EACH | 62.34 | 57.81 | 55.07 | 50.64 | 49.53 |
| 33 33 0102 | Sodium Hydroxide Solution, 700 Lb Drummed Liquid | EACH | 163.04 | 158.51 | 155.77 | 151.34 | 150.23 |
| 33 33 0103 | Sodium Hydroxide, 500 Lb Container, Beads | EACH | 234.06 | 229.53 | 226.79 | 222.36 | 221.25 |
| 33 33 0104 | Sodium Hydroxide, 100 Lb Container, Flakes | EACH | 89.90 | 85.37 | 82.63 | 78.20 | 77.09 |
| 33 33 0105 | Sodium Hydroxide, 400 Lb Container, Flakes | EACH | 169.60 | 169.60 | 169.60 | 169.60 | 169.60 |
| 33 33 0106 | Magnesium Hydroxide Slurry, 55 Gallon Drums, < 40 Drums | EACH | 249.77 | 249.77 | 249.77 | 249.77 | 249.77 |

**Environmental Remediation: Assemblies Cost Book**

*Page 3-302*

# Cost Data for Remediation

## Soil Washing

| Assembly | Description | Unit | Unit Cost by Safety Level | | | | |
|---|---|---|---|---|---|---|---|
| | | | A | B | C | D | E |
| | **Capital Costs** | | | | | | |
| 33 33 0107 | Magnesium Hydroxide Slurry, 55 Gallon Drums, 40 - 71 Drums | EACH | 234.26 | 234.26 | 234.26 | 234.26 | 234.26 |
| 33 33 0108 | Magnesium Hydroxide Slurry, 55 Gallon Drums, 72 Drums | EACH | 107.33 | 107.33 | 107.33 | 107.33 | 107.33 |
| 33 33 0109 | Magnesium Hydroxide, Bulk Quantity, 20 Tons | EACH | 4,452 | 4,452 | 4,452 | 4,452 | 4,452 |
| 33 33 0110 | Sulfuric Acid Solution, 220 Lb Drummed Liquid | EACH | 45.38 | 40.85 | 38.11 | 33.68 | 32.57 |
| 33 33 0111 | Sulfuric Acid Solution, 750 Lb Drummed Liquid | EACH | 93.61 | 89.08 | 86.34 | 81.91 | 80.80 |
| 33 33 0112 | Soda Ash, 50 Lb Bag, Powdered | EACH | 28.42 | 23.89 | 21.15 | 16.72 | 15.61 |
| 33 33 0113 | Soda Ash, 100 Lb Bag, Powdered | EACH | 41.14 | 36.61 | 33.87 | 29.44 | 28.33 |
| 33 33 0114 | Quicklime, 1/4" Nominal Granules, Bulk Quantity | TON | 96.49 | 96.49 | 96.48 | 96.48 | 96.47 |
| 33 33 0115 | Quicklime, 3/4" Nominal Granules, Bulk Quantity | TON | 96.49 | 96.49 | 96.48 | 96.48 | 96.47 |
| 33 33 0116 | Quicklime, Combination 1/4" & 3/4" Granules, Bulk Quantity | TON | 96.49 | 96.49 | 96.48 | 96.48 | 96.47 |
| 33 33 0117 | Hydrated Lime, Powdered, Bulk | TON | 97.55 | 97.55 | 97.54 | 97.54 | 97.53 |
| 33 33 0118 | Hydrated Lime, Powdered, 50 Lb Bag | EACH | 25.76 | 21.23 | 18.49 | 14.06 | 12.94 |
| 33 33 0119 | Aluminum Sulfate (Alum), Bulk | TON | 240.07 | 240.06 | 240.06 | 240.05 | 240.05 |
| 33 33 0120 | Ferric Sulfate, Bulk | TON | 274.04 | 274.04 | 274.03 | 274.03 | 274.02 |
| 33 33 0121 | Ferric Chloride, Bulk | TON | 241.18 | 241.18 | 241.17 | 241.17 | 241.16 |
| 33 33 0122 | Montmorillonite Clay Flocculant Aid, Bulk | LB | 0.14 | 0.13 | 0.13 | 0.12 | 0.12 |
| 33 33 0123 | Sodium Hectorite Clay Flocculant Aid, Bulk | LB | 0.17 | 0.17 | 0.16 | 0.15 | 0.15 |
| 33 33 0124 | Super Dispersible Sodium Flocculant Aid, Bulk | LB | 0.19 | 0.19 | 0.18 | 0.18 | 0.17 |
| 33 33 0125 | Alkylbenzene Sulfonate Surfactant, Bulk | LB | 1.16 | 1.16 | 1.16 | 1.16 | 1.16 |
| 33 33 0127 | Non-ionic Polymeric Flocculant | LB | 3.01 | 3.01 | 3.01 | 3.01 | 3.01 |
| 33 33 0171 | Hydrogen Peroxide, 50% Solution, 500 Lb Drums | EACH | 667.80 | 667.80 | 667.80 | 667.80 | 667.80 |
| 33 42 0101 | Electrical Charge | KWH | 0.06 | 0.06 | 0.06 | 0.06 | 0.06 |
| 33 42 0302 | Process Water, Supplied by Water Line | KGAL | 2.56 | 2.56 | 2.56 | 2.56 | 2.56 |

**Environmental Remediation: Assemblies Cost Book**

# Cost Data for Remediation
## Solidification/Stabilization

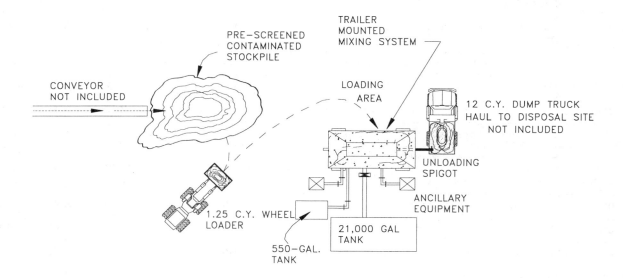

### General:
*Ex situ solidification/stabilization is a treatment technology in which chemical reagents are mixed with waste to make use of complex chemical and physical reactions to improve physical properties and reduce contaminant solubility, toxicity, and/or mobility. Solidification/stabilization is a viable treatment for contaminated materials when the constituents cannot be treated, recovered, or destroyed by other methods because of technical or economic limitations.*

### Typical Treatment Train:
*Excavation, transportation of wastes to process site, transportation to and from mixing area, drum removal and staging, disposal, sampling and analysis.*

### Common Cost Components:
1. Structural slab
2. Holding tank
3. Equipment purchase or rental: conveyor, hopper, mixing unit, pumps, drum shredder
4. Batch processing: load, mix, and unload
5. Chemical additives: water, proprietary chemical binders, portland cement, fly ash, cement kiln dust, hydrated lime, bitumen, activated carbon
6. Handling of nonhomogeneous materials

### Other Cost Considerations:
*Bench/pilot testing, personal protective equipment (PPE), backfill excavated area.*

**Environmental Remediation: Assemblies Cost Book**

# Cost Data for Remediation

## Solidification/Stabilization

### Unit Cost by Safety Level

| Assembly | Description | Unit | A | B | C | D | E |
|---|---|---|---|---|---|---|---|
| | **Capital Costs** | | | | | | |
| 17 03 0220 | 910, 1.25 CY, Wheel Loader | HOUR | 130.20 | 104.06 | 87.23 | 61.63 | 55.86 |
| 17 03 0221 | 916, 1.5 CY, Wheel Loader | HOUR | 114.32 | 90.83 | 76.64 | 53.69 | 47.92 |
| 17 03 0222 | 926, 2.0 CY, Wheel Loader | HOUR | 154.30 | 124.14 | 103.30 | 73.68 | 67.91 |
| 17 03 0223 | 950, 3.0 CY, Wheel Loader | HOUR | 180.10 | 145.64 | 120.50 | 86.58 | 80.81 |
| 17 03 0224 | 966, 4.0 CY, Wheel Loader | HOUR | 211.58 | 171.88 | 141.48 | 102.32 | 96.55 |
| 17 03 0225 | 980, 5.25 CY, Wheel Loader | HOUR | 298.84 | 244.38 | 199.68 | 145.79 | 139.74 |
| 17 03 0226 | 988, 7.0 CY, Wheel Loader | HOUR | 384.90 | 316.10 | 257.05 | 188.82 | 182.77 |
| 17 03 0227 | 992, 13.5 CY, Wheel Loader | HOUR | 648.92 | 536.11 | 433.06 | 320.83 | 314.78 |
| 17 03 0235 | Crawler-mounted, 5.5 CY, Koehring 1266, Hydraulic Excavator | HOUR | 491.70 | 404.92 | 328.27 | 242.08 | 235.80 |
| 17 03 0285 | 12 CY, Dump Truck | HOUR | 119.55 | 95.96 | 80.05 | 56.91 | 52.15 |
| 17 03 0431 | 580K, 1.0 CY, Backhoe with Front-end Loader | HOUR | 127.72 | 101.78 | 85.60 | 60.23 | 54.18 |
| 17 03 0432 | Bobcat with Backhoe | MONTH | 1,500 | 1,500 | 1,500 | 1,500 | 1,500 |
| 18 02 0320 | 4" Structural Slab on Grade | SF | 6.45 | 5.45 | 4.98 | 4.00 | 3.67 |
| 18 02 0321 | 6" Structural Slab on Grade | SF | 6.98 | 5.95 | 5.47 | 4.48 | 4.14 |
| 18 02 0322 | 8" Structural Slab on Grade | SF | 9.75 | 8.38 | 7.71 | 6.38 | 5.95 |
| 18 02 0323 | 10" Structural Slab on Grade | SF | 10.56 | 9.13 | 8.43 | 7.04 | 6.58 |
| 18 02 0324 | 12" Structural Slab on Grade | SF | 11.57 | 10.04 | 9.30 | 7.82 | 7.34 |
| 18 02 0341 | 6" Unreinforced Slab on Grade | SF | 4.87 | 4.14 | 3.79 | 3.08 | 2.85 |
| 18 02 0342 | 8" Unreinforced Slab on Grade | SF | 6.45 | 5.54 | 5.07 | 4.19 | 3.93 |
| 18 02 0343 | 10" Unreinforced Slab on Grade | SF | 6.97 | 6.04 | 5.56 | 4.66 | 4.38 |
| 18 02 0344 | 12" Unreinforced Slab on Grade | SF | 7.64 | 6.67 | 6.16 | 5.22 | 4.93 |
| 19 04 0401 | 550 Gallon, Stainless Steel Aboveground Wastewater Holding Tank, Rental | MONTH | 320.65 | 320.65 | 320.65 | 320.65 | 320.65 |
| 19 04 0408 | 21,000 Gallon Steel, Open Top, Tank Rental | MONTH | 1,124 | 1,124 | 1,124 | 1,124 | 1,124 |
| 20 99 9901 | 200 KW Diesel Generator, 3 Phase | EACH | 48,691 | 46,436 | 45,434 | 43,253 | 42,477 |
| 33 01 0462 | Truck Scale Rental | MONTH | 2,862 | 2,862 | 2,862 | 2,862 | 2,862 |
| 33 10 0118 | R60 Rough Terrain Forklift, 6,000 Lb @ 24" LC | HOUR | 114.98 | 91.38 | 77.08 | 54.02 | 48.25 |
| 33 13 1120 | Pretreatment System Operational Labor Cost | DAY | 217.57 | 217.57 | 217.57 | 217.57 | 217.57 |
| 33 13 2401 | Weekly Rental, Portable Drum Shredder, 60 - 100 Drums/Hour | WEEK | 11,408 | 11,408 | 11,408 | 11,408 | 11,408 |
| 33 13 2402 | Daily Rental, Portable Drum Shredder, 60 - 100 Drums/Hour | DAY | 2,282 | 2,282 | 2,282 | 2,282 | 2,282 |
| 33 13 2403 | Monthly Rental, Portable Drum Shredder, 60 - 100 Drums/Hour | MONTH | 45,633 | 45,633 | 45,633 | 45,633 | 45,633 |

**Environmental Remediation: Assemblies Cost Book**

*Page 3-305*

1998 by ECHOS. All rights reserved

# Cost Data for Remediation

## Solidification/Stabilization

| Assembly | Description | Unit | Unit Cost by Safety Level | | | | |
|---|---|---|---|---|---|---|---|
| | | | A | B | C | D | E |
| | **Capital Costs** | | | | | | |
| 33 13 2404 | Shredder Mobilize/DeMobilize | EACH | 1,700 | 1,700 | 1,700 | 1,700 | 1,700 |
| 33 13 2405 | Electric Shredder, Minimum Rental Charge, 50 - 75 Tons/Hour | EACH | 41,750 | 41,750 | 41,750 | 41,750 | 41,750 |
| 33 13 2406 | Electric Shredder, Additional Monthly Rental, 50 - 75 Tons/Hour | MONTH | 18,875 | 18,875 | 18,875 | 18,875 | 18,875 |
| 33 13 2407 | Electric Shredder, Purchase, 35 Tons/Hour Output | EACH | 271,283 | 271,283 | 271,283 | 271,283 | 271,283 |
| 33 13 2408 | Portable Drum Shredder, 60 - 100 Drums/Hour Output | EACH | 251,900 | 251,900 | 251,900 | 251,900 | 251,900 |
| 33 15 0401 | Cement Kiln Dust (Bulk) | TON | 5.26 | 5.26 | 5.26 | 5.26 | 5.26 |
| 33 15 0402 | Fly Ash, Class C (Bulk) | TON | 14.84 | 14.84 | 14.84 | 14.84 | 14.84 |
| 33 15 0403 | Fly Ash, Class F (Bulk) | TON | 15.64 | 15.64 | 15.64 | 15.64 | 15.64 |
| 33 15 0404 | Bottom Ash (Bulk) | TON | 2.83 | 2.83 | 2.83 | 2.83 | 2.83 |
| 33 15 0405 | Portland Cement Type I (Bulk) | TON | 77.73 | 77.73 | 77.73 | 77.73 | 77.73 |
| 33 15 0406 | Portland Cement Type K (Bulk) | TON | 103.88 | 103.88 | 103.88 | 103.88 | 103.88 |
| 33 15 0407 | Lime (Crushed Limestone) Bulk | TON | 5.30 | 5.30 | 5.30 | 5.30 | 5.30 |
| 33 15 0408 | Urrichem Proprietary Additive (Bulk) | TON | 980.00 | 980.00 | 980.00 | 980.00 | 980.00 |
| 33 15 0409 | Bitumen (Bulk) | TON | 40.28 | 40.28 | 40.28 | 40.28 | 40.28 |
| 33 15 0410 | Chloranan Proprietary Additive (Bulk) | TON | 636.00 | 636.00 | 636.00 | 636.00 | 636.00 |
| 33 15 0411 | P4 Proprietary Reagent (Bulk) | TON | 725.00 | 725.00 | 725.00 | 725.00 | 725.00 |
| 33 15 0412 | P27 Proprietary Reagent (Bulk) | TON | 275.00 | 275.00 | 275.00 | 275.00 | 275.00 |
| 33 15 0413 | Activated Carbon, Granular or Powdered Form (Bulk) | TON | 3,604 | 3,604 | 3,604 | 3,604 | 3,604 |
| 33 15 0414 | Sodium Silicate (Bulk) | TON | 275.60 | 275.60 | 275.60 | 275.60 | 275.60 |
| 33 15 0415 | Tank Truck Standby Time for Solidification/Stabilization Unit | HOUR | 21.80 | 18.16 | 14.53 | 10.90 | 10.90 |
| 33 15 0417 | 7.5 HP Sludge Pump, 1" Maximum Particle Size, Rental | MONTH | 950.00 | 950.00 | 950.00 | 950.00 | 950.00 |
| 33 15 0418 | 1 CY Plywood Boxes | EACH | 79.07 | 66.27 | 60.70 | 48.32 | 43.83 |
| 33 15 0420 | Operational Labor for Process Equipment | HOUR | 101.08 | 77.92 | 68.00 | 45.61 | 37.40 |
| 33 15 0421 | Bulk Chemical Transport (40,000 Lb Truckload) | EACH | 2,350 | 2,350 | 2,350 | 2,350 | 2,350 |
| 33 15 0422 | 2 CY Mixing System | MONTH | 1,577 | 1,577 | 1,577 | 1,577 | 1,577 |
| 33 15 0423 | 10 CY Mixing System | MONTH | 4,690 | 4,690 | 4,690 | 4,690 | 4,690 |
| 33 15 0425 | High-pressure Water System for 2 CY Waste Mixer, 28 Gallon | EACH | 2,866 | 2,866 | 2,866 | 2,866 | 2,866 |
| 33 15 0426 | Nonpressurized Water System for 10 CY Waste Mixer | EACH | 1,948 | 1,948 | 1,948 | 1,948 | 1,948 |
| 33 15 0427 | Barrel Loader for 2 CY Waste Mixer | EACH | 5,485 | 5,485 | 5,485 | 5,485 | 5,485 |
| 33 15 0428 | Belt Feeder for 10 CY Mixer, 13' Long | EACH | 9,947 | 9,947 | 9,947 | 9,947 | 9,947 |
| 33 15 0429 | 50 CY/Hour Decumulative Batch Plant, 8 CY Capacity | EACH | 185,553 | 185,553 | 185,553 | 185,553 | 185,553 |

## Environmental Remediation: Assemblies Cost Book

# Cost Data for Remediation

## Solidification/Stabilization

| Assembly | Description | Unit | Unit Cost by Safety Level A | B | C | D | E |
|----------|-------------|------|------|------|------|------|------|
| | **Capital Costs** | | | | | | |
| 33 15 0430 | Dust Collection with 2 HP Blower and Controls | EACH | 7,240 | 7,240 | 7,240 | 7,240 | 7,240 |
| 33 15 0431 | Water Pump, 3" Self-priming with 10 HP Motor | EACH | 7,314 | 7,314 | 7,314 | 7,314 | 7,314 |
| 33 15 0432 | Radial Stacking Conveyor with 2 CY Hopper, 55' Long | EACH | 48,961 | 48,961 | 48,961 | 48,961 | 48,961 |
| 33 15 0433 | 5 CY Waste Mixer | MONTH | 1,542 | 1,542 | 1,542 | 1,542 | 1,542 |
| 33 15 0434 | 15 CY Waste Mixer | MONTH | 4,666 | 4,666 | 4,666 | 4,666 | 4,666 |
| 33 15 0435 | Solidification/Stabilization Ancillary Equipment | EACH | 7,100 | 7,100 | 7,100 | 7,100 | 7,100 |
| 33 15 0436 | Mobilize/DeMobilize of Solidification/Stabilization Equipment | LS | 32,851 | 25,323 | 22,100 | 14,823 | 12,155 |
| 33 15 0437 | Maintenance of Solidification/Stabilization Unit | YEAR | 14,354 | 11,064 | 9,656 | 6,477 | 5,311 |
| 33 17 0816 | 3,000 PSI Pressure Washer, 4.5 GPM | EACH | 8,868 | 8,868 | 8,868 | 8,868 | 8,868 |
| 33 17 0823 | Operation of Pressure Washer, Including Water, Soap, Electricity, Labor | HOUR | 79.38 | 63.31 | 56.43 | 40.91 | 35.21 |
| 33 18 8401 | 41.5' Automatic Conveyor, 45 FPM, Horizontal 24" Belt, Center Drive | EACH | 8,054 | 7,271 | 6,882 | 6,121 | 5,877 |
| 33 18 8402 | 61.5' Automatic Conveyor, 45 FPM, Horizontal 24" Belt, Center Drive | EACH | 10,682 | 9,637 | 9,119 | 8,105 | 7,779 |
| 33 18 8403 | 34' Automatic Inclined Conveyor, 25 Degree, 24" Belt, Loader/End Idler | EACH | 9,284 | 8,239 | 7,721 | 6,707 | 6,381 |
| 33 18 8501 | Refuse Bottom Hopper, Aluminized Steel, 18" Diameter | EACH | 1,334 | 1,251 | 1,215 | 1,135 | 1,105 |
| 33 18 8601 | 4' x 12' Single-tray Vibrating Screening Unit, with Motor & Accessories | EACH | 16,360 | 15,705 | 15,362 | 14,725 | 14,531 |
| 33 18 8602 | 5' x 16' Double-tray Vibrating Screening Unit, with Motor & Accessories | EACH | 25,858 | 25,105 | 24,711 | 23,979 | 23,757 |
| 33 18 8603 | 6' x 20' Triple-tray Vibrating Screening Unit, with Motor & Accessories | EACH | 32,164 | 30,853 | 30,168 | 28,894 | 28,506 |
| 33 18 8604 | 7' x 24' Triple-tray Vibrating Screening Unit, with Motor & Accessories | EACH | 39,668 | 37,805 | 36,831 | 35,020 | 34,469 |
| 33 19 0109 | Dredging Sludge < 1,000' to Loading Area | DAY | 3,048 | 2,457 | 2,040 | 1,459 | 1,351 |
| 33 19 0110 | Dredging Sludge > 1,000' to Loading Area | DAY | 4,339 | 3,533 | 2,901 | 2,105 | 1,997 |
| 33 19 9921 | DOT Steel Drum, 55 Gallon | EACH | 65.19 | 65.19 | 65.19 | 65.19 | 65.19 |
| 33 26 0502 | 2" Polyethylene (SDR 21) Piping | LF | 11.71 | 9.24 | 8.18 | 5.79 | 4.91 |
| 33 26 0503 | 3" Polyethylene (SDR 21) Piping | LF | 12.96 | 10.32 | 9.12 | 6.56 | 5.67 |
| 33 26 0504 | 4" Polyethylene (SDR 21) Piping | LF | 16.79 | 13.35 | 11.80 | 8.48 | 7.31 |
| 33 26 0507 | 10" Polyethylene (SDR 21) Piping | LF | 47.02 | 38.71 | 34.53 | 26.46 | 23.90 |
| 33 29 0103 | 50 GPM, 100' Head, 3 HP, Centrifugal Pump | EACH | 2,658 | 2,500 | 2,432 | 2,279 | 2,223 |
| 33 29 0106 | 100 GPM, 150' Head, 7.5 HP, Centrifugal Pump | EACH | 3,419 | 3,191 | 3,093 | 2,873 | 2,793 |
| 33 29 0109 | 10 HP, 250 GPM, Centrifugal Pump | EACH | 3,032 | 2,805 | 2,707 | 2,487 | 2,407 |

**Environmental Remediation: Assemblies Cost Book**

*Page 3-307*

1998 by ECHOS. All rights reserved

# Cost Data for Remediation

## Solidification/Stabilization

| Assembly | Description | Unit | Unit Cost by Safety Level | | | | |
|---|---|---|---|---|---|---|---|
| | | | A | B | C | D | E |
| | **Capital Costs** | | | | | | |
| 33 29 0110 | 15 HP, 300 GPM, Centrifugal Pump | EACH | 3,441 | 3,160 | 3,040 | 2,770 | 2,670 |
| 33 29 0116 | 100 HP, 3000 GPM, Centrifugal Pump | EACH | 15,575 | 14,046 | 13,338 | 11,857 | 11,348 |
| 33 33 0117 | Hydrated Lime, Powdered, Bulk | TON | 97.55 | 97.55 | 97.54 | 97.54 | 97.53 |
| 33 42 0101 | Electrical Charge | KWH | 0.06 | 0.06 | 0.06 | 0.06 | 0.06 |
| 33 42 0201 | Diesel Fuel | GAL | 1.31 | 1.31 | 1.31 | 1.31 | 1.31 |
| 33 42 0301 | Water | KGAL | 8.39 | 8.39 | 8.39 | 8.39 | 8.39 |

# Cost Data for Remediation
## Solvent Extraction

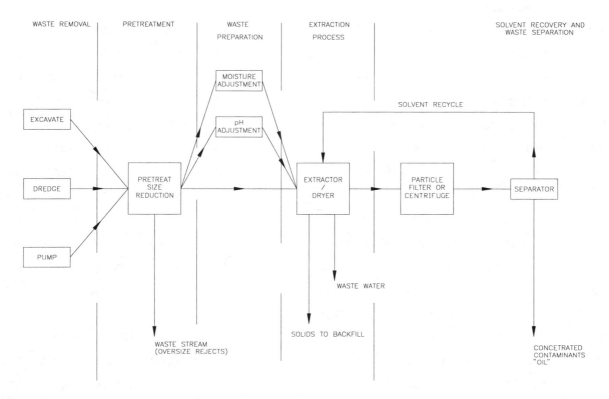

### General:

Solvent extraction is an ex situ process where contaminated sediment, soil, or sludge is mixed with a solvent to separate the contaminant from its existing matrix. When solvent extraction is used for site remediation, the solvent concentrates the contaminants; thus reducing the quantity of hazardous waste. The resultant waste stream may be suitable for either recycle or reuse, or it may require additional treatment and disposal. Solvent extraction is a relatively new technology and the capital costs are significant.

### Typical Treatment Train:

Excavation, transportation of wastes to treatment area, handling of effluent (transportation and disposal), sampling and analysis.

### Common Cost Components:

1. Process equipment mobilization, assembly, start up, decontamination, demobilization. Note: process units are available in three sizes:
   - 1,500 cubic yards per month
   - 6,000 cubic yards per month
   - 18,000 cubic yards per month
2. Site preparation, based on size of process unit
3. Pretreatment equipment
4. Operational labor

### Other Cost Considerations:

Access road, structural slab capable of supporting 250 to 400 pounds per square foot, storage tanks, containment berms.

**Environmental Remediation: Assemblies Cost Book**

# Cost Data for Remediation

## Solvent Extraction

|  |  | | Unit Cost by Safety Level | | | | |
|---|---|---|---|---|---|---|---|
| **Assembly** | **Description** | **Unit** | **A** | **B** | **C** | **D** | **E** |
|  | **Capital Costs** |  |  |  |  |  |  |
| 33 12 0201 | 1,500 CY/Month Unit, Site Preparation Charge | EACH | 125,000 | 125,000 | 125,000 | 125,000 | 125,000 |
| 33 12 0202 | 6,000 CY/Month Unit - Site Preparation Charge | EACH | 251,000 | 251,000 | 251,000 | 251,000 | 251,000 |
| 33 12 0203 | 18,000 CY/Month Unit - Site Preparation Charge | EACH | 415,000 | 415,000 | 415,000 | 415,000 | 415,000 |
| 33 12 0204 | 1,500 CY/Month Unit - Mobilize and Assemble | EACH | 166,500 | 166,500 | 166,500 | 166,500 | 166,500 |
| 33 12 0205 | 6,000 CY/Month Unit - Mobilize and Assemble | EACH | 495,000 | 495,000 | 495,000 | 495,000 | 495,000 |
| 33 12 0206 | 18,000 CY/Month Unit - Mobilize and Assemble | EACH | 120,750 | 120,750 | 120,750 | 120,750 | 120,750 |
| 33 12 0207 | 1,500 CY/Month Unit - Start-up Charge | EACH | 33,800 | 33,800 | 33,800 | 33,800 | 33,800 |
| 33 12 0208 | 6,000 CY/Month Unit - Start-up Charge | EACH | 66,475 | 66,475 | 66,475 | 66,475 | 66,475 |
| 33 12 0209 | 18,000 CY/Month Unit - Start-up Charge | EACH | 132,500 | 132,500 | 132,500 | 132,500 | 132,500 |
| 33 12 0210 | 1,500 CY/Month Unit - Decontaminate/DeMobilize | EACH | 75,300 | 75,300 | 75,300 | 75,300 | 75,300 |
| 33 12 0211 | 6,000 CY/Month Unit - Decontaminate/DeMobilize | EACH | 150,500 | 150,500 | 150,500 | 150,500 | 150,500 |
| 33 12 0212 | 18,000 CY/Month Unit - Decontaminate/DeMobilize | EACH | 255,000 | 255,000 | 255,000 | 255,000 | 255,000 |
| 33 12 0213 | 1,500 CY/Month Unit - Process Equipment Rental | MONTH | 327,773 | 273,182 | 218,592 | 164,001 | 164,001 |
| 33 12 0214 | 6,000 CY/Month Unit - Process Equipment Rental | MONTH | 723,909 | 603,340 | 482,771 | 362,202 | 362,202 |
| 33 12 0215 | 18,000 CY/Month Unit - Process Equipment Rental | MONTH | 1,493,347 | 1,244,621 | 995,895 | 747,169 | 747,169 |
| 33 12 0216 | 1,500 CY/Month Unit - Process Labor | MONTH | 155,750 | 120,074 | 104,802 | 70,318 | 57,674 |
| 33 12 0217 | 6,000 CY/Month Unit - Process Labor | MONTH | 288,717 | 222,584 | 194,273 | 130,350 | 106,911 |
| 33 12 0218 | 18,000 CY/Month Unit - Process Labor | MONTH | 288,717 | 222,584 | 194,273 | 130,350 | 106,911 |
| 33 12 0219 | 1,500 CY/Month Unit - Consumables | CY | 33.96 | 33.96 | 33.96 | 33.96 | 33.96 |
| 33 12 0220 | 6,000 CY/Month Unit - Consumables | CY | 22.02 | 22.02 | 22.02 | 22.02 | 22.02 |
| 33 12 0221 | 18,000 CY/Month Unit - Consumables | CY | 18.90 | 18.90 | 18.90 | 18.90 | 18.90 |
| 33 12 0222 | 1,500 CY/Month Unit - Pretreatment Unit | EACH | 55,463 | 55,463 | 55,463 | 55,463 | 55,463 |
| 33 12 0223 | 6,000 CY/Month Unit - Pretreatment Unit | EACH | 69,690 | 69,690 | 69,690 | 69,690 | 69,690 |
| 33 12 0224 | 18,000 CY/Month Unit - Pretreatment Unit | EACH | 104,625 | 104,625 | 104,625 | 104,625 | 104,625 |

**Environmental Remediation: Assemblies Cost Book**

# Cost Data for Remediation
## Storage Tank Installation

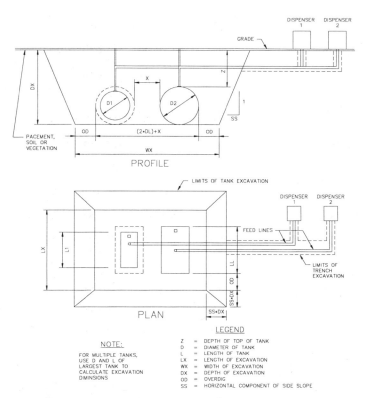

### General:

The costs presented here are appropriate for (1) new installation of either aboveground or belowground fuel storage tanks and (2) replacement of existing tanks (reference the UST Closure section for removal costs). When calculating quantities for excavation, foundation slab, cover, etc., consider contiguous excavation, slab, etc. for multiple tanks.

### Typical Treatment Train:

UST closure, site preparation, groundwater monitoring wells, sampling and analysis.

### Common Cost Components:

1. Tanks: single- or double-walled steel or fiberglass
2. Feed lines: single- or double-walled steel or fiberglass piping
3. Belowground tanks: cover demolition, excavation, tank and feed line installation, backfill, concrete or asphalt covers, leak detection system
4. Aboveground tanks: tank installation, foundation slab, feed line installation (above or belowground), containment walls

### Other Cost Considerations:

1. Tanks: single- or double-walled steel or fiberglass
2. Feed lines: single- or double-walled steel or fiberglass piping
3. Belowground tanks: cover demolition, excavation, tank and feed line installation, backfill, concrete or asphalt covers, leak detection system
4. Aboveground tanks: tank installation, foundation slab, feed line installation (above or belowground), containment walls

**Environmental Remediation: Assemblies Cost Book**

# Cost Data for Remediation

## Storage Tank Installation

| Assembly | Description | Unit | Unit Cost by Safety Level | | | | |
|---|---|---|---|---|---|---|---|
| | | | A | B | C | D | E |
| | **Capital Costs** | | | | | | |
| 17 02 0201 | Demolish Bituminous Road with Power Equipment | CY | 46.17 | 36.60 | 30.96 | 21.62 | 19.19 |
| 17 02 0203 | Demolish Bituminous Pavement with Air Equipment | CY | 80.15 | 63.38 | 53.76 | 37.41 | 32.98 |
| 17 02 0205 | Demolish Unreinforced Concrete to 6" Thick with Air Equipment | CY | 126.90 | 100.35 | 85.12 | 59.24 | 52.23 |
| 17 02 0206 | Demolish Mesh Reinforced Concrete to 6" Thick with Air Equipment | CY | 183.47 | 145.09 | 123.07 | 85.65 | 75.51 |
| 17 02 0207 | Demolish Rod Reinforced Concrete to 6" Thick with Air Equipment | CY | 199.06 | 157.42 | 133.53 | 92.92 | 81.92 |
| 17 02 0208 | Demolish Mesh Reinforced Concrete to 6" Thick with Power Equipment | CY | 96.73 | 76.68 | 64.87 | 45.30 | 40.20 |
| 17 02 0209 | Demolish Rod Reinforced Concrete to 6" Thick with Power Equipment | CY | 106.91 | 84.75 | 71.70 | 50.07 | 44.43 |
| 17 02 0210 | Demolish Unreinforced Concrete 7" to 24" Thick with Power Equipment | CY | 258.44 | 204.87 | 173.31 | 121.03 | 107.40 |
| 17 02 0211 | Demolish Reinforced Concrete 7" to 24" Thick with Power Equipment | CY | 356.37 | 282.51 | 238.98 | 166.90 | 148.10 |
| 17 03 0276 | 1 CY, Crawler-mounted, Hydraulic Excavator | CY | 7.74 | 6.25 | 5.18 | 3.71 | 3.44 |
| 17 03 0277 | 2 CY, Crawler-mounted, Hydraulic Excavator | CY | 5.16 | 4.19 | 3.45 | 2.49 | 2.35 |
| 17 03 0278 | 3 CY, Crawler-mounted, Hydraulic Excavator | CY | 3.67 | 2.95 | 2.46 | 1.75 | 1.61 |
| 17 03 0279 | 4 CY, Crawler-mounted, Hydraulic Excavator | CY | 3.00 | 2.42 | 2.01 | 1.44 | 1.33 |
| 17 03 0418 | Delivered & Dumped, Backfill with Stone | BCY | 23.83 | 23.26 | 22.84 | 22.28 | 22.18 |
| 17 03 0901 | Steel Sheeting, Pull & Salvage, to 15' | SF | 13.65 | 11.21 | 9.59 | 7.20 | 6.70 |
| 17 03 0902 | Steel Sheeting, Pull & Salvage, to 20' | SF | 14.11 | 11.61 | 9.95 | 7.50 | 6.99 |
| 17 03 0903 | Steel Sheeting, Pull & Salvage, to 25' | SF | 12.32 | 10.20 | 8.79 | 6.72 | 6.28 |
| 17 03 0904 | Steel Sheeting, Pull & Salvage, to 40' | SF | 14.52 | 12.05 | 10.40 | 7.98 | 7.46 |
| 17 03 1002 | 2" Diameter Contractor's Trash Pump, 75 GPM | DAY | 50.20 | 47.01 | 45.64 | 42.55 | 41.42 |
| 18 02 0301 | Asphalt Pavement - 10" Subgrade, 9" Base, 1 1/2" Topping | SY | 61.11 | 50.94 | 43.58 | 33.57 | 31.83 |
| 18 02 0321 | 6" Structural Slab on Grade | SF | 6.98 | 5.95 | 5.47 | 4.48 | 4.14 |
| 18 02 0322 | 8" Structural Slab on Grade | SF | 9.75 | 8.38 | 7.71 | 6.38 | 5.95 |
| 18 02 0324 | 12" Structural Slab on Grade | SF | 11.57 | 10.04 | 9.30 | 7.82 | 7.34 |
| 18 02 0331 | 6" Mesh Reinforced Slab on Grade | SF | 6.01 | 5.10 | 4.67 | 3.79 | 3.49 |
| 18 02 0341 | 6" Unreinforced Slab on Grade | SF | 4.87 | 4.14 | 3.79 | 3.08 | 2.85 |
| 18 04 0205 | CIP Walls Form & Strip (4 Uses) | SF | 8.86 | 6.99 | 6.19 | 4.38 | 3.73 |
| 20 06 0101 | 3 - 9 Lb Magnesium Anodes, Cathodic Protection Point | EACH | 953.82 | 778.29 | 693.26 | 523.02 | 466.94 |
| 20 06 0102 | 3 - 17 Lb Magnesium Anodes, Cathodic Protection Point | EACH | 1,364 | 1,121 | 1,002 | 766.32 | 688.94 |
| 20 06 0103 | 3 - 32 Lb Magnesium Anodes, Cathodic Protection Point | EACH | 1,845 | 1,539 | 1,390 | 1,092 | 995.23 |

## Environmental Remediation: Assemblies Cost Book

# Cost Data for Remediation

## Storage Tank Installation

| Assembly | Description | Unit | A | B | C | D | E |
|---|---|---|---|---|---|---|---|
| | **Capital Costs** | | | | | | |
| 20 06 0104 | 3 - 48 Lb Magnesium Anodes, Cathodic Protection Point | EACH | 2,585 | 2,166 | 1,960 | 1,553 | 1,420 |
| 33 09 0701 | CIP Reinforced Concrete Containment Wall | CY | 279.72 | 243.62 | 224.28 | 189.16 | 178.78 |
| 33 10 9601 | 550 Gallon Single-wall Steel UST | EACH | 2,663 | 2,354 | 2,222 | 1,923 | 1,814 |
| 33 10 9602 | 1,000 Gallon Single-wall Steel UST | EACH | 2,985 | 2,643 | 2,497 | 2,166 | 2,045 |
| 33 10 9603 | 2,000 Gallon Single-wall Steel UST | EACH | 4,885 | 4,382 | 4,142 | 3,654 | 3,491 |
| 33 10 9604 | 3,000 Gallon Single-wall Steel UST | EACH | 5,715 | 5,212 | 4,972 | 4,484 | 4,321 |
| 33 10 9605 | 5,000 Gallon Single-wall Steel UST | EACH | 6,693 | 6,050 | 5,742 | 5,117 | 4,909 |
| 33 10 9606 | 8,000 Gallon Single-wall Steel UST | EACH | 10,296 | 9,565 | 9,215 | 8,505 | 8,269 |
| 33 10 9607 | 10,000 Gallon Single-wall Steel UST | EACH | 11,686 | 10,881 | 10,496 | 9,716 | 9,456 |
| 33 10 9608 | 12,000 Gallon Single-wall Steel UST | EACH | 13,430 | 12,536 | 12,108 | 11,241 | 10,952 |
| 33 10 9609 | 15,000 Gallon Single-wall Steel UST | EACH | 16,092 | 15,086 | 14,605 | 13,630 | 13,304 |
| 33 10 9610 | 20,000 Gallon Single-wall Steel UST | EACH | 16,719 | 15,431 | 14,815 | 13,567 | 13,150 |
| 33 10 9611 | 30,000 Gallon Single-wall Steel UST | EACH | 23,914 | 22,305 | 21,535 | 19,974 | 19,454 |
| 33 10 9612 | 550 Gallon Double-wall Steel UST | EACH | 3,280 | 3,119 | 3,042 | 2,886 | 2,834 |
| 33 10 9613 | 1,000 Gallon Double-wall Steel UST | EACH | 15,449 | 14,443 | 13,962 | 12,987 | 12,661 |
| 33 10 9614 | 2,000 Gallon Double-wall Steel UST | EACH | 5,023 | 4,775 | 4,656 | 4,416 | 4,336 |
| 33 10 9615 | 3,000 Gallon Double-wall Steel UST | EACH | 8,275 | 7,997 | 7,865 | 7,596 | 7,506 |
| 33 10 9616 | 5,000 Gallon Double-wall Steel UST | EACH | 11,096 | 10,291 | 9,906 | 9,125 | 8,865 |
| 33 10 9617 | 8,000 Gallon Double-wall Steel UST | EACH | 13,567 | 12,647 | 12,207 | 11,315 | 11,018 |
| 33 10 9618 | 10,000 Gallon Double-wall Steel UST | EACH | 15,449 | 14,443 | 13,962 | 12,987 | 12,661 |
| 33 10 9619 | 12,000 Gallon Double-wall Steel UST | EACH | 21,756 | 20,638 | 20,103 | 19,019 | 18,658 |
| 33 10 9620 | 15,000 Gallon Double-wall Steel UST | EACH | 22,920 | 21,658 | 21,055 | 19,832 | 19,424 |
| 33 10 9621 | 20,000 Gallon Double-wall Steel UST | EACH | 27,936 | 26,326 | 25,556 | 23,995 | 23,475 |
| 33 10 9622 | 30,000 Gallon Double-wall Steel UST | EACH | 38,941 | 36,929 | 35,966 | 34,015 | 33,365 |
| 33 10 9623 | 550 Gallon Single-wall Fiberglass UST | EACH | 2,721 | 2,411 | 2,278 | 1,977 | 1,867 |
| 33 10 9624 | 1,000 Gallon Single-wall Fiberglass UST | EACH | 3,205 | 2,863 | 2,717 | 2,387 | 2,266 |
| 33 10 9625 | 2,000 Gallon Single-wall Fiberglass UST | EACH | 4,747 | 4,244 | 4,003 | 3,515 | 3,353 |
| 33 10 9626 | 3,000 Gallon Single-wall Fiberglass UST | EACH | 5,361 | 4,858 | 4,617 | 4,129 | 3,967 |
| 33 10 9627 | 5,000 Gallon Single-wall Fiberglass UST | EACH | 6,861 | 6,241 | 5,945 | 5,345 | 5,145 |
| 33 10 9628 | 8,000 Gallon Single-wall Fiberglass UST | EACH | 7,944 | 7,212 | 6,862 | 6,153 | 5,916 |
| 33 10 9629 | 10,000 Gallon Single-wall Fiberglass UST | EACH | 8,891 | 8,086 | 7,701 | 6,921 | 6,661 |
| 33 10 9630 | 12,000 Gallon Single-wall Fiberglass UST | EACH | 9,973 | 9,078 | 8,651 | 7,784 | 7,494 |
| 33 10 9631 | 15,000 Gallon Single-wall Fiberglass UST | EACH | 12,869 | 11,863 | 11,382 | 10,406 | 10,081 |

*Unit Cost by Safety Level*

**Environmental Remediation: Assemblies Cost Book**

*Page 3-313*

1998 by ECHOS. All rights reserved

# Cost Data for Remediation

## Storage Tank Installation

| Assembly | Description | Unit | \(Unit Cost by Safety Level\) A | B | C | D | E |
|---|---|---|---|---|---|---|---|
| | **Capital Costs** | | | | | | |
| 33 10 9632 | 20,000 Gallon Single-wall Fiberglass UST | EACH | 16,830 | 15,488 | 14,847 | 13,546 | 13,112 |
| 33 10 9633 | 30,000 Gallon Single-wall Fiberglass UST | EACH | 38,392 | 36,782 | 36,013 | 34,452 | 33,931 |
| 33 10 9634 | 550 Gallon Double-wall Fiberglass UST | EACH | 3,797 | 3,636 | 3,559 | 3,403 | 3,351 |
| 33 10 9635 | 1,000 Gallon Double-wall Fiberglass UST | EACH | 5,187 | 4,986 | 4,890 | 4,695 | 4,630 |
| 33 10 9636 | 2,000 Gallon Double-wall Fiberglass UST | EACH | 6,316 | 6,072 | 5,955 | 5,719 | 5,640 |
| 33 10 9637 | 3,000 Gallon Double-wall Fiberglass UST | EACH | 8,079 | 7,835 | 7,719 | 7,482 | 7,403 |
| 33 10 9638 | 5,000 Gallon Double-wall Fiberglass UST | EACH | 12,169 | 11,364 | 10,979 | 10,199 | 9,938 |
| 33 10 9639 | 8,000 Gallon Double-wall Fiberglass UST | EACH | 13,540 | 12,645 | 12,218 | 11,351 | 11,061 |
| 33 10 9640 | 10,000 Gallon Double-wall Fiberglass UST | EACH | 19,259 | 18,253 | 17,772 | 16,797 | 16,471 |
| 33 10 9641 | 12,000 Gallon Double-wall Fiberglass UST | EACH | 18,395 | 17,389 | 16,908 | 15,932 | 15,607 |
| 33 10 9642 | 15,000 Gallon Double-wall Fiberglass UST | EACH | 23,231 | 21,889 | 21,248 | 19,947 | 19,513 |
| 33 10 9643 | 20,000 Gallon Double-wall Fiberglass UST | EACH | 29,667 | 28,057 | 27,288 | 25,727 | 25,206 |
| 33 10 9644 | 30,000 Gallon Double-wall Fiberglass UST | EACH | 45,408 | 43,396 | 42,434 | 40,483 | 39,832 |
| 33 10 9645 | Tank Fittings | EACH | 1,309 | 1,084 | 982.55 | 764.72 | 688.39 |
| 33 10 9646 | Overfill Spill Container | EACH | 687.16 | 592.78 | 550.06 | 458.70 | 426.68 |
| 33 10 9647 | 30" Diameter Manhole at Grade | EACH | 623.95 | 557.80 | 521.66 | 457.26 | 438.66 |
| 33 10 9648 | Turbine Enclosure with Secondary Containment Collar | EACH | 1,980 | 1,980 | 1,980 | 1,980 | 1,980 |
| 33 10 9649 | Pump, Cast-iron Close Coupling, 2 HP, 50 GPM | EACH | 1,398 | 1,284 | 1,236 | 1,126 | 1,085 |
| 33 10 9650 | Control Panel, 1 - 8 Single-wall Tanks, Monitoring System | EACH | 3,320 | 3,320 | 3,320 | 3,320 | 3,320 |
| 33 10 9651 | Control Panel, 1 - 4 Double-wall Tanks, Monitoring System | EACH | 2,995 | 2,995 | 2,995 | 2,995 | 2,995 |
| 33 10 9652 | Control Panel, 1 - 8 Double-wall Tanks, Monitoring System | EACH | 3,946 | 3,946 | 3,946 | 3,946 | 3,946 |
| 33 10 9653 | Tank Leak Detection Probe | EACH | 1,226 | 1,226 | 1,226 | 1,226 | 1,226 |
| 33 10 9654 | Secondary Containment Collar Probe | EACH | 146.87 | 146.87 | 146.87 | 146.87 | 146.87 |
| 33 10 9655 | Interstitial Leak Detection Probe | EACH | 268.20 | 268.20 | 268.20 | 268.20 | 268.20 |
| 33 10 9656 | 550 Gallon Single-wall Steel Aboveground Tank | EACH | 1,598 | 1,390 | 1,301 | 1,099 | 1,026 |
| 33 10 9657 | 1,000 Gallon Single-wall Steel Aboveground Tank | EACH | 2,118 | 1,844 | 1,727 | 1,463 | 1,366 |
| 33 10 9658 | 2,000 Gallon Single-wall Steel Aboveground Tank | EACH | 3,457 | 2,997 | 2,777 | 2,332 | 2,183 |
| 33 10 9659 | 3,000 Gallon Single-wall Steel Aboveground Tank | EACH | 4,433 | 3,959 | 3,733 | 3,274 | 3,121 |
| 33 10 9660 | 5,000 Gallon Single-wall Steel Aboveground Tank | EACH | 6,079 | 5,494 | 5,214 | 4,646 | 4,457 |
| 33 10 9661 | 8,000 Gallon Single-wall Steel Aboveground Tank | EACH | 7,876 | 7,205 | 6,884 | 6,234 | 6,017 |
| 33 10 9662 | 10,000 Gallon Single-wall Steel Aboveground Tank | EACH | 9,147 | 8,415 | 8,065 | 7,356 | 7,119 |
| 33 10 9663 | 12,000 Gallon Single-wall Steel Aboveground Tank | EACH | 10,460 | 9,655 | 9,270 | 8,490 | 8,230 |
| 33 10 9664 | 15,000 Gallon Single-wall Steel Aboveground Tank | EACH | 11,591 | 10,697 | 10,269 | 9,402 | 9,113 |

## Environmental Remediation: Assemblies Cost Book

# Cost Data for Remediation

## Storage Tank Installation

| Assembly | Description | Unit | A | B | C | D | E |
|---|---|---|---|---|---|---|---|
| | | | \multicolumn{5}{c}{Unit Cost by Safety Level} | | | | |
| | **Capital Costs** | | | | | | |
| 33 10 9665 | 20,000 Gallon Single-wall Steel Aboveground Tank | EACH | 14,594 | 13,445 | 12,895 | 11,780 | 11,408 |
| 33 10 9666 | 30,000 Gallon Single-wall Steel Aboveground Tank | EACH | 6,427 | 5,085 | 4,444 | 3,143 | 2,709 |
| 33 10 9667 | Steel Dike for 550 Gallon Tank, Uncoated | EACH | 919.80 | 919.80 | 919.80 | 919.80 | 919.80 |
| 33 10 9668 | Steel Dike for 1,000 Gallon Tank, Uncoated | EACH | 1,546 | 1,546 | 1,546 | 1,546 | 1,546 |
| 33 10 9669 | Steel Dike for 2,000 Gallon Tank, Uncoated | EACH | 2,814 | 2,814 | 2,814 | 2,814 | 2,814 |
| 33 10 9670 | Steel Dike for 3,000 Gallon Tank, Uncoated | EACH | 3,807 | 3,807 | 3,807 | 3,807 | 3,807 |
| 33 10 9671 | Steel Dike for 5,000 Gallon Tank, Uncoated | EACH | 4,972 | 4,972 | 4,972 | 4,972 | 4,972 |
| 33 10 9672 | Steel Dike for 8,000 Gallon Tank, Uncoated | EACH | 6,718 | 6,718 | 6,718 | 6,718 | 6,718 |
| 33 10 9673 | Steel Dike for 10,000 Gallon Tank, Uncoated | EACH | 7,815 | 7,815 | 7,815 | 7,815 | 7,815 |
| 33 10 9674 | Steel Dike for 12,000 Gallon Tank, Uncoated | EACH | 8,315 | 8,315 | 8,315 | 8,315 | 8,315 |
| 33 10 9675 | Steel Dike for 15,000 Gallon Tank, Uncoated | EACH | 9,674 | 9,674 | 9,674 | 9,674 | 9,674 |
| 33 10 9676 | Steel Dike for 20,000 Gallon Tank, Uncoated | EACH | 12,573 | 12,573 | 12,573 | 12,573 | 12,573 |
| 33 10 9677 | 500 Barrel Fuel Oil Tank, 15' Diameter x 16' H, Cone Roof, Including Foundation | EACH | 84,414 | 73,073 | 67,214 | 56,194 | 52,796 |
| 33 10 9678 | 2,000 Barrel Fuel Oil Tank, 25' Diameter x 24' H, Cone Roof, Including Foundation | EACH | 157,565 | 134,652 | 122,815 | 100,549 | 93,685 |
| 33 10 9679 | 5,000 Barrel Fuel Oil Tank, 40' Diameter x 24' H, Cone Roof, Including Foundation | EACH | 220,519 | 190,175 | 174,499 | 145,012 | 135,922 |
| 33 10 9680 | 12,000 Barrel Fuel Oil Tank, 60' Diameter x 24' H, Cone Roof, Including Foundation | EACH | 343,088 | 293,188 | 267,410 | 218,920 | 203,971 |
| 33 10 9681 | 20,000 Barrel Fuel Oil Tank, 60' Diameter x 40' H, Cone Roof, Including Foundation | EACH | 499,244 | 431,198 | 396,047 | 329,924 | 309,540 |
| 33 10 9682 | 24,000 Barrel Fuel Oil Tank, 70' Diameter x 32' H, Cone Roof, Including Foundation | EACH | 535,481 | 463,046 | 425,627 | 355,238 | 333,538 |
| 33 10 9683 | 36,000 Barrel Fuel Oil Tank, 80' Diameter x 40' H, Cone Roof, Including Foundation | EACH | 682,685 | 585,055 | 534,620 | 439,748 | 410,501 |
| 33 10 9684 | 45,000 Barrel Fuel Oil Tank, 90' Diameter x 40' H, Cone Roof, Including Foundation | EACH | 822,428 | 704,244 | 643,192 | 528,347 | 492,942 |
| 33 10 9685 | 56,000 Barrel Fuel Oil Tank, 100' Diameter x 40' H, Cone Roof, with Foundation | EACH | 977,395 | 837,052 | 764,552 | 628,174 | 586,130 |
| 33 10 9686 | 64,000 Barrel Fuel Oil Tank, 120' Diameter x 32' H, Cone Roof, with Foundation | EACH | 1,123,447 | 963,054 | 880,198 | 724,337 | 676,287 |
| 33 10 9687 | 80,000 Barrel Fuel Oil Tank, 120' Diameter x 40' H, Cone Roof, with Foundation | EACH | 1,381,494 | 1,177,359 | 1,071,905 | 873,536 | 812,382 |
| 33 10 9688 | 88,000 Barrel Fuel Oil Tank, 140' Diameter x 32' H, Cone Roof, with Foundation | EACH | 1,549,654 | 1,345,518 | 1,240,065 | 1,041,696 | 980,542 |
| 33 10 9689 | 110,000 Barrel Fuel Oil Tank, 140' Diameter x 40' H, Cone Roof, with Foundation | EACH | 1,619,516 | 1,415,381 | 1,309,927 | 1,111,558 | 1,050,404 |
| 33 10 9690 | 143,000 Barrel Fuel Oil Tank, 160' Diameter x 40' H, Cone Roof, with Foundation | EACH | 1,838,330 | 1,613,781 | 1,497,782 | 1,279,576 | 1,212,307 |

**Environmental Remediation: Assemblies Cost Book**

# Cost Data for Remediation

## Storage Tank Installation

| Assembly | Description | Unit | A | B | C | D | E |
|---|---|---|---|---|---|---|---|
| | **Capital Costs** | | | | | | |
| 33 10 9691 | 180,000 Barrel Fuel Oil Tank, 180' Diameter x 40' H, Cone Roof, with Foundation | EACH | 2,290,384 | 2,009,698 | 1,864,699 | 1,591,942 | 1,507,855 |
| 33 10 9692 | 224,000 Barrel Fuel Oil Tank, 200' Diameter x 40' H, Cone Roof, with Foundation | EACH | 2,948,911 | 2,574,663 | 2,381,331 | 2,017,655 | 1,905,539 |
| 33 10 9693 | Tank Tightness Testing, <= 12,000 Gallon Tank | EACH | 300.00 | 300.00 | 300.00 | 300.00 | 300.00 |
| 33 10 9694 | Tank Tightness Testing, 12,001 - 29,999 Gallon Tank | EACH | 475.00 | 475.00 | 475.00 | 475.00 | 475.00 |
| 33 10 9695 | Tank Tightness Testing, >= 30,000 Gallon Tank | EACH | 650.00 | 650.00 | 650.00 | 650.00 | 650.00 |
| 33 23 2203 | 8" x 7.5" Manhole Cover | EACH | 343.19 | 285.78 | 240.65 | 183.96 | 176.35 |
| 33 26 0102 | 2" Carbon Steel Piping | LF | 10.50 | 8.72 | 7.95 | 6.23 | 5.60 |
| 33 26 0104 | 4" Carbon Steel Piping | LF | 23.20 | 19.86 | 18.43 | 15.21 | 14.02 |
| 33 26 0645 | (2", 4") Carbon Steel Double-wall Piping | EACH | 33.70 | 28.58 | 26.39 | 21.44 | 19.62 |
| 33 27 0425 | 2" Bronze Gate Valve | EACH | 125.01 | 107.16 | 99.52 | 82.26 | 75.94 |
| 33 27 0502 | 2" Steel Tee | EACH | 68.33 | 54.79 | 48.99 | 35.91 | 31.11 |
| 33 27 0504 | (2", 4") Tee, Carbon Steel | EACH | 254.31 | 216.17 | 199.84 | 162.98 | 149.46 |
| 33 27 0512 | 2" Steel 90-degree Elbow | EACH | 46.11 | 36.96 | 33.05 | 24.21 | 20.97 |
| 33 27 0514 | (2", 4") 90-degree Elbow, Carbon Steel | EACH | 180.91 | 152.29 | 140.03 | 112.37 | 102.23 |
| 33 27 0522 | 2" Steel 45-degree Elbow | EACH | 46.75 | 37.60 | 33.68 | 24.84 | 21.60 |
| 33 27 0524 | (2", 4") 45-degree Elbow, Carbon Steel | EACH | 185.37 | 156.75 | 144.50 | 116.83 | 106.69 |
| 33 27 0532 | 2" Steel Coupling | EACH | 47.17 | 38.02 | 34.11 | 25.27 | 22.03 |
| 33 27 0534 | (2", 4") Coupling, Carbon Steel | EACH | 182.34 | 153.72 | 141.47 | 113.80 | 103.66 |
| 33 27 0544 | 4" x 2" Eccentric Reducer, Steel | EACH | 326.23 | 259.50 | 229.31 | 164.72 | 142.07 |
| 33 27 0602 | 2" Tee, Fiberglass Reinforced | EACH | 131.59 | 109.76 | 100.41 | 79.31 | 71.57 |
| 33 27 0604 | (2", 4") Tee, Fiberglass Reinforced | EACH | 279.06 | 235.39 | 216.70 | 174.49 | 159.01 |
| 33 27 0612 | 2" 90-degree Elbow, Fiberglass Reinforced | EACH | 126.48 | 104.65 | 95.30 | 74.20 | 66.46 |
| 33 27 0614 | (2", 4") 90-degree Elbow, Fiberglass Reinforced | EACH | 258.24 | 214.57 | 195.88 | 153.67 | 138.19 |
| 33 27 0622 | 2" 45-degree Elbow, Fiberglass Reinforced | EACH | 126.48 | 104.65 | 95.30 | 74.20 | 66.46 |
| 33 27 0624 | (2", 4") 45-degree Elbow, Fiberglass Reinforced | EACH | 258.24 | 214.57 | 195.88 | 153.67 | 138.19 |
| 33 27 0632 | 2" Coupling, Fiberglass Reinforced | EACH | 102.87 | 81.04 | 71.69 | 50.59 | 42.85 |
| 33 27 0634 | (2", 4") Coupling, Fiberglass Reinforced | EACH | 221.99 | 178.32 | 159.63 | 117.42 | 101.94 |
| 33 27 0642 | 2" Termination Reducer, Fiberglass Reinforced | EACH | 109.33 | 87.50 | 78.15 | 57.05 | 49.31 |
| 33 27 0644 | (2", 4") Termination Reducer, Fiberglass Reinforced | EACH | 223.39 | 179.72 | 161.03 | 118.82 | 103.34 |
| 33 27 0902 | 2" Fiberglass Reinforced Piping | EACH | 12.46 | 10.27 | 9.34 | 7.23 | 6.46 |
| 33 27 0904 | (2", 4") Fiberglass Reinforced Double-wall Piping | LF | 25.86 | 21.50 | 19.63 | 15.41 | 13.86 |

**Environmental Remediation: Assemblies Cost Book**

*Page 3-316*

# Cost Data for Remediation
## Thermal & Catalytic Oxidation

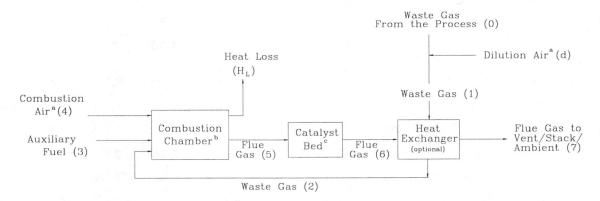

(a) When required.
(b) Referred to as preheat chamber in the case of catalytic incinerators.
(c) Included only in catalytic incinerators.

Diagram of a Thermal/Catalytic Incineration System
(Numbers indicate locations in the system)

### General:

Environmental remediation processes, such as soil vapor extraction or air stripping, will generate gas waste streams containing volatile organic compounds (VOCs). These gas streams are characterized by high humidity levels and by high initial concentrations of VOCs, which decay rapidly to lower values and asymptotically approach a steady state concentration. Vapor emission oxidation is a process where oxygen and organics react under high temperatures to produce carbon dioxide, water vapor and in some cases acidic gases (such as hydrochloric acid). Oxidation systems are relatively simple devices capable of achieving destruction efficiencies of 98 percent or greater.

### Typical Treatment Train:

Environmental remediation processes that generate gas waste streams containing VOCs, such as soil vapor extraction or air stripping; sampling and analysis

### Common Cost Components:

1. Thermal and catalytic oxidation process unit  2. Auxiliary fuel (natural gas, propane)  3. Electrical usage  4. Structural slab  5. Catalyst (fixed bed catalytic oxidation only)  6. Operational labor  7. Operations and maintenance

### Other Cost Considerations:

Electrical distribution, access road, gas distribution.

---

**Environmental Remediation: Assemblies Cost Book**

# Cost Data for Remediation

## Thermal & Catalytic Oxidation

| Assembly | Description | Unit | A | B | C | D | E |
|---|---|---|---|---|---|---|---|
| | **Capital Costs** | | | | | | |
| 18 02 0322 | 8" Structural Slab on Grade | SF | 9.75 | 8.38 | 7.71 | 6.38 | 5.95 |
| 33 04 0145 | Crane,22Tn Slf Cntnd,Hyd,Rough Trrn,4WD,for Ordnnc Rmvl | HR | 326.84 | 260.74 | 219.02 | 154.34 | 139.23 |
| 33 07 0401 | 150 SCFM Simple Thermal Oxidizer | EACH | 35,251 | 34,543 | 34,218 | 33,533 | 33,295 |
| 33 07 0402 | 250 SCFM Simple Thermal Oxidizer | EACH | 37,501 | 36,793 | 36,468 | 35,783 | 35,545 |
| 33 07 0403 | 500 SCFM Simple Thermal Oxidizer | EACH | 49,901 | 49,193 | 48,868 | 48,183 | 47,945 |
| 33 07 0404 | 1,000 SCFM Simple Thermal Oxidizer | EACH | 69,223 | 68,162 | 67,675 | 66,647 | 66,291 |
| 33 07 0405 | 5,000 SCFM Simple Thermal Oxidizer | EACH | 260,223 | 259,162 | 258,675 | 257,647 | 257,291 |
| 33 07 0406 | 150 SCFM Recuperative Thermal Oxidation Unit | EACH | 48,351 | 47,643 | 47,318 | 46,633 | 46,395 |
| 33 07 0407 | 250 SCFM Recuperative Thermal Oxidation Unit | EACH | 50,901 | 50,193 | 49,868 | 49,183 | 48,945 |
| 33 07 0408 | 500 SCFM Recuperative Thermal Oxidation Unit | EACH | 63,901 | 63,193 | 62,868 | 62,183 | 61,945 |
| 33 07 0409 | 1,000 SCFM Recuperative Thermal Oxidation Unit | EACH | 85,723 | 84,662 | 84,175 | 83,147 | 82,791 |
| 33 07 0410 | 5,000 SCFM Recuperative Thermal Oxidation Unit | EACH | 387,823 | 386,762 | 386,275 | 385,247 | 384,891 |
| 33 07 0414 | 2,500 SCFM Regenerative Thermal Oxidation Unit | EACH | 145,223 | 144,162 | 143,675 | 142,647 | 142,291 |
| 33 07 0415 | 5,000 SCFM Regenerative Thermal Oxidation Unit | EACH | 181,271 | 180,210 | 179,723 | 178,696 | 178,340 |
| 33 07 0416 | 5 SCFM Flameless Thermal Oxidation Unit | EACH | 54,251 | 53,543 | 53,218 | 52,533 | 52,295 |
| 33 07 0417 | 100 SCFM Flameless Thermal Oxidation Unit | EACH | 105,351 | 104,643 | 104,318 | 103,633 | 103,395 |
| 33 07 0418 | 1,000 SCFM Flameless Thermal Oxidation Unit | EACH | 151,723 | 150,662 | 150,175 | 149,147 | 148,791 |
| 33 07 0419 | 5,000 SCFM Flameless Thermal Oxidation Unit | EACH | 288,223 | 287,162 | 286,675 | 285,647 | 285,291 |
| 33 07 0420 | 1,000 SCFM Recuperative Flameless Thermal Unit | EACH | 173,223 | 172,162 | 171,675 | 170,647 | 170,291 |
| 33 07 0421 | 5,000 SCFM Recuperative Flameless Thermal Unit | EACH | 315,023 | 313,962 | 313,475 | 312,447 | 312,091 |
| 33 07 0441 | 100 SCFM Fixed Bed Catalytic Unit | EACH | 41,451 | 40,743 | 40,418 | 39,733 | 39,495 |
| 33 07 0442 | 150 SCFM Fixed Bed Catalytic Unit | EACH | 41,751 | 41,043 | 40,718 | 40,033 | 39,795 |
| 33 07 0443 | 250 SCFM Fixed Bed Catalytic Unit | EACH | 45,151 | 44,443 | 44,118 | 43,433 | 43,195 |
| 33 07 0444 | 500 SCFM Fixed Bed Catalytic Unit | EACH | 64,873 | 63,812 | 63,325 | 62,297 | 61,941 |
| 33 07 0445 | 1,000 SCFM Fixed Bed Catalytic Unit | EACH | 94,723 | 93,662 | 93,175 | 92,147 | 91,791 |
| 33 07 0446 | 5,000 SCFM Fixed Bed Catalytic Unit | EACH | 113,723 | 112,662 | 112,175 | 111,147 | 110,791 |
| 33 07 0447 | Precious Metal Catalyst | SCF | 4,000 | 4,000 | 4,000 | 4,000 | 4,000 |
| 33 07 0451 | 100 SCFM Recuperative Fixed Bed Catalytic | EACH | 54,151 | 53,443 | 53,118 | 52,433 | 52,195 |
| 33 07 0452 | 150 SCFM Recuperative Fixed Bed Catalytic | EACH | 55,151 | 54,443 | 54,118 | 53,433 | 53,195 |
| 33 07 0453 | 250 SCFM Recuperative Fixed Bed Catalytic | EACH | 58,651 | 57,943 | 57,618 | 56,933 | 56,695 |
| 33 07 0454 | 500 SCFM Recuperative Fixed Bed Catalytic | EACH | 77,351 | 76,643 | 76,318 | 75,633 | 75,395 |
| 33 07 0455 | 1,000 SCFM Recuperative Fixed Bed Catalytic | EACH | 84,138 | 82,424 | 81,605 | 79,944 | 79,390 |
| 33 07 0456 | 2,000 SCFM Recuperative Fixed Bed Catalytic | EACH | 106,323 | 104,610 | 103,791 | 102,129 | 101,575 |

**Environmental Remediation: Assemblies Cost Book**

# Cost Data for Remediation

## Thermal & Catalytic Oxidation

| Assembly | Description | Unit | Unit Cost by Safety Level A | B | C | D | E |
|---|---|---|---|---|---|---|---|
| | **Capital Costs** | | | | | | |
| 33 07 0457 | 3,000 SCFM Recuperative Fixed Bed Catalytic | EACH | 130,150 | 128,004 | 126,979 | 124,899 | 124,205 |
| 33 07 0458 | 4,000 SCFM Recuperative Fixed Bed Catalytic | EACH | 155,948 | 153,803 | 152,777 | 150,697 | 150,004 |
| 33 07 0459 | 5,000 SCFM Recuperative Fixed Bed Catalytic | EACH | 158,121 | 155,976 | 154,950 | 152,870 | 152,177 |
| 33 07 0461 | Remote Monitoring Unit | EACH | 3,850 | 3,850 | 3,850 | 3,850 | 3,850 |
| 33 10 0118 | R60 Rough Terrain Forklift, 6,000 Lb @ 24" LC | HOUR | 114.98 | 91.38 | 77.08 | 54.02 | 48.25 |
| 33 13 2916 | Natural Gas Usage, per 1,000 CF | MCF | 5.50 | 5.50 | 5.50 | 5.50 | 5.50 |
| 33 13 2917 | Propane | GAL | 1.28 | 1.28 | 1.28 | 1.28 | 1.28 |
| 33 23 0102 | 4" PVC, Schedule 40, Well Casing | LF | 26.98 | 22.38 | 18.77 | 14.24 | 13.63 |
| 33 27 0114 | 4" PVC, 90 Degree, Elbow | EACH | 101.72 | 80.57 | 71.51 | 51.07 | 43.57 |
| 33 42 0101 | Electrical Charge | KWH | 0.06 | 0.06 | 0.06 | 0.06 | 0.06 |
| | **Operations & Maintenance** | | | | | | |
| 33 07 0447 | Precious Metal Catalyst | SCF | 4,000 | 4,000 | 4,000 | 4,000 | 4,000 |
| 33 13 2916 | Natural Gas Usage, per 1,000 CF | MCF | 5.50 | 5.50 | 5.50 | 5.50 | 5.50 |
| 33 13 2917 | Propane | GAL | 1.28 | 1.28 | 1.28 | 1.28 | 1.28 |
| 33 42 0101 | Electrical Charge | KWH | 0.06 | 0.06 | 0.06 | 0.06 | 0.06 |

**Environmental Remediation: Assemblies Cost Book**

# Cost Data for Remediation
## Transportation

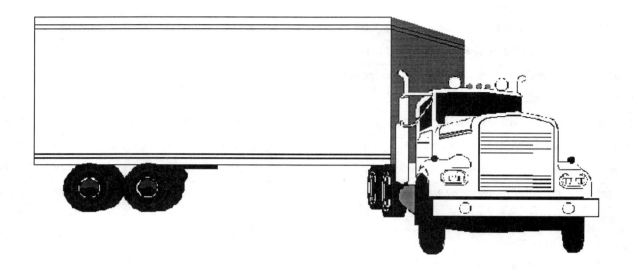

### General:
*Off-site transportation of material, either hazardous or nonhazardous, is regulated by the Department of Transportation. The type of waste transported can have a significant impact on the shipping cost. Transportation of hazardous waste is more expensive than nonhazardous waste due to cargo liability, regulatory restraints, and equipment requirements.*

### Typical Treatment Train:
*Waste pretreatment and/or staging, unloading wastes after transport, sampling and analysis.*

### Common Cost Components:
1. Mileage charge: based on type of waste (explosive, hazardous, radioactive)
2. Waste consolidation, containerization
3. Loading
4. Containers, drums, overpacks
5. Miscellaneous: rail tariff, demurrage, manifest discrepancies, etc.

### Other Cost Considerations:
*Truck wash-out, and grubbing, electrical distribution.*

# Cost Data for Remediation

## Transportation

| Assembly | Description | Unit | Unit Cost by Safety Level A | B | C | D | E |
|---|---|---|---|---|---|---|---|
| | **Capital Costs** | | | | | | |
| 17 03 0220 | 910, 1.25 CY, Wheel Loader | HOUR | 130.20 | 104.06 | 87.23 | 61.63 | 55.86 |
| 17 03 0221 | 916, 1.5 CY, Wheel Loader | HOUR | 114.32 | 90.83 | 76.64 | 53.69 | 47.92 |
| 17 03 0222 | 926, 2.0 CY, Wheel Loader | HOUR | 154.30 | 124.14 | 103.30 | 73.68 | 67.91 |
| 17 03 0224 | 966, 4.0 CY, Wheel Loader | HOUR | 211.58 | 171.88 | 141.48 | 102.32 | 96.55 |
| 17 03 0226 | 988, 7.0 CY, Wheel Loader | HOUR | 384.90 | 316.10 | 257.05 | 188.82 | 182.77 |
| 17 03 0227 | 992, 13.5 CY, Wheel Loader | HOUR | 648.92 | 536.11 | 433.06 | 320.83 | 314.78 |
| 17 03 1002 | 2" Diameter Contractor's Trash Pump, 75 GPM | DAY | 50.20 | 47.01 | 45.64 | 42.55 | 41.42 |
| 17 03 1003 | 3" Diameter Contractor's Trash Pump, 150 GPM | DAY | 61.72 | 57.05 | 55.05 | 50.54 | 48.88 |
| 17 03 1004 | 4" Diameter Contractor's Trash Pump, 300 GPM | DAY | 80.30 | 74.23 | 71.63 | 65.77 | 63.61 |
| 33 10 0102 | Consolidate Drums into Tanker Truck | GAL | 0.94 | 0.74 | 0.63 | 0.44 | 0.38 |
| 33 10 0103 | Consolidate Drums into Dump Truck | CY | 155.31 | 122.89 | 104.17 | 72.55 | 64.06 |
| 33 10 0109 | 250 Gallon Bulk Container, Polyethylene with Metal Case | EACH | 749.95 | 749.95 | 749.95 | 749.95 | 749.95 |
| 33 10 0110 | 350 Gallon Bulk Container, Polyethylene | EACH | 294.33 | 294.33 | 294.33 | 294.33 | 294.33 |
| 33 17 0809 | Triple-rinse 55 Gallon Drums | EACH | 14.07 | 10.92 | 9.46 | 6.41 | 5.36 |
| 33 19 0103 | Load Drums on Disposal Vehicle | EACH | 6.21 | 4.95 | 4.16 | 2.93 | 2.64 |
| 33 19 0104 | Load Bulk Liquid/Sludge Waste Into 55 Gallon Drums, Drums Separate | EACH | 17.53 | 13.51 | 11.79 | 7.91 | 6.48 |
| 33 19 0105 | Load Bulk Solid Waste Into 55 Gallon Drums, Drums Separate | EACH | 62.31 | 48.03 | 41.92 | 28.12 | 23.06 |
| 33 19 0106 | Recontainerize Drums, Not Including Salvage Drums | EACH | 105.53 | 84.84 | 70.65 | 50.34 | 46.31 |
| 33 19 0108 | Tanker Pumping Equipment to Load Liquid | HOUR | 46.88 | 46.88 | 46.88 | 46.88 | 46.88 |
| 33 19 0109 | Dredging Sludge < 1,000' to Loading Area | DAY | 3,048 | 2,457 | 2,040 | 1,459 | 1,351 |
| 33 19 0110 | Dredging Sludge > 1,000' to Loading Area | DAY | 4,339 | 3,533 | 2,901 | 2,105 | 1,997 |
| 33 19 0111 | Vacuum Truck on Board Loader | HOUR | 32.13 | 32.13 | 32.13 | 32.13 | 32.13 |
| 33 19 0112 | Load Bulk Liquid/Sludge Waste Into 350 Gallon Container | EACH | 111.46 | 85.92 | 74.98 | 50.29 | 41.24 |
| 33 19 0113 | Load Bulk Liquid/Sludge Waste Into 250 Gallon Container | EACH | 79.67 | 61.41 | 53.60 | 35.95 | 29.48 |
| 33 19 0114 | Load Bulk Liquid/Sludge Waste Into 110 Gallon Drum | EACH | 35.05 | 27.02 | 23.58 | 15.82 | 12.97 |
| 33 19 0115 | Load Bulk Liquid/Sludge Waste Into 95 Gallon Drum | EACH | 30.27 | 23.33 | 20.36 | 13.66 | 11.20 |
| 33 19 0116 | Load Bulk Liquid/Sludge Waste Into 85 Gallon Drum | EACH | 27.09 | 20.89 | 18.23 | 12.23 | 10.02 |
| 33 19 0117 | Load Bulk Liquid/Sludge Waste Into 83 Gallon Drum | EACH | 26.46 | 20.40 | 17.80 | 11.94 | 9.79 |
| 33 19 0118 | Load Bulk Liquid/Sludge Waste Into 35 Gallon Drum | EACH | 11.15 | 8.59 | 7.50 | 5.03 | 4.12 |
| 33 19 0119 | Load Bulk Liquid/Sludge Waste Into 30 Gallon Drum | EACH | 9.57 | 7.38 | 6.44 | 4.32 | 3.54 |
| 33 19 0120 | Load Bulk Liquid/Sludge Waste Into 20 Gallon Drum | EACH | 6.38 | 4.92 | 4.29 | 2.88 | 2.36 |
| 33 19 0121 | Load Bulk Liquid/Sludge Waste Into 15 Gallon Drum | EACH | 4.78 | 3.69 | 3.22 | 2.16 | 1.77 |

**Environmental Remediation: Assemblies Cost Book**

# Cost Data for Remediation

## Transportation

| Assembly | Description | Unit | Unit Cost by Safety Level A | B | C | D | E |
|---|---|---|---|---|---|---|---|
| | **Capital Costs** | | | | | | |
| 33 19 0122 | Load Bulk Liquid/Sludge Waste Into 10 Gallon Drum | EACH | 3.19 | 2.46 | 2.15 | 1.44 | 1.18 |
| 33 19 0123 | Load Bulk Solid Waste Into 110 Gallon Drum | EACH | 124.62 | 96.06 | 83.84 | 56.23 | 46.11 |
| 33 19 0124 | Load Bulk Solid Waste Into 95 Gallon Drum | EACH | 107.61 | 82.95 | 72.39 | 48.56 | 39.82 |
| 33 19 0125 | Load Bulk Solid Waste Into 85 Gallon Drum | EACH | 96.33 | 74.26 | 64.81 | 43.47 | 35.64 |
| 33 19 0126 | Load Bulk Solid Waste Into 83 Gallon Drum | EACH | 94.09 | 72.53 | 63.30 | 42.46 | 34.81 |
| 33 19 0127 | Load Bulk Solid Waste Into 35 Gallon Drum | EACH | 39.63 | 30.55 | 26.66 | 17.88 | 14.66 |
| 33 19 0128 | Load Bulk Solid Waste Into 30 Gallon Drum | EACH | 34.02 | 26.23 | 22.89 | 15.35 | 12.59 |
| 33 19 0129 | Load Bulk Solid Waste Into 20 Gallon Drum | EACH | 22.68 | 17.48 | 15.26 | 10.23 | 8.39 |
| 33 19 0130 | Load Bulk Solid Waste Into 15 Gallon Drum | EACH | 17.01 | 13.11 | 11.44 | 7.68 | 6.29 |
| 33 19 0131 | Load Bulk Solid Waste Into 10 Gallon Drum | EACH | 11.34 | 8.74 | 7.63 | 5.12 | 4.20 |
| 33 19 0132 | Load Bulk Liquid/Sludge into HazMax | EACH | 64.37 | 49.62 | 43.30 | 29.05 | 23.82 |
| 33 19 0133 | Load Bulk Solid Waste into HazMax | EACH | 228.87 | 176.42 | 153.97 | 103.27 | 84.68 |
| 33 19 0209 | Dump Truck Transportation Hazardous Waste Minimum Charge | EACH | 650.00 | 650.00 | 650.00 | 650.00 | 650.00 |
| 33 19 0210 | Dump Truck Transportation Hazardous Waste 200 - 299 Miles | MILE | 2.17 | 2.17 | 2.17 | 2.17 | 2.17 |
| 33 19 0211 | Dump Truck Transportation Hazardous Waste 300 - 399 Miles | MILE | 1.97 | 1.97 | 1.97 | 1.97 | 1.97 |
| 33 19 0212 | Dump Truck Transportation Hazardous Waste 400 - 499 Miles | MILE | 1.84 | 1.84 | 1.84 | 1.84 | 1.84 |
| 33 19 0213 | Dump Truck Transportation Hazardous Waste 500 - 599 Miles | MILE | 1.76 | 1.76 | 1.76 | 1.76 | 1.76 |
| 33 19 0214 | Dump Truck Transportation Hazardous Waste 600 - 699 Miles | MILE | 1.74 | 1.74 | 1.74 | 1.74 | 1.74 |
| 33 19 0215 | Dump Truck Transportation Hazardous Waste 700 - 799 Miles | MILE | 1.70 | 1.70 | 1.70 | 1.70 | 1.70 |
| 33 19 0216 | Dump Truck Transportation Hazardous Waste 800 - 899 Miles | MILE | 1.68 | 1.68 | 1.68 | 1.68 | 1.68 |
| 33 19 0217 | Dump Truck Transportation Hazardous Waste 900 - 999 Miles | MILE | 1.67 | 1.67 | 1.67 | 1.67 | 1.67 |
| 33 19 0218 | Dump Truck Transportation Hazardous Waste 1000+ Miles | MILE | 1.65 | 1.65 | 1.65 | 1.65 | 1.65 |
| 33 19 0219 | Dump Truck Transportation Explosive Waste Minimum Charge | EACH | 650.00 | 650.00 | 650.00 | 650.00 | 650.00 |
| 33 19 0220 | Dump Truck Transportation Explosive Waste 200 - 299 Miles | MILE | 2.17 | 2.17 | 2.17 | 2.17 | 2.17 |
| 33 19 0221 | Dump Truck Transportation Explosive Waste 300 - 399 Miles | MILE | 1.97 | 1.97 | 1.97 | 1.97 | 1.97 |
| 33 19 0222 | Dump Truck Transportation Explosive Waste 400 - 499 Miles | MILE | 1.84 | 1.84 | 1.84 | 1.84 | 1.84 |

**Environmental Remediation: Assemblies Cost Book**

# Cost Data for Remediation

## Transportation

| Assembly | Description | Unit | A | B | C | D | E |
|---|---|---|---|---|---|---|---|
| | **Capital Costs** | | | | | | |
| 33 19 0223 | Dump Truck Transportation Explosive Waste 500 - 599 Miles | MILE | 1.76 | 1.76 | 1.76 | 1.76 | 1.76 |
| 33 19 0224 | Dump Truck Transportation Explosive Waste 600 - 699 Miles | MILE | 1.74 | 1.74 | 1.74 | 1.74 | 1.74 |
| 33 19 0225 | Dump Truck Transportation Explosive Waste 700 - 799 Miles | MILE | 1.70 | 1.70 | 1.70 | 1.70 | 1.70 |
| 33 19 0226 | Dump Truck Transportation Explosive Waste 800 - 899 Miles | MILE | 1.68 | 1.68 | 1.68 | 1.68 | 1.68 |
| 33 19 0227 | Dump Truck Transportation Explosive Waste 900 - 999 Miles | MILE | 1.67 | 1.67 | 1.67 | 1.67 | 1.67 |
| 33 19 0228 | Dump Truck Transportation Explosive Waste 1,000+ Miles | MILE | 1.65 | 1.65 | 1.65 | 1.65 | 1.65 |
| 33 19 0229 | Van Trailer Transportation Hazardous Waste Minimum Charge | EACH | 732.33 | 732.33 | 732.33 | 732.33 | 732.33 |
| 33 19 0230 | Van Trailer Transportation Hazardous Waste 200 - 299 Miles | MILE | 2.48 | 2.48 | 2.48 | 2.48 | 2.48 |
| 33 19 0231 | Van Trailer Transportation Hazardous Waste 300 - 399 Miles | MILE | 2.35 | 2.35 | 2.35 | 2.35 | 2.35 |
| 33 19 0232 | Van Trailer Transportation Hazardous Waste 400 - 499 Miles | MILE | 2.27 | 2.27 | 2.27 | 2.27 | 2.27 |
| 33 19 0233 | Van Trailer Transportation Hazardous Waste 500 - 599 Miles | MILE | 2.22 | 2.22 | 2.22 | 2.22 | 2.22 |
| 33 19 0234 | Van Trailer Transportation Hazardous Waste 600 - 699 Miles | MILE | 2.20 | 2.20 | 2.20 | 2.20 | 2.20 |
| 33 19 0235 | Van Trailer Transportation Hazardous Waste 700 - 799 Miles | MILE | 2.16 | 2.16 | 2.16 | 2.16 | 2.16 |
| 33 19 0236 | Van Trailer Transportation Hazardous Waste 800 - 899 Miles | MILE | 2.14 | 2.14 | 2.14 | 2.14 | 2.14 |
| 33 19 0237 | Van Trailer Transportation Hazardous Waste 900 - 999 Miles | MILE | 2.13 | 2.13 | 2.13 | 2.13 | 2.13 |
| 33 19 0238 | Van Trailer Transportation Hazardous Waste 1,000 - 1,099 Miles | MILE | 2.11 | 2.11 | 2.11 | 2.11 | 2.11 |
| 33 19 0239 | Van Trailer Transportation Hazardous Waste 1,100 - 1,199 Miles | MILE | 2.09 | 2.09 | 2.09 | 2.09 | 2.09 |
| 33 19 0240 | Van Trailer Transportation Hazardous Waste 1,200+ Miles | MILE | 2.07 | 2.07 | 2.07 | 2.07 | 2.07 |
| 33 19 0241 | Van Trailer Transportation Explosive Waste Minimum Charge | EACH | 650.00 | 650.00 | 650.00 | 650.00 | 650.00 |
| 33 19 0242 | Van Trailer Transportation Explosive Waste 200 - 299 Miles | MILE | 2.17 | 2.17 | 2.17 | 2.17 | 2.17 |
| 33 19 0243 | Van Trailer Transportation Explosive Waste 300 - 399 Miles | MILE | 1.97 | 1.97 | 1.97 | 1.97 | 1.97 |
| 33 19 0244 | Van Trailer Transportation Explosive Waste 400 - 499 Miles | MILE | 1.84 | 1.84 | 1.84 | 1.84 | 1.84 |
| 33 19 0245 | Van Trailer Transportation Explosive Waste 500 - 599 Miles | MILE | 1.76 | 1.76 | 1.76 | 1.76 | 1.76 |
| 33 19 0246 | Van Trailer Transportation Explosive Waste 600 - 699 Miles | MILE | 1.74 | 1.74 | 1.74 | 1.74 | 1.74 |

**Environmental Remediation: Assemblies Cost Book**

# Cost Data for Remediation

## Transportation

| Assembly | Description | Unit | A | B | C | D | E |
|---|---|---|---|---|---|---|---|
| | **Capital Costs** | | | | | | |
| 33 19 0247 | Van Trailer Transportation Explosive Waste 700 - 799 Miles | MILE | 1.70 | 1.70 | 1.70 | 1.70 | 1.70 |
| 33 19 0248 | Van Trailer Transportation Explosive Waste 800 - 899 Miles | MILE | 1.68 | 1.68 | 1.68 | 1.68 | 1.68 |
| 33 19 0249 | Van Trailer Transportation Explosive Waste 900 - 999 Miles | MILE | 1.67 | 1.67 | 1.67 | 1.67 | 1.67 |
| 33 19 0250 | Van Trailer Transportation Explosive Waste 1,000 - 1,099 Miles | MILE | 1.65 | 1.65 | 1.65 | 1.65 | 1.65 |
| 33 19 0251 | Van Trailer Transportation Explosive Waste 1,100 - 1,199 Miles | MILE | 1.63 | 1.63 | 1.63 | 1.63 | 1.63 |
| 33 19 0252 | Van Trailer Transportation Explosive Waste 1,200+ Miles | MILE | 1.61 | 1.61 | 1.61 | 1.61 | 1.61 |
| 33 19 0253 | Tanker Trailer Transport Hazardous Waste Minimum Charge | EACH | 732.33 | 732.33 | 732.33 | 732.33 | 732.33 |
| 33 19 0254 | Tanker Trailer Transport Hazardous Waste 200 - 299 Miles | MILE | 2.48 | 2.48 | 2.48 | 2.48 | 2.48 |
| 33 19 0255 | Tanker Trailer Transport Hazardous Waste 300 - 399 Miles | MILE | 2.35 | 2.35 | 2.35 | 2.35 | 2.35 |
| 33 19 0256 | Tanker Trailer Transport Hazardous Waste 400 - 499 Miles | MILE | 2.27 | 2.27 | 2.27 | 2.27 | 2.27 |
| 33 19 0257 | Tanker Trailer Transport Hazardous Waste 500 - 599 Miles | MILE | 2.22 | 2.22 | 2.22 | 2.22 | 2.22 |
| 33 19 0258 | Tanker Trailer Transport Hazardous Waste 600 - 699 Miles | MILE | 2.20 | 2.20 | 2.20 | 2.20 | 2.20 |
| 33 19 0259 | Tanker Trailer Transport Hazardous Waste 700 - 799 Miles | MILE | 2.16 | 2.16 | 2.16 | 2.16 | 2.16 |
| 33 19 0260 | Tanker Trailer Transport Hazardous Waste 800 - 899 Miles | MILE | 2.14 | 2.14 | 2.14 | 2.14 | 2.14 |
| 33 19 0261 | Tanker Trailer Transport Hazardous Waste 900 - 999 Miles | MILE | 2.13 | 2.13 | 2.13 | 2.13 | 2.13 |
| 33 19 0262 | Tanker Trailer Transport Hazardous Waste 1,000+ Miles | MILE | 2.11 | 2.11 | 2.11 | 2.11 | 2.11 |
| 33 19 0263 | Tanker Trailer Transport Explosive Waste Minimum Charge | EACH | 650.00 | 650.00 | 650.00 | 650.00 | 650.00 |
| 33 19 0264 | Tanker Trailer Transport Explosive Waste 200 - 299 Miles | MILE | 2.17 | 2.17 | 2.17 | 2.17 | 2.17 |
| 33 19 0265 | Tanker Trailer Transport Explosive Waste 300 - 399 Miles | MILE | 1.97 | 1.97 | 1.97 | 1.97 | 1.97 |
| 33 19 0266 | Tanker Trailer Transport Explosive Waste 400 - 499 Miles | MILE | 1.84 | 1.84 | 1.84 | 1.84 | 1.84 |
| 33 19 0267 | Tanker Trailer Transport Explosive Waste 500 - 599 Miles | MILE | 1.76 | 1.76 | 1.76 | 1.76 | 1.76 |
| 33 19 0268 | Tanker Trailer Transport Explosive Waste 600 - 699 Miles | MILE | 1.74 | 1.74 | 1.74 | 1.74 | 1.74 |
| 33 19 0269 | Tanker Trailer Transport Explosive Waste 700 - 799 Miles | MILE | 1.70 | 1.70 | 1.70 | 1.70 | 1.70 |
| 33 19 0270 | Tanker Trailer Transport Explosive Waste 800 - 899 Miles | MILE | 1.68 | 1.68 | 1.68 | 1.68 | 1.68 |
| 33 19 0271 | Tanker Trailer Transport Explosive Waste 900 - 999 Miles | MILE | 1.67 | 1.67 | 1.67 | 1.67 | 1.67 |
| 33 19 0272 | Tanker Trailer Transport Explosive Waste 1,000+ Miles | MILE | 1.65 | 1.65 | 1.65 | 1.65 | 1.65 |
| 33 19 0273 | Vacuum Trailer Transport Hazardous Waste Minimum Charge | EACH | 732.33 | 732.33 | 732.33 | 732.33 | 732.33 |
| 33 19 0274 | Vacuum Trailer Transport Hazardous Waste 200 - 299 Miles | MILE | 2.27 | 2.27 | 2.27 | 2.27 | 2.27 |
| 33 19 0275 | Vacuum Trailer Transport Hazardous Waste 300 - 399 Miles | MILE | 2.21 | 2.21 | 2.21 | 2.21 | 2.21 |

**Environmental Remediation: Assemblies Cost Book**

# Cost Data for Remediation

## Transportation

| Assembly | Description | Unit | Unit Cost by Safety Level A | B | C | D | E |
|---|---|---|---|---|---|---|---|
| | **Capital Costs** | | | | | | |
| 33 19 0276 | Vacuum Trailer Transport Hazardous Waste 400 - 899 Miles | MILE | 2.16 | 2.16 | 2.16 | 2.16 | 2.16 |
| 33 19 0277 | Vacuum Trailer Transport Hazardous Waste 900+ Miles | MILE | 2.12 | 2.12 | 2.12 | 2.12 | 2.12 |
| 33 19 0278 | Vacuum Trailer Transport Explosive Waste Minimum Charge | EACH | 776.27 | 776.27 | 776.27 | 776.27 | 776.27 |
| 33 19 0279 | Vacuum Trailer Transport Explosive Waste 200 - 299 Miles | MILE | 2.41 | 2.41 | 2.41 | 2.41 | 2.41 |
| 33 19 0280 | Vacuum Trailer Transport Explosive Waste 300 - 399 Miles | MILE | 2.34 | 2.34 | 2.34 | 2.34 | 2.34 |
| 33 19 0281 | Vacuum Trailer Transport Explosive Waste 400 - 899 Miles | MILE | 2.29 | 2.29 | 2.29 | 2.29 | 2.29 |
| 33 19 0282 | Vacuum Trailer Transport Explosive Waste 900+Miles | MILE | 2.25 | 2.25 | 2.25 | 2.25 | 2.25 |
| 33 19 0283 | Radioactive Truck Haul Over 500 Miles | MILE | 3.39 | 3.39 | 3.39 | 3.39 | 3.39 |
| 33 19 0284 | Radioactive Truck Haul 251 - 500 Miles | MILE | 2.93 | 2.93 | 2.93 | 2.93 | 2.93 |
| 33 19 0285 | Radioactive Truck Haul 101 - 250 Miles | MILE | 2.45 | 2.45 | 2.45 | 2.45 | 2.45 |
| 33 19 0286 | Radioactive Truck Haul 0 - 100 Miles | MILE | 1,000 | 1,000 | 1,000 | 1,000 | 1,000 |
| 33 19 0287 | Rail Boxcar | CWT | 3.18 | 3.18 | 3.18 | 3.18 | 3.18 |
| 33 19 0288 | Rail Tanker | CWT | 4.24 | 4.24 | 4.24 | 4.24 | 4.24 |
| 33 19 0289 | Rail Gondola | CWT | 5.30 | 5.30 | 5.30 | 5.30 | 5.30 |
| 33 19 0290 | Rail Flatbed | CWT | 6.36 | 6.36 | 6.36 | 6.36 | 6.36 |
| 33 19 0311 | Truck Washout | EACH | 150.00 | 150.00 | 150.00 | 150.00 | 150.00 |
| 33 19 0318 | Manifest Discrepancies | EACH | 50.00 | 50.00 | 50.00 | 50.00 | 50.00 |
| 33 19 0327 | Demurrage Vacuum, Dump, and Tanker | HOUR | 60.67 | 60.67 | 60.67 | 60.67 | 60.67 |
| 33 19 0328 | Demurrage Van Trailer | HOUR | 67.50 | 67.50 | 67.50 | 67.50 | 67.50 |
| 33 19 0329 | Overnight Demurrage | EACH | 226.50 | 226.50 | 226.50 | 226.50 | 226.50 |
| 33 19 0330 | Maximum 24 Hour Demurrage | EACH | 530.00 | 530.00 | 530.00 | 530.00 | 530.00 |
| 33 19 0331 | Additional Stops <= 3 | EACH | 62.50 | 62.50 | 62.50 | 62.50 | 62.50 |
| 33 19 0332 | Additional Stops > 3 | EACH | 100.00 | 100.00 | 100.00 | 100.00 | 100.00 |
| 33 19 0333 | Tanker Trailer Drop (Daily) | DAY | 117.50 | 117.50 | 117.50 | 117.50 | 117.50 |
| 33 19 0334 | Tanker Trailer Drop (Monthly) | MONTH | 107.50 | 107.50 | 107.50 | 107.50 | 107.50 |
| 33 19 0335 | Dump Trailer Drop (Daily) | DAY | 15.00 | 15.00 | 15.00 | 15.00 | 15.00 |
| 33 19 0336 | Dump Trailer Drop (Monthly) | MONTH | 300.00 | 300.00 | 300.00 | 300.00 | 300.00 |
| 33 19 0337 | Van Trailer Drop (Daily) | DAY | 36.25 | 36.25 | 36.25 | 36.25 | 36.25 |
| 33 19 0338 | Van Trailer Drop (Monthly) | MONTH | 307.63 | 307.63 | 307.63 | 307.63 | 307.63 |
| 33 19 0401 | 55 Gallon 17C Closed Head Steel Drum | EACH | 60.06 | 60.06 | 60.06 | 60.06 | 60.06 |
| 33 19 0402 | 55 Gallon 17H Open Head Steel Drum | EACH | 68.30 | 68.30 | 68.30 | 68.30 | 68.30 |
| 33 19 0403 | 55 Gallon 17H Closed Head Steel Drum | EACH | 59.88 | 59.88 | 59.88 | 59.88 | 59.88 |

**Environmental Remediation: Assemblies Cost Book**

# Cost Data for Remediation

## Transportation

| Assembly | Description | Unit | A | B | C | D | E |
|---|---|---|---|---|---|---|---|
| | **Capital Costs** | | | | | | |
| 33 19 0404 | 20 Gallon 17C Open Head Steel Drum | EACH | 47.28 | 47.28 | 47.28 | 47.28 | 47.28 |
| 33 19 0405 | 20 Gallon 17E Open Head Steel Drum | EACH | 56.08 | 56.08 | 56.08 | 56.08 | 56.08 |
| 33 19 0406 | 55 Gallon Stainless Steel DOT 5C Closed Head Seamless 16 Gauge (1 - 9) | EACH | 627.49 | 627.49 | 627.49 | 627.49 | 627.49 |
| 33 19 0407 | 55 Gallon Stainless Steel DOT 5C Closed Head Seamless 16 Gauge (10 - 24) | EACH | 660.54 | 660.54 | 660.54 | 660.54 | 660.54 |
| 33 19 0408 | 55 Gallon Stainless Steel DOT 5C Closed Head Seamless 16 Gauge (25 - 49) | EACH | 660.54 | 660.54 | 660.54 | 660.54 | 660.54 |
| 33 19 0409 | 55 Gallon Stainless Steel DOT 5C Closed Head Seamless 16 Gauge (50+) | EACH | 660.54 | 660.54 | 660.54 | 660.54 | 660.54 |
| 33 19 0410 | 55 Gallon Stainless Steel DOT 5C Closed Head 16 Gauge (1 - 9) | EACH | 665.14 | 665.14 | 665.14 | 665.14 | 665.14 |
| 33 19 0411 | 55 Gallon Stainless Steel DOT 5C Closed Head 16 Gauge (10 - 24) | EACH | 665.14 | 665.14 | 665.14 | 665.14 | 665.14 |
| 33 19 0412 | 55 Gallon Stainless Steel DOT 5C Closed Head 16 Gauge (25 - 49) | EACH | 665.14 | 665.14 | 665.14 | 665.14 | 665.14 |
| 33 19 0413 | 55 Gallon Stainless Steel DOT 5C Closed Head 16 Gauge (50+) | EACH | 665.14 | 665.14 | 665.14 | 665.14 | 665.14 |
| 33 19 0414 | 55 Gallon Stainless Steel DOT 17E Closed Head 18 Gauge (1 - 9) | EACH | 558.08 | 558.08 | 558.08 | 558.08 | 558.08 |
| 33 19 0415 | 55 Gallon Stainless Steel DOT 17E Closed Head 18 Gauge (10 - 24) | EACH | 558.08 | 558.08 | 558.08 | 558.08 | 558.08 |
| 33 19 0416 | 55 Gallon Stainless Steel DOT 17E Closed Head 18 Gauge (25 - 49) | EACH | 558.08 | 558.08 | 558.08 | 558.08 | 558.08 |
| 33 19 0417 | 55 Gallon Stainless Steel DOT 17E Closed Head 18 Gauge (50+) | EACH | 558.08 | 558.08 | 558.08 | 558.08 | 558.08 |
| 33 19 0418 | 30 Gallon Stainless Steel DOT 5C Closed Head 16 Gauge (1 - 9) | EACH | 560.93 | 560.93 | 560.93 | 560.93 | 560.93 |
| 33 19 0419 | 30 Gallon Stainless Steel DOT 5C Closed Head 16 Gauge (10 - 24) | EACH | 560.93 | 560.93 | 560.93 | 560.93 | 560.93 |
| 33 19 0420 | 30 Gallon Stainless Steel DOT 5C Closed Head 16 Gauge (25 - 49) | EACH | 560.93 | 560.93 | 560.93 | 560.93 | 560.93 |
| 33 19 0421 | 30 Gallon Stainless Steel DOT 5C Closed Head 16 Gauge (50+) | EACH | 560.93 | 560.93 | 560.93 | 560.93 | 560.93 |
| 33 19 0422 | 55 Gallon Stainless Steel DOT 5C Seamless Open Head 16 Gauge (1 - 9) | EACH | 788.41 | 788.41 | 788.41 | 788.41 | 788.41 |
| 33 19 0423 | 55 Gallon Stainless Steel DOT 5C Seamless Open Head 16 Gauge (10 - 24) | EACH | 788.41 | 788.41 | 788.41 | 788.41 | 788.41 |
| 33 19 0424 | 55 Gallon Stainless Steel DOT 5C Seamless Open Head 16 Gauge (25 - 49) | EACH | 788.41 | 788.41 | 788.41 | 788.41 | 788.41 |
| 33 19 0425 | 55 Gallon Stainless Steel DOT 5C Seamless Open Head 16 Gauge (50+) | EACH | 788.41 | 788.41 | 788.41 | 788.41 | 788.41 |

**Environmental Remediation: Assemblies Cost Book**

*Page 3-326*

# Cost Data for Remediation

## Transportation

| Assembly | Description | Unit | A | B | C | D | E |
|---|---|---|---|---|---|---|---|
| | **Capital Costs** | | | | | | |
| 33 19 0426 | 55 Gallon Stainless Steel DOT 5C Open Head 16 Gauge (1 - 9) | EACH | 607.83 | 607.83 | 607.83 | 607.83 | 607.83 |
| 33 19 0427 | 55 Gallon Stainless Steel DOT 5C Open Head 16 Gauge (10 - 24) | EACH | 607.83 | 607.83 | 607.83 | 607.83 | 607.83 |
| 33 19 0428 | 55 Gallon Stainless Steel DOT 5C Open Head 16 Gauge (25 - 49) | EACH | 607.83 | 607.83 | 607.83 | 607.83 | 607.83 |
| 33 19 0429 | 55 Gallon Stainless Steel DOT 5C Open Head 16 Gauge (50+) | EACH | 607.83 | 607.83 | 607.83 | 607.83 | 607.83 |
| 33 19 0430 | 55 Gallon Reconditioned Steel 17H (1 - 9) | EACH | 52.96 | 52.96 | 52.96 | 52.96 | 52.96 |
| 33 19 0431 | 56 Gallon Reconditioned Steel 17H (10 - 24) | EACH | 49.25 | 49.25 | 49.25 | 49.25 | 49.25 |
| 33 19 0432 | 57 Gallon Reconditioned Steel 17H (25 - 49) | EACH | 45.40 | 45.40 | 45.40 | 45.40 | 45.40 |
| 33 19 0433 | 58 Gallon Reconditioned Steel 17H (50+) | EACH | 44.69 | 44.69 | 44.69 | 44.69 | 44.69 |
| 33 19 0434 | 110 Gallon Steel Overpack 16 Gauge | EACH | 130.38 | 130.38 | 130.38 | 130.38 | 130.38 |
| 33 19 0435 | 85 Gallon Steel Overpack 16 Gauge | EACH | 57.06 | 57.06 | 57.06 | 57.06 | 57.06 |
| 33 19 0436 | 83 Gallon Steel Overpack 16 Gauge | EACH | 55.12 | 55.12 | 55.12 | 55.12 | 55.12 |
| 33 19 0501 | 35 Gallon Polyethylene Closed Head | EACH | 35.37 | 35.37 | 35.37 | 35.37 | 35.37 |
| 33 19 0502 | 30 Gallon Polyethylene Closed Head | EACH | 32.44 | 32.44 | 32.44 | 32.44 | 32.44 |
| 33 19 0503 | 20 Gallon Polyethylene Closed Head | EACH | 23.78 | 23.78 | 23.78 | 23.78 | 23.78 |
| 33 19 0504 | 15 Gallon Polyethylene Closed Head | EACH | 20.32 | 20.32 | 20.32 | 20.32 | 20.32 |
| 33 19 0505 | 10 Gallon DOT 21C-250 Closed Head PET/Poly Lined | EACH | 21.39 | 21.39 | 21.39 | 21.39 | 21.39 |
| 33 19 0506 | 20 Gallon 20" Diameter DOT 21C-250 Closed Head PET/Poly Lined | EACH | 17.38 | 17.38 | 17.38 | 17.38 | 17.38 |
| 33 19 0507 | 20 Gallon 23" Diameter DOT 21C-250 Closed Head PET/Poly Lined | EACH | 20.86 | 20.86 | 20.86 | 20.86 | 20.86 |
| 33 19 0508 | 35 Gallon Polyethylene Open Head | EACH | 41.39 | 41.39 | 41.39 | 41.39 | 41.39 |
| 33 19 0509 | 30 Gallon Polyethylene Open Head | EACH | 36.02 | 36.02 | 36.02 | 36.02 | 36.02 |
| 33 19 0510 | 20 Gallon Polyethylene Open Head | EACH | 30.48 | 30.48 | 30.48 | 30.48 | 30.48 |
| 33 19 0511 | 15 Gallon Polyethylene Open Head | EACH | 23.14 | 23.14 | 23.14 | 23.14 | 23.14 |
| 33 19 0512 | 95 Gallon Poly Overpack DOT E9618 | EACH | 127.20 | 127.20 | 127.20 | 127.20 | 127.20 |
| 33 19 0513 | 85 Gallon Poly Overpack DOT E9775 | EACH | 98.21 | 98.21 | 98.21 | 98.21 | 98.21 |
| 33 19 0603 | 55 Gallon Fiberglass DOT-21C (1 - 9) | EACH | 33.16 | 33.16 | 33.16 | 33.16 | 33.16 |
| 33 19 0604 | 55 Gallon Fiberglass DOT-21C (10 - 24) | EACH | 31.87 | 31.87 | 31.87 | 31.87 | 31.87 |
| 33 19 0605 | 55 Gallon Fiberglass DOT-21C (25 - 49) | EACH | 30.10 | 30.10 | 30.10 | 30.10 | 30.10 |
| 33 19 0606 | 55 Gallon Fiberglass DOT-21C (50 - 99) | EACH | 29.46 | 29.46 | 29.46 | 29.46 | 29.46 |
| 33 19 0607 | 55 Gallon Fiberglass DOT-21C (100+) | EACH | 28.51 | 28.51 | 28.51 | 28.51 | 28.51 |
| 33 19 0608 | 30 Gallon Fiberglass DOT-21C (1 - 9) | EACH | 26.71 | 26.71 | 26.71 | 26.71 | 26.71 |

## Environmental Remediation: Assemblies Cost Book

# Cost Data for Remediation

## Transportation

| Assembly | Description | Unit | Unit Cost by Safety Level A | B | C | D | E |
|---|---|---|---|---|---|---|---|
| | **Capital Costs** | | | | | | |
| 33 19 0609 | 30 Gallon Fiberglass DOT-21C (10 - 24) | EACH | 25.77 | 25.77 | 25.77 | 25.77 | 25.77 |
| 33 19 0610 | 30 Gallon Fiberglass DOT-21C (25 - 49) | EACH | 24.70 | 24.70 | 24.70 | 24.70 | 24.70 |
| 33 19 0611 | 30 Gallon Fiberglass DOT-21C (50+) | EACH | 24.22 | 24.22 | 24.22 | 24.22 | 24.22 |
| 33 19 0616 | HazMax Box 1 CY with Liner (1 - 9) | EACH | 83.55 | 83.55 | 83.55 | 83.55 | 83.55 |
| 33 19 0617 | HazMax Box 1 CY with Liner (10 - 24) | EACH | 80.37 | 80.37 | 80.37 | 80.37 | 80.37 |
| 33 19 0618 | HazMax Box 1 CY with Liner (25 - 49) | EACH | 76.13 | 76.13 | 76.13 | 76.13 | 76.13 |
| 33 19 0619 | HazMax Box 1 CY with Liner (50+) | EACH | 67.86 | 67.86 | 67.86 | 67.86 | 67.86 |
| 33 19 0701 | Tall 55 Gallon Drum for 250 Lb UO2 | EACH | 1,203 | 1,203 | 1,203 | 1,203 | 1,203 |
| 33 19 0702 | 5,020 Lb, 30" Cylinder Enriched UF6 | EACH | 3,633 | 3,633 | 3,633 | 3,633 | 3,633 |
| 33 19 0703 | 27,560 Lb, Cylinder UF6 | EACH | 8,289 | 8,289 | 8,289 | 8,289 | 8,289 |
| 33 19 0704 | 55 Gallon Drum for 2 5-Gallon Pails UO2 | EACH | 867.61 | 867.61 | 867.61 | 867.61 | 867.61 |
| 33 19 0705 | 5" 55 Lb UF6 Cylinder Overpack | EACH | 2,247 | 2,247 | 2,247 | 2,247 | 2,247 |
| 33 19 0706 | 8" 250 Lb UF6 Cylinder Overpack | EACH | 4,452 | 4,452 | 4,452 | 4,452 | 4,452 |
| 33 19 0707 | 12" 430 Lb UF6 Cylinder Overpack | EACH | 5,003 | 5,003 | 5,003 | 5,003 | 5,003 |
| 33 19 0708 | 30" 2.5 Ton UF6 Cylinder Overpack | EACH | 14,374 | 14,374 | 14,374 | 14,374 | 14,374 |
| 33 19 0709 | 48" 10 Ton UF6 Cylinder Overpack | EACH | 66,804 | 66,804 | 66,804 | 66,804 | 66,804 |
| 33 19 0710 | 48" 14 Ton UF6 Cylinder Overpack | EACH | 111,353 | 111,353 | 111,353 | 111,353 | 111,353 |
| 33 19 0711 | 14 - 55 Gallon Drum Cask <= 20 R/Hour (Daily Rent) | DAY | 283.55 | 283.55 | 283.55 | 283.55 | 283.55 |
| 33 19 0712 | 14 - 55 Gallon Drum Cask <= 20 R/Hour (Monthly Rental) | MONTH | 7,780 | 7,780 | 7,780 | 7,780 | 7,780 |
| 33 19 0801 | 55 Gallon, 15 Mil, Corrugated | EACH | 8.73 | 8.73 | 8.73 | 8.73 | 8.73 |
| 33 19 0802 | 55 Gallon, 10 Mil, Corrugated | EACH | 8.62 | 8.62 | 8.62 | 8.62 | 8.62 |
| 33 19 0803 | 55 Gallon, 6 Mil | EACH | 8.82 | 8.82 | 8.82 | 8.82 | 8.82 |
| 33 19 0804 | 55 Gallon, 90 Mil High-density Polyethylene Radioactive Waste | EACH | 95.14 | 95.14 | 95.14 | 95.14 | 95.14 |
| 33 19 0805 | 85 Gallon, 8 Mil | EACH | 4.69 | 4.69 | 4.69 | 4.69 | 4.69 |
| 33 19 0806 | 32' Dump Truck, 4 Mil | EACH | 14.72 | 14.72 | 14.72 | 14.72 | 14.72 |
| 33 19 0807 | 32' Dump Truck, 6 Mil | EACH | 22.05 | 22.05 | 22.05 | 22.05 | 22.05 |
| 33 19 0808 | 39' Dump Truck, 72" Deep, 4 Mil | EACH | 17.41 | 17.41 | 17.41 | 17.41 | 17.41 |
| 33 19 0809 | 39' Dump Truck, 96" Deep, 4 Mil | EACH | 21.61 | 21.61 | 21.61 | 21.61 | 21.61 |
| 33 19 0810 | 40' Dump Truck, 96" Deep, 8 Mil | EACH | 21.86 | 21.86 | 21.86 | 21.86 | 21.86 |
| 33 19 0811 | 40 to 44' Van, 96" Deep, 8 Mil | EACH | 56.64 | 56.64 | 56.64 | 56.64 | 56.64 |
| 33 19 0812 | 50 to 54' Rail Gondola, 10 Mil | EACH | 112.81 | 112.81 | 112.81 | 112.81 | 112.81 |
| 33 19 7280 | 30 Gallon, 17C, Open | EACH | 57.64 | 57.64 | 57.64 | 57.64 | 57.64 |
| 33 19 7281 | 30 Gallon, 17H, Open | EACH | 57.64 | 57.64 | 57.64 | 57.64 | 57.64 |

## Environmental Remediation: Assemblies Cost Book

# Cost Data for Remediation

## Transportation

| Assembly | Description | Unit | Unit Cost by Safety Level | | | | |
|---|---|---|---|---|---|---|---|
| | | | A | B | C | D | E |
| | **Capital Costs** | | | | | | |
| 33 19 7282 | 30 Gallon, 17E, Open | EACH | 59.31 | 59.31 | 59.31 | 59.31 | 59.31 |
| 33 19 9921 | DOT Steel Drum, 55 Gallon | EACH | 65.19 | 65.19 | 65.19 | 65.19 | 65.19 |
| 33 19 9922 | Polyethylene Closed Head Drum, 55 Gallon | EACH | 43.63 | 43.63 | 43.63 | 43.63 | 43.63 |
| 33 19 9923 | Polyethylene Open Head Drum, 55 Gallon | EACH | 49.90 | 49.90 | 49.90 | 49.90 | 49.90 |

**Environmental Remediation: Assemblies Cost Book**

# Cost Data for Remediation
## Ultraviolet Oxidation

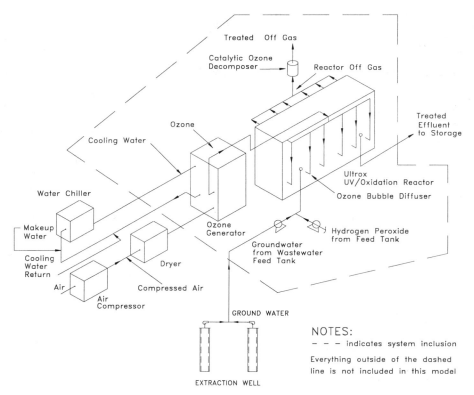

### General:

The oxidation and reduction mechanism is a well understood type of chemical reaction. The ultraviolet (UV) oxidation process uses this mechanism, although the complete degradation pathways are unknown. Processes that generate hydroxyl radicals which include UV photolyzed oxidation are generally categorized as advanced oxidation processes (AOPs). This technology is intended to be used to develop cost estimates for the AOP group that utilizes UV light to enhance generation of hydroxyl radicals. It should not be used to supplement or replace treatability studies and/or pilot plants.

### Typical Treatment Train:

This model includes minimal pretreatment equipment for the removal of suspended solids. Additional pretreatment, such as media filtration, coagulation/flocculation, and pH adjustment, should be estimated using other remediation methods; transportation and disposal of treated wastewater, treatment residuals, and gas emissions; sampling and analysis.

### Common Cost Components:

1. Assemble/shakedown ultraviolet system
2. Ultraviolet system
3. Pretreatment equipment
4. Fugitive emission control system  5. Piping (influent, effluent)
6. Electrical usage
7. Operations and maintenance

### Other Cost Considerations:

Access road; clearing and grubbing; excavation, cut and fill; fencing; loading and hauling; electrical distribution; water distribution.

**Environmental Remediation: Assemblies Cost Book**

# Cost Data for Remediation

## Ultraviolet Oxidation

| Assembly | | | Description | Unit | Unit Cost by Safety Level | | | | |
|---|---|---|---|---|---|---|---|---|---|
| | | | | | A | B | C | D | E |
| | | | **Capital Costs** | | | | | | |
| 17 | 01 | 0105 | Heavy Brush with Average Grub, Medium - Heavy Trees, Clearing | ACRE | 1,738 | 1,408 | 1,162 | 837.39 | 785.18 |
| 18 | 02 | 0320 | 4" Structural Slab on Grade | SF | 6.45 | 5.45 | 4.98 | 4.00 | 3.67 |
| 18 | 02 | 0322 | 8" Structural Slab on Grade | SF | 9.75 | 8.38 | 7.71 | 6.38 | 5.95 |
| 18 | 02 | 0323 | 10" Structural Slab on Grade | SF | 10.56 | 9.13 | 8.43 | 7.04 | 6.58 |
| 18 | 02 | 0324 | 12" Structural Slab on Grade | SF | 11.57 | 10.04 | 9.30 | 7.82 | 7.34 |
| 19 | 01 | 0202 | 1", Class 200, PVC Piping | LF | 7.48 | 5.84 | 5.07 | 3.48 | 2.93 |
| 19 | 01 | 0204 | 2", Class 200, PVC Piping | LF | 8.86 | 6.96 | 6.07 | 4.23 | 3.60 |
| 19 | 01 | 0206 | 3", Class 200, PVC Piping | LF | 11.81 | 9.34 | 8.20 | 5.80 | 4.98 |
| 19 | 01 | 0207 | 4", Class 200, PVC Piping | LF | 12.33 | 9.86 | 8.72 | 6.32 | 5.50 |
| 19 | 01 | 0208 | 6", Class 200, PVC Piping | LF | 14.95 | 12.20 | 10.93 | 8.27 | 7.36 |
| 19 | 01 | 0212 | 8", Class 150, PVC Piping | LF | 18.36 | 15.27 | 13.84 | 10.85 | 9.82 |
| 19 | 02 | 0101 | 4", Class 50, Bell & Spigot Sanitary Sewer, Cast-iron Pipe | LF | 8.49 | 7.49 | 7.05 | 6.08 | 5.72 |
| 19 | 02 | 0102 | 5", Class 50, Bell & Spigot Sanitary Sewer, Cast-iron Pipe | LF | 11.80 | 10.41 | 9.81 | 8.47 | 7.98 |
| 19 | 02 | 0103 | 6", Class 50, Bell & Spigot Sanitary Sewer, Cast-iron Pipe | LF | 13.25 | 11.79 | 11.16 | 9.75 | 9.23 |
| 19 | 02 | 0104 | 8", Class 50, Bell & Spigot Sanitary Sewer, Cast-iron Pipe | LF | 18.98 | 17.26 | 16.52 | 14.86 | 14.25 |
| 19 | 02 | 0105 | 10", Class 50, Bell & Spigot Sanitary Sewer, Cast-iron Pipe | LF | 27.66 | 25.51 | 24.59 | 22.51 | 21.75 |
| 19 | 02 | 0106 | 12", Class 50, Bell & Spigot Sanitary Sewer, Cast-iron Pipe | LF | 36.36 | 34.21 | 33.29 | 31.21 | 30.45 |
| 19 | 02 | 0107 | 15", Class 50, Bell & Spigot Sanitary Sewer, Cast-iron Pipe | LF | 51.76 | 49.01 | 47.83 | 45.17 | 44.19 |
| 19 | 04 | 0401 | 550 Gallon, Stainless Steel Aboveground Wastewater Holding Tank, Rental | MONTH | 320.65 | 320.65 | 320.65 | 320.65 | 320.65 |
| 19 | 04 | 0402 | 630 Gallon, Polyethylene Aboveground Wastewater Holding Tank, Rental | MONTH | 320.65 | 320.65 | 320.65 | 320.65 | 320.65 |
| 19 | 04 | 0403 | 4,000 Gallon Polyethylene Wastewater Tank, Rental | MONTH | 545.90 | 545.90 | 545.90 | 545.90 | 545.90 |
| 19 | 04 | 0404 | 4,000 Gallon, Polyethylene Trailer-mounted Wastewater Holding Tank, Rental | MONTH | 1,124 | 1,124 | 1,124 | 1,124 | 1,124 |
| 19 | 04 | 0405 | 6,000 Gallon, Polyethylene Aboveground Wastewater Holding Tank, Rental | MONTH | 641.30 | 641.30 | 641.30 | 641.30 | 641.30 |
| 19 | 04 | 0406 | 21,000 Gallon Steel Wastewater Holding Tank, Rental | MONTH | 1,219 | 1,219 | 1,219 | 1,219 | 1,219 |
| 19 | 04 | 0408 | 21,000 Gallon Steel, Open Top, Tank Rental | MONTH | 1,124 | 1,124 | 1,124 | 1,124 | 1,124 |
| 19 | 04 | 0418 | 55 Gallon Nalgene Horizontal XLPE Tank without legs | EACH | 295.00 | 295.00 | 295.00 | 295.00 | 295.00 |
| 19 | 04 | 0419 | 110 Gallon Nalgene Horizontal XLPE Tank without legs | EACH | 410.00 | 410.00 | 410.00 | 410.00 | 410.00 |
| 19 | 04 | 0420 | 200 Gallon Nalgene Horizontal XLPE Tank without legs | EACH | 510.00 | 510.00 | 510.00 | 510.00 | 510.00 |
| 19 | 04 | 0421 | 300 Gallon Nalgene Horizontal XLPE Tank without legs | EACH | 555.00 | 555.00 | 555.00 | 555.00 | 555.00 |
| 19 | 04 | 0422 | 500 Gallon Nalgene Horizontal XLPE Tank without legs | EACH | 695.00 | 695.00 | 695.00 | 695.00 | 695.00 |

**Environmental Remediation: Assemblies Cost Book**

# Cost Data for Remediation

## Ultraviolet Oxidation

| Assembly | Description | Unit | Unit Cost by Safety Level | | | | |
|---|---|---|---|---|---|---|---|
| | | | A | B | C | D | E |
| | **Capital Costs** | | | | | | |
| 19 04 0423 | 1,000 Gallon Nalgene Horizontal XLPE Tank without legs | EACH | 1,495 | 1,495 | 1,495 | 1,495 | 1,495 |
| 19 04 0424 | 1,650 Gallon Nalgene Horizontal XLPE Tank without legs | EACH | 1,995 | 1,995 | 1,995 | 1,995 | 1,995 |
| 19 04 0425 | 2,500 Gallon Nalgene Horizontal XLPE Tank without legs | EACH | 3,250 | 3,250 | 3,250 | 3,250 | 3,250 |
| 19 04 0426 | 55 Gallon Tank Fiberglass Saddle | EACH | 295.00 | 295.00 | 295.00 | 295.00 | 295.00 |
| 19 04 0427 | 110 Gallon Tank Fiberglass Saddle | EACH | 410.00 | 410.00 | 410.00 | 410.00 | 410.00 |
| 19 04 0428 | 200 Gallon Tank Fiberglass Saddle | EACH | 585.00 | 585.00 | 585.00 | 585.00 | 585.00 |
| 19 04 0429 | 300 Gallon Tank Fiberglass Saddle | EACH | 995.00 | 995.00 | 995.00 | 995.00 | 995.00 |
| 19 04 0430 | 500 Gallon Tank Fiberglass Saddle | EACH | 1,995 | 1,995 | 1,995 | 1,995 | 1,995 |
| 19 04 0431 | 60 Gallon Nalgene Horizontal XLPE Tank with legs | EACH | 531.94 | 487.69 | 467.40 | 424.55 | 409.70 |
| 19 04 0432 | 125 Gallon Nalgene Horizontal XLPE Tank with legs | EACH | 636.94 | 592.69 | 572.40 | 529.55 | 514.70 |
| 19 04 0433 | 225 Gallon Nalgene Horizontal XLPE Tank with legs | EACH | 791.94 | 747.69 | 727.40 | 684.55 | 669.70 |
| 19 04 0434 | 300 Gallon Nalgene Horizontal XLPE Tank with legs | EACH | 811.94 | 767.69 | 747.40 | 704.55 | 689.70 |
| 19 04 0435 | 500 Gallon Nalgene Horizontal XLPE Tank with legs | EACH | 916.94 | 872.69 | 852.40 | 809.55 | 794.70 |
| 19 04 0436 | 1,575 Gallon Conical Bottom Vertical XLPE Tank | EACH | 2,441 | 2,386 | 2,361 | 2,307 | 2,288 |
| 19 04 0437 | 2,200 Gallon Conical Bottom Vertical XLPE Tank | EACH | 2,886 | 2,831 | 2,806 | 2,752 | 2,733 |
| 19 04 0438 | 2,600 Gallon Conical Bottom Vertical XLPE Tank | EACH | 3,496 | 3,441 | 3,416 | 3,362 | 3,343 |
| 19 04 0439 | 3,000 Gallon Conical Bottom Vertical XLPE Tank | EACH | 3,945 | 3,879 | 3,849 | 3,784 | 3,762 |
| 19 04 0440 | 4,200 Gallon Conical Bottom Vertical XLPE Tank | EACH | 5,595 | 5,529 | 5,499 | 5,434 | 5,412 |
| 19 04 0441 | 6,000 Gallon Conical Bottom Vertical XLPE Tank | EACH | 7,795 | 7,729 | 7,699 | 7,634 | 7,612 |
| 19 04 0442 | 8,000 Gallon Conical Bottom Vertical XLPE Tank | EACH | 10,495 | 10,429 | 10,399 | 10,334 | 10,312 |
| 19 04 0443 | 1,575 Gallon Conical Tank Stand | EACH | 1,495 | 1,495 | 1,495 | 1,495 | 1,495 |
| 19 04 0444 | 2,200 Gallon Conical Tank Stand | EACH | 1,695 | 1,695 | 1,695 | 1,695 | 1,695 |
| 19 04 0445 | 2,600 Gallon Conical Tank Stand | EACH | 1,795 | 1,795 | 1,795 | 1,795 | 1,795 |
| 19 04 0446 | 3,000 Gallon Conical Tank Stand | EACH | 1,795 | 1,795 | 1,795 | 1,795 | 1,795 |
| 19 07 0120 | 1 1/4", 60 PSI, Polyethylene Pipe | LF | 5.29 | 4.21 | 3.72 | 2.68 | 2.32 |
| 19 07 0121 | 1 1/2", 60 PSI, Polyethylene Pipe | LF | 5.38 | 4.30 | 3.81 | 2.77 | 2.41 |
| 19 07 0122 | 2", 60 PSI, Polyethylene Pipe | LF | 6.05 | 4.85 | 4.31 | 3.15 | 2.75 |
| 19 07 0123 | 3" Polyethylene 40' Joints, 60 PSI | LF | 7.90 | 6.47 | 5.82 | 4.42 | 3.94 |
| 19 07 0124 | 4" Polyethylene 40' Joints, 60 PSI | LF | 19.10 | 15.75 | 13.90 | 10.64 | 9.71 |
| 19 07 0125 | 6" Polyethylene 40' Joints, 60 PSI | LF | 25.85 | 22.22 | 20.22 | 16.68 | 15.67 |
| 19 07 0126 | 8" Polyethylene 40' Joints, 60 PSI | LF | 35.37 | 31.01 | 28.61 | 24.37 | 23.16 |
| 33 12 0801 | Cavitation System Treatability Study | EACH | 2,495 | 2,495 | 2,495 | 2,495 | 2,495 |
| 33 12 0802 | Peroxide Treatability/Pilot Plant Study | EACH | 6,100 | 6,100 | 6,100 | 6,100 | 6,100 |

## Environmental Remediation: Assemblies Cost Book

*Page 3-332*

# Cost Data for Remediation

## Ultraviolet Oxidation

| Assembly | Description | Unit | A | B | C | D | E |
|---|---|---|---|---|---|---|---|
| | **Capital Costs** | | | | | | |
| 33 12 0803 | Peroxide System Mob/Assembly/Shakedown | EA | 8,320 | 8,320 | 8,320 | 8,320 | 8,320 |
| 33 12 0804 | Ozone System On-site Operator Training | EACH | 7,838 | 6,042 | 5,273 | 3,537 | 2,900 |
| 33 12 0805 | Operator Health and Safety Course | EACH | 2,699 | 2,161 | 1,931 | 1,410 | 1,219 |
| 33 12 0806 | Supervisor Health and Safety Course | EACH | 843.78 | 664.17 | 587.27 | 413.66 | 350.00 |
| 33 12 0807 | Cavitation Permit & Regulatory Fee | EACH | 20,000 | 20,000 | 20,000 | 20,000 | 20,000 |
| 33 12 0808 | Cavitation System Demobilization & Disposal | EACH | 83,566 | 67,302 | 56,986 | 41,070 | 37,384 |
| 33 12 0809 | 5 KW High Intensity Ultraviolet, H2O2 Capital Equipment | EACH | 59,866 | 56,865 | 54,962 | 52,025 | 51,345 |
| 33 12 0810 | 10 KW High Intensity Ultraviolet, H2O2 Capital Equipment | EACH | 69,150 | 65,789 | 63,657 | 60,368 | 59,606 |
| 33 12 0811 | 15 KW High Intensity Ultraviolet, H2O2 Capital Equipment | EACH | 76,214 | 72,507 | 70,156 | 66,528 | 65,688 |
| 33 12 0812 | 20 KW High Intensity Ultraviolet, H2O2 Capital Equipment | EACH | 83,442 | 79,376 | 76,796 | 72,817 | 71,896 |
| 33 12 0813 | 25 KW High Intensity Ultraviolet, H2O2 Capital Equipment | EACH | 90,808 | 86,386 | 83,580 | 79,252 | 78,250 |
| 33 12 0814 | 30 KW High Intensity Ultraviolet, H2O2 Capital Equipment | EACH | 97,862 | 93,105 | 90,088 | 85,434 | 84,356 |
| 33 12 0815 | 35 KW High Intensity Ultraviolet, H2O2 Capital Equipment | EACH | 105,028 | 99,935 | 96,705 | 91,721 | 90,567 |
| 33 12 0816 | 40 KW High Intensity Ultraviolet, H2O2 Capital Equipment | EACH | 112,447 | 106,967 | 103,491 | 98,128 | 96,886 |
| 33 12 0817 | 45 KW High Intensity Ultraviolet, H2O2 Capital Equipment | EACH | 119,508 | 113,712 | 110,036 | 104,365 | 103,052 |
| 33 12 0818 | 50 KW High Intensity Ultraviolet, H2O2 Capital Equipment | EACH | 126,758 | 120,609 | 116,709 | 110,692 | 109,299 |
| 33 12 0819 | 60 KW High Intensity Ultraviolet, H2O2 Capital Equipment | EACH | 140,951 | 134,138 | 129,816 | 123,148 | 121,604 |
| 33 12 0820 | 70 KW High Intensity Ultraviolet, H2O2 Capital Equipment | EACH | 155,477 | 147,952 | 143,179 | 135,814 | 134,109 |
| 33 12 0821 | 80 KW High Intensity Ultraviolet, H2O2 Capital Equipment | EACH | 170,044 | 161,778 | 156,535 | 148,447 | 146,574 |
| 33 12 0822 | 90 KW High Intensity Ultraviolet, H2O2 Capital Equipment | EACH | 184,699 | 175,696 | 169,985 | 161,174 | 159,134 |
| 33 12 0823 | 100 KW High Intensity Ultraviolet, H2O2 Capital Equipment | EACH | 199,130 | 189,434 | 183,284 | 173,795 | 171,598 |
| 33 12 0824 | 125 KW High Intensity Ultraviolet, H2O2 Capital Equipment | EACH | 235,263 | 223,804 | 216,536 | 205,322 | 202,725 |
| 33 12 0825 | 150 KW High Intensity Ultraviolet, H2O2 Capital Equipment | EACH | 271,725 | 258,457 | 250,041 | 237,057 | 234,050 |
| 33 12 0826 | 175 KW High Intensity Ultraviolet, H2O2 Capital Equipment | EACH | 299,025 | 285,757 | 277,341 | 264,357 | 261,350 |
| 33 12 0827 | 200 KW High Intensity Ultraviolet, H2O2 Capital Equipment | EACH | 344,052 | 327,246 | 316,585 | 300,139 | 296,330 |
| 33 12 0828 | 225 KW High Intensity Ultraviolet, H2O2 Capital Equipment | EACH | 380,502 | 361,829 | 349,984 | 331,709 | 327,478 |
| 33 12 0829 | 250 KW High Intensity Ultraviolet, H2O2 Capital Equipment | EACH | 415,502 | 395,335 | 382,543 | 362,806 | 358,236 |
| 33 12 0830 | 10 GPM Low Energy Cavitation System Capital Equipment | EACH | 54,321 | 50,501 | 48,079 | 44,341 | 43,475 |
| 33 12 0831 | 25 GPM Low Energy Cavitation System Capital Equipment | EACH | 73,328 | 68,235 | 65,005 | 60,021 | 58,867 |
| 33 12 0832 | 10 GPM High Energy Cavitation System Capital Equipment | EACH | 55,121 | 51,301 | 48,879 | 45,141 | 44,275 |
| 33 12 0833 | 25 GPM High Energy Cavitation System Capital Equipment | EACH | 73,728 | 68,635 | 65,405 | 60,421 | 59,267 |
| 33 12 0834 | 20 GPM Ultraviolet Reactor, Ozone and Peroxide Capital Equipment | EACH | 73,364 | 69,657 | 67,306 | 63,678 | 62,838 |

**Environmental Remediation: Assemblies Cost Book**

# Cost Data for Remediation

## Ultraviolet Oxidation

| Assembly | Description | Unit | Unit Cost by Safety Level A | B | C | D | E |
|---|---|---|---|---|---|---|---|
| | **Capital Costs** | | | | | | |
| 33 12 0835 | 100 GPM Ultraviolet Reactor, Ozone and Peroxide Capital Equipment | EACH | 163,944 | 155,941 | 150,865 | 143,033 | 141,219 |
| 33 12 0836 | 250 GPM Ultraviolet Reactor, Ozone and Peroxide Capital Equipment | EACH | 85,725 | 72,457 | 64,041 | 51,057 | 48,050 |
| 33 12 0837 | Cavitation System Peroxide Consumption | LB | 0.43 | 0.43 | 0.43 | 0.43 | 0.43 |
| 33 12 0838 | 20 GPM Ozone System Consumables | WK | 250.00 | 250.00 | 250.00 | 250.00 | 250.00 |
| 33 12 0839 | 100 GPM Ozone System Consumables | WK | 400.00 | 400.00 | 400.00 | 400.00 | 400.00 |
| 33 12 0840 | 250 GPM Ozone System Consumables | WK | 500.00 | 500.00 | 500.00 | 500.00 | 500.00 |
| 33 12 0845 | Ozone System Demobilization | EACH | 14,031 | 11,199 | 9,402 | 6,630 | 5,988 |
| 33 12 0847 | Peroxide System Sodium Hydroxide | KGAL | 0.49 | 0.49 | 0.49 | 0.49 | 0.49 |
| 33 12 0848 | Fugitive Emission Control System | EACH | 26,121 | 24,886 | 24,102 | 22,893 | 22,613 |
| 33 12 0849 | Fugitive Emission System Maintenance | MONTH | 2,996 | 2,309 | 2,015 | 1,352 | 1,108 |
| 33 12 0850 | 60 watt Ultraviolet Source Low Intensity Lamp | EACH | 95.00 | 95.00 | 95.00 | 95.00 | 95.00 |
| 33 12 0851 | 7.5 KW Ultraviolet Source High Intensity Lamp | EACH | 150.00 | 150.00 | 150.00 | 150.00 | 150.00 |
| 33 12 0852 | 20 KW Ultraviolet Source High Intensity Lamp | EACH | 425.00 | 425.00 | 425.00 | 425.00 | 425.00 |
| 33 12 0853 | 5 KW Ultraviolet Source High Intensity Lamp | EACH | 125.00 | 125.00 | 125.00 | 125.00 | 125.00 |
| 33 12 0854 | 15 KW Ultraviolet Source High Intensity Lamp | EACH | 300.00 | 300.00 | 300.00 | 300.00 | 300.00 |
| 33 12 0855 | 10 GPM Cavitation System Maintenance | EACH | 1,100 | 1,100 | 1,100 | 1,100 | 1,100 |
| 33 12 0856 | 25 GPM Cavitation System Maintenance | EACH | 1,500 | 1,500 | 1,500 | 1,500 | 1,500 |
| 33 12 0860 | Ozone System Operational/Sampling Labor | HR | 84.98 | 65.51 | 57.17 | 38.35 | 31.44 |
| 33 12 0861 | Ozone System Lamp Replacement (Installed) | EACH | 76.63 | 76.63 | 76.63 | 76.63 | 76.63 |
| 33 12 0862 | 20 GPM Ozone System Maintenance & Repair | EACH | 11,867 | 9,233 | 8,106 | 5,560 | 4,627 |
| 33 12 0863 | 100 GPM Ozone System Maintenance & Repair | EACH | 60,218 | 46,830 | 41,099 | 28,159 | 23,415 |
| 33 12 0864 | 250 GPM Ozone System Maintenance & Repair | EACH | 120,536 | 93,761 | 82,298 | 56,418 | 46,929 |
| 33 12 0865 | Peroxide System Sulfuric Acid Consumption | KGAL | 0.15 | 0.15 | 0.15 | 0.15 | 0.15 |
| 33 12 0866 | System Operational/Sampling Labor | HR | 56.66 | 43.67 | 38.11 | 25.56 | 20.96 |
| 33 12 0867 | Peroxide System Demobilizaton | EACH | 35,000 | 35,000 | 35,000 | 35,000 | 35,000 |
| 33 13 0107 | 5 GPM Polypropylene (PP) Filter Chamber | EACH | 61.43 | 59.89 | 59.24 | 57.76 | 57.21 |
| 33 13 0108 | 5 GPM Polypropylene/Styrene Filter Chamber | EACH | 61.43 | 59.89 | 59.24 | 57.76 | 57.21 |
| 33 13 0109 | 5 GPM Nylon Filter Chamber | EACH | 76.75 | 75.21 | 74.56 | 73.07 | 72.53 |
| 33 13 0110 | 15 GPM CPVC Filter Chamber | EACH | 994.67 | 989.02 | 986.60 | 981.15 | 979.15 |
| 33 13 0111 | 30 GPM CPVC Filter Chamber | EACH | 1,084 | 1,078 | 1,076 | 1,070 | 1,068 |
| 33 13 0112 | 45 GPM CPVC Filter Chamber | EACH | 1,197 | 1,191 | 1,189 | 1,183 | 1,181 |
| 33 13 0113 | 25 GPM Polypropylene Filter Chamber | EACH | 1,245 | 1,240 | 1,237 | 1,232 | 1,230 |

**Environmental Remediation: Assemblies Cost Book**

*Page 3-334*

1998 by ECHOS. All rights reserved

# Cost Data for Remediation

## Ultraviolet Oxidation

| Assembly | Description | Unit | Unit Cost by Safety Level | | | | |
|---|---|---|---|---|---|---|---|
| | | | A | B | C | D | E |
| | **Capital Costs** | | | | | | |
| 33 13 0114 | 42 GPM Polypropylene Filter Chamber | EACH | 1,318 | 1,312 | 1,310 | 1,304 | 1,302 |
| 33 13 0115 | 60 GPM Polypropylene Filter Chamber | EACH | 1,358 | 1,353 | 1,350 | 1,345 | 1,343 |
| 33 13 0116 | 0 - 50 GPM Cartridge Filter Equipment | EACH | 2,014 | 1,992 | 1,982 | 1,960 | 1,952 |
| 33 13 0117 | 50 - 100 GPM Cartridge Filter Equipment | EACH | 3,930 | 3,907 | 3,898 | 3,876 | 3,868 |
| 33 13 0118 | Replacement Cartridge Filter up to 100 GPM | EACH | 198.26 | 196.73 | 196.07 | 194.59 | 194.04 |
| 33 19 7202 | Landfill Hazardous Solid Waste, 55 Gallon Drum | EACH | 87.67 | 87.67 | 87.67 | 87.67 | 87.67 |
| 33 26 0101 | 1" Carbon Steel Piping | LF | 6.46 | 5.28 | 4.77 | 3.63 | 3.21 |
| 33 26 0102 | 2" Carbon Steel Piping | LF | 10.50 | 8.72 | 7.95 | 6.23 | 5.60 |
| 33 26 0103 | 3" Carbon Steel Piping | LF | 18.02 | 15.20 | 13.99 | 11.26 | 10.26 |
| 33 26 0104 | 4" Carbon Steel Piping | LF | 23.20 | 19.86 | 18.43 | 15.21 | 14.02 |
| 33 26 0105 | 6" Carbon Steel Piping | LF | 36.37 | 30.24 | 27.46 | 21.52 | 19.44 |
| 33 26 0106 | 8" Carbon Steel Piping | LF | 40.77 | 34.08 | 31.15 | 24.68 | 22.36 |
| 33 26 0107 | 10" Carbon Steel Piping | LF | 54.40 | 46.05 | 42.38 | 34.30 | 31.39 |
| 33 26 0201 | 1" Stainless Steel Piping, Schedule 40, Threaded | LF | 18.67 | 16.12 | 15.03 | 12.56 | 11.65 |
| 33 26 0202 | 2" Stainless Steel Piping, Schedule 40, Threaded | LF | 33.47 | 29.21 | 27.39 | 23.28 | 21.77 |
| 33 26 0203 | 3" Stainless Steel Piping, Schedule 40, Threaded | LF | 57.00 | 50.61 | 47.88 | 41.71 | 39.45 |
| 33 26 0204 | 4" Stainless Steel Piping, Schedule 40, Threaded | LF | 80.15 | 71.72 | 68.12 | 59.98 | 57.00 |
| 33 26 0205 | 6" Stainless Steel Piping, Schedule 10, Type 316 | LF | 78.58 | 70.01 | 66.14 | 57.84 | 54.94 |
| 33 26 0206 | 8" Stainless Steel Piping, Schedule 40, Welded | LF | 197.22 | 183.32 | 176.89 | 163.42 | 158.80 |
| 33 26 0601 | (1 1/2", 3") Stainless Steel Double-wall Piping, with Fittings | LF | 134.61 | 117.05 | 109.53 | 92.56 | 86.33 |
| 33 26 0602 | (2", 4") Stainless Steel Double-wall Piping, with Fittings | LF | 185.98 | 163.46 | 153.81 | 132.04 | 124.05 |
| 33 26 0603 | (2 1/2", 4") Stainless Steel Double-wall Piping, with Fittings | LF | 206.86 | 182.17 | 171.60 | 147.74 | 138.99 |
| 33 26 0604 | (4", 6") Stainless Steel Double-wall Piping, with Fittings | LF | 339.64 | 280.79 | 254.18 | 197.21 | 177.23 |
| 33 26 0621 | (1 1/2", 3") PVC Double-wall Piping, with Fittings | LF | 33.15 | 27.05 | 24.43 | 18.53 | 16.36 |
| 33 26 0622 | (2", 4") PVC Double-wall Piping, with Fittings | LF | 46.22 | 37.98 | 34.43 | 26.46 | 23.55 |
| 33 26 0623 | (2 1/2", 4") PVC Double-wall Piping, with Fittings | LF | 49.22 | 40.39 | 36.61 | 28.07 | 24.94 |
| 33 26 0624 | (4", 6") PVC Double-wall Piping, with Fittings | LF | 70.61 | 58.00 | 52.59 | 40.40 | 35.93 |
| 33 26 0641 | (1 1/2", 3") Carbon Steel Double-wall Piping, with Fittings | LF | 48.16 | 40.64 | 37.42 | 30.15 | 27.48 |
| 33 26 0642 | (2", 4") Carbon Steel Double-wall Piping, with Fittings | LF | 64.71 | 55.69 | 51.83 | 43.11 | 39.91 |
| 33 26 0643 | (2 1/2", 4") Carbon Steel Double-wall Piping, with Fittings | LF | 70.81 | 61.02 | 56.83 | 47.37 | 43.90 |
| 33 26 0644 | (4", 6") Carbon Steel Double-wall Piping, with Fittings | LF | 114.39 | 100.33 | 94.32 | 80.73 | 75.75 |
| 33 29 0102 | 10 GPM, 1/2 HP, Centrifugal Pump | EACH | 1,063 | 952.32 | 905.07 | 798.39 | 759.27 |
| 33 29 0103 | 50 GPM, 100' Head, 3 HP, Centrifugal Pump | EACH | 2,658 | 2,500 | 2,432 | 2,279 | 2,223 |

## Environmental Remediation: Assemblies Cost Book

*Page 3-335*

1998 by ECHOS. All rights reserved

# Cost Data for Remediation

## Ultraviolet Oxidation

| Assembly | Description | Unit | Unit Cost by Safety Level | | | | |
|---|---|---|---|---|---|---|---|
| | | | A | B | C | D | E |
| | **Capital Costs** | | | | | | |
| 33 29 0104 | 75 GPM, 100' Head, 5 HP, Centrifugal Pump | EACH | 2,946 | 2,754 | 2,672 | 2,487 | 2,419 |
| 33 29 0105 | 5 HP, 100 GPM, Centrifugal Pump | EACH | 2,475 | 2,301 | 2,227 | 2,059 | 1,998 |
| 33 29 0108 | 10 HP, 200 GPM, Centrifugal Pump | EACH | 2,941 | 2,714 | 2,616 | 2,396 | 2,316 |
| 33 29 0110 | 15 HP, 300 GPM, Centrifugal Pump | EACH | 3,441 | 3,160 | 3,040 | 2,770 | 2,670 |
| 33 29 0111 | 20 HP, 500 GPM, Centrifugal Pump | EACH | 4,544 | 4,241 | 4,111 | 3,818 | 3,710 |
| 33 29 0112 | 30 HP, 750 GPM, Centrifugal Pump | EACH | 5,213 | 4,808 | 4,635 | 4,244 | 4,100 |
| 33 29 0113 | 40 HP, 1050 GPM, Centrifugal Pump | EACH | 5,655 | 5,199 | 5,004 | 4,564 | 4,403 |
| 33 29 0143 | 15 GPM, Centrifugal Pump, 6' Head, 1/8 HP, Bronze | EACH | 602.28 | 557.16 | 537.84 | 494.23 | 478.23 |
| 33 41 0101 | Pump & Motor Maintenance/Repair | EACH | 865.17 | 671.48 | 581.58 | 393.96 | 329.64 |
| 33 42 0101 | Electrical Charge | KWH | 0.06 | 0.06 | 0.06 | 0.06 | 0.06 |
| | **Operations & Maintenance** | | | | | | |
| 19 04 0401 | 550 Gallon, Stainless Steel Aboveground Wastewater Holding Tank, Rental | MONTH | 320.65 | 320.65 | 320.65 | 320.65 | 320.65 |
| 19 04 0402 | 630 Gallon, Polyethylene Aboveground Wastewater Holding Tank, Rental | MONTH | 320.65 | 320.65 | 320.65 | 320.65 | 320.65 |
| 19 04 0403 | 4,000 Gallon Polyethylene Wastewater Tank, Rental | MONTH | 545.90 | 545.90 | 545.90 | 545.90 | 545.90 |
| 19 04 0404 | 4,000 Gallon, Polyethylene Trailer-mounted Wastewater Holding Tank, Rental | MONTH | 1,124 | 1,124 | 1,124 | 1,124 | 1,124 |
| 19 04 0405 | 6,000 Gallon, Polyethylene Aboveground Wastewater Holding Tank, Rental | MONTH | 641.30 | 641.30 | 641.30 | 641.30 | 641.30 |
| 19 04 0406 | 21,000 Gallon Steel Wastewater Holding Tank, Rental | MONTH | 1,219 | 1,219 | 1,219 | 1,219 | 1,219 |
| 19 04 0408 | 21,000 Gallon Steel, Open Top, Tank Rental | MONTH | 1,124 | 1,124 | 1,124 | 1,124 | 1,124 |
| 33 12 0808 | Cavitation System Demobilization & Disposal | EACH | 83,566 | 67,302 | 56,986 | 41,070 | 37,384 |
| 33 12 0837 | Cavitation System Peroxide Consumption | LB | 0.43 | 0.43 | 0.43 | 0.43 | 0.43 |
| 33 12 0838 | 20 GPM Ozone System Consumables | WK | 250.00 | 250.00 | 250.00 | 250.00 | 250.00 |
| 33 12 0839 | 100 GPM Ozone System Consumables | WK | 400.00 | 400.00 | 400.00 | 400.00 | 400.00 |
| 33 12 0840 | 250 GPM Ozone System Consumables | WK | 500.00 | 500.00 | 500.00 | 500.00 | 500.00 |
| 33 12 0845 | Ozone System Demobilization | EACH | 14,031 | 11,199 | 9,402 | 6,630 | 5,988 |
| 33 12 0847 | Peroxide System Sodium Hydroxide | KGAL | 0.49 | 0.49 | 0.49 | 0.49 | 0.49 |
| 33 12 0849 | Fugitive Emission System Maintenance | MONTH | 2,996 | 2,309 | 2,015 | 1,352 | 1,108 |
| 33 12 0850 | 60 watt Ultraviolet Source Low Intensity Lamp | EACH | 95.00 | 95.00 | 95.00 | 95.00 | 95.00 |
| 33 12 0851 | 7.5 KW Ultraviolet Source High Intensity Lamp | EACH | 150.00 | 150.00 | 150.00 | 150.00 | 150.00 |
| 33 12 0852 | 20 KW Ultraviolet Source High Intensity Lamp | EACH | 425.00 | 425.00 | 425.00 | 425.00 | 425.00 |
| 33 12 0853 | 5 KW Ultraviolet Source High Intensity Lamp | EACH | 125.00 | 125.00 | 125.00 | 125.00 | 125.00 |
| 33 12 0854 | 15 KW Ultraviolet Source High Intensity Lamp | EACH | 300.00 | 300.00 | 300.00 | 300.00 | 300.00 |

**Environmental Remediation: Assemblies Cost Book**

# Cost Data for Remediation

## Ultraviolet Oxidation

| Assembly | Description | Unit | Unit Cost by Safety Level |  |  |  |  |
|---|---|---|---|---|---|---|---|
|  |  |  | A | B | C | D | E |
|  | **Operations & Maintenance** |  |  |  |  |  |  |
| 33 12 0855 | 10 GPM Cavitation System Maintenance | EACH | 1,100 | 1,100 | 1,100 | 1,100 | 1,100 |
| 33 12 0856 | 25 GPM Cavitation System Maintenance | EACH | 1,500 | 1,500 | 1,500 | 1,500 | 1,500 |
| 33 12 0860 | Ozone System Operational/Sampling Labor | HR | 84.98 | 65.51 | 57.17 | 38.35 | 31.44 |
| 33 12 0861 | Ozone System Lamp Replacement (Installed) | EACH | 76.63 | 76.63 | 76.63 | 76.63 | 76.63 |
| 33 12 0862 | 20 GPM Ozone System Maintenance & Repair | EACH | 11,867 | 9,233 | 8,106 | 5,560 | 4,627 |
| 33 12 0863 | 100 GPM Ozone System Maintenance & Repair | EACH | 60,218 | 46,830 | 41,099 | 28,159 | 23,415 |
| 33 12 0864 | 250 GPM Ozone System Maintenance & Repair | EACH | 120,536 | 93,761 | 82,298 | 56,418 | 46,929 |
| 33 12 0865 | Peroxide System Sulfuric Acid Consumption | KGAL | 0.15 | 0.15 | 0.15 | 0.15 | 0.15 |
| 33 12 0866 | System Operational/Sampling Labor | HR | 56.66 | 43.67 | 38.11 | 25.56 | 20.96 |
| 33 12 0867 | Peroxide System Demobilizaton | EACH | 35,000 | 35,000 | 35,000 | 35,000 | 35,000 |
| 33 13 0118 | Replacement Cartridge Filter up to 100 GPM | EACH | 198.26 | 196.73 | 196.07 | 194.59 | 194.04 |
| 33 41 0101 | Pump & Motor Maintenance/Repair | EACH | 865.17 | 671.48 | 581.58 | 393.96 | 329.64 |
| 33 42 0101 | Electrical Charge | KWH | 0.06 | 0.06 | 0.06 | 0.06 | 0.06 |

**Environmental Remediation: Assemblies Cost Book**

*Page 3-337*

# Cost Data for Remediation
## UST Closure

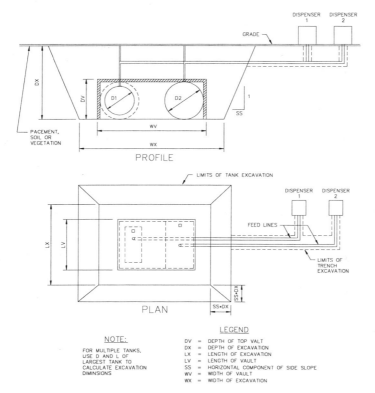

### General:

Permanent closure of an underground storage tank (UST) system can be accomplished either through disposal in-place or by physical removal. Tank removal is the preferred method of closure since it involves the removal of a contaminant source. Disposal in-place is appropriate when removal of the tanks may damage or undermine adjacent structures or when the presence of underground utilities render removal impractical.

### Typical Treatment Train:

Remediation of the excavated soil and/or groundwater, tank disposal, tank replacement, sampling and analysis.

### Common Cost Components:

1. Tank removal: excavate, pump remaining liquid, purge tanks of vapors, decontaminate tanks, remove tank and feed lines, demolish concrete vault (if present)
2. Dispose in-place: excavate to top of tank, pump remaining liquid, purge tanks of vapors, fill tanks and any containment vaults with inert substance
3. Demolition of existing cover and construction of replacement cover
4. Sidewall protection
5. Backfill
6. Monitoring wells

### Other Cost Considerations:

Load and haul, transportation.

**Environmental Remediation: Assemblies Cost Book**

# Cost Data for Remediation

## UST Closure

| Assembly | Description | Unit | Unit Cost by Safety Level A | B | C | D | E |
|---|---|---|---|---|---|---|---|
| | **Capital Costs** | | | | | | |
| 17 02 0201 | Demolish Bituminous Road with Power Equipment | CY | 46.17 | 36.60 | 30.96 | 21.62 | 19.19 |
| 17 02 0203 | Demolish Bituminous Pavement with Air Equipment | CY | 80.15 | 63.38 | 53.76 | 37.41 | 32.98 |
| 17 02 0205 | Demolish Unreinforced Concrete to 6" Thick with Air Equipment | CY | 126.90 | 100.35 | 85.12 | 59.24 | 52.23 |
| 17 02 0206 | Demolish Mesh Reinforced Concrete to 6" Thick with Air Equipment | CY | 183.47 | 145.09 | 123.07 | 85.65 | 75.51 |
| 17 02 0207 | Demolish Rod Reinforced Concrete to 6" Thick with Air Equipment | CY | 199.06 | 157.42 | 133.53 | 92.92 | 81.92 |
| 17 02 0208 | Demolish Mesh Reinforced Concrete to 6" Thick with Power Equipment | CY | 96.73 | 76.68 | 64.87 | 45.30 | 40.20 |
| 17 02 0209 | Demolish Rod Reinforced Concrete to 6" Thick with Power Equipment | CY | 106.91 | 84.75 | 71.70 | 50.07 | 44.43 |
| 17 02 0210 | Demolish Unreinforced Concrete 7" to 24" Thick with Power Equipment | CY | 258.44 | 204.87 | 173.31 | 121.03 | 107.40 |
| 17 02 0211 | Demolish Reinforced Concrete 7" to 24" Thick with Power Equipment | CY | 356.37 | 282.51 | 238.98 | 166.90 | 148.10 |
| 17 02 0610 | Piping Demolition | LF | 15.80 | 12.45 | 10.60 | 7.34 | 6.41 |
| 17 03 0277 | 2 CY, Crawler-mounted, Hydraulic Excavator | CY | 5.16 | 4.19 | 3.45 | 2.49 | 2.35 |
| 17 03 0278 | 3 CY, Crawler-mounted, Hydraulic Excavator | CY | 3.67 | 2.95 | 2.46 | 1.75 | 1.61 |
| 17 03 0279 | 4 CY, Crawler-mounted, Hydraulic Excavator | CY | 3.00 | 2.42 | 2.01 | 1.44 | 1.33 |
| 17 03 0415 | Backfill with Excavated Material | CY | 7.50 | 5.97 | 5.12 | 3.63 | 3.21 |
| 17 03 0422 | Unclassified Fill, 6" Lifts, On-Site | CY | 14.23 | 11.65 | 9.60 | 7.05 | 6.72 |
| 17 03 0423 | Unclassified Fill, 6" Lifts, Off-Site | CY | 11.08 | 9.81 | 8.83 | 7.58 | 7.40 |
| 17 03 0901 | Steel Sheeting, Pull & Salvage, to 15' | SF | 13.65 | 11.21 | 9.59 | 7.20 | 6.70 |
| 17 03 0902 | Steel Sheeting, Pull & Salvage, to 20' | SF | 14.11 | 11.61 | 9.95 | 7.50 | 6.99 |
| 17 03 0903 | Steel Sheeting, Pull & Salvage, to 25' | SF | 12.32 | 10.20 | 8.79 | 6.72 | 6.28 |
| 17 03 0904 | Steel Sheeting, Pull & Salvage, to 40' | SF | 14.52 | 12.05 | 10.40 | 7.98 | 7.46 |
| 17 03 1001 | Wellpoint for Trench, Install & Remove <500', 1 Month | LFHD | 55.06 | 55.06 | 55.06 | 55.06 | 55.06 |
| 17 03 1002 | 2" Diameter Contractor's Trash Pump, 75 GPM | DAY | 50.20 | 47.01 | 45.64 | 42.55 | 41.42 |
| 17 03 1003 | 3" Diameter Contractor's Trash Pump, 150 GPM | DAY | 61.72 | 57.05 | 55.05 | 50.54 | 48.88 |
| 17 03 1004 | 4" Diameter Contractor's Trash Pump, 300 GPM | DAY | 80.30 | 74.23 | 71.63 | 65.77 | 63.61 |
| 18 02 0301 | Asphalt Pavement - 10" Subgrade, 9" Base, 1 1/2" Topping | SY | 61.11 | 50.94 | 43.58 | 33.57 | 31.83 |
| 18 02 0321 | 6" Structural Slab on Grade | SF | 6.98 | 5.95 | 5.47 | 4.48 | 4.14 |
| 18 02 0322 | 8" Structural Slab on Grade | SF | 9.75 | 8.38 | 7.71 | 6.38 | 5.95 |
| 18 02 0324 | 12" Structural Slab on Grade | SF | 11.57 | 10.04 | 9.30 | 7.82 | 7.34 |
| 18 02 0331 | 6" Mesh Reinforced Slab on Grade | SF | 6.01 | 5.10 | 4.67 | 3.79 | 3.49 |

**Environmental Remediation: Assemblies Cost Book**

*Page 3-339*

1998 by ECHOS. All rights reserved

# Cost Data for Remediation

## UST Closure

| Assembly | Description | Unit | A | B | C | D | E |
|---|---|---|---|---|---|---|---|
| | **Capital Costs** | | | | | | |
| 18 02 0341 | 6" Unreinforced Slab on Grade | SF | 4.87 | 4.14 | 3.79 | 3.08 | 2.85 |
| 18 05 0402 | Seeding, Vegetative Cover | ACRE | 1,874 | 1,823 | 1,791 | 1,741 | 1,729 |
| 18 05 0405 | Sodding, Vegetative Cover | ACRE | 25,697 | 22,064 | 20,238 | 16,710 | 15,589 |
| 33 01 0101 | Mobilize/DeMobilize Drilling Rig & Crew | LS | 4,949 | 4,031 | 3,308 | 2,401 | 2,280 |
| 33 02 0303 | Organic Vapor Analyzer Rental, per Day | DAY | 184.30 | 184.30 | 184.30 | 184.30 | 184.30 |
| 33 10 0202 | Tank Purging with Dry Ice | KGAL | 58.05 | 45.00 | 39.03 | 26.40 | 22.01 |
| 33 10 0203 | Pump <=220 Gallon Liquid from Tank | GAL | 1.79 | 1.40 | 1.21 | 0.82 | 0.70 |
| 33 10 0204 | Pump 221 - 500 Gallon Liquid from Tank | GAL | 0.85 | 0.67 | 0.57 | 0.40 | 0.36 |
| 33 10 0205 | Pump 501 - 3,000 Gallon Liquid from Tank | GAL | 0.77 | 0.62 | 0.51 | 0.37 | 0.34 |
| 33 10 0206 | Pump >3000 Gallon Liquid from Tank | GAL | 0.48 | 0.39 | 0.32 | 0.23 | 0.22 |
| 33 10 0207 | Drain/Flush Liquids in Pipes, 1 Pipe | EACH | 192.52 | 180.27 | 175.02 | 163.18 | 158.83 |
| 33 10 9501 | Fill Tank/Vault with Sand | GAL | 0.46 | 0.38 | 0.32 | 0.24 | 0.23 |
| 33 10 9502 | Remove Steel/Fiberglass UST | EACH | 1,999 | 1,620 | 1,337 | 963.48 | 903.35 |
| 33 10 9503 | Concrete Tank/Vault Demolition, to 6" Thick Mesh Reinforced/Power | CY | 96.73 | 76.68 | 64.87 | 45.30 | 40.20 |
| 33 10 9504 | Concrete Tank/Vault Demolition, to 6" Thick Rod Reinforced/Power | CY | 106.91 | 84.75 | 71.70 | 50.07 | 44.43 |
| 33 17 0801 | Decontaminate Light Equipment | EACH | 189.11 | 146.65 | 127.13 | 86.01 | 71.80 |
| 33 17 0802 | Decontaminate Medium Equipment | EACH | 378.22 | 293.30 | 254.27 | 172.03 | 143.59 |
| 33 17 0803 | Decontaminate Heavy Equipment | EACH | 560.32 | 434.51 | 376.69 | 254.86 | 212.73 |
| 33 17 0804 | Decontaminate Tank By Steam Cleaning, 50 SF/Hour | SF | 2.51 | 1.95 | 1.69 | 1.15 | 0.97 |
| 33 17 0805 | Decontaminate Tank By High-pressure Wash, 40 SF/Hour | SF | 4.49 | 3.54 | 3.01 | 2.09 | 1.83 |
| 33 17 0806 | Decontaminate Tank By Sandblasting, 40 SF/Hour | SF | 2.52 | 1.99 | 1.74 | 1.23 | 1.06 |
| 33 23 0101 | 2" PVC, Schedule 40, Well Casing | LF | 17.45 | 14.39 | 11.98 | 8.95 | 8.55 |
| 33 23 0201 | 2" PVC, Schedule 40, Well Screen | LF | 23.47 | 19.52 | 16.42 | 12.52 | 12.00 |
| 33 23 1101 | Hollow-stem Auger, 8" Outside Diameter Borehole for 2" Well | LF | 89.98 | 73.28 | 60.15 | 43.66 | 41.45 |
| 33 23 1111 | Well Development Equipment Rental | WEEK | 478.91 | 462.93 | 456.09 | 440.65 | 434.98 |
| 33 23 1122 | Move Rig/Equipment Around Site | EACH | 154.66 | 125.95 | 103.39 | 75.04 | 71.24 |
| 33 23 1126 | Furnish 55 Gallon Drum for Drill Cuttings & Development Water | EACH | 65.19 | 65.19 | 65.19 | 65.19 | 65.19 |
| 33 23 2402 | Teflon Bailer, 3/4" Outside Diameter x 3', 180 cc | EACH | 227.74 | 227.74 | 227.74 | 227.74 | 227.74 |

**Environmental Remediation: Assemblies Cost Book**

*Page 3-340*

1998 by ECHOS. All rights reserved

# Cost Data for Remediation
## Well Drilling and Installation

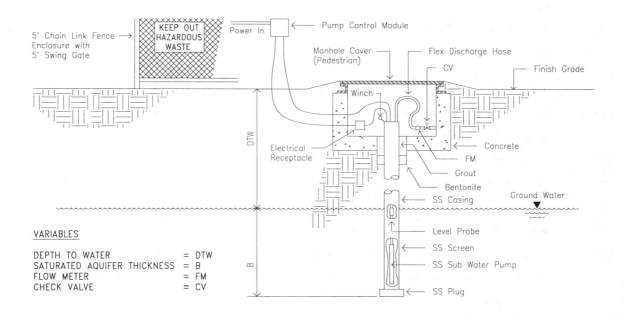

### General:

This section provides for the cost of drilling and installing vertical, slant, or horizontal wells. It does not include costs for technology-specific components, such as pumps, vacuum extraction blowers, or bailers. The only material costs in this model are well completion materials. Technology-specific components are addressed in the appropriate remediation sections.

### Typical Treatment Train:

This technology may be used alone or in conjunction with other remediation methods to estimate costs for the installation and operation of a subsurface remediation system employing slant or horizontal wells.

### Common Cost Components:

1. Mobilization/demobilization of drill rig and crew
2. Drilling, decontamination of equipment
3. Screen, casing, plug, grout, etc.
4. Well development
5. Soil sample collection during borehole advancement (vertical wells only)
6. Containment of drill cuttings

### Other Cost Considerations:

Trenching in sand, rock, or soil with boulders; pumps; vacuum extraction blowers; bailers; professional field labor; access road; clearing and grubbing; fencing.

**Environmental Remediation: Assemblies Cost Book**

# Cost Data for Remediation

## Well Drilling and Installation

| Assembly | Description | Unit | Unit Cost by Safety Level |  |  |  |  |
|---|---|---|---|---|---|---|---|
|  |  |  | A | B | C | D | E |
|  | **Capital Costs** |  |  |  |  |  |  |
| 17 03 0259 | Cat 225, 1.5 CY, Soil/Sand, Trenching | CY | 1.84 | 1.49 | 1.23 | 0.88 | 0.82 |
| 17 03 0263 | Cat 235, 2.5 CY, Soil/Sand, 10' - 20' Deep Trench Box, Trench | CY | 5.70 | 4.75 | 4.07 | 3.14 | 2.98 |
| 17 03 0273 | Pull Trench Box, Cat 235, 2.5 CY, Soil/Sand, Trenching | CY | 5.31 | 4.37 | 3.68 | 2.75 | 2.59 |
| 17 03 0401 | 950, 3.00 CY, Backfill with Excavated Material | CY | 2.54 | 2.05 | 1.70 | 1.22 | 1.14 |
| 17 03 0511 | Compact Soil with Vibrating Plate | CY | 7.20 | 5.57 | 4.84 | 3.26 | 2.70 |
| 17 03 0515 | Compact With Pogosticks | CY | 5.91 | 4.58 | 3.98 | 2.68 | 2.23 |
| 17 03 0901 | Steel Sheeting, Pull & Salvage, to 15' | SF | 13.65 | 11.21 | 9.59 | 7.20 | 6.70 |
| 17 03 0902 | Steel Sheeting, Pull & Salvage, to 20' | SF | 14.11 | 11.61 | 9.95 | 7.50 | 6.99 |
| 17 03 0903 | Steel Sheeting, Pull & Salvage, to 25' | SF | 12.32 | 10.20 | 8.79 | 6.72 | 6.28 |
| 17 03 0912 | Cat 235, 2.5 CY, Soil/Sand, < 15' Trench in Sheeting | CY | 4.02 | 3.25 | 2.69 | 1.93 | 1.80 |
| 17 03 0913 | Cat 235, 2.5 CY, Soil/Sand, 15' - 25' Trench in Sheeting | CY | 6.03 | 4.87 | 4.04 | 2.90 | 2.70 |
| 33 01 0101 | Mobilize/DeMobilize Drilling Rig & Crew | LS | 4,949 | 4,031 | 3,308 | 2,401 | 2,280 |
| 33 02 0303 | Organic Vapor Analyzer Rental, per Day | DAY | 184.30 | 184.30 | 184.30 | 184.30 | 184.30 |
| 33 02 0405 | Monitoring Well Slug Testing Equipment Rental | WEEK | 552.08 | 552.08 | 552.08 | 552.08 | 552.08 |
| 33 02 0601 | Drilling 2.5" Diameter Soil Borings, No Sampling | LF | 21.74 | 18.75 | 16.71 | 13.78 | 13.18 |
| 33 13 2320 | Installation Using Chain Trencher, Depth <= 4' | CY | 2.78 | 2.18 | 1.86 | 1.29 | 1.12 |
| 33 17 0801 | Decontaminate Light Equipment | EACH | 189.11 | 146.65 | 127.13 | 86.01 | 71.80 |
| 33 17 0802 | Decontaminate Medium Equipment | EACH | 378.22 | 293.30 | 254.27 | 172.03 | 143.59 |
| 33 17 0803 | Decontaminate Heavy Equipment | EACH | 560.32 | 434.51 | 376.69 | 254.86 | 212.73 |
| 33 17 0808 | Decontaminate Rig, Augers, Screen (Rental Equipment) | DAY | 205.34 | 205.34 | 205.34 | 205.34 | 205.34 |
| 33 17 0811 | Decontaminate Trenching Equipment | EACH | 3,835 | 3,345 | 3,029 | 2,549 | 2,441 |
| 33 23 0101 | 2" PVC, Schedule 40, Well Casing | LF | 17.45 | 14.39 | 11.98 | 8.95 | 8.55 |
| 33 23 0102 | 4" PVC, Schedule 40, Well Casing | LF | 26.98 | 22.38 | 18.77 | 14.24 | 13.63 |
| 33 23 0103 | 6" PVC, Schedule 40, Well Casing | LF | 27.07 | 22.66 | 19.19 | 14.84 | 14.25 |
| 33 23 0104 | 8" PVC, Schedule 40, Well Casing | LF | 34.50 | 29.27 | 25.15 | 19.98 | 19.29 |
| 33 23 0105 | 10" PVC, Schedule 40, Well Casing | LF | 38.28 | 33.04 | 28.92 | 23.75 | 23.06 |
| 33 23 0106 | 12" PVC, Schedule 40, Well Casing | LF | 44.32 | 39.08 | 34.97 | 29.80 | 29.10 |
| 33 23 0111 | 2" PVC, Schedule 80, Well Casing | LF | 17.78 | 14.72 | 12.31 | 9.28 | 8.88 |
| 33 23 0112 | 4" PVC, Schedule 80, Well Casing | LF | 28.55 | 23.95 | 20.34 | 15.81 | 15.20 |
| 33 23 0113 | 6" PVC, Schedule 80, Well Casing | LF | 47.52 | 40.17 | 34.40 | 27.14 | 26.17 |
| 33 23 0115 | 8" PVC, Schedule 80, Well Casing | LF | 61.99 | 52.81 | 45.58 | 36.51 | 35.30 |
| 33 23 0116 | 10" PVC, Schedule 80, Well Casing | LF | 65.25 | 56.06 | 48.84 | 39.77 | 38.56 |
| 33 23 0117 | 12" PVC, Schedule 80, Well Casing | LF | 68.60 | 59.41 | 52.19 | 43.12 | 41.91 |

## Environmental Remediation: Assemblies Cost Book

*Page 3-342*

1998 by ECHOS. All rights reserved

# Cost Data for Remediation

## Well Drilling and Installation

| Assembly | Description | Unit | A | B | C | D | E |
|---|---|---|---|---|---|---|---|
| | **Capital Costs** | | | | | | |
| 33 23 0121 | 2" Stainless Steel, Well Casing | LF | 31.75 | 28.14 | 25.30 | 21.73 | 21.25 |
| 33 23 0122 | 4" Stainless Steel, Well Casing | LF | 47.66 | 43.06 | 39.45 | 34.92 | 34.31 |
| 33 23 0123 | 6" Stainless Steel Well Casing, 5' Sections, Flush Threaded | LF | 370.73 | 311.96 | 265.77 | 207.74 | 199.94 |
| 33 23 0124 | 6" Stainless Steel Well Casing, 10' Sections, Flush Threaded | LF | 348.53 | 289.76 | 243.57 | 185.54 | 177.74 |
| 33 23 0125 | 8" Stainless Steel Well Casing, 5' Sections, Flush Threaded | LF | 526.42 | 452.94 | 395.18 | 322.61 | 312.87 |
| 33 23 0126 | 8" Stainless Steel Well Casing, 10' Sections, Flush Threaded | LF | 478.02 | 404.54 | 346.78 | 274.21 | 264.47 |
| 33 23 0127 | 10" Stainless Steel Well Casing, 5' Sections, Flush Threaded | LF | 533.65 | 460.17 | 402.41 | 329.84 | 320.10 |
| 33 23 0128 | 10" Stainless Steel Well Casing, 10' Sections, Flush Threaded | LF | 512.82 | 439.34 | 381.58 | 309.01 | 299.27 |
| 33 23 0129 | 12" Stainless Steel Well Casing, 5' Sections, Flush Threaded | LF | 561.57 | 488.09 | 430.33 | 357.76 | 348.02 |
| 33 23 0130 | 12" Stainless Steel Well Casing, 10' Sections, Flush Threaded | LF | 533.82 | 460.34 | 402.58 | 330.01 | 320.27 |
| 33 23 0131 | 10" Stainless Steel Well Casing, 5' Sections, Welded Ring | LF | 482.01 | 408.53 | 350.77 | 278.20 | 268.46 |
| 33 23 0132 | 10" Stainless Steel Well Casing, 10' Sections, Welded Ring | LF | 459.46 | 385.98 | 328.22 | 255.65 | 245.91 |
| 33 23 0133 | 12" Stainless Steel Well Casing, 5' Sections, Welded Ring | LF | 519.88 | 446.40 | 388.64 | 316.07 | 306.33 |
| 33 23 0134 | 12" Stainless Steel Well Casing, 10' Sections, Welded Ring | LF | 484.79 | 411.31 | 353.55 | 280.98 | 271.24 |
| 33 23 0135 | 2" PVC, Schedule 40, Horizontal Well Casing, Material Only | LF | 0.95 | 0.95 | 0.95 | 0.95 | 0.95 |
| 33 23 0136 | 4" PVC, Schedule 40, Horizontal Well Casing, Material Only | LF | 2.23 | 2.23 | 2.23 | 2.23 | 2.23 |
| 33 23 0137 | 6" PVC, Schedule 40, Horizontal Well Casing, Material Only | LF | 5.52 | 5.52 | 5.52 | 5.52 | 5.52 |
| 33 23 0138 | 8" PVC, Schedule 40, Horizontal Well Casing, Material Only | LF | 11.04 | 11.04 | 11.04 | 11.04 | 11.04 |
| 33 23 0139 | 2" Stainless Steel, Horizontal Well Casing, Material Only | LF | 10.41 | 10.41 | 10.41 | 10.41 | 10.41 |
| 33 23 0140 | 4" Stainless Steel, Horizontal Well Casing, Material Only | LF | 22.91 | 22.91 | 22.91 | 22.91 | 22.91 |
| 33 23 0141 | 6" Diameter, Stainless Steel, Horizontal Well Casing, Material Only | LF | 54.10 | 54.10 | 54.10 | 54.10 | 54.10 |
| 33 23 0142 | 8" Diameter, Stainless Steel, Horizontal Well Casing, Material Only | LF | 130.50 | 130.50 | 130.50 | 130.50 | 130.50 |
| 33 23 0143 | 2" High-density Polyethylene Horizontal Well Casing, Material Only | LF | 1.28 | 1.28 | 1.28 | 1.28 | 1.28 |
| 33 23 0144 | 4" High-density Polyethylene Horizontal Well Casing, Material Only | LF | 5.11 | 5.11 | 5.11 | 5.11 | 5.11 |
| 33 23 0145 | 6" High-density Polyethylene Horizontal Well Casing, Material Only | LF | 12.77 | 12.77 | 12.77 | 12.77 | 12.77 |
| 33 23 0146 | 8" High-density Polyethylene Horizontal Well Casing, Material Only | LF | 20.44 | 20.44 | 20.44 | 20.44 | 20.44 |
| 33 23 0147 | 2" PVC, Schedule 40, Well Casing, Slant Well | LF | 22.40 | 18.42 | 15.29 | 11.36 | 10.83 |

**Environmental Remediation: Assemblies Cost Book**

# Cost Data for Remediation

## Well Drilling and Installation

| Assembly | Description | Unit | A | B | C | D | E |
|---|---|---|---|---|---|---|---|
| | **Capital Costs** | | | | | | |
| 33 23 0148 | 4" PVC, Schedule 40, Well Casing, Slant Well | LF | 34.40 | 28.43 | 23.74 | 17.84 | 17.05 |
| 33 23 0150 | 2" PVC, Schedule 80, Well Casing, Slant Well | LF | 22.73 | 18.75 | 15.62 | 11.69 | 11.16 |
| 33 23 0151 | 4" PVC, Schedule 80, Well Casing, Slant Well | LF | 35.97 | 30.00 | 25.31 | 19.41 | 18.62 |
| 33 23 0153 | 2" Stainless Steel Well Casing, Slant Well | LF | 31.86 | 27.88 | 24.75 | 20.82 | 20.29 |
| 33 23 0154 | 4" Stainless Steel Well Casing, Slant Well | LF | 55.08 | 49.11 | 44.42 | 38.52 | 37.73 |
| 33 23 0201 | 2" PVC, Schedule 40, Well Screen | LF | 23.47 | 19.52 | 16.42 | 12.52 | 12.00 |
| 33 23 0202 | 4" PVC, Schedule 40, Well Screen | LF | 38.16 | 32.00 | 27.16 | 21.09 | 20.27 |
| 33 23 0203 | 6" PVC, Schedule 40, Well Screen | LF | 47.13 | 39.78 | 34.01 | 26.75 | 25.78 |
| 33 23 0204 | 8" PVC, Schedule 40, Well Screen | LF | 62.02 | 52.84 | 45.61 | 36.54 | 35.33 |
| 33 23 0205 | 10" PVC, Schedule 40, Well Screen | LF | 67.86 | 58.67 | 51.45 | 42.38 | 41.17 |
| 33 23 0206 | 12" PVC, Schedule 40, Well Screen | LF | 70.91 | 61.73 | 54.50 | 45.43 | 44.22 |
| 33 23 0211 | 2" PVC, Schedule 80, Well Screen | LF | 18.70 | 15.64 | 13.23 | 10.20 | 9.80 |
| 33 23 0212 | 4" PVC, Schedule 80, Well Screen | LF | 30.01 | 25.41 | 21.80 | 17.27 | 16.66 |
| 33 23 0213 | 6" PVC, Schedule 80, Well Screen | LF | 49.72 | 42.37 | 36.60 | 29.34 | 28.37 |
| 33 23 0214 | 8" PVC, Schedule 80, Well Screen | LF | 65.91 | 56.73 | 49.50 | 40.43 | 39.22 |
| 33 23 0215 | 10" PVC, Schedule 80, Well Screen | LF | 73.71 | 64.53 | 57.30 | 48.23 | 47.02 |
| 33 23 0216 | 12" PVC, Schedule 80, Well Screen | LF | 77.41 | 68.22 | 61.00 | 51.93 | 50.72 |
| 33 23 0221 | 2" Stainless Steel, Well Screen | LF | 26.91 | 23.84 | 21.44 | 18.41 | 18.01 |
| 33 23 0222 | 4" Stainless Steel, Well Screen | LF | 47.66 | 43.06 | 39.45 | 34.92 | 34.31 |
| 33 23 0223 | 6" Stainless Steel Well Screen, 5' Sec, Flush Threaded | LF | 366.05 | 307.28 | 261.09 | 203.06 | 195.26 |
| 33 23 0224 | 6" Stainless Steel Well Screen, 10' Sec, Flush Threaded | LF | 355.96 | 297.19 | 251.00 | 192.97 | 185.17 |
| 33 23 0225 | 8" Stainless Steel Well Screen, 5' Sec, Flush Threaded | LF | 462.22 | 388.74 | 330.98 | 258.41 | 248.67 |
| 33 23 0226 | 8" Stainless Steel Well Screen, 10' Sec, Flush Threaded | LF | 449.32 | 375.84 | 318.08 | 245.51 | 235.77 |
| 33 23 0227 | 10" Stainless Steel Well Screen, 5' Sec, Flush Threaded | LF | 496.10 | 422.62 | 364.86 | 292.29 | 282.55 |
| 33 23 0228 | 10" Stainless Steel Well Screen, 10' Sec, Flush Threaded | LF | 475.48 | 402.00 | 344.24 | 271.67 | 261.93 |
| 33 23 0229 | 12" Stainless Steel Well Screen, 5' Sec, Flush Threaded | LF | 519.16 | 445.68 | 387.92 | 315.35 | 305.61 |
| 33 23 0230 | 12" Stainless Steel Well Screen, 10' Sec, Flush Threaded | LF | 490.96 | 417.48 | 359.72 | 287.15 | 277.41 |
| 33 23 0231 | 10" Dia, SS Well Screen, 5' Sec, Welded Ring | LF | 436.05 | 355.88 | 291.43 | 212.17 | 202.43 |
| 33 23 0232 | 10" Stainless Steel Well Screen, 10' Sec, Welded Ring | LF | 465.38 | 391.90 | 334.14 | 261.57 | 251.83 |
| 33 23 0233 | 12" Stainless Steel Well Screen, 5' Sec, Welded Ring | LF | 506.81 | 433.33 | 375.57 | 303.00 | 293.26 |
| 33 23 0234 | 12" Stainless Steel Well Screen, 10' Sec, Welded Ring | LF | 484.80 | 411.32 | 353.56 | 280.99 | 271.25 |
| 33 23 0235 | 4" SDR 17, High-density Polyethylene, Well Screen | LF | 10.00 | 10.00 | 10.00 | 10.00 | 10.00 |
| 33 23 0236 | 6" SDR 17, High-density Polyethylene, Well Screen | LF | 12.00 | 12.00 | 12.00 | 12.00 | 12.00 |

**Environmental Remediation: Assemblies Cost Book**

*Page 3-344*

# Cost Data for Remediation

## Well Drilling and Installation

| | | | | Unit Cost by Safety Level | | | | |
|---|---|---|---|---|---|---|---|---|
| Assembly | Description | Unit | A | B | C | D | E |
| | **Capital Costs** | | | | | | |
| 33 23 0237 | 2" PVC, Schedule 40, Horizontal Well Screen, Material Only | LF | 1.70 | 1.70 | 1.70 | 1.70 | 1.70 |
| 33 23 0238 | 4" PVC, Schedule 40, Horizontal Well Screen, Material Only | LF | 3.73 | 3.73 | 3.73 | 3.73 | 3.73 |
| 33 23 0239 | 6" PVC, Schedule 40, Horizontal Well Screen, Material Only | LF | 7.54 | 7.54 | 7.54 | 7.54 | 7.54 |
| 33 23 0240 | 8" PVC, Schedule 40, Horizontal Well Screen, Material Only | LF | 12.53 | 12.53 | 12.53 | 12.53 | 12.53 |
| 33 23 0241 | 2" Stainless Steel Horizontal Well Screen, Material Only | LF | 10.41 | 10.41 | 10.41 | 10.41 | 10.41 |
| 33 23 0242 | 4" Stainless Steel Horizontal Well Screen, Material Only | LF | 22.91 | 22.91 | 22.91 | 22.91 | 22.91 |
| 33 23 0243 | 6" Diameter, Stainless Steel Horizontal Well Screen, Material Only | LF | 49.42 | 49.42 | 49.42 | 49.42 | 49.42 |
| 33 23 0244 | 8" Diameter, Stainless Steel Horizontal Well Screen, Material Only | LF | 66.30 | 66.30 | 66.30 | 66.30 | 66.30 |
| 33 23 0245 | 2" High-density Polyethylene, Horizontal Well Screen, Material Only | LF | 4.47 | 4.47 | 4.47 | 4.47 | 4.47 |
| 33 23 0246 | 4" High-density Polyethylene, Horizontal Well Screen, Material Only | LF | 10.60 | 10.60 | 10.60 | 10.60 | 10.60 |
| 33 23 0247 | 6" High-density Polyethylene, Horizontal Well Screen, Material Only | LF | 22.98 | 22.98 | 22.98 | 22.98 | 22.98 |
| 33 23 0248 | 8" High-density Polyethylene Horizontal Well Screen, Material Only | LF | 30.65 | 30.65 | 30.65 | 30.65 | 30.65 |
| 33 23 0249 | 2" PVC, Schedule 40, Horizontal Prepack, Material Only | LF | 19.16 | 19.16 | 19.16 | 19.16 | 19.16 |
| 33 23 0250 | 4" PVC, Schedule 40, Horizontal Prepack, Material Only | LF | 35.76 | 35.76 | 35.76 | 35.76 | 35.76 |
| 33 23 0251 | 6" PVC, Schedule 40, Horizontal Prepack, Material Only | LF | 51.08 | 51.08 | 51.08 | 51.08 | 51.08 |
| 33 23 0253 | 2" Stainless Steel, Horizontal Prepack, Material Only | LF | 63.86 | 63.86 | 63.86 | 63.86 | 63.86 |
| 33 23 0254 | 4" Stainless Steel, Horizontal Prepack, Material Only | LF | 108.56 | 108.56 | 108.56 | 108.56 | 108.56 |
| 33 23 0255 | 6" Stainless Steel, Horizontal Prepack, Material Only | LF | 223.50 | 223.50 | 223.50 | 223.50 | 223.50 |
| 33 23 0256 | 2" PVC, Schedule 40, Well Screen, Slant Well | LF | 29.37 | 24.23 | 20.20 | 15.13 | 14.44 |
| 33 23 0257 | 2" High-density Polyethylene, Horizontal Prepack, Material Only | LF | 19.16 | 19.16 | 19.16 | 19.16 | 19.16 |
| 33 23 0258 | 4" High-density Polyethylene, Horizontal Prepack, Material Only | LF | 35.76 | 35.76 | 35.76 | 35.76 | 35.76 |
| 33 23 0259 | 6" High-density Polyethylene, Horizontal Prepack, Material Only | LF | 57.47 | 57.47 | 57.47 | 57.47 | 57.47 |
| 33 23 0260 | 2" PVC, Schedule 80, Well Screen, Slant Well | LF | 48.27 | 40.08 | 33.63 | 25.54 | 24.45 |
| 33 23 0261 | 4" PVC, Schedule 40, Well Screen, Slant Well | LF | 70.53 | 58.96 | 49.86 | 38.44 | 36.90 |
| 33 23 0262 | 4" PVC, Schedule 80, Well Screen, Slant Well | LF | 47.39 | 39.49 | 33.29 | 25.49 | 24.44 |
| 33 23 0264 | 2" Stainless Steel Well Screen, Slant Well | LF | 38.08 | 32.94 | 28.91 | 23.84 | 23.15 |
| 33 23 0265 | 4" Stainless Steel Well Screen, Slant Well | LF | 65.45 | 57.55 | 51.35 | 43.55 | 42.50 |
| 33 23 0301 | 2" PVC, Well Plug | EACH | 29.37 | 24.77 | 21.16 | 16.63 | 16.02 |

**Environmental Remediation: Assemblies Cost Book**

# Cost Data for Remediation

## Well Drilling and Installation

| Assembly | Description | Unit | \multicolumn{5}{c}{Unit Cost by Safety Level} |
|---|---|---|---|---|---|---|---|
| | | | A | B | C | D | E |
| | **Capital Costs** | | | | | | |
| 33 23 0302 | 4" PVC, Well Plug | EACH | 55.65 | 48.91 | 43.62 | 36.97 | 36.07 |
| 33 23 0303 | 6" PVC, Well Plug | EACH | 112.63 | 101.15 | 92.13 | 80.79 | 79.27 |
| 33 23 0304 | 8" PVC, Well Plug | EACH | 132.70 | 119.58 | 109.26 | 96.31 | 94.57 |
| 33 23 0305 | 10" PVC, Well Plug | EACH | 156.31 | 141.01 | 128.97 | 113.85 | 111.82 |
| 33 23 0306 | 12" PVC, Well Plug | EACH | 197.87 | 179.50 | 165.06 | 146.92 | 144.48 |
| 33 23 0311 | 2" Stainless Steel, Well Plug | EACH | 83.24 | 74.06 | 66.83 | 57.76 | 56.55 |
| 33 23 0312 | 4" Stainless Steel, Well Plug | EACH | 112.24 | 102.03 | 94.01 | 83.93 | 82.58 |
| 33 23 0313 | 6" Stainless Steel, Well Plug | EACH | 573.04 | 481.18 | 408.98 | 318.27 | 306.09 |
| 33 23 0314 | 8" Stainless Steel, Well Plug | EACH | 702.53 | 597.56 | 515.04 | 411.38 | 397.46 |
| 33 23 0315 | 10" Stainless Steel, Well Plug | EACH | 891.17 | 768.70 | 672.43 | 551.48 | 535.25 |
| 33 23 0316 | 4" High-density Polyethylene, Well Plug | EACH | 70.64 | 70.64 | 70.64 | 70.64 | 70.64 |
| 33 23 0317 | 6" High-density Polyethylene, Well Plug | EACH | 86.08 | 86.08 | 86.08 | 86.08 | 86.08 |
| 33 23 0325 | 12" Stainless Steel, Well Plug | EACH | 1,023 | 879.90 | 767.56 | 626.43 | 607.48 |
| 33 23 0326 | 2" PVC Plug for Horizontal Well, Material Only | EACH | 9.24 | 9.24 | 9.24 | 9.24 | 9.24 |
| 33 23 0327 | 4" PVC Plug for Horizontal Well, Material Only | EACH | 29.33 | 29.33 | 29.33 | 29.33 | 29.33 |
| 33 23 0328 | 6" PVC Plug for Horizontal Well, Material Only | LF | 50.77 | 50.77 | 50.77 | 50.77 | 50.77 |
| 33 23 0329 | 8" PVC Plug for Horizontal Well, Material Only | EACH | 62.00 | 62.00 | 62.00 | 62.00 | 62.00 |
| 33 23 0330 | 2" Stainless Steel Plug for Horizontal, Material Only | EACH | 26.82 | 26.82 | 26.82 | 26.82 | 26.82 |
| 33 23 0331 | 4" Stainless Steel Plug for Horizontal, Material Only | EACH | 54.91 | 54.91 | 54.91 | 54.91 | 54.91 |
| 33 23 0332 | 6" Stainless Steel Plug for Horizontal, Material Only | LF | 102.17 | 102.17 | 102.17 | 102.17 | 102.17 |
| 33 23 0333 | 8" Stainless Steel Plug for Horizontal, Material Only | EACH | 212.00 | 212.00 | 212.00 | 212.00 | 212.00 |
| 33 23 0334 | 2" High-density Polyethylene Plug for Horizontal Well, Material Only | EACH | 6.39 | 6.39 | 6.39 | 6.39 | 6.39 |
| 33 23 0335 | 4" High-density Polyethylene Plug for Horizontal Well, Material Only | EACH | 29.37 | 29.37 | 29.37 | 29.37 | 29.37 |
| 33 23 0336 | 6" High-density Polyethylene Plug for Horizontal Well, Material Only | LF | 70.24 | 70.24 | 70.24 | 70.24 | 70.24 |
| 33 23 0337 | 8" High-density Polyethylene Plug for Horizontal Well, Material Only | EACH | 114.94 | 114.94 | 114.94 | 114.94 | 114.94 |
| 33 23 0338 | 2" PVC, Schedule 40, Well Plug, Slant Well | EACH | 41.41 | 35.44 | 30.75 | 24.85 | 24.06 |
| 33 23 0339 | 4" PVC, Schedule 40, Well Plug, Slant Well | EACH | 76.52 | 67.76 | 60.87 | 52.23 | 51.06 |
| 33 23 0341 | 2" Stainless Steel Well Plug, Slant Well | EACH | 91.17 | 79.22 | 69.84 | 58.04 | 56.46 |
| 33 23 0342 | 4" Stainless Steel Well Plug, Slant Well | EACH | 126.41 | 113.14 | 102.71 | 89.60 | 87.84 |
| 33 23 1101 | Hollow-stem Auger, 8" Outside Diameter Borehole for 2" Well | LF | 89.98 | 73.28 | 60.15 | 43.66 | 41.45 |

---

**Environmental Remediation: Assemblies Cost Book**

# Cost Data for Remediation

## Well Drilling and Installation

| Assembly | Description | Unit | Unit Cost by Safety Level | | | | |
|---|---|---|---|---|---|---|---|
| | | | A | B | C | D | E |
| | **Capital Costs** | | | | | | |
| 33 23 1102 | Hollow-stem Auger, 11" Outside Diameter Borehole for 4" Well | LF | 109.98 | 89.57 | 73.52 | 53.36 | 50.66 |
| 33 23 1103 | Hollow-stem Auger, 13 3/4" Outside Diameter Borehole for 6" Well | LF | 141.40 | 115.16 | 94.53 | 68.61 | 65.13 |
| 33 23 1105 | Hollow-stem Auger, 16" Outside Diameter Borehole for 8" Well | LF | 164.97 | 134.35 | 110.28 | 80.05 | 75.99 |
| 33 23 1106 | Split Spoon Sample, 2" x 24", During Drilling | EACH | 46.67 | 46.67 | 46.67 | 46.67 | 46.67 |
| 33 23 1107 | Coring (Fluid or Air) | LF | 102.14 | 88.16 | 78.67 | 64.96 | 62.17 |
| 33 23 1109 | Hollow-stem Auger, 8" Outside Diameter Borehole, Slant Well | LF | 38.31 | 38.31 | 38.31 | 38.31 | 38.31 |
| 33 23 1111 | Well Development Equipment Rental | WEEK | 478.91 | 462.93 | 456.09 | 440.65 | 434.98 |
| 33 23 1121 | Standby for Drilling | EACH | 618.63 | 503.81 | 413.56 | 300.17 | 284.95 |
| 33 23 1122 | Move Rig/Equipment Around Site | EACH | 154.66 | 125.95 | 103.39 | 75.04 | 71.24 |
| 33 23 1125 | Concrete Coring (Minimum Charge) | DAY | 2,475 | 2,015 | 1,654 | 1,201 | 1,140 |
| 33 23 1126 | Furnish 55 Gallon Drum for Drill Cuttings & Development Water | EACH | 65.19 | 65.19 | 65.19 | 65.19 | 65.19 |
| 33 23 1128 | Jackhammer Rental, per Day | DAY | 166.03 | 166.03 | 166.03 | 166.03 | 166.03 |
| 33 23 1129 | Concrete Saw Rental, per Day | DAY | 102.00 | 102.00 | 102.00 | 102.00 | 102.00 |
| 33 23 1151 | Mud Drilling, 4" Diameter Borehole | LF | 73.65 | 59.98 | 49.23 | 35.73 | 33.92 |
| 33 23 1152 | Mud Drilling, 6" Diameter Borehole | LF | 81.01 | 65.98 | 54.16 | 39.31 | 37.31 |
| 33 23 1153 | Mud Drilling, 8" Diameter Borehole | LF | 88.38 | 71.97 | 59.08 | 42.88 | 40.71 |
| 33 23 1154 | Mud Drilling, 10" Diameter Borehole | LF | 100.65 | 81.97 | 67.29 | 48.84 | 46.36 |
| 33 23 1155 | Mud Drilling, 12" Diameter Borehole | LF | 112.92 | 91.97 | 75.49 | 54.79 | 52.01 |
| 33 23 1156 | Mud Drilling, 15" Diameter Borehole | LF | 137.47 | 111.96 | 91.90 | 66.71 | 63.32 |
| 33 23 1157 | Bucket Auger, 18" Diameter Borehole | LF | 45.98 | 45.98 | 45.98 | 45.98 | 45.98 |
| 33 23 1161 | Air Rotary 6" Borehole in Unconsolidated | LF | 73.65 | 59.98 | 49.23 | 35.73 | 33.92 |
| 33 23 1162 | Air Rotary 8" Borehole in Unconsolidated | LF | 122.74 | 99.96 | 82.06 | 59.56 | 56.54 |
| 33 23 1163 | Air Rotary 10" Borehole in Unconsolidated | LF | 220.94 | 179.93 | 147.70 | 107.20 | 101.77 |
| 33 23 1164 | Air Rotary 12" Borehole in Unconsolidated | LF | 294.59 | 239.91 | 196.93 | 142.94 | 135.69 |
| 33 23 1165 | Air Rotary 16" Borehole in Unconsolidated | LF | 392.78 | 319.88 | 262.58 | 190.59 | 180.92 |
| 33 23 1168 | Air Rotary 6" Borehole in Consolidated | LF | 58.92 | 47.98 | 39.39 | 28.59 | 27.14 |
| 33 23 1169 | Air Rotary 8" Borehole in Consolidated | LF | 147.29 | 119.96 | 98.47 | 71.47 | 67.85 |
| 33 23 1170 | Air Rotary 10" Borehole in Consolidated | LF | 294.59 | 239.91 | 196.93 | 142.94 | 135.69 |
| 33 23 1171 | Air Rotary 12" Borehole in Consolidated | LF | 613.72 | 499.81 | 410.28 | 297.79 | 282.69 |
| 33 23 1172 | Air Rotary 4" Borehole, Slant Well | LF | 34.49 | 34.49 | 34.49 | 34.49 | 34.49 |

## Environmental Remediation: Assemblies Cost Book

*Page 3-347*

1998 by ECHOS. All rights reserved

# Cost Data for Remediation

## Well Drilling and Installation

| Assembly | Description | Unit | Unit Cost by Safety Level A | B | C | D | E |
|---|---|---|---|---|---|---|---|
| | **Capital Costs** | | | | | | |
| 33 23 1181 | Trenching for Horizontal Well Installation | LF | 31.93 | 31.93 | 31.93 | 31.93 | 31.93 |
| 33 23 1185 | Standby for Trenching | EACH | 383.54 | 334.53 | 302.86 | 254.87 | 244.12 |
| 33 23 1187 | Mobilize/DeMobilize Trencher/Crew | EACH | 38,354 | 33,453 | 30,286 | 25,487 | 24,412 |
| 33 23 1201 | Mobilize/Demobilize Directional Drill Rig | EACH | 14,000 | 14,000 | 14,000 | 14,000 | 14,000 |
| 33 23 1203 | Fluid Drilling to 400' Length, Consolidated, Continuous | LF | 44.70 | 44.70 | 44.70 | 44.70 | 44.70 |
| 33 23 1204 | Mud Drilling, 400 - 1,200' Length, Consolidated, Continuous | LF | 75.35 | 75.35 | 75.35 | 75.35 | 75.35 |
| 33 23 1205 | Mud Drilling, > 1,200' Length, Consolidated, Continuous | LF | 116.22 | 116.22 | 116.22 | 116.22 | 116.22 |
| 33 23 1206 | Fluid Drilling to 400' Length, Consolidated, Blind Well | LF | 60.03 | 60.03 | 60.03 | 60.03 | 60.03 |
| 33 23 1207 | Mud Drilling, 400 - 1,200' Length, Consolidated, Blind Well | LF | 102.17 | 102.17 | 102.17 | 102.17 | 102.17 |
| 33 23 1208 | Mud Drilling, > 1,200' Length, Consolidated, Blind Well | LF | 155.80 | 155.80 | 155.80 | 155.80 | 155.80 |
| 33 23 1209 | Fluid Drilling to 400' Length, Unconsolidated, Continuous | LF | 57.47 | 57.47 | 57.47 | 57.47 | 57.47 |
| 33 23 1210 | Mud Drilling, 400 - 1,200' Length, Unconsolidated, Continuous | LF | 90.68 | 90.68 | 90.68 | 90.68 | 90.68 |
| 33 23 1211 | Mud Drilling, > 1,200' Length, Unconsolidated, Continuous | LF | 140.48 | 140.48 | 140.48 | 140.48 | 140.48 |
| 33 23 1401 | 2" Screen, Filter Pack | LF | 16.49 | 13.88 | 11.84 | 9.27 | 8.92 |
| 33 23 1402 | 4" Screen, Filter Pack | LF | 29.10 | 24.50 | 20.89 | 16.36 | 15.75 |
| 33 23 1403 | 6" Screen, Filter Pack | LF | 42.19 | 35.53 | 30.29 | 23.72 | 22.83 |
| 33 23 1404 | Gravel Pack, 8" Well | LF | 29.47 | 23.93 | 19.82 | 14.36 | 13.48 |
| 33 23 1405 | Gravel Pack, 10" Well | LF | 43.84 | 35.59 | 29.45 | 21.33 | 20.02 |
| 33 23 1406 | Gravel Pack, 12" Well | LF | 54.65 | 44.40 | 36.79 | 26.70 | 25.07 |
| 33 23 1407 | Gravel Pack for Horizontal Well Installation | CF | 0.84 | 0.80 | 0.78 | 0.74 | 0.73 |
| 33 23 1408 | 2" Screen, Filter Pack, Slant Well | LF | 24.41 | 21.03 | 18.37 | 15.03 | 14.59 |
| 33 23 1409 | 4" Screen, Filter Pack, Slant Well | LF | 43.11 | 37.14 | 32.45 | 26.55 | 25.76 |
| 33 23 1502 | Surface Pad, Concrete, 4' x 4' x 4" | EACH | 24.21 | 21.82 | 20.74 | 18.43 | 17.61 |
| 33 23 1504 | Surface Pad, Concrete, 2' x 2' x 4" | EACH | 6.05 | 5.45 | 5.18 | 4.61 | 4.40 |
| 33 23 1811 | 2" Well, Portland Cement Grout | LF | 0.92 | 0.92 | 0.92 | 0.92 | 0.92 |
| 33 23 1812 | 4" Well, Portland Cement Grout | LF | 1.38 | 1.38 | 1.38 | 1.38 | 1.38 |
| 33 23 1813 | 6" Well, Portland Cement Grout | LF | 7.80 | 7.80 | 7.80 | 7.80 | 7.80 |
| 33 23 1814 | 8" Well, Portland Cement Grout | LF | 11.02 | 11.02 | 11.02 | 11.02 | 11.02 |
| 33 23 1815 | 10" Well, Portland Cement Grout | LF | 12.85 | 12.85 | 12.85 | 12.85 | 12.85 |
| 33 23 1816 | 12" Well, Portland Cement Grout | LF | 14.69 | 14.69 | 14.69 | 14.69 | 14.69 |
| 33 23 1820 | Grout Continuous Borehole | CF | 4.59 | 4.59 | 4.59 | 4.59 | 4.59 |
| 33 23 2101 | 2" Well, Bentonite Seal | EACH | 63.00 | 52.67 | 44.55 | 34.34 | 32.97 |
| 33 23 2102 | 4" Well, Bentonite Seal | EACH | 157.54 | 131.70 | 111.39 | 85.87 | 82.44 |

**Environmental Remediation: Assemblies Cost Book**

*Page 3-348*

# Cost Data for Remediation

## Well Drilling and Installation

| Assembly | Description | Unit | \multicolumn{5}{c}{Unit Cost by Safety Level} |
|---|---|---|---|---|---|---|---|
| | | | A | B | C | D | E |
| | **Capital Costs** | | | | | | |
| 33 23 2103 | 6" Well, Bentonite Seal | EACH | 252.01 | 210.68 | 178.18 | 137.37 | 131.89 |
| 33 23 2105 | 8" Well, Bentonite Seal | EACH | 346.60 | 289.75 | 245.06 | 188.92 | 181.39 |
| 33 23 2106 | 10" Well, Bentonite Seal | EACH | 441.13 | 368.78 | 311.90 | 240.45 | 230.86 |
| 33 23 2107 | 12" Well, Bentonite Seal | EACH | 535.66 | 447.80 | 378.74 | 291.98 | 280.33 |
| 33 23 2108 | 2" Well, Bentonite Seal, Slant Well | EACH | 84.05 | 70.61 | 60.05 | 46.78 | 45.00 |
| 33 23 2109 | 4" Well, Bentonite Seal, Slant Well | EACH | 195.66 | 166.46 | 143.50 | 114.66 | 110.79 |
| 33 23 2204 | Hazardous Area, Pedestrian Load, Well Protection | EACH | 6,261 | 5,266 | 4,505 | 3,524 | 3,380 |
| 33 23 2205 | Hazardous Area, Traffic Load, Well Protection | EACH | 6,266 | 5,271 | 4,510 | 3,529 | 3,384 |
| 33 23 2206 | Restricted Area, Well Protection (with 4 Posts & Explosionproof Receptacle) | EACH | 1,264 | 1,133 | 1,077 | 949.88 | 903.62 |
| 33 23 2213 | 8" x 7.5" Locking Manhole Cover, Watertight | EACH | 345.53 | 288.12 | 242.99 | 186.30 | 178.69 |
| 33 23 2214 | 12" x 7.5" Locking Manhole Cover, Watertight | EACH | 459.57 | 383.02 | 322.86 | 247.27 | 237.12 |
| 33 23 2215 | 4" x 4" x 5' Steel Protective Cover, Lockable | EACH | 539.95 | 448.09 | 375.89 | 285.18 | 273.00 |
| 33 23 2216 | 6" x 6" x 5' Steel Protective Cover, Lockable | EACH | 560.94 | 469.08 | 396.88 | 306.17 | 293.99 |
| 33 23 2217 | 8" x 8" x 5' Steel Protective Cover, Lockable | EACH | 620.41 | 528.55 | 456.35 | 365.64 | 353.46 |
| 33 23 2301 | 5' Guard Posts, Cast Iron, Concrete Fill | EACH | 98.83 | 83.39 | 76.77 | 61.84 | 56.37 |
| 33 26 0801 | 2" Slotted PVC Pipe | LF | 5.64 | 4.83 | 4.48 | 3.69 | 3.40 |
| 33 26 0802 | 4" Slotted PVC Pipe | LF | 8.67 | 7.65 | 7.22 | 6.23 | 5.87 |
| 33 26 0803 | 6" Slotted PVC Pipe | LF | 14.25 | 12.89 | 12.31 | 10.99 | 10.51 |
| 33 27 0112 | 2" PVC, 90 Degree, Elbow | EACH | 41.36 | 32.22 | 28.30 | 19.46 | 16.22 |
| 33 27 0114 | 4" PVC, 90 Degree, Elbow | EACH | 101.72 | 80.57 | 71.51 | 51.07 | 43.57 |
| 33 27 0115 | 6" PVC, 90 Degree, Elbow | EACH | 135.44 | 111.27 | 100.92 | 77.56 | 68.99 |
| 33 27 0116 | 8" PVC, 90 Degree, Elbow | EACH | 200.21 | 172.01 | 159.94 | 132.68 | 122.68 |
| 33 27 0144 | 8" Flange Assemblies, PVC, Schedule 40 | EACH | 31.29 | 31.29 | 31.29 | 31.29 | 31.29 |
| 33 27 0153 | 8" x 6" Reducer, PVC Schedule 40 | EACH | 22.98 | 22.98 | 22.98 | 22.98 | 22.98 |
| 33 27 0155 | 8" x 4" Reducer, PVC Schedule 40 | EACH | 67.28 | 67.28 | 67.28 | 67.28 | 67.28 |
| 33 27 0211 | 2" Stainless Steel, 90 Degree, Elbow | EACH | 114.28 | 93.45 | 84.53 | 64.39 | 57.01 |
| 33 27 0212 | 4" Stainless Steel, 90 Degree, Elbow | EACH | 391.27 | 338.50 | 315.91 | 264.90 | 246.20 |
| 33 27 0213 | 6" Stainless Steel, 90 Degree, Elbow | EACH | 425.65 | 376.17 | 355.00 | 307.18 | 289.64 |
| 33 27 0352 | 4" SDR-17 Flange Assemblies | EACH | 34.09 | 34.09 | 34.09 | 34.09 | 34.09 |
| 33 27 0353 | 6" SDR-17 Flange Assemblies | EACH | 71.20 | 71.20 | 71.20 | 71.20 | 71.20 |

**Environmental Remediation: Assemblies Cost Book**

# Cost Data for Remediation

## Access Roads

**General:**

Site accessibility is often a major concern in any site remediation project involving the use of heavy equipment. When a site is not accessible to equipment and vehicles, construction of an access road may be required. The access road may be one or two lanes, constructed of dirt, gravel, or asphalt. If the access road is constructed in the contaminated zone, the excavated material must be treated and disposed.

**Common Cost Components:**

1. Clearing
2. Excavation for road base, ditches, culverts
3. Subgrade: natural soil, compacted, or soil stabilization mixtures
4. Base: asphalt, cement stabilized coarse aggregate mixture, or graded, crushed gravel
5. Pavement

# Cost Data for Site Work

## Access Roads

| Assembly | Description | Unit | A | B | C | D | E |
|---|---|---|---|---|---|---|---|
| | **Capital Costs** | | | | | | |
| 17 01 0106 | Heavy Brush, Light Trees, Clear, Grub, Haul | ACRE | 11,338 | 9,081 | 7,594 | 5,382 | 4,905 |
| 17 01 0107 | Medium Brush, Medium Trees, Clear, Grub, Haul | ACRE | 13,478 | 10,783 | 9,029 | 6,389 | 5,807 |
| 17 01 0108 | Light Brush, Heavy Trees, Clear, Grub, Haul | ACRE | 17,877 | 14,287 | 11,977 | 8,462 | 7,669 |
| 17 01 0113 | Heavy Brush & Trees, Wet Clearing | ACRE | 2,455 | 1,969 | 1,644 | 1,167 | 1,067 |
| 17 03 0103 | Rough Grading, 14G, 1 Pass | SY | 2.06 | 1.68 | 1.38 | 1.00 | 0.94 |
| 17 03 0108 | Fine Grading, 130G, 2 Passes | SY | 0.48 | 0.39 | 0.32 | 0.23 | 0.22 |
| 17 03 0202 | Ditch Excavation, Normal Soil, Haul Spoil 1 Mile | CY | 9.27 | 7.49 | 6.20 | 4.45 | 4.14 |
| 17 03 0203 | Roadway Soil Excavation, with Scraper, Load & Haul Spoil | CY | 9.78 | 7.86 | 6.55 | 4.67 | 4.30 |
| 17 03 0204 | Roadway Sand Excavation, with Scraper, Load & Haul Spoil | CY | 18.35 | 14.79 | 12.28 | 8.78 | 8.13 |
| 17 03 0315 | Roadway Rock Excavator Borrow Subgrade (5 Miles) & Spread | CY | 15.44 | 12.54 | 10.32 | 7.47 | 7.04 |
| 17 03 0501 | Compact Subgrade, 2 Lifts | CY | 0.84 | 0.68 | 0.56 | 0.40 | 0.37 |
| 17 03 0504 | Compact Borrowed Subgrade, 4 Lifts | BCY | 5.43 | 5.27 | 5.15 | 4.99 | 4.97 |
| 17 03 0506 | Compact Sand Subgrade (Wet & 2 Passes) | SY | 0.98 | 0.80 | 0.66 | 0.47 | 0.44 |
| 17 03 0601 | Lime Stabilization, Dry | BCY | 34.46 | 29.87 | 25.87 | 21.32 | 20.95 |
| 17 03 0602 | Cement Stabilization, 6% | BCY | 13.58 | 12.66 | 11.94 | 11.03 | 10.90 |
| 18 01 0102 | Gravel, Delivered & Dumped | CY | 24.33 | 22.93 | 22.03 | 20.66 | 20.36 |
| 18 01 0103 | Gravel (90%) & Sand Base (10%), with Calcium Chloride 3/4 - 1 Lb/CY | CY | 25.70 | 24.15 | 23.16 | 21.65 | 21.31 |
| 18 01 0105 | Asphalt, Stabilized Base Course | CY | 30.79 | 30.01 | 29.41 | 28.64 | 28.54 |
| 18 01 0310 | Prime Coat | SY | 0.55 | 0.52 | 0.51 | 0.49 | 0.48 |
| 18 01 0312 | Asphalt Wearing Course, 1 Pass (Line Item Includes 5% Waste) | TON | 83.61 | 72.83 | 66.20 | 55.67 | 53.10 |
| 19 03 0401 | 18' Complete, 24" Corrugated Metal Pipe Culvert with Headwalls | EACH | 6,021 | 5,084 | 4,522 | 3,606 | 3,374 |

**Environmental Remediation: Assemblies Cost Book**

# Cost Data for Site Work

## Access Roads

| Assembly | Description | Unit | Unit Cost by Safety Level | | | | |
|----------|-------------|------|------|------|------|------|------|
| | | | A | B | C | D | E |
| 19 03 0402 | 34' Complete, 24" Corrugated Metal Pipe Culvert with Headwalls | EACH | 6,746 | 5,734 | 5,135 | 4,147 | 3,892 |

# Cost Data for Remediation
## Arterial Roads/Divided Highways

**General:**

This section provides for the cost of installing a primary two-lane, four-lane, or divided four-lane arterial road with a gravel, asphalt, or concrete finished surface. Note: If cut and fill is required, refer to the Excavation, Cut and Fill section. All waste materials generated from the road construction process are assumed to be uncontaminated. Refer to the appropriate remediation sections for handling of contaminated material.

**Common Cost Components:**

1. Clearing
2. Excavation for road base, ditches, culverts
3. Subgrade: natural soil, compacted, or soil stabilization mixtures
4. Base: asphalt stabilized, cement stabilized, or unstabilized gravel
5. Pavement
6. Pavement markings, signage, fencing

# Cost Data for Site Work
## Arterial Roads/Divided Highways

| Assembly | Description | Unit | A | B | C | D | E |
|---|---|---|---|---|---|---|---|
| | **Capital Costs** | | | | | | |
| 17 01 0106 | Heavy Brush, Light Trees, Clear, Grub, Haul | ACRE | 11,338 | 9,081 | 7,594 | 5,382 | 4,905 |
| 17 01 0107 | Medium Brush, Medium Trees, Clear, Grub, Haul | ACRE | 13,478 | 10,783 | 9,029 | 6,389 | 5,807 |
| 17 01 0108 | Light Brush, Heavy Trees, Clear, Grub, Haul | ACRE | 17,877 | 14,287 | 11,977 | 8,462 | 7,669 |
| 17 03 0103 | Rough Grading, 14G, 1 Pass | SY | 2.06 | 1.68 | 1.38 | 1.00 | 0.94 |
| 17 03 0107 | Fine Grading, 120G, 2 Passes | SY | 0.42 | 0.34 | 0.28 | 0.20 | 0.19 |
| 17 03 0108 | Fine Grading, 130G, 2 Passes | SY | 0.48 | 0.39 | 0.32 | 0.23 | 0.22 |
| 17 03 0202 | Ditch Excavation, Normal Soil, Haul Spoil 1 Mile | CY | 9.27 | 7.49 | 6.20 | 4.45 | 4.14 |
| 17 03 0203 | Roadway Soil Excavation, with Scraper, Load & Haul Spoil | CY | 9.78 | 7.86 | 6.55 | 4.67 | 4.30 |
| 17 03 0204 | Roadway Sand Excavation, with Scraper, Load & Haul Spoil | CY | 18.35 | 14.79 | 12.28 | 8.78 | 8.13 |
| 17 03 0205 | Curb/Sidewalk Excavation & Backfill, 27% Haul Spoil, 1 Mile | CY | 38.69 | 30.82 | 25.93 | 18.23 | 16.37 |
| 17 03 0281 | Borrow Subgrade, Load & Haul 5 Miles, Spreading | CY | 11.41 | 9.21 | 7.63 | 5.48 | 5.10 |
| 17 03 0315 | Roadway Rock Excavator Borrow Subgrade (5 Miles) & Spread | CY | 15.44 | 12.54 | 10.32 | 7.47 | 7.04 |
| 17 03 0316 | Roadway Rock Excavation & Haul (5 Miles), No Borrow | CY | 11.45 | 9.32 | 7.66 | 5.55 | 5.26 |
| 17 03 0501 | Compact Subgrade, 2 Lifts | CY | 0.84 | 0.68 | 0.56 | 0.40 | 0.37 |
| 17 03 0504 | Compact Borrowed Subgrade, 4 Lifts | BCY | 5.43 | 5.27 | 5.15 | 4.99 | 4.97 |
| 17 03 0506 | Compact Sand Subgrade (Wet & 2 Passes) | SY | 0.98 | 0.80 | 0.66 | 0.47 | 0.44 |
| 17 03 0512 | Compact Roadside Fill, Hand & Machine | CY | 36.63 | 29.26 | 24.54 | 17.33 | 15.68 |
| 17 03 0601 | Lime Stabilization, Dry | BCY | 34.46 | 29.87 | 25.87 | 21.32 | 20.95 |
| 17 03 0602 | Cement Stabilization, 6% | BCY | 13.58 | 12.66 | 11.94 | 11.03 | 10.90 |
| 18 01 0101 | Cement-stabilized Base | CY | 41.04 | 36.91 | 33.71 | 29.62 | 29.04 |
| 18 01 0102 | Gravel, Delivered & Dumped | CY | 24.33 | 22.93 | 22.03 | 20.66 | 20.36 |

**Environmental Remediation: Assemblies Cost Book**

*Page 4-3*

1998 by ECHOS. All rights reserved

## Cost Data for Site Work

### Arterial Roads/Divided Highways

| Assembly | Description | Unit | A | B | C | D | E |
|---|---|---|---|---|---|---|---|
| 18 01 0104 | Asphalt, Intermediate Course (Line Item Includes 5% Waste) | TON | 76.50 | 65.72 | 59.09 | 48.56 | 45.99 |
| 18 01 0105 | Asphalt, Stabilized Base Course | CY | 30.79 | 30.01 | 29.41 | 28.64 | 28.54 |
| 18 01 0201 | Concrete Curb, 6" x 6" | LF | 2.78 | 2.38 | 2.21 | 1.83 | 1.70 |
| 18 01 0202 | Concrete Curb & Gutter, 6" x 24", Formed | LF | 20.78 | 16.61 | 14.83 | 10.80 | 9.32 |
| 18 01 0205 | Curb Inlet Frame Grate & Box, 525 Lb | EACH | 6,004 | 4,987 | 4,534 | 3,550 | 3,200 |
| 18 01 0301 | 6" Concrete Pavement, Formed, 2 Lanes | SY | 31.94 | 28.59 | 27.12 | 23.88 | 22.72 |
| 18 01 0302 | 8" Concrete Pavement, Formed, 2 Lanes | SY | 35.65 | 32.19 | 30.67 | 27.33 | 26.12 |
| 18 01 0305 | 6" Concrete Pavement, Slipform, Integral Curb, 2 Lanes | SY | 29.62 | 26.75 | 25.46 | 22.68 | 21.70 |
| 18 01 0306 | 8" Concrete Pavement, Slipform, Integral Curb, 2 Lanes | SY | 33.19 | 30.24 | 28.91 | 26.05 | 25.04 |
| 18 01 0307 | 10" Concrete Pavement, Slipform, Integral Curb, 2 Lanes | SY | 37.14 | 34.03 | 32.61 | 29.60 | 28.55 |
| 18 01 0310 | Prime Coat | SY | 0.55 | 0.52 | 0.51 | 0.49 | 0.48 |
| 18 01 0311 | Tack Coat | SY | 0.70 | 0.64 | 0.60 | 0.53 | 0.52 |
| 18 01 0312 | Asphalt Wearing Course, 1 Pass (Line Item Includes 5% Waste) | TON | 83.61 | 72.83 | 66.20 | 55.67 | 53.10 |
| 18 01 0401 | Crosswalk, Stop Lines, Per Lane, Intersection Painting | EACH | 109.66 | 87.46 | 77.96 | 56.50 | 48.63 |
| 18 01 0402 | Turn Lane, Per Lane, Intersection Painting | EACH | 45.82 | 36.69 | 31.42 | 22.52 | 20.12 |
| 18 01 0403 | Arrows, Per Lane, Intersection Painting | EACH | 27.59 | 22.73 | 20.65 | 15.96 | 14.24 |
| 18 01 0405 | No Pass Stripe, Yellow | LF | 0.19 | 0.17 | 0.16 | 0.13 | 0.13 |
| 18 01 0406 | Centerline Stripe, White | LF | 2.80 | 2.75 | 2.72 | 2.67 | 2.66 |
| 18 01 0407 | Edge Stripe, Yellow | LF | 0.19 | 0.17 | 0.16 | 0.13 | 0.13 |
| 18 01 0410 | Street Signs, Average | EACH | 86.60 | 70.54 | 63.66 | 48.13 | 42.44 |
| 18 01 0411 | Traffic Signs & Posts, Average | EACH | 90.91 | 74.85 | 67.97 | 52.44 | 46.75 |
| 18 01 0501 | Guardrail, Single Rail, Wood Posts | LF | 34.02 | 28.59 | 24.84 | 19.51 | 18.47 |
| 18 01 0502 | Guardrail, Single Rail, Wood Posts, Ends | EACH | 230.17 | 199.53 | 178.37 | 148.27 | 142.40 |
| 18 04 0110 | Barbed-wire Fencing, 3-Strand | LF | 1.36 | 1.36 | 1.36 | 1.36 | 1.36 |
| 19 03 0403 | 52' Complete, 24" Corrugated Metal Pipe Culvert with Headwalls | EACH | 7,562 | 6,466 | 5,825 | 4,756 | 4,474 |
| 19 03 0404 | 76' Complete, 24" Corrugated Metal Pipe Culvert with Headwalls | EACH | 8,650 | 7,443 | 6,745 | 5,567 | 5,251 |

**Unit Cost by Safety Level**

---

**Environmental Remediation: Assemblies Cost Book**

# Cost Data for Remediation

## Bridges

### General:

This section provides the costs for the following types of bridges: cast-in-place concrete T-beam, precast I-beam sections, precast box sections, concrete and steel composite, and timber-laminated deck. Note: For bridges constructed in conjunction with an access road to a remediation site, it is assumed that all road and bridge construction will occur outside of the contaminated zone.

### Common Cost Components:

1. Footing excavation, backfill, and compaction
2. Abutment grading
3. Area cleanup
4. Stabilized base at abutment roadway, bituminous pavement (timber laminated deck), guardrails, footings, wood piles (timber laminated deck), bent caps, columns, abutments, decks and beams, retaining walls at abutments, approach slabs, parapets, expansion joints, bearing pads
5. Slope protection
6. Seeding and sediment control
7. Fertilizer

# Cost Data for Site Work

## Bridges

| Assembly | Description | Unit | Unit Cost by Safety Level A | B | C | D | E |
|---|---|---|---|---|---|---|---|
| | **Capital Costs** | | | | | | |
| 17 03 0102 | Rough Grading, 12G, 1 Pass | SY | 1.28 | 1.04 | 0.85 | 0.62 | 0.59 |
| 17 03 0107 | Fine Grading, 120G, 2 Passes | SY | 0.42 | 0.34 | 0.28 | 0.20 | 0.19 |
| 17 03 0201 | Excavation, Spoil to Side | CY | 1.85 | 1.48 | 1.24 | 0.88 | 0.80 |
| 17 03 0282 | Soil, 5 Miles, Dump Truck, Load/Haul Spoil From Trench | CY | 6.97 | 5.60 | 4.66 | 3.33 | 3.06 |
| 17 03 0405 | 950, 3.00 CY, Delivered & Dumped, Backfill with Sand | CY | 38.31 | 35.03 | 32.95 | 29.74 | 29.00 |
| 17 03 0416 | Backfill, Large Spread Footing Excavated Material, 950, 3.0 CY | CY | 5.63 | 4.55 | 3.77 | 2.71 | 2.53 |
| 17 03 0508 | Compact, Footing Excavator, Excavated Material Backfill | CY | 4.36 | 3.37 | 2.93 | 1.97 | 1.64 |
| 17 03 0511 | Compact Soil with Vibrating Plate | CY | 7.20 | 5.57 | 4.84 | 3.26 | 2.70 |
| 17 03 0513 | Spread Dumped Borrow & Compact with Roller | CY | 0.84 | 0.68 | 0.56 | 0.40 | 0.37 |
| 17 04 0101 | General Area Cleanup | ACRE | 640.88 | 509.59 | 429.63 | 301.34 | 269.54 |
| 18 01 0101 | Cement-stabilized Base | CY | 41.04 | 36.91 | 33.71 | 29.62 | 29.04 |
| 18 01 0312 | Asphalt Wearing Course, 1 Pass (Line Item Includes 5% Waste) | TON | 83.61 | 72.83 | 66.20 | 55.67 | 53.10 |
| 18 01 0501 | Guardrail, Single Rail, Wood Posts | LF | 34.02 | 28.59 | 24.84 | 19.51 | 18.47 |
| 18 01 0502 | Guardrail, Single Rail, Wood Posts, Ends | EACH | 230.17 | 199.53 | 178.37 | 148.27 | 142.40 |
| 18 05 0201 | Security Fence, Temporary | LF | 10.84 | 8.88 | 7.66 | 5.74 | 5.28 |
| 18 05 0402 | Seeding, Vegetative Cover | ACRE | 1,874 | 1,823 | 1,791 | 1,741 | 1,729 |
| 18 05 0405 | Sodding, Vegetative Cover | ACRE | 25,697 | 22,064 | 20,238 | 16,710 | 15,589 |
| 18 05 0408 | Fertilizer, Hydro Spread | ACRE | 241.03 | 207.54 | 182.57 | 149.58 | 144.30 |
| 18 06 0101 | Form, Straight Beam Bottoms | SF | 12.24 | 9.72 | 8.62 | 6.18 | 5.29 |
| 18 06 0102 | Form, Straight Beam Sides | SF | 7.94 | 6.32 | 5.62 | 4.06 | 3.49 |

**Environmental Remediation: Assemblies Cost Book**

*Page 4-5*

1998 by ECHOS. All rights reserved

# Cost Data for Site Work

## Bridges

| Assembly | Description | Unit | A | B | C | D | E |
|---|---|---|---|---|---|---|---|
| 18 06 0103 | Rebar, CIP Beams | LB | 1.17 | 0.98 | 0.88 | 0.69 | 0.63 |
| 18 06 0104 | Pour & Cure, CIP Beams | CY | 162.22 | 144.21 | 134.73 | 117.22 | 111.92 |
| 18 06 0105 | Form, Beam Bracing Diaphragms | SF | 7.96 | 6.36 | 5.66 | 4.10 | 3.54 |
| 18 06 0106 | Rebar, Bracing Diaphragms | LB | 1.17 | 0.98 | 0.88 | 0.69 | 0.63 |
| 18 06 0107 | Pour & Cure, Bracing Diaphragms | CY | 207.46 | 181.03 | 167.07 | 141.37 | 133.64 |
| 18 06 0108 | Form Deck, 3 Uses | SF | 6.98 | 5.46 | 4.81 | 3.34 | 2.81 |
| 18 06 0109 | Rebar, Bridge Deck | LB | 1.00 | 0.84 | 0.75 | 0.60 | 0.55 |
| 18 06 0110 | Pour & Cure, Deck | CY | 180.58 | 158.16 | 147.10 | 125.35 | 118.31 |
| 18 06 0111 | Bush Hammer Finish | SF | 4.07 | 3.15 | 2.75 | 1.85 | 1.52 |
| 18 06 0112 | Armor Joints | LF | 122.02 | 121.06 | 120.64 | 119.70 | 119.37 |
| 18 06 0113 | Beam Bearing Pads (Elastomeric) | SF | 139.80 | 119.24 | 110.43 | 90.56 | 83.27 |
| 18 06 0120 | Precast I Beam, Type I, 1' 4" x 2' 4" Deep | LF | 159.58 | 146.54 | 138.18 | 125.42 | 122.53 |
| 18 06 0125 | Precast Box Beam, 3' x 2' 9" Deep | LF | 677.20 | 626.42 | 593.85 | 544.14 | 532.86 |
| 18 06 0126 | Precast Box Beam, 4' x 2' 9" Deep | LF | 828.28 | 770.25 | 733.02 | 676.21 | 663.33 |
| 18 06 0127 | Pour & Cure Concrete Deck Topping | CY | 228.60 | 196.59 | 180.86 | 149.81 | 139.72 |
| 18 06 0128 | Shear Key, High Strength, Nonshrink Grout | CY | 133.66 | 119.35 | 111.40 | 97.47 | 93.53 |
| 18 06 0129 | 1" Bituminous Fiber Expansion Joint | LF | 4.33 | 3.78 | 3.54 | 3.01 | 2.82 |
| 18 06 0130 | 1/2" x 1" Poured Rubber Joint | LF | 1.53 | 1.24 | 1.12 | 0.84 | 0.74 |
| 18 06 0131 | Structural Steel Beams, Rolled Shapes, A36, 2 - 6" Studs @ 3" | TON | 3,828 | 3,252 | 2,957 | 2,397 | 2,222 |
| 18 06 0132 | Structural Steel Beam Bracing | TON | 2,616 | 2,402 | 2,287 | 2,080 | 2,019 |
| 18 06 0133 | Rocker/Fixed Bearing Assembly for Steel Beams, Complete | LB | 4.55 | 4.45 | 4.40 | 4.31 | 4.28 |
| 18 06 0135 | Treated Timber Piles/Abutments, 12" Butt, 8" Tip | LF | 23.87 | 20.53 | 18.40 | 15.13 | 14.38 |
| 18 06 0136 | Treated Timber Piles (Bentcaps), 12" Butt, 8" Tip | LF | 23.87 | 20.53 | 18.40 | 15.13 | 14.38 |
| 18 06 0137 | Heavy Construction Timber, Treated with Misc Blocking | BF | 3.57 | 3.24 | 3.09 | 2.77 | 2.66 |
| 18 06 0138 | Bolts, Washers, & Nuts, Installed | LB | 22.29 | 21.68 | 21.42 | 20.84 | 20.63 |
| 18 06 0139 | Spikes & Nails, Installed | LB | 22.20 | 18.41 | 16.80 | 13.14 | 11.80 |
| 18 06 0140 | Shear Connectors, Dowels & Shear Plates | LB | 12.63 | 10.91 | 10.17 | 8.50 | 7.89 |
| 18 06 0141 | Heavy Construction Timber (Decks), Treated with Misc Blocking | BF | 5.42 | 4.51 | 4.01 | 3.13 | 2.88 |
| 18 06 0142 | Heavy Construction Timber (Beams), Treated with Misc Blocking | BF | 3.57 | 3.24 | 3.09 | 2.77 | 2.66 |
| 18 06 0145 | Large Spread Footing, Edge Form & Strip, 4 Uses | SF | 6.22 | 5.05 | 4.54 | 3.41 | 3.00 |
| 18 06 0146 | Footing, Rebar | LB | 1.17 | 0.98 | 0.88 | 0.69 | 0.63 |
| 18 06 0147 | Large Spread Footing, Pour & Cure Concrete | CY | 138.75 | 125.05 | 117.78 | 104.46 | 100.47 |

**Environmental Remediation: Assemblies Cost Book**

# Cost Data for Site Work

## Bridges

| Assembly | Description | Unit | Unit Cost by Safety Level | | | | |
|----------|-------------|------|------|------|------|------|------|
| | | | A | B | C | D | E |
| 18 06 0148 | Column Forms, Fiberglass, 24" Round | LF | 21.80 | 18.74 | 17.43 | 14.48 | 13.39 |
| 18 06 0149 | Columns, Rebar, Spiral | LB | 1.49 | 1.29 | 1.18 | 0.98 | 0.92 |
| 18 06 0150 | Columns, Pour & Finish Concrete | CY | 332.66 | 275.61 | 250.56 | 195.38 | 175.55 |
| 18 06 0151 | Form, Bentcap Bottom | SF | 12.24 | 9.72 | 8.62 | 6.18 | 5.29 |
| 18 06 0152 | Form, Bentcap Sides | SF | 7.15 | 5.72 | 5.10 | 3.72 | 3.22 |
| 18 06 0153 | Bentcap, Rebar | LB | 1.00 | 0.84 | 0.75 | 0.60 | 0.55 |
| 18 06 0154 | Bentcap, Pour & Cure | CY | 177.81 | 155.76 | 143.75 | 122.29 | 116.07 |
| 18 06 0155 | Edgeforms, 10" Approach Slab, 2 Uses | LF | 9.56 | 7.55 | 6.69 | 4.75 | 4.03 |
| 18 06 0156 | Welded Wire Mesh, Approach Slab, 6 x 6 x 4/4, 58 Lb/SQ | SQ | 85.68 | 71.30 | 65.14 | 51.25 | 46.15 |
| 18 06 0157 | Pour & Cure, Approach Slab | CY | 162.78 | 143.89 | 134.50 | 116.17 | 110.28 |
| 18 06 0158 | Parapet, Form | SF | 5.57 | 4.37 | 3.85 | 2.69 | 2.26 |
| 18 06 0159 | Parapet, Rebar | LB | 0.85 | 0.73 | 0.68 | 0.56 | 0.52 |
| 18 06 0160 | Parapet, Pour & Cure | CY | 200.92 | 176.58 | 163.96 | 140.30 | 133.03 |

**Environmental Remediation: Assemblies Cost Book**

# Cost Data for Remediation
## Cleanup and Landscaping

**General:**

Many environmental remediation projects include cleanup and landscaping as a final step. If the cleanup or landscaping occurs prior to the complete removal or treatment of contaminants, any debris or excavated soil must be treated and/or disposed.

**Common Cost Components:**

1. Cleanup: general area cleanup, removal of any on-site debris, loading and hauling of debris, and pavement sweeping
2. Landscaping: area preparation, slope protection, seeding or sodding, fertilizing, watering, mowing

# Cost Data for Site Work

## Cleanup and Landscaping

| Assembly | Description | Unit | Unit Cost by Safety Level | | | | |
|---|---|---|---|---|---|---|---|
| | | | A | B | C | D | E |
| | **Cleanup** | | | | | | |
| 17 04 0101 | General Area Cleanup | ACRE | 640.88 | 509.59 | 429.63 | 301.34 | 269.54 |
| 17 04 0101 | General Area Cleanup | ACRE | 640.88 | 509.59 | 429.63 | 301.34 | 269.54 |
| 17 04 0102 | Pavement Sweeping, Machine | SY | 0.02 | 0.02 | 0.01 | 0.01 | 0.01 |
| 17 04 0102 | Pavement Sweeping, Machine | SY | 0.02 | 0.02 | 0.01 | 0.01 | 0.01 |
| 17 04 0103 | Load & Haul Debris, 5 Miles, Dumptruck | CY | 8.09 | 6.53 | 5.41 | 3.88 | 3.61 |
| 17 04 0103 | Load & Haul Debris, 5 Miles, Dumptruck | CY | 8.09 | 6.53 | 5.41 | 3.88 | 3.61 |
| 18 05 0101 | Area Preparation, 67% Level & 33% Slope | ACRE | 141.24 | 114.14 | 94.51 | 67.84 | 63.22 |
| 18 05 0101 | Area Preparation, 67% Level & 33% Slope | ACRE | 141.24 | 114.14 | 94.51 | 67.84 | 63.22 |
| 18 05 0201 | Security Fence, Temporary | LF | 10.84 | 8.88 | 7.66 | 5.74 | 5.28 |
| 18 05 0201 | Security Fence, Temporary | LF | 10.84 | 8.88 | 7.66 | 5.74 | 5.28 |
| 18 05 0401 | Seeding, 67% Level & 33% Slope, Hydroseeding | ACRE | 626.53 | 562.96 | 521.02 | 458.72 | 445.32 |
| 18 05 0401 | Seeding, 67% Level & 33% Slope, Hydroseeding | ACRE | 626.53 | 562.96 | 521.02 | 458.72 | 445.32 |
| 18 05 0402 | Seeding, Vegetative Cover | ACRE | 1,874 | 1,823 | 1,791 | 1,741 | 1,729 |
| 18 05 0402 | Seeding, Vegetative Cover | ACRE | 1,874 | 1,823 | 1,791 | 1,741 | 1,729 |
| 18 05 0404 | Sodding, Average CONUS | ACRE | 25,588 | 21,976 | 20,165 | 16,658 | 15,542 |
| 18 05 0404 | Sodding, Average CONUS | ACRE | 25,588 | 21,976 | 20,165 | 16,658 | 15,542 |
| 18 05 0405 | Sodding, Vegetative Cover | ACRE | 25,697 | 22,064 | 20,238 | 16,710 | 15,589 |
| 18 05 0405 | Sodding, Vegetative Cover | ACRE | 25,697 | 22,064 | 20,238 | 16,710 | 15,589 |
| 18 05 0408 | Fertilizer, Hydro Spread | ACRE | 241.03 | 207.54 | 182.57 | 149.58 | 144.30 |
| 18 05 0408 | Fertilizer, Hydro Spread | ACRE | 241.03 | 207.54 | 182.57 | 149.58 | 144.30 |
| 18 05 0409 | Fertilize, 800 Lbs/Acre, Push Rotary | ACRE | 140.58 | 118.67 | 105.24 | 83.82 | 78.56 |
| 18 05 0409 | Fertilize, 800 Lbs/Acre, Push Rotary | ACRE | 140.58 | 118.67 | 105.24 | 83.82 | 78.56 |
| 18 05 0410 | Fertilize, 800 Lbs/Acre, Spray from Truck | ACRE | 107.45 | 90.70 | 78.22 | 61.72 | 59.08 |
| 18 05 0410 | Fertilize, 800 Lbs/Acre, Spray from Truck | ACRE | 107.45 | 90.70 | 78.22 | 61.72 | 59.08 |
| 18 05 0411 | Crushed Limestone, pH Adjustment, 800 Lbs/Acre | ACRE | 156.71 | 134.79 | 121.37 | 99.94 | 94.68 |

**Environmental Remediation: Assemblies Cost Book**

*Page 4-8*

1998 by ECHOS. All rights reserved

# Cost Data for Site Work

## Cleanup and Landscaping

| Assembly | Description | Unit | Unit Cost by Safety Level | | | | |
|---|---|---|---|---|---|---|---|
| | | | A | B | C | D | E |
| 18 05 0411 | Crushed Limestone, pH Adjustment, 800 Lbs/Acre | ACRE | 156.71 | 134.79 | 121.37 | 99.94 | 94.68 |
| 18 05 0412 | Purchase & Spread Dry Granular Limestone for pH Control | SY | 0.05 | 0.04 | 0.04 | 0.03 | 0.03 |
| 18 05 0412 | Purchase & Spread Dry Granular Limestone for pH Control | SY | 0.05 | 0.04 | 0.04 | 0.03 | 0.03 |
| 18 05 0413 | Watering with 3,000-Gallon Tank Truck, per Pass | ACRE | 111.81 | 90.08 | 75.73 | 54.43 | 49.85 |
| 18 05 0413 | Watering with 3,000-Gallon Tank Truck, per Pass | ACRE | 111.81 | 90.08 | 75.73 | 54.43 | 49.85 |
| 18 05 0415 | Mowing | ACRE | 64.67 | 51.13 | 43.38 | 30.18 | 26.60 |
| 18 05 0415 | Mowing | ACRE | 64.67 | 51.13 | 43.38 | 30.18 | 26.60 |

**Environmental Remediation: Assemblies Cost Book**

# Cost Data for Remediation
## Clear and Grub

**General:**

Clearing and grubbing are often required to prepare a site for remediation activity. Note: the loading, hauling, and disposal costs included in this section assume that all soil and cover material are uncontaminated. Refer to the appropriate remediation sections for handling of contaminated material.

**Common Cost Components:**

1. Clearing: removal of vegetation such as trees, shrubs, and brush.
2. Grubbing: removal of stumps, roots, and debris from the soil by using dozers or other heavy equipment
3. Soil stripping: removal of topsoil that contains unsuitable organic material

# Cost Data for Site Work

## Clear and Grub

| Assembly | Description | Unit | Unit Cost by Safety Level | | | | |
|---|---|---|---|---|---|---|---|
| | | | A | B | C | D | E |
| | **Capital Costs** | | | | | | |
| 17 01 0101 | Light Brush without Grub, Clearing | ACRE | 139.11 | 110.49 | 93.27 | 65.31 | 58.24 |
| 17 01 0102 | Medium Brush without Grub, Clearing | ACRE | 330.00 | 264.79 | 220.99 | 157.03 | 143.77 |
| 17 01 0104 | Heavy Brush without Grub, Clearing | ACRE | 653.47 | 530.18 | 437.04 | 315.52 | 296.84 |
| 17 01 0110 | Light, without Grub D5 LGP, Wet Clearing | ACRE | 1,255 | 997.24 | 841.35 | 589.58 | 526.47 |
| 17 01 0111 | Medium, without Grub D7 LGP, Wet Clearing | ACRE | 1,189 | 949.65 | 796.81 | 562.33 | 508.59 |
| 17 01 0112 | Heavy, without Grub D7 LGP, Wet Clearing | ACRE | 2,455 | 1,969 | 1,644 | 1,167 | 1,067 |
| 17 01 0201 | <= 6" Tree Removal, Hand | EACH | 235.55 | 188.90 | 157.75 | 112.00 | 102.40 |
| 17 01 0202 | > 6" and <= 12" Tree Removal, Hand | EACH | 353.32 | 283.35 | 236.62 | 168.01 | 153.60 |
| 17 01 0203 | > 12" and <= 24" Tree Removal, Hand | EACH | 471.10 | 377.79 | 315.50 | 224.01 | 204.79 |
| 17 01 0204 | > 24" and <= 36" Tree Removal, Hand | EACH | 706.64 | 566.69 | 473.25 | 336.01 | 307.19 |
| 17 01 0210 | Clear Trees to 6" Diameter with D8 Cat | EACH | 12.03 | 9.85 | 8.04 | 5.88 | 5.66 |
| 17 01 0211 | Clear Trees to 12" Diameter with D8 Cat | EACH | 22.45 | 18.39 | 15.00 | 10.98 | 10.56 |
| 17 01 0212 | Clear Trees to 24" Diameter with D8 Cat | EACH | 33.68 | 27.58 | 22.50 | 16.46 | 15.84 |
| 17 01 0213 | Clear Trees to 36" Diameter with D8 Cat | EACH | 67.36 | 55.17 | 45.00 | 32.93 | 31.67 |
| 17 01 0301 | <= 6" Stump Removal, with Excavator | EACH | 107.29 | 87.01 | 71.76 | 51.77 | 48.66 |
| 17 01 0302 | > 6" and <= 12" Stump Removal, with Excavator | EACH | 134.11 | 108.76 | 89.70 | 64.72 | 60.82 |
| 17 01 0303 | > 12" and <= 24" Stump Removal, with Excavator | EACH | 178.81 | 145.02 | 119.60 | 86.29 | 81.10 |
| 17 01 0304 | > 24" and <= 36" Stump Removal, with Excavator | EACH | 268.22 | 217.52 | 179.39 | 129.43 | 121.64 |
| 17 01 0310 | < 6" Wet Stump Removal, with LGP D7 | EACH | 130.68 | 106.64 | 87.34 | 63.58 | 60.65 |
| 17 01 0311 | > 6" < 12" Wet Stump Removal, with LGP D7 | EACH | 163.35 | 133.30 | 109.18 | 79.47 | 75.81 |
| 17 01 0312 | > 12" < 24" Wet Stump Removal, with LGP D7 | EACH | 217.80 | 177.74 | 145.57 | 105.96 | 101.08 |
| 17 01 0313 | > 24" < 36" Wet Stump Removal, with LGP D7 | EACH | 326.70 | 266.61 | 218.35 | 158.95 | 151.61 |
| 17 01 0314 | <= 6" Stump Removal, with D8 | EACH | 8.42 | 6.90 | 5.63 | 4.12 | 3.96 |
| 17 01 0315 | > 6" and <= 12" Stump Removal, with D8 | EACH | 13.47 | 11.03 | 9.00 | 6.59 | 6.33 |

**Environmental Remediation: Assemblies Cost Book**

# Cost Data for Site Work

## Clear and Grub

| Assembly | Description | Unit | Unit Cost by Safety Level A | B | C | D | E |
|---|---|---|---|---|---|---|---|
| 17 01 0316 | > 12" and <= 24" Stump Removal, with D8 | EACH | 134.73 | 110.34 | 90.01 | 65.86 | 63.34 |
| 17 01 0317 | > 24" and <= 36" Stump Removal, with D8 | EACH | 336.82 | 275.85 | 225.01 | 164.64 | 158.36 |
| 17 01 0401 | Light Brush without Grub, Chipping | ACRE | 2,513 | 2,015 | 1,683 | 1,195 | 1,092 |
| 17 01 0402 | Medium Brush without Grub, Chipping | ACRE | 3,231 | 2,591 | 2,164 | 1,536 | 1,405 |
| 17 01 0403 | Heavy Brush without Grub, Chipping | ACRE | 4,524 | 3,628 | 3,030 | 2,151 | 1,967 |
| 17 01 0501 | Dozer 105 HP D5, Grubbing & Stacking | CY | 13.44 | 10.84 | 8.99 | 6.44 | 5.98 |
| 17 01 0502 | Dozer 200 HP D7, Grubbing & Stacking | CY | 5.84 | 4.77 | 3.91 | 2.84 | 2.70 |
| 17 01 0510 | Dozer 105 HP D5 LGP, Wet Grubbing & Stacking | CY | 13.44 | 10.84 | 8.99 | 6.44 | 5.98 |
| 17 01 0511 | Dozer 200 HP D7 LGP, Wet Grubbing & Stacking | CY | 5.84 | 4.77 | 3.91 | 2.84 | 2.70 |
| 17 01 0701 | Light Burning | ACRE | 4,137 | 3,796 | 3,580 | 3,247 | 3,169 |
| 17 01 0702 | Medium Burning | ACRE | 5,769 | 5,331 | 5,053 | 4,624 | 4,525 |
| 17 01 0703 | Heavy Burning | ACRE | 7,537 | 6,923 | 6,535 | 5,934 | 5,794 |

**Environmental Remediation: Assemblies Cost Book**

*Page 4-11*

1998 by ECHOS. All rights reserved

# Cost Data for Remediation
## Communications

**General:**

When underground telephone service is needed for a site, this section can be used to determine the amount of underground ducts and wiring required for communications capabilities.

**Common Cost Components:**

1. Excavation, backfill
2. Conduit
3. Concrete encasement for ductbank
4. Communications cable
5. Manhole
6. Seal slab
7. Dewatering pump

# Cost Data for Site Work
## Communications

| Assembly | Description | Unit | \_Unit Cost by Safety Level\_ A | B | C | D | E |
|---|---|---|---|---|---|---|---|
| | **Capital Costs** | | | | | | |
| 17 03 0259 | Cat 225, 1.5 CY, Soil/Sand, Trenching | CY | 1.84 | 1.49 | 1.23 | 0.88 | 0.82 |
| 17 03 0261 | Cat 225, 1.5 CY, Soil/Sand with Boulders, Trenching | CY | 3.91 | 3.15 | 2.62 | 1.87 | 1.74 |
| 17 03 0306 | Cat 225, 1.5 CY, Rock, No Haul or Borrow, Trenching | BCY | 149.48 | 122.48 | 106.81 | 80.48 | 73.45 |
| 17 03 0401 | 950, 3.00 CY, Backfill with Excavated Material | CY | 2.54 | 2.05 | 1.70 | 1.22 | 1.14 |
| 17 03 0405 | 950, 3.00 CY, Delivered & Dumped, Backfill with Sand | CY | 38.31 | 35.03 | 32.95 | 29.74 | 29.00 |
| 17 03 0410 | 950, Delivered & Dumped, Backfill with Cement Stabilized Base Material | CY | 46.93 | 43.66 | 41.58 | 38.37 | 37.63 |
| 17 03 0420 | Backfill Trench, Borrow Material, Delivered & Dumped Only | CY | 9.86 | 8.74 | 7.93 | 6.83 | 6.63 |
| 17 03 0511 | Compact Soil with Vibrating Plate | CY | 7.20 | 5.57 | 4.84 | 3.26 | 2.70 |
| 17 03 0515 | Compact With Pogosticks | CY | 5.91 | 4.58 | 3.98 | 2.68 | 2.23 |
| 17 03 0516 | Compact with 50% Pogosticks, 50% Hand Roller | CY | 4.24 | 3.30 | 2.85 | 1.94 | 1.63 |
| 17 03 1002 | 2" Diameter Contractor's Trash Pump, 75 GPM | DAY | 50.20 | 47.01 | 45.64 | 42.55 | 41.42 |
| 18 04 0203 | Pour & Cure Concrete, Continuous Footing | CY | 140.13 | 126.22 | 118.87 | 105.35 | 101.29 |
| 20 02 0601 | 2" Steel Conduit | LF | 12.33 | 11.02 | 10.46 | 9.19 | 8.73 |
| 20 02 0602 | 3" Steel Conduit | LF | 22.17 | 19.98 | 19.04 | 16.93 | 16.16 |
| 20 02 0603 | 4" Steel Conduit | LF | 31.60 | 28.53 | 27.22 | 24.25 | 23.16 |
| 20 02 0604 | 6" Steel Conduit | LF | 72.02 | 62.75 | 58.79 | 49.83 | 46.55 |
| 20 02 0610 | 2" PVC Conduit | LF | 4.85 | 3.86 | 3.44 | 2.49 | 2.15 |
| 20 02 0611 | 3" PVC Conduit | LF | 8.30 | 6.66 | 5.95 | 4.37 | 3.78 |
| 20 02 0612 | 4" PVC Conduit | LF | 11.68 | 9.38 | 8.40 | 6.17 | 5.36 |
| 20 02 0613 | 6" PVC Conduit | LF | 20.78 | 16.83 | 15.13 | 11.31 | 9.90 |
| 20 02 0615 | Concrete Encasement for Ductbank | CY | 231.39 | 194.67 | 178.27 | 142.74 | 130.15 |
| 20 04 0101 | 27 Pair No. 22 AWG Wire, Communication Cable | LF | 3.04 | 2.57 | 2.36 | 1.90 | 1.74 |
| 20 04 0102 | 51 Pair No. 22 AWG Wire, Communication Cable | LF | 4.47 | 3.84 | 3.58 | 2.97 | 2.75 |

**Environmental Remediation: Assemblies Cost Book**

# Cost Data for Site Work

## Communications

| Assembly | Description | Unit | Unit Cost by Safety Level | | | | |
|---|---|---|---|---|---|---|---|
| | | | A | B | C | D | E |
| 20 04 0103 | 100 Pair No. 22 AWG Wire, Communication Cable | LF | 8.45 | 7.10 | 6.52 | 5.22 | 4.74 |
| 20 04 0105 | Electrical & Communications Manhole 10.5' Square x 10' Deep, Cable Tray | EACH | 33,997 | 28,212 | 24,944 | 19,307 | 17,747 |

# Cost Data for Remediation
## Demolition, Bridges

**General:**

The demolition section can be used to calculate the cost of demolishing bridges of masonry, concrete, or steel construction. Note: If the debris is hazardous, costs for transportation and disposal should be estimated using cost data included with the remediation technologies.

**Common Cost Components:**

1. Demolition using non-explosive methods

Note: Refer to the Load and Haul section for loading, hauling, and disposal costs for nonhazardous materials

# Cost Data for Site Work

## Demolition, Bridges

| Assembly | Description | Unit | Unit Cost by Safety Level | | | | |
|---|---|---|---|---|---|---|---|
| | | | A | B | C | D | E |
| | **Capital Costs** | | | | | | |
| 17 02 0101 | Multilevel, Steel, Nonexplosive, Building Demolition | CF | 0.14 | 0.11 | 0.10 | 0.07 | 0.06 |
| 17 02 0102 | Multilevel, Concrete, Nonexplosive, Building Demolition | CF | 0.20 | 0.16 | 0.13 | 0.09 | 0.08 |
| 17 02 0103 | Multilevel, Masonry, Nonexplosive, Building Demolition | CF | 0.15 | 0.12 | 0.10 | 0.07 | 0.06 |
| 17 02 0105 | Single-level, Steel, Nonexplosive, Building Demolition | CF | 0.28 | 0.23 | 0.19 | 0.14 | 0.12 |
| 17 02 0106 | Single-level, Concrete, Nonexplosive, Building Demolition | CF | 0.37 | 0.30 | 0.25 | 0.18 | 0.16 |
| 17 02 0107 | Single-level, Masonry, Nonexplosive, Building Demolition | CF | 0.28 | 0.23 | 0.19 | 0.14 | 0.12 |
| 17 02 0108 | Single-level, Wood, Nonexplosive, Building Demolition | CF | 0.28 | 0.23 | 0.19 | 0.14 | 0.12 |

**Environmental Remediation: Assemblies Cost Book**

*Page 4-14*

1998 by ECHOS. All rights reserved

# Cost Data for Remediation
## Demolition, Catch Basins/Manholes

### General:

The demolition -- catch basins/manholes section provides the cost for either abandonment or demolition of catch basins and manholes. Note: If the debris is hazardous, costs for transportation and disposal should be estimated using cost data included with the remediation technologies.

### Common Cost Components:

1. Demolition using non-explosive methods

Note: Refer to the Load and Haul section for loading, hauling, and disposal costs for nonhazardous materials

## Cost Data for Site Work

### Demolition, Catch Basins/Manholes

| Assembly | Description | Unit | Unit Cost by Safety Level | | | | |
|---|---|---|---|---|---|---|---|
| | | | A | B | C | D | E |
| | **Capital Costs** | | | | | | |
| 17 02 0305 | Abandon Catch Basin/Manhole | CY | 53.63 | 42.22 | 35.99 | 24.89 | 21.68 |
| 17 02 0306 | Remove Catch Basin/Manhole | CY | 93.85 | 73.89 | 62.99 | 43.55 | 37.94 |
| 17 03 0420 | Backfill Trench, Borrow Material, Delivered & Dumped Only | CY | 9.86 | 8.74 | 7.93 | 6.83 | 6.63 |

# Cost Data for Remediation
## Demolition, Curbs

**General:**                                    **Common Cost Components:**

## Cost Data for Site Work

### Demolition, Curbs

| Assembly | Description | Unit | Unit Cost by Safety Level | | | | |
|---|---|---|---|---|---|---|---|
| | | | A | B | C | D | E |
| | **Capital Costs** | | | | | | |
| 17 02 0220 | Demolish Bituminous Curbs | LF | 2.66 | 2.09 | 1.78 | 1.23 | 1.07 |
| 17 02 0221 | Demolish Unreinforced Concrete Curbs | LF | 7.26 | 5.71 | 4.87 | 3.37 | 2.93 |
| 17 02 0222 | Demolish Reinforced Concrete Curbs | LF | 9.90 | 7.79 | 6.64 | 4.59 | 4.00 |
| 17 02 0223 | Demolish Granite Curbs | LF | 6.60 | 5.19 | 4.43 | 3.06 | 2.67 |

**Environmental Remediation: Assemblies Cost Book**

# Cost Data for Remediation
## Demolition, Fencing

### General:

The demolition section can be used to calculate the cost of demolishing fences including wire, chain link, wood, and masonry fences. Note: If the debris is hazardous, costs for transportation and disposal should be estimated using cost data included with the remediation technologies.

### Common Cost Components:

1. Demolition using non-explosive methods

Note: Refer to the Load and Haul section for loading, hauling, and disposal costs for nonhazardous materials

# Cost Data for Site Work

## Demolition, Fencing

| Assembly | Description | Unit | Unit Cost by Safety Level | | | | |
|---|---|---|---|---|---|---|---|
| | | | A | B | C | D | E |
| | **Capital Costs** | | | | | | |
| 17 02 0225 | Remove Chain-link Fence | LF | 3.62 | 2.79 | 2.43 | 1.63 | 1.34 |
| 17 02 0228 | Remove 3-strand Barbed-wire Fence | LF | 2.82 | 2.18 | 1.90 | 1.28 | 1.06 |
| 17 02 0229 | Remove 5-strand Barbed-wire Fence | LF | 4.03 | 3.12 | 2.71 | 1.83 | 1.51 |
| 17 02 0231 | Remove Wood Fence (To 6' High), Maximum | LF | 4.24 | 3.27 | 2.85 | 1.91 | 1.57 |

# Cost Data for Remediation
## Demolition, Pavements

**General:**

Demolition of pavements may be necessary for installation of new roadway, installation of utility lines, and some subsurface remediation projects when the contamination is located below a paved surface. Note: If the debris is hazardous, costs for transportation and disposal should be estimated using cost data included with the remediation technologies.

**Common Cost Components:**

1. Pavement demolition using power or air equipment

Note: Refer to the Load and Haul section for loading, hauling, and disposal costs for nonhazardous materials

# Cost Data for Site Work

## Demolition, Pavements

| Assembly | Description | Unit | Unit Cost by Safety Level | | | | |
|---|---|---|---|---|---|---|---|
| | | | A | B | C | D | E |
| | **Capital Costs** | | | | | | |
| 17 02 0201 | Demolish Bituminous Road with Power Equipment | CY | 46.17 | 36.60 | 30.96 | 21.62 | 19.19 |
| 17 02 0202 | Demolish Bituminous Driveway with Power Equipment | CY | 50.16 | 39.76 | 33.63 | 23.49 | 20.84 |
| 17 02 0203 | Demolish Bituminous Pavement with Air Equipment | CY | 80.15 | 63.38 | 53.76 | 37.41 | 32.98 |
| 17 02 0205 | Demolish Unreinforced Concrete to 6" Thick with Air Equipment | CY | 126.90 | 100.35 | 85.12 | 59.24 | 52.23 |
| 17 02 0206 | Demolish Mesh Reinforced Concrete to 6" Thick with Air Equipment | CY | 183.47 | 145.09 | 123.07 | 85.65 | 75.51 |
| 17 02 0207 | Demolish Rod Reinforced Concrete to 6" Thick with Air Equipment | CY | 199.06 | 157.42 | 133.53 | 92.92 | 81.92 |
| 17 02 0208 | Demolish Mesh Reinforced Concrete to 6" Thick with Power Equipment | CY | 96.73 | 76.68 | 64.87 | 45.30 | 40.20 |
| 17 02 0209 | Demolish Rod Reinforced Concrete to 6" Thick with Power Equipment | CY | 106.91 | 84.75 | 71.70 | 50.07 | 44.43 |
| 17 02 0210 | Demolish Unreinforced Concrete 7" to 24" Thick with Power Equipment | CY | 258.44 | 204.87 | 173.31 | 121.03 | 107.40 |
| 17 02 0211 | Demolish Reinforced Concrete 7" to 24" Thick with Power Equipment | CY | 356.37 | 282.51 | 238.98 | 166.90 | 148.10 |

**Environmental Remediation: Assemblies Cost Book**

# Cost Data for Remediation
## Demolition, Pipes

### General:

Many environmental remediation projects, especially those involving groundwater treatment systems, involve the installation of underground piping. The piping is commonly removed upon completion of the project. Note: Demolition and removal of contaminated or hazardous materials should be estimated using remediation cost data.

### Common Cost Components:

1. Removal of pipe
2. Backfill

Note: Costs for excavation, loading, hauling, and disposal are included elsewhere in this book.

# Cost Data for Site Work

## Demolition, Pipes

| Assembly | Description | Unit | Unit Cost by Safety Level | | | | |
| --- | --- | --- | --- | --- | --- | --- | --- |
| | | | A | B | C | D | E |
| | **Capital Costs** | | | | | | |
| 17 02 0601 | Remove 4" Diameter Steel-welded Connections, Not Including Excavation | LF | 15.80 | 12.45 | 10.60 | 7.34 | 6.41 |
| 17 02 0602 | Remove 10" Diameter Steel-welded Connections, Not Including Excavation | LF | 29.34 | 23.12 | 19.69 | 13.63 | 11.90 |
| 17 02 0603 | Remove 12" Diameter Concrete Pipe, Not Including Excavation | LF | 13.69 | 10.79 | 9.19 | 6.36 | 5.55 |
| 17 02 0604 | Remove 15" Diameter Concrete Pipe, Not Including Excavation | LF | 16.43 | 12.95 | 11.03 | 7.63 | 6.66 |
| 17 02 0605 | Remove 24" Diameter Concrete Pipe, Not Including Excavation | LF | 20.54 | 16.18 | 13.78 | 9.54 | 8.33 |
| 17 02 0606 | Remove 36" Diameter Concrete Pipe, Not Including Excavation | LF | 27.38 | 21.58 | 18.38 | 12.72 | 11.11 |
| 17 03 0259 | Cat 225, 1.5 CY, Soil/Sand, Trenching | CY | 1.84 | 1.49 | 1.23 | 0.88 | 0.82 |
| 17 03 0401 | 950, 3.00 CY, Backfill with Excavated Material | CY | 2.54 | 2.05 | 1.70 | 1.22 | 1.14 |
| 17 03 0420 | Backfill Trench, Borrow Material, Delivered & Dumped Only | CY | 9.86 | 8.74 | 7.93 | 6.83 | 6.63 |
| 17 03 0511 | Compact Soil with Vibrating Plate | CY | 7.20 | 5.57 | 4.84 | 3.26 | 2.70 |
| 17 03 1002 | 2" Diameter Contractor's Trash Pump, 75 GPM | DAY | 50.20 | 47.01 | 45.64 | 42.55 | 41.42 |

# Cost Data for Remediation

## Demolition, Sidewalks

**General:**

Demolition of sidewalks may be necessary for installation of utility lines and other survey and right-of-way repair. Note: If the debris is hazardous, costs for transportation and disposal should be estimated using remediation cost data.

**Common Cost Components:**

1. Sidewalk demolition: bituminous, concrete, brick

Note: Refer to the Load and Haul section for loading, hauling, and disposal costs for nonhazardous materials

# Cost Data for Site Work

## Demolition, Sidewalks

| Assembly | Description | Unit | Unit Cost by Safety Level | | | | |
|---|---|---|---|---|---|---|---|
| | | | A | B | C | D | E |
| | **Capital Costs** | | | | | | |
| 17 02 0215 | Demolish Bituminous Sidewalk | CY | 154.36 | 121.53 | 103.59 | 71.64 | 62.41 |
| 17 02 0216 | Demolish Unreinforced Concrete Sidewalk | CY | 122.48 | 96.43 | 82.20 | 56.84 | 49.52 |
| 17 02 0217 | Demolish Mesh Reinforced Concrete Sidewalk | CY | 130.64 | 102.86 | 87.68 | 60.63 | 52.82 |
| 17 02 0218 | Demolish Brick-Set-in-Mortar Sidewalk | CY | 105.93 | 83.40 | 71.09 | 49.16 | 42.83 |

**Environmental Remediation: Assemblies Cost Book**

# Cost Data for Remediation

## Excavation, Cut and Fill

**General:**

This section provides costs for estimating machine hours required to excavate (cut) and place embankment (fill). Five types of equipment (dozer, scraper, excavator, track loader, wheel loader) are included that are commonly used for various excavation tasks, in addition to two methods of rock removal (ripping and blasting). This section can also be used to calculate excavation and removal to spoil piles. Note: Loading of spoil and hauling off site should be estimated using the Load and Haul section.

**Common Cost Components:**

1. Excavation equipment, including operator costs
2. Blasting, ripping
3. Compaction
4. Borrow soil

## Cost Data for Site Work

### Excavation, Cut and Fill

| Assembly | Description | Unit | A | B | C | D | E |
|---|---|---|---|---|---|---|---|
| | **Capital Costs** | | | | | | |
| 17 03 0206 | D3 with A-blade Bulldozer | HOUR | 151.96 | 122.19 | 101.74 | 72.51 | 66.74 |
| 17 03 0207 | D4 with A-blade Bulldozer | HOUR | 153.56 | 123.53 | 102.80 | 73.31 | 67.54 |
| 17 03 0208 | D5 with A-blade Bulldozer | HOUR | 181.44 | 146.76 | 121.39 | 87.25 | 81.48 |
| 17 03 0209 | D6 with A-blade Bulldozer | HOUR | 208.40 | 169.01 | 139.38 | 100.57 | 94.52 |
| 17 03 0210 | D7 with U-blade Bulldozer | HOUR | 296.12 | 242.11 | 197.86 | 144.43 | 138.38 |
| 17 03 0215 | 931, 1.0 CY, Track Loader | HOUR | 112.08 | 88.96 | 75.15 | 52.57 | 46.80 |
| 17 03 0216 | 943, 1.5 CY, Track Loader | HOUR | 170.50 | 137.43 | 114.12 | 81.62 | 75.57 |
| 17 03 0217 | 953, 2.0 CY, Track Loader | HOUR | 195.08 | 158.13 | 130.48 | 94.07 | 88.30 |
| 17 03 0218 | 963, 2.5 CY, Track Loader | HOUR | 280.58 | 229.38 | 187.48 | 136.82 | 131.05 |
| 17 03 0219 | 973, 3.75 CY, Track Loader | HOUR | 357.58 | 293.33 | 238.84 | 175.16 | 169.11 |
| 17 03 0220 | 910, 1.25 CY, Wheel Loader | HOUR | 130.20 | 104.06 | 87.23 | 61.63 | 55.86 |
| 17 03 0221 | 916, 1.5 CY, Wheel Loader | HOUR | 114.32 | 90.83 | 76.64 | 53.69 | 47.92 |
| 17 03 0222 | 926, 2.0 CY, Wheel Loader | HOUR | 154.30 | 124.14 | 103.30 | 73.68 | 67.91 |
| 17 03 0223 | 950, 3.0 CY, Wheel Loader | HOUR | 180.10 | 145.64 | 120.50 | 86.58 | 80.81 |
| 17 03 0224 | 966, 4.0 CY, Wheel Loader | HOUR | 211.58 | 171.88 | 141.48 | 102.32 | 96.55 |
| 17 03 0225 | 980, 5.25 CY, Wheel Loader | HOUR | 298.84 | 244.38 | 199.68 | 145.79 | 139.74 |
| 17 03 0226 | 988, 7.0 CY, Wheel Loader | HOUR | 384.90 | 316.10 | 257.05 | 188.82 | 182.77 |
| 17 03 0227 | 992, 13.5 CY, Wheel Loader | HOUR | 648.92 | 536.11 | 433.06 | 320.83 | 314.78 |
| 17 03 0230 | Crawler-mounted, 1.0 CY, 215 Hydraulic Excavator | HOUR | 210.40 | 170.50 | 140.73 | 101.43 | 95.15 |
| 17 03 0231 | Crawler-mounted, 1.25 CY, 225 Hydraulic Excavator | HOUR | 210.40 | 170.50 | 140.73 | 101.43 | 95.15 |
| 17 03 0232 | Crawler-mounted, 2.0 CY, 235 Hydraulic Excavator | HOUR | 330.90 | 270.92 | 221.07 | 161.68 | 155.40 |
| 17 03 0233 | Crawler-mounted, 3.125 CY, 245 Hydraulic Excavator | HOUR | 363.36 | 297.97 | 242.71 | 177.91 | 171.63 |

**Environmental Remediation: Assemblies Cost Book**

*Page 4-21*

1998 by ECHOS. All rights reserved

# Cost Data for Site Work

## Excavation, Cut and Fill

| Assembly | Description | Unit | Unit Cost by Safety Level | | | | |
|---|---|---|---|---|---|---|---|
| | | | A | B | C | D | E |
| 17 03 0234 | Crawler-mounted, 4.0 CY, Koehring 1166 Hydraulic Excavator | HOUR | 363.36 | 297.97 | 242.71 | 177.91 | 171.63 |
| 17 03 0235 | Crawler-mounted, 5.5 CY, Koehring 1266, Hydraulic Excavator | HOUR | 491.70 | 404.92 | 328.27 | 242.08 | 235.80 |
| 17 03 0236 | Scraper - Standard, 15 CY - 621 with D8L Dozer | HOUR | 569.73 | 468.78 | 380.40 | 280.19 | 272.40 |
| 17 03 0237 | Scraper - Standard, 22 CY - 631 with D9L Dozers | HOUR | 755.52 | 622.44 | 504.37 | 372.17 | 362.87 |
| 17 03 0238 | Scraper - Standard, 34 CY - 651 with D9L Dozers | HOUR | 652.54 | 536.62 | 435.72 | 320.68 | 311.38 |
| 17 03 0240 | Scraper - Elevating, 11 CY - 613 | HOUR | 181.08 | 146.07 | 121.19 | 86.77 | 80.49 |
| 17 03 0241 | Scraper - Elevating, 22 CY - 623 | HOUR | 280.40 | 228.83 | 187.40 | 136.43 | 130.15 |
| 17 03 0243 | Scraper - Tandem, 14 CY - 627 with D8L Dozer | HOUR | 392.35 | 320.97 | 262.15 | 191.50 | 183.71 |
| 17 03 0244 | Scraper - Tandem, 21 CY - 637 with D9L Dozer | HOUR | 555.75 | 457.13 | 371.08 | 273.19 | 265.40 |
| 17 03 0245 | Scraper - Tandem, 32 CY - 657 with D9L Dozers | HOUR | 762.00 | 627.84 | 508.69 | 375.41 | 366.11 |
| 17 03 0246 | Scraper - Tandem Elevating, 34 CY - with D9L Dozer | HOUR | 655.76 | 540.47 | 437.76 | 323.20 | 315.41 |
| 17 03 0301 | 2 1/2" Bore @ 25' with Track Drill | LF | 4.78 | 3.89 | 3.20 | 2.31 | 2.19 |
| 17 03 0302 | 5 1/2" Bore @ 25' with Track Drill | LF | 21.97 | 17.72 | 14.79 | 10.61 | 9.79 |
| 17 03 0303 | Ammonium Nitrate Explosive, Loaded, Packed & Blown | LB | 9.78 | 8.59 | 8.05 | 6.90 | 6.50 |
| 17 03 0310 | D8 with U-blade & Single-shank Ripper, Bulldozer | HOUR | 396.60 | 325.85 | 264.85 | 194.67 | 188.62 |
| 17 03 0311 | D9 with U-blade & Single-shank Ripper, Bulldozer | HOUR | 455.94 | 375.30 | 304.41 | 224.34 | 218.29 |
| 17 03 0312 | D10 with U-blade & Single-shank Ripper, Bulldozer | HOUR | 476.58 | 392.50 | 318.17 | 234.66 | 228.61 |
| 17 03 0424 | Unclassified Fill, Delivered, Off-Site | CY | 4.03 | 4.03 | 4.03 | 4.03 | 4.03 |
| 17 03 0517 | Spread/Compact Large Areas, 6" Lifts, D8 & Towed Sheepsfoot | CY | 1.42 | 1.15 | 0.95 | 0.68 | 0.64 |
| 17 03 0518 | Spread/Compact Large Areas, 8" Lifts, D8 & Towed Sheepsfoot | CY | 0.79 | 0.64 | 0.53 | 0.38 | 0.36 |

# Cost Data for Remediation

## Excavation, Trench/Channel

**General:**

Channel excavation includes relocation of a creek or stream, usually because it flows through a right-of-way. Draglines are typically used in swampy areas or where the underlying soil is too soft to support a front-end loader or shovel. Clamshells are generally used in trenching where sheeting of sides is required and it is necessary to dig between braces and to great depths. Excavators are typically used for trenching or borrow excavation on dry, stable ground.

**Common Cost Components:**

1. Excavation equipment, including operator costs

Note: Excavation of contaminated materials should be estimated using remediation cost data.

## Cost Data for Site Work

### Excavation, Trench/Channel

| Assembly | Description | Unit | A | B | C | D | E |
|---|---|---|---|---|---|---|---|
| | **Capital Costs** | | | | | | |
| 17 03 0230 | Crawler-mounted, 1.0 CY, 215 Hydraulic Excavator | HOUR | 210.40 | 170.50 | 140.73 | 101.43 | 95.15 |
| 17 03 0231 | Crawler-mounted, 1.25 CY, 225 Hydraulic Excavator | HOUR | 210.40 | 170.50 | 140.73 | 101.43 | 95.15 |
| 17 03 0232 | Crawler-mounted, 2.0 CY, 235 Hydraulic Excavator | HOUR | 330.90 | 270.92 | 221.07 | 161.68 | 155.40 |
| 17 03 0233 | Crawler-mounted, 3.125 CY, 245 Hydraulic Excavator | HOUR | 363.36 | 297.97 | 242.71 | 177.91 | 171.63 |
| 17 03 0234 | Crawler-mounted, 4.0 CY, Koehring 1166 Hydraulic Excavator | HOUR | 363.36 | 297.97 | 242.71 | 177.91 | 171.63 |
| 17 03 0235 | Crawler-mounted, 5.5 CY, Koehring 1266, Hydraulic Excavator | HOUR | 491.70 | 404.92 | 328.27 | 242.08 | 235.80 |
| 17 03 0247 | 1.0 CY Dragline, with 40' Boom | HOUR | 271.10 | 217.12 | 181.58 | 128.69 | 117.26 |
| 17 03 0248 | 1.5 CY Dragline, with 60' Boom | HOUR | 367.88 | 297.77 | 246.10 | 177.08 | 165.65 |
| 17 03 0249 | 2.5 CY Dragline, with 50' Boom | HOUR | 353.00 | 285.37 | 236.18 | 169.64 | 158.21 |
| 17 03 0250 | 4 CY Dragline, with 95' Boom | HOUR | 418.84 | 340.24 | 280.08 | 202.56 | 191.13 |
| 17 03 0251 | 1 CY Clamshell, with 40' Boom | HOUR | 281.10 | 225.45 | 188.25 | 133.69 | 122.26 |
| 17 03 0252 | 1.5 CY Clamshell, with 60' Boom | HOUR | 381.74 | 309.32 | 255.34 | 184.01 | 172.58 |
| 17 03 0253 | 2.5 CY Clamshell, with 50' Boom | HOUR | 368.94 | 298.65 | 246.81 | 177.61 | 166.18 |

**Environmental Remediation: Assemblies Cost Book**

# Cost Data for Remediation

## Fencing

**General:**

Fencing is often required at remediation sites for public protection. Barbed wire, gates, and warning signs may also be required.

**Common Cost Components:**

1. Installation of fence, including labor and materials
2. Gates
3. Signs
4. Barbed wire

## Cost Data for Site Work

### Fencing

| Assembly | Description | Unit | Unit Cost by Safety Level | | | | |
|---|---|---|---|---|---|---|---|
| | | | A | B | C | D | E |
| | **Capital Costs** | | | | | | |
| 18 04 0101 | Security Fence, 10' Galvanized with 3 Strands Barbed Wire | LF | 66.09 | 58.86 | 54.27 | 47.20 | 45.57 |
| 18 04 0102 | Security Fence, with 1' x 1' Grade Beam, 10' Galvanized with 3 Strands Barbed Wire | LF | 69.37 | 61.98 | 57.31 | 50.08 | 48.39 |
| 18 04 0103 | Privacy Fence, 6' High, Wood | LF | 19.11 | 16.35 | 15.16 | 12.50 | 11.52 |
| 18 04 0105 | Boundary Fence, 5' Galvanized | LF | 10.28 | 8.46 | 7.32 | 5.54 | 5.11 |
| 18 04 0106 | 5' Galvanized Chain-link Fence | LF | 19.90 | 19.20 | 18.91 | 18.24 | 18.00 |
| 18 04 0107 | 6' Galvanized Chain-link Fence | LF | 21.07 | 20.35 | 20.04 | 19.35 | 19.09 |
| 18 04 0108 | 7' Galvanized Chain-link Fence | LF | 27.68 | 26.93 | 26.61 | 25.88 | 25.62 |
| 18 04 0110 | Barbed-wire Fencing, 3-Strand | LF | 1.36 | 1.36 | 1.36 | 1.36 | 1.36 |
| 18 04 0111 | Galvanized Barbed Wire, 3-Strand | LF | 1.81 | 1.45 | 1.27 | 0.92 | 0.81 |
| 18 04 0115 | Swing Gates, Complete | EACH | 1,122 | 1,024 | 960.82 | 864.14 | 841.88 |
| 18 04 0116 | 5' Swing Gate, 12' Double | EACH | 780.94 | 725.61 | 697.51 | 643.78 | 626.90 |
| 18 04 0117 | 6' Swing Gate, 12' Double | EACH | 896.03 | 836.09 | 805.65 | 747.44 | 729.16 |
| 18 04 0118 | 7' Swing Gate, 12' Double | EACH | 329.09 | 329.09 | 329.09 | 329.09 | 329.09 |
| 18 04 0501 | Hazardous Waste Signing | EACH | 86.60 | 70.54 | 63.66 | 48.13 | 42.44 |

**Environmental Remediation: Assemblies Cost Book**

# Cost Data for Remediation
## Gas Distribution

**General:**

A natural gas supply is sometimes required at a remediation site. The costs presented here can be used to estimate the construction of a main distribution line to the site. Removal and disposal of the excavated soil is addressed in the Load and Haul section.

**Common Cost Components:**

1. Trenching
2. Pipe
3. Backfill
4. Seal slab
5. Trench box
6. Dewatering pump
7. Wellpoints

## Cost Data for Site Work

### Gas Distribution

| Assembly | Description | Unit | Unit Cost by Safety Level |  |  |  |  |
|---|---|---|---|---|---|---|---|
|  |  |  | A | B | C | D | E |
|  | **Capital Costs** |  |  |  |  |  |  |
| 17 03 0259 | Cat 225, 1.5 CY, Soil/Sand, Trenching | CY | 1.84 | 1.49 | 1.23 | 0.88 | 0.82 |
| 17 03 0260 | Cat 225, 1.5 CY, Soil/Sand, 10' - 20' Deep Trench Box, Trench | CY | 7.22 | 6.04 | 5.20 | 4.04 | 3.82 |
| 17 03 0261 | Cat 225, 1.5 CY, Soil/Sand with Boulders, Trenching | CY | 3.91 | 3.15 | 2.62 | 1.87 | 1.74 |
| 17 03 0272 | Pull Trench Box, Cat 225, 1.5 CY, Soil/Sand, Trenching | CY | 6.67 | 5.49 | 4.65 | 3.48 | 3.27 |
| 17 03 0306 | Cat 225, 1.5 CY, Rock, No Haul or Borrow, Trenching | BCY | 149.48 | 122.48 | 106.81 | 80.48 | 73.45 |
| 17 03 0401 | 950, 3.00 CY, Backfill with Excavated Material | CY | 2.54 | 2.05 | 1.70 | 1.22 | 1.14 |
| 17 03 0405 | 950, 3.00 CY, Delivered & Dumped, Backfill with Sand | CY | 38.31 | 35.03 | 32.95 | 29.74 | 29.00 |
| 17 03 0410 | 950, Delivered & Dumped, Backfill with Cement Stabilized Base Material | CY | 46.93 | 43.66 | 41.58 | 38.37 | 37.63 |
| 17 03 0420 | Backfill Trench, Borrow Material, Delivered & Dumped Only | CY | 9.86 | 8.74 | 7.93 | 6.83 | 6.63 |
| 17 03 0511 | Compact Soil with Vibrating Plate | CY | 7.20 | 5.57 | 4.84 | 3.26 | 2.70 |
| 17 03 0515 | Compact With Pogosticks | CY | 5.91 | 4.58 | 3.98 | 2.68 | 2.23 |
| 17 03 0516 | Compact with 50% Pogosticks, 50% Hand Roller | CY | 4.24 | 3.30 | 2.85 | 1.94 | 1.63 |
| 17 03 1001 | Wellpoint for Trench, Install & Remove <500', 1 Month | LFHD | 55.06 | 55.06 | 55.06 | 55.06 | 55.06 |
| 17 03 1002 | 2" Diameter Contractor's Trash Pump, 75 GPM | DAY | 50.20 | 47.01 | 45.64 | 42.55 | 41.42 |
| 18 04 0203 | Pour & Cure Concrete, Continuous Footing | CY | 140.13 | 126.22 | 118.87 | 105.35 | 101.29 |
| 19 07 0101 | 1" Black Steel Pipe, Welded T & C Schedule 40 | LF | 5.70 | 5.00 | 4.68 | 4.00 | 3.75 |
| 19 07 0102 | 2" Black Steel Pipe, Welded T & C Schedule 40 | LF | 7.65 | 6.76 | 6.37 | 5.51 | 5.21 |
| 19 07 0103 | 3" Black Steel Pipe, Welded T & C Schedule 40 | LF | 7.83 | 6.82 | 6.37 | 5.39 | 5.04 |
| 19 07 0104 | 4" Black Steel Pipe, Welded T & C Schedule 40 | LF | 9.70 | 8.57 | 8.06 | 6.97 | 6.58 |
| 19 07 0105 | 5" Black Steel Pipe, Welded T & C Schedule 40 | LF | 13.96 | 12.73 | 12.18 | 10.99 | 10.57 |
| 19 07 0106 | 6" Black Steel Pipe, Welded T & C Schedule 40 | LF | 14.65 | 13.20 | 12.47 | 11.06 | 10.61 |
| 19 07 0107 | 8" Black Steel Pipe, Welded T & C Schedule 40 | LF | 18.68 | 17.22 | 16.49 | 15.09 | 14.64 |
| 19 07 0120 | 1 1/4", 60 PSI, Polyethylene Pipe | LF | 5.29 | 4.21 | 3.72 | 2.68 | 2.32 |

**Environmental Remediation: Assemblies Cost Book**

# Cost Data for Site Work

## Gas Distribution

| Assembly | Description | Unit | Unit Cost by Safety Level | | | | |
|---|---|---|---|---|---|---|---|
| | | | A | B | C | D | E |
| 19 07 0121 | 1 1/2", 60 PSI, Polyethylene Pipe | LF | 5.38 | 4.30 | 3.81 | 2.77 | 2.41 |
| 19 07 0122 | 2", 60 PSI, Polyethylene Pipe | LF | 6.05 | 4.85 | 4.31 | 3.15 | 2.75 |
| 19 07 0123 | 3" Polyethylene 40' Joints, 60 PSI | LF | 7.90 | 6.47 | 5.82 | 4.42 | 3.94 |
| 19 07 0124 | 4" Polyethylene 40' Joints, 60 PSI | LF | 19.10 | 15.75 | 13.90 | 10.64 | 9.71 |
| 19 07 0125 | 6" Polyethylene 40' Joints, 60 PSI | LF | 25.85 | 22.22 | 20.22 | 16.68 | 15.67 |
| 19 07 0126 | 8" Polyethylene 40' Joints, 60 PSI | LF | 35.37 | 31.01 | 28.61 | 24.37 | 23.16 |
| 20 06 0101 | 3 - 9 Lb Magnesium Anodes, Cathodic Protection Point | EACH | 953.82 | 778.29 | 693.26 | 523.02 | 466.94 |
| 20 06 0102 | 3 - 17 Lb Magnesium Anodes, Cathodic Protection Point | EACH | 1,364 | 1,121 | 1,002 | 766.32 | 688.94 |
| 20 06 0103 | 3 - 32 Lb Magnesium Anodes, Cathodic Protection Point | EACH | 1,845 | 1,539 | 1,390 | 1,092 | 995.23 |
| 20 06 0104 | 3 - 48 Lb Magnesium Anodes, Cathodic Protection Point | EACH | 2,585 | 2,166 | 1,960 | 1,553 | 1,420 |

**Environmental Remediation: Assemblies Cost Book**

# Cost Data for Remediation
## Heating/Cooling Distribution System

**General:**

When a facility requires heating and cooling from a central plant source, this section will provide the cost for overhead or buried distribution systems. The overhead system includes three types of support frames: wood, galvanized steel, and aluminum. There are two methods of underground distribution: direct buried and precast concrete trench sections. Note: The model assumes no clearing required, ready access, and right-of-way availability.

**Common Cost Components:**

1. Trenching
2. Pipe
3. Backfill
4. Seal slab
5. Trench box
6. Dewatering pump

# Cost Data for Site Work

## Heating/Cooling Distribution System

| Assembly | Description | Unit | Unit Cost by Safety Level | | | | |
|---|---|---|---|---|---|---|---|
| | | | A | B | C | D | E |
| | **Capital Costs** | | | | | | |
| 17 03 0262 | Cat 235, 2.5 CY, Soil/Sand, Trenching | CY | 2.01 | 1.62 | 1.35 | 0.97 | 0.90 |
| 17 03 0265 | Cat 245, 3.0 CY, Soil/Sand, Trenching | CY | 2.78 | 2.23 | 1.86 | 1.33 | 1.22 |
| 17 03 0273 | Pull Trench Box, Cat 235, 2.5 CY, Soil/Sand, Trenching | CY | 5.31 | 4.37 | 3.68 | 2.75 | 2.59 |
| 17 03 0307 | Cat 235, 2.0 CY, Rock, No Haul or Borrow, Trenching | BCY | 133.78 | 109.18 | 95.00 | 71.01 | 64.55 |
| 17 03 0401 | 950, 3.00 CY, Backfill with Excavated Material | CY | 2.54 | 2.05 | 1.70 | 1.22 | 1.14 |
| 17 03 0402 | 966, 4.00 CY, Backfill with Excavated Material | CY | 1.70 | 1.38 | 1.13 | 0.82 | 0.77 |
| 17 03 0405 | 950, 3.00 CY, Delivered & Dumped, Backfill with Sand | CY | 38.31 | 35.03 | 32.95 | 29.74 | 29.00 |
| 17 03 0406 | 966, 4.00 CY, Delivered & Dumped, Backfill with Sand | CY | 37.79 | 34.61 | 32.60 | 29.49 | 28.76 |
| 17 03 0410 | 950, Delivered & Dumped, Backfill with Cement Stabilized Base Material | CY | 46.93 | 43.66 | 41.58 | 38.37 | 37.63 |
| 17 03 0411 | 966, Delivered & Dumped, Backfill with Cement Stabilized Sand | CY | 46.41 | 43.23 | 41.23 | 38.12 | 37.39 |
| 17 03 0420 | Backfill Trench, Borrow Material, Delivered & Dumped Only | CY | 9.86 | 8.74 | 7.93 | 6.83 | 6.63 |
| 17 03 0511 | Compact Soil with Vibrating Plate | CY | 7.20 | 5.57 | 4.84 | 3.26 | 2.70 |
| 17 03 0515 | Compact With Pogosticks | CY | 5.91 | 4.58 | 3.98 | 2.68 | 2.23 |
| 17 03 0516 | Compact with 50% Pogosticks, 50% Hand Roller | CY | 4.24 | 3.30 | 2.85 | 1.94 | 1.63 |
| 17 03 1002 | 2" Diameter Contractor's Trash Pump, 75 GPM | DAY | 50.20 | 47.01 | 45.64 | 42.55 | 41.42 |
| 18 04 0203 | Pour & Cure Concrete, Continuous Footing | CY | 140.13 | 126.22 | 118.87 | 105.35 | 101.29 |
| 19 05 0101 | Pre-Insulated Steel Pipe, 1 1/4" Supply/Return HTHW W/H | LF | 34.27 | 30.27 | 27.98 | 24.07 | 23.00 |
| 19 05 0102 | Pre-Insulated Steel Pipe, 1 1/2" Supply/Return HTHW W/H | LF | 35.90 | 31.83 | 29.52 | 25.55 | 24.45 |
| 19 05 0103 | Pre-Insulated Steel Pipe, 2" Supply/Return HTHW W/H | LF | 40.72 | 35.99 | 33.27 | 28.65 | 27.40 |
| 19 05 0104 | Pre-Insulated Steel Pipe, 2 1/2" Supply/Return HTHW W/H | LF | 43.89 | 38.67 | 35.64 | 30.54 | 29.18 |
| 19 05 0105 | Pre-Insulated Steel Pipe, 3" Supply/Return HTHW W/H | LF | 48.83 | 42.74 | 39.23 | 33.29 | 31.69 |
| 19 05 0106 | Pre-Insulated Steel Pipe, 4" Supply/Return HTHW W/H | LF | 59.96 | 52.09 | 47.48 | 39.80 | 37.78 |

**Environmental Remediation: Assemblies Cost Book**

# Cost Data for Site Work

## Heating/Cooling Distribution System

| Assembly | Description | Unit | A | B | C | D | E |
|---|---|---|---|---|---|---|---|
| 19 05 0107 | Pre-Insulated Steel Pipe, 6" Supply/Return HTHW W/H | LF | 78.94 | 69.39 | 63.72 | 54.40 | 51.99 |
| 19 05 0108 | Pre-Insulated Steel Pipe, 8" Supply/Return HTHW W/H | LF | 99.14 | 88.02 | 81.47 | 70.62 | 67.79 |
| 19 05 0109 | Pre-Insulated Steel Pipe, 10" Supply/Return HTHW W/H | LF | 124.40 | 111.93 | 104.68 | 92.52 | 89.29 |
| 19 05 0201 | Field Insulated Steel Pipe, 1 1/4" Supply, 3/4" Return Steam W/H | LF | 40.42 | 33.90 | 31.10 | 24.80 | 22.50 |
| 19 05 0202 | Field Insulated Steel Pipe, 1 1/2" Supply, 1" Return Steam W/H | LF | 44.07 | 36.99 | 33.96 | 27.12 | 24.61 |
| 19 05 0203 | Field Insulated Steel Pipe, 2" Supply, 1" Return Steam W/H | LF | 46.95 | 39.44 | 36.22 | 28.97 | 26.31 |
| 19 05 0204 | Field Insulated Steel Pipe, 2 1/2" Supply, 1 1/2"Return Steam W/H | LF | 54.77 | 46.22 | 42.55 | 34.28 | 31.26 |
| 19 05 0205 | Field Insulated Steel Pipe, 3" Supply, 2" Return Steam W/H | LF | 64.40 | 54.29 | 49.95 | 40.17 | 36.59 |
| 19 05 0206 | Field Insulated Steel Pipe, 4" Supply, 2" Return Steam W/H | LF | 72.68 | 61.66 | 56.94 | 46.29 | 42.40 |
| 19 05 0207 | Field Insulated Steel Pipe, 6" Supply, 2 1/2" Return Steam W/H | LF | 94.43 | 80.39 | 74.37 | 60.80 | 55.83 |
| 19 05 0208 | Field Insulated Steel Pipe, 8" Supply, 3" Return Steam W/H | LF | 103.14 | 87.29 | 80.50 | 65.18 | 59.57 |
| 19 05 0209 | Field Insulated Steel Pipe, 10" Supply, 4" Return Steam W/H | LF | 129.77 | 110.85 | 102.74 | 84.45 | 77.75 |
| 19 05 0301 | Pre-Insulated Steel Pipe, 1 1/4" Supply/Return HTHW Direct Burial | LF | 27.84 | 25.22 | 23.53 | 20.97 | 20.39 |
| 19 05 0302 | Pre-Insulated Steel Pipe, 1 1/2" Supply/Return HTHW Direct Burial | LF | 28.95 | 26.38 | 24.72 | 22.20 | 21.63 |
| 19 05 0303 | Pre-Insulated Steel Pipe, 2" Supply/Return HTHW Direct Burial | LF | 33.38 | 30.21 | 28.17 | 25.07 | 24.38 |
| 19 05 0304 | Pre-Insulated Steel Pipe, 2 1/2" Supply/Return HTHW Direct Burial | LF | 36.18 | 32.60 | 30.29 | 26.79 | 26.00 |
| 19 05 0305 | Pre-Insulated Steel Pipe, 3" Supply/Return HTHW Direct Burial | LF | 39.61 | 35.49 | 32.83 | 28.80 | 27.89 |
| 19 05 0306 | Pre-Insulated Steel Pipe, 4" Supply/Return HTHW Direct Burial | LF | 49.47 | 43.81 | 40.16 | 34.62 | 33.37 |
| 19 05 0307 | Pre-Insulated Steel Pipe, 6" Supply/Return HTHW Direct Burial | LF | 67.38 | 60.18 | 55.53 | 48.47 | 46.89 |
| 19 05 0308 | Pre-Insulated Steel Pipe, 8" Supply/Return HTHW Direct Burial | LF | 84.42 | 76.20 | 70.89 | 62.84 | 61.04 |
| 19 05 0309 | Pre-Insulated Steel Pipe, 10" Supply/Return HTHW Direct Burial | LF | 104.59 | 95.84 | 90.19 | 81.62 | 79.70 |
| 19 05 0401 | Field Insulated Steel Pipe, 1 1/4" Supply, 3/4" Return Steam Direct Burial | LF | 34.20 | 29.02 | 26.81 | 21.81 | 19.97 |
| 19 05 0402 | Field Insulated Steel Pipe, 1 1/2" Supply, 1" Return Steam Direct Burial | LF | 37.48 | 31.83 | 29.40 | 23.94 | 21.93 |
| 19 05 0403 | Field Insulated Steel Pipe, 2" Supply, 1" Return Steam Direct Burial | LF | 40.17 | 34.12 | 31.52 | 25.67 | 23.53 |
| 19 05 0404 | Field Insulated Steel Pipe, 2 1/2" Supply, 1 1/2"Return Steam Direct Burial | LF | 47.43 | 40.46 | 37.47 | 30.73 | 28.26 |

**Environmental Remediation: Assemblies Cost Book**

*Page 4-28*

1998 by ECHOS. All rights reserved

# Cost Data for Site Work

## Heating/Cooling Distribution System

| Assembly | Description | Unit | A | B | C | D | E |
|---|---|---|---|---|---|---|---|
| 19 05 0405 | Field Insulated Steel Pipe, 3" Supply, 2" Return Steam Direct Burial | LF | 55.59 | 47.36 | 43.84 | 35.88 | 32.97 |
| 19 05 0406 | Field Insulated Steel Pipe, 4" Supply, 2" Return Steam Direct Burial | LF | 63.23 | 54.22 | 50.36 | 41.66 | 38.46 |
| 19 05 0407 | Field Insulated Steel Pipe, 6" Supply, 2 1/2" Return Steam Direct Burial | LF | 83.53 | 71.75 | 66.71 | 55.33 | 51.16 |
| 19 05 0408 | Field Insulated Steel Pipe, 8" Supply, 3" Return Steam Direct Burial | LF | 90.80 | 77.46 | 71.75 | 58.87 | 54.14 |
| 19 05 0409 | Field Insulated Steel Pipe, 10" Supply, 4" Return Steam Direct Burial | LF | 113.88 | 98.08 | 91.31 | 76.03 | 70.43 |
| 19 05 0501 | 2' x 2' (Inside), Precast Trench Box | LF | 44.12 | 35.74 | 31.08 | 22.91 | 20.61 |
| 19 05 0502 | 2' 6" x 2' (Inside), Precast Trench Box | LF | 46.84 | 38.02 | 33.11 | 24.52 | 22.09 |
| 19 05 0503 | 3' x 2' (Inside), Precast Trench Box | LF | 49.81 | 40.49 | 35.31 | 26.24 | 23.68 |
| 19 05 0504 | 3' 6" x 2' (Inside), Precast Trench Box | LF | 53.03 | 43.17 | 37.69 | 28.08 | 25.37 |
| 19 05 0505 | 4' x 2' (Inside), Precast Trench Box | LF | 56.58 | 46.10 | 40.28 | 30.07 | 27.19 |
| 19 05 0506 | 5' x 2' (Inside), Precast Trench Box | LF | 60.85 | 49.68 | 43.46 | 32.57 | 29.50 |
| 19 05 0507 | 6' x 2' (Inside), Precast Trench Box | LF | 65.59 | 53.61 | 46.95 | 35.29 | 31.99 |
| 19 05 9901 | 2 - 20' Wood Supports for Overhead Heat Distribution - Complete | EACH | 2,603 | 2,151 | 1,934 | 1,496 | 1,351 |
| 19 05 9902 | 2 - 30' Wood Supports for Overhead Heat Distribution - Complete | EACH | 2,729 | 2,257 | 2,029 | 1,572 | 1,420 |
| 19 05 9905 | 2 - 20' Galvanized Steel Supports for Overhead Heat Distribution - Complete | EACH | 4,701 | 4,194 | 3,955 | 3,464 | 3,298 |
| 19 05 9906 | 2 - 30' Galvanized Steel Supports for Overhead Heat Distribution - Complete | EACH | 5,693 | 5,103 | 4,823 | 4,251 | 4,059 |
| 19 05 9910 | 2 - 20' Aluminum Supports for Overhead Heat Distribution - Complete | EACH | 3,870 | 3,397 | 3,175 | 2,716 | 2,560 |
| 19 05 9911 | 2 - 30' Aluminum Supports for Overhead Heat Distribution - Complete | EACH | 5,890 | 5,333 | 5,070 | 4,531 | 4,349 |
| 19 06 0101 | Pre-Insulated Steel Pipe, 1 1/4" Supply/Return Chilled Water W/H | LF | 34.27 | 30.27 | 27.98 | 24.07 | 23.00 |
| 19 06 0102 | Pre-Insulated Steel Pipe, 1 1/2" Supply/Return Chilled Water W/H | LF | 35.90 | 31.83 | 29.52 | 25.55 | 24.45 |
| 19 06 0103 | Pre-Insulated Steel Pipe, 2" Supply/Return Chilled Water W/H | LF | 40.72 | 35.99 | 33.27 | 28.65 | 27.40 |
| 19 06 0104 | Pre-Insulated Steel Pipe, 2 1/2" Supply/Return Chilled Water W/H | LF | 43.89 | 38.67 | 35.64 | 30.54 | 29.18 |
| 19 06 0105 | Pre-Insulated Steel Pipe, 3" Supply/Return Chilled Water W/H | LF | 48.83 | 42.74 | 39.23 | 33.29 | 31.69 |
| 19 06 0106 | Pre-Insulated Steel Pipe, 4" Supply/Return Chilled Water W/H | LF | 59.96 | 52.09 | 47.48 | 39.80 | 37.78 |
| 19 06 0107 | Pre-Insulated Steel Pipe, 6" Supply/Return Chilled Water W/H | LF | 78.94 | 69.39 | 63.72 | 54.40 | 51.99 |

**Environmental Remediation: Assemblies Cost Book**

*Page 4-29*

1998 by ECHOS. All rights reserved

# Cost Data for Site Work

## Heating/Cooling Distribution System

| Assembly | Description | Unit | A | B | C | D | E |
|---|---|---|---|---|---|---|---|
| | | | **Unit Cost by Safety Level** | | | | |
| 19 06 0108 | Pre-Insulated Steel Pipe, 8" Supply/Return Chilled Water W/H | LF | 99.14 | 88.02 | 81.47 | 70.62 | 67.79 |
| 19 06 0109 | Pre-Insulated Steel Pipe, 10" Supply/Return Chilled Water W/H | LF | 124.40 | 111.93 | 104.68 | 92.52 | 89.29 |
| 19 06 0201 | Pre-Insulated Steel Pipe, 1 1/4" Supply/Return Chilled Water Direct Burial | LF | 27.84 | 25.22 | 23.53 | 20.97 | 20.39 |
| 19 06 0202 | Pre-Insulated Steel Pipe, 1 1/2" Supply/Return Chilled Water Direct Burial | LF | 28.95 | 26.38 | 24.72 | 22.20 | 21.63 |
| 19 06 0203 | Pre-Insulated Steel Pipe, 2" Supply/Return Chilled Water Direct Burial | LF | 33.38 | 30.21 | 28.17 | 25.07 | 24.38 |
| 19 06 0204 | Pre-Insulated Steel Pipe, 2 1/2" Supply/Return Chilled Water Direct Burial | LF | 36.18 | 32.60 | 30.29 | 26.79 | 26.00 |
| 19 06 0205 | Pre-Insulated Steel Pipe, 3" Supply/Return Chilled Water Direct Burial | LF | 39.61 | 35.49 | 32.83 | 28.80 | 27.89 |
| 19 06 0206 | Pre-Insulated Steel Pipe, 4" Supply/Return Chilled Water Direct Burial | LF | 49.47 | 43.81 | 40.16 | 34.62 | 33.37 |
| 19 06 0207 | Pre-Insulated Steel Pipe, 6" Supply/Return Chilled Water Direct Burial | LF | 67.38 | 60.18 | 55.53 | 48.47 | 46.89 |
| 19 06 0208 | Pre-Insulated Steel Pipe, 8" Supply/Return Chilled Water Direct Burial | LF | 84.42 | 76.20 | 70.89 | 62.84 | 61.04 |
| 19 06 0209 | Pre-Insulated Steel Pipe, 10" Supply/Return Chilled Water Direct Burial | LF | 104.59 | 95.84 | 90.19 | 81.62 | 79.70 |
| 19 06 0301 | 2' x 2' (Inside), Precast Trench Box | LF | 44.12 | 35.74 | 31.08 | 22.91 | 20.61 |
| 19 06 0302 | 2' 6" x 2' (Inside), Precast Trench Box | LF | 46.84 | 38.02 | 33.11 | 24.52 | 22.09 |
| 19 06 0303 | 3' x 2' (Inside), Precast Trench Box | LF | 49.81 | 40.49 | 35.31 | 26.24 | 23.68 |
| 19 06 0304 | 3' 6" x 2' (Inside), Precast Trench Box | LF | 53.03 | 43.17 | 37.69 | 28.08 | 25.37 |
| 19 06 0305 | 4' x 2' (Inside), Precast Trench Box | LF | 56.58 | 46.10 | 40.28 | 30.07 | 27.19 |
| 19 06 0306 | 5' x 2' (Inside), Precast Trench Box | LF | 60.85 | 49.68 | 43.46 | 32.57 | 29.50 |
| 19 06 0307 | 6' x 2' (Inside), Precast Trench Box | LF | 65.59 | 53.61 | 46.95 | 35.29 | 31.99 |

**Environmental Remediation: Assemblies Cost Book**

# Cost Data for Remediation

## Lighting, Interstate, Roadway, Parking

**General:**

This section provides the capital cost of illuminating areas for vehicle/pedestrian usage. It is assumed that all fixtures and appurtenances are permanent and will utilize local utility power. Applicable building codes should be consulted for variances that may affect the quantities and total cost.

**Common Cost Components:**

1. Trenching
2. Backfill
3. Wire
4. Poles
5. Fixtures
6. Conduit
7. Transformers

# Cost Data for Site Work

## Lighting, Interstate, Roadway, Parking

| Assembly | Description | Unit | Unit Cost by Safety Level | | | | |
|---|---|---|---|---|---|---|---|
| | | | A | B | C | D | E |
| | **Capital Costs** | | | | | | |
| 17 03 0255 | Trenching to 48" Deep, Including Backfill & Compaction | CY | 10.66 | 8.37 | 7.16 | 4.93 | 4.26 |
| 17 03 0257 | Cat 215, 1.0 CY, Soil, Shallow, Trenching | CY | 2.42 | 1.95 | 1.62 | 1.16 | 1.07 |
| 17 03 0282 | Soil, 5 Miles, Dump Truck, Load/Haul Spoil From Trench | CY | 6.97 | 5.60 | 4.66 | 3.33 | 3.06 |
| 17 03 0401 | 950, 3.00 CY, Backfill with Excavated Material | CY | 2.54 | 2.05 | 1.70 | 1.22 | 1.14 |
| 17 03 0417 | Delivered & Dumped - Hand, Backfill with Sand | CY | 82.86 | 70.10 | 62.85 | 50.41 | 47.00 |
| 17 03 0511 | Compact Soil with Vibrating Plate | CY | 7.20 | 5.57 | 4.84 | 3.26 | 2.70 |
| 17 03 0515 | Compact With Pogosticks | CY | 5.91 | 4.58 | 3.98 | 2.68 | 2.23 |
| 20 02 0501 | 1/C #4 Copper Grounded 600V Direct Burial, Wire | LF | 1.50 | 1.27 | 1.18 | 0.96 | 0.88 |
| 20 02 0502 | 1/C #6 Copper Grounded 600V Direct Burial, Wire | LF | 1.14 | 0.95 | 0.87 | 0.69 | 0.62 |
| 20 02 0503 | 3/C #6, W/#6 Grounded 5 KV Direct Burial, Wire | LF | 9.43 | 8.11 | 7.46 | 6.18 | 5.76 |
| 20 02 0504 | 3/c Underground, 15 KV Direct Burial, #1 with Bare Grounded, Wire | LF | 19.95 | 18.36 | 17.58 | 16.05 | 15.55 |
| 20 02 0505 | 2 - #2 Underground, 600V Direct Burial, Wire | LF | 6.10 | 5.44 | 5.11 | 4.47 | 4.26 |
| 20 02 0508 | #10 Insulated Strand Wire (THW) | LF | 0.90 | 0.74 | 0.68 | 0.53 | 0.47 |
| 20 03 0101 | 4160V/480V, 1, 37.5 KVA Transformer, Oil, Pad Mounted | EACH | 2,817 | 2,601 | 2,505 | 2,296 | 2,221 |
| 20 03 0102 | 5/15 KV - 277/480V, 75 KVA Transformer, Oil, Pad Mounted | EACH | 11,019 | 10,571 | 10,372 | 9,939 | 9,785 |
| 20 03 0301 | 30' Area Lighting Pole | EACH | 2,168 | 2,044 | 1,983 | 1,862 | 1,823 |
| 20 03 0302 | 30' Galvanized Steel Pole (Single Arm) with Base | EACH | 2,579 | 2,352 | 2,244 | 2,024 | 1,950 |
| 20 03 0303 | 20' Brushed Aluminum Pole (Single Arm) with Base | EACH | 1,656 | 1,494 | 1,418 | 1,260 | 1,207 |
| 20 03 0601 | 150W High Pressure Sodium Fixture | EACH | 586.61 | 519.47 | 490.73 | 425.83 | 402.04 |
| 20 03 0602 | 400W High Intensity Sodium Fixture | EACH | 1,123 | 1,069 | 1,046 | 994.37 | 975.35 |
| 20 03 9901 | 1" Rigid Steel Conduit | LF | 6.85 | 5.56 | 5.00 | 3.76 | 3.30 |
| 20 03 9903 | Concrete Backfill Around Conduit | CY | 105.94 | 99.08 | 95.52 | 88.86 | 86.81 |

**Environmental Remediation: Assemblies Cost Book**

*Page 4-31*

1998 by ECHOS. All rights reserved

# Cost Data for Remediation

## Load and Haul

**General:**

The costs presented here can be used to estimate the costs of loading stockpiled material onto highway or off-highway trucks, transporting the material to a disposal facility, and disposing of the materials.

Note: If the materials are hazardous, the costs should be estimated using remediation data such as transportation and disposal options.

**Common Cost Components:**

1. Loaders
2. Trucks (highway or off-highway)
3. Dump charges

Note: These costs are often used in conjunction with other site work efforts, such as road construction, gas distribution, and demolition.

## Cost Data for Site Work

### Load and Haul

| Assembly | Description | Unit | Unit Cost by Safety Level | | | | |
|---|---|---|---|---|---|---|---|
| | | | A | B | C | D | E |
| | **Capital Costs** | | | | | | |
| 17 02 0401 | Dump Charges | CY | 20.00 | 20.00 | 20.00 | 20.00 | 20.00 |
| 17 03 0215 | 931, 1.0 CY, Track Loader | HOUR | 112.08 | 88.96 | 75.15 | 52.57 | 46.80 |
| 17 03 0216 | 943, 1.5 CY, Track Loader | HOUR | 170.50 | 137.43 | 114.12 | 81.62 | 75.57 |
| 17 03 0217 | 953, 2.0 CY, Track Loader | HOUR | 195.08 | 158.13 | 130.48 | 94.07 | 88.30 |
| 17 03 0218 | 963, 2.5 CY, Track Loader | HOUR | 280.58 | 229.38 | 187.48 | 136.82 | 131.05 |
| 17 03 0219 | 973, 3.75 CY, Track Loader | HOUR | 357.58 | 293.33 | 238.84 | 175.16 | 169.11 |
| 17 03 0220 | 910, 1.25 CY, Wheel Loader | HOUR | 130.20 | 104.06 | 87.23 | 61.63 | 55.86 |
| 17 03 0221 | 916, 1.5 CY, Wheel Loader | HOUR | 114.32 | 90.83 | 76.64 | 53.69 | 47.92 |
| 17 03 0222 | 926, 2.0 CY, Wheel Loader | HOUR | 154.30 | 124.14 | 103.30 | 73.68 | 67.91 |
| 17 03 0223 | 950, 3.0 CY, Wheel Loader | HOUR | 180.10 | 145.64 | 120.50 | 86.58 | 80.81 |
| 17 03 0224 | 966, 4.0 CY, Wheel Loader | HOUR | 211.58 | 171.88 | 141.48 | 102.32 | 96.55 |
| 17 03 0225 | 980, 5.25 CY, Wheel Loader | HOUR | 298.84 | 244.38 | 199.68 | 145.79 | 139.74 |
| 17 03 0226 | 988, 7.0 CY, Wheel Loader | HOUR | 384.90 | 316.10 | 257.05 | 188.82 | 182.77 |
| 17 03 0227 | 992, 13.5 CY, Wheel Loader | HOUR | 648.92 | 536.11 | 433.06 | 320.83 | 314.78 |
| 17 03 0284 | 8 CY, Dump Truck | HOUR | 142.67 | 115.22 | 95.47 | 68.47 | 63.71 |
| 17 03 0285 | 12 CY, Dump Truck | HOUR | 119.55 | 95.96 | 80.05 | 56.91 | 52.15 |
| 17 03 0287 | 20 CY, Semi Dump | HOUR | 161.11 | 130.59 | 107.76 | 77.69 | 72.93 |
| 17 03 0288 | 26 CY, Semi Dump | HOUR | 149.99 | 121.33 | 100.35 | 72.13 | 67.37 |
| 17 03 0289 | 32 CY, Semi Dump | HOUR | 188.91 | 153.76 | 126.29 | 91.59 | 86.83 |
| 17 03 0291 | 20 CY, Bottom Dump | HOUR | 140.41 | 113.34 | 93.96 | 67.34 | 62.58 |
| 17 03 0293 | 18 CY, Bottom Dump, 27 Ton | HOUR | 169.81 | 137.84 | 113.56 | 82.04 | 77.28 |
| 17 03 0294 | 30 CY, Bottom Dump, 30 Ton | HOUR | 155.03 | 125.53 | 103.71 | 74.65 | 69.89 |
| 17 03 0295 | 35 Ton, 769, Off-highway Truck | HOUR | 0.00 | 0.00 | 0.00 | 0.00 | 0.00 |

**Environmental Remediation: Assemblies Cost Book**

# Cost Data for Site Work

## Load and Haul

| Assembly | Description | Unit | Unit Cost by Safety Level | | | | |
|---|---|---|---|---|---|---|---|
| | | | A | B | C | D | E |
| 17 03 0296 | 50 Ton, 773, Off-highway Truck | HOUR | 398.85 | 328.71 | 266.25 | 196.56 | 191.80 |
| 17 03 0297 | 85 Ton, 777, Off-highway Truck | HOUR | 446.47 | 368.39 | 298.00 | 220.37 | 215.61 |

# Cost Data for Remediation
## Materials Plant

### General:

This section provides hourly cost for materials plants, which process either sand or crushed rock, and the loaders necessary to load the plants. A screen/wash plant is used for washing and screening sand to a desired gradation. A crusher is used for crushing rock to be used as crushed aggregate.

### Common Cost Components:

1. Materials plant equipment, including operator costs

# Cost Data for Site Work

## Materials Plant

| Assembly | Description | Unit | Unit Cost by Safety Level | | | | |
|---|---|---|---|---|---|---|---|
| | | | A | B | C | D | E |
| | **Capital Costs** | | | | | | |
| 17 03 9901 | Washing/Screening Plant (125 Tons/Hour) Including 966 | HOUR | 625.69 | 500.62 | 419.14 | 296.62 | 269.61 |
| 17 03 9902 | Crusher (200 Tons/Hour) Including 950 & 953 Loaders | HOUR | 760.64 | 613.39 | 509.08 | 364.34 | 337.72 |

**Environmental Remediation: Assemblies Cost Book**

# Cost Data for Remediation
## Overhead Electrical Distribution

**General:**

Electrical power is frequently required at remediation sites for the operation of pumps and treatment equipment.

**Common Cost Components:**

1. Poles
2. Wire

# Cost Data for Site Work

## Overhead Electrical Distribution

| Assembly | Description | Unit | A | B | C | D | E |
|---|---|---|---|---|---|---|---|
| | **Capital Costs** | | | | | | |
| 20 02 0101 | Pole-mounted Transformer, 15 KV - 480/277 3 Phase | EACH | 13,899 | 13,043 | 12,663 | 11,835 | 11,541 |
| 20 02 0103 | Pole-mounted Capacitors, 6 KW | EACH | 810.53 | 730.56 | 691.14 | 613.54 | 588.41 |
| 20 02 0301 | 1/0 ACSR Conductor | LF | 1.45 | 1.19 | 1.07 | 0.82 | 0.73 |
| 20 02 0302 | 4/0 ACSR Conductor | LF | 2.18 | 1.82 | 1.65 | 1.30 | 1.18 |
| 20 02 0305 | 477.0 ACSR Conductor | LF | 12.48 | 10.07 | 8.89 | 6.55 | 5.80 |
| 20 02 0306 | 795.0 ACSR Conductor | LF | 20.91 | 16.94 | 14.99 | 11.14 | 9.89 |
| 20 02 0310 | 1/C #2 Aluminum, Bare, Wire | LF | 1.32 | 1.07 | 0.96 | 0.71 | 0.63 |
| 20 02 0401 | 30' Class 3 Treated Power Pole | EACH | 630.82 | 527.53 | 476.61 | 376.38 | 343.92 |
| 20 02 0402 | 35' Class 3 Treated Power Pole | EACH | 867.84 | 731.25 | 663.92 | 531.38 | 488.46 |
| 20 02 0403 | 40' Class 3 Treated Power Pole | EACH | 950.41 | 800.17 | 726.10 | 580.31 | 533.10 |
| 20 02 0404 | 45' Class 3 Treated Power Pole | EACH | 1,053 | 885.18 | 802.68 | 640.28 | 587.69 |
| 20 02 0405 | 50' Class 3 Treated Power Pole | EACH | 1,172 | 984.56 | 891.99 | 709.74 | 650.73 |
| 20 02 0406 | 55' Class 3 Treated Power Pole | EACH | 1,324 | 1,118 | 1,016 | 815.32 | 750.40 |
| 20 02 0407 | 60' Class 3 Treated Power Pole | EACH | 1,539 | 1,312 | 1,200 | 979.59 | 908.21 |
| 20 02 0408 | 65' Class 3 Treated Power Pole | EACH | 1,736 | 1,488 | 1,366 | 1,126 | 1,048 |
| 20 02 0409 | 70' Class 3 Treated Power Pole | EACH | 1,978 | 1,715 | 1,585 | 1,330 | 1,247 |
| 20 02 0410 | 75' Class 3 Treated Power Pole | EACH | 2,212 | 1,931 | 1,792 | 1,519 | 1,431 |
| 20 02 0420 | Straight-line Structure, 5 KV Pole Top | EACH | 706.49 | 571.09 | 504.35 | 372.97 | 330.43 |
| 20 02 0421 | Straight-line Structure, 15 KV Pole Top | EACH | 722.43 | 587.04 | 520.30 | 388.92 | 346.37 |
| 20 02 0422 | Straight-line Structure, 35 KV Pole Top | EACH | 1,331 | 1,108 | 998.06 | 781.53 | 711.42 |
| 20 02 0423 | Straight-line Structure, 70 KV Pole Top | EACH | 1,811 | 1,523 | 1,381 | 1,101 | 1,010 |
| 20 02 0430 | Terminal Structure, 5 KV Pole Top | EACH | 2,720 | 2,206 | 1,953 | 1,454 | 1,293 |
| 20 02 0431 | Terminal Structure, 15 KV Pole Top | EACH | 2,806 | 2,293 | 2,039 | 1,541 | 1,379 |
| 20 02 0432 | Terminal Structure, 35 KV Pole Top | EACH | 3,575 | 2,947 | 2,637 | 2,028 | 1,831 |
| 20 02 0433 | Terminal Structure, 70 KV Pole Top | EACH | 4,744 | 3,929 | 3,527 | 2,736 | 2,480 |
| 20 02 0506 | 3/C #2 Underground, 600V Direct Burial, Wire | LF | 7.67 | 6.23 | 5.61 | 4.21 | 3.70 |
| 20 02 0545 | 5 KV, 1/0 to 4/0 Conductor, Terminations & Splicing | EACH | 794.85 | 703.86 | 664.91 | 576.96 | 544.72 |

**Environmental Remediation: Assemblies Cost Book**

*Page 4-35*

1998 by ECHOS. All rights reserved

# Cost Data for Site Work

## Overhead Electrical Distribution

| Assembly | Description | Unit | Unit Cost by Safety Level | | | | |
|---|---|---|---|---|---|---|---|
| | | | A | B | C | D | E |
| 20 02 0546 | 15 KV, 1/0 to 4/0 Conductor, Terminations & Splicing | EACH | 856.33 | 679.41 | 603.67 | 432.66 | 369.96 |
| 20 02 0547 | 35 KV, 1/0 to 4/0 Conductor, Terminations & Splicing | EACH | 1,500 | 1,206 | 1,080 | 795.91 | 691.73 |
| 20 02 0549 | 70 KV, #2 to 4/0 Conductor, Terminations & Splicing | EACH | 1,532 | 1,238 | 1,112 | 827.84 | 723.65 |
| 20 02 0550 | 5 KV, 3/0 to 500 MCM Conductor, Terminations & Splicing | EACH | 930.86 | 771.63 | 703.46 | 549.55 | 493.12 |
| 20 02 0551 | 5 KV, 400 to 700 MCM Conductor, Terminations & Splicing | EACH | 1,289 | 1,107 | 1,029 | 853.33 | 788.84 |
| 20 03 9901 | 1" Rigid Steel Conduit | LF | 6.85 | 5.56 | 5.00 | 3.76 | 3.30 |
| 20 03 9902 | 4" Rigid Steel Conduit | LF | 23.92 | 20.24 | 18.67 | 15.12 | 13.82 |

**Environmental Remediation: Assemblies Cost Book**

# Cost Data for Remediation

## Parking Lots

**General:**

Parking lots to support remediation projects are typically made of asphalt or gravel and may include storm drains and piping. Because a parking lot is typically separate from the contaminated site, the trench spoil and debris can be disposed of as nonhazardous waste.

**Common Cost Components:**

1. Clearing and grubbing
2. Earthwork
3. Compaction
4. Trenching
5. Pavements, curbs
6. Drains, pipe
7. Striping, parking bumpers

# Cost Data for Site Work

## Parking Lots

| Assembly | Description | Unit | Unit Cost by Safety Level | | | | |
|---|---|---|---|---|---|---|---|
| | | | A | B | C | D | E |
| | **Capital Costs** | | | | | | |
| 17 01 0106 | Heavy Brush, Light Trees, Clear, Grub, Haul | ACRE | 11,338 | 9,081 | 7,594 | 5,382 | 4,905 |
| 17 01 0107 | Medium Brush, Medium Trees, Clear, Grub, Haul | ACRE | 13,478 | 10,783 | 9,029 | 6,389 | 5,807 |
| 17 01 0108 | Light Brush, Heavy Trees, Clear, Grub, Haul | ACRE | 17,877 | 14,287 | 11,977 | 8,462 | 7,669 |
| 17 03 0102 | Rough Grading, 12G, 1 Pass | SY | 1.28 | 1.04 | 0.85 | 0.62 | 0.59 |
| 17 03 0107 | Fine Grading, 120G, 2 Passes | SY | 0.42 | 0.34 | 0.28 | 0.20 | 0.19 |
| 17 03 0203 | Roadway Soil Excavation, with Scraper, Load & Haul Spoil | CY | 9.78 | 7.86 | 6.55 | 4.67 | 4.30 |
| 17 03 0204 | Roadway Sand Excavation, with Scraper, Load & Haul Spoil | CY | 18.35 | 14.79 | 12.28 | 8.78 | 8.13 |
| 17 03 0256 | Cat 215, 1.0 CY, Sand, Shallow, Trenching | CY | 1.99 | 1.60 | 1.33 | 0.95 | 0.88 |
| 17 03 0257 | Cat 215, 1.0 CY, Soil, Shallow, Trenching | CY | 2.42 | 1.95 | 1.62 | 1.16 | 1.07 |
| 17 03 0305 | Cat 215, 1.0 CY, Rock, No Haul or Borrow, Trenching | BCY | 176.44 | 145.32 | 127.05 | 96.67 | 88.70 |
| 17 03 0501 | Compact Subgrade, 2 Lifts | CY | 0.84 | 0.68 | 0.56 | 0.40 | 0.37 |
| 17 03 0510 | Dry Roll Gravel, Steel Roller | SY | 1.22 | 0.96 | 0.82 | 0.57 | 0.49 |
| 17 03 0515 | Compact With Pogosticks | CY | 5.91 | 4.58 | 3.98 | 2.68 | 2.23 |
| 17 03 0601 | Lime Stabilization, Dry | BCY | 34.46 | 29.87 | 25.87 | 21.32 | 20.95 |
| 17 03 0602 | Cement Stabilization, 6% | BCY | 13.58 | 12.66 | 11.94 | 11.03 | 10.90 |
| 18 01 0102 | Gravel, Delivered & Dumped | CY | 24.33 | 22.93 | 22.03 | 20.66 | 20.36 |
| 18 01 0103 | Gravel (90%) & Sand Base (10%), with Calcium Chloride 3/4 - 1 Lb/CY | CY | 25.70 | 24.15 | 23.16 | 21.65 | 21.31 |
| 18 01 0104 | Asphalt, Intermediate Course (Line Item Includes 5% Waste) | TON | 76.50 | 65.72 | 59.09 | 48.56 | 45.99 |
| 18 01 0201 | Concrete Curb, 6" x 6" | LF | 2.78 | 2.38 | 2.21 | 1.83 | 1.70 |
| 18 01 0202 | Concrete Curb & Gutter, 6" x 24", Formed | LF | 20.78 | 16.61 | 14.83 | 10.80 | 9.32 |
| 18 01 0205 | Curb Inlet Frame Grate & Box, 525 Lb | EACH | 6,004 | 4,987 | 4,534 | 3,550 | 3,200 |
| 18 01 0310 | Prime Coat | SY | 0.55 | 0.52 | 0.51 | 0.49 | 0.48 |

**Environmental Remediation: Assemblies Cost Book**

# Cost Data for Site Work

## Parking Lots

| Assembly | Description | Unit | Unit Cost by Safety Level | | | | |
|---|---|---|---|---|---|---|---|
| | | | A | B | C | D | E |
| 18 01 0311 | Tack Coat | SY | 0.70 | 0.64 | 0.60 | 0.53 | 0.52 |
| 18 01 0312 | Asphalt Wearing Course, 1 Pass (Line Item Includes 5% Waste) | TON | 83.61 | 72.83 | 66.20 | 55.67 | 53.10 |
| 18 01 0401 | Crosswalk, Stop Lines, Per Lane, Intersection Painting | EACH | 109.66 | 87.46 | 77.96 | 56.50 | 48.63 |
| 18 01 0403 | Arrows, Per Lane, Intersection Painting | EACH | 27.59 | 22.73 | 20.65 | 15.96 | 14.24 |
| 18 01 0410 | Street Signs, Average | EACH | 86.60 | 70.54 | 63.66 | 48.13 | 42.44 |
| 18 01 0411 | Traffic Signs & Posts, Average | EACH | 90.91 | 74.85 | 67.97 | 52.44 | 46.75 |
| 18 02 0203 | 26" x 26", 5' Deep Area Drain with Grate | EACH | 3,425 | 2,959 | 2,752 | 2,302 | 2,141 |
| 18 02 0204 | 27" x 20", 5' Deep Area Drain with Grate | EACH | 3,346 | 2,880 | 2,673 | 2,223 | 2,062 |
| 18 02 0205 | 24" Diameter, 5' Deep Area Drain with Grate | EACH | 3,131 | 2,673 | 2,470 | 2,027 | 1,869 |
| 18 02 0401 | Parking Space Striping | EACH | 13.65 | 10.58 | 9.27 | 6.30 | 5.21 |
| 18 02 0402 | Handicap Symbol, Painted | EACH | 39.31 | 30.46 | 26.67 | 18.11 | 14.97 |
| 18 02 0403 | Handicap Parking Sign, with Post | EACH | 86.60 | 70.54 | 63.66 | 48.13 | 42.44 |
| 18 02 0501 | Precast Parking Stops | EACH | 34.69 | 30.20 | 27.91 | 23.55 | 22.18 |
| 19 03 0112 | 8" Corrugated Metal Pipe, Plain | LF | 8.69 | 7.91 | 7.51 | 6.75 | 6.51 |
| 19 03 0113 | 10" Corrugated Metal Pipe, Plain | LF | 9.99 | 9.21 | 8.81 | 8.05 | 7.81 |
| 19 03 0114 | 12" Corrugated Metal Pipe, Plain | LF | 11.23 | 10.38 | 9.96 | 9.13 | 8.87 |
| 19 03 0115 | 15" Corrugated Metal Pipe, Plain | LF | 13.37 | 12.43 | 11.95 | 11.04 | 10.74 |
| 19 03 0116 | 18" Corrugated Metal Pipe, Plain | LF | 16.76 | 15.33 | 14.61 | 13.22 | 12.77 |
| 19 03 0117 | 24" Corrugated Metal Pipe, Plain | LF | 24.55 | 22.64 | 21.68 | 19.82 | 19.23 |
| 19 03 0158 | 8" Extra Strength, Nonreinforced Concrete Pipe | LF | 12.86 | 10.79 | 9.75 | 7.74 | 7.10 |
| 19 03 0159 | 10" Extra Strength, Nonreinforced Concrete Pipe | LF | 16.79 | 14.59 | 13.49 | 11.36 | 10.68 |
| 19 03 0160 | 12" Extra Strength, Nonreinforced Concrete Pipe | LF | 17.97 | 15.63 | 14.46 | 12.20 | 11.48 |
| 19 03 0161 | 15" Extra Strength, Nonreinforced Concrete Pipe | LF | 19.95 | 17.36 | 16.06 | 13.55 | 12.74 |
| 19 03 0162 | 18" Extra Strength, Nonreinforced Concrete Pipe | LF | 24.93 | 22.06 | 20.62 | 17.83 | 16.94 |
| 19 03 0163 | 21" Extra Strength, Nonreinforced Concrete Pipe | LF | 30.32 | 27.45 | 26.01 | 23.22 | 22.33 |
| 19 03 0164 | 24" Extra Strength, Nonreinforced Concrete Pipe | LF | 39.69 | 36.49 | 34.88 | 31.77 | 30.78 |
| 19 03 0174 | 12" Reinforced Concrete Pipe, Class 4, with Gaskets | LF | 14.33 | 13.02 | 12.29 | 11.01 | 10.65 |
| 19 03 0175 | 15" Reinforced Concrete Pipe, Class 4, with Gaskets | LF | 16.47 | 14.84 | 13.94 | 12.36 | 11.91 |
| 19 03 0176 | 18" Reinforced Concrete Pipe, Class 4, with Gaskets | LF | 23.67 | 21.05 | 19.60 | 17.05 | 16.33 |
| 19 03 0178 | 24" Reinforced Concrete Pipe, Class 4, with Gaskets | LF | 34.58 | 31.22 | 29.36 | 26.09 | 25.17 |

**Environmental Remediation: Assemblies Cost Book**

*Page 4-38*

1998 by ECHOS. All rights reserved

# Cost Data for Remediation
## Railroad Tracks and Crossings

**General:**

The costs for railroad spurs and/or railroad track crossings are included in this section. Note: Transportation and disposal of contaminated or hazardous materials should be estimated using remediation cost data.

**Common Cost Components:**

1. Railroad track: track bed preparation (i.e., subgrade, sub-ballast, ballast, track), tracks, ties, turnouts, culverts, stops, bumpers
2. Crossings: culverts, markings, signs, road crossing material

Note: If cut and fill is required, the costs should be estimated using Excavation, Cut and Fill cost data.

# Cost Data for Site Work
## Railroad Tracks and Crossings

| Assembly | Description | Unit | A | B | C | D | E |
|---|---|---|---|---|---|---|---|
| | **Capital Costs** | | | | | | |
| 17 01 0106 | Heavy Brush, Light Trees, Clear, Grub, Haul | ACRE | 11,338 | 9,081 | 7,594 | 5,382 | 4,905 |
| 17 01 0107 | Medium Brush, Medium Trees, Clear, Grub, Haul | ACRE | 13,478 | 10,783 | 9,029 | 6,389 | 5,807 |
| 17 01 0108 | Light Brush, Heavy Trees, Clear, Grub, Haul | ACRE | 17,877 | 14,287 | 11,977 | 8,462 | 7,669 |
| 17 03 0103 | Rough Grading, 14G, 1 Pass | SY | 2.06 | 1.68 | 1.38 | 1.00 | 0.94 |
| 17 03 0107 | Fine Grading, 120G, 2 Passes | SY | 0.42 | 0.34 | 0.28 | 0.20 | 0.19 |
| 17 03 0108 | Fine Grading, 130G, 2 Passes | SY | 0.48 | 0.39 | 0.32 | 0.23 | 0.22 |
| 17 03 0202 | Ditch Excavation, Normal Soil, Haul Spoil 1 Mile | CY | 9.27 | 7.49 | 6.20 | 4.45 | 4.14 |
| 17 03 0315 | Roadway Rock Excavator Borrow Subgrade (5 Miles) & Spread | CY | 15.44 | 12.54 | 10.32 | 7.47 | 7.04 |
| 17 03 0501 | Compact Subgrade, 2 Lifts | CY | 0.84 | 0.68 | 0.56 | 0.40 | 0.37 |
| 17 03 0504 | Compact Borrowed Subgrade, 4 Lifts | BCY | 5.43 | 5.27 | 5.15 | 4.99 | 4.97 |
| 17 03 0506 | Compact Sand Subgrade (Wet & 2 Passes) | SY | 0.98 | 0.80 | 0.66 | 0.47 | 0.44 |
| 17 03 0510 | Dry Roll Gravel, Steel Roller | SY | 1.22 | 0.96 | 0.82 | 0.57 | 0.49 |
| 17 03 0601 | Lime Stabilization, Dry | BCY | 34.46 | 29.87 | 25.87 | 21.32 | 20.95 |
| 17 03 0602 | Cement Stabilization, 6% | BCY | 13.58 | 12.66 | 11.94 | 11.03 | 10.90 |
| 18 01 0102 | Gravel, Delivered & Dumped | CY | 24.33 | 22.93 | 22.03 | 20.66 | 20.36 |
| 18 01 0404 | Railroad Crossing Markings, Stop Line, Per Lane | EACH | 70.62 | 56.33 | 50.21 | 36.40 | 31.34 |
| 18 01 0411 | Traffic Signs & Posts, Average | EACH | 90.91 | 74.85 | 67.97 | 52.44 | 46.75 |
| 18 06 0201 | Ballast | CY | 101.23 | 85.43 | 75.56 | 60.11 | 56.43 |
| 18 06 0202 | Gravel (90%) & Sand Base (10%); with Calcium Chloride 3/4 - 1 Lb/CY | CY | 25.70 | 24.15 | 23.16 | 21.65 | 21.31 |
| 18 06 0210 | 110 Lb #8 Turnout to New Track | EACH | 27,796 | 25,599 | 24,585 | 22,458 | 21,726 |
| 18 06 0211 | 115 Lb #8 Turnout to New Track | EACH | 31,196 | 28,999 | 27,985 | 25,858 | 25,126 |
| 18 06 0212 | 132 Lb #8 Turnout to New Track | EACH | 44,796 | 42,599 | 41,585 | 39,458 | 38,726 |

**Environmental Remediation: Assemblies Cost Book**

# Cost Data for Site Work

## Railroad Tracks and Crossings

| Assembly | Description | Unit | Unit Cost by Safety Level | | | | |
|---|---|---|---|---|---|---|---|
| | | | A | B | C | D | E |
| 18 06 0215 | 110 Lb #8 Turnout to Existing Track | EACH | 26,887 | 24,690 | 23,676 | 21,549 | 20,817 |
| 18 06 0216 | 115 Lb #8 Turnout to Existing Track | EACH | 30,650 | 28,453 | 27,439 | 25,312 | 24,580 |
| 18 06 0217 | 132 Lb #8 Turnout to Existing Track | EACH | 33,136 | 30,939 | 29,925 | 27,798 | 27,066 |
| 18 06 0220 | Railroad Car Wheel Stops | EACH | 330.91 | 305.85 | 291.08 | 266.63 | 260.25 |
| 18 06 0221 | Railroad Car Bumpers, Standard | EACH | 3,711 | 3,360 | 3,153 | 2,811 | 2,721 |
| 18 06 0222 | Railroad Car Bumpers, Heavy Duty | EACH | 6,106 | 5,755 | 5,548 | 5,206 | 5,116 |
| 18 06 0223 | Hand-throw Derailer with Standard Timbers, Open Stand & Target | EACH | 1,469 | 1,341 | 1,266 | 1,142 | 1,109 |
| 18 06 0224 | Railroad Crossing Precast Concrete Inserts, 12' Wide | EACH | 2,054 | 1,935 | 1,871 | 1,755 | 1,721 |
| 18 06 0225 | Railroad Crossing Molded Rubber with Headers, 12' Wide | EACH | 4,187 | 4,068 | 4,004 | 3,888 | 3,854 |
| 18 06 0230 | New 110 Lb Track | LF | 24.77 | 23.69 | 23.18 | 22.13 | 21.77 |
| 18 06 0231 | New 115 Lb Track | LF | 28.41 | 27.33 | 26.82 | 25.77 | 25.41 |
| 18 06 0232 | New 132 Lb Track | LF | 31.91 | 30.83 | 30.32 | 29.27 | 28.91 |
| 18 06 0240 | 110 Lb Angle Bar with Bolts and Washers | PAIR | 110.96 | 103.96 | 99.58 | 92.73 | 91.10 |
| 18 06 0241 | 115 Lb Angle Bar with Bolts and Washers | PAIR | 120.96 | 113.96 | 109.58 | 102.73 | 101.10 |
| 18 06 0242 | 132 Lb Angle Bar with Bolts and Washers | PAIR | 130.96 | 123.96 | 119.58 | 112.73 | 111.10 |
| 18 06 0250 | Crossties with 110 Lb Tie Plates and Spikes | EACH | 130.93 | 112.93 | 101.69 | 84.08 | 79.89 |
| 18 06 0251 | Crossties with 115 Lb Tie Plates and Spikes | EACH | 130.93 | 112.93 | 101.69 | 84.08 | 79.89 |
| 18 06 0252 | Crossties with 132 Lb Tie Plates and Spikes | EACH | 132.80 | 114.80 | 103.56 | 85.95 | 81.76 |
| 19 03 0402 | 34' Complete, 24" Corrugated Metal Pipe Culvert with Headwalls | EACH | 6,746 | 5,734 | 5,135 | 4,147 | 3,892 |
| 19 03 0403 | 52' Complete, 24" Corrugated Metal Pipe Culvert with Headwalls | EACH | 7,562 | 6,466 | 5,825 | 4,756 | 4,474 |
| 19 03 0404 | 76' Complete, 24" Corrugated Metal Pipe Culvert with Headwalls | EACH | 8,650 | 7,443 | 6,745 | 5,567 | 5,251 |

**Environmental Remediation: Assemblies Cost Book**

# Cost Data for Remediation

## Restriping Roadways/Parking Lots

**General:**

Restriping is the painting, either by machine or by hand, of traffic control markings on resurfaced roadways/parking lots, or the touching up of existing painted markings. This section provides a method for estimating the cost of cleaning and repainting pavements.

**Common Cost Components:**

1. Pavement sweeping
2. Intersection painting: crosswalks, turn lanes, arrows
3. Striping: no pass, centerline, edge, parking space

# Cost Data for Site Work

## Restriping Roadways/Parking Lots

| Assembly | Description | Unit | A | B | C | D | E |
|---|---|---|---|---|---|---|---|
| | **Capital Costs** | | | | | | |
| 17 04 0102 | Pavement Sweeping, Machine | SY | 0.02 | 0.02 | 0.01 | 0.01 | 0.01 |
| 18 01 0401 | Crosswalk, Stop Lines, Per Lane, Intersection Painting | EACH | 109.66 | 87.46 | 77.96 | 56.50 | 48.63 |
| 18 01 0402 | Turn Lane, Per Lane, Intersection Painting | EACH | 45.82 | 36.69 | 31.42 | 22.52 | 20.12 |
| 18 01 0403 | Arrows, Per Lane, Intersection Painting | EACH | 27.59 | 22.73 | 20.65 | 15.96 | 14.24 |
| 18 01 0405 | No Pass Stripe, Yellow | LF | 0.19 | 0.17 | 0.16 | 0.13 | 0.13 |
| 18 01 0406 | Centerline Stripe, White | LF | 2.80 | 2.75 | 2.72 | 2.67 | 2.66 |
| 18 01 0407 | Edge Stripe, Yellow | LF | 0.19 | 0.17 | 0.16 | 0.13 | 0.13 |
| 18 02 0401 | Parking Space Striping | EACH | 13.65 | 10.58 | 9.27 | 6.30 | 5.21 |
| 18 02 0402 | Handicap Symbol, Painted | EACH | 39.31 | 30.46 | 26.67 | 18.11 | 14.97 |

*Unit Cost by Safety Level*

**Environmental Remediation: Assemblies Cost Book**

# Cost Data for Remediation

## Resurfacing Roadways/Parking Lots

**General:**

Resurfacing is the installation of a new surface course over the existing pavement. Resurfacing may be required after years of service where the existing surface has deteriorated so badly that routine maintenance measures are not cost effective or adequate. The existing surface can be used as a base for the new top layer.

**Common Cost Components:**

1. Prepare existing surface: scarifying, rough grading, fine grading, pavement sweeping
2. Resurface roadway/parking lot: prime coat, tack coat, asphalt wearing course, asphalt intermediate course
3. Pavement markings: center line, edge, no pass, crosswalk/stop line, turn lane, arrows, parking stripes

## Cost Data for Site Work

### Resurfacing Roadways/Parking Lots

| Assembly | Description | Unit | Unit Cost by Safety Level | | | | |
|---|---|---|---|---|---|---|---|
| | | | A | B | C | D | E |
| | **Capital Costs** | | | | | | |
| 17 01 0520 | Scarifying, 12G - 2 Pass | SY | 0.45 | 0.36 | 0.30 | 0.22 | 0.20 |
| 17 01 0521 | Scarifying, 14G - 2 Pass | SY | 0.38 | 0.31 | 0.26 | 0.19 | 0.18 |
| 17 03 0102 | Rough Grading, 12G, 1 Pass | SY | 1.28 | 1.04 | 0.85 | 0.62 | 0.59 |
| 17 03 0103 | Rough Grading, 14G, 1 Pass | SY | 2.06 | 1.68 | 1.38 | 1.00 | 0.94 |
| 17 03 0106 | Fine Grading, 12G, 2 Passes | SY | 0.62 | 0.50 | 0.42 | 0.30 | 0.28 |
| 17 03 0108 | Fine Grading, 130G, 2 Passes | SY | 0.48 | 0.39 | 0.32 | 0.23 | 0.22 |
| 17 03 0510 | Dry Roll Gravel, Steel Roller | SY | 1.22 | 0.96 | 0.82 | 0.57 | 0.49 |
| 17 04 0102 | Pavement Sweeping, Machine | SY | 0.02 | 0.02 | 0.01 | 0.01 | 0.01 |
| 18 01 0102 | Gravel, Delivered & Dumped | CY | 24.33 | 22.93 | 22.03 | 20.66 | 20.36 |
| 18 01 0103 | Gravel (90%) & Sand Base (10%), with Calcium Chloride 3/4 - 1 Lb/CY | CY | 25.70 | 24.15 | 23.16 | 21.65 | 21.31 |
| 18 01 0104 | Asphalt, Intermediate Course (Line Item Includes 5% Waste) | TON | 76.50 | 65.72 | 59.09 | 48.56 | 45.99 |
| 18 01 0310 | Prime Coat | SY | 0.55 | 0.52 | 0.51 | 0.49 | 0.48 |
| 18 01 0311 | Tack Coat | SY | 0.70 | 0.64 | 0.60 | 0.53 | 0.52 |
| 18 01 0312 | Asphalt Wearing Course, 1 Pass (Line Item Includes 5% Waste) | TON | 83.61 | 72.83 | 66.20 | 55.67 | 53.10 |
| 18 01 0401 | Crosswalk, Stop Lines, Per Lane, Intersection Painting | EACH | 109.66 | 87.46 | 77.96 | 56.50 | 48.63 |
| 18 01 0402 | Turn Lane, Per Lane, Intersection Painting | EACH | 45.82 | 36.69 | 31.42 | 22.52 | 20.12 |
| 18 01 0403 | Arrows, Per Lane, Intersection Painting | EACH | 27.59 | 22.73 | 20.65 | 15.96 | 14.24 |
| 18 01 0405 | No Pass Stripe, Yellow | LF | 0.19 | 0.17 | 0.16 | 0.13 | 0.13 |
| 18 01 0406 | Centerline Stripe, White | LF | 2.80 | 2.75 | 2.72 | 2.67 | 2.66 |
| 18 01 0407 | Edge Stripe, Yellow | LF | 0.19 | 0.17 | 0.16 | 0.13 | 0.13 |
| 18 02 0401 | Parking Space Striping | EACH | 13.65 | 10.58 | 9.27 | 6.30 | 5.21 |
| 18 02 0402 | Handicap Symbol, Painted | EACH | 39.31 | 30.46 | 26.67 | 18.11 | 14.97 |
| 18 02 0501 | Precast Parking Stops | EACH | 34.69 | 30.20 | 27.91 | 23.55 | 22.18 |

---

**Environmental Remediation: Assemblies Cost Book**

# Cost Data for Remediation

## Retaining Wall, Cast-In-Place Concrete

**General:**

Retaining structures hold back soil or other loose material and prevent its assuming the natural angle of repose at locations where an abrupt change in elevation occurs.

**Common Cost Components:**

1. Excavation for footing
2. Retaining wall, cast-in-place (CIP) concrete
3. Backfill, compaction of backfill
4. Finish of exposed wall surface
5. Grading of exterior side of wall

# Cost Data for Site Work

## Retaining Wall, Cast-In-Place Concrete

| Assembly | Description | Unit | A | B | C | D | E |
|---|---|---|---|---|---|---|---|
| | **Capital Costs** | | | | | | |
| 17 03 0105 | Fine Grading, Hand | SY | 4.21 | 3.24 | 2.83 | 1.90 | 1.56 |
| 17 03 0107 | Fine Grading, 120G, 2 Passes | SY | 0.42 | 0.34 | 0.28 | 0.20 | 0.19 |
| 17 03 0257 | Cat 215, 1.0 CY, Soil, Shallow, Trenching | CY | 2.42 | 1.95 | 1.62 | 1.16 | 1.07 |
| 17 03 0282 | Soil, 5 Miles, Dump Truck, Load/Haul Spoil From Trench | CY | 6.97 | 5.60 | 4.66 | 3.33 | 3.06 |
| 17 03 0401 | 950, 3.00 CY, Backfill with Excavated Material | CY | 2.54 | 2.05 | 1.70 | 1.22 | 1.14 |
| 17 03 0516 | Compact with 50% Pogosticks, 50% Hand Roller | CY | 4.24 | 3.30 | 2.85 | 1.94 | 1.63 |
| 18 04 0201 | Continuous Footing, Edge Form, 4 Uses | SF | 5.52 | 4.48 | 4.03 | 3.03 | 2.67 |
| 18 04 0202 | Footing, Rebar | LB | 1.17 | 0.98 | 0.88 | 0.69 | 0.63 |
| 18 04 0203 | Pour & Cure Concrete, Continuous Footing | CY | 140.13 | 126.22 | 118.87 | 105.35 | 101.29 |
| 18 04 0205 | CIP Walls Form & Strip (4 Uses) | SF | 8.86 | 6.99 | 6.19 | 4.38 | 3.73 |
| 18 04 0206 | Reinforced Steel, Retaining Wall | LB | 0.96 | 0.81 | 0.73 | 0.59 | 0.55 |
| 18 04 0207 | Pour & Cure Concrete, Retaining Wall | CY | 173.85 | 154.04 | 143.64 | 124.39 | 118.56 |
| 18 04 0208 | Bush Hammer Finish | SF | 4.07 | 3.15 | 2.75 | 1.85 | 1.52 |
| 18 04 0210 | Keyway | LF | 1.53 | 1.22 | 1.09 | 0.79 | 0.68 |

**Environmental Remediation: Assemblies Cost Book**

# Cost Data for Remediation
## Sanitary Sewer

**General:**

This section provides the cost for estimating a gravity flow sanitary sewer system. Since it is a gravity flow system, costs are not included for sewage pumps, lift stations, and force mains. In addition, costs for influent/effluent sampling and wastewater disposal fees are not included. The Discharge to POTW section should be used for estimating the cost of discharging contaminated aqueous wastes such as contaminated groundwater, leachate, or surface runoff from a remediation site to a publicly owned treatment works (POTW). Note: Transportation and disposal of contaminated or hazardous excavated materials should be estimated using remediation cost data.

**Common Cost Components:**

1. Trenching
2. Pipe
3. Trench box
4. Backfill, compaction
5. Wellpoints
6. Manholes

Note: Refer to the Load and Haul section for loading, hauling, and disposal costs for nonhazardous materials.

# Cost Data for Site Work

## Sanitary Sewer

| | | | | Unit Cost by Safety Level | | | | |
|---|---|---|---|---|---|---|---|---|
| Assembly | Description | Unit | A | B | C | D | E |
| | **Capital Costs** | | | | | | |
| 17 03 0259 | Cat 225, 1.5 CY, Soil/Sand, Trenching | CY | 1.84 | 1.49 | 1.23 | 0.88 | 0.82 |
| 17 03 0260 | Cat 225, 1.5 CY, Soil/Sand, 10' - 20' Deep Trench Box, Trench | CY | 7.22 | 6.04 | 5.20 | 4.04 | 3.82 |
| 17 03 0261 | Cat 225, 1.5 CY, Soil/Sand with Boulders, Trenching | CY | 3.91 | 3.15 | 2.62 | 1.87 | 1.74 |
| 17 03 0262 | Cat 235, 2.5 CY, Soil/Sand, Trenching | CY | 2.01 | 1.62 | 1.35 | 0.97 | 0.90 |
| 17 03 0263 | Cat 235, 2.5 CY, Soil/Sand, 10' - 20' Deep Trench Box, Trench | CY | 5.70 | 4.75 | 4.07 | 3.14 | 2.98 |
| 17 03 0264 | Cat 235, 2.5 CY, Soil/Sand with Boulders, Trenching | CY | 4.28 | 3.45 | 2.86 | 2.05 | 1.91 |
| 17 03 0265 | Cat 245, 3.0 CY, Soil/Sand, Trenching | CY | 2.78 | 2.23 | 1.86 | 1.33 | 1.22 |
| 17 03 0266 | Cat 245, 3.0 CY, Soil/Sand, 10' - 20' Deep Trench Box, Trench | CY | 11.21 | 9.18 | 7.78 | 5.79 | 5.39 |
| 17 03 0267 | Cat 245, 3.0 CY, Soil/Sand with Boulders, Trenching | CY | 5.18 | 4.17 | 3.47 | 2.47 | 2.28 |
| 17 03 0272 | Pull Trench Box, Cat 225, 1.5 CY, Soil/Sand, Trenching | CY | 6.67 | 5.49 | 4.65 | 3.48 | 3.27 |
| 17 03 0273 | Pull Trench Box, Cat 235, 2.5 CY, Soil/Sand, Trenching | CY | 5.31 | 4.37 | 3.68 | 2.75 | 2.59 |
| 17 03 0274 | Pull Trench Box, Cat 245, 3.0 CY, Soil/Sand, Trenching | CY | 10.79 | 8.76 | 7.36 | 5.37 | 4.97 |
| 17 03 0306 | Cat 225, 1.5 CY, Rock, No Haul or Borrow, Trenching | BCY | 149.48 | 122.48 | 106.81 | 80.48 | 73.45 |
| 17 03 0307 | Cat 235, 2.0 CY, Rock, No Haul or Borrow, Trenching | BCY | 133.78 | 109.18 | 95.00 | 71.01 | 64.55 |
| 17 03 0308 | Cat 245, 3.0 CY, Rock, No Haul or Borrow, Trenching | BCY | 125.50 | 101.91 | 88.44 | 65.44 | 59.17 |
| 17 03 0401 | 950, 3.00 CY, Backfill with Excavated Material | CY | 2.54 | 2.05 | 1.70 | 1.22 | 1.14 |
| 17 03 0402 | 966, 4.00 CY, Backfill with Excavated Material | CY | 1.70 | 1.38 | 1.13 | 0.82 | 0.77 |
| 17 03 0405 | 950, 3.00 CY, Delivered & Dumped, Backfill with Sand | CY | 38.31 | 35.03 | 32.95 | 29.74 | 29.00 |
| 17 03 0406 | 966, 4.00 CY, Delivered & Dumped, Backfill with Sand | CY | 37.79 | 34.61 | 32.60 | 29.49 | 28.76 |

**Environmental Remediation: Assemblies Cost Book**

# Cost Data for Site Work

## Sanitary Sewer

| Assembly | Description | Unit | A | B | C | D | E |
|---|---|---|---|---|---|---|---|
| 17 03 0410 | 950, Delivered & Dumped, Backfill with Cement Stabilized Base Material | CY | 46.93 | 43.66 | 41.58 | 38.37 | 37.63 |
| 17 03 0411 | 966, Delivered & Dumped, Backfill with Cement Stabilized Sand | CY | 46.41 | 43.23 | 41.23 | 38.12 | 37.39 |
| 17 03 0420 | Backfill Trench, Borrow Material, Delivered & Dumped Only | CY | 9.86 | 8.74 | 7.93 | 6.83 | 6.63 |
| 17 03 0511 | Compact Soil with Vibrating Plate | CY | 7.20 | 5.57 | 4.84 | 3.26 | 2.70 |
| 17 03 0515 | Compact With Pogosticks | CY | 5.91 | 4.58 | 3.98 | 2.68 | 2.23 |
| 17 03 0516 | Compact with 50% Pogosticks, 50% Hand Roller | CY | 4.24 | 3.30 | 2.85 | 1.94 | 1.63 |
| 17 03 1001 | Wellpoint for Trench, Install & Remove <500', 1 Month | LFHD | 55.06 | 55.06 | 55.06 | 55.06 | 55.06 |
| 17 03 1002 | 2" Diameter Contractor's Trash Pump, 75 GPM | DAY | 50.20 | 47.01 | 45.64 | 42.55 | 41.42 |
| 18 04 0203 | Pour & Cure Concrete, Continuous Footing | CY | 140.13 | 126.22 | 118.87 | 105.35 | 101.29 |
| 19 02 0101 | 4", Class 50, Bell & Spigot Sanitary Sewer, Cast-iron Pipe | LF | 8.49 | 7.49 | 7.05 | 6.08 | 5.72 |
| 19 02 0102 | 5", Class 50, Bell & Spigot Sanitary Sewer, Cast-iron Pipe | LF | 11.80 | 10.41 | 9.81 | 8.47 | 7.98 |
| 19 02 0103 | 6", Class 50, Bell & Spigot Sanitary Sewer, Cast-iron Pipe | LF | 13.25 | 11.79 | 11.16 | 9.75 | 9.23 |
| 19 02 0104 | 8", Class 50, Bell & Spigot Sanitary Sewer, Cast-iron Pipe | LF | 18.98 | 17.26 | 16.52 | 14.86 | 14.25 |
| 19 02 0105 | 10", Class 50, Bell & Spigot Sanitary Sewer, Cast-iron Pipe | LF | 27.66 | 25.51 | 24.59 | 22.51 | 21.75 |
| 19 02 0106 | 12", Class 50, Bell & Spigot Sanitary Sewer, Cast-iron Pipe | LF | 36.36 | 34.21 | 33.29 | 31.21 | 30.45 |
| 19 02 0107 | 15", Class 50, Bell & Spigot Sanitary Sewer, Cast-iron Pipe | LF | 51.76 | 49.01 | 47.83 | 45.17 | 44.19 |
| 19 02 0110 | 4" Extra-strength Vitrified Clay Pipe, Class 200, Premium Joints | LF | 17.47 | 14.14 | 12.39 | 9.15 | 8.17 |
| 19 02 0111 | 5" Extra-strength Vitrified Clay Pipe, Class 200, Premium Joints | LF | 18.99 | 15.45 | 13.59 | 10.14 | 9.10 |
| 19 02 0112 | 6" Extra-strength Vitrified Clay Pipe, Class 200, Premium Joints | LF | 20.06 | 16.40 | 14.48 | 10.92 | 9.84 |
| 19 02 0113 | 8" Extra-strength Vitrified Clay Pipe, Class 200, Premium Joints | LF | 22.16 | 18.37 | 16.38 | 12.70 | 11.58 |
| 19 02 0114 | 10" Extra-strength Vitrified Clay Pipe, Class 200, Premium Joints | LF | 27.02 | 22.79 | 20.57 | 16.46 | 15.22 |
| 19 02 0115 | 12" Extra-strength Vitrified Clay Pipe, Class 200, Premium Joints | LF | 28.25 | 23.86 | 21.55 | 17.28 | 15.98 |
| 19 02 0116 | 15" Extra-strength Vitrified Clay Pipe, Class 200, Premium Joints | LF | 39.29 | 34.29 | 31.67 | 26.81 | 25.34 |
| 19 02 0117 | 18" Extra-strength Vitrified Clay Pipe, Class 200, Premium Joints | LF | 50.85 | 45.06 | 42.02 | 36.40 | 34.70 |
| 19 02 0118 | 24" Extra-strength Vitrified Clay Pipe, Class 200, Premium Joints | LF | 85.93 | 78.09 | 73.96 | 66.33 | 64.03 |
| 19 02 0119 | 30" Extra-strength Vitrified Clay Pipe, Class 200, Premium Joints | LF | 103.01 | 92.02 | 86.25 | 75.57 | 72.34 |
| 19 02 0120 | 36" Extra-strength Vitrified Clay Pipe, Class 200, Premium Joints | LF | 143.76 | 130.03 | 122.81 | 109.46 | 105.42 |

**Environmental Remediation: Assemblies Cost Book**

# Cost Data for Site Work

## Sanitary Sewer

| Assembly | Description | Unit | Unit Cost by Safety Level | | | | |
|---|---|---|---|---|---|---|---|
| | | | A | B | C | D | E |
| 19 02 0125 | 4" PVC Pipe Sanitary | LF | 13.31 | 10.99 | 9.71 | 7.45 | 6.80 |
| 19 02 0126 | 6" PVC Pipe Sanitary | LF | 15.68 | 13.19 | 11.82 | 9.39 | 8.70 |
| 19 02 0127 | 8" PVC Pipe Sanitary | LF | 17.03 | 14.43 | 13.00 | 10.47 | 9.74 |
| 19 02 0128 | 10" PVC Pipe Sanitary | LF | 22.78 | 19.45 | 17.70 | 14.46 | 13.48 |
| 19 02 0129 | 12" PVC Pipe Sanitary | LF | 26.69 | 23.26 | 21.45 | 18.11 | 17.11 |
| 19 02 0130 | 15" PVC Pipe Sanitary | LF | 42.35 | 36.56 | 33.52 | 27.90 | 26.20 |
| 19 02 0201 | Precast, CIP Base, 4' Diameter, 6' Deep, Manhole | EACH | 1,459 | 1,276 | 1,179 | 1,001 | 947.36 |
| 19 02 0202 | Precast, CIP Base, 4' Diameter, 8' Deep, Manhole | EACH | 1,957 | 1,697 | 1,557 | 1,304 | 1,229 |
| 19 02 0203 | Precast, CIP Base, 4' Diameter, 12' Deep, Manhole | EACH | 2,835 | 2,459 | 2,257 | 1,891 | 1,783 |
| 20 06 0101 | 3 - 9 Lb Magnesium Anodes, Cathodic Protection Point | EACH | 953.82 | 778.29 | 693.26 | 523.02 | 466.94 |
| 20 06 0102 | 3 - 17 Lb Magnesium Anodes, Cathodic Protection Point | EACH | 1,364 | 1,121 | 1,002 | 766.32 | 688.94 |
| 20 06 0103 | 3 - 32 Lb Magnesium Anodes, Cathodic Protection Point | EACH | 1,845 | 1,539 | 1,390 | 1,092 | 995.23 |
| 20 06 0104 | 3 - 48 Lb Magnesium Anodes, Cathodic Protection Point | EACH | 2,585 | 2,166 | 1,960 | 1,553 | 1,420 |

**Environmental Remediation: Assemblies Cost Book**

# Cost Data for Remediation
## Sidewalks

**General:**

Sidewalks are commonly used in urban areas where pedestrian traffic is allowed within the right-of-way. They are also provided along some short stretches of rural highways for use by schoolchildren

**Common Cost Components:**

1. Clearing, grubbing, hauling
2. Excavation, backfill, compaction
3. Sidewalk: concrete (wire mesh reinforced and formed), bituminous, brick

# Cost Data for Site Work
## Sidewalks

| Assembly | Description | Unit | Unit Cost by Safety Level | | | | |
|----------|-------------|------|------|------|------|------|------|
|          |             |      | A    | B    | C    | D    | E    |
|          | **Capital Costs** | | | | | | |
| 17 01 0106 | Heavy Brush, Light Trees, Clear, Grub, Haul | ACRE | 11,338 | 9,081 | 7,594 | 5,382 | 4,905 |
| 17 01 0107 | Medium Brush, Medium Trees, Clear, Grub, Haul | ACRE | 13,478 | 10,783 | 9,029 | 6,389 | 5,807 |
| 17 01 0108 | Light Brush, Heavy Trees, Clear, Grub, Haul | ACRE | 17,877 | 14,287 | 11,977 | 8,462 | 7,669 |
| 17 03 0105 | Fine Grading, Hand | SY | 4.21 | 3.24 | 2.83 | 1.90 | 1.56 |
| 17 03 0205 | Curb/Sidewalk Excavation & Backfill, 27% Haul Spoil, 1 Mile | CY | 38.69 | 30.82 | 25.93 | 18.23 | 16.37 |
| 17 03 0417 | Delivered & Dumped - Hand, Backfill with Sand | CY | 82.86 | 70.10 | 62.85 | 50.41 | 47.00 |
| 17 03 0511 | Compact Soil with Vibrating Plate | CY | 7.20 | 5.57 | 4.84 | 3.26 | 2.70 |
| 18 03 0301 | Standard 4" Sidewalk with Mesh, Formed | SF | 4.39 | 3.62 | 3.28 | 2.53 | 2.26 |
| 18 03 0302 | 2" Thick Bituminous Sidewalk | SF | 0.86 | 0.74 | 0.69 | 0.58 | 0.54 |
| 18 03 0303 | Brick Sidewalk with Sand Joints (4.5 Bricks/SF) | SF | 13.82 | 11.33 | 10.27 | 7.86 | 6.98 |
| 18 03 0304 | Standard 5" Sidewalk with Mesh, Formed | SF | 4.67 | 3.88 | 3.53 | 2.77 | 2.49 |
| 18 03 0305 | Standard 6" Sidewalk with Mesh, Formed | SF | 4.90 | 4.10 | 3.75 | 2.98 | 2.70 |
| 18 03 0306 | 2 1/2" Thick Bituminous Sidewalk | SF | 1.06 | 0.92 | 0.85 | 0.71 | 0.66 |
| 18 03 0307 | Brick Sidewalk with Sand Joints (7.2 Bricks/SF) | SF | 19.22 | 15.31 | 13.63 | 9.85 | 8.47 |

# Cost Data for Remediation
## Sprinkler System

### General:
When a landscaping irrigation system is required, this section can be used to estimate the cost of installation.

### Common Cost Components:
1. Trenching
2. Backfill, compaction
3. Pipe
4. Sprinkler heads: full circle, semi-circle, shrub/tree
5. Appurtenances: control box, gate valve, reducer, testing and inspection of system

# Cost Data for Site Work
## Sprinkler System

| Assembly | Description | Unit | Unit Cost by Safety Level | | | | |
|---|---|---|---|---|---|---|---|
| | | | A | B | C | D | E |
| | **Capital Costs** | | | | | | |
| 17 03 0255 | Trenching to 48" Deep, Including Backfill & Compaction | CY | 10.66 | 8.37 | 7.16 | 4.93 | 4.26 |
| 18 05 0701 | Full Circle Sprinkler Head, 30' Diameter | EACH | 94.41 | 76.34 | 68.61 | 51.14 | 44.74 |
| 18 05 0702 | Semicircle Sprinkler Head, 20' Diameter | EACH | 43.12 | 34.09 | 30.22 | 21.49 | 18.29 |
| 18 05 0703 | Shrub/Tree Sprinkler Head | EACH | 40.45 | 31.42 | 27.55 | 18.82 | 15.62 |
| 18 05 0704 | Full Circle Sprinkler Head, 80' Diameter | EACH | 302.56 | 287.17 | 280.57 | 265.69 | 260.23 |
| 18 05 0705 | Control Box | EACH | 1,598 | 1,461 | 1,402 | 1,270 | 1,221 |
| 18 05 0706 | 2 1/2", Cast-iron Body, Gate Valve | EACH | 257.59 | 239.88 | 232.30 | 215.19 | 208.91 |
| 18 05 0707 | 4" Reducer | EACH | 80.38 | 64.15 | 57.20 | 41.51 | 35.76 |
| 18 05 0710 | Testing & Inspection of Sprinkler System | LS | 856.76 | 660.42 | 576.36 | 386.59 | 317.00 |
| 19 01 0201 | 3/4", Class 200, PVC Piping | LF | 7.45 | 5.81 | 5.04 | 3.45 | 2.90 |
| 19 01 0202 | 1", Class 200, PVC Piping | LF | 7.48 | 5.84 | 5.07 | 3.48 | 2.93 |
| 19 01 0204 | 2", Class 200, PVC Piping | LF | 8.86 | 6.96 | 6.07 | 4.23 | 3.60 |
| 19 01 0205 | 2 1/2", Class 200, PVC Piping | LF | 9.03 | 7.13 | 6.24 | 4.40 | 3.77 |

**Environmental Remediation: Assemblies Cost Book**

# Cost Data for Remediation
## Storm Sewer

**General:**

Storm sewers may be required to divert surface runoff from remediation sites. If the runoff is contaminated, the costs included in the Discharge to POTW are more appropriate.

**Common Cost Components:**

1. Trenching
2. Pipe
3. Trench box
4. Backfill, compaction
5. Wellpoints
6. Inlets, drains, manholes

# Cost Data for Site Work
## Storm Sewer

| Assembly | Description | Unit | Unit Cost by Safety Level | | | | |
|---|---|---|---|---|---|---|---|
| | | | A | B | C | D | E |
| | **Capital Costs** | | | | | | |
| 17 03 0259 | Cat 225, 1.5 CY, Soil/Sand, Trenching | CY | 1.84 | 1.49 | 1.23 | 0.88 | 0.82 |
| 17 03 0260 | Cat 225, 1.5 CY, Soil/Sand, 10' - 20' Deep Trench Box, Trench | CY | 7.22 | 6.04 | 5.20 | 4.04 | 3.82 |
| 17 03 0261 | Cat 225, 1.5 CY, Soil/Sand with Boulders, Trenching | CY | 3.91 | 3.15 | 2.62 | 1.87 | 1.74 |
| 17 03 0262 | Cat 235, 2.5 CY, Soil/Sand, Trenching | CY | 2.01 | 1.62 | 1.35 | 0.97 | 0.90 |
| 17 03 0263 | Cat 235, 2.5 CY, Soil/Sand, 10' - 20' Deep Trench Box, Trench | CY | 5.70 | 4.75 | 4.07 | 3.14 | 2.98 |
| 17 03 0264 | Cat 235, 2.5 CY, Soil/Sand with Boulders, Trenching | CY | 4.28 | 3.45 | 2.86 | 2.05 | 1.91 |
| 17 03 0265 | Cat 245, 3.0 CY, Soil/Sand, Trenching | CY | 2.78 | 2.23 | 1.86 | 1.33 | 1.22 |
| 17 03 0266 | Cat 245, 3.0 CY, Soil/Sand, 10' - 20' Deep Trench Box, Trench | CY | 11.21 | 9.18 | 7.78 | 5.79 | 5.39 |
| 17 03 0267 | Cat 245, 3.0 CY, Soil/Sand with Boulders, Trenching | CY | 5.18 | 4.17 | 3.47 | 2.47 | 2.28 |
| 17 03 0268 | Koehring 1166, 4.0 CY, Soil/Sand, Trenching | CY | 2.27 | 1.83 | 1.52 | 1.09 | 1.01 |
| 17 03 0269 | Koehring 1166, 4.0 CY, Soil/Sand,10' - 20' Deep Trench Box, Trench | CY | 6.44 | 5.28 | 4.46 | 3.32 | 3.11 |
| 17 03 0270 | Koehring 1166, 4.0 CY, Soil/Sand with Boulders, Trenching | CY | 3.00 | 2.42 | 2.01 | 1.44 | 1.33 |
| 17 03 0272 | Pull Trench Box, Cat 225, 1.5 CY, Soil/Sand, Trenching | CY | 6.67 | 5.49 | 4.65 | 3.48 | 3.27 |
| 17 03 0273 | Pull Trench Box, Cat 235, 2.5 CY, Soil/Sand, Trenching | CY | 5.31 | 4.37 | 3.68 | 2.75 | 2.59 |
| 17 03 0274 | Pull Trench Box, Cat 245, 3.0 CY, Soil/Sand, Trenching | CY | 10.79 | 8.76 | 7.36 | 5.37 | 4.97 |
| 17 03 0275 | Pull Trench Box, Koehring 1166, 4.0 CY, Soil/Sand, Trench | CY | 6.22 | 5.06 | 4.23 | 3.09 | 2.88 |
| 17 03 0306 | Cat 225, 1.5 CY, Rock, No Haul or Borrow, Trenching | BCY | 149.48 | 122.48 | 106.81 | 80.48 | 73.45 |
| 17 03 0307 | Cat 235, 2.0 CY, Rock, No Haul or Borrow, Trenching | BCY | 133.78 | 109.18 | 95.00 | 71.01 | 64.55 |
| 17 03 0308 | Cat 245, 3.0 CY, Rock, No Haul or Borrow, Trenching | BCY | 125.50 | 101.91 | 88.44 | 65.44 | 59.17 |
| 17 03 0309 | Koehring 1166, 4.0 CY, Rock, No Haul or Borrow, Trench | BCY | 117.46 | 95.22 | 82.61 | 60.93 | 54.97 |
| 17 03 0401 | 950, 3.00 CY, Backfill with Excavated Material | CY | 2.54 | 2.05 | 1.70 | 1.22 | 1.14 |
| 17 03 0402 | 966, 4.00 CY, Backfill with Excavated Material | CY | 1.70 | 1.38 | 1.13 | 0.82 | 0.77 |

**Environmental Remediation: Assemblies Cost Book**

# Cost Data for Site Work

## Storm Sewer

| Assembly | Description | Unit | A | B | C | D | E |
|---|---|---|---|---|---|---|---|
| 17 03 0403 | 980, 5.25 CY, Backfill with Excavated Material | CY | 1.89 | 1.54 | 1.26 | 0.92 | 0.88 |
| 17 03 0405 | 950, 3.00 CY, Delivered & Dumped, Backfill with Sand | CY | 38.31 | 35.03 | 32.95 | 29.74 | 29.00 |
| 17 03 0406 | 966, 4.00 CY, Delivered & Dumped, Backfill with Sand | CY | 37.79 | 34.61 | 32.60 | 29.49 | 28.76 |
| 17 03 0407 | 980, 5.25 CY, Delivered & Dumped, Backfill with Sand | CY | 37.76 | 34.60 | 32.58 | 29.49 | 28.79 |
| 17 03 0410 | 950, Delivered & Dumped, Backfill with Cement Stabilized Base Material | CY | 46.93 | 43.66 | 41.58 | 38.37 | 37.63 |
| 17 03 0411 | 966, Delivered & Dumped, Backfill with Cement Stabilized Sand | CY | 46.41 | 43.23 | 41.23 | 38.12 | 37.39 |
| 17 03 0412 | 980, Delivered & Dumped, Backfill with Cement Stabilized Sand | CY | 46.39 | 43.23 | 41.21 | 38.12 | 37.41 |
| 17 03 0420 | Backfill Trench, Borrow Material, Delivered & Dumped Only | CY | 9.86 | 8.74 | 7.93 | 6.83 | 6.63 |
| 17 03 0511 | Compact Soil with Vibrating Plate | CY | 7.20 | 5.57 | 4.84 | 3.26 | 2.70 |
| 17 03 0515 | Compact With Pogosticks | CY | 5.91 | 4.58 | 3.98 | 2.68 | 2.23 |
| 17 03 0516 | Compact with 50% Pogosticks, 50% Hand Roller | CY | 4.24 | 3.30 | 2.85 | 1.94 | 1.63 |
| 17 03 1001 | Wellpoint for Trench, Install & Remove <500', 1 Month | LFHD | 55.06 | 55.06 | 55.06 | 55.06 | 55.06 |
| 17 03 1002 | 2" Diameter Contractor's Trash Pump, 75 GPM | DAY | 50.20 | 47.01 | 45.64 | 42.55 | 41.42 |
| 17 03 1003 | 3" Diameter Contractor's Trash Pump, 150 GPM | DAY | 61.72 | 57.05 | 55.05 | 50.54 | 48.88 |
| 17 03 1004 | 4" Diameter Contractor's Trash Pump, 300 GPM | DAY | 80.30 | 74.23 | 71.63 | 65.77 | 63.61 |
| 18 02 0201 | Area Inlets, Precast, 4' Deep | EACH | 1,225 | 1,033 | 937.09 | 750.68 | 691.01 |
| 18 02 0202 | Area Drains with Grates, 6' Deep | EACH | 3,527 | 2,997 | 2,763 | 2,250 | 2,066 |
| 18 02 0203 | 26" x 26", 5' Deep Area Drain with Grate | EACH | 3,425 | 2,959 | 2,752 | 2,302 | 2,141 |
| 18 02 0204 | 27" x 20", 5' Deep Area Drain with Grate | EACH | 3,346 | 2,880 | 2,673 | 2,223 | 2,062 |
| 18 02 0205 | 24" Diameter, 5' Deep Area Drain with Grate | EACH | 3,131 | 2,673 | 2,470 | 2,027 | 1,869 |
| 18 04 0203 | Pour & Cure Concrete, Continuous Footing | CY | 140.13 | 126.22 | 118.87 | 105.35 | 101.29 |
| 19 02 0201 | Precast, CIP Base, 4' Diameter, 6' Deep, Manhole | EACH | 1,459 | 1,276 | 1,179 | 1,001 | 947.36 |
| 19 02 0202 | Precast, CIP Base, 4' Diameter, 8' Deep, Manhole | EACH | 1,957 | 1,697 | 1,557 | 1,304 | 1,229 |
| 19 02 0203 | Precast, CIP Base, 4' Diameter, 12' Deep, Manhole | EACH | 2,835 | 2,459 | 2,257 | 1,891 | 1,783 |
| 19 03 0101 | 8" Corrugated Metal Pipe, Bituminous Coated & Paved | LF | 9.71 | 8.90 | 8.49 | 7.70 | 7.45 |
| 19 03 0102 | 10" Corrugated Metal Pipe, Bituminous Coated & Paved | LF | 11.21 | 10.40 | 9.99 | 9.20 | 8.95 |
| 19 03 0103 | 12" Corrugated Metal Pipe, Bituminous Coated & Paved | LF | 13.07 | 12.20 | 11.77 | 10.92 | 10.66 |
| 19 03 0104 | 15" Corrugated Metal Pipe, Bituminous Coated & Paved | LF | 15.83 | 14.86 | 14.38 | 13.44 | 13.14 |
| 19 03 0105 | 18" Corrugated Metal Pipe, Bituminous Coated & Paved | LF | 17.63 | 16.66 | 16.18 | 15.24 | 14.94 |
| 19 03 0106 | 24" Corrugated Metal Pipe, Bituminous Coated & Paved | LF | 31.83 | 30.33 | 29.59 | 28.14 | 27.68 |
| 19 03 0107 | 30" Corrugated Metal Pipe, Bituminous Coated & Paved | LF | 38.98 | 37.02 | 36.03 | 34.13 | 33.52 |
| 19 03 0108 | 36" Corrugated Metal Pipe, Bituminous Coated & Paved | LF | 50.01 | 47.69 | 46.39 | 44.12 | 43.48 |

**Environmental Remediation: Assemblies Cost Book**

# Cost Data for Site Work

## Storm Sewer

| Assembly | Description | Unit | A | B | C | D | E |
|---|---|---|---|---|---|---|---|
| | | | **Unit Cost by Safety Level** | | | | |
| 19 03 0109 | 48" Corrugated Metal Pipe, Bituminous Coated & Paved | LF | 63.29 | 60.59 | 59.08 | 56.45 | 55.70 |
| 19 03 0110 | 60" Corrugated Metal Pipe, Bituminous Coated & Paved | LF | 91.75 | 88.53 | 86.73 | 83.59 | 82.71 |
| 19 03 0111 | 72" Corrugated Metal Pipe, Bituminous Coated & Paved | LF | 133.03 | 128.37 | 125.78 | 121.24 | 119.96 |
| 19 03 0112 | 8" Corrugated Metal Pipe, Plain | LF | 8.69 | 7.91 | 7.51 | 6.75 | 6.51 |
| 19 03 0113 | 10" Corrugated Metal Pipe, Plain | LF | 9.99 | 9.21 | 8.81 | 8.05 | 7.81 |
| 19 03 0114 | 12" Corrugated Metal Pipe, Plain | LF | 11.23 | 10.38 | 9.96 | 9.13 | 8.87 |
| 19 03 0115 | 15" Corrugated Metal Pipe, Plain | LF | 13.37 | 12.43 | 11.95 | 11.04 | 10.74 |
| 19 03 0116 | 18" Corrugated Metal Pipe, Plain | LF | 16.76 | 15.33 | 14.61 | 13.22 | 12.77 |
| 19 03 0117 | 24" Corrugated Metal Pipe, Plain | LF | 24.55 | 22.64 | 21.68 | 19.82 | 19.23 |
| 19 03 0118 | 30" Corrugated Metal Pipe, Plain | LF | 30.25 | 28.34 | 27.38 | 25.52 | 24.93 |
| 19 03 0119 | 36" Corrugated Metal Pipe, Plain | LF | 45.22 | 42.95 | 41.69 | 39.48 | 38.86 |
| 19 03 0120 | 48" Corrugated Metal Pipe, Plain | LF | 58.89 | 56.27 | 54.81 | 52.26 | 51.54 |
| 19 03 0121 | 60" Corrugated Metal Pipe, Plain | LF | 87.46 | 84.30 | 82.54 | 79.46 | 78.59 |
| 19 03 0122 | 18" x 11" (15" Equivalent) Corrugated Steel Pipe Arch, Coated & Paved | LF | 16.38 | 15.41 | 14.93 | 13.99 | 13.69 |
| 19 03 0123 | 22" x 13" (18" Equivalent) Corrugated Steel Pipe Arch, Coated & Paved | LF | 25.43 | 24.46 | 23.98 | 23.04 | 22.74 |
| 19 03 0124 | 29" x 18" (24" Equivalent) Corrugated Steel Pipe Arch, Coated & Paved | LF | 30.83 | 29.33 | 28.59 | 27.14 | 26.68 |
| 19 03 0125 | 36" x 22" (30" Equivalent) Corrugated Steel Pipe Arch, Coated & Paved | LF | 46.48 | 44.52 | 43.53 | 41.63 | 41.02 |
| 19 03 0126 | 43" x 27" (36" Equivalent) Corrugated Steel Pipe Arch, Coated & Paved | LF | 50.51 | 48.19 | 46.89 | 44.62 | 43.98 |
| 19 03 0127 | 50" x 31" (42" Equivalent) Corrugated Steel Pipe Arch, Coated & Paved | LF | 57.89 | 55.27 | 53.81 | 51.26 | 50.54 |
| 19 03 0128 | 58" x 36" (48" Equivalent) Corrugated Steel Pipe Arch, Coated & Paved | LF | 64.39 | 61.77 | 60.31 | 57.76 | 57.04 |
| 19 03 0130 | 72", 3 Gauge, Sectional Plate Pipe | LF | 561.40 | 500.52 | 463.95 | 404.48 | 389.42 |
| 19 03 0131 | 78", 3 Gauge, Sectional Plate Pipe | LF | 605.18 | 544.30 | 507.73 | 448.26 | 433.20 |
| 19 03 0132 | 84", 3 Gauge, Sectional Plate Pipe | LF | 671.56 | 606.36 | 567.71 | 504.06 | 487.61 |
| 19 03 0133 | 90", 3 Gauge, Sectional Plate Pipe | LF | 684.73 | 619.53 | 580.88 | 517.23 | 500.78 |
| 19 03 0134 | 96", 3 Gauge, Sectional Plate Pipe | LF | 742.58 | 673.25 | 632.72 | 565.07 | 547.23 |
| 19 03 0135 | 108", 3 Gauge, Sectional Plate Pipe | LF | 1,161 | 1,031 | 953.77 | 826.47 | 793.56 |
| 19 03 0136 | 120", 3 Gauge, Sectional Plate Pipe | LF | 1,222 | 1,091 | 1,014 | 886.80 | 853.89 |
| 19 03 0137 | 132", 3 Gauge, Sectional Plate Pipe | LF | 2,024 | 1,763 | 1,609 | 1,354 | 1,288 |
| 19 03 0138 | 144", 3 Gauge, Sectional Plate Pipe | LF | 2,024 | 1,763 | 1,609 | 1,354 | 1,288 |
| 19 03 0139 | 72", 5 Gauge, Sectional Plate Pipe | LF | 524.63 | 463.75 | 427.18 | 367.71 | 352.65 |

**Environmental Remediation: Assemblies Cost Book**

# Cost Data for Site Work

## Storm Sewer

| Assembly | Description | Unit | Unit Cost by Safety Level A | B | C | D | E |
|---|---|---|---|---|---|---|---|
| 19 03 0140 | 78", 5 Gauge, Sectional Plate Pipe | LF | 563.73 | 502.85 | 466.28 | 406.81 | 391.75 |
| 19 03 0141 | 84", 5 Gauge, Sectional Plate Pipe | LF | 562.06 | 501.18 | 464.61 | 405.14 | 390.08 |
| 19 03 0142 | 90", 5 Gauge, Sectional Plate Pipe | LF | 609.83 | 544.63 | 505.98 | 442.33 | 425.88 |
| 19 03 0143 | 96", 5 Gauge, Sectional Plate Pipe | LF | 620.56 | 555.36 | 516.71 | 453.06 | 436.61 |
| 19 03 0144 | 108", 5 Gauge, Sectional Plate Pipe | LF | 802.11 | 732.78 | 692.25 | 624.60 | 606.76 |
| 19 03 0145 | 120", 5 Gauge, Sectional Plate Pipe | LF | 1,110 | 988.19 | 915.05 | 796.13 | 766.00 |
| 19 03 0146 | 132", 5 Gauge, Sectional Plate Pipe | LF | 1,340 | 1,210 | 1,132 | 1,005 | 972.13 |
| 19 03 0147 | 144", 5 Gauge, Sectional Plate Pipe | LF | 1,328 | 1,198 | 1,120 | 992.97 | 960.06 |
| 19 03 0148 | 72", 8 Gauge, Sectional Plate Pipe | LF | 447.06 | 386.18 | 349.61 | 290.14 | 275.08 |
| 19 03 0149 | 78", 8 Gauge, Sectional Plate Pipe | LF | 486.73 | 425.85 | 389.28 | 329.81 | 314.75 |
| 19 03 0150 | 84", 8 Gauge, Sectional Plate Pipe | LF | 518.40 | 457.52 | 420.95 | 361.48 | 346.42 |
| 19 03 0151 | 90", 8 Gauge, Sectional Plate Pipe | LF | 545.23 | 484.35 | 447.78 | 388.31 | 373.25 |
| 19 03 0152 | 96", 8 Gauge, Sectional Plate Pipe | LF | 550.40 | 489.52 | 452.95 | 393.48 | 378.42 |
| 19 03 0153 | 108", 8 Gauge, Sectional Plate Pipe | LF | 733.90 | 668.70 | 630.05 | 566.40 | 549.95 |
| 19 03 0154 | 120", 8 Gauge, Sectional Plate Pipe | LF | 1,038 | 916.02 | 842.88 | 723.96 | 693.83 |
| 19 03 0155 | 132", 8 Gauge, Sectional Plate Pipe | LF | 1,170 | 1,048 | 974.88 | 855.96 | 825.83 |
| 19 03 0156 | 144", 8 Gauge, Sectional Plate Pipe | LF | 1,224 | 1,102 | 1,029 | 909.96 | 879.83 |
| 19 03 0157 | 6" Extra Strength, Nonreinforced Concrete Pipe | LF | 10.78 | 8.92 | 7.98 | 6.17 | 5.59 |
| 19 03 0158 | 8" Extra Strength, Nonreinforced Concrete Pipe | LF | 12.86 | 10.79 | 9.75 | 7.74 | 7.10 |
| 19 03 0159 | 10" Extra Strength, Nonreinforced Concrete Pipe | LF | 16.79 | 14.59 | 13.49 | 11.36 | 10.68 |
| 19 03 0160 | 12" Extra Strength, Nonreinforced Concrete Pipe | LF | 17.97 | 15.63 | 14.46 | 12.20 | 11.48 |
| 19 03 0165 | 12" Reinforced Concrete Pipe, Class 3, with Gaskets | LF | 14.33 | 13.02 | 12.29 | 11.01 | 10.65 |
| 19 03 0166 | 15" Reinforced Concrete Pipe, Class 3, with Gaskets | LF | 16.47 | 14.84 | 13.94 | 12.36 | 11.91 |
| 19 03 0167 | 18" Reinforced Concrete Pipe, Class 3, with Gaskets | LF | 23.67 | 21.05 | 19.60 | 17.05 | 16.33 |
| 19 03 0168 | 24" Reinforced Concrete Pipe, Class 3, with Gaskets | LF | 34.58 | 31.22 | 29.36 | 26.09 | 25.17 |
| 19 03 0169 | 30" Reinforced Concrete Pipe, Class 3, with Gaskets | LF | 44.77 | 40.58 | 38.25 | 34.17 | 33.02 |
| 19 03 0170 | 36" Reinforced Concrete Pipe, Class 3, with Gaskets | LF | 61.34 | 56.10 | 53.19 | 48.09 | 46.65 |
| 19 03 0171 | 48" Reinforced Concrete Pipe, Class 3, with Gaskets | LF | 88.98 | 82.77 | 79.32 | 73.27 | 71.56 |
| 19 03 0172 | 60" Reinforced Concrete Pipe, Class 3, with Gaskets | LF | 127.64 | 119.58 | 115.09 | 107.24 | 105.03 |
| 19 03 0173 | 72" Reinforced Concrete Pipe, Class 3, with Gaskets | LF | 163.82 | 155.43 | 150.77 | 142.61 | 140.30 |
| 19 03 0174 | 12" Reinforced Concrete Pipe, Class 4, with Gaskets | LF | 14.33 | 13.02 | 12.29 | 11.01 | 10.65 |
| 19 03 0175 | 15" Reinforced Concrete Pipe, Class 4, with Gaskets | LF | 16.47 | 14.84 | 13.94 | 12.36 | 11.91 |
| 19 03 0176 | 18" Reinforced Concrete Pipe, Class 4, with Gaskets | LF | 23.67 | 21.05 | 19.60 | 17.05 | 16.33 |
| 19 03 0177 | 21" Reinforced Concrete Pipe, Class 4, with Gaskets | LF | 27.08 | 24.38 | 22.87 | 20.24 | 19.50 |

**Environmental Remediation: Assemblies Cost Book**

*Page 4-53*

# Cost Data for Site Work

## Storm Sewer

| Assembly | Description | Unit | A | B | C | D | E |
|---|---|---|---|---|---|---|---|
| | | | **Unit Cost by Safety Level** | | | | |
| 19 03 0178 | 24" Reinforced Concrete Pipe, Class 4, with Gaskets | LF | 34.58 | 31.22 | 29.36 | 26.09 | 25.17 |
| 19 03 0179 | 27" Reinforced Concrete Pipe, Class 4, with Gaskets | LF | 39.57 | 35.85 | 33.77 | 30.15 | 29.12 |
| 19 03 0180 | 30" Reinforced Concrete Pipe, Class 4, with Gaskets | LF | 44.77 | 40.58 | 38.25 | 34.17 | 33.02 |
| 19 03 0181 | 36" Reinforced Concrete Pipe, Class 4, with Gaskets | LF | 61.34 | 56.10 | 53.19 | 48.09 | 46.65 |
| 19 03 0182 | 12" Reinforced Concrete Pipe, Class 5, with Gaskets | LF | 14.33 | 13.02 | 12.29 | 11.01 | 10.65 |
| 19 03 0183 | 15" Reinforced Concrete Pipe, Class 5, with Gaskets | LF | 16.47 | 14.84 | 13.94 | 12.36 | 11.91 |
| 19 03 0184 | 18" Reinforced Concrete Pipe, Class 5, with Gaskets | LF | 23.67 | 21.05 | 19.60 | 17.05 | 16.33 |
| 19 03 0185 | 21" Reinforced Concrete Pipe, Class 5, with Gaskets | LF | 27.08 | 24.38 | 22.87 | 20.24 | 19.50 |
| 19 03 0186 | 24" Reinforced Concrete Pipe, Class 5, with Gaskets | LF | 34.58 | 31.22 | 29.36 | 26.09 | 25.17 |
| 19 03 0187 | 27" Reinforced Concrete Pipe, Class 5, with Gaskets | LF | 39.57 | 35.85 | 33.77 | 30.15 | 29.12 |
| 19 03 0188 | 30" Reinforced Concrete Pipe, Class 5, with Gaskets | LF | 44.77 | 40.58 | 38.25 | 34.17 | 33.02 |
| 19 03 0189 | 36" Reinforced Concrete Pipe, Class 5, with Gaskets | LF | 61.34 | 56.10 | 53.19 | 48.09 | 46.65 |
| 19 03 0190 | 42" Reinforced Concrete Pipe, Class 5, with Gaskets | LF | 72.24 | 67.00 | 64.09 | 58.99 | 57.55 |
| 19 03 0191 | 48" Reinforced Concrete Pipe, Class 5, with Gaskets | LF | 88.98 | 82.77 | 79.32 | 73.27 | 71.56 |
| 19 03 0192 | 60" Reinforced Concrete Pipe, Class 5, with Gaskets | LF | 127.64 | 119.58 | 115.09 | 107.24 | 105.03 |
| 19 03 0193 | 72" Reinforced Concrete Pipe, Class 5, with Gaskets | LF | 163.82 | 155.43 | 150.77 | 142.61 | 140.30 |
| 19 03 9901 | 8" Corrugated Metal Pipe, Bituminous Coated, End Section | EACH | 286.55 | 236.80 | 211.85 | 163.55 | 148.17 |
| 19 03 9902 | 10" Corrugated Metal Pipe, Bituminous Coated, End Section | EACH | 344.44 | 282.25 | 251.06 | 190.68 | 171.47 |
| 19 03 9903 | 12" Corrugated Metal Pipe, Bituminous Coated, End Section | EACH | 377.83 | 303.20 | 265.78 | 193.32 | 170.26 |
| 19 03 9904 | 15" Corrugated Metal Pipe, Bituminous Coated, End Section | EACH | 421.76 | 338.83 | 297.25 | 216.74 | 191.12 |
| 19 03 9905 | 18" Corrugated Metal Pipe, Bituminous Coated, End Section | EACH | 489.16 | 395.88 | 349.09 | 258.52 | 229.70 |
| 19 03 9906 | 24" Corrugated Metal Pipe, Bituminous Coated, End Section | EACH | 579.12 | 472.50 | 419.04 | 315.53 | 282.59 |
| 19 03 9907 | 30" Corrugated Metal Pipe, Bituminous Coated, End Section | EACH | 699.13 | 584.31 | 526.73 | 415.26 | 379.78 |
| 19 03 9908 | 36" Corrugated Metal Pipe, Bituminous Coated, End Section | EACH | 827.89 | 703.50 | 641.13 | 520.36 | 481.93 |
| 19 03 9909 | 48" Corrugated Metal Pipe, Bituminous Coated, End Section | EACH | 1,273 | 1,123 | 1,049 | 903.64 | 857.52 |
| 19 03 9910 | 60" Corrugated Metal Pipe, Bituminous Coated, End Section | EACH | 1,853 | 1,667 | 1,573 | 1,392 | 1,334 |
| 19 03 9911 | 72" Corrugated Metal Pipe, Bituminous Coated, End Section | EACH | 2,200 | 1,987 | 1,880 | 1,673 | 1,607 |
| 19 03 9915 | 8" Corrugated Metal Pipe, End Section | EACH | 140.25 | 120.08 | 109.97 | 90.38 | 84.15 |
| 19 03 9916 | 10" Corrugated Metal Pipe, End Section | EACH | 146.25 | 126.08 | 115.97 | 96.38 | 90.15 |

**Environmental Remediation: Assemblies Cost Book**

*Page 4-54*

1998 by ECHOS. All rights reserved

# Cost Data for Site Work

## Storm Sewer

| Assembly | Description | Unit | A | B | C | D | E |
|---|---|---|---|---|---|---|---|
| | | | **Unit Cost by Safety Level** | | | | |
| 19 03 9917 | 12" Corrugated Metal Pipe, End Section | EACH | 160.52 | 139.20 | 128.51 | 107.81 | 101.22 |
| 19 03 9918 | 15" Corrugated Metal Pipe, End Section | EACH | 164.52 | 143.20 | 132.51 | 111.81 | 105.22 |
| 19 03 9919 | 18" Corrugated Metal Pipe, End Section | EACH | 186.28 | 161.40 | 148.93 | 124.77 | 117.09 |
| 19 03 9920 | 24" Corrugated Metal Pipe, End Section | EACH | 240.53 | 210.68 | 195.71 | 166.73 | 157.50 |
| 19 03 9921 | 30" Corrugated Metal Pipe, End Section | EACH | 340.53 | 310.68 | 295.71 | 266.73 | 257.50 |
| 19 03 9922 | 36" Corrugated Metal Pipe, End Section | EACH | 468.67 | 431.35 | 412.64 | 376.41 | 364.88 |
| 19 03 9923 | 48" Corrugated Metal Pipe, End Section | EACH | 887.55 | 837.80 | 812.85 | 764.55 | 749.17 |
| 19 03 9924 | 60" Corrugated Metal Pipe, End Section | EACH | 1,491 | 1,417 | 1,379 | 1,307 | 1,284 |
| 19 03 9930 | 12" Flared-end Reinforced Concrete Pipe, Class 3 | EACH | 309.07 | 281.43 | 267.57 | 240.73 | 232.19 |
| 19 03 9931 | 15" Flared-end Reinforced Concrete Pipe, Class 3 | EACH | 298.20 | 268.35 | 253.38 | 224.40 | 215.17 |
| 19 03 9932 | 18" Flared-end Reinforced Concrete Pipe, Class 3 | EACH | 343.20 | 313.35 | 298.38 | 269.40 | 260.17 |
| 19 03 9933 | 24" Flared-end Reinforced Concrete Pipe, Class 3 | EACH | 432.34 | 395.02 | 376.31 | 340.08 | 328.55 |
| 19 03 9934 | 30" Flared-end Reinforced Concrete Pipe, Class 3 | EACH | 526.30 | 484.84 | 464.05 | 423.79 | 410.98 |
| 19 03 9935 | 36" Flared-end Reinforced Concrete Pipe, Class 3 | EACH | 747.55 | 697.80 | 672.85 | 624.55 | 609.17 |
| 19 03 9936 | 48" Flared-end Reinforced Concrete Pipe, Class 3 | EACH | 1,237 | 1,143 | 1,097 | 1,006 | 977.20 |
| 19 03 9937 | 60" Flared-end Reinforced Concrete Pipe, Class 3 | EACH | 1,983 | 1,833 | 1,759 | 1,614 | 1,568 |
| 19 03 9938 | 72" Flared-end Reinforced Concrete Pipe, Class 3 | EACH | 3,094 | 2,845 | 2,720 | 2,479 | 2,402 |
| 19 03 9940 | 12" Flared-end Reinforced Concrete Pipe, Class 4 | EACH | 309.07 | 281.43 | 267.57 | 240.73 | 232.19 |
| 19 03 9941 | 15" Flared-end Reinforced Concrete Pipe, Class 4 | EACH | 298.20 | 268.35 | 253.38 | 224.40 | 215.17 |
| 19 03 9942 | 18" Flared-end Reinforced Concrete Pipe, Class 4 | EACH | 343.20 | 313.35 | 298.38 | 269.40 | 260.17 |
| 19 03 9943 | 21" Flared-end Reinforced Concrete Pipe, Class 4 | EACH | 501.65 | 467.73 | 450.72 | 417.78 | 407.30 |
| 19 03 9944 | 24" Flared-end Reinforced Concrete Pipe, Class 4 | EACH | 432.34 | 395.02 | 376.31 | 340.08 | 328.55 |
| 19 03 9945 | 27" Flared-end Reinforced Concrete Pipe, Class 4 | EACH | 613.15 | 573.87 | 554.17 | 516.04 | 503.90 |
| 19 03 9946 | 30" Flared-end Reinforced Concrete Pipe, Class 4 | EACH | 526.30 | 484.84 | 464.05 | 423.79 | 410.98 |
| 19 03 9947 | 36" Flared-end Reinforced Concrete Pipe, Class 4 | EACH | 747.55 | 697.80 | 672.85 | 624.55 | 609.17 |
| 19 03 9950 | 12" Flared-end Reinforced Concrete Pipe, Class 5 | EACH | 309.07 | 281.43 | 267.57 | 240.73 | 232.19 |
| 19 03 9951 | 15" Flared-end Reinforced Concrete Pipe, Class 5 | EACH | 298.20 | 268.35 | 253.38 | 224.40 | 215.17 |
| 19 03 9952 | 18" Flared-end Reinforced Concrete Pipe, Class 5 | EACH | 343.20 | 313.35 | 298.38 | 269.40 | 260.17 |
| 19 03 9953 | 21" Flared-end Reinforced Concrete Pipe, Class 5 | EACH | 501.65 | 467.73 | 450.72 | 417.78 | 407.30 |
| 19 03 9954 | 24" Flared-end Reinforced Concrete Pipe, Class 5 | EACH | 432.34 | 395.02 | 376.31 | 340.08 | 328.55 |
| 19 03 9955 | 27" Flared-end Reinforced Concrete Pipe, Class 5 | EACH | 613.15 | 573.87 | 554.17 | 516.04 | 503.90 |
| 19 03 9956 | 30" Flared-end Reinforced Concrete Pipe, Class 5 | EACH | 526.30 | 484.84 | 464.05 | 423.79 | 410.98 |
| 19 03 9957 | 36" Flared-end Reinforced Concrete Pipe, Class 5 | EACH | 747.55 | 697.80 | 672.85 | 624.55 | 609.17 |
| 19 03 9958 | 42" Flared-end Reinforced Concrete Pipe, Class 5 | EACH | 1,001 | 926.70 | 889.28 | 816.82 | 793.76 |

# Cost Data for Site Work

## Storm Sewer

| Assembly | Description | Unit | Unit Cost by Safety Level | | | | |
|---|---|---|---|---|---|---|---|
| | | | A | B | C | D | E |
| 19 03 9959 | 48" Flared-end Reinforced Concrete Pipe, Class 5 | EACH | 1,237 | 1,143 | 1,097 | 1,006 | 977.20 |
| 19 03 9960 | 60" Flared-end Reinforced Concrete Pipe, Class 5 | EACH | 1,983 | 1,833 | 1,759 | 1,614 | 1,568 |
| 19 03 9961 | 72" Flared-end Reinforced Concrete Pipe, Class 5 | EACH | 3,094 | 2,845 | 2,720 | 2,479 | 2,402 |
| 20 06 0101 | 3 - 9 Lb Magnesium Anodes, Cathodic Protection Point | EACH | 953.82 | 778.29 | 693.26 | 523.02 | 466.94 |
| 20 06 0102 | 3 - 17 Lb Magnesium Anodes, Cathodic Protection Point | EACH | 1,364 | 1,121 | 1,002 | 766.32 | 688.94 |
| 20 06 0103 | 3 - 32 Lb Magnesium Anodes, Cathodic Protection Point | EACH | 1,845 | 1,539 | 1,390 | 1,092 | 995.23 |
| 20 06 0104 | 3 - 48 Lb Magnesium Anodes, Cathodic Protection Point | EACH | 2,585 | 2,166 | 1,960 | 1,553 | 1,420 |

**Environmental Remediation: Assemblies Cost Book**

# Cost Data for Remediation
## Structures, Culverts

**General:**

cast-in-place (CIP) concrete double barrel or large corrugated metal pipe arch (CMPA) culverts are typically used under roadways where they cross stormwater drainage ditches and creeks. Note: It is assumed that all surface runoff passing through the culverts is uncontaminated and that all excavated soil is uncontaminated. Costs for transportation and disposal of contaminated excavated soil should be estimated using remediation cost data.

**Common Cost Components:**

1. Trenching
2. Backfill, compaction
3. Wellpoints
4. Culverts: cast-in-place (CIP) concrete or corrugated metal pipe arch

Note: Refer to the Load and Haul section for loading, hauling, and disposal costs for nonhazardous materials.

# Cost Data for Site Work
## Structures, Culverts

| Assembly | Description | Unit | Unit Cost by Safety Level | | | | |
| --- | --- | --- | --- | --- | --- | --- | --- |
| | | | A | B | C | D | E |
| | **Capital Costs** | | | | | | |
| 17 03 0268 | Koehring 1166, 4.0 CY, Soil/Sand, Trenching | CY | 2.27 | 1.83 | 1.52 | 1.09 | 1.01 |
| 17 03 0269 | Koehring 1166, 4.0 CY, Soil/Sand,10' - 20' Deep Trench Box, Trench | CY | 6.44 | 5.28 | 4.46 | 3.32 | 3.11 |
| 17 03 0270 | Koehring 1166, 4.0 CY, Soil/Sand with Boulders, Trenching | CY | 3.00 | 2.42 | 2.01 | 1.44 | 1.33 |
| 17 03 0275 | Pull Trench Box, Koehring 1166, 4.0 CY, Soil/Sand, Trench | CY | 6.22 | 5.06 | 4.23 | 3.09 | 2.88 |
| 17 03 0309 | Koehring 1166, 4.0 CY, Rock, No Haul or Borrow, Trench | BCY | 117.46 | 95.22 | 82.61 | 60.93 | 54.97 |
| 17 03 0402 | 966, 4.00 CY, Backfill with Excavated Material | CY | 1.70 | 1.38 | 1.13 | 0.82 | 0.77 |
| 17 03 0403 | 980, 5.25 CY, Backfill with Excavated Material | CY | 1.89 | 1.54 | 1.26 | 0.92 | 0.88 |
| 17 03 0406 | 966, 4.00 CY, Delivered & Dumped, Backfill with Sand | CY | 37.79 | 34.61 | 32.60 | 29.49 | 28.76 |
| 17 03 0407 | 980, 5.25 CY, Delivered & Dumped, Backfill with Sand | CY | 37.76 | 34.60 | 32.58 | 29.49 | 28.79 |
| 17 03 0412 | 980, Delivered & Dumped, Backfill with Cement Stabilized Sand | CY | 46.39 | 43.23 | 41.21 | 38.12 | 37.41 |
| 17 03 0420 | Backfill Trench, Borrow Material, Delivered & Dumped Only | CY | 9.86 | 8.74 | 7.93 | 6.83 | 6.63 |
| 17 03 0510 | Dry Roll Gravel, Steel Roller | SY | 1.22 | 0.96 | 0.82 | 0.57 | 0.49 |
| 17 03 0511 | Compact Soil with Vibrating Plate | CY | 7.20 | 5.57 | 4.84 | 3.26 | 2.70 |
| 17 03 0515 | Compact With Pogosticks | CY | 5.91 | 4.58 | 3.98 | 2.68 | 2.23 |
| 17 03 0516 | Compact with 50% Pogosticks, 50% Hand Roller | CY | 4.24 | 3.30 | 2.85 | 1.94 | 1.63 |
| 17 03 0901 | Steel Sheeting, Pull & Salvage, to 15' | SF | 13.65 | 11.21 | 9.59 | 7.20 | 6.70 |
| 17 03 0902 | Steel Sheeting, Pull & Salvage, to 20' | SF | 14.11 | 11.61 | 9.95 | 7.50 | 6.99 |
| 17 03 0903 | Steel Sheeting, Pull & Salvage, to 25' | SF | 12.32 | 10.20 | 8.79 | 6.72 | 6.28 |
| 17 03 0904 | Steel Sheeting, Pull & Salvage, to 40' | SF | 14.52 | 12.05 | 10.40 | 7.98 | 7.46 |
| 17 03 1001 | Wellpoint for Trench, Install & Remove <500', 1 Month | LFHD | 55.06 | 55.06 | 55.06 | 55.06 | 55.06 |
| 17 03 1002 | 2" Diameter Contractor's Trash Pump, 75 GPM | DAY | 50.20 | 47.01 | 45.64 | 42.55 | 41.42 |
| 18 04 0203 | Pour & Cure Concrete, Continuous Footing | CY | 140.13 | 126.22 | 118.87 | 105.35 | 101.29 |

**Environmental Remediation: Assemblies Cost Book**

*Page 4-57*

1998 by ECHOS. All rights reserved

# Cost Data for Site Work

## Structures, Culverts

| Assembly | Description | Unit | Unit Cost by Safety Level | | | | |
|---|---|---|---|---|---|---|---|
| | | | A | B | C | D | E |
| 18 04 0205 | CIP Walls Form & Strip (4 Uses) | SF | 8.86 | 6.99 | 6.19 | 4.38 | 3.73 |
| 18 06 0129 | 1" Bituminous Fiber Expansion Joint | LF | 4.33 | 3.78 | 3.54 | 3.01 | 2.82 |
| 18 06 0146 | Footing, Rebar | LB | 1.17 | 0.98 | 0.88 | 0.69 | 0.63 |
| 19 03 0122 | 18" x 11" (15" Equivalent) Corrugated Steel Pipe Arch, Coated & Paved | LF | 16.38 | 15.41 | 14.93 | 13.99 | 13.69 |
| 19 03 0123 | 22" x 13" (18" Equivalent) Corrugated Steel Pipe Arch, Coated & Paved | LF | 25.43 | 24.46 | 23.98 | 23.04 | 22.74 |
| 19 03 0124 | 29" x 18" (24" Equivalent) Corrugated Steel Pipe Arch, Coated & Paved | LF | 30.83 | 29.33 | 28.59 | 27.14 | 26.68 |
| 19 03 0125 | 36" x 22" (30" Equivalent) Corrugated Steel Pipe Arch, Coated & Paved | LF | 46.48 | 44.52 | 43.53 | 41.63 | 41.02 |
| 19 03 0126 | 43" x 27" (36" Equivalent) Corrugated Steel Pipe Arch, Coated & Paved | LF | 50.51 | 48.19 | 46.89 | 44.62 | 43.98 |
| 19 03 0127 | 50" x 31" (42" Equivalent) Corrugated Steel Pipe Arch, Coated & Paved | LF | 57.89 | 55.27 | 53.81 | 51.26 | 50.54 |
| 19 03 0128 | 58" x 36" (48" Equivalent) Corrugated Steel Pipe Arch, Coated & Paved | LF | 64.39 | 61.77 | 60.31 | 57.76 | 57.04 |
| 19 03 0129 | 72" x 44" (58" Equivalent) Corrugated Steel Pipe Arch, Coated & Paved | LF | 75.39 | 72.77 | 71.31 | 68.76 | 68.04 |
| 19 03 0410 | Continuous Footing, Edge Form, 4 Uses | SF | 5.52 | 4.48 | 4.03 | 3.03 | 2.67 |
| 19 03 0411 | Continuous Footing, Pour Concrete | CY | 93.52 | 87.98 | 85.48 | 80.11 | 78.22 |
| 19 03 0415 | CIP Wall Rebar | LB | 0.96 | 0.81 | 0.73 | 0.59 | 0.55 |
| 19 03 0416 | Pour & Finish Concrete (1-Side), Retaining Wall | CY | 173.85 | 154.04 | 143.64 | 124.39 | 118.56 |
| 19 03 0419 | Slab on Grade, Edge Form, 4 Uses | LF | 5.72 | 4.56 | 4.06 | 2.95 | 2.54 |
| 19 03 0420 | Slab on Grade, Rebar | LB | 1.17 | 0.98 | 0.88 | 0.69 | 0.63 |
| 19 03 0421 | Pour & Cure Concrete, Slab on Grade | CY | 172.33 | 152.06 | 142.70 | 123.07 | 116.31 |
| 19 03 0424 | Top Slab Cover, Form & Strip, 4 Uses | SF | 8.40 | 6.63 | 5.87 | 4.16 | 3.53 |
| 19 03 0425 | Elevated Slab Rebar | LB | 1.00 | 0.84 | 0.75 | 0.60 | 0.55 |
| 19 03 0426 | Pour & Cure Top Slab Cover | CY | 258.72 | 221.06 | 203.02 | 166.50 | 154.34 |
| 19 03 0429 | Bush Hammer Finish | SF | 4.07 | 3.15 | 2.75 | 1.85 | 1.52 |
| 19 03 0501 | Lean Concrete Riprap/Headwall/Splash Slab, No Reinforcement | SF | 8.90 | 7.44 | 6.55 | 5.12 | 4.77 |

**Environmental Remediation: Assemblies Cost Book**

# Cost Data for Remediation
## Treatment Plants/Lift Stations

### General:

Treatment plants/lift stations are used to treat and transport fluid. The treatment plants and lift stations included in this section are all packaged plants/stations; thus, the costs are the installed costs for the entire operational system. Note: This section assumes that all wastewater is nonhazardous. The Discharge to POTW section addresses costs normally associated with discharging contaminated effluent from a remediation site to a sewage treatment facility or publicly owned treatment works (POTW), It is also important to note that the Discharge to POTW section includes costs for lift stations.

### Common Cost Components:

1. Packaged sewage lift stations including wet well for lift stations
2. Packaged sewage treatment plants
3. Packaged water treatment plants

# Cost Data for Site Work

## Treatment Plants/Lift Stations

| Assembly | Description | Unit | Unit Cost by Safety Level | | | | |
|---|---|---|---|---|---|---|---|
| | | | A | B | C | D | E |
| | **Capital Costs** | | | | | | |
| 19 01 0801 | 360 GPD, Packaged Water Treatment Plant | EACH | 5,000 | 5,000 | 5,000 | 5,000 | 5,000 |
| 19 01 0802 | 720 GPD, Packaged Water Treatment Plant, Skid Mounted | EACH | 6,250 | 6,250 | 6,250 | 6,250 | 6,250 |
| 19 01 0803 | 2,880 GPD, Packaged Water Treatment Plant | EACH | 8,631 | 8,631 | 8,631 | 8,631 | 8,631 |
| 19 01 0804 | 6,480 GPD, Packaged Water Treatment Plant | EACH | 12,330 | 12,330 | 12,330 | 12,330 | 12,330 |
| 19 01 0805 | 12,240 GPD, Packaged Water Treatment Plant | EACH | 16,900 | 16,900 | 16,900 | 16,900 | 16,900 |
| 19 01 0806 | 24,480 GPD, Packaged Water Treatment Plant | EACH | 19,835 | 19,835 | 19,835 | 19,835 | 19,835 |
| 19 01 0807 | 36,000 GPD, Packaged Water Treatment Plant | EACH | 24,427 | 24,427 | 24,427 | 24,427 | 24,427 |
| 19 02 0301 | 10,000 GPD (7 GPM) Lift Station | EACH | 2,828 | 2,659 | 2,587 | 2,423 | 2,363 |
| 19 02 0302 | 25,000 GPD (18 GPM) Lift Station | EACH | 3,539 | 3,313 | 3,217 | 2,998 | 2,918 |
| 19 02 0303 | 50,000 GPD (35 GPM) Lift Station | EACH | 4,960 | 4,621 | 4,476 | 4,149 | 4,029 |
| 19 02 0304 | 100,000 GPD (70 GPM) Lift Station | EACH | 7,842 | 7,509 | 7,366 | 7,043 | 6,925 |
| 19 02 0305 | 200,000 GPD Packaged Lift Station | EACH | 133,007 | 123,412 | 118,907 | 109,609 | 106,455 |
| 19 02 0306 | 500,000 GPD Packaged Lift Station | EACH | 157,190 | 144,431 | 138,440 | 126,076 | 121,881 |
| 19 02 0307 | 800,000 GPD Packaged Lift Station | EACH | 188,049 | 172,867 | 165,739 | 151,027 | 146,036 |
| 19 02 0311 | 12' x 36" Diameter Reinforced Concrete Pipe Wet Well for Lift Station | EACH | 9,840 | 8,577 | 7,816 | 6,581 | 6,270 |
| 19 02 0312 | 24' x 60" Diameter Reinforced Concrete Pipe Wet Well for Lift Station | EACH | 39,142 | 33,897 | 30,681 | 25,554 | 24,296 |
| 19 02 0401 | 200,000 GPD, Packaged Sewage Treatment Plant | EACH | 341,000 | 341,000 | 341,000 | 341,000 | 341,000 |
| 19 02 0402 | 500,000 GPD, Packaged Sewage Treatment Plant | EACH | 812,500 | 812,500 | 812,500 | 812,500 | 812,500 |
| 19 02 0403 | 800,000 GPD, Packaged Sewage Treatment Plant | EACH | 1,260,000 | 1,260,000 | 1,260,000 | 1,260,000 | 1,260,000 |

**Environmental Remediation: Assemblies Cost Book**

# Cost Data for Remediation
## Underground Electrical Distribution

**General:**

Electrical power is frequently required at remediation sites for the operation of pumps and treatment equipment. This section provides costs for either concrete encased, direct buried, or residential primary power distribution.

**Common Cost Components:**

1. Trenching
2. Backfill, compaction
3. Wire
4. Cable
5. Conduit
6. Manholes
7. Light fixtures, poles
8. Transformers

## Cost Data for Site Work
### Underground Electrical Distribution

| Assembly | Description | Unit | Unit Cost by Safety Level | | | | |
|---|---|---|---|---|---|---|---|
| | | | A | B | C | D | E |
| | **Capital Costs** | | | | | | |
| 17 03 0255 | Trenching to 48" Deep, Including Backfill & Compaction | CY | 10.66 | 8.37 | 7.16 | 4.93 | 4.26 |
| 17 03 0259 | Cat 225, 1.5 CY, Soil/Sand, Trenching | CY | 1.84 | 1.49 | 1.23 | 0.88 | 0.82 |
| 17 03 0261 | Cat 225, 1.5 CY, Soil/Sand with Boulders, Trenching | CY | 3.91 | 3.15 | 2.62 | 1.87 | 1.74 |
| 17 03 0262 | Cat 235, 2.5 CY, Soil/Sand, Trenching | CY | 2.01 | 1.62 | 1.35 | 0.97 | 0.90 |
| 17 03 0264 | Cat 235, 2.5 CY, Soil/Sand with Boulders, Trenching | CY | 4.28 | 3.45 | 2.86 | 2.05 | 1.91 |
| 17 03 0306 | Cat 225, 1.5 CY, Rock, No Haul or Borrow, Trenching | BCY | 149.48 | 122.48 | 106.81 | 80.48 | 73.45 |
| 17 03 0307 | Cat 235, 2.0 CY, Rock, No Haul or Borrow, Trenching | BCY | 133.78 | 109.18 | 95.00 | 71.01 | 64.55 |
| 17 03 0401 | 950, 3.00 CY, Backfill with Excavated Material | CY | 2.54 | 2.05 | 1.70 | 1.22 | 1.14 |
| 17 03 0405 | 950, 3.00 CY, Delivered & Dumped, Backfill with Sand | CY | 38.31 | 35.03 | 32.95 | 29.74 | 29.00 |
| 17 03 0410 | 950, Delivered & Dumped, Backfill with Cement Stabilized Base Material | CY | 46.93 | 43.66 | 41.58 | 38.37 | 37.63 |
| 17 03 0420 | Backfill Trench, Borrow Material, Delivered & Dumped Only | CY | 9.86 | 8.74 | 7.93 | 6.83 | 6.63 |
| 17 03 0511 | Compact Soil with Vibrating Plate | CY | 7.20 | 5.57 | 4.84 | 3.26 | 2.70 |
| 17 03 0515 | Compact With Pogosticks | CY | 5.91 | 4.58 | 3.98 | 2.68 | 2.23 |
| 17 03 0516 | Compact with 50% Pogosticks, 50% Hand Roller | CY | 4.24 | 3.30 | 2.85 | 1.94 | 1.63 |
| 17 03 1002 | 2" Diameter Contractor's Trash Pump, 75 GPM | DAY | 50.20 | 47.01 | 45.64 | 42.55 | 41.42 |
| 20 02 0105 | 12.47 KV - 120/240V, 1, 50 KV, Oil, Pad-mounted Transformer | EACH | 3,798 | 3,543 | 3,429 | 3,182 | 3,094 |
| 20 02 0506 | 3/C #2 Underground, 600V Direct Burial, Wire | LF | 7.67 | 6.23 | 5.61 | 4.21 | 3.70 |
| 20 02 0507 | 3/C Underground, 15 KV Direct Burial, #6 with Bare Grounded Wire | LF | 10.62 | 10.14 | 9.94 | 9.48 | 9.31 |
| 20 02 0511 | 5 KV, 3/0, Shielded Cable, Copper | LF | 5.90 | 5.24 | 4.92 | 4.27 | 4.07 |
| 20 02 0512 | 5 KV, 4/0, Shielded Cable, Copper | LF | 6.64 | 5.90 | 5.53 | 4.81 | 4.58 |
| 20 02 0513 | 5 KV, 350 MCM, Shielded Cable, Copper | LF | 9.17 | 8.19 | 7.70 | 6.74 | 6.43 |
| 20 02 0514 | 5 KV, 500 MCM, Shielded Cable, Copper | LF | 11.52 | 10.36 | 9.79 | 8.66 | 8.30 |

---

**Environmental Remediation: Assemblies Cost Book**

*Page 4-60*

1998 by ECHOS. All rights reserved

# Cost Data for Site Work

## Underground Electrical Distribution

|  |  |  |  | Unit Cost by Safety Level | | | | |
|---|---|---|---|---|---|---|---|---|
| Assembly | Description | Unit | A | B | C | D | E |
| 20 02 0521 | 15 KV, 3/0, Shielded Cable, Copper | LF | 6.91 | 6.18 | 5.83 | 5.12 | 4.89 |
| 20 02 0522 | 15 KV, 350 MCM, Shielded Cable, Copper | LF | 9.07 | 8.29 | 7.91 | 7.15 | 6.91 |
| 20 02 0526 | 35 KV, 3/0, Shielded Cable, Copper | LF | 11.91 | 10.64 | 10.04 | 8.81 | 8.39 |
| 20 02 0528 | 70 KV, 350 MCM, Shielded Cable, Copper | LF | 36.18 | 34.73 | 34.05 | 32.65 | 32.17 |
| 20 02 0531 | 5 KV, 3/0, Nonshielded Cable, Copper | LF | 6.51 | 5.79 | 5.44 | 4.75 | 4.52 |
| 20 02 0532 | 5 KV, 4/0, Nonshielded Cable, Copper | LF | 6.76 | 6.03 | 5.67 | 4.97 | 4.74 |
| 20 02 0533 | 5 KV, 350 MCM, Nonshielded Cable, Copper | LF | 8.02 | 7.27 | 6.89 | 6.16 | 5.92 |
| 20 02 0534 | 5 KV, 500 MCM, Nonshielded Cable, Copper | LF | 11.27 | 10.48 | 10.09 | 9.32 | 9.07 |
| 20 02 0541 | 15 KV, 3/0, Nonshielded Cable, Copper | LF | 8.42 | 7.69 | 7.33 | 6.63 | 6.40 |
| 20 02 0542 | 15 KV, 350 MCM, Nonshielded Cable, Copper | LF | 11.36 | 10.59 | 10.20 | 9.45 | 9.20 |
| 20 02 0545 | 5 KV, 1/0 to 4/0 Conductor, Terminations & Splicing | EACH | 794.85 | 703.86 | 664.91 | 576.96 | 544.72 |
| 20 02 0546 | 15 KV, 1/0 to 4/0 Conductor, Terminations & Splicing | EACH | 856.33 | 679.41 | 603.67 | 432.66 | 369.96 |
| 20 02 0547 | 35 KV, 1/0 to 4/0 Conductor, Terminations & Splicing | EACH | 1,500 | 1,206 | 1,080 | 795.91 | 691.73 |
| 20 02 0548 | 70 KV, 350 MCM Conductor, Terminations & Splicing | EACH | 5,384 | 4,986 | 4,816 | 4,431 | 4,290 |
| 20 02 0550 | 5 KV, 3/0 to 500 MCM Conductor, Terminations & Splicing | EACH | 930.86 | 771.63 | 703.46 | 549.55 | 493.12 |
| 20 02 0551 | 5 KV, 400 to 700 MCM Conductor, Terminations & Splicing | EACH | 1,289 | 1,107 | 1,029 | 853.33 | 788.84 |
| 20 02 0552 | 5 KV, 750 MCM Conductor, Terminations & Splicing | EACH | 1,916 | 1,704 | 1,613 | 1,408 | 1,333 |
| 20 02 0553 | 5 KV, 1,000 MCM Conductor, Terminations & Splicing | EACH | 3,370 | 3,051 | 2,915 | 2,607 | 2,494 |
| 20 02 0555 | 15 KV, 3/0 to 4/0 Conductor, Terminations & Splicing | EACH | 1,032 | 819.81 | 728.92 | 523.71 | 448.46 |
| 20 02 0601 | 2" Steel Conduit | LF | 12.33 | 11.02 | 10.46 | 9.19 | 8.73 |
| 20 02 0602 | 3" Steel Conduit | LF | 22.17 | 19.98 | 19.04 | 16.93 | 16.16 |
| 20 02 0603 | 4" Steel Conduit | LF | 31.60 | 28.53 | 27.22 | 24.25 | 23.16 |
| 20 02 0604 | 6" Steel Conduit | LF | 72.02 | 62.75 | 58.79 | 49.83 | 46.55 |
| 20 02 0610 | 2" PVC Conduit | LF | 4.85 | 3.86 | 3.44 | 2.49 | 2.15 |
| 20 02 0611 | 3" PVC Conduit | LF | 8.30 | 6.66 | 5.95 | 4.37 | 3.78 |
| 20 02 0612 | 4" PVC Conduit | LF | 11.68 | 9.38 | 8.40 | 6.17 | 5.36 |
| 20 02 0613 | 6" PVC Conduit | LF | 20.78 | 16.83 | 15.13 | 11.31 | 9.90 |
| 20 02 0615 | Concrete Encasement for Ductbank | CY | 231.39 | 194.67 | 178.27 | 142.74 | 130.15 |
| 20 02 0616 | Electrical & Communication Manhole, 10.5' Square x 10.0' Deep, with Cable Tray | EACH | 33,997 | 28,212 | 24,944 | 19,307 | 17,747 |
| 20 03 0303 | 20' Brushed Aluminum Pole (Single Arm) with Base | EACH | 1,656 | 1,494 | 1,418 | 1,260 | 1,207 |
| 20 03 0601 | 150W High Pressure Sodium Fixture | EACH | 586.61 | 519.47 | 490.73 | 425.83 | 402.04 |

**Environmental Remediation: Assemblies Cost Book**

*Page 4-61*

1998 by ECHOS. All rights reserved

# Cost Data for Remediation
## Water Distribution

**General:**

Water distribution may be required for potable or irrigation water at remediation sites.

Note: Piping for transportation of hazardous substances or piping installed in a contaminated zone should be estimated using remediation cost data.

**Common Cost Components:**

1. Trenching
2. Trench box
3. Pipe
4. Backfill, compaction
5. Wellpoints

## Cost Data for Site Work

### Water Distribution

| Assembly | Description | Unit | Unit Cost by Safety Level | | | | |
|---|---|---|---|---|---|---|---|
| | | | A | B | C | D | E |
| | **Capital Costs** | | | | | | |
| 17 03 0259 | Cat 225, 1.5 CY, Soil/Sand, Trenching | CY | 1.84 | 1.49 | 1.23 | 0.88 | 0.82 |
| 17 03 0260 | Cat 225, 1.5 CY, Soil/Sand, 10' - 20' Deep Trench Box, Trench | CY | 7.22 | 6.04 | 5.20 | 4.04 | 3.82 |
| 17 03 0261 | Cat 225, 1.5 CY, Soil/Sand with Boulders, Trenching | CY | 3.91 | 3.15 | 2.62 | 1.87 | 1.74 |
| 17 03 0262 | Cat 235, 2.5 CY, Soil/Sand, Trenching | CY | 2.01 | 1.62 | 1.35 | 0.97 | 0.90 |
| 17 03 0263 | Cat 235, 2.5 CY, Soil/Sand, 10' - 20' Deep Trench Box, Trench | CY | 5.70 | 4.75 | 4.07 | 3.14 | 2.98 |
| 17 03 0264 | Cat 235, 2.5 CY, Soil/Sand with Boulders, Trenching | CY | 4.28 | 3.45 | 2.86 | 2.05 | 1.91 |
| 17 03 0272 | Pull Trench Box, Cat 225, 1.5 CY, Soil/Sand, Trenching | CY | 6.67 | 5.49 | 4.65 | 3.48 | 3.27 |
| 17 03 0273 | Pull Trench Box, Cat 235, 2.5 CY, Soil/Sand, Trenching | CY | 5.31 | 4.37 | 3.68 | 2.75 | 2.59 |
| 17 03 0306 | Cat 225, 1.5 CY, Rock, No Haul or Borrow, Trenching | BCY | 149.48 | 122.48 | 106.81 | 80.48 | 73.45 |
| 17 03 0307 | Cat 235, 2.0 CY, Rock, No Haul or Borrow, Trenching | BCY | 133.78 | 109.18 | 95.00 | 71.01 | 64.55 |
| 17 03 0401 | 950, 3.00 CY, Backfill with Excavated Material | CY | 2.54 | 2.05 | 1.70 | 1.22 | 1.14 |
| 17 03 0405 | 950, 3.00 CY, Delivered & Dumped, Backfill with Sand | CY | 38.31 | 35.03 | 32.95 | 29.74 | 29.00 |
| 17 03 0410 | 950, Delivered & Dumped, Backfill with Cement Stabilized Base Material | CY | 46.93 | 43.66 | 41.58 | 38.37 | 37.63 |
| 17 03 0420 | Backfill Trench, Borrow Material, Delivered & Dumped Only | CY | 9.86 | 8.74 | 7.93 | 6.83 | 6.63 |
| 17 03 0511 | Compact Soil with Vibrating Plate | CY | 7.20 | 5.57 | 4.84 | 3.26 | 2.70 |
| 17 03 0515 | Compact With Pogosticks | CY | 5.91 | 4.58 | 3.98 | 2.68 | 2.23 |
| 17 03 0516 | Compact with 50% Pogosticks, 50% Hand Roller | CY | 4.24 | 3.30 | 2.85 | 1.94 | 1.63 |
| 17 03 1001 | Wellpoint for Trench, Install & Remove <500', 1 Month | LFHD | 55.06 | 55.06 | 55.06 | 55.06 | 55.06 |
| 17 03 1002 | 2" Diameter Contractor's Trash Pump, 75 GPM | DAY | 50.20 | 47.01 | 45.64 | 42.55 | 41.42 |
| 18 04 0203 | Pour & Cure Concrete, Continuous Footing | CY | 140.13 | 126.22 | 118.87 | 105.35 | 101.29 |
| 19 01 0201 | 3/4", Class 200, PVC Piping | LF | 7.45 | 5.81 | 5.04 | 3.45 | 2.90 |
| 19 01 0202 | 1", Class 200, PVC Piping | LF | 7.48 | 5.84 | 5.07 | 3.48 | 2.93 |
| 19 01 0203 | 1 1/2", Class 200, PVC Piping | LF | 7.59 | 5.95 | 5.18 | 3.59 | 3.04 |

**Environmental Remediation: Assemblies Cost Book**

# Cost Data for Site Work

## Water Distribution

| Assembly | Description | Unit | A | B | C | D | E |
|---|---|---|---|---|---|---|---|
| | | | \multicolumn{5}{c}{Unit Cost by Safety Level} |
| 19 01 0204 | 2", Class 200, PVC Piping | LF | 8.86 | 6.96 | 6.07 | 4.23 | 3.60 |
| 19 01 0205 | 2 1/2", Class 200, PVC Piping | LF | 9.03 | 7.13 | 6.24 | 4.40 | 3.77 |
| 19 01 0206 | 3", Class 200, PVC Piping | LF | 11.81 | 9.34 | 8.20 | 5.80 | 4.98 |
| 19 01 0207 | 4", Class 200, PVC Piping | LF | 12.33 | 9.86 | 8.72 | 6.32 | 5.50 |
| 19 01 0208 | 6", Class 200, PVC Piping | LF | 14.95 | 12.20 | 10.93 | 8.27 | 7.36 |
| 19 01 0210 | 4", Class 150, PVC Piping | LF | 12.42 | 9.95 | 8.81 | 6.41 | 5.59 |
| 19 01 0211 | 6", Class 150, PVC Piping | LF | 14.84 | 12.09 | 10.82 | 8.16 | 7.25 |
| 19 01 0212 | 8", Class 150, PVC Piping | LF | 18.36 | 15.27 | 13.84 | 10.85 | 9.82 |
| 19 01 0213 | 10", Class 150, PVC Piping | LF | 22.90 | 19.37 | 17.73 | 14.32 | 13.14 |
| 19 01 0214 | 12", Class 150, PVC Piping | LF | 28.75 | 24.63 | 22.72 | 18.73 | 17.36 |
| 19 01 0220 | 4", Class 250, Mechanical Joint, Ductile Iron Pipe | LF | 15.69 | 13.75 | 12.77 | 10.88 | 10.28 |
| 19 01 0221 | 6", Class 250, Mechanical Joint, Ductile Iron Pipe | LF | 17.56 | 15.44 | 14.37 | 12.32 | 11.66 |
| 19 01 0222 | 8", Class 250, Mechanical Joint, Ductile Iron Pipe | LF | 19.64 | 17.52 | 16.45 | 14.40 | 13.74 |
| 19 01 0223 | 10", Class 250, Mechanical Joint, Ductile Iron Pipe | LF | 26.80 | 24.21 | 22.91 | 20.40 | 19.59 |
| 19 01 0224 | 12", Class 250, Mechanical Joint, Ductile Iron Pipe | LF | 28.67 | 26.08 | 24.78 | 22.27 | 21.46 |
| 19 01 0225 | 14", Class 250, Mechanical Joint, Ductile Iron Pipe | LF | 36.05 | 33.31 | 31.93 | 29.27 | 28.42 |
| 19 01 0226 | 16", Class 250, Mechanical Joint, Ductile Iron Pipe | LF | 39.55 | 36.81 | 35.43 | 32.77 | 31.92 |
| 19 01 0227 | 18", Class 250, Mechanical Joint, Ductile Iron Pipe | LF | 43.32 | 39.99 | 38.32 | 35.08 | 34.05 |
| 19 01 0228 | 20", Class 250, Mechanical Joint, Ductile Iron Pipe | LF | 50.58 | 46.34 | 44.22 | 40.10 | 38.79 |
| 19 01 0229 | 24", Class 250, Mechanical Joint, Ductile Iron Pipe | LF | 105.89 | 92.16 | 84.94 | 71.59 | 67.55 |
| 19 01 0235 | 4", Class 250, Push-on Joint, Ductile Iron Pipe | LF | 14.73 | 12.87 | 11.93 | 10.12 | 9.54 |
| 19 01 0236 | 6", Class 250, Push-on Joint, Ductile Iron Pipe | LF | 16.00 | 14.06 | 13.08 | 11.19 | 10.59 |
| 19 01 0237 | 8", Class 250, Push-on Joint, Ductile Iron Pipe | LF | 18.62 | 16.68 | 15.70 | 13.81 | 13.21 |
| 19 01 0238 | 10", Class 250, Push-on Joint, Ductile Iron Pipe | LF | 23.02 | 20.68 | 19.51 | 17.25 | 16.53 |
| 19 01 0239 | 12", Class 250, Push-on Joint, Ductile Iron Pipe | LF | 26.76 | 24.42 | 23.25 | 20.99 | 20.27 |
| 19 01 0240 | 14", Class 250, Push-on Joint, Ductile Iron Pipe | LF | 30.78 | 28.32 | 27.09 | 24.71 | 23.95 |
| 19 01 0241 | 16", Class 250, Push-on Joint, Ductile Iron Pipe | LF | 33.73 | 31.27 | 30.04 | 27.66 | 26.90 |
| 19 01 0242 | 18", Class 250, Push-on Joint, Ductile Iron Pipe | LF | 41.02 | 38.11 | 36.65 | 33.82 | 32.91 |
| 19 01 0243 | 20", Class 250, Push-on Joint, Ductile Iron Pipe | LF | 47.49 | 44.06 | 42.25 | 38.91 | 37.91 |
| 19 01 0244 | 24", Class 250, Push-on Joint, Ductile Iron Pipe | LF | 57.52 | 53.59 | 51.53 | 47.72 | 46.56 |
| 19 01 0250 | 4", Schedule 80 Steel Pipe, Plain End, Tar Coated & Wrapped | LF | 11.65 | 10.52 | 10.01 | 8.92 | 8.53 |
| 19 01 0251 | 6", Schedule 80 Steel Pipe, Plain End, Tar Coated & Wrapped | LF | 16.60 | 15.15 | 14.42 | 13.01 | 12.56 |

**Environmental Remediation: Assemblies Cost Book**

# Cost Data for Site Work

## Water Distribution

| Assembly | Description | Unit | Unit Cost by Safety Level | | | | |
|----------|-------------|------|------|------|------|------|------|
| | | | A | B | C | D | E |
| 19 01 0252 | 8", Schedule 80 Steel Pipe, Plain End, Tar Coated & Wrapped | LF | 20.63 | 19.17 | 18.44 | 17.04 | 16.59 |
| 19 01 0253 | 12", Schedule 80 Steel Pipe, Plain End, Tar Coated & Wrapped | LF | 29.35 | 27.17 | 26.07 | 23.96 | 23.29 |
| 19 01 0254 | 16", Schedule 80 Steel Pipe, Plain End, Tar Coated & Wrapped | LF | 39.10 | 36.17 | 34.81 | 31.98 | 31.01 |
| 19 01 0255 | 20", Schedule 80 Steel Pipe, Plain End, Tar Coated & Wrapped | LF | 57.92 | 53.54 | 51.50 | 47.24 | 45.79 |
| 19 01 0256 | 24", Schedule 80 Steel Pipe, Plain End, Tar Coated & Wrapped | LF | 77.51 | 72.03 | 69.48 | 64.16 | 62.34 |
| 19 01 0260 | 1", Schedule 40 Steel Pipe, Plain End, Tar Coated & Wrapped | LF | 7.65 | 6.95 | 6.63 | 5.95 | 5.70 |
| 19 01 0261 | 2", Schedule 40 Steel Pipe, Plain End, Tar Coated & Wrapped | LF | 9.60 | 8.71 | 8.32 | 7.46 | 7.16 |
| 19 01 0262 | 3", Schedule 40 Steel Pipe, Plain End, Tar Coated & Wrapped | LF | 9.78 | 8.77 | 8.32 | 7.34 | 6.99 |
| 19 01 0263 | 4", Schedule 40 Steel Pipe, Plain End, Tar Coated & Wrapped | LF | 11.65 | 10.52 | 10.01 | 8.92 | 8.53 |
| 19 01 0264 | 5", Schedule 40 Steel Pipe, Plain End, Tar Coated & Wrapped | LF | 15.91 | 14.68 | 14.13 | 12.94 | 12.52 |
| 19 01 0265 | 6", Schedule 40 Steel Pipe, Plain End, Tar Coated & Wrapped | LF | 16.60 | 15.15 | 14.42 | 13.01 | 12.56 |
| 19 01 0266 | 8", Schedule 40 Steel Pipe, Plain End, Tar Coated & Wrapped | LF | 20.63 | 19.17 | 18.44 | 17.04 | 16.59 |
| 19 01 0267 | 10", Schedule 40 Steel Pipe, Plain End, Tar Coated & Wrapped | LF | 29.41 | 27.23 | 26.14 | 24.02 | 23.35 |
| 19 01 0268 | 12", Schedule 40 Steel Pipe, Plain End, Tar Coated & Wrapped | LF | 29.35 | 27.17 | 26.07 | 23.96 | 23.29 |
| 19 01 0269 | 14", Schedule 40 Steel Pipe, Plain End, Tar Coated & Wrapped | LF | 35.52 | 32.62 | 31.16 | 28.34 | 27.45 |
| 19 01 0270 | 16", Schedule 40 Steel Pipe, Plain End, Tar Coated & Wrapped | LF | 39.10 | 36.17 | 34.81 | 31.98 | 31.01 |
| 19 01 0271 | 18", Schedule 40 Steel Pipe, Plain End, Tar Coated & Wrapped | LF | 47.57 | 43.92 | 42.22 | 38.67 | 37.46 |
| 19 01 0272 | 20", Schedule 40 Steel Pipe, Plain End, Tar Coated & Wrapped | LF | 57.92 | 53.54 | 51.50 | 47.24 | 45.79 |
| 19 01 0273 | 24", Schedule 40 Steel Pipe, Plain End, Tar Coated & Wrapped | LF | 77.51 | 72.03 | 69.48 | 64.16 | 62.34 |
| 20 06 0101 | 3 - 9 Lb Magnesium Anodes, Cathodic Protection Point | EACH | 953.82 | 778.29 | 693.26 | 523.02 | 466.94 |
| 20 06 0102 | 3 - 17 Lb Magnesium Anodes, Cathodic Protection Point | EACH | 1,364 | 1,121 | 1,002 | 766.32 | 688.94 |
| 20 06 0103 | 3 - 32 Lb Magnesium Anodes, Cathodic Protection Point | EACH | 1,845 | 1,539 | 1,390 | 1,092 | 995.23 |
| 20 06 0104 | 3 - 48 Lb Magnesium Anodes, Cathodic Protection Point | EACH | 2,585 | 2,166 | 1,960 | 1,553 | 1,420 |

**Environmental Remediation: Assemblies Cost Book**

# Cost Data for Remediation

## Water Storage Tanks

### General:

Water storage tanks are used for providing storage capability for a specified area. If topography will not allow a surface reservoir, a standpipe or elevated tank must be constructed. Elevated storage tanks are also used to increase pressure for a water distribution system. This section provides the cost for either elevated tanks or tanks at grade.

### Common Cost Components:

1. Installation of water storage tanks, either elevated or at grade

## Cost Data for Site Work

### Water Storage Tanks

| Assembly | Description | Unit | Unit Cost by Safety Level | | | | |
|---|---|---|---|---|---|---|---|
| | | | A | B | C | D | E |
| | **Capital Costs** | | | | | | |
| 19 01 0301 | 50,000 Gallon Water Tank, Prestressed Concrete @ Grade | EACH | 62,212 | 51,872 | 46,828 | 36,798 | 33,516 |
| 19 01 0302 | 100,000 Gallon Water Tank, Prestressed Concrete @ Grade | EACH | 124,406 | 106,829 | 98,253 | 81,202 | 75,623 |
| 19 01 0303 | 500,000 Gallon Water Tank, Prestressed Concrete @ Grade | EACH | 300,260 | 260,801 | 241,550 | 203,271 | 190,748 |
| 19 01 0304 | 1,000,000 Gallon Water Tank, Prestressed Concrete @ Grade | EACH | 541,560 | 467,195 | 430,914 | 358,774 | 335,172 |
| 19 01 0310 | 5,000 Gallon Water Tank, Steel, Horizontal @ Grade | EACH | 15,945 | 13,359 | 12,252 | 9,752 | 8,835 |
| 19 01 0311 | 10,000 Gallon Water Tank, Steel, Horizontal @ Grade | EACH | 18,756 | 16,170 | 15,063 | 12,563 | 11,646 |
| 19 01 0312 | 20,000 Gallon Water Tank, Steel, Horizontal @ Grade | EACH | 28,844 | 24,853 | 23,144 | 19,286 | 17,871 |
| 19 01 0313 | 35,000 Gallon Water Tank, Steel, Horizontal @ Grade | EACH | 45,913 | 40,740 | 38,525 | 33,525 | 31,692 |
| 19 01 0320 | 250,000 Gallon Water Tank, Steel Stand Pipe | EACH | 137,432 | 130,000 | 126,818 | 119,634 | 117,000 |
| 19 01 0321 | 500,000 Gallon Water Tank, Steel Stand Pipe | EACH | 273,216 | 266,750 | 263,982 | 257,732 | 255,440 |
| 19 01 0325 | 5,000 Gallon Water Tank, Elevated Steel on Towers | EACH | 83,956 | 67,760 | 58,753 | 42,978 | 38,523 |
| 19 01 0326 | 10,000 Gallon Water Tank, Elevated Steel on Towers | EACH | 87,179 | 70,983 | 61,976 | 46,201 | 41,746 |
| 19 01 0327 | 20,000 Gallon Water Tank, Elevated Steel on Towers | EACH | 91,310 | 75,114 | 66,107 | 50,332 | 45,877 |
| 19 01 0328 | 35,000 Gallon Water Tank, Elevated Steel on Towers | EACH | 110,017 | 91,230 | 80,782 | 62,482 | 57,315 |
| 19 01 0330 | 50,000 Gallon Water Tanks, Elevated (100' +) | EACH | 415,467 | 360,699 | 332,407 | 279,186 | 262,779 |
| 19 01 0331 | 100,000 Gallon Water Tanks, Elevated (100' +) | EACH | 539,152 | 471,107 | 435,956 | 369,833 | 349,448 |
| 19 01 0332 | 250,000 Gallon Water Tanks, Elevated (100' +) | EACH | 760,741 | 663,111 | 612,676 | 517,804 | 488,557 |
| 19 01 0333 | 500,000 Gallon Water Tanks, Elevated (100' +) | EACH | 1,191,135 | 1,041,436 | 964,103 | 818,633 | 773,787 |
| 19 01 0334 | 750,000 Gallon Water Tanks, Elevated (100' +) | EACH | 1,726,703 | 1,502,154 | 1,386,155 | 1,167,950 | 1,100,680 |
| 19 01 0335 | 1,000,000 Gallon Water Tanks, Elevated (100' +) | EACH | 2,095,879 | 1,815,193 | 1,670,194 | 1,397,437 | 1,313,350 |

# Cost Data for Contractor Costs

## Contractor Costs / General Conditions

**General:**

The total cost of a contract is the sum of direct costs, general conditions, overhead and profit. Direct costs include all of the costs which can be directly attributed to a particular item of work or activity. The prime contractor's direct cost also includes the total subcontractor's price including overhead and profit. General conditions costs include field related items that are required to execute a contract but are not directly attributable to a technology or restoration task. Overhead and profit are the final costs. For multiple-site projects, general conditions items, overhead, and profit are typically estimated for the overall project rather than for individual sites.

**Common Cost Components:**

1. Supervision
2. Job trailers
3. Storage trailers
4. Portable toilets
5. Temporary plants
6. Personal protective equipment (PPE)
7. Travel and per diem
8. Permits, taxes, insurance, bonds
9. Decontamination facilities
10. Professional labor personnel, such as project managers, engineers, hydrogeologists, industrial hygienists, and health and safety engineers.

## Cost Data for Site Work

### Contractor Costs / General Conditions

| Assembly | Description | Unit | Unit Cost Safety Level E |
|---|---|---|---|
| 33 01 0102 | Van or Pickup Rental | DAY | 32.69 |
| 33 01 0201 | Mobilize Crew, >= 500 Miles, per Person | EACH | 357.75 |
| 33 01 0202 | Per Diem | DAY | 104.94 |
| 33 01 0203 | Mobilize Crew, 250 Miles, per Person | EACH | 178.88 |
| 33 01 0204 | Mobilize Crew, 100 Miles, per Person | EACH | 71.55 |
| 33 01 0205 | Mobilize Crew, 50 Miles, per Person | EACH | 53.66 |
| 33 01 0206 | Mobilize Crew, Local, per Person | EACH | 17.89 |
| 33 01 0402 | Supplied Air | EACH | 784.88 |
| 33 01 0403 | Breathing Apparatus (SCBA) | EACH | 2557 |
| 33 01 0404 | Level "C" Breathing Apparatus (Full Face Respirator) | EACH | 175.58 |
| 33 01 0405 | Level "C" Breathing Apparatus (Half-face Respirator) | EACH | 23.06 |
| 33 01 0406 | Level "C" Respirator Cartridges | EACH | 7.91 |
| 33 01 0407 | Basic Level "A" Encapsulation Suit | EACH | 500.13 |
| 33 01 0408 | Butyl Level "A" Suit | EACH | 1882 |
| 33 01 0409 | PVC Level "A" Suit | EACH | 613.87 |
| 33 01 0410 | Teflon/Nomex Level "A" Suit | EACH | 477.29 |
| 33 01 0411 | Viton Level "A" Suit | EACH | 4116 |

**Environmental Remediation: Assemblies Cost Book**

# Cost Data for Site Work

## Contractor Costs / General Conditions

| Assembly | | | Description | Unit | Unit Cost Safety Level E |
|---|---|---|---|---|---|
| 33 | 01 | 0412 | Basic Level "B" Suit | EACH | 162.77 |
| 33 | 01 | 0413 | Reusable PVC Boots | PAIR | 12.35 |
| 33 | 01 | 0414 | Reusable Neoprene Chemical-resistant Boots | PAIR | 34.87 |
| 33 | 01 | 0415 | Hard Hat | EACH | 6.34 |
| 33 | 01 | 0416 | Reusable Gloves (Butyl) | PAIR | 20.37 |
| 33 | 01 | 0417 | Reusable Gloves (Viton) for Chlorin/Aromatic Solvents | PAIR | 0.31 |
| 33 | 01 | 0418 | Reusable Gloves (PVC/Nitrile) | PAIR | 0.31 |
| 33 | 01 | 0419 | Reusable Gloves (Neoprene) | PAIR | 0.36 |
| 33 | 01 | 0420 | Two-way Radio (Reusable) | EACH | 37.24 |
| 33 | 01 | 0421 | Disposable Boot Covers (Tyvek) | PAIR | 4.91 |
| 33 | 01 | 0422 | Briefs (Disposable) | EACH | 0.14 |
| 33 | 01 | 0423 | Disposable Gloves (Latex) | PAIR | 0.15 |
| 33 | 01 | 0424 | Disposable Coveralls (Tyvek/Polycoated) | EACH | 5.55 |
| 33 | 01 | 0425 | Disposable Coveralls (Tyvek) | EACH | 3.83 |
| 33 | 01 | 0426 | Face Shield (Reusable) | EACH | 13.12 |
| 33 | 01 | 0427 | Safety Goggles | EACH | 1.91 |
| 33 | 01 | 0428 | Safety Glasses | EACH | 5.57 |
| 33 | 01 | 0429 | Disposable Ear Plugs | PAIR | 0.11 |
| 33 | 01 | 0430 | Aluminized Outer Cover Suit for Level "A" | EACH | 1845 |
| 33 | 01 | 0431 | Emergency Escape Breathing Apparatus (5 Minute) | EACH | 421.30 |
| 99 | 01 | 0101 | Site Project Manager - Minimum Cost | MWK | 1150 |
| 99 | 01 | 0102 | Site Project Manager - Average Cost | MWK | 1285 |
| 99 | 01 | 0103 | Site Project Manager - Maximum Cost | MWK | 1450 |
| 99 | 01 | 0201 | Superintendent - Minimum Cost | MWK | 1095 |
| 99 | 01 | 0202 | Superintendent - Average Cost | MWK | 1210 |
| 99 | 01 | 0203 | Superintendent - Maximum Cost | MWK | 1370 |
| 99 | 01 | 0301 | Clerk - Average Cost | MWK | 255.20 |
| 99 | 04 | 0101 | Temporary Office 20' x 8' | MONTH | 179.00 |
| 99 | 04 | 0102 | Temporary Office 32' x 8' | MONTH | 214.00 |

**Environmental Remediation: Assemblies Cost Book**

*Page 5-2*

1998 by ECHOS. All rights reserved

# Cost Data for Site Work

## Contractor Costs / General Conditions

| Assembly | | | Description | Unit | Unit Cost Safety Level E |
|---|---|---|---|---|---|
| 99 | 04 | 0103 | Temporary Office 50' x 10' | MONTH | 314.00 |
| 99 | 04 | 0104 | Temporary Office 50' x 12' | MONTH | 360.00 |
| 99 | 04 | 0201 | Temporary Storage Trailer 16' x 8' | MONTH | 70.00 |
| 99 | 04 | 0202 | Temporary Storage Trailer 28' x 10' | MONTH | 95.00 |
| 99 | 04 | 0301 | Security Fencing 5' Chain-Link | LF | 5.42 |
| 99 | 04 | 0302 | Security Fencing 6' Chain-Link | LF | 5.72 |
| 99 | 04 | 0401 | Construction Signs | SF | 10.11 |
| 99 | 04 | 0501 | Portable Toilets - Chemical | MONTH | 71.67 |
| 99 | 04 | 1101 | Construction Photographs | SET | 255.20 |
| 99 | 04 | 1102 | Time Lapse Video Equipment & Tapes, Per Month | MONTH | 319.63 |
| 99 | 04 | 1201 | Surveying - 2-man Crew | DAY | 598.80 |
| 99 | 04 | 1202 | Surveying - 3-man Crew | DAY | 760.00 |
| 99 | 04 | 1203 | Surveying - 4-man Crew | DAY | 921.20 |
| 99 | 04 | 1301 | Uniformed Watchman - Minimum | HOUR | 7.80 |
| 99 | 04 | 1302 | Uniformed Watchman - Maximum | HOUR | 14.15 |

**Environmental Remediation: Assemblies Cost Book**

# Cost Data for Contractor Costs
## Other General Conditions

| Assembly | Description | Guidelines |
|---|---|---|
| 99 03 0101 | Builder's Risk Insurance, Standard | 0.1 - 1.0% of Construction Costs |
| 99 03 0102 | Contractor's Pollution Liability Insurance | 1.0 - 5.0% of Construction Costs |
| 99 03 0301 | Public Liability Insurance | Typically 3% of Construction Costs (2.0 - 10.0% of Construction Costs) |
| 99 03 0501 | Performance Bond | Typically 3% of Construction Costs (2.0 - 10.0% of Construction Costs) |
| 99 03 0601 | Permits | Location/Project Specific |
| 99 03 0602 | Local Fees | Location/Project Specific |
| 99 03 0603 | State Permits | Location/Project Specific |
| 99 03 0604 | Federal Permits | Location/Project Specific |
| 99 03 0701 | Special Builder's Insurance | Location/Project Specific |
| 99 05 0101 | Sales Tax | Typically Applied to Equipment & Materials only |
| 99 05 0401 | Special Labor Tax | Location/Project Specific |
| 99 06 0101 | Personnel Mobilization | 0.5 - 1.0% of Construction Costs |
| 99 06 0201 | Equipment Mobilization | 0.15 - 3.0% of Construction Costs |
| 99 06 0301 | Off-Site Construction Camp | Location/Project Specific |
| 99 06 0401 | Construction Plant | Location/Project Specific |
| 99 06 0402 | Construction Plant Operational Expenses | Location/Project Specific |
| 99 06 0501 | Demobilization | 0.15 - 3.0% of Construction Costs |
| 99 07 0101 | Off-Site Construction Camp Operations | Location/Project Specific |

**Environmental Remediation: Assemblies Cost Book**

# Cost Data for Contractor Costs

## State Sales Tax

| State | Tax (%) | State | Tax (%) |
|---|---|---|---|
| Alabama | 4.000 | New Hampshire | 0.000 |
| Alaska | 0.000 | New Jersey | 6.000 |
| Arizona | 5.000 | New Mexico | 5.000 |
| Arkansas | 4.500 | New York | 4.000 |
| California | 7.250 | North Carolina | 4.000 |
| Colorado | 3.000 | North Dakota | 5.000 |
| Connecticut | 6.000 | Ohio | 5.000 |
| Delaware | 0.000 | Oklahoma | 4.500 |
| District of Columbia | 5.750 | Oregon | 0.000 |
| Florida | 6.000 | Pennsylvania | 6.000 |
| Georgia | 6.000 | Rhode Island | 7.000 |
| Hawaii | 4.000 | South Carolina | 5.000 |
| Idaho | 5.000 | South Dakota | 4.000 |
| Illinois | 6.250 | Tennessee | 6.000 |
| Indiana | 5.000 | Texas | 6.250 |
| Iowa | 5.000 | Utah | 6.125 |
| Kansas | 4.900 | Vermont | 5.000 |
| Kentucky | 6.000 | Virginia | 4.500 |
| Louisiana | 4.000 | Washington | 6.500 |
| Maine | 6.000 | West Virginia | 6.000 |
| Maryland | 5.000 | Wisconsin | 5.000 |
| Massachusetts | 5.000 | Wyoming | 4.000 |
| Michigan | 6.000 | | |
| Minnesota | 6.500 | | |
| Mississippi | 7.000 | | |
| Missouri | 4.225 | | |
| Montana | 0.000 | | |
| Nebraska | 5.000 | | |
| Nevada | 6.750 | | |

**Environmental Remediation: Assemblies Cost Book**

# Notes

# Notes

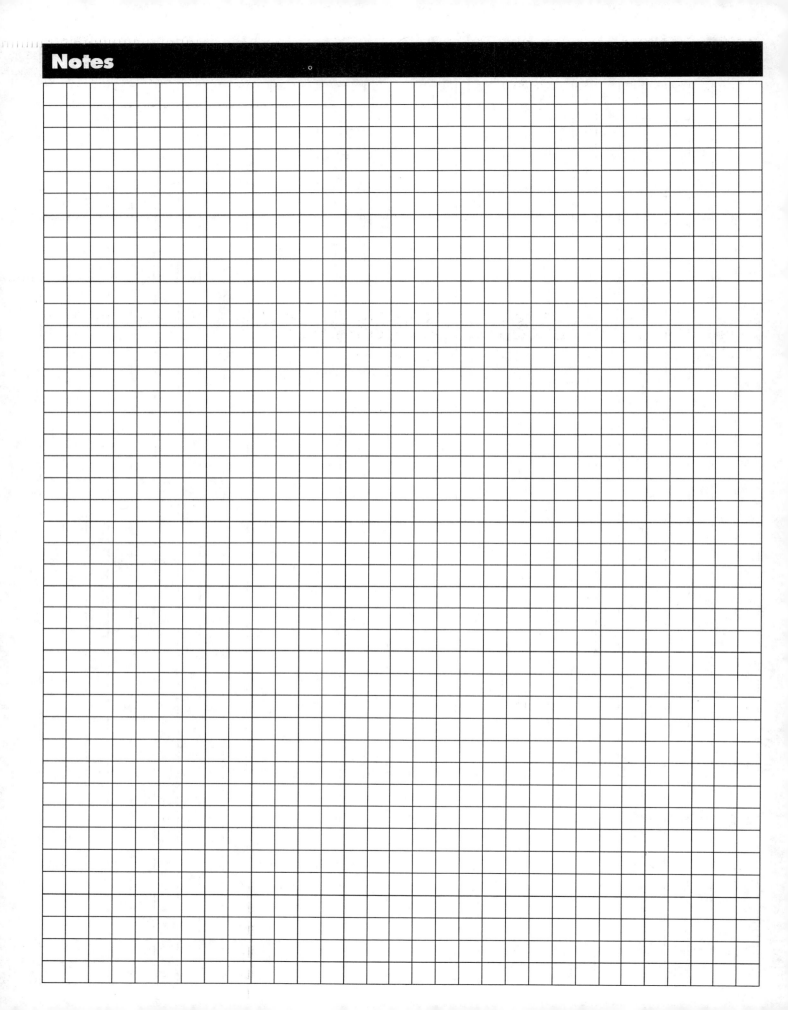

# R.S. Means Company, Inc. . . a tradition of excellence in Construction Cost Information and Services since 1942.

## Table of Contents

Annual Cost Guides, Page 2
Reference Books, Page 6
Seminars, Page 11
Consulting Services, Page 13
Electronic Data, Page 14
Order Form, Page 16

## Book Selection Guide

The following table provides definitive information on the content of each cost data publication. The number of lines of data provided in each unit price or assemblies division, as well as the number of reference tables and crews is listed for each book. The presence of other elements such as an historical cost index, city cost indexes, square foot models or cross-referenced index is also indicated. You can use the table to help select the Means' book that has the quantity and type of information you most need in your work.

| Unit Cost Divisions | Building Construction Costs | Mechanical | Electrical | Repair & Remodel. | Square Foot | Site Work Landsc. | Assemblies | Interior | Concrete Masonry | Open Shop | Heavy Construc. | Resi-dential | Light Commercial | Facil. Construc. | Plumbing | Western Construction Costs |
|---|---|---|---|---|---|---|---|---|---|---|---|---|---|---|---|---|
| 1 | 1033 | 521 | 519 | 702 | | 945 | | 436 | 969 | 1032 | 1045 | 372 | 556 | 1480 | 592 | 1030 |
| 2 | 3366 | 1384 | 451 | 2158 | | 5281 | | 1134 | 1594 | 3315 | 5169 | 1131 | 1206 | 4925 | 1636 | 3346 |
| 3 | 1502 | 91 | 68 | 747 | | 1321 | | 188 | 1932 | 1478 | 1449 | 262 | 206 | 1342 | 43 | 1480 |
| 4 | 874 | 27 | 0 | 666 | | 805 | | 633 | 1227 | 856 | 709 | 317 | 402 | 1134 | 0 | 851 |
| 5 | 1284 | 170 | 165 | 461 | | 787 | | 457 | 653 | 1247 | 1087 | 267 | 293 | 1347 | 93 | 1264 |
| 6 | 1370 | 96 | 83 | 1275 | | 475 | | 1287 | 328 | 1339 | 593 | 1521 | 1429 | 1367 | 59 | 1727 |
| 7 | 1378 | 60 | 9 | 1304 | | 484 | | 539 | 458 | 1377 | 364 | 824 | 1014 | 1370 | 70 | 1377 |
| 8 | 1837 | 60 | 0 | 1861 | | 339 | | 1746 | 717 | 1812 | 9 | 1094 | 1151 | 1967 | 0 | 1837 |
| 9 | 1639 | 50 | 0 | 1484 | | 186 | | 1734 | 322 | 1637 | 114 | 1311 | 1380 | 1833 | 50 | 1629 |
| 10 | 973 | 56 | 30 | 546 | | 206 | | 804 | 195 | 973 | 0 | 244 | 446 | 971 | 249 | 973 |
| 11 | 1019 | 261 | 200 | 586 | | 53 | | 896 | 34 | 1002 | 16 | 115 | 234 | 1016 | 231 | 1001 |
| 12 | 353 | 0 | 0 | 47 | | 227 | | 1533 | 30 | 344 | 0 | 73 | 68 | 1536 | 0 | 344 |
| 13 | 881 | 413 | 201 | 154 | | 464 | | 356 | 105 | 879 | 325 | 120 | 133 | 966 | 239 | 930 |
| 14 | 365 | 43 | 0 | 262 | | 36 | | 330 | 0 | 365 | 40 | 9 | 18 | 363 | 17 | 363 |
| 15 | 2724 | 15477 | 703 | 2302 | | 2022 | | 1772 | 80 | 2643 | 2327 | 986 | 1613 | 13058 | 11712 | 2600 |
| 16 | 1547 | 798 | 10589 | 1121 | | 1098 | | 1346 | 62 | 1562 | 1034 | 704 | 1275 | 10170 | 705 | 1489 |
| 17 | 459 | 369 | 368 | 1 | | 0 | | 0 | 0 | 460 | 0 | 0 | 0 | 459 | 369 | 459 |
| **Totals** | **22604** | **19876** | **13386** | **15677** | | **14729** | | **15191** | **8706** | **22321** | **14281** | **9350** | **11424** | **45304** | **16065** | **22700** |

| Assembly Divisions | Building Construction Costs | Mechanical | Electrical | Repair & Remodel. | Square Foot | Site Work Landsc. | Assemblies | Interior | Concrete Masonry | Open Shop | Heavy Construc. | Resi-dential | Light Commercial | Facil. Construc. | Plumbing | Western Construction Costs |
|---|---|---|---|---|---|---|---|---|---|---|---|---|---|---|---|---|
| 1 | | 0 | 0 | 202 | 131 | 738 | 775 | 0 | 689 | | 752 | 844 | 131 | 89 | 0 | |
| 2 | | 0 | 0 | 40 | 33 | 35 | 48 | 0 | 49 | | 0 | 589 | 32 | 0 | 0 | |
| 3 | | 0 | 0 | 445 | 1114 | 0 | 3064 | 228 | 1245 | | 0 | 1549 | 806 | 174 | 0 | |
| 4 | | 0 | 0 | 713 | 1353 | 0 | 3175 | 255 | 1273 | | 0 | 2047 | 1171 | 26 | 0 | |
| 5 | | 0 | 0 | 289 | 231 | 0 | 465 | 0 | 0 | | 0 | 1082 | 227 | 28 | 0 | |
| 6 | | 0 | 0 | 1006 | 857 | 0 | 1308 | 1736 | 152 | | 0 | 1084 | 749 | 333 | 0 | |
| 7 | | 0 | 0 | 42 | 89 | 0 | 179 | 159 | 0 | | 0 | 723 | 33 | 71 | 0 | |
| 8 | | 2255 | 149 | 998 | 1699 | 0 | 2725 | 925 | 0 | | 0 | 1694 | 1197 | 1114 | 2091 | |
| 9 | | 0 | 1347 | 355 | 357 | 0 | 1284 | 307 | 0 | | 0 | 242 | 358 | 319 | 0 | |
| 10 | | 0 | 0 | 0 | 0 | 0 | 0 | 0 | 0 | | 0 | 0 | 0 | 0 | 0 | |
| 11 | | 0 | 0 | 365 | 465 | 0 | 724 | 174 | 0 | | 0 | 0 | 466 | 222 | 0 | |
| 12 | | 501 | 160 | 539 | 84 | 2392 | 699 | 0 | 698 | | 541 | 0 | 84 | 119 | 850 | |
| **Totals** | | **2756** | **1656** | **4994** | **6413** | **3165** | **14446** | **3784** | **4106** | | **1293** | **9854** | **5254** | **2495** | **2941** | |

| Reference Section | Building Construction Costs | Mechanical | Electrical | Repair & Remodel. | Square Foot | Site Work Landsc. | Assemblies | Interior | Concrete Masonry | Open Shop | Heavy Construc. | Resi-dential | Light Commercial | Facil. Construc. | Plumbing | Western Construction Costs |
|---|---|---|---|---|---|---|---|---|---|---|---|---|---|---|---|---|
| Tables | 156 | 46 | 85 | 70 | 4 | 87 | 223 | 60 | 84 | 156 | 56 | 51 | 74 | 88 | 49 | 158 |
| Models | | | | | 102 | | | | | | | 32 | 43 | | | |
| Crews | 397 | 397 | 397 | 397 | | 397 | | 397 | 397 | 397 | 397 | 397 | 397 | 397 | 397 | 397 |
| City Cost Indexes | yes | yes | yes | yes | yes | yes | yes | yes | yes | yes | yes | yes | yes | yes | yes | yes |
| Historical Cost Indexes | yes | yes | yes | yes | yes | yes | yes | yes | yes | yes | yes | no | yes | yes | yes | yes |
| Index | yes | yes | yes | yes | no | yes | yes | yes | yes | yes | yes | yes | yes | yes | yes | yes |

# Annual Cost Guides

## Means Building Construction Cost Data 1998

### Available in Both Softbound and Looseleaf Editions

*The "Bible" of the industry comes in the standard softcover edition or the looseleaf edition.*

Many customers enjoy the convenience and flexibility of the looseleaf binder, which increases the usefulness of *Means Building Construction Cost Data 1998* by making it easy to add and remove pages. You can insert your own cost information pages, so everything is in one place. Copying pages for faxing is easier also. Whichever edition you prefer, softbound or the convenient looseleaf edition, you'll get *The DI Quarterly/Change Notice* newsletter at no extra cost.

$84.95 per copy, Softbound
Catalog No. 60018

$109.95 per copy, Looseleaf
Catalog No. 61018

## Means Building Construction Cost Data 1998

Offers you unchallenged unit price reliability in an easy-to-use arrangement. Whether used for complete, finished estimates or for periodic checks, it supplies more cost facts better and faster than any comparable source. Over 21,000 unit prices for 1998. The City Cost Indexes now cover over 930 areas, for indexing to any project location in North America. Order and get *The DI Quarterly/Change Notice* newsletter sent to you FREE. You'll have year-long access to the Means Estimating **Hotline** FREE with your subscription. Expert assistance when using Means data is just a phone call away.

$84.95 per copy
Over 650 pages, illustrated, available Oct. 1997
Catalog No. 60018

## Means Building Construction Cost Data 1998

**Metric Version**

The Federal Government has stated that all federal construction projects must now use metric documentation. The *Metric Version* of *Means Building Construction Cost Data 1998* is presented in metric measurements covering all construction areas. Don't miss out on these billion dollar opportunities. Make the switch to metric today.

$89.95 per copy
Over 650 pages, illustrated, available Nov. 1997
Catalog No. 63018

# Annual Cost Guides

## Means Mechanical Cost Data 1998
- HVAC  • Controls

Total unit and systems price guidance for mechanical construction... materials, parts, fittings, and complete labor cost information. Includes prices for piping, heating, air conditioning, ventilation, and all related construction.

Plus new 1998 unit costs for:
- Over 3000 installed HVAC/controls assemblies
- "On Site" Location Factors for close to 1,000 cities and towns in the U.S. and Canada
- Crews, labor and equipment

$87.95 per copy
Over 600 pages, illustrated, available Oct. 1997
Catalog No. 60028

## Means Plumbing Cost Data 1998

Comprehensive unit prices and assemblies for plumbing, irrigation systems, commercial and residential fire protection, point-of-use water heaters, and the latest approved materials. This publication and its companion, *Means Mechanical Cost Data*, provide full-range cost estimating coverage for all the mechanical trades.

$87.95 per copy
Over 500 pages, illustrated, available Oct. 1997
Catalog No. 60218

## Means Electrical Cost Data 1998

Pricing information for every part of electrical cost planning: More than 17,000 unit and systems costs with design tables; clear specifications and drawings; engineering guides and illustrated estimating procedures; complete labor-hour and materials costs for better scheduling and procurement; the latest electrical products and construction methods.
- (NEW FOR 1998) Underground Marking Tape and Polyethylene Pull Rope costs
- A Variety of Special Electrical Systems including Cathodic Protection
- Costs for maintenance, demolition, HVAC/mechanical, specialties, equipment, and more

$87.95 per copy
Over 450 pages, illustrated, available Oct. 1997
Catalog No. 60038

## Means Electrical Change Order Cost Data 1998

You are provided with electrical unit prices exclusively for pricing change orders—based on the recent, direct experience of contractors and suppliers. Analyze and check your own change order estimates against the experience others have had doing the same work. It also covers productivity analysis and change order cost justifications. With useful information for calculating the effects of change orders and dealing with their administration.

$89.95 per copy
Over 440 pages, available Oct. 1997
Catalog No. 60238

## Means Facilities Maintenance & Repair Cost Data 1998

Published in a looseleaf format, *Means Facilities Maintenance & Repair Cost Data* gives you a complete system to manage and plan your facility repair and maintenance costs and budget efficiently. Guidelines for auditing a facility and developing an annual maintenance plan. Budgeting is included, along with reference tables on cost and management and information on frequency and productivity of maintenance operations.

The only nationally recognized source of maintenance and repair costs. Developed in cooperation with the Army Corps of Engineers.

$199.95 per copy
Over 600 pages, illustrated, available Dec. 1997
Catalog No. 60308

## Means Square Foot Costs 1998

*It's Accurate and Easy To Use!*
- **NEW 1998 price information**, based on nationwide figures from suppliers, estimators, labor experts and contractors.
- **NEW "How-to-Use" Sections**, with **better, clearer examples** of commercial, residential, industrial, and institutional structures.
- **NEW**, more realistic graphics, offering true-to-life illustrations of building projects.
- **NEW**, more extensive information on using square foot cost data, including **sample estimates** and **alternate pricing methods.**

$97.95 per copy
Over 460 pages, illustrated, available Nov. 1997
Catalog No. 60058

# Annual Cost Guides

## Means Repair & Remodeling Cost Data 1998
**Commercial/Residential**

You can use this valuable tool to estimate commercial and residential renovation and remodeling.

**Includes:** New 1998 costs for hundreds of unique methods, materials and conditions that only come up in repair and remodeling. PLUS:
- 1998 unit costs for over 15,000 construction components
- 1998 installed costs for over 4,000 assemblies.
- 1998 costs for 300+ construction crews
- Over 930 "On Site" localization factors for the U.S. and Canada.

$79.95 per copy
Over 600 pages, illustrated, available Oct. 1997
Catalog No. 60048

## Means Facilities Construction Cost Data 1998

For the maintenance and construction of commercial, industrial, municipal, and institutional properties. Costs are shown for new and remodeling construction and are broken down into materials, labor, equipment, overhead, and profit. Special emphasis is given to sections on mechanical, electrical, furnishings, site work, building maintenance, finish work, and demolition. More than 40,000 unit costs plus assemblies and reference sections are included.

$209.95 per copy
Over 1100 pages, illustrated, available Nov. 1997
Catalog No. 60208

## Means Residential Cost Data 1998

Now contains square foot costs for 30 basic home models with the look of today—plus hundreds of custom additions and modifications you can quote right off the page. With costs for the 100 residential systems you're most likely to use in the year ahead. Complete with blank estimating forms, sample estimates and step-by-step instructions.

$74.95 per copy
Over 550 pages, illustrated, available Dec. 1997
Catalog No. 60178

## Means Light Commercial Cost Data 1998

Specifically addresses the light commercial market, which is an increasingly specialized niche in the industry. Aids you, the owner/designer/contractor, in preparing all types of estimates, from budgets to detailed bids. Includes new advances in methods and materials. Assemblies section allows you to evaluate alternatives in the early stages of design/planning.

Over 10,000 unit costs for 1998 ensure you have the prices you need... when you need them.

$76.95 per copy
Over 600 pages, illustrated, available Nov. 1997
Catalog No. 60188

## Means Assemblies Cost Data 1998

*Means Assemblies Cost Data 1998* takes the guesswork out of preliminary or conceptual estimates. Now you don't have to try to calculate the assembled cost by working up individual components costs. We've done all the work for you.

Presents detailed illustrations, descriptions, specifications and costs for every conceivable building assembly—240 types in all—arranged in the easy-to-use UniFormat system. Each illustrated "assembled" cost includes a complete grouping of materials and associated installation costs including the installing contractor's overhead and profit.

$139.95 per copy
Over 570 pages, illustrated, available Oct. 1997
Catalog No. 60068

## Means Site Work & Landscape Cost Data 1998

*Means Site Work & Landscape Cost Data 1998* is organized to assist you in all your estimating needs. Hundreds of fact-filled pages help you make accurate cost estimates efficiently.

**New for 1998!**
- New or expanded demolition features— ceilings, doors, electrical, flooring, HVAC, millwork, plumbing, roofing, walls and windows
- State-of-the-art segmental retaining walls
- Flywheel trenching costs and details
- Expanded Wells section
- Thousands of landscape materials, flowers, shrubs and trees

$89.95 per copy
Over 570 pages, illustrated, available Nov. 1997
Catalog No. 60288

# Annual Cost Guides

## Means Open Shop Building Construction Cost Data 1998

The latest costs for accurate budgeting and estimating of new commercial and residential construction... renovation work... change orders... cost engineering. *Means Open Shop BCCD* will assist you to...
- Develop benchmark prices for change orders
- Plug gaps in preliminary estimates, budgets
- Estimate complex projects
- Substantiate invoices on contracts
- Price ADA-related renovations

$86.95 per copy
Over 650 pages, illustrated, available Nov. 1997
Catalog No. 60158

## Means Heavy Construction Cost Data 1998

A comprehensive guide to heavy construction costs. Includes costs for highly specialized projects such as tunnels, dams, highways, airports, and waterways. Information on different labor rates, equipment, and material costs is included. Has unit price costs, systems costs, and numerous reference tables for costs and design. Valuable not only to contractors and civil engineers, but also to government agencies and city/town engineers.

$89.95 per copy
Over 450 pages, illustrated, available Dec. 1997
Catalog No. 60168

## Means Building Construction Cost Data 1998
### Western Edition

This regional edition provides more precise cost information for western North America. Labor rates are based on union rates from 13 western states and western Canada. Included are western practices and materials not found in our national edition: tilt-up concrete walls, glu-lam structural systems, specialized timber construction, seismic restraints, landscape and irrigation systems.

$89.95 per copy
Over 600 pages, illustrated, available Nov. 1997
Catalog No. 60228

## Means Heavy Construction Cost Data 1998
### Metric Version

Make sure you have the Means industry standard metric costs for the federal, state, municipal and private marketplace. With thousands of up-to-date metric unit prices in tables by CSI standard divisions. Supplies you with assemblies costs using the metric standard for reliable cost projections in the design stage of your project. Helps you determine sizes, material amounts, and has tips for handling metric estimates.

$89.95 per copy
Over 450 pages, illustrated, available Dec. 1997
Catalog No. 63168

## Means Construction Cost Indexes 1998

Who knows what 1998 holds? What materials and labor costs will change unexpectedly? By how much?
- Breakdowns for 305 major cities.
- National averages for 30 key cities.
- Expanded five major city indexes.
- Historical construction cost indexes.

$198.00 per year/$49.50 individual quarters
Catalog No. 60148

## Means Interior Cost Data 1998

Provides you with prices and guidance needed to make accurate interior work estimates. Contains costs on materials, equipment, hardware, custom installations, furnishings, labor costs . . . every cost factor for new and remodel commercial and industrial interior construction, plus more than 50 reference tables. For contractors, facility managers, owners.

**Newly expanded information on office furnishings.**

$83.95 per copy
Over 550 pages, illustrated, available Nov. 1997
Catalog No. 60098

## Means Concrete & Masonry Cost Data 1998

Provides you with cost facts for virtually all concrete/masonry estimating needs, from complicated form work to various sizes and face finishes of brick and block, all in great detail. The comprehensive unit cost section contains more than 15,000 selected entries. The assemblies cost section is illustrated with isometric drawings. A detailed reference section supplements the cost data.

$77.95 per copy
Over 520 pages, illustrated, available Dec. 1997
Catalog No. 60118

## Means Labor Rates for the Construction Industry 1998

Complete information for estimating labor costs, making comparisons and negotiating wage rates by trade for over 300 cities (United States and Canada), 46 construction trades listed by local union number in each city, historical wage rates included for comparison. No similar book is available through the trade.

**New: Each city chart now lists the county and is alphabetically arranged with handy visual flip tabs for quick reference.**

$189.95 per copy
Over 330 pages, available Dec. 1997
Catalog No. 60128

# Reference Books

## Means Environmental Remediation Estimating Methods
By Richard R. Rast  **NEW TITLE**

The first-ever guide to estimating any size environmental remediation project... anywhere in the country.

Field-tested guidelines for estimating 50 standard remediation technologies. This resource will help you: prepare preliminary budgets, develop detailed estimates, compare costs, select solutions, estimate liability, review quotes, and negotiate settlements.

A valuable support tool for the *Environmental Restoration* cost data books.

$99.95 per copy
Over 600 pages, illustrated, Hardcover
Catalog No. 64777

## Cyberplaces: The Internet Guide for Architects, Engineers & Contractors
By Paul Doherty  **NEW TITLE**

Internet applications for business and project management. Includes book, CD-ROM and Web site.

**The CD-ROM** offers built-in links to Web sites, tours of captured sites, FREE browser software and a document workshop, plus a test for Continuing Education Credits. **The Web Site** keeps the book current with updates on new technologies, interactive workshops, links to new tools and sites, and reports from top firms.

See Page 14 for full-page description

$59.95 per copy
Over 700 pages, illustrated, Softcover
Catalog No. 67317

## Value Engineering: Practical Applications
...For Design, Construction, Maintenance & Operations  **NEW TITLE**

By Alphonse Dell'Isola, P.E., the recognized leading authority on Value Engineering

A tool for immediate application—for engineers, architects, facility managers, owners and contractors. Includes: Making the Case for VE—The Management Briefing, Integrating VE into Planning and Budgeting, Conducting Life Cycle Costing, Integrating VE into the Design Process, and Using VE Methodology in Design Review and Consultant Selection, Case Studies (corporate, commercial, hospital, industrial and civil), and an expertly organized VE Workbook.

$79.95 per copy
Over 450 pages, illustrated, Hardcover
Catalog No. 67319

## HVAC: Design Criteria, Options, Selection
*Expanded Second Edition*

By William H. Rowe III, AIA, PE

*Now including Indoor Air Quality, CFC Removal, Energy Efficient Systems and Special Systems by Building Type.*

The book that helps you solve a wide range of HVAC system design and selection problems effectively and economically. Gives you explanations of the latest ASHRAE standards.

$84.95 per copy
Over 600 pages, illustrated, Hardcover
Catalog No. 67306

## The ADA in Practice
*(Revised, expanded edition of the award-winning\* New ADA: Compliance & Costs)*

By Deborah S. Kearney, PhD

Helps you meet and budget for the requirements of the Americans with Disabilities Act. Shows you how to do the job right, by understanding what the law actually requires and allows. Includes an objective "authoritative buyers guide" for 70 ADA-compliant products with specs and purchasing information. With sample evaluation forms and illustrations of ADA-compliant products.

\*Winner of the "Distinguished Author of the Year" award from the International Facility Managers Association.

$72.95 per copy
Over 600 pages, illustrated, Softcover
Catalog No. 67147A

## Means ADA Compliance Pricing Guide

Gives you detailed cost estimates for budgeting modification projects, including estimates for each of 260 alternates. You get the 75 most commonly needed modifications for ADA compliance, each an assembly estimate with detailed cost breakdown including materials, labor hours and contractor's overhead. With 3,000 additional ADA compliance-related unit cost line items and costs easily adjusted to 900 cities and towns.

$72.95 per copy
Over 350 pages, illustrated, Softcover
Catalog No. 67310

# Reference Books

## Cost Planning & Estimating for Facilities Maintenance

In this unique book, a team of facilities management authorities shares their expertise at:
- Evaluating and budgeting operations
- Maintaining & repairing key building components
- Applying *Means Facilities Maintenance & Repair Cost Data* to your estimating

With the special maintenance requirements of 10 different building types.

$82.95 per copy
Over 475 pages, Hardcover
Catalog No. 67314

## Facilities Planning & Relocation

**By David D. Owen**

An A–Z working guide—complete with the checklists, schematic diagrams and questionnaires to ensure the success of every office relocation. Complete with a step-by-step manual, 100-page technical reference section, over 50 reproducible forms in hard copy and on computer diskettes.
Winner of the International Facility Managers Assoc. "Distinguished Author of the Year" award.

$109.95 per copy, Textbook-384 pages,
Forms Binder-137 pages, illustrated, Hardcover
Catalog No. 67301

## Means Facilities Maintenance Standards

*Unique features of this one-of-a-kind working guide for facilities maintenance*

A working encyclopedia that points the way to solutions to every kind of maintenance and repair dilemma. With a labor-hours section to provide productivity figures for over 180 maintenance tasks. Included are ready-to-use forms, checklists, worksheets and comparisons, as well as analysis of materials systems and remedies for deterioration and wear.

$159.95 per copy, 600 pages, 205 tables, checklists and diagrams, Hardcover
Catalog No. 67246

## Maintenance Management Audit

At a time when many companies and institutions are reorganizing or downsizing, this annual audit program is essential for every organization in need of a proper assessment of its maintenance operation. The forms presented in this easy-to-use workbook allow managers to identify and correct problems, enhance productivity, and impact the bottom line.

New reduced price

Now $32.48 per copy, limited quantity
125 pages, spiral bound, illustrated, Hardcover
Catalog No. 67299

## The Facilities Manager's Reference

**By Harvey H. Kaiser, PhD**

The tasks and tools the facility manager needs to accomplish the organization's objectives, and develop individual and staff skills. Includes Facilities and Property Management, Administrative Control, Planning and Operations, Support Services, and a complete building audit with forms and instructions, widely used by facilities managers nationwide.

$86.95 per copy, over 250 pages,
with prototype forms and graphics, Hardcover
Catalog No. 67264

## Facilities Maintenance Management

**By Gregory H. Magee, PE**

Now you can get successful management methods and techniques for all aspects of facilities maintenance. This comprehensive reference explains and demonstrates successful management techniques for all aspects of maintenance, repair and improvements for buildings, machinery, equipment and grounds. Plus, guidance for outsourcing and managing internal staffs.

$86.95 per copy
Over 280 pages with illustrations, Hardcover
Catalog No. 67249

## Understanding Building Automation Systems

- Direct Digital Control
- Security/Access Control
- Energy Management
- Life Safety
- Lighting

**By Reinhold A. Carlson, PE & Robert Di Giandomenico**

The authors, leading authorities on the design and installation of these systems, describe the major building systems in both an overview with estimating and selection criteria, and in system configuration-level detail.

$79.95 per copy
Over 200 pages, illustrated, Hardcover
Catalog No. 67284

## Planning and Managing Interior Projects

**By Carol E. Farren**

Expert, up-to-date guidance for managing interior installation projects. Includes: project phases; winning client support; space planning and design; budgeting, bidding and purchasing, and more.
For interior designers, architects, facilities professionals.

$76.95 per copy
Over 330 pages, illustrated, Hardcover
Catalog No. 67245

# Reference Books

## Basics For Builders: Plan Reading & Material Takeoff
By Wayne J. DelPico

*For Residential and Light Commercial Construction*

A valuable tool for understanding plans and specs, and accurately calculating material quantities. Step-by-step instructions and takeoff procedures based on a full set of working drawings.

$35.95 per copy
Over 420 pages, Softcover
Catalog No. 67307

## Means Illustrated Construction Dictionary Unabridged Edition

Written in contractor's language, the information is adaptable for report writing, specifications or just intelligent discussion. Contains over 17,000 construction terms, words, phrases, acronyms and abbreviations, slang, regional terminology, and hundreds of illustrations.

$99.95 per copy
Over 700 pages, Hardcover
Catalog No. 67292

## Superintending for Contractors:
How to Bring Jobs in On-time, On-Budget
By Paul J. Cook

This book examines the complex role of the superintendent/field project manager, and provides guidelines for the efficient organization of this job. Includes administration of contracts, change orders, purchase orders, and more.

$35.95 per copy
Over 220 pages, illustrated, Softcover
Catalog No. 67233

## Means Forms for Contractors
For general and specialty contractors

Means editors have created and collected the most needed forms as requested by contractors of various-sized firms and specialties. This book covers all project phases. Includes sample correspondence and personnel administration.
Full-size forms on durable paper for photocopying or reprinting.

$79.95 per copy
Over 400 pages, three-ring binder, more than 80 forms
Catalog No. 67288

## Bidding for Contractors:
How to Make Bids that Make Money
By Paul J. Cook

The author shares the benefit of his more than 30 years of experience in construction project management, providing contractors with the tools they need to develop competitive bids.

**New reduced price**

Now $17.98 per copy, limited quantity
Over 225 pages with graphics, Softcover
Catalog No. 67180

## Means Heavy Construction Handbook

Informed guidance for planning, estimating and performing today's heavy construction projects. Provides expert advice on every aspect of heavy construction work including hazardous waste remediation and estimating. To assist planning, estimating, performing, or overseeing work.

$74.95 per copy
Over 430 pages, illustrated, Hardcover
Catalog No. 67148

## Estimating for Contractors:
How to Make Estimates that Win Jobs
By Paul J. Cook

*Estimating for Contractors* is a reference that will be used over and over, whether to check a specific estimating procedure, or to take a complete course in estimating.

$35.95 per copy
Over 225 pages, illustrated, Softcover
Catalog No. 67160

## HVAC Systems Evaluation
By Harold R. Colen, PE

You get direct comparisons of how each type of system works, with the relative costs of installation, operation, maintenance and applications by type of building. With requirements for hooking up electrical power to HVAC components. Contains experienced advice for repairing operational problems in existing HVAC systems, ductwork, fans, cooling coils, and much more!

$84.95 per copy
Over 500 pages, illustrated, Hardcover
Catalog No. 67281

## Quantity Takeoff for Contractors:
How to Get Accurate Material Counts
By Paul J. Cook

Contractors who are new to material takeoffs or want to be sure they are using the best techniques will find helpful information in this book, organized by CSI MasterFormat division.

**New reduced price**

Now $17.48 per copy, limited quantity
Over 250 pages, illustrated, Softcover
Catalog No. 67262

## Roofing: Design Criteria, Options, Selection
By R.D. Herbert, III

This book is required reading for those who specify, install or have to maintain roofing systems. It covers all types of roofing technology and systems. You'll get the facts needed to intelligently evaluate and select both traditional and new roofing systems.

$62.95 per copy
Over 225 pages with illustrations, Hardcover
Catalog No. 67253

# Reference Books

## Means Estimating Handbook

This comprehensive reference covers a full spectrum of technical data for estimating, with information on sizing, productivity, equipment requirements, codes, design standards and engineering factors.
*Means Estimating Handbook* will help you: evaluate architectural plans and specifications, prepare accurate quantity takeoffs, prepare estimates from conceptual to detailed, and evaluate change orders.

$99.95 per copy
Over 900 pages, Hardcover
Catalog No. 67276

## Means Repair and Remodeling Estimating Third Edition
### By Edward B. Wetherill & R.S. Means

Focuses on the unique problems of estimating renovations of existing structures. It helps you determine the true costs of remodeling through careful evaluation of architectural details and a site visit.
**New section on disaster restoration costs.**

$69.95 per copy
Over 450 pages, illustrated, Hardcover
Catalog No. 67265A

## Successful Estimating Methods:
### From Concept to Bid
### By John D. Bledsoe, PhD, PE

A highly practical, all-in-one guide to the tips and practices of today's successful estimator. Presents techniques for all types of estimates, *and* advanced topics such as life cycle cost analysis, value engineering, and automated estimating.
Estimate spreadsheets available at Means Web site.

$64.95 per copy
Over 300 pages, illustrated, Hardcover
Catalog No. 67287

## Means Productivity Standards for Construction
### Expanded Edition (Formerly Man-Hour Standards)

Here is the working encyclopedia of labor productivity information for construction professionals, with labor requirements for thousands of construction functions in CSI MasterFormat.
Completely updated, with over 3,000 new work items.

$159.95 per copy
Over 800 pages, Hardcover
Catalog No. 67236A

## Means Electrical Estimating Methods Second Edition

Expanded version includes sample estimates and cost information in keeping with the latest version of the CSI MasterFormat. Contains new coverage of Fiber Optic and Uninterruptible Power Supply electrical systems, broken down by components and explained in detail. A practical companion to *Means Electrical Cost Data*.

$62.95 per copy
Over 325 pages, Hardcover
Catalog No. 67230A

## Means Mechanical Estimating
### Second Edition

This guide assists you in making a review of plans, specs and bid packages with suggestions for takeoff procedures, listings, substitutions and pre-bid scheduling. Includes suggestions for budgeting labor and equipment usage. Compares materials and construction methods to allow you to select the best options for your job.

$64.95 per copy
Over 350 pages, illustrated, Hardcover
Catalog No. 67294

## Means Scheduling Manual
### Third Edition
### By F. William Horsley

Fast, convenient expertise for keeping your scheduling skills right in step with today's cost-conscious times. Covers bar charts, PERT, precedence and CPM scheduling methods. Now updated to include computer applications.

$62.95 per copy
Over 200 pages, spiral-bound, Softcover
Catalog No. 67291

## Means Graphic Construction Standards

*Means Graphic Construction Standards* bridges the gap between design and actual construction methods. With illustrations of unit assemblies, systems and components, you can see quickly which construction methods work best to meet design, budget and time objectives.

$124.95 per copy
Over 540 pages, illustrated, Hardcover
Catalog No. 67210

## Means Square Foot Estimating Methods Second Edition
### By Billy J. Cox and F. William Horsley

Proven techniques for conceptual and design-stage cost planning. Steps you through the square foot cost process, demonstrating faster, better ways to relate the design to the budget. Now updated to the latest version of UniFormat.

$69.95 per copy
Over 300 pages, illustrated, Hardcover
Catalog No. 67145A

## The Building Professional's Guide to Contract Documents
### By Waller S. Poage, AIA, CSI, CCS

Includes latest changes to AIA prototype contracts. Directions for preparing specifications and technical requirements . . . description writing, proprietary and performance specs, standards. Guidance on owner, designer and contractor liability.

$64.95 per copy
Over 400 pages, illustrated, Hardcover
Catalog No. 67261

# Reference Books

### Unit Price Estimating Methods
2nd Edition
$59.95 per copy
Catalog No. 67303

### Legal Reference for Design & Construction
By Charles R. Heuer, Esq., AIA
$109.95, Now $54.98 per copy, limited quantity
Catalog No. 67266

### Plumbing Estimating
By Joseph J. Galeno & Sheldon T. Greene
$59.95 per copy
Catalog No. 67283

### Structural Steel Estimating
By S. Paul Bunea, PhD
$79.95 per copy
Catalog No. 67241

### Landscape Estimating
2nd Edition
By Sylvia H. Fee
$64.95 per copy
Catalog No. 67295

### Business Management for Contractors
How to Make Profits in Today's Market
By Paul J. Cook
$35.95, Now $17.98 per copy, limited quantity
Catalog No. 67250

### Understanding Legal Aspects of Design/Build
By Timothy R. Twomey, Esq., AIA
$79.95 per copy
Catalog No. 67259

### Construction Paperwork
An Efficient Management System
By J. Edward Grimes
$52.95 per copy
Catalog No. 67268

### Contractor's Business Handbook
By Michael S. Milliner
$42.95, Now $21.48 per copy, limited quantity
Catalog No. 67255

### Basics for Builders: How to Survive and Prosper in Construction
By Thomas N. Frisby
$34.95 per copy
Catalog No. 67273

### Successful Interior Projects Through Effective Contract Documents
By Joel Downey & Patricia K. Gilbert
$69.95 per copy
Catalog No. 67313

### Project Planning & Control for Construction
By David R. Pierce, Jr.
$64.95 per copy
Catalog No. 67247

### Illustrated Construction Dictionary, Condensed
$59.95 per copy
Catalog No. 67282

### Managing Construction Purchasing
By John G. McConville, CCC, CPE
$62.95, Now $31.48 per copy, limited quantity
Catalog No. 67302

### Hazardous Material & Hazardous Waste
By Francis J. Hopcroft, PE, David L. Vitale, M. Ed., & Donald L. Anglehart, Esq.
$89.95, Now $44.98 per copy, limited quantity
Catalog No. 67258

### Fundamentals of the Construction Process
By Kweku K. Bentil, AIC
$69.95, Now $34.98 per copy, limited quantity
Catalog No. 67260

### Construction Delays
By Theodore J. Trauner, Jr., PE, PP
$69.95 per copy
Catalog No. 67278

### Basics for Builders: Framing & Rough Carpentry
By Scot Simpson
$24.95 per copy
Catalog No. 67298

### Interior Home Improvement Costs
5th Edition
$19.95 per copy
Catalog No. 67308A

### Exterior Home Improvement Costs
5th Edition
$19.95 per copy
Catalog No. 67309A

### Concrete Repair and Maintenance Illustrated
By Peter H. Emmons
$64.95 per copy
Catalog No. 67146

### How to Estimate with Metric Units
$49.95, Now $24.98 per copy, limited quantity
Catalog No. 67304

# Seminars

## How to Develop Facility Assessment Programs

This two-day program concentrates on the management process required for planning, conducting, and documenting the physical condition and functional adequacy of buildings and other facilities. Regular facilities condition inspections are one of the facility department's most important duties. However, gathering reliable data hinges on the design of the overall program used to identify and gauge deferred maintenance requirements. Knowing where to look . . . and reporting results effectively are the keys. This seminar is designed to give the facility executive essential steps for conducting facilities inspection programs.

**Inspection Program Requirements** • Where is the deficiency? • What is the nature of the problem? • How can it be remedied? • How much will it cost in labor, equipment and materials? • When should it be accomplished? • Who is best suited to do the work?

**Note: Because of its management focus, this course will not address trade practices and procedures.**

## Mechanical and Electrical Estimating

This seminar is tailored to fit the needs of those seeking to develop or improve their skills and to have a better understanding of how mechanical and electrical estimates are prepared during the conceptual, planning, budgeting and bidding stages. Learn how to avoid costly omissions and overlaps between these two interrelated specialties by preparing complete and thorough cost estimates for both trades. Featured are order of magnitude, assemblies, and unit price estimating. In combination with the use of **Means Mechanical Cost Data**, **Means Plumbing Cost Data** and **Means Electrical Cost Data**, this seminar will ensure more accurate and complete Mechanical/Electrical estimates for both unit price and preliminary estimating procedures.

## Square Foot Cost Estimating

Learn how to make better preliminary estimates with a limited amount of budget and design information. You will benefit from examples of a wide range of systems estimates with specifications limited to building use requirements, budget, building codes, and type of building. And yet, with minimal information, you will obtain a remarkable degree of accuracy.

Workshop sessions will provide you with model square foot estimating problems and other skill-building exercises. The exclusive Means building assemblies square foot cost approach shows how to make very reliable estimates using "bare bones" budget and design information.

## Facilities Maintenance and Repair Estimating

With our Facilities Maintenance and Repair Estimating seminar, you'll learn how to plan, budget, and estimate the cost of ongoing and preventive maintenance and repair for all your buildings and grounds. Based on R.S. Means' groundbreaking cost estimating book, this two-day seminar will show you how to decide to either contract out or retain maintenance and repair work in-house. In addition, you'll learn how to prepare budgets and schedules that help cut down on unplanned and costly emergency repair projects. Facilities Maintenance and Repair Estimating crystallizes what facilities professionals have learned over the years, but never had time to organize or document. This program covers a variety of maintenance and repair projects, from underground storage tank removal, roof repair and maintenance, exterior wall renovations, and energy source conversions, to service upgrades and estimating energy-saving alternatives.

## Repair and Remodeling Estimating

Repair and remodeling work is becoming increasingly competitive as more professionals enter the market. Recycling existing buildings can pose difficult estimating problems. Labor costs, energy use concerns, building codes, and the limitations of working with an existing structure place enormous importance on the development of accurate estimates. Using the exclusive techniques associated with Means' widely acclaimed **Repair & Remodeling Cost Data**, this seminar sorts out and discusses solutions to the problems of building alteration estimating. Attendees will receive two intensive days of eye-opening methods for handling virtually every kind of repair and remodeling situation . . . from demolition and removal to final restoration.

## Unit Price Estimating

This seminar shows how today's advanced estimating techniques and cost information sources can be used to develop more reliable unit price estimates for projects of any size. It demonstrates how to organize data, use plans efficiently, and avoid embarrassing errors by using better methods of checking.

You'll get down-to-earth help and easy-to-apply guidance for:

• making maximum use of construction cost information sources
• organizing estimating procedures in order to save time and reduce mistakes
• sorting out and identifying unusual job requirements to improve estimating accuracy.

## Scheduling and Project Management

This seminar helps you successfully establish project priorities, develop realistic schedules, and apply today's advanced management techniques to your construction projects. Hands-on exercises familiarize participants with network approaches such as the Critical Path Method. Special emphasis is placed on cost control, including use of computer-based systems. Through this seminar you'll perfect your scheduling and management skills, ensuring completion of your projects *on time* and *within budget*. Includes hands-on application of **Means Scheduling Manual** and **Means Building Construction Cost Data**.

## Managing Facilities Construction and Maintenance

In you're involved in new facility construction, renovation or maintenance projects and are concerned about getting quality work done on time and on or below budget, in Means' seminar **Managing Facilities Construction and Maintenance** you'll learn management techniques needed to effectively plan, organize, control and get the most out of your limited facilities resources.

Learn how to develop budgets, reduce expenditures and check productivity. With the knowledge gained in this course, you'll be better prepared to successfully sell accurate project budgets, timing and manpower needs to senior management . . . plus understand how to evaluate the impact of today's facility decisions on tomorrow's budget.

# Call 1-800-448-8182 for more information

# Seminars

## 1998 Means Seminar Schedule

| Location | Dates |
|---|---|
| Las Vegas, NV | March 9-12 |
| Washington, DC | April 20-23 |
| Denver, CO | May 4-7 |
| Los Angeles, CA | May 18-21 |
| San Francisco, CA | June 1-4 |
| New Orleans, LA | June 15-18 |
| Washington, DC | September 14-17 |
| Orlando, FL | November 16-19 |

## Registration Information

**How to Register** Register by phone today! Means toll-free number for making reservations is: **1-800-448-8182**

**Individual Seminar Registration Fee $845** To register by mail, complete the registration form and return with your full fee (or minimum deposit of $300/person for one 2-day seminar, or minimum deposit of $475/person for two 2-day seminars) to: Seminar Division, R.S. Means Company, Inc., 100 Construction Plaza, P.O. Box 800, Kingston, MA 02364-0800.

**Federal Government Pricing** All federal government employees save 25% off regular seminar price. Other promotional discounts cannot be combined with Federal Government discount.

**Team Discount Program** Two to four seminar registrations: $745 per person—Five or more seminar registrations: $695 per person—Ten or more seminar registrations: Call for pricing.

**Consecutive Seminar Offer** One individual signing up for two separate courses at the same location during the designated time period pays only $1,345. You get the second course for only $500 (**a 40% discount**). Payment must be received at least ten days prior to seminar dates to confirm attendance.

**Refunds** Cancellations will be accepted up to ten days prior to the seminar start. There are no refunds for cancellations postmarked later than ten working days prior to the first day of the seminar. A $150 processing fee will be charged for all cancellations. Written notice or telegram is required for all cancellations. Substitutions can be made at any time before the session starts. **No-shows are subject to the full seminar fee.**

**AIA Continuing Education** R.S. Means is registered with the AIA Continuing Education System (AIA/CES) and is committed to developing quality learning activities in accordance with the CES criteria. R.S. Means seminars meet the AIA/CES criteria for Quality Level 2. AIA members will receive (28) learning units (LUs) for each two day R.S. Means Course.

**Daily Course Schedule** The first day of each seminar session begins at 8:30 A.M. and ends at 4:30 P.M. The second day is 8:00 A.M.–4:00 P.M. Participants are urged to bring a hand-held calculator since many actual problems will be worked out in each session.

**Continental Breakfast** Your registration includes the cost of a continental breakfast, a morning coffee break, and an afternoon cola break. These informal segments will allow you to discuss topics of mutual interest with other members of the seminar. (You are free to make your own lunch and dinner arrangements.)

**Hotel/Transportation Arrangements** R.S. Means has arranged to hold a block of rooms at each hotel hosting a seminar. To take advantage of special group rates when making your reservation be sure to mention that you are attending the Means Seminar. You are of course free to stay at the lodging place of your choice. (**Hotel reservations and transportation arrangements should be made directly by seminar attendees.**)

**Important** Class sizes are limited, so please register as soon as possible.

## Registration Form     Call 1-800-448-8182, ext. 701 to register or FAX 1-617-585-7466

Please register the following people for the Means Construction Seminars as shown here. Full payment or deposit is enclosed, and we understand that we must make our own hotel reservations if overnight stays are necessary.

☐ Full payment of $ _____ enclosed.

☐ Deposit of $ _____ enclosed.

Balance Due is $ _____
U.S. Funds

Name of Registrant(s)
(To appear on certificate of completion)

_____

_____

_____

Firm Name _____

Address _____

City/State/Zip _____

Telephone No. _____ Fax No. _____

E-Mail Address _____

Charge our registration(s) to: ☐ MasterCard ☐ VISA ☐ American Express ☐ Discover

Account No. _____ Exp. Date _____

Cardholder's Signature _____

Seminar Name                City                Dates

_____

_____

**Please mail check to: R.S. MEANS COMPANY, INC., 100 Construction Plaza, P.O. Box 800, Kingston, MA 02364-0800 USA**

# Consulting Services Group

## *Proven Solutions for Managing the Costs of Construction and Facility Operations*

Business and government leaders go to great lengths today to control costs and increase return on their construction-related activities. But they seldom have an opportunity to achieve optimum success. No matter what the activity . . . planning and design, new construction, facilities management, or introducing new building technologies . . . it comes down to the same problem: To control costs one must accurately predict them. R.S. Means Consulting Services Group has a proven record of success helping businesses meet this challenge while greatly increasing opportunity to maximize return on construction investments. With its extensive, highly specialized experience understanding both construction costs and relational database technology, Means Consulting Services Group offers an unparalleled opportunity for businesses to realize dramatic improvement in their construction/facilities cost control and valuation programs.

## Database Management Solutions

- **Custom Database Development**—Means expertise in construction cost engineering and database management can be put to work creating customized cost databases and applications.
- **Data Licensing & Integration**—To enhance applications dealing with construction, any segment of Means vast database can be licensed for use, and harnessed in a format compatible with a previously developed proprietary system.
- **Cost Modeling**—Pre-built custom cost models provide organizations that expend countless hours estimating repetitive work with a systematic time-saving estimating solution.
- **Database Auditing & Maintenance**—For clients with in-house data, Means can help organize it, and by linking it with Means database, fill in any gaps that exist and provide necessary updates to maintain current and relevant proprietary cost data.

## Cost Planning Solutions

- **Estimating Service**—Means expertise is available to perform construction cost estimates, as well as to develop baseline schedules and establish management plans for projects of all sizes and types. Conceptual, budget and detailed estimates are available.
- **Benchmarking**—Means can run baseline estimates on existing project estimates. Gauging estimating accuracy and identifying inefficiencies improves the success ratio, precision and productivity of estimates.
- **Project Feasibility Studies**—The Consulting Services Group can assist in the review and clarification of the most sound and practical construction approach in terms of time, cost and use.
- **Professional Review/Expert Witness**—Means is available to provide opinions of value and to supply expert interpretations or testimony in the resolution of construction cost claims, litigation and mediation.

## Training and Development Solutions

- **Core Curriculum**—Means educational programs, delivered on-site, are designed to sharpen professional skills and to maximize effective use of cost estimating and management tools. On-site training cuts down on travel expenses and time away from the office. The broad curriculum covers such topics as repair, new construction, and conceptual estimating; scheduling and project management; facilities management; metrication; and delivery order contracting (DOC) methods.
- **Custom Curriculum**—Means can custom-tailor courses to meet the specific needs and requirements of clients. The goal is to simultaneously boost skills and broaden cost estimating and management knowledge while focusing on applications that bring immediate benefits to unique operations, challenges, or markets.
- **Development Programs**—In addition to custom curricula, Means can work with a client's Human Resources Department or with individual operating units to create programs consistent with long-term employee development objectives.

**For more information and a copy of our capabilities brochure, please call
1-800-448-8182 and ask for the Consulting Services Group, or reach us at http://www.rsmeans.com**

*Earn up to 60 continuing education credits*

Book with CD-ROM $59.95
Catalog No. 67317

## Three-Part Guide Includes Book, CD-ROM, and Continually Updated Web Site

*Written by Paul Doherty, a leading authority on Internet and Electronic Information Management*

With a CD-ROM and its own frequently updated Web site, *Cyberplaces* is a new concept in book publishing that will bring AEC professionals up to speed on Internet and intranet technologies . . . and keep them on the forefront with updates on evolving tools and applications.

Specifically for architects, engineers and contractors . . . from getting started to advanced applications, and ingenious new uses for familiar Internet tools that will reduce project administration time and cost, while vastly improving project and in-house communications.

**The CD-ROM:** Interactive review questions give you credits toward association education requirements. Plus free software, direct links to "best of" Web sites, and complete captured Web sites to explore without going on-line.

**The Web Site:** Keeps the book current with updates on technology and applications, with interactive workshops.

### What people are saying about Cyberplaces and the author:

"Paul Doherty is not only well-informed, he is a member of the inner circle of information technology strategists in the AEC industry."

John Young, CAD Manager, Gap, Inc.

"Readers will take this book and turn it into new types of value for their clients."

John R. Sorrenti, FAIA, President, JRS Architects & National AIA Vice President

RSMeans
*A Construction Market Data Group Company*

# MeansData™

CONSTRUCTION COSTS FOR SOFTWARE APPLICATIONS
**Your construction estimating software is only as good as your cost data.**

## Software Integration

A proven construction cost database is a mandatory part of any estimating package. We have linked MeansData™ directly into the industry's leading software applications. The following list of software providers can offer you MeansData™ as an added feature for their estimating systems. Visit them on-line at ***http://www.rsmeans.com/demo/*** for more information and free demos. Or call their numbers listed below.

**AEC DATA SYSTEMS**
1-800-659-9001

**ARMSTRONG
& ASSOCIATES**
Reserve Study Systems
1-808-263-7732

**BSD**
Building System Design
1-800-875-0047

**CDCI**
Construction Data Controls, Inc.
1-800-285-3929

**CMS**
Computerized Micro Solutions
1-800-255-7407

**CONAC GROUP**
1-604-273-3463

**CONSTRUCTIVE
COMPUTING, Inc.**
1-800-456-2113

**DATAQUIRE**
1-401-253-8969

**EAGLE POINT**
1-800-678-6565

**ESTIMATING
SYSTEMS, Inc.**
1-800-967-8572

**G2, Inc.**
1-800-657-6312

**GEAC COMMERCIAL
SYSTEMS, Inc.**
1-800-851-1115

**GRANTLUN
CORPORATION**
1-602-897-7750

**IQ BENECO**
1-801-565-1122

**MC$^2$**
Management Computer
Controls
1-800-225-5622

**PRISM COMPUTER
CORPORATION**
Facility Management Software
1-800-774-7622

**PDA CONSTRUCTION**
Handheld Cost Estimating
1-909-653-5878

**SANDERS
SOFTWARE, Inc.**
1-404-934-8423

**STN, Inc.**
Workline Maintenance Systems
1-800-321-1969

**TIMBERLINE SOFTWARE
CORP.**
1-800-628-6583

**TMA SYSTEMS, Inc.**
Facility Management Software
1-918-494-2890

**US COST, Inc.**
1-800-955-1385

**VERTIGRAPH, Inc.**
1-800-989-4243

**WINESTIMATOR, Inc.**
1-800-950-2374

**DemoSource™** **One-stop shopping for the latest cost estimating software for just $19.95.** This evaluation tool includes product literature and demo diskettes for ten or more estimating systems, all of which link to MeansData™. **Call 1-800-334-3509 to order.**

## FOR MORE INFORMATION ON ELECTRONIC PRODUCTS CALL
## 1-800-448-8182 OR FAX 1-800-632-6732.

MeansData™ is a registered trademark of R.S. Means Co., Inc., *A **Construction Market Data Group** Company.*

# ORDER TOLL FREE 1-800-334-3509
# OR FAX 1-800-632-6732.

# 1998 Order Form

| Qty. | Book No. | COST ESTIMATING BOOKS | Unit Price | Total |
|---|---|---|---|---|
| | 60068 | Assemblies Cost Data 1998 | $139.95 | |
| | 60018 | Building Construction Cost Data 1998 | 84.95 | |
| | 61018 | Building Const. Cost Data-Looseleaf Ed. 1998 | 109.95 | |
| | 63018 | Building Const. Cost Data-Metric Version 1998 | 89.95 | |
| | 60228 | Building Const. Cost Data-Western Ed. 1998 | 89.95 | |
| | 60118 | Concrete & Masonry Cost Data 1998 | 77.95 | |
| | 50140 | Construction Cost Indexes 1998 | 198.00 | |
| | 60148A | Construction Cost Index-January 1998 | 49.50 | |
| | 60148B | Construction Cost Index-April 1998 | 49.50 | |
| | 60148C | Construction Cost Index-July 1998 | 49.50 | |
| | 60148D | Construction Cost Index-October 1998 | 49.50 | |
| | 60318 | Contr. Pricing Guide: Framing/Carpentry | 34.95 | |
| | 60338 | Contr. Pricing Guide: Resid. Detailed | 36.95 | |
| | 60328 | Contr. Pricing Guide: Resid. Sq. Ft. | 39.95 | |
| | 64028 | ECHOS Assemblies Cost Book 1998 | 149.95 | |
| | 64018 | ECHOS Unit Cost Book 1998 | 99.95 | |
| | 60238 | Electrical Change Order Cost Data 1998 | 89.95 | |
| | 60038 | Electrical Cost Data 1998 | 87.95 | |
| | 60208 | Facilities Construction Cost Data 1998 | 209.95 | |
| | 60308 | Facilities Maintenance & Repair Cost Data 1998 | 199.95 | |
| | 60168 | Heavy Construction Cost Data 1998 | 89.95 | |
| | 63168 | Heavy Const. Cost Data-Metric Version 1998 | 89.95 | |
| | 60098 | Interior Cost Data 1998 | 83.95 | |
| | 60128 | Labor Rates for the Const. Industry 1998 | 189.95 | |
| | 60188 | Light Commercial Cost Data 1998 | 76.95 | |
| | 60028 | Mechanical Cost Data 1998 | 87.95 | |
| | 60158 | Open Shop Building Const. Cost Data 1998 | 86.95 | |
| | 60218 | Plumbing Cost Data 1998 | 87.95 | |
| | 60048 | Repair and Remodeling Cost Data 1998 | 79.95 | |
| | 60178 | Residential Cost Data 1998 | 74.95 | |
| | 60288 | Site Work & Landscape Cost Data 1998 | 89.95 | |
| | 60058 | Square Foot Costs 1998 | 97.95 | |
| | | REFERENCE BOOKS | | |
| | 67147A | ADA in Practice | 72.95 | |
| | 67310 | ADA Pricing Guide | 72.95 | |
| | 67298 | Basics for Builders: Framing & Rough Carpentry | 24.95 | |
| | 67273 | Basics for Builders: How to Survive and Prosper | 34.95 | |
| | 67307 | Basics for Builders: Plan Reading & Takeoff | 35.95 | |
| | 67180 | Bidding for Contractors | 17.98 | |
| | 67261 | Building Profess. Guide to Contract Documents | 64.95 | |
| | 67250 | Business Management for Contractors | 17.98 | |
| | 67146 | Concrete Repair & Maintenance Illustrated | 64.95 | |
| | 67278 | Construction Delays | 69.95 | |
| | 67268 | Construction Paperwork | 52.95 | |
| | 67255 | Contractor's Business Handbook | 21.48 | |
| | 67314 | Cost Planning & Est. for Facil. Maint. | 82.95 | |
| | 67317 | Cyberplaces: The Internet Guide for A/E/C | 59.95 | |
| | 67230A | Electrical Estimating Methods-2nd Ed. | 62.95 | |
| | 64777 | Environmental Remediation Est. Methods | 99.95 | |
| | 67160 | Estimating for Contractors | 35.95 | |
| | 67276 | Estimating Handbook | 99.95 | |
| | 67249 | Facilities Maintenance Management | 86.95 | |

| Qty. | Book No. | REFERENCE BOOKS (Con't) | Unit Price | Total |
|---|---|---|---|---|
| | 67246 | Facilities Maintenance Standards | 159.95 | |
| | 67264 | Facilities Manager's Reference | 86.95 | |
| | 67301 | Facilities Planning & Relocation | 109.95 | |
| | 67288 | Forms for Contractors | 79.95 | |
| | 67260 | Fundamentals of the Construction Process | 34.98 | |
| | 67210 | Graphic Construction Standards | 124.95 | |
| | 67258 | Hazardous Material & Hazardous Waste | 44.98 | |
| | 67148 | Heavy Construction Handbook | 74.95 | |
| | 67308A | Home Improvement Costs-Interior Projects | 19.95 | |
| | 67309A | Home Improvement Costs-Exterior Projects | 19.95 | |
| | 67304 | How to Estimate with Metric Units | 24.98 | |
| | 67306 | HVAC: Design Criteria, Options, Select.-2nd Ed. | 84.95 | |
| | 67281 | HVAC Systems Evaluation | 84.95 | |
| | 67282 | Illustrated Construction Dictionary, Condensed | 59.95 | |
| | 67292 | Illustrated Construction Dictionary, Unabridged | 99.95 | |
| | 67295 | Landscape Estimating-2nd Ed. | 64.95 | |
| | 67266 | Legal Reference for Design & Construction | 54.98 | |
| | 67299 | Maintenance Management Audit | 32.48 | |
| | 67302 | Managing Construction Purchasing | 31.48 | |
| | 67294 | Mechanical Estimating-2nd Ed. | 64.95 | |
| | 67245 | Planning and Managing Interior Projects | 76.95 | |
| | 67283 | Plumbing Estimating Methods | 59.95 | |
| | 67236A | Productivity Standards for Constr.-3rd Ed. | 159.95 | |
| | 67247 | Project Planning & Control | 64.95 | |
| | 67262 | Quantity Takeoff for Contractors | 17.48 | |
| | 67265A | Repair & Remodeling Estimating-3rd Ed. | 69.95 | |
| | 67253 | Roofing: Design Criteria, Options, Selection | 62.95 | |
| | 67291 | Scheduling Manual-3rd Ed. | 62.95 | |
| | 67145A | Square Foot Estimating Methods-2nd Ed. | 69.95 | |
| | 67241 | Structural Steel Estimating | 79.95 | |
| | 67287 | Successful Estimating Methods | 64.95 | |
| | 67313 | Successful Interior Projects | 69.95 | |
| | 67233 | Superintending for Contractors | 35.95 | |
| | 67284 | Understanding Building Automation Systems | 79.95 | |
| | 67259 | Understanding Legal Aspects of Design/Build | 79.95 | |
| | 67303 | Unit Price Estimating Methods-2nd Ed. | 59.95 | |
| | 67319 | Value Engineering: Practical Applications | 79.95 | |

|  |  |
|---|---|
| MA residents add 5% state sales tax | |
| Shipping & Handling** | |
| Total (U.S. Funds)* | |

Prices are subject to change and are for U.S. delivery only. *Canadian customers may call for current prices. **Shipping & handling charges: Add 6.5% of total order for check and credit card payments. Add 9% of total order for invoiced orders.

**Send Order To:**     **ADDV-1001**

Name (Please Print) _____

Company _____

☐ **Company**
☐ **Home**     Address _____

_____

City/State/Zip _____

Phone # _____ P.O. # _____

**(Must accompany all orders being billed)**

Mail To: **R.S. Means Company, Inc.**, P.O. Box 800, Kingston, MA 02364-0800